"十二五"普通高等教育本科国家级规划教材
国家级精品课程双语教材
高等学校交通运输与工程类专业规划教材

Principle of Structural Design
结构设计原理

(第3版)

Jianren Zhang, Xiaoyan Liu

张建仁　刘小燕　**主编**

Outline(内容提要)

This book is a bilingual textbook a national level excellent course textbook. This book was edited based on the requirements of teaching advisory committee of civil engineering or road, bridge and crossing engineering professions and in accordance to the latest standards of P. R. of China and provisions of Ministry of Transportation. This book compressively introduces the mechanical behaviors of design principles of reinforced concrete structures, prestressed concrete structures and masonry structures. The second edition of this book has been much revised based on the advices of authors and many teachers who teach this subject sufficiently.

This book can be used as a bilingual textbook not only for civil engineering or road, bridge and crossing engineering professions in advanced schools, but also for overseas students who major in the related professions. Moreover, the book can be used as a reference for engineers home and abroad who engage in the bridge design and management field.

本书为国家级精品课程教材,且为双语教材。本书根据高等学校土木工程、道路桥梁与渡河工程专业结构设计原理课程的教学要求,参照中华人民共和国国家和交通行业现行标准与规范,对公路桥涵所用的钢筋混凝土结构、预应力混凝土结构、圬工结构的各种基本受力构件进行全面介绍。本书再版是在使用实践、充分听取任课教师意见的基础上进行的。

本书可以作为高等学校土木工程、道路桥梁与渡河工程专业的双语教材使用,也可供相关专业的外国留学生使用,同时也可供从事桥梁设计与管理的国内外专业技术人员参考。

图书在版编目(CIP)数据

结构设计原理 = Principle of Structural Design:英文 / 张建仁主编. — 3 版. — 北京:人民交通出版社股份有限公司,2019.9
 ISBN 978-7-114-14985-6

Ⅰ.①结… Ⅱ.①张… Ⅲ.①建筑结构—结构设计—高等学校—教材—英文 Ⅳ.①TU318

中国版本图书馆 CIP 数据核字(2019)第 195138 号

"十二五"普通高等教育本科国家级规划教材
国家级精品课程双语教材
高等学校交通运输与工程类专业规划教材

书　　名:	Principle of Structural Design
	结构设计原理(第3版)
著 作 者:	张建仁　刘小燕
责任编辑:	岑　瑜　赵晓雪
责任校对:	孙国靖　龙　雪　扈　婕
责任印制:	张　凯
出版发行:	人民交通出版社股份有限公司
地　　址:	(100011)北京市朝阳区安定门外外馆斜街3号
网　　址:	http://www.ccpress.com.cn
销售电话:	(010)59757973
总 经 销:	人民交通出版社股份有限公司发行部
经　　销:	各地新华书店
印　　刷:	北京鑫正大印刷有限公司
开　　本:	787×1092　1/16
印　　张:	40
字　　数:	920 千
版　　次:	2011 年 9 月　第 1 版
	2013 年 3 月　第 2 版
	2019 年 9 月　第 3 版
印　　次:	2019 年 9 月　第 3 版　第 1 次印刷　共 4 次印刷
书　　号:	ISBN 978-7-114-14985-6
定　　价:	85.00 元

(有印刷、装订质量问题的图书由本公司负责调换)

Preface for the third edition

  Reinforced concrete is widely used to many structures in infrastructure engineering as a good performance architecture material. As for civil engineering technicians, it is the basic requirements to master the design principles of reinforced concrete structures, prestressed concrete structures and masonry structures. The book was supplemented and revised accordingly based on the requirements of teaching advisory committee of civil engineering or road and bridge professional committee and in accordance to the latest standards of P. R. of China and provisions of Ministry of Transportation, such as ***General Code for Design of Highway Bridges and Culverts(JTG D60—2015)***, ***Code for Design of Reinforced Concrete and Prestressed Concrete Highway Bridges and Culverts(JTG 3362—2018)***.

  This book was completed on the basis of the author's long-term teaching practice and the creation of the "principle of structural design" national superior course, high-quality resource-sharing course, and the full absorption of the essence of excellent Chinese and English textbooks at home and abroad. The content layout is from simplicity to complexity, step by step, and focusing on combining theory and practice. Symbols and units of measurement are in accordance with international standards.

  There are many contributors who undertake to revise this book as: Dr. Jianxin Peng contributes the General Introduction, Chapters 1, 2 and 3; Dr. Naiwei Lu contributes the Chapters 4 and 6; Dr. Yafei Ma contributes the Chapters 7 and 10; Dr. Yigang Lv contributes the Chapter 9; Dr. Ming Yuan contributes the Chapters 11 and 12; and Professor Xiaoyan Liu contributes the final compilation and edit.

There are some unavoidable errors in this book due to the limited time and authors' abilities. Any corrections by readers are welcome.

Jianren Zhang, Xiaoyan Liu
August 20, 2019

第3版前言

钢筋混凝土作为一种优质建筑材料被广泛应用于土木工程的各种结构中,作为土木工程专业的技术人员,掌握钢筋混凝土结构、预应力混凝土结构、圬工结构的设计原理是专业的基本要求。本书基于高等院校土木工程专业、道路桥梁与渡河工程专业教学指导委员会的要求,按照中华人民共和国国家标准和交通运输行业标准与最新设计规范《公路桥涵设计通用规范》(JTG D60—2015),《公路钢筋混凝土及预应力混凝土桥涵设计规范》(JTG 3362—2018),对教材进行了新增补充和相应的修订。

本书是在作者总结长期教学实践经验和创建"结构设计原理"国家精品课程、精品资源共享课程、充分吸取国内外优秀中英文教材精华的基础上完成的,在内容编排上,由浅入深,循序渐进,注重理论联系实际。符号和计量单位按国际标准。

参加修订的人员及分工为:彭建新博士修改总论和第1、2、3章,鲁乃唯博士修改第4、6章,马亚飞博士修改第7、10章,吕毅刚博士修改第9章,袁明博士修改第11、12章。刘小燕教授负责统稿。

由于编写时间短促,编者水平有限,书中错误在所难免,恳请读者批评指正。

张建仁　刘小燕
2019 年 8 月 20 日

Reprint preface for the second edition

Reinforced concrete is widely used to many structures in infrastructure engineering as a good performance architecture material. As for civil engineering technicians, it is the basic requirements to master the design principles of reinforced concrete structures, prestressed concrete structures and masonry structures. The book was edited based on the requirements of teaching advisory committee of civil engineering or road, bridge and crossing engineering professions and in accordance to the latest standards of P. R. of China and provisions of Ministry of Transportation, such as ***General Code for Design of Highway Bridges and Culverts** (**JTG D60—2004**), **Code for Design of Reinforced Concrete and Prestressed Concrete Highway Bridges and Culverts** (**JTG D62—2004**)* and ***Code for Design of Masonry Highway Bridges and Culverts** (**JTG D61—2005**)*. This book is a bilingual textbook adapted to related professional technicians.

The books have been sold out due to readers' choose. The book was selected as national planning teaching material, and has been revised based on teachers' opinions. Series of diagrams corresponding to English ones also have been supplemented in Chinese chapters. It is convenient for teaching.

There are many contributors who undertake to edit this book as: Dr. Jianxin Peng contributes the General Introduction, Chapters 1, 2 and 3, Dr. Xiaoxia Xu and Dr. Dan Xiao contribute the Chapters 4, 5 and 8, Dr. Lei Wang contributes the Chapters 6 and 7, Dr. Hui Peng contributes the Chapters 9 and 10, Professor Xiaoyan Liu and Dr. Ming Yuan contribute the Chapters 11, 12 and 13, and Dr. Xinfeng Yin contributes the Chapters 14, 15 and 16.

There are some unavoidable errors in this book due to the limit time and authors' abilities. Any corrections by readers are welcome.

Jianren Zhang, Xiaoyan Liu

03/2015

第2版重印前言

钢筋混凝土作为一种优质的建筑材料被广泛应用于土木工程的各种结构中，作为土木工程专业的技术人员，掌握钢筋混凝土结构、预应力混凝土结构、圬工结构的设计原理是专业的基本要求。本书基于高等院校土木工程专业、道路桥梁工程专业教学指导委员会的要求，按照中华人民共和国国家标准和现行的交通运输部行业标准与最新设计规范《公路桥涵设计规范》(JTG D60—2004)、《公路钢筋混凝土及预应力混凝土桥涵设计规范》(JTG D62—2004)、《公路圬工桥涵设计规范》(JTG D61—2005)进行编写，便于相关的专业技术人员使用。

承蒙读者的厚爱，本书第二版第1次印刷已经用完，又经过一年多的实践，教材被评为"十二五"普通高等教育本科国家级规划教材，在充分听取任课教师意见的基础上进行了修改，补充了与英文相对应的中文表格，这样更便于中文教学。

参加修改的人员及分工为：总论，第1、2、3章由彭建新撰写；第4、5、8章由徐晓霞、肖丹撰写；第6、7章由王磊撰写；第9、10章由彭晖撰写；第11、12、13章由刘小燕、袁明撰写；第14、15、16章由殷新锋撰写。

由于编者的水平有限，书中错误在所难免，恳请读者批评指正。

张建仁　刘小燕
2015 年 3 月

Preface for the second edition

Reinforced concrete is widely used to many structures in infrastructure engineering as a good performance architecture material. As for civil engineering technicians, it is the basic requirements to master the design principles of reinforced concrete structures, prestressed concrete structures and masonry structures. The book was edited based on the requirements of teaching advisory committee of civil engineering or road, bridge and crossing engineering professions and in accordance to the latest standards of P. R. of China and provisions of Ministry of Transportation, such as *General Code for Design of Highway Bridges and Culverts* (*JTG D60—2004*), *Code for Design of Reinforced Concrete and Prestressed Concrete Highway Bridges and Culverts* (*JTG D62—2004*) and *Code for Design of Masonry Highway Bridges and Culverts* (*JTG D61—2005*).

The second edition of this book has been much revised based on the advices of authors and many teachers who teach this subject. An example of T-section has been added in the Chapter 3. Series of diagrams corresponding to English ones also have been supplemented in Chinese chapters. Then it is more convenient for Chinese teaching.

There are many contributors who undertake to edit this book as: Prof. Jianren Zhang and Dr. Jianxin Peng contribute the General Introduction, Chapters 1, 2 and 3; Dr. Xiaoxia Xu and Dr. Dan Xiao contribute the Chapters 4, 5 and 8; Dr. Lei Wang contributes the Chapters 6 and 7; Professor, Xiaoyan Liu and Dr. Hui Peng contribute the Chapters 9 and 10; Dr. Ming Yuan and

Prof. Xiaoyan Liu contribute the Chapters 11, 12 and 13; and Dr. Xinfeng Yin contributes the Chapters 14, 15 and 16.

There are some unavoidable errors in this book due to the limited time and authors' abilities. Any corrections by readers are welcome.

Jianren Zhang
02/2013

第2版前言

钢筋混凝土作为一种优质的建筑材料被广泛应用于土木工程的各种结构中，作为土木工程专业的技术人员，掌握钢筋混凝土结构、预应力混凝土结构、圬工结构的设计原理是专业的基本要求。本书基于高等院校土木工程专业、道路桥梁工程专业教学指导委员会的要求，按照中华人民共和国国家标准和现行的交通运输部行业标准与最新设计规范《公路桥涵设计通用规范》(JTG D60—2004)、《公路钢筋混凝土及预应力混凝土桥涵设计规范》(JTG D62—2004)、《公路圬工桥涵设计规范》(JTG D61—2005)进行编写。

本书再版是作者在经过一年多的使用实践、充分听取任课教师意见的基础上进行的。第三章补充了T形截面的计算实例，中文部分增加了与英文相对应的图，这样更便于中文教学。

参加编写的人员及分工为：总论、第1、2、3章由张建仁和彭建新撰写，第4、5、8章由徐晓霞和肖丹撰写，第6、7章由王磊撰写，第9、10章由刘小燕和彭晖撰写，第11、12、13章由袁明和刘小燕撰写，第14、15、16章由殷新锋撰写。

由于编者的水平有限，书中错误在所难免，恳请读者批评指正。

张建仁

2013 年 2 月

Preface for the first edition

Reinforced concrete is widely used to many structures in infrastructure engineering as a good performance architecture material. As for civil engineering technicians, it is the basic requirements to master the design principles of reinforced concrete structures, prestressed concrete structures and masonry structures.

The book was edited based on the requirements of teaching advisory committee of civil engineering or road, bridge and crossing engineering professions and in accordance to the latest standards of P. R. of China and provisions of Ministry of Transportation, such as ***General Code for Design of Highway Bridges and Culverts*** (***JTG D60—2004***), ***Code for Design of Reinforced Concrete and Prestressed Concrete Highway Bridges and Culverts*** (***JTG D62—2004***) and ***Code for Design of Masonry Highway Bridges and Culverts*** (***JTG D61—2005***). This book is a bilingual textbook adapted to related professional technicians.

The book was completed based on long-term teaching practices of authors, National Advanced Course of the Principles of Structural Design and existed Chinese and English textbooks all over the world. This book has characteristics of clear concepts and concise content. All symbols and units conform to the national standards.

There are many contributors who undertake to edit this book as: Prof. Zhang Jianren and Dr. Peng Jianxin contribute the General Introduction, Chapters 1, 2 and 3, Dr. Xu Xiaoxia and Dr. Xiao Dan contribute the Chapters 4, 5 and 8, Dr. Wang Lei contributes the Chapters 6 and 7, Dr. Peng Hui contribute the Chapters 9 and 10, Dr. Yuan Ming and Prof. Liu Xiaoyan contribute the Chapters 11, 12 and 13, and Dr. Yin Xinfeng contributes the Chapters 14, 15 and 16.

The professional help provided by Prof. Steve-C. S. Cai, who is a professor of Civil Engineering at Louisiana State University, USA, is gratefully appreciated. The English revisions by Prof. Xiong Lijun, College of Foreign Language, and Associate Prof. Wang Yan, Department of International Communications, Changsha University of Science & Technology, are much acknowledged.

There are some unavoidable errors in this book due to the limited time and authors' abilities. Any corrections by readers are welcome.

Jianren Zhang
12/2010

第1版前言

钢筋混凝土作为一种优质的建筑材料被广泛应用于土木工程的各种结构中,作为土木工程专业的技术人员,掌握钢筋混凝土结构、预应力混凝土结构、圬工结构的设计原理是专业的基本要求。本书基于高等院校土木工程专业、道路桥梁工程专业教学指导委员会的要求,按照中华人民共和国国家标准和现行的交通运输部行业标准与最新设计规范《公路桥涵设计通用规范》(JTG D60—2004)、《公路钢筋混凝土及预应力混凝土桥涵设计规范》(JTG D62—2004)、《公路圬工桥涵设计规范》(JTG D61—2005)进行编写的,便于相关的专业技术人员使用。

本书是在作者长期教学实践和创建"结构设计原理"国家精品课程、充分汲取国内外优秀中英文教材精华的基础上完成的,在内容编排上,由浅入深,循序渐进,注重理论联系实际。符号和计量单位按国家标准的规定使用。

参加编写的人员及分工为:总论、第1、2、3章由张建仁和彭建新撰写,第4、5、8章由徐晓霞和肖丹撰写,第6、7章由王磊撰写,第9、10章由彭晖撰写,第11、12、13章由袁明和刘小燕撰写,第14、15、16章由殷新锋撰写。

在书稿的撰写过程中,得到了美国路易斯安那州大学蔡春声教授的指导,得到了长沙理工大学外语部熊丽君教授、国际交流处王燕副教授的审核,特此表示衷心感谢。

由于编写时间短促和编者的水平有限,书中错误在所难免,恳请读者批评指正。

张建仁

2010 年 12 月

Contents

General Introduction 1
  0.1 Introduction 1
  0.2 Basic concepts 2
  0.3 Characteristics of the course and guidance for how to learn the course 3

PART 1 Reinforced Concrete Structures

Chapter 1 Basic Concepts of Reinforced Concrete Structures and Physical and Mechanical Properties of Materials 7
  1.1 Basic concepts of reinforced concrete structures 7
  1.2 Concrete 9
  1.3 Steel reinforcement 18
  1.4 Bonding between steel reinforcement and concrete 23

Chapter 2 Structural Probabilistic Design Method and Principles 26
  2.1 Introduction 26
  2.2 Basic concepts of probabilistic limit state design method 29
  2.3 Calculation principles of the *CHBC* 32
  2.4 Values of material strength 33
  2.5 Actions effects and combination of action effects 35

Chapter 3 Calculation of Load-carrying Capacity of Flexural Member 40
  3.1 Basic concept of flexural members 40

3.2　Construction and shape of cross-section for flexural members　41

3.3　Failure mode and mechanical behavior of whole process for flexural members　48

3.4　Basic calculation principles of load-carrying capacity of normal section for flexural member　54

3.5　Singly-reinforced rectangular section for flexural members　59

3.6　Doubly-reinforced flexural members with rectangular section　65

3.7　T-shaped flexural members　70

Chapter 4　Calculation of Inclined Section's Load-carrying Capacity of Flexural Members　79

4.1　Inclined section's performance characteristics and failure forms of flexural members　79

4.2　Factors affecting the shear capacity of diagonal section　83

4.3　The calculation of inclined section's shear capacity of flexural members　84

4.4　Inclined section's flexural capacity of flexural members　88

4.5　Load-carrying capacity checking of the whole beam and the detailing requirements　90

Chapter 5　Calculation of Load-carrying Capacity of Torsion Members　100

5.1　Overview　100

5.2　Failure characteristics and capacity calculation of pure torsion members　101

5.3　The bearing capacity of rectangular section member under the action of bending, shear and torsion　107

5.4　Torsion member of T-shaped and I-shaped cross-section　111

5.5　Torsion member with box section　112

5.6　Construction requirements　113

Chapter 6　Calculation of Strength of Axially Loaded Members　115

6.1　Introduction　115

6.2　Calculation on axially loaded members with longitudinal bars and tied stirrups　116

6.3　Axially loaded members with longitudinal bars and spiral stirrups　120

Chapter 7　Calculation of Strength of Eccentrically Loaded Members　125

7.1　Failure modes and mechanical characteristics　126

7.2　Buckling of eccentrically loaded members　128

7.3　Eccentrically loaded rectangular member ……………………………………… 130

7.4　Eccentrically loaded members with I-shaped and T-shaped sections ………… 142

7.5　Eccentrically loaded circular member ……………………………………… 146

Chapter 8　Calculation of Load-carrying Capacity of Tensile Members ………… 151

8.1　Overview …………………………………………………………………… 151

8.2　Calculation of load-carrying capacity of axial tensile members …………… 152

8.3　Calculation of strength of eccentric tensile members ……………………… 152

Chapter 9　Calculation of Stress, Cracking and Deflection of Reinforced Concrete Flexural Members ……………………………………………… 155

9.1　Introduction ………………………………………………………………… 155

9.2　Transformed section ………………………………………………………… 157

9.3　Checking of stress …………………………………………………………… 160

9.4　Checking of cracks and crack width of flexural members ………………… 161

9.5　Deformation (deflection) checking of flexural members …………………… 163

9.6　Durability of concrete structure ……………………………………………… 171

Chapter 10　Local Compression ………………………………………………… 174

10.1　The mode and mechanism of failure under local compression …………… 175

10.2　Enhancement coefficient of concrete strength for local compression ……… 177

10.3　Calculation of local compression zone …………………………………… 179

PART 2　Prestressed Concrete Structures

Chapter 11　Basic Concepts and Materials of Prestressed Concrete Structures ……… 185

11.1　Introduction ………………………………………………………………… 185

11.2　Methods and equipments for prestressing construction …………………… 187

11.3　Materials of prestressed concrete structures ……………………………… 193

Chapter 12　Design and Calculation of Prestressed Concrete Flexural Members …… 195

12.1　Mechanical phases and calculation characteristics ………………………… 195

12.2　Calculation of ultimate load-carrying capacity …………………………… 199

12.3　Calculation of tension control stress and loss of prestressing ……………… 202

12.4　Calculation and checking of stress for the flexural member of prestressed concrete ……………………………………………………………………… 211

12.5 Checking of anti-cracking ····· 218

12.6 Calculation of deformation and setting of pre-camber ····· 220

12.7 Calculation of the end in the anchorage zone ····· 222

12.8 Design of simply-supported prestressed concrete beams ····· 224

12.9 Example for calculation of the hollow-core deck of prestressed concrete ····· 233

Chapter 13 Partially Prestressed Concrete Flexural Members ····· 272

13.1 Concepts and characteristics of PPC members ····· 272

13.2 Classification and mechanic characteristic of PPC structure ····· 273

13.3 The design calculation for category B of PPC ····· 274

13.4 Design of PPC flexural member crack allowed ····· 282

Chapter 14 Calculation of the Flexural Members of Unbonded Prestressed Concrete ····· 287

14.1 Flexural strength of unbonded prestressed concrete members ····· 288

14.2 Calculation of flexural members of unbonded partially prestressed concrete ····· 291

14.3 Design of a cross-section of a flexural member of unbonded partially prestressed concrete member ····· 294

14.4 Structure of flexural members of unbonded partially prestressed concrete ····· 296

PART 3 Masonry Structures

Chapter 15 Basic Concepts and Materials of Masonry Structures ····· 303

15.1 Basic concepts of masonry structures ····· 303

15.2 Materials of the masonry ····· 304

15.3 The strength and displacement of the masonry ····· 307

Chapter 16 Calculation of Load-carrying Capacity of Masonry Structures ····· 310

16.1 Principles of calculation ····· 310

16.2 Load-carrying capacity of the compressed member ····· 311

16.3 Sectional local compression, bended, and shear calculations of the load-carrying capacity ····· 315

Appendix ····· 317

References ····· 330

General Introduction

0.1 Introduction

Human beings took advantage of soil, stone, wood and other natural materials for construction activities in earlier age.

Civil engineering has experienced three breakthroughs in human history.

First breakthrough: 11th-3rd century BC, brick and tile, compressive human-made materials were invented. This was an era of widely use of wood and brick (stone) structures, such as the Great Wall, the Pyramids and Metrical.

Second breakthrough: the 17th century cast iron, the 19th century wrought iron, and the mid-19th century high-quality steel (tensile and compression resistant material) appeared. This was an era of steel works, for example, 300m high Eiffel Tower, steel structures with a main Cross of 512m of the Fox Channel.

Third breakthrough: the 19-20th century, reinforced concrete structures occurred. After the invention of steel and cement, until this century, the reinforced concrete structure calculation theory, and stress loss and anchorage of prestressed concrete structure had been solved. This dominated in the field of civil engineering.

"Principle of Structural Design" is an important and technically basic course, which studies structural elements combined bridge engineering after learning "mechanics of materials" and "road construction materials". Therefore, it is a very practical subject.

0.2 Basic concepts

1) Component

Its function is to carry loading and to play a role as skeleton. According to mechanical characteristics, typical components include flexural members (beams and plates), compression members, tensile members and torsion members, etc.

2) Structure

Structure means a space or plane system consists of several components made of construction materials. For example, beam of a bridge, pier and foundation compose of bridge's load-bearing system, which is called structure.

3) Structural category

(1) According to the construction materials, structures are divided into concrete structures, masonry structures, steel structures and wood structures. Concrete structures include plain concrete structure, reinforced concrete structure, prestressed concrete structure, and concrete-steel composite structure.

(2) Based on the use and function of structures, structures include bridge structures, building structures, underground structures, hydraulic structures, and special structures. Bridge structures can be divided into following types according to mechanical characteristics of load-carrying components as beam bridge(flexure), arch bridge(compression), rigid frame(beam and column combination), suspension bridge(in tension) and cable-stayed bridge(combined stress).

4) Reinforced concrete structures

Reinforced concrete structure, consisting of steel reinforcement and concrete, is a kind of architecture component that takes advantage of the merits of steel reinforcement and concrete.

Its advantages are:

(1) Local resources;

(2) Low cost;

(3) Durability;

(4) Fire resistance;

(5) Good adaptability.

Its disadvantages are:

(1) Large self-weight;

(2) Construction impacted by season;

(3) Cracking;

(4) Not suitable for high-strength material.

Reinforced concrete structure is mainly applied to small and medium span beams, plates, piers, arches and towers.

5) Prestressed concrete structures

Prestressed concrete structure is a kind of structure that a stress is imposed to concrete to improve stress condition of concrete. Its advantages include use of high-strength material, light weight, capacity of large span and good cracking resistance. Its disadvantages are complex workmanship and more equipments required. It is mainly applied to the bridge span that is more than 50m.

6) Masonry structures

Masonry structure consists of brick and stone such as masonry block according to certain rules of construction materials made of the structure. Its advantages are a wide source of materials, high compressive strength, durable and simple construction. Its drawbacks are that it is only suitable for compressive members.

7) Steel structures

The advantages of steel structures are light weight, high strength, material uniformity, large elastic modulus, and ideal elastic material. In addition, the construction is convenient and fast. The disadvantages are high cost and poor durability, regular maintenance, and poor fire resistance. Steel structure is mainly suitable for long-span steel bridge superstructures and light steel buildings.

0.3 Characteristics of the course and guidance for how to learn the course

1) Main tasks of this course

(1) To learn basic concepts and theories of reinforced concrete and prestressed concrete structures;

(2) To learn the principles of design and calculation of various engineering materials;

(3) To master working performance, failure modes, computation theories, design and construction requirements and other issues under various loading condition of reinforced concrete structures, prestressed concrete structures, masonry structures and steel structures. Through course design project, the entire beam design is mastered, and the project provides a basis for graduation design and future engineering design.

2) Features of the course

(1) The course is an interim curriculum, transiting from the basic courses to technical courses. Research methods inherit from previous courses but differ each other.

(2) Reinforced concrete design method is based on experimental investigation. Design method of reinforced concrete members depends heavily on experimental tests, and we should pay attention to application scopes and conditions of theoretical analysis.

(3) Design has a character of variation. Structural design should obey principles of safety,

adaptability, economy and reasonability. When a load is given, the same design elements have various answers. A reasonable selection is made based on comprehensive analysis and comparison.

(4) For this course, design process and steps of analysis are needed to understand.

(5) Learning how to use design specifications.

The following codes are associated with this course:

①*General Code for Design of Highway Bridge and Culverts*(*JTG D60—2015*).

②*Code for Design of Reinforced Concrete and Prestressed Concrete Highway Bridges and Culverts*(*JTG 3362—2018*). This is abbreviated by the **CHBC**.

③*Code for Design of Masonry Highway Bridges and Culverts*(*JTG D61—2005*).

④*Uniform Standard for Engineering Structural Reliability*(*GB 50153—2008*).

⑤ *Uniform Standard for Highway Engineering Structural Reliability* (*GB/T 50283—1999*).

⑥**CEB-FIP Model Code 2010**.

The mentioned codes are necessary to be understood during the study.

3) Problems needed to be solved

(1) Selection of structural materials;

(2) Selection of section type;

(3) Pre-design sectional dimensions;

(4) Reinforcement Computing;

(5) Checking the strength conditions, the stiffness, the stability and the cracking resistance.

4) Ways to learn the course

(1) Learning contents on class well, and grasping internal relations and nuances of content system.

(2) Combining theory with practice, and addressing on testing, experience and meeting design standard requirements.

(3) Understanding the material properties, because the structure features are determined by the material properties.

(4) Structural design features in diversity, and possible to search optimal design strategy.

(5) Familiar with the provisions and addressing construction requirements.

PART 1 | Reinforced Concrete Structures

Chapter 1
Basic Concepts of Reinforced Concrete Structures and Physical and Mechanical Properties of Materials

1.1 Basic concepts of reinforced concrete structures

1) Definition of reinforced concrete structures

Concrete is a non-homogeneous material with high compressive strength and very low tensile strength. However, steel is a kind of material with very high tensile and compressive strength, being able to bear the tensile stress. The combination of steel and concrete is used for working together. Concrete is good in pressure, however poor in tensile. Steel rebar bears tension. The combination will be able to use their strengths fully, and hence produce a reinforced concrete structure. Therefore, reinforced concrete structure is composed of concrete configured by the ordinary mechanical steel rebars or skeleton of reinforcement.

2) Example

The following example illustrate the stress state of a pure concrete member and a reinforced concrete structure.

Example: a span is 4m, a concentrated load is placed at the middle of the test beam, beam section size is 200mm × 300mm. The grade of concrete is C25. The details are shown in Figure 1-1.

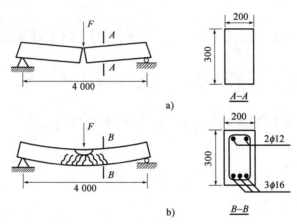

Figure 1-1　Concrete and reinforced concrete beams(unit: mm)

(1) Experimental results

①As shown in Figure 1-1a), for concrete beams, load P is 8kN and controlled by the concrete tensile strength and failure mode is brittle;

②As shown in Figure 1-1b), the ultimate load of reinforced concrete beams P is 36kN, and the steel is in tension reaching yielding and concrete is in compression. The failure mode is plastic.

(2) Experimental findings

The combination of steel and concrete improves the bearing capacity and the mechanical properties of the structure.

3) Reasons for working together of steel and concrete

The reasons for working together of steel reinforcement and concrete are as follows:

(1) Good bonding between the concrete and steel. Hence the two can be reliably combined into an integrity, which deform well under load together to complete the structural function.

(2) Temperature expansion coefficient of rebar $\alpha_{st} = 1.2 \times 10^{-5}$, and temperature expansion coefficient of concrete $\alpha_{ct} = (1.0 \sim 1.5) \times 10^{-5}$. They have similar temperature coefficient of linear expansion, so there is no greater temperature stress and relative deformation and destruction of bond strength because of no temperature changes.

(3) Sufficient cover thickness. Reinforcement in concrete can be protected against corrosion so it can be ensured that concrete can work with reinforcement in alkaline environment.

4) The main advantages and disadvantages of RC structures

(1) Advantages

①Using local resources and saving reinforced concrete materials;

②Good durability and fire resistance;
③Good modulability, easy implementation of structural types;
④Good structural integrity for casting in site or prefabricated, enough stiffness.

(2) Disadvantages

①High self-weight (improvement measures: use of lightweight aggregate concrete);
②Poor crack resistance (improvement measures: prestressed concrete);
③Construction influenced by climatic conditions;
④Difficult strengthening and reconstruction, poor performance of heating and noise.

As mentioned above, it is realized that RC structures have advantages and disadvantages, and so structural engineers may use its advantages in design practices. In engineering light aggregate is used for reducing self-weight and so enhance structural span.

1.2 Concrete

1) Introduction

Concrete is made from cement, stone, sand, and water, which behaviors a certain mix of different grades. Concrete quality is high strength, workability (easy bleeding), durability, economy and so on. Its performance characteristics are: ①compressive strength is high and tensile strength is low; ②many factors affect its strength, such as: cement grades, aggregate properties, concrete age, production methods, curing conditions and test methods; ③concrete compressive strength is conditional—the compressive strength of concrete depends on the lateral deformation constraints.

2) Concrete strength

Strength of concrete has three indicators: cubic compressive strength f_{cu}, axial compressive strength f_c, and tensile strength f_t. There are various factors affecting its strength indices, including: physical performance of materials, concrete mix, curing environment, construction methods, the shape and size of specimen, test methods, loading conditions and the loading model of the sample.

(1) The cubic compressive strength (f_{cu})—basic strength index

Concrete cubic compressive strength is the basic strength index, making the standard cure method (temperature 20℃ ±3℃, relative humidity of not less than 90% curing 28d) of side length of 150mm cube specimens. Standard Test Method is that (test surface is lubricantly coated, the whole cross-section is in pressure, and load speed is 0.15~0.25 MPa/s) concrete cubic compressive strength is measured in the 28d age with a 95% guarantee rate of compressive strength (in MPa unit), using the symbol of f_{cu}. The factors affecting cube strength are sample size and test methods.

$$f_{cu}(150) = 0.95f_{cu}(100) \brace f_{cu}(150) = 1.05f_{cu}(200)} \qquad (1\text{-}1)$$

Concrete strength changes from C25 to C80 divided into 12 grades. The medium incremental is 5MPa. Concrete lower than C50 is named as ordinary strength concrete. C50 and above C50 is high strength concrete.

According to the **CHBC**, for RC structures, strength should not be less than C25. When steel types with HRB400 or RRB400-class, its strength should not be less than C30. But for prestressed concrete members, strength should not be less than C40.

It is worthy to mention that concrete cylinder compressive strength is often used abroad, such as the United States, Japan and Europe Concrete Association using diameter of 6 inches (152mm) and height of 12 inches (305mm) standard cylinder compressive strength of specimens as a index of axial compressive strength, denoted by f'_c. For concrete lowering than C60, concrete cylinder compressive strength and the standard value of f_{cu} have a relation, which can be converted as

$$f'_c = 0.79 f_{cu,k} \qquad (1\text{-}2)$$

When $f_{cu,k}$ exceeds 60MPa with an increase of compressive strength, the ratio of f'_c and $f_{cu,k}$ (that is, the coefficient in the formula) increase. For CEB-FIP code, for C60 concrete the ratio is 0.833 and for C70 concrete ratio is 0.857. For C80 concrete the ratio is 0.875.

(2) Axial compressive strength (f_c)

Axial compressive strength is able to reflect mainly the compressive strength of concrete structure, which is 150mm × 150mm × 300mm prism as the standard sample measured compressive strength. The standard value of axial compressive strength (f_{ck}) has a relationship with standard value of cubic compressive strength ($f_{cu,k}$), and so the relationship is

$$f_{ck} = 0.88 a f_{cu,k} \qquad (1\text{-}3)$$

Where for C50 concrete and below, $a = 0.76$. For C55 to C80 concrete, $a = 0.78 \sim 0.82$.

(3) Axial tensile strength (f_t)

Concrete tensile strength is much smaller than the compressive strength of concrete with 1/8 ~ 1/18 of the compressive strength. Axial tensile strength is also a basic strength index. The main test methods are:

①Face-to-face tensioned test method.

As shown in Figure 1-2, specimen is a cylinder with 100mm × 100mm × 500mm. Reinforcing bars are placed at geometric centroid of members, and tensile force is acted by using tools to tension steel rebars. When the central specimen destroys and transverse concrete cracks, the average axial tensile stress of damaged cross section shall be the average axial tensile strength f_t.

②Splitting test method.

As shown in Figure 1-3, 150mm cylinder are used as a standard specimen of concrete to measure splitting tensile strength, and splitting tensile strength of concrete f_{ts} can be measured as

$$f_{ts} = \frac{2F}{\pi A} = 0.637 \frac{F}{A} \qquad (1\text{-}4)$$

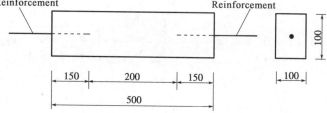

Figure 1-2 Concrete tensile strength of test specimens (unit: mm)

Where f_{ts} is the concrete splitting tensile strength (MPa); F is the splitting fracture load; A is split area of the sample surface (mm^2).

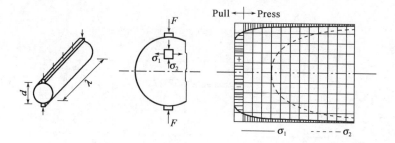

Figure 1-3 Splitting test

It is noted that measured value of concrete splitting tensile strength which converted into axial tensile strength should be multiplied using the conversion factor of 0.9 using the above test methods, that is $f_t = 0.9 f_{ts}$.

The standard value of axial tensile strength (f_{tk}) has a conversion relationship with standard value of cubic compressive strength ($f_{cu,k}$), and it is

$$f_{tk} = 0.88 \times 0.395 f_{cu,k}^{0.55} (1 - 1.645 \delta_f)^{0.45} \qquad (1\text{-}5)$$

Where δ_f is coefficient of variation of cubic compressive strength.

(4) Concrete strength under complex stress

①Bi-directional normal stress.

As shown in Figure 1-4, When σ_1 and σ_2 (compression-compression) act, concrete strength increases. When σ_1 and σ_2 (tension-compression) act, concrete strength decreases. When σ_1 and σ_2 (tension-tension) act, concrete strength remains unchanged.

②Normal stress and shear stress.

As shown Figure 1-5, when $\sigma/f_c < (0.5 \sim 0.7)$, the concrete shear strength increases with an increase of compressive stress. When $\sigma/f_c > (0.5 \sim 0.7)$, the concrete shear strength decreases with an increase of compressive stress.

When the compressive stress is about $0.6 f_c$, the shear strength increases to the maximum value, and compressive stress continues to increase significantly. Due to the development of internal cracks, the shear strength will decrease with an increase of compressive stress. For compression-shear element, the ratio should be considered.

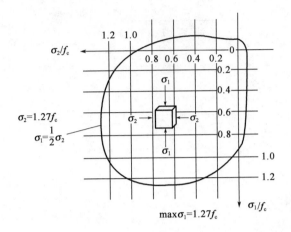

Figure 1-4 Concrete strength curves under biaxial stress

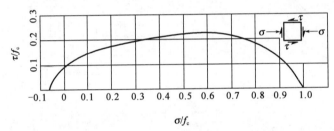

Figure 1-5 Strength curve under combination of normal stress and shear stress

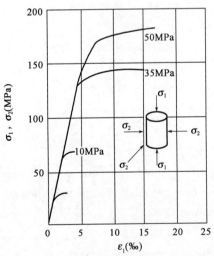

Figure 1-6 Three-compressed concrete under compression

③Concrete triaxial compression.

As shown in Figure 1-6, for the three-compressed concrete, its compressive strength is

$$f'_{cc} = f'_c + k'\sigma_2 \qquad (1\text{-}6)$$

Where σ_2 is the lateral compressive stress, and f'_c is the cylinder compressive strength.

Application of the three-compressed concrete exists concrete-filled steel tubular concrete and reinforced concrete column with screwed stirrups (Figure 1-7).

3) Deformation of concrete

(1) Characteristics of deformation of concrete

Deformation factors of concrete include loading methods, loading time, temperature, humidity, test size, shape and strength of concrete. The main deformation is the load-induced deformation and volume deformation. Deformation includes deformation under monotonic loading under monotonous and short-term load, long-term deformation under loading, and repeated deformation under loading. Volume deformation is related to shrinkage deformation and deformation caused by temperature changes.

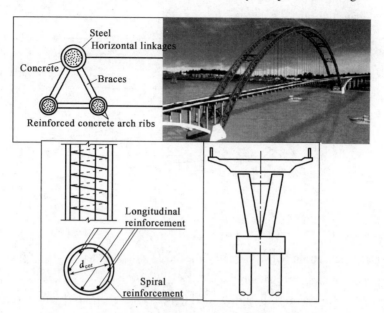

Figure 1-7   Application in steel tube, close with spiral reinforcement

(2) Concrete deformation performance under monotonous and short-term load

The experimental conditions of concrete under monotonous and short-term load are: a. uniaxial compression; b. loading under equal strain rate, etc.; c. test specimen and the high-elastic component attached to the specimen under in compression.

Concrete stress-strain test σ-ε curve is shown in Figure 1-8.

In Figure 1-8, three characteristic values of concrete stress-strain curves can be presented as: ① f_c is the axial compressive strength; ② ε_{c0} corresponds to the peak strain, $\varepsilon_{c0} = 0.002$; ③ ε_{cu} corresponds to concrete ultimate compressive strain, $\varepsilon_{cu} = 3.0 \times 10^{-3}$, based on the **CHBC**.

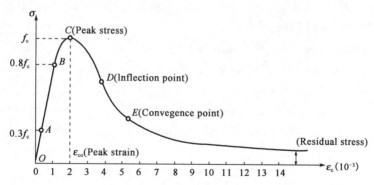

Figure 1-8   Compressive stress-strain curves of concrete

Concrete loading process is divided into three phases, namely: OA stage, elastic phase: $\sigma < 0.3f_c$; AB stage, plastic stage (crack stabilization phase): $\sigma = 0.3f_c \sim 0.8f_c$; BC phase, unstable phase of cracking: $\sigma = 0.8f_c \sim 1.0f_c$.

Many factors that impact axial compression stress-strain curves of concrete include:

①Concrete strength. As shown in Figure 1-9, when low-grade concrete is in compression, the ductility is better than high-grade concrete.

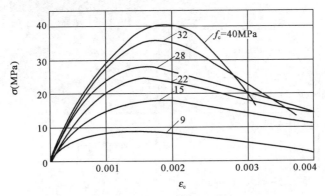

Figure 1-9 Concrete stress-strain curves of different strength grades

②Strain rate. when strain rate is small, the peak stress f_c decreases. When ε_{c0} increases, curve slopes descend significantly slowly.

③The lateral confinement. As shown in Figure 1-10, there is friction force between the bearing plates of testing machine and the upper and lower surfaces of the specimen. When damage occurs, a truncated square pyramid is formed along two pairs of top overlay (the measured strength is high). When lubricant is smeared between the upper and lower surfaces of the specimen and bearing plates of test machine, the failure mode, shown in Figure 1-10c), shows the cracking failure (measured strength is lower and descending phase can not be measured.).

④Test technology and conditions. Equal strain loading model should be used. If equal stress loading is used, it is difficult to measure the descending segment of the curve. Test machine stiffness impacts the descending phase. If the testing machine stiffness is insufficient, it is very difficult to detect the descending stress-strain curve. Strain gauge measurement also influences the descending curve. A higher strain gauge measurement will result in a steeper curve. However the smaller standard gauge will have slower slope.

(3) Concrete elastic modulus and deformation modulus

Figure 1-11 is the three types of presentation of concrete elastic modulus.

①The origin elastic modulus.

The origin elastic modulus of concrete is origin of the tangent of the compressive stress-strain curve. Namely, the origin elastic modulus is the slope of tangent. It can be expressed as

$$E'_c = \frac{\sigma}{\varepsilon_{ce}} = \tan\alpha_0 \tag{1-7}$$

②Tangent modulus.

Tangent modulus is the tangent of a location σ_c from the stress-strain curve of concrete. The slope of tangent corresponds to the tangent modulus. It can be expressed as

$$E''_c = \frac{d\sigma}{d\varepsilon} \tag{1-8}$$

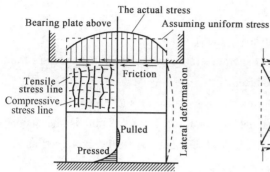 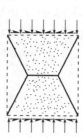 

a) Without grease lubricant test methods   b) Failure state   c) Bearing plate and the specimen between the upper and lower surfaces coated with grease lubricant

Figure 1-10 Concrete strength testing

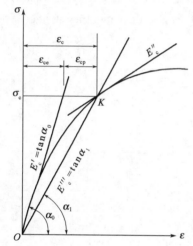

Figure 1-11 Representation of concrete elastic modulus

③Secant modulus.

Secant modulus of concrete is secant connecting origin O of the stress-strain curve and the point K on the curve. Concrete stress σ_c at location of K point is $0.5f_c$. Then the slope of secant (OK) shall be deformation modulus of concrete. It is also known as the secant modulus or elastic plastic modulus, which can be calculated as

$$E'''_c = \tan\alpha_1 = \frac{\sigma_c}{\varepsilon_c} \qquad (1\text{-}9)$$

Elastic modulus of concrete in tension and in compression are equal.

Concrete shear modulus can be expressed as

$$G_c = \frac{E_c}{2(1+\mu_c)} \qquad (1\text{-}10)$$

Where μ_c is the coefficient of lateral deformation of concrete (Poisson ratio). When $\mu_c = 0.2$, $G_c = 0.4E_c$.

Chinese code provides determination of concrete elastic modulus E_c as: the standard size of 150mm × 150mm × 300mm prism specimens. At first, it is loaded to $\sigma_c = 0.5f_c$, and then it is unloaded to zero. The number of repeatedly loading and unloading is 5 to 10 times. As the concrete is not a flexible material, when unloading to zero stress, there is residual deformation. With the number of loading increases, the stress-strain curve tends to become more stable and essentially linear. The slope of the straight line is concrete elastic modulus as obtained in Equation (1-11). The results showed that the elastic modulus measured according to the above method is slightly lower than the elastic modulus from the origin tangent to the stress-strain curve

$$E_c = \frac{10^5}{2.2 + \dfrac{34.74}{f_{cu,k}}} \qquad (1\text{-}11)$$

(4) Concrete uniaxial compressive stress-strain mathematic model

In order to do non-linear analysis of reinforced concrete structures, the non-linear stress-strain relationship is used and so several models are introduced.

As shown in Figure 1-12, a wide range of descriptions at home and abroad of concrete uniaxial compressive stress-strain curves of the model are that the rising section is the second parabola segment and the descending segment is oblique line segment. It can be expressed as

$$\left.\begin{array}{ll} \text{Rising section:} & \sigma = f_c \left[\dfrac{2\varepsilon}{\varepsilon_0} - \left(\dfrac{\varepsilon}{\varepsilon_0}\right)^2 \right], \ 0 \leqslant \varepsilon \leqslant \varepsilon_0 \\ \text{Descending segment:} & \sigma = f_c \left(1 - 0.15 \dfrac{\varepsilon - \varepsilon_0}{\varepsilon_u - \varepsilon_0}\right), \ \varepsilon_0 \leqslant \varepsilon \leqslant \varepsilon_u \end{array}\right\} \qquad (1\text{-}12)$$

Where the concrete peak strain $\varepsilon_0 = 0.002$ and ultimate compressive strain $\varepsilon_{cu} = 0.0038$.

As shown in Figure 1-13, the model based on the Chinese code is relatively simple. The ascending section is quadratic parabola and the decreased section is horizontal straight line. The model can be expressed as

$$\left.\begin{array}{ll} \text{Ascending section:} & \sigma_c = f_c \left[1 - \left(1 - \dfrac{\varepsilon_{cu}}{\varepsilon_0}\right)^n \right], \ \varepsilon \leqslant \varepsilon_0 \\ \text{Decreased section:} & \sigma_c = f_c, \ \varepsilon_0 \leqslant \varepsilon \leqslant \varepsilon_{cu} \end{array}\right\} \qquad (1\text{-}13)$$

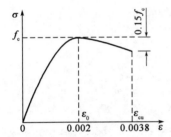
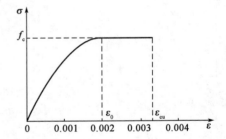

Figure 1-12 Concrete stress-strain curve　　　Figure 1-13 Concrete stress-strain curve in the **CHBC**

Where n, ε_0 and ε_{cu} are

$$n = 2 - \frac{1}{60}(f_{cu,k} - 50) \leqslant 2.0 \qquad (1\text{-}14)$$

$$\varepsilon_0 = 0.002 + 0.5(f_{cu,k} - 50) \times 10^{-5} \leqslant 0.002 \qquad (1\text{-}15)$$

$$\varepsilon_{cu} = 0.0033 - (f_{cu,k} - 50) \times 10^{-5} \leqslant 0.0033 \qquad (1\text{-}16)$$

(5) Creep deformation ε_c of concrete under long-term load

The deformation of concrete will increase with time under the long-term load. In the case of constant stress, the strain of concrete continues to increase with time. This phenomenon is known as creep shown in Figure 1-14.

Creep is a comprehensive result of various factors under long-term effect of loading, including that the water gel is gradually pressed out, viscous flow of cement is gradually occurred, micro-gap gradually closes, internal crystal gradually slides and micro cracks begin to generate and develop.

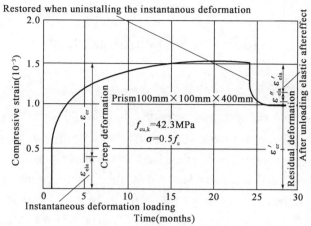

Figure 1-14 Concrete creep curve

Many factors influence concrete creep, mainly includes: ①internal factors, including the concrete composition, age, etc. The earlier age, the greater creep deformation, so prestress technology should be avoided earlier; ②environment condition refers to the curing condition and the environmental conditions (temperature, humidity, etc.). A higher humidity has less creep, however the higher temperature and lower humidity will result in the greater creep; ③ stress conditions, when the compressive stress σ is less than $0.5f_c$, the creep and stress have a linear relationship. When the compressive stress σ varies in range $(0.5 \sim 0.8)f_c$, the creep growth is faster than stress growth. This condition is called non-linear creep. When the compressive stress $\sigma > 0.8f_c$, the non-linear creep of concrete is often not convergent. It can be explained that when the stress is too large, a sharp increase in creep will lead to concrete failure. For prestressed concrete members a higher prestress force is too dangerous.

Concrete creep influences structural mechanical behaviors, and mainly reflected as: ①increase in the deformation of component; ②redistribution of sectional stress for a determinate structure; ③producing redundant for indeterminate structures; ④ prestress loss happening to prestressed concrete structures.

(6) Volume deformation—concrete non-load deformation (shrinkage)

The property that concrete volume decreases over time during the concrete setting and hardening of the physical and chemical processes is called shrinkage. Shrinkage of concrete includes the following two main reasons: ①the initial hardening cement producing solidification process in volume change (chemical shrinkage, its volume shrinkage); ② later shrinkage in concrete, which is mainly caused by free water evaporation (physical shrinkage, dryness).

The main factors that affect shrinkage are: ①the composition and the ratio of concrete; ②the curing conditions, environmental temperature and humidity, and all impact factors that keep moisture in concrete of members; ③the body surface ratio of members, where the smaller ratio will result in the greater shrinkage.

Shrinkage of concrete impacts the structure mainly as: ①members are cracked before loaded; ②prestress loss occurs in prestressed structures; ③ the second internal force is produced for indeterminate structures.

1.3 Steel reinforcement

1) Basic concepts

Steel is a kind of material with high strength, good ductility and weldability. It has better bonding properties with concrete materials. There are many varieties and types of reinforcement.

Reinforcement can be divided into carbon steel and ordinary low-alloy steel by the chemical ingredients. The carbon steel, including low carbon steel (carbon content of 0.25% or less), middle carbon steel (carbon content of 0.25%-0.6%), and high carbon steel (carbon content greater than 0.6%). Ordinary low-alloy steel refers to a small amount of low-alloy steel elements for enhancing its strength and ductility (for example: 20MnSi, 20MnSiV).

By production methods classification, reinforcement can be divided into: ①hot-rolled steel; ②heat treatment steel; ③cold-manufacture steel. The cold-manufacture steel rebar includes cold drawn steel, cold-rolled steel, ribbed steel and cold rolled twisted bars. However, due to poor ductility of cold drawn steel, now it is less used. If it is used in the project, engineers should comply with the special provisions.

Based on the shape feature classification, hot-rolled steel is divided into round bar (HPB300) and hot-rolled ribbed steel bar (HRB400, HRBF400, RRB400, HRB500), as shown in Figure 1-15. For ordinary steel, it can be selected as HPB300, HRB400, HRBF400, HRB500 rolled steel rebar. The prestressing strand and wire (Figure 1-16) should be used in prestressed concrete. For small and medium components, the vertical or horizontal reinforcing bar can also use mill and ribbed steel rebar. Surface forms of hot-rolled steel are listed in Table 1-1.

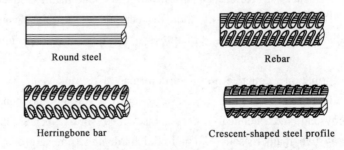

Figure 1-15 Surface forms of hot-rolled steel

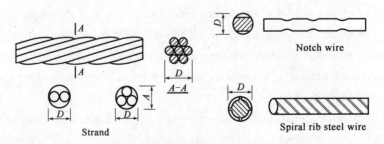

Figure 1-16 Prestressing steel

Hot-rolled reinforced bar   Table 1-1

Steel	Symbol	Steel	Symbol
HPB300($d = 6 \sim 22$)	Φ	HRBF400($d = 6 \sim 50$)	Φ^F
HRB500($d = 6 \sim 50$)	Φ	HRB400($d = 6 \sim 50$)	Φ

Based on the mechanical properties, reinforcement has different categories as: ①mild steel, appearing significant yielding platform (hot-rolled steel, cold-drawn steel); ② hard steel: no significant yielding platform (high strength carbon steel wire, steel cutter line).

2) Mechanical properties of steel rebar (strength and deformation)

As shown in Figure 1-17, for significant yielding platform of mild steel reinforcement with the stress-strain curves, the curve is divided into three stages as: ①oa phase, the elastic stage; ②$abcd$ phase, yielding stage; ③fd phase, strengthening stage; ④de stage, necking stage.

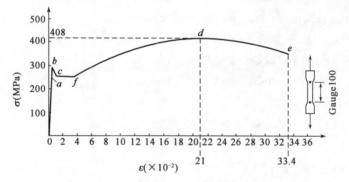

Figure 1-17 Reinforcement stress-strain curves with yielding platform

The stress-strain curve has three points, namely: ① proportional strength, σ_a; ② yield strength, σ_b; ③ultimate strength, σ_d.

Steel reinforcement has two physical indices and two mechanical indices.

(1) Two strength indices

①Point b, yield strength, σ_b, is the main basis for design and calculation of reinforced concrete structures, because the stress of steel reinforcement reaches the yield strength without increasing the load but strain continues to grow. This makes reinforced concrete crack too broadly, the structure has too large deformation and the structure does not work. Generally, it is taken lower limit value of yielding as yield strength.

②Point d, ultimate tensile strength, σ_d, which is an actual failure strength of material. It is used to assess large amount of deformation measured after the tensile strength of reinforcement, and can not be the basis of calculations. Parameters in Equation (1-17) are defined as a ratio of yield strength of RC structures and the ultimate tensile strength, and can indicate the potential reliability of the structure. The ratio is smaller, the structure has higher reliability. But small the ratio results in low utilization rate of steel. In practice, in order to ensure comprehensive strength properties of reinforcement, when the quality of steel bars is in the test, its ultimate tensile of reinforcements has to be ensured and inspection standards should be satisfied. Especially for structural seismic design, because the structure is needed to resist large deformation, the strengthening phase of reinforcements should be taken into accounts.

$$\text{Ratio} = \frac{\text{Yielding strength}}{\text{Ultimate tensile strength}} \qquad (1\text{-}17)$$

(2) Two plastic indices

①Elongation rate, also known as elongation. Point e corresponding to the abscissa (measurement gauge refers to 5 times or 10 times the bar diameter). So elongation rate is expressed as

$$\delta = \frac{l_2 - l_1}{l_1} \qquad (1\text{-}18)$$

Where l_1 and l_2 are initial and deformation lengths of reinforcing bar in a range of gauge, respectively.

②Another index is the cold bending property of steel rebar, which is a performance of non-cracking, fraction and brittle failure in the process of cold-manufacture. Through the cold bending test (bending center of the diameter and the angle of cold bending), whether or not the specimen surfaces crack or fracture is used to reflect the plastic property.

Stress-strain curve of steel reinforcement without significant yielding platform is shown in Figure 1-18. Its characteristics are high mechanical strength, poor ductility and brittleness. Its eigenvalue of stress-strain curve is $\sigma_{0.2}$. Usually it is taken as 85% of ultimate strength.

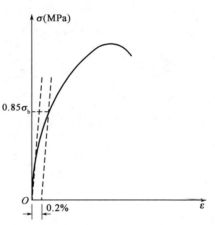

Figure 1-18 Stress-strain curve of steel reinforcement without significant yielding platform

3) Stress-strain mathematic model

Several common stress-strain mathematic model are presented as follows.

(1) Bilinear model

Bilinear model for a description of full elastic-plastic behavior is applicable to low strength steel rebar with longer yielding platform. This model simplifies stress-strain curve to Figure 1-19a), comprising of two lines. Mathematical expression of bilinear model is as

When $\varepsilon_s < \varepsilon_y$, $\left. \begin{array}{l} \sigma_s = E_s \varepsilon_s, \ E_s = \dfrac{f_y}{\varepsilon_y} \\ \sigma_s = f_y \end{array} \right\} \qquad (1\text{-}19)$

When $\varepsilon_y < \varepsilon_s < \varepsilon_{s,h}$,

(2) Three-line model

As shown in Figure 1-19b), three-line model for a description of full elastic-plastic plus hardening is applicable to mild steel with a short yield platform and can describe the strain hardening occurs immediately after yielding (stress hardening) of steel reinforcement. The stress above the yield strain can be correctly estimated. The three-line model can be mathematically expressed as

When $\varepsilon_s \leqslant \varepsilon_y$, $\varepsilon_y \leqslant \varepsilon_s \leqslant \varepsilon_{s,h}$, the expression has a same type as Equation (1-19);

When $\varepsilon_{s,h} \leq \varepsilon_s \leq \varepsilon_{s,u}$, the expression is

$$f_s = f_y + (\varepsilon_s - \varepsilon_{s,h})\tan\theta', \text{ where } \tan\theta' = E'_s = 0.01E_s \quad (1\text{-}20)$$

(3) Bi-slant model

As shown in Figure 1-19c), bi-slant model for a description of elastic-plastic can describe the stress-strain curve of high-strength steel reinforcement or wires without significant yield platform. The bi-slant model can be mathematically expressed as

When $\varepsilon_s < \varepsilon_y$, $\quad \sigma_s = E_s\varepsilon_s, \ E_s = \dfrac{f_y}{\varepsilon_y}$

When $\varepsilon_y \leq \varepsilon_s \leq \varepsilon_{s,u}$, $\quad \sigma_s = f_y + (\varepsilon_s - \varepsilon_y)\tan\theta''$ $\quad (1\text{-}21)$

Where $\tan\theta'' = E''_s = \dfrac{f_{s,u} - f_y}{\varepsilon_{s,u} - \varepsilon_y}$.

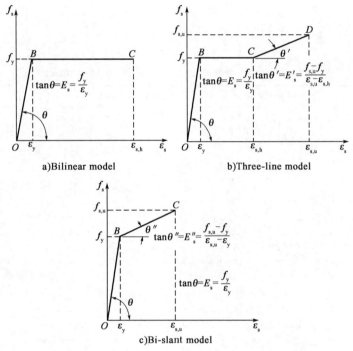

Figure 1-19 Stress-strain mathematic models of steel reinforcement

4) Joints, hook and bend of steel reinforcement

As shown in Figure 1-20, joints of steel reinforced have three types including welded joints, mechanical connectors and binding joints. Welded joints include butt welding and lap welding. Mechanical connectors include pressed sleeve connectors and screwed connectors. Binding connectors mainly refer to the bar diameters of binding joints. Generally its bar diameter is no greater than 28 mm and diameter of compressive steel reinforcement for member in compression is less than 32mm. For lap length of steel reinforcement, in tension zone it should be less than 25% of steel area and in compression zone it should be less than 50%.

As shown in Figure 1-21, different specifications have different lap lengths of binding joints. According to the ***CHBC***, longitudinal lap lengths of binding joints for tensile steel reinforcement are specified in Table 1-2. The longitudinal lap length of binding joints for steel reinforcement in compression zone should be taken as 0.7 times the lap length of binding joint in tension steel reinforcement.

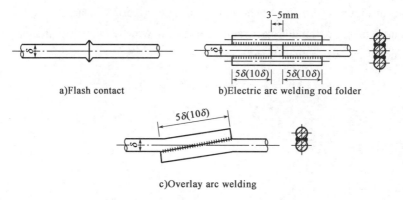

Figure 1-20 Welded joints for steel reinforcement

Note: Figures in brackets are for the side seam.

The minimum anchorage length of the reinforcement Table 1-2

Steel	HPB300				HRB400, HRBF400, RRB400			HRB500		
Concrete	C25	C30	C35	≥C40	C30	C35	≥C40	C30	C35	≥C40
Pressure bar (Straight side)	$45d$	$40d$	$38d$	$35d$	$30d$	$28d$	$25d$	$35d$	$33d$	$30d$
Tension bar — Straight side	—	—	—	—	$35d$	$33d$	$30d$	$45d$	$43d$	$40d$
Tension bar — Hook end	$40d$	$35d$	$33d$	$30d$	$30d$	$28d$	$25d$	$35d$	$33d$	$30d$

Notes: ① d is the diameter of steel bar.

②For the anchorage ribs and the anchorage lengths of the tension ribs with equivalent diameter $d_e \leq 28\text{mm}$, the equivalent generation diameter shall be determined according to the table value, and each single reinforcement of the ribs may be cut at the same anchoring end point; $d_e > 28\text{mm}$ tension ribs, each single reinforcing bar in the ribs should start from the anchoring point, and the anchoring length of the single reinforcing bar specified in the table is 1.3 times, which is stepped and then cut off, that is, starting from the starting point of anchoring. An anchor length is 1.3 times a single bar; the second extends 2.6 times the anchor length of a single bar; and the third extends 3.9 times the anchor length of a single bar.

③When epoxy-coated steel bars are used, the minimum anchorage length of the tensile reinforcement should be increased by 25%.

④When the concrete is susceptible to disturbance during solidification, the anchor length should be increased by 25%.

⑤When the end of the tensile bar is a hook, the length of the anchor is the length of the projection including the hook.

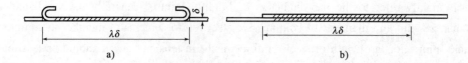

Figure 1-21 Binding lap joints for steel reinforcement

As shown in Figure 1-22, the hook and bend of steel reinforcement in tension are needed to meet the end set that is at semi-circular hook of mild steel reinforcement. For ribbed steel reinforcement rectangular-shaped hook and inside of bending section, diameter D shall not be smaller than $20d$.

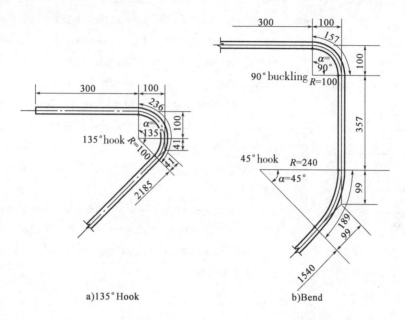

Figure 1-22 Schematic of hook and bend of steel reinforcement (unit: mm)

5) Requirements of RC structures to steel reinforcements

For the strength demand of steel reinforcement, a smaller yielding-strength ratio of a structure will have a higher reliability. But if it is too small, utilization rate of steel is low. In order to have sufficient deformation capacity, steel reinforcement shall be required to have a certain plastic ability. Elongation and cold-drawn steel performance are key indicators to check quality of materials of steel reinforcement. Weldability is an important indicator for inspection performance of welded joints of the steel reinforcement. Good weldability should satisfy requirements of steel welding cracks and excessive deformations. The fire resistance of hot-rolled steel reinforcement is best. The fine resistance of cold-rolled steel reinforcement is the second. The fire resistance of prestressing steel is the worst. Structural design engineers should pay attention to the thickness of concrete to meet fire resistance requirements of a component. In order to ensure steel reinforcement and concrete working together, enough bond force is needed.

1.4 Bonding between steel reinforcement and concrete

Bond strength is an interaction force at interface between steel and concrete. Bond stress is shear stress along the contact surface between concrete and steel reinforcement. The interaction can pass stress and conformity between steel reinforcement and concrete. Bond stress varies along the length of the steel reinforcement. On the contrary, there is no bond stress for unchangeable stress. Bond strength is maximum average bond stress when bond fails (steel rebar is pulled out or concrete is split).

Bond behavior between steel and concrete plays roles as: ① To resist slide of steel reinforcement, which is a basis for working together of two materials; ② To conduct a finite element analysis of RC structures, where bond-slip relationship is needed; ③ Degradation bonding effect, which influences resistances of fatigue and earthquake for structures.

Composition of bonding forces include chemical cohesive force, friction force and mechanical bite force. Chemical cohesive force is chemical adsorption at surface of steel reinforcement and concrete. Friction force is produced by concrete shrinkage. Mechanical bite force is produced by the uneven surface of steel reinforcement and concrete.

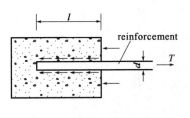

Figure 1-23   Pull-out test

An example of pull-out test is analyzing bond stress, and the testing equipment is shown in Figure 1-23.

At the early time of loading, stress of steel reinforcement at loaded end is

$$\sigma_s = \frac{F}{A_s}, \quad \varepsilon_s = \frac{\sigma_s}{E_s} \qquad (1\text{-}22)$$

Steel stress at end of specimen is

$$\sigma_s = 0, \quad \varepsilon_s = 0 \qquad (1\text{-}23)$$

Bond stress τ induced by strain difference passes tensile stress of steel reinforcement to concrete. Stress of steel reinforcement decreases, however stress of concrete increases. This behavior is extended to the length of l where $\varepsilon_c = \varepsilon_s$ and $\tau = 0$.

Test results are shown in Figure 1-24. Therefore, we can see that bond stress distribution is curve and bond stress distributions of round bar and ribbed bar are different. Bond strength between ribbed steel and concrete is much higher than that between the round rebar and concrete. Based on the statistical results, for the round bar, the average bond stress $\bar{\tau} = 1.5 \sim 3.5\text{MPa}$; for the ribbed bar, the average bond stress $\bar{\tau} = 2.5 \sim 6.0\text{MPa}$.

For round steel reinforcement, once concrete and steel reinforcement slip, the friction force is provided by mechanical bite force due to uneven surfaces of steel reinforcement. Failure mode is shear failure and steel reinforcement is pulled out from concrete. Failure occurs on the contact surface between concrete and steel. Measures that are applied to avoid bond failure are that the end hook and adequate anchorage length (compression, it can't do hook) are used. The anchorage length can be found in provision.

For ribbed steel reinforcement, before slip happens, there is only chemical adhesion force. Once concrete and steel reinforcement slip, the bonding force is provided by the friction force and mechanical bite force due to uneven surfaces of steel reinforcement. Failure mode has two types, including shear failure (high cover and loop stirrup) and splitting bond failure (low cover and no loop stirrup). Measures that are applied to avoid bond failure are the end hook, adequate anchorage length are used, and there is enough cover.

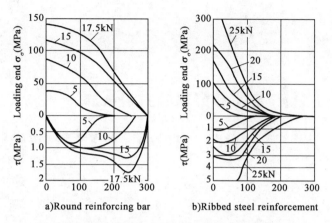

Figure 1-24 Bond stress distributions of steel reinforcement

Factors affecting bond strength between steel and concrete include concrete strength grades, positions of steel reinforcements when concrete pouring, the clear distance between steel reinforcements, concrete cover and surface shape of steel reinforcement. In order to ensure enough bond strength between steel reinforcement and concrete, spacing and concrete cover are needed to meet the requirements of code including enough anchorage length of steel reinforcements.

Chapter 2
Structural Probabilistic Design Method and Principles

2.1 Introduction

Structure is designed to meet the functional requirements, that is to say, the role of the external effects on the structure is compared with resistance of structure itself. This is applied to meet requirements of safety, durability, environmental protection, economy and aethetics for structural design.

Structural design has a long history of development. From Galileo to today, structural design possibly has about 300 years. The structural design has undergone various developments, which can be summarized as two aspects: a. from the structural design theory, it was developed from the elasticity theory to probability-based limit state theory; b. from its method, the evolution of design is from the fixed value design to the probabilistic design.

At present, calculation theory and method for structural design include allowable stress method, fracture phase of design method, multi-factor limit state design method and reliability-based probabilistic limit state method.

Allowable stress method is that the stress of a section meets or exceeds the maximum allowable stress of the material under external loading. Then the component fails or destroyed, that is to say, satisfying

$$\sigma \leqslant [\sigma] = \frac{\text{strength of materials}}{\text{factor of safety}} \qquad (2\text{-}1)$$

For RC members,

Steel Reinforcement:
$$\sigma_s \leqslant [\sigma_s] = \frac{f_s}{k_s} \qquad (2\text{-}2a)$$

Concrete:
$$\sigma_c \leqslant [\sigma_c] = \frac{f_c}{k_c} \qquad (2\text{-}2b)$$

Where k_c and k_s are the safety factors, respectively; $[\sigma_c]$ and $[\sigma_s]$ are the allowable stress of concrete and steel rebar, respectively.

As shown in Figure 2-1, the allowable stress design method is based on elasticity theory to consider design index for a RC member. The following assumptions are as follows: ①elastic assumption, steel and concrete are elastic materials; ②plane section assumption, cross section remains before and after deformation; ③ignoring tensile ability of concrete.

Allowable stress method has the following design characteristics:

(1) The safety factor k is greater than 1. A higher k will have a higher safety potentials, however resulting in more materials.

(2) There is no consideration of diversity of structure. For a structure, on the one hand, it is needed to consider the load-carrying capacity; on the other hand, it is needed to consider serviceability.

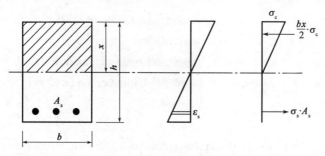

Figure 2-1 Allowable stress method diagram for a RC element

(3) Safety factor k is determined on empirical judgment rather than scientific basis.

Allowable stress method is adaptive to non-linear shaped structures (such as the large volume of dam, spatial shell structure, etc). The specification did not give a clear formula, and elasticity method is still more practical method.

Facture phase of design method was developed in the 1930's. A cross-section internal force reaches a certain limit internal force under external loading, and component fails (damages). Taking flexural members as an example, its expression is

$$M \leqslant M_u/k \qquad (2\text{-}3)$$

Where M is the computed moment of a section; M_u is the ultimate moment that cross-section bears; and k is the safety factor.

As shown in Figure 2-2, for RC structures, for example, this design should satisfy the following assumptions: a. the failure phase of component is taken as basis for calculation; b. without regard to tension; c. distribution of concrete stress in compression zone is curve, and equivalent rectangular stress block is used when calculated.

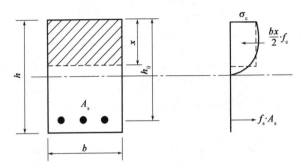

Figure 2-2 Schematic of failure phase for RC structures

The features of this design are:

(1) considering the full usage of material plasticity and strength, the ultimate load is confirmed by experimental testing, and structural safety index is relatively clear;

(2) determination of safety factor depends on experience and it is a constant;

(3) without consideration of the structural diversity.

The shortcomings of this design are that the stress distribution and displacement can not be appropriately predicted because of the use of the limit equilibrium theory.

Multi-factor limit state design is that the limit state of a component includes not only the limit state of load capacity, but also the serviceability limit state of deflection (deformation) and crack width. This covers structural safety and applicability.

For the ultimate limit state, due to loads and materials having variabilities, no single factor but many coefficients expressions are adopted. Ultimate load-carrying capacity of multi-state expression coefficient is

$$M(\sum n_i q_{ik}) \leq m M_u(k_s f_s, k_c f_c, a \cdots) \tag{2-4}$$

Where q_{ik} is standard load or effect; n_i is the corresponding overloading factor; f_s and f_c are the strength of steel and concrete; k_s and k_c are the corresponding homogeneous material coefficients; m is the coefficient of working conditions, and a is the geometrical property of a cross section.

It should be noted that material strength was obtained based on more than 95% guaranteed rate, and it is lower percentiles value. Load value is also based on statistics of various loads if possible according to more than 95% guarantee rate, which is lower percentiles value. Material strength coefficient and load factor are based upon the engineering experience. Load variation coefficients have different values.

Multi-factor limit state design is characterized by the selection of the safety factor which has changed from purely empirical to partially probability statistics. However, in essential this method is a semi-empirical semi-probabilistic approach.

The ***CHBC*** (JTG 3362—2018) uses a polynomial expression and single-factor limit state design.

Reliability-based probabilistic limit state design is based on reliability theory proposed by American scholar professor A. M. Freadentbal in 1940s. The reliability theory made a great progress in 1960s to 1970s. Until 1970s, structural reliability theory has been practically used in the field of civil engineering on the international and the theory has been gradually developed.

In China, in the middle of 1970s, researchers began to study reliability theory. At the late 1980s, structural reliability theory was used as practical method in the field of building structure. The following series of national standards have been published, such as: ***Uniform Standard for Engineering Structural Reliability*** (*GB 50153—92*) and ***Uniform Standard for Highway Engineering Structural Reliability*** (*GB/T 50283—1999*).

Structural probabilistic design method was divided into three levels in terms of development process, namely: ①level Ⅰ, semi-probability design method, some parameters that influence structural reliability were statistically analyzed and combined with experience, introducing some of empirical coefficients and so structural reliability can't be quantitatively estimated; ②level Ⅱ, approximate probabilistic design method, using probability theory and mathematical statistics, structural reliability of a component, structure and a cross section was relatively approximately estimated. The relation of variable and time is ignored or simplified in an analysis and non-linear limit state equation is linearized; ③level Ⅲ, whole probability design, the design is going on research. The ***CHBC*** is at the stage of level Ⅱ, which is approximate probability limit state design.

2.2 Basic concepts of probabilistic limit state design method

1) Functional requirements for structure design

According to the probabilistic limit state design, structural functional requirements are: a. Safety, known as strength requirements. Structure should withstand all kinds of loads occurring during normal construction and normal service, and imposed deformation and constrained deformation; b. Applicability. Structure has good working performance under normal service; c. Durability. Structure has enough durability in specified time under normal operation and normal maintenance; d. Stability. After accidental loads and incidents, the structure can still maintain stable, not collapses.

2) Structural reliability

Reliability refers to capability that structures completes scheduled function in the specified time (design reference period) and the specified conditions (with refer to normal design, construction and service of structure), and consists of safety (strength and stability), applicability and durability. Structural reliability refers to quantitative description of reliability using probability method.

Specified time refers to design reference period. It is determined according to ***Uniform Standard for Engineering Structural Reliability*** (*GB 50153—2008*). Generally, design reference period for bridge structure is 100 years. Structural design service period is associated with structural life, but not identical. When the structural service life exceeds designed service life, it does not mean that the structure is scrapped, but its reliability decreases. Specified condition refers to specified normal construction, design and operation.

3) Structural limit state

When the entire structure or part of the structure exceeding a particular state can not meet the design need of functional requirements, this particular state is called limit state function. Limit state is divided into the ultimate limit state, serviceability limit state and failure-safety limit state. The ultimate limit state corresponds that element or structure reaches the maximum load capacity or is inappropriate to bear deformation, which can be presented as: ①the whole structure or part of structure loses balance as a rigid body; ②structural components or connect joints fail in exceeding material strength; ③the structure becomes a mobile system; ④stability of structural members is lost; ⑤deformation is too large, and structure can't continue to carry load and use. The serviceability limit state mainly corresponds that a structure or component reaches the normal service or a specified limit value of durability. The specified performance is shown as: ①affecting the normal use due to appearance of deformation; ②affecting the normal use due to local damage of durability; ③affecting the normal use due to shock; ④affecting the normal use due to other specified state. Failure-safety limit state refers to the local damage caused by accident, and the remaining part may not cause the state of progressive collapse. According to the ***CHBC***, safety-fracture limit state is included in ultimate limit state.

4) Structural failure and reliability index

In a reliability analysis, structural performance function is used to describe structural limit state. There are n random variables (X_1, X_2, $\cdots$, X_n) affecting structural reliability, and structural limit state equation Z is composed of load effect S and structural resistance R

$$Z = Z(X_1, X_2, \cdots, X_n) = g(R, S) = R - S = \begin{cases} > 0, \text{ Reliable state} \\ = 0, \text{ Limit state} \\ < 0, \text{ Failure state} \end{cases} \quad (2\text{-}5)$$

Where R is the structural resistance corresponding to capacity of bearing loading and deformation; S is the load effect (action), referring to action or load effects.

Action is the reason of structural internal force, deformation, stress and strain, and is divided into direct action and indirect action. Direct action is loads are applied to structures directly (concentrated load and distributed load), such as weight, vehicle load, etc. Indirect action causes structural deformation and restraint deformation. For example, the imposed deformation means that structure is forced to be deformed, such as uniform settlement of base, and seismic. Constrained deformation is shrinkage and expansion of structural materials, and the deformation is restrained by the structural bearing or node to produce indirect deformation, such as concrete shrinkage, steel welding, atmospheric temperature changes.

Failure is part of the structure or structure can not meet the requirements of designed specified function, which meet or exceed a limit value corresponding to carrying capacity limit state or serviceability limit state. Failure probability is a probability that structure fails, and can be expressed as

Structural failure probability: $$p_f = p(z < 0) \quad (2\text{-}6)$$
Structural reliable probability: $$p_r = 1 - p_f \quad (2\text{-}7)$$

Reliability index (β) of a structure is used to measure structural reliability as

$$\beta = \frac{m_Z}{\sigma_Z} = \frac{m_R - m_S}{\sqrt{\sigma_R^2 + \sigma_S^2}} \quad (2\text{-}8)$$

Where m_Z is the mean of function Z, and σ_Z is the standard deviation of function Z. m_R and m_S are the means of resistance and load effect; σ_R and σ_S are the standard deviations of resistance and load effect.

When the R and S are normal distributions, as shown on Figure 2-3, failure probability can be derived

$$P_f = \Phi(-\beta) \quad (2\text{-}9)$$

P_f and β establish relationship one by one. A larger P_f will result in a smaller β, which means less reliable structures.

Target reliability index is used as a reliability index for the basis of structural design, and it is determined by "calibration method", experience of engineering and economic optimization principles. Calibration method uses the reliability calculation method based on the statistical parameters of the basic variables and probability distribution type, and reveals the reliability index implied in the past **CHBC**, and then takes it as a basis for determining target reliability index. This approach in general admits design experience and reliability level in the past **CHBC**, taking into account objective statistical analysis from reality.

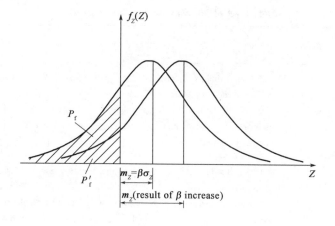

Figure 2-3 Reliability index β and the mean m_Z diagram

Based on *Uniform Standard for Highway Engineering Structural Reliability*(*GB/T 50283—1999*), structural target reliability indices are listed in Table 2-1.

Target reliability index of highway bridges　　　　　　　　　　Table 2-1

Safety levels Damage types of structural component	1	2	3
Ductile failure	4.7	4.2	3.7
Brittle fracture	5.2	4.7	4.2

2.3　Calculation principles of the *CHBC*

1) Basic Concepts

The *CHBC* provides three conditions of structural design: persistent state, transient state and accidental state. Persistent state means that bridge bears self-weight and vehicle load is long durable action. The situation is bridge service stage. It is needed to consider ultimate limit state and serviceability limit state for design. Transient state is that bridges carry temporary load during construction, and the condition corresponds to the bridge construction phase, considering ultimate limit state design. Accidental state is accidental situation during bridge service (state corresponding to earthquakes situation), considering ultimate limit state design.

Bridge and culvert structural damage according to the severity of consequences, can be divided into: Safety level 1, Safety level 2 and Safety level 3, as shown in Table 2-2.

2) Expressions of limit state design

Ultimate limit state design is established based on the plastic theory, and the design principle is that the design value of most maximum combination of actions (basic combination) must be less than or equal to the design value of structural resistance.

Safety level of highway bridges and culverts　　　　　　　　　　Table 2-2

Safety level	Damage effects	Bridge types	Structural important Factors
1	Very serious	1) Super long-span bridges, long-span bridges, and medium bridges located at all-rated highway; 2) Small bridges located at high-speed highway, first rate highway, second-rate highway, military highway, and highway that has busy traffic near city.	1.1
2	Serious	1) Small bridges located at three-rated highway and four-rated highway; 2) Culverts located at high-speed highway, first rate highway, second-rate highway, military highway, and highway that has busy traffic near city.	1.0
3	Not serious	Culverts located at three-rated highway and four-rated highway.	0.9

The basic expression of ultimate limit state is

$$\gamma_0 S_d \leqslant R = R(f_d, a_d) \qquad (2\text{-}10)$$

Where γ_0 is the importance factor, $\gamma_0 = 1.1$, 1.0 and 0.9, which corresponds to structural safety level 1, 2 and 3, respectively; f_d is the design value of strength of material; a_d are the design value of geometrical parameters.

When there is no reliable data, a_d, the standard value of geometrical parameter is adopted, namely specified value in design documents.

Design principle of persistent state (serviceability limit state design) is based on the elastic theory or elastic plastic theory, and uses short-term action combinations and long-term action combinations or short-term action combinations considering the impact of long-term action combinations of components, and stress, crack width and deflection are checked. The calculated value should not be more than the prescribed value in the **CHBC**. The expression for the ultimate limit state design is

$$S \leqslant c_1 \qquad (2\text{-}11)$$

Where S is the combined design value of serviceability limit state; c_1 is the limit value for a component reaching normal use, such as deformation, crack width and cross sectional stress.

Design principles of stress expressions of persistent and transient conditions are very complex, and different components have different design principles. For a reinforced concrete and prestressed concrete members, stress are induced by its self-weight and construction loads in fabrication, and transportation and installation stages based on transient condition shall not exceed the limit. For design of flexural prestressed concrete members based on persistent conditions, stress in service stage should not exceed limit values. To sum up, structural stress does not exceed limit values. Design expressions of limit state design for persistent and transient conditions is as

$$S \leqslant c_2 \qquad (2\text{-}12)$$

Where S is the standard value of actions (considering the impact of vehicle load factor). A load combination factor is not considered when loads are combined; c_2 is the structural function (stress) limit value.

2.4 Values of material strength

1) Principles of values for material strength

Standard value of material strength is a characteristic value of material strength, which is determined by standard test methods according to mathematical statistics using probability distribution by 0.05 percentile values.

Principle of specifying value is to meet that the overall strength of the material measured of specified quality should not be less than 95% of the guaranteed rate. The basic expression is

$$f_k = f_m(1 - 1.645\delta_f) \qquad (2\text{-}13)$$

Where f_m is the mean for the material strength, and δ_f is the variation coefficient for the strength of the material.

The design value of material strength is that standard value of material strength divides by partial factors of material performance, and expressed as

$$f_s = \frac{f_k}{\gamma} \tag{2-14}$$

Different materials have different values. For example, for concrete materials, $\gamma_c = 1.45$. For steel reinforcement, $\gamma_s = 1.2$. For tensile high strength steel wire and strand, $\gamma_s = 1.47$.

2) Standard and design values of concrete strength

Standard values of concrete cubic compressive strength $f_{cu,k}$ are obtained from the standard method of production and curing of the side lengths of 150mm cube specimens, in 28 days, measured by the standard test method with 95% of the guaranteed rate. Based on the **CHBC** for reinforced concrete components, concrete grade is not lower than C25, when HRB400, HRBF400 and RRB400 reinforcement rebars are used, concrete grade is not lower than C30. For prestressed concrete components, the concrete grade is not lower than C40.

Standard values of concrete compressive strength f_{ck} are assumed that variation coefficient of axial compressive strength of concrete is similar to that concrete cubic compressive strength. The standard value of concrete compressive strength is determined as

$$f_{ck} = f_{c.m}(1 - 1.645\delta_f) = 0.88\alpha_{c1}\alpha_{c2}f_{cu,m}(1 - 1.645\delta_f) = 0.88\alpha_{c1}\alpha_{c2}f_{cu,k} \tag{2-15}$$

Where $f_{c,m}$ and $f_{cu,m}$ are the means of concrete compressive strength and cube compressive strength, respectively; α_{c1} is the ratio of axial compressive strength of concrete and cube compressive strength. For lower than C50, α_{c1} is 0.76, and for C80 α_{c1} is taken as 0.82, and linear interpolation is used between concrete C50 and C80; α_{c2} is the reduction factor for the brittleness of concrete. For C40 concrete, α_{c2} is taken as 1.0. For C80 concrete, α_{c2} is taken as 0.87, and linear interpolation is used between concrete C40 and C80.

Standard values of concrete tensile strength are assumed that the variation coefficient of axial tensile strength of concrete is similar to that of cubic compressive strength, which can be expressed as

$$f_{tk} = 0.348\alpha_{c2}(f_{cu,k})^{0.55}(1 - 1.645\delta_f)^{0.45} \tag{2-16}$$

Where f_{tk} and $f_{cu,k}$ are the standard values of concrete tensile strength and concrete cubic compressive strength, respectively. According to the **CHBC**, partial factor is taken as 1.45 for concrete compressive strength and axial tensile strength. The design value is obtained from a standard value dividing the corresponding partial factor. (Close to the safety level corresponding to target reliability index considering brittle fracture of a component.)

3) Standard and design values of steel reinforcements strength

(1) Standard value

In national standards, standard value of yield strength of steel reinforcement is set as limit value for checking production as the guaranteed rate being not less than 95%.

For significant yield platform of hot-rolled steel, standard value of tensile strength of steel reinforcement is taken as standard value of yield. For no significant yield platform of steel reinforcement, standard value of tensile strength of steel wire and strand is taken as 0.85 σ_b with guaranteed rate being not less than 95% (σ_b is the ultimate tensile strength specified in national standard) using as conditional yield strength for design (corresponding to stress of reinforcement with residual strain of 0.2% of the tensile strength).

Based on the ***CHBC***, for the twisted steel reinforcement and hot-rolled steel reinforcement, partial factor of material performance is 1.20. For the steel strand and wire, partial factor of material properties is 1.47.

(2) Design value

The design value of tensile strength of steel reinforcement is that its standard value of tensile strength divides partial factor of 1.20 or 1.47. Design value of compressive strength of steel reinforcement is determined by $f'_{sd} = \varepsilon'_s E'_s$ or $f'_{pd} = \varepsilon'_p E'_p$. E_s and E_p are the elastic modulus of hot-rolled steel reinforcement and steel wire; ε'_s and ε'_p are the compressive strains of the similar types. $\varepsilon'_s(\varepsilon'_p)$ is equal to 0.002. f'_{sd} (f'_{pd}) should not be greater than the design value of the corresponding tensile strength of steel rebars.

2.5 Action effects and combination of action effects

Action is the reason that structure induces internal force, deformation, stress and strain. Actions are divided into: permanent action, variable action and accidental action. Permanent action is an action that does not change over time or changes are negligible compared with mean value during structural service period. Variable action is an action that varies over time and when compared to mean value changes can't be ignored during the service time. Accidental action is an action with low probability, and once the event happens its value is huge in short duration.

1) Representative value of action

When designing highway bridges and culverts, different representative effects are adopted for different effects according to the following regulations:

(1) The representative value of permanent action is its standard value. The permanent value is standard value and can be determined by statistics, calculations, and comprehensive analysis of engineering experience.

(2) Representative values of the variable action include standard values, combined values, frequency values, and quasi-permanent values. The combined value is the action value that makes the over-probability of the combined action effect coincide with the over-probability of the effect of the standard value of the action alone period; or the combined action value of the structure with the specified reliable index. It can be expressed by a combination of the value coefficient a and the reduction of the action standard value. The frequency value is the value of the total time that is exceeded the design reference period or the frequency that is exceeded is limited to the value within the specified frequency. It can be expressed by the reduction of the action standard value by

the frequency value coefficient. The quasi-permanent value is the value of the total time that is exceeded during the design base period as a percentage of the design base period. It can be expressed by the reduction of the effective standard value by the quasi-permanent value coefficient.

(3) The accidental effect takes design value as its representative value, which can be accordingly determined by historical records, observations and tests, combined with comprehensive analysis of engineering experience, and can also be determined according to the special regulations of relevant standards.

(4) The representative value of the seismic action is standard value. The standard value of seismic action should be determined in accordance with the current *Code for Seismic Engineering of Highway Engineering* (*JTG B02*).

(5) The design value of the action should be the standard value or combined value of the action multiplied by the corresponding action component coefficient. The value of the partial coefficient can be found in the related Regulations (2015).

When designing of highway bridges and culverts it should take into account the effects that may occur at the same time on the structure. The combination of load capacity limit state and serviceability limit state should be designed according to the following principles:

(1) Combinations are only made at the same time when the structure or structural members need to be checked for different directions of force, they should be calculated with the most unfavorable combination effects in different directions.

(2) When the occurrence of variable effect has a favorable effect on a structure or structural members, the action should not take account into the combination. When the effect is impossible to simultaneously occur or the effect of the combination probability is pretty low, the participation combination should be not considered in accordance with Table 4.1.4.

2) Combination of action effects

When the highway bridge and culvert structure is designed according to the limit state of the bearing capacity, the basic combination of the effect of the permanent design condition and the short-term design condition should be adopted. The accidental design should adopt the accidental combination of the action, and the seismic combination should be applied to the seismic design.

(1) Limit state action effect

① Basic combination

For the limit state of the capacity, the "Bridge code" stipulates that the basic combination is

$$S_{ud} = \gamma_0 S \left(\sum_{i=1}^{m} \gamma_{G_j} G_{ik}, \gamma_{Q1} \gamma_{L1} Q_{1k}, \psi_c \sum_{j=2}^{n} \gamma_{Lj} \gamma_{Qj} Q_{jk} \right) \qquad (2\text{-}17)$$

Or

$$S_{ud} = \gamma_0 S \left(\sum_{i=1}^{m} G_{id}, Q_{1d}, \sum_{j=2}^{n} Q_{jd} \right) \qquad (2\text{-}18)$$

Where γ_0 is the importance coefficient of the bridge structure, which is adopted according to the structural design safety level. For highway bridges, where the safety level is 1, 2, and 3, γ_0 are

1.1, 1.0, and 0.9 respectively; γ_{Gi} is the partial coefficient of the ith permanent effect. When the permanent action effect (structural gravity and pre-stress) is unfavorable to the structural bearing capacity, $\gamma_G = 1.2$. When the carrying capacity of the structure is favorable, the value of the partial coefficient γ_G is 1.0, and the partial coefficients of other permanent effects are detailed in the "Bridge code". γ_{Q1} is the partial coefficient of vehicle loads (including automobile impact force, centrifugal force). When the lane load is calculated, $\gamma_{Q1} = 1.4$; using the vehicle load calculation period, the partial coefficient $\gamma_{Q1} = 1.8$. When a variable effect exceeds the vehicle load effect in the combination, the effect replaces the vehicle load, and its partial coefficient $\gamma_{Q1} = 1.4$. For a structure or device designed to withstand an effect, the sub-factor for this effect is taken as $\gamma_{Q1} = 1.4$. Q_{1k} and Q_{1d} are standard values and design values for vehicle loads (including automotive impact, centrifugal force); γ_{Qj} is the partial coefficient of the jth variable action in addition to the vehicle loads (including automobile impact force, centrifugal force) and wind load in the action combination. However, the partial coefficient of wind load is $\gamma_{Qj} = 1.1$; Q_{jk} and Q_{jd} are the standard and design values of the jth variable action in addition to the vehicle loads (including automobile impact force, centrifugal force) in the action combination. ψ_c is the combined value coefficient of the variable action except the vehicle loads (including the automobile impact force, centrifugal force) in the action combination, taking $\psi_c = 0.75$; γ_{Lj} is the jth variable action structure design using the age load adjustment factor, $\gamma_{Lj} = 1.0$ generally. When the action and action effects can be considered in a linear relationship, the effect design value S_{ud} of the basic combination can be calculated by adding the action effect.

②Accidental combination

The permanent action standard value is combined with a variable representative value and an accidental design value; the variable action occurring at the same time as the accidental effect can be used according to the observation data and engineering experience value.

The effect design value of the accidental combination can be calculated as

$$S_{ad} = S\left(\sum_{i=1}^{m} G_{ik}, A_d, (\psi_{fl} \text{ or } \psi_{ql})Q_{1k}, \sum_{j=2}^{n} \psi_{qj}Q_{jk}\right) \qquad (2\text{-}19)$$

Where S_{ad} is the effect design value of the accidental combination of the load capacity limit state; A_d is the design value of the accidental effect; ψ_{fl} is the frequency coefficient of the vehicle loads (including the impact force of the car, centrifugal force), taking $\psi_{fl} = 0.7$; When a variable effect in the combination exceeds the effect of the vehicle loads, the variable effect replaces the vehicle loads, and the crowd load $\psi_{fl} = 1.0$, wind load $\psi_{fl} = 0.75$, temperature gradient action $\psi_{fl} = 0.8$, other effects $\psi_q = 1.0$; $\psi_{ql}Q_{1k}$ and $\psi_{qj}Q_{jk}$ are the quasi-permanent values of the first and jth variable effects.

③Seismic combination

For bridge engineering, the bearing capacity of the occasional combination of seismic action under the limit state of bearing capacity is

$$\gamma_0\left(\sum_{i=1}^{m}\gamma_{Gi}S_{GiK} + \sum_{j=1}^{n}S_{QjK} + Q_e\right) \leqslant R(\gamma_f, f_K, \gamma_a, \alpha_K) \qquad (2\text{-}20)$$

Where γ_0 is the structural importance coefficient; S_{GiK} is the ith permanent effect; S_{QjK} is a certain magnitude effect of the jth variable action that may act simultaneously with the seismic action; Q_e is the seismic effect; γ_{Gi} is the permanent action component coefficient. The specific values are calculated in the current **Code for Design of Highway Completion Bridge and Culvert** (***JTG D61***) and the relevant provisions of **Design Specification for Highway Reinforced Concrete and Prestressed Concrete Bridges and Culverts** (***JTG D62***). γ_f is a partial coefficient of structural materials and geotechnical properties; γ_a is a partial coefficient of the structural or component geometry parameters; f_K is the standard value of material and geotechnical properties. α_K is the standard value of the geometric parameter.

(2) Serviceability action effect

The normal use limit state should be based on different design requirements, using the frequency combination or quasi-permanent combination of effects, and the effect combination expression is

Frequent combination effects:

$$S_{fd} = S\left(\sum_{i=1}^{m} G_{ik}, \psi_{f1} Q_{1k}, \sum_{j=2}^{n} \psi_{qj} Q_{jk}\right) \qquad (2\text{-}21)$$

Where S_{fd} is the effect design value of the combination of frequency effects; G_{ik} is the ith permanent standard value. For the car load (excluding the car impact force) frequency factor, it is 0.7. When the action and action effects can be considered in a linear relationship, the effect design value S_{fd} of the basic combination can be calculated by adding the action effect.

Quasi-permanent combination effect:

$$S_{qd} = S\left(\sum_{i=1}^{m} G_{ik}, \sum_{j=1}^{n} \psi_{qj} Q_{jk}\right) \qquad (2\text{-}22)$$

Where S_{qd} is the effect design value that acts as a quasi-permanent combination; ψ_{qj} is the quasi-permanent coefficient of the vehicle load (excluding the impact of the car), taking 0.4. When the action and action effects can be considered in a linear relationship, the effect design value S_{qd} of the basic combination can be calculated by adding the action effect.

3) Examples

A simply supported reinforced concrete bridge beam bears self-weight, vehicle action and pedestrian action, respectively. The standard values of bending moment at 1/4 span are: for bending moment by self-weight, $M_{Gk} = 512$ kN·m; for bending moment by vehicle action, $M_{Q1k} = 425$ kN·m. Impact factor $1 + \mu = 1.19$. For bending moment by pedestrain action $M_{Q2k} = 38.5$ kN·m. Structural importance factor $\gamma_0 = 1.0$. It is needed to calculate effect of combination for structural design.

(1) Basic combination of the ultimate limit state

$$\gamma_0 = 1.0, \gamma_{G1} = 1.2, \gamma_{Q1} = 1.4, \gamma_{Q2} = 1.4, \psi_c = 0.75, \gamma_{L1} = \gamma_{L2} = 1.4$$

①Basic equation is (Adding action effect)

$$M_{ud} = \gamma_0 \left(\sum_{i=1}^{m} \gamma_{G_i} M_{Gik} + \gamma_{Q1} \gamma_{L1} M_{Q1k} + \psi_c \sum_{j=2}^{n} \gamma_{Lj} \gamma_{Qj} M_{Qjk}\right)$$

$$= 1.0 \times (1.2 \times 512 + 1.4 \times 1.4 \times 425 + 0.75 \times 1.4 \times 1.4 \times 38.5)$$

$$= 1503.995 (\text{kN} \cdot \text{m})$$

②Casual combination

$$M_{ud} = \sum_{i=1}^{m} M_{Gik} + \psi_{f1} M_{Q1k} + \sum_{j=2}^{n} \psi_{qj} M_{Qjk}$$
$$= 512 + 0.7 \times 425 + 1.0 \times 38.5$$
$$= 848 (\text{kN} \cdot \text{m})$$

(2) Combination of effects during normal use limit state design

①Effect frequency combination effect:

$$M_{Q1k} = 425/(1+\mu) = 357.1, \psi_{f1} = 0.7, \psi_{q2} = 1.0$$

The combination of effect frequency combination effect design values are

$$M_{fd} = \sum_{i=1}^{m} M_{Gik} + \psi_{f1} M_{Q1k} + \sum_{j=2}^{n} \psi_{qj} M_{Qjk}$$
$$= 512 + 0.7 \times 357.1 + 1.0 \times 38.5 = 800.47 (\text{kN} \cdot \text{m})$$

②Quasi-permanent combination effect:

$$M_{Q1k} = 425/(1+\mu) = 357.1, \psi_{q1} = \psi_{q2} = 0.4$$
$$M_{qd} = \sum_{i=1}^{m} G_{ik} + \sum_{j=1}^{n} \psi_{qj} M_{Qjk}$$
$$= 512 + 0.4 \times 357.1 + 0.4 \times 38.5 = 670.24 (\text{kN} \cdot \text{m})$$

Chapter 3
Calculation of Load-carrying Capacity of Flexural Member

3.1 Basics concepts of flexural members

A flexural member is a component that bears bending moment, shear interaction and a negligible axial force on a cross-section (Figure 3-1). Reinforced concrete beams and plates are typical flexural members in civil engineering which are widely applied to bridge engineering, such as small and medium span beams or slab bridge structures in the load-carrying beams and plates, sidewalk panels, lane decks (Figure 3-2).

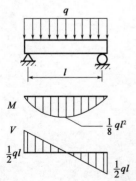

Figure 3-1 Schematic diagram of flexural members

The cross-section of flexural members bears the bending moment M and shear V under load. Therefore the design of flexural members, generally, should meet the following two aspects of requirements:

(1) Due to a bending moment M, a component may fail along a normal cross section (vertical line with orthogonal to surface in plane section of the beam or plate) (Figure 3-3), and so calculation of load-carrying capacity of normal cross-section of a flexural member is needed.

Chapter 3 · Calculation of Load-carrying Capacity of Flexural Member

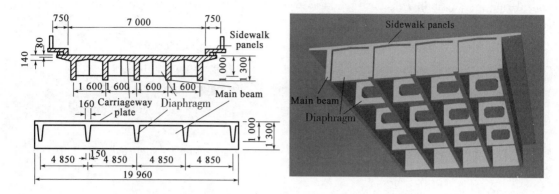

Figure 3-2　Schematic of RC T-beam bridge of flexural members(unit: mm)

(2) Under a bending moment M and shear V together, a member may fail along an inclined cross-section under a range in shear and compression (vertical line or the surface with the oblique plane of a beam or plate) (Figure 3-4). Hence, calculation of load-carrying capacity of an oblique section is needed.

This chapter focuses on calculations of load-carrying capacity of normal cross-section for reinforced concrete beams and plates. The aims are: ① structural design, and it is to determine and design the area and longitudinal steel reinforcement of reinforced concrete beams and plates based on combined design value M_d, and ② structural check.

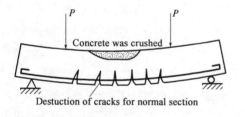

Figure 3-3　Failure mode of a normal section

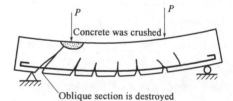

Figure 3-4　Failure mode of an oblique section

3.2　Construction and shape of cross-section for flexural members

1) Construction requirements

Construction requirements are based on specifications or lessons learned. Some of the restrictions on the component size, material strength, grade, variety, steel quantity, layout location, spacing, diameter, connectivity, etc. are made.

Structural design involves many problems. Some problems are non-essential, but can't be ignored. Some problems have not yet been figured out, and can't be solved with a computational theory and quantitatively determined. All the problems that can't be reflected by design formula and identified quantitatively, according to engineering experience and scientific research, considering the possibility of construction as well as requirements of technical and economic, are summarized as construction requirements as:

(1) To offset theoretical disadvantages and effects of uncertain factors.

(2) Construction requirements, such as: requirements of minimum thickness of plate.

(3) Engineering experience, such as: spacing of distribution reinforcement.

(4) Other technical and economic requirements, such as: materials consumption.

The roles of the construction requirements to structural design are:

(1) To provide a reference for preliminary size of a component, such as $h/l = 1/10 \sim 1/18$ and $h/b = 2 \sim 4$;

(2) To be complementary with computation;

(3) To reflect features of actual project design.

2) Section type and size

(1) Section types

The section types of reinforced concrete beams (slabs) are shown in Figure 3-5.

①Types of beams sections, such as rectangular-shaped, T-shaped, I-shaped, box-shaped and inverted T-shaped, etc.

②Types of plates sections, such as flat deck, corrugated and porous plates (the most common aype).

(2) Size requirements

①Size requirements of slab

a. For integrated cast-in-site slab, section width is large [Figure 3-5a)], but a unit width (e.g. 1m for the unit) of rectangular cross-section is used for calculation.

b. For prefabricated slab, slab widths are controlled in 1.0 ~ 1.5m, in order to meet requirements of transportation and hoisting. As the construction conditions are perfect, not only rectangular solid plate is used [Figure 3-5b)], but also more complex shapes with rectangular cross-section hollow plate are applied to reduce self-weight [Figure 3-5c)].

c. Slab thickness h is controlled by the largest bending moment in control cross section and bending stiffness of slab, and needs to meet construction requirements of slab. In order to ensure quality and durability requirements, the **CHBC** provides that minimum thickness of a variety of slabs. For the sidewalk slab (integrated cast-in-site) the minimum thickness should not be less than 80mm, and for the sidewalk slab (precasted) it is 60mm. The top and bottom slab thickness for hollow deck should not be less than 80mm.

②Size requirements of beam

a. Width of cast-in-site rectangular beam b is often taken as 120mm, 150mm, 180mm, 200mm, 220mm and 250mm. Beam width increases of 50mm when beam height $h \leqslant 800$mm, and increases by 100mm when beam high $h > 800$mm.

Ratio of height to width of rectangular beam h/b is generally taken as $2.0 \sim 2.5$.

b. For prefabricated T-shaped beam, the ratio of beam height h and span l (called high-span ratio) is generally defined as $h/l = 1/11 \sim 1/16$, and a larger span often has a smaller ratio.

Width b of beam rib is often taken to be 150 ~ 180mm based on main reinforcement arrangement and shear requirements in beam.

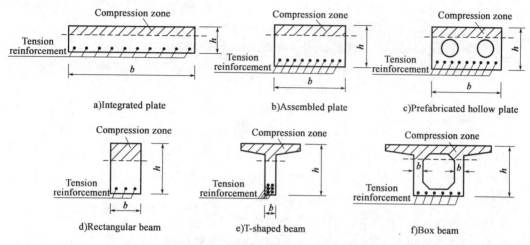

Figure 3-5 Cross-section types of bridge flexural members

c. Thickness of T-shaped cantilever beam end flange is not less than 100mm, and thickness of beam flange is not less than ten percent of beam height.

3) Steel reinforcement layout of flexural members

(1) Basic concepts

①Singly-reinforced and doubly-reinforced sections

When reinforced concrete beam (slab) bears bending moment, section above the neutral axis is in compression while it is in tension below the neutral axis. Longitudinal tensile reinforcement is placed at the tensile region of beam (slab). This component is called singly-reinforced flexural members. If steel reinforcement is also placed in compression zone of cross-section, this component is called doubly-reinforced flexural members.

②Reinforcement ratio $\rho(\%)$

Reinforcement ratio is usually used to measure how much reinforcement is applied on the cross-section. For rectangular cross-sections and T-shaped cross-section, the tensile steel reinforcement ratio $\rho(\%)$ is expressed as

$$\rho = \frac{A_s}{bh_0} \tag{3-1}$$

Where A_s is the cross-sectional area of longitudinal tensile reinforcement; b is the width of rectangular beam or web width of T-beam; h_0 is the effective height of cross-section (Figure 3-6), $h_0 = h - a_s$, where h is the beam height, and a_s is a distance from the center of the longitudinal tensile reinforcement to the tensile edge of a cross-section.

③Concrete cover

As shown in Figure 3-6, the parameter c is known as the concrete cover. Concrete cover is a concrete layer with sufficient thickness, and taken as the shortest distance from the edge of reinforcement to the surface of member. Concrete cover is used to protect the steel reinforcement

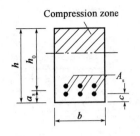

Figure 3-6 Calculation graph of reinforcement ratio ρ

from being affected directly by atmospheric corrosion, chloride, and other environmental factors, and also to ensure a good bond between steel rebars and concrete. It is an important indicator of durability design of reinforced concrete structures.

(2) Steel reinforcement of slabs

①Based on mechanical features, slabs are classified as:

a. Cantilever slab;

b. The slab with four side supports: Based on the mechanical behavior, the slab is composed of one-way slab and two-way slab.

For one-way slab: l_2(long side)/l_1(short side)$\geqslant 2$, the main bar is in one direction.

For two-way slab: $l_2/l_1 < 2$, the main bars are in two directions.

c. The slab with two-side support: one-way slab.

②The type and role of steel reinforcement.

a. Main reinforcement (longitudinal tensile steel reinforcement).

One-way slab steel reinforcement is placed in the tension zone along the slab span direction (short side direction) of the slab. The amount of steel reinforcement is determined by calculation, and needs to meet construction requirements. The diameter of main reinforcement should not be less than 10 mm (lane slab) or 8mm (pavement slab). The main reinforcement located at beam rib is bent according to angle 30° ~ 45° along the central axis line of slab (1/4 to 1/6) of calculation span. However, for main rebars passing the supports without being bent, the number of main reinforcement is not less than three per unit meter slab and should not be less than 1/4 of cross-section of main reinforcements.

At locations of midspan of simply-supported slab and support of continuous slab, the spacing of main rebar inside slab is not larger than 200 mm.

The minimum concrete cover c of steel reinforcement of lane slab (Figure 3-7) should not be less than the nominal diameter of steel rebars and meet the requirements of minimum thickness.

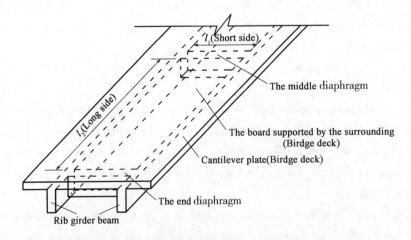

Figure 3-7 Schematic of peripheral supporting bridge slab and cantilever bridge deck

b. Distribution reinforcement.

Distribution reinforcement is connection reinforcement that is set at a certain distance for connection on main reinforcement, and its amount is determined by provisions calculation. The role of the distribution reinforcement is to improve mechanism of main reinforcement, fix main reinforcement, and resist stress induced by temperature and shrinkage of concrete.

Distribution reinforcement should be placed at the top of the main reinforcement (Figure 3-8). According to the **CHBC**, the bar diameter of distribution reinforcement in lane slab is not less than 8 mm, the spacing should not be more than 200 mm, and cross-sectional area should not be less than 0.1% of cross-sectional area of slab. At the location of bending of steel bars, the distribution reinforcement should be set. The bar diameter of distribution reinforcement in sidewalks should be not less than 6mm, and the distance should not exceed 200mm.

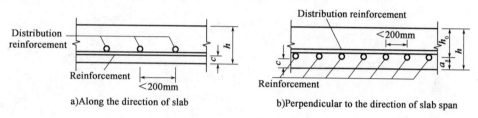

Figure 3-8   Reinforcement in one-way lane slab

For four sides supported two-way slab, two directions of concrete slab (the long side of slab and the short side of slab) both bear bending moments, and so the slab should be set the main reinforcement along both directions.

Prefabricated slab are widely used in the assembly slab. Lane deck of bridge slab is composed of several pieces of precast decks using matching seam filled with concrete to be connected together. From structural analysis of mechanical performance, under the load, it is not the integrated slab width of two directions bearing load, but a series of narrow slab beam that one direction bears load. Concrete between matching-seam of various slabs (known as the concrete hinge) bears shear force, and is also called prefabricated beam slab. Therefore, the steel reinforcement layout of precast slab is similar to that of rectangular beam.

(3) Steel rebars in beam

①Skeleton reinforcement covers binding steel skeleton (Figure 3-9) and welded steel skeleton (Figure 3-10). Binding skeleton is that vertical bars and horizontal steel are fabricated into a spatial steel skeleton (Figure 3-9). The skeleton is that the first longitudinal tensile reinforcement (main reinforcement), bent or inclined steel bars and skeleton reinforcement are welded into plane skeleton, and then stirrup is used to weld plane skeleton into a spatial steel reinforcement skeleton. Figure 3-10 shows a welded plane skeleton.

②Category of steel reinforcement

There are longitudinal tensile steel reinforcements, bent or inclined steel reinforcements, stirrups, skeleton reinforcements and longitudinal steel rebars in a beam.

a. Main reinforcement (longitudinal reinforcement), divided into the main tensile and

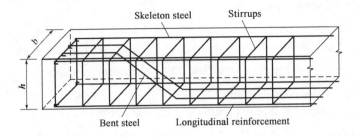

Figure 3-9 Binding steel skeleton

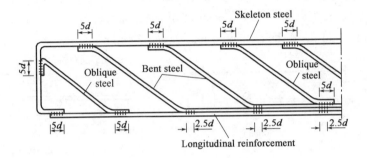

Figure 3-10 Schematic of welded steel skeleton

compressive steel reinforcement, helps concrete in tension and compression to improve the flexural load-capacity of beam. Main longitudinal reinforcement is determined by calculation of load capacity, and meets construction requirements. Bar diameter is normally in the range of 12~32 mm, and ≤40mm.

Arrangement rules for the main steel reinforcement of beam are arranged in a layer as possible to reduce the layers of main reinforcement (to increase arm and so save steel rebar). For binding skeleton, the layers of main bar is not more than 3 layers. Concrete cover at the bottom is larger than bar diameter. When arranged into two or more levels, general principle of arrangement of the upper and lower steel should be aligned and symmetrical.

The minimum concrete cover of steel rebar shall not be less than the nominal diameter of steel, and should meet specifications. For example, when a bridge located at the I types of environmental conditions, the main steel bar (nominal bar diameter d) in reinforced concrete beams and the bottom concrete cover, and the main bar nearest beam side and its concrete cover c (Figure 3-11) should not be less than the nominal bar diameter d and 30mm. When concrete cover in tensile reinforcement zone is greater than 50mm, steel reinforcement grid should be placed with a bar diameter of no less than 6mm and space of no more than 100mm. The minimum concrete cover of steel reinforcement grid is 25mm.

For the spacing of steel reinforcement skeleton, when the layers of reinforcement is three or below three, the clear transverse distance of main bars and the spacing between vertical layers should not be less than 30mm and not less than the bar diameter; while layers of reinforcement are above three, it is not less than 40mm and not less than 1.25 times bar diameter.

For welded steel reinforcement skeleton, there is no gap between the vertical multi-layer main steel reinforcements, and the number of layers is generally not more than 6 layers. Spacing requirements of welded skeleton is shown in Figure 3-11.

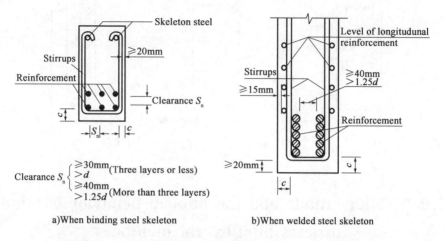

a) When binding steel skeleton   b) When welded steel skeleton

Figure 3-11 Spacing of main steel rebars and concrete cover of beams

The number of the main bars that pass the supports should not be less than 2 and its area should be not less than 20% of area of main tensile reinforcement.

b. Stirrup is determined by the load-carrying capacity of inclined section, meets construction requirements, and must be set within the beam. Role of stirrup is to enhance the shear capacity of beams, to form skeleton with longitudinal reinforcement, and to fix the location of the main bar. The bar diameter of stirrup $d \geq 8$mm and is greater than 1/4 of bar diameter of the main steel reinforcement. Spacing can be found in specifications.

The forms of stirrups include opened ones, closed ones, four limbs, two limbs, and one limb (Figure 3-12). For the limb number: single limb is generally not used, double limbs uses generally in single hoop, and four limbs in the hoop in tensile reinforcement on each layer adopt more than 5 layers or the hoop reinforcement in compression each layer adopt more than 3 layers.

c. The number and layout of oblique tendons (bent reinforcement), as shown in Figure 3-10, are determined by the load-carrying capacity of inclined section, and meet construction requirements. Bending steel reinforcement in beam is required based on the location and angle of bend to the upper beam and meets the requirements of the anchorage of steel reinforcement, and inclined bar is designed to set the diagonal reinforcement. Bending angle is generally 45°.

d. Skeleton steel is construction steel reinforcement. Based on requirement of construction, its function is to fix stirrup and to bind main steel reinforcement and stirrup to be a skeleton. Bar diameter is usually between 10 ~ 14mm.

e. Longitudinal and horizontal steel rebars are construction steel reinforcement. Its layout is designed according to construction requirements. Its role is to resist stress induced by thermal and concrete shrinkage and to prevent concrete cracks induced by shrinkage and temperature changes in the concrete. Bar diameter varies from 6mm to 8mm, and when the beam is high, longitudinal steel reinforcement is arranged along the both sides of beam ribs and the lateral side of stirrups. Area is $(0.001 \sim 0.002)bh$. Spacing in a tension zone in the web should not be greater than the width b and should not be greater than 200mm, and in compression zone it should not exceed 300mm.

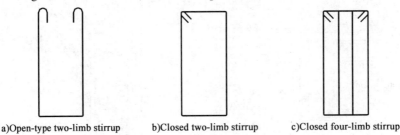

a) Open-type two-limb stirrup   b) Closed two-limb stirrup   c) Closed four-limb stirrup

Figure 3-12   Forms of stirrups

3.3 Failure mode and mechanical behavior of whole process for flexural members

1) Experimental research

Reinforced concrete is composed of composite materials with different physical and mechanical properties. It is non-homogeneous, non-elastic material and is not consistent with Hooke Law after the loading reaching to some extent (σ and ε are not proportional). Theoretical and experimental results are very different, so the calculation of reinforced concrete must be established based on an experimental investigation.

In order to address the mechanical behavior and deformation of normal cross-section of beam under load, as shown in Figure 3-13, a simply-supported reinforced concrete beam with a span of 1.8m is used as a test beam. The size of cross-section of a longitudinal beam is $b \times h = 100\text{mm} \times 160\text{mm}$.

(1) Experimental purposes

To understand the failure process of reinforced concrete beams, deformation features and mechanical behavior of the cross section under limit load are tested to develop strength calculation formula of RC beams.

(2) Introduction of test

The longitudinal bar diameter is $2\phi10$. Measured prism compressive strength of concrete beam $f_c = 20.2\text{MPa}$, and the measured tensile strength of longitudinal reinforced $f_s = 395\text{MPa}$.

A tested beam is a singly reinforced rectangular beam, using hydraulic jacks to impose two concentrated loads F, and diagram of bending moment and shear force is shown in Figure 3-13. The shear is zero (ignoring beam weight) in CD section of the beam, while the bending moment is a constant, known as the "pure bending" section, which is the main target of experimental research. The concentrated force F is measured by force sensor. Deflection is measured with dial indicator, placing at the mid-span of the test beam under the E point. Concrete strain is measured using a hand-holding strain gauge of 200mm, arranging measured points of a, b, c, d and e along the height of beam. It is needed to record the applied force, deflection and strain, and observe cracking.

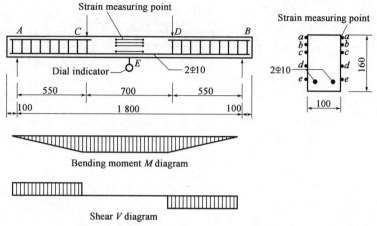

Figure 3-13   Layout of a tested beam (unit: mm)

2) Mechanical features of the tested beam

(1) F-w curve

From F-w curve of the tested beam (Figure 3-14), it can be seen that there are two characteristic points and the mechanism and the deformation process of the beam are divided into three phases.

①Phase Ⅰ: Overall working phase without cracks in beams.

②Phase Ⅱ: Working with cracks.

③Phase Ⅲ: Failure phase, cracks propagate sharply. Steel stress maintains the level of yield strength.

Three characteristic points, for example: the end of phase Ⅰ (presented by: Ⅰ$_a$), cracks will appear soon; the end of phase Ⅱ (presented by Ⅱ$_a$), longitudinal steel reinforcement yields; and the end of phase Ⅲ (indicated with Ⅲ$_a$), concrete in compression zone of beam is crushed, and the entire cross-section of beam is damaged.

(2) Stress-strain curve

Figure 3-15 shows sectional average strain of concrete under various loads and corresponding stress distribution. As shown in Figure 3-15, with an increase of loading the strain of concrete is also increasing, but the strain diagram is basically a symmetrical triangle. It is also found that the neutral axis gradually increases as the load increases.

Sectional stress must be calculated from stress-strain relationship (Hooke Law). The stress in Figure 3-15b) is calculated based on the measured value as indicated in Figure 3-15a) (a, b, c, d, e measured points) and the stress-strain relationship in Figure 3-16 along the section from the top to the bottom.

(3) Mechanical behavior of each phase

The stress and deformation of three phases for the tested beam have the following features.

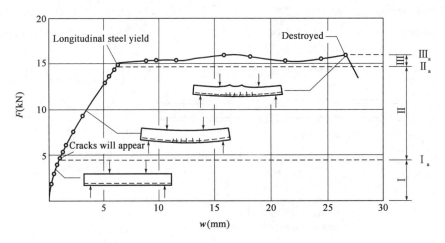

Figure 3-14 Load-deflection (F-w) diagram of the tested beam

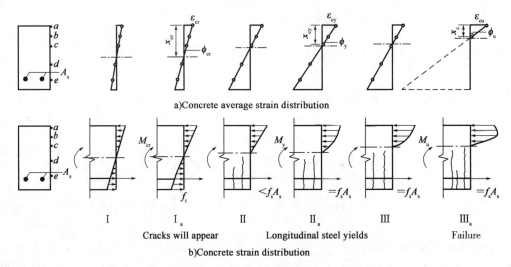

Figure 3-15 Strain and stress schematic of all phases of the tested beam

For phase I: whole cross section of concrete beam works, and concrete compressive stress and tensile stress of concrete basically follow triangular distributions. Longitudinal steel carries tensile stress. Concrete is in the elastic stage, that is to say, concrete stress is proportional to strain.

End of phase I: The stress of concrete in compression zone is still in a triangular distribution. However, the plastic deformation of concrete in tensile zone develops rapidly, and tensile strain grows quickly. According to the stress-strain curve in Figure 3-16c), stress diagram of concrete in tensile zone appears a curve shape. Meanwhile, tensile strain of concrete in tensile zone is nearly ultimate tensile strain, and the tensile stress reaches the tensile strength of concrete. This means cracks will appear and the corresponding bending moment imposed on the beam is M_{cr}.

For phase II: Bending moment under load reaches M_{cr}, the weakest section of tensile strength of concrete on the beams shows cracks. At this time, there are cracks in the cross section, and concrete in cracked sections quits working. This results in tension carried by concrete transferring to the steel reinforcements, and so there is significant stress redistribution. Tensile stress of steel reinforcement increases with the increase of load. Concrete compressive stress appears no longer a triangle, forming slightly in a bent and curved shape, and neutral axis moves upward.

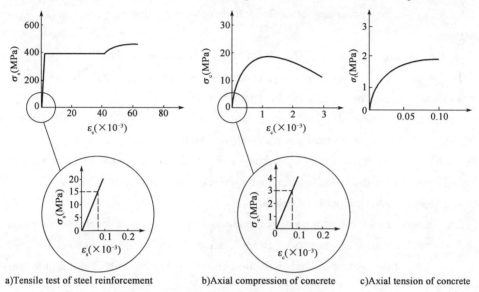

a) Tensile test of steel reinforcement b) Axial compression of concrete c) Axial tension of concrete

Figure 3-16 Stress-strain relationship of material

End of phase II: Tensile strain of steel reinforcement reaches a value that steel rebar yields, which indicates that steel stress reaches its yield strength.

For phase III: the tensile strain of steel reinforcement increases rapidly, but its tensile stress generally remains a level of yield strength (for obvious yield platform). Then, crack propagates rapidly, and the neutral axis goes up continually. Concrete compression zone reduces, and concrete compressive stress increases. The compressive stress diagram appears in a fully curved shape.

End of phase III: The concrete compressive strain in top zone reaches its ultimate compressive strain, and compressive stress block shows a curved shape. Moreover, the maximum compressive stress is not located at the top edge but on the edge of the concrete section of the compressive strain block, with a little down from the top edge. This is determined by the concrete stress-strain diagram in compression. The compressive strength of concrete in compressive zone exhausts, and horizontal cracks appear at both sides in a certain section. Then concrete is crushed, and beam fails. At this stage, the tensile stress of steel reinforcement remains in the level of yield strength.

(4) Mechanical behavior of appropriately-reinforced beam

Through experimental research, the mechanical characteristics of singly-reinforced concrete beams can be properly drawn:

①Comparison with homogeneous elastic beams.

Homogeneous elastic beams: M is proportional to σ ($\sigma = M/I$). At the location of neutral axis, the shape of stress and strain diagrams (straight line distribution) keeps unchangeable, and only quantitative changes happen.

Reinforced concrete beams: the value of σ changes as M increases. At location of the neutral axis, stress diagram changes. The beam works with cracks at most of the stages. M is not proportional to the σ and w (rigidity EI decreases).

②Cross-section remains plane after flexural deformation (that is consistent with assumption of plane section).

Concrete ultimate compressive strain is taken as 0.003 in the **CHBC**.

③The neutral axis increases with an increase of loading, which is resulted by static equilibrium under stress change and cracking condition.

For phase Ⅱ, the neutral axis changes a little, and the arm of internal force keeps basically unchangeable. Increasing σ_s is used for resisting the external bending moments.

For phase Ⅲ, sectional stress is unchangeable. In order to maintain cross-sectional static equilibrium, the arm of the internal force increases and the neutral axis moves up.

④The characteristics of plastic damage.

Failure characteristic of appropriately reinforced beam is that steel reinforcement yields firstly, and then concrete is crushed. Before failure, steel reinforcement has large plastic elongation, cracks fully develop, and deflection increases dramatically with significant damage characteristics.

These characteristics reflect two basic aspects of the mechanical properties of concrete structures, namely, the tensile strength of concrete is much smaller than the compressive strength, and cracks appear under small load induced. Concrete is an elastic-plastic material. When the stress exceeds a certain limit, plastic deformation will occur.

It can be obtained that the calculation basis of reinforced concrete beams as:

①Point $Ⅰ_a$ can be used as the basis for calculating crack capacity of flexural member;

②Phase Ⅱ is used as a basis for deformation and crack propagation over service time;

③Point $Ⅲ_a$ can be used as the basis for calculating the load-carrying capacity of limit state.

3) Failure modes of flexural member

Reinforced concrete flexural beam has two failure properties: one is plastic failure (ductile failure), which refers to the structure or components have significant deformation or other signs before failure [Figure 3-17a)], and the other is brittle failure, which refers to the structure or components have no significant deformation or other signs before failure [Figure 3-17b), c)]. Based on the experimental testing, failure mode of reinforced concrete beam is dependent on reinforcement ratio ρ, steel strength grade and concrete strength grade. For the commonly used hot-rolled steel and ordinary concrete, the main failure mode is affected by reinforcement ratio ρ. Therefore, in accordance with steel reinforcement in concrete beam and corresponding failure characteristics, it is expected that there are three failure forms of cross-section.

(1) Appropriately-reinforced beam—plastic failure (ductile failure)

Tensile steel reinforcement of beam firstly reaches yield strength, and the stress remains stable while the strain increases significantly until compressive strain of concrete at the edge of the

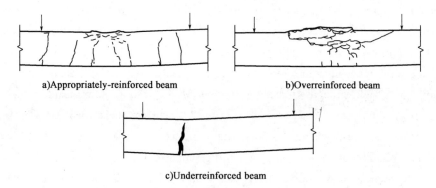

a) Appropriately-reinforced beam b) Overreinforced beam

c) Underreinforced beam

Figure 3-17 Failure modes of beam

compression zone reaches its limit. Longitudinal and horizontal cracks in compression zone appear and then concrete crushes. Before failure of a beam, crack of beam propagates rapidly with large deflection, and cross-section of a beam has a great plastic deformation. Then the beam has obvious sign of damage which is plastic failure.

The sectional curvature ϕ of a flexural member is an indicator of comprehensive strain, stiffness, and deformation capacity, and its expression is $\phi = \dfrac{\varepsilon_c}{\xi_i h_0}$ (Figure 3-15), where ε_c is the concrete strain of the edge of a section; h_0 is the effective height of a cross-section; ξ_i is the relative compression height, while the depth of compression zone is $x_i = \xi_i h_0$. As shown in Figure 3-15, ϕ_y is the sectional curvature when steel yields, ϕ_u is the ultimate curvature when the beam fails. The sharp increase of ε_c and small $\xi_i h_0$ make ϕ_u much greater than ϕ_y. That is to say, a higher $\phi_u - \phi_y$ indicates lower component stiffness, resulting in deformation increasing but better ability to bear deformation. This is called ductility. Ductility is an important feature for a component bearing earthquake and impact load.

(2) Overreinforced beam (ρ is large or when high strength steel is used) —brittle failure

When sectional steel ratio ρ of a beam increases, the steel stress increases slowly. Concrete stress in compressive zone grows rapidly. If ρ is larger, the bending moment M_y corresponding to longitudinal steel yielding is close to M_u referring to bending moment of beam failure. It means that phase Ⅲ becomes shorter. When the ρ increases to make $M_y = M_u$, it means the tension steel yields and compression concrete crushes simultaneously. This is known as balance failure or boundary failure, and the corresponding ρ value is called the maximum reinforcement ratio ρ_{max}.

When the actual reinforcement ratio $\rho > \rho_{max}$, the compression concrete of the beam is crushed, while the tension stress of steel reinforcement has not yet reached its yield strength. Before failure, deflection and the section curvature of beam have no obvious turning points. The crack in tension zone is not wide and its extension is not high. There is no obvious warning sign of failure. Damage happens suddenly where it is brittle failure. This failure is also called overreinforced beam failure.

The failure of overreinforced beam is to use up the concrete compressive strength, and the tensile strength of steel has not been fully used. Therefore, M_u corresponding to failure of overreinforced beam is independent on steel reinforcement but depends on the concrete compressive strength.

(3) Underreinforced beam (ρ is small)—brittle failure

When the reinforcement ratio ρ of a beam is small, after beam tension concrete cracks, steel stress is close to its yield strength. Cracking moment M_{cr} is close to the moment M_y corresponding to tensile steel yielding. It means phase Ⅱ is short. When ρ reduces to $M_{cr} = M_y$, once cracks occur, reinforcement stress reaches yield strength immediately, and the reinforcement ratio is called minimum reinforcement ratio ρ_{min}. When the actual reinforcement ratio $\rho < \rho_{min}$, once beam tension concrete cracks, tensile steel reaches yielding. Steel stress quickly enters the strengthening phase through yielding platform, and there is only a concentrated crack in beam. Crack is wide and develops to a certain height along the beam. At the same time the compression concrete has not been crushed, however crack is not very wide and deflection is not very small. Steel reinforcement is even pulled off. As the beam fails suddenly, it is a brittle failure. Beam that behaves this way is called the underreinforced beam.

Flexural load-carrying capacity of underreinforced beams depends on the tensile strength of concrete, and in bridge engineering it is not allowed to be used.

In summary, the failure characteristics of normal cross section of flexural member vary with reinforcement ratio, and features are: ①When reinforcement ratio is too small, the component failure strength depends on the concrete tensile strength and cross sectional size, and failure appears brittle; ②For excessive reinforcement, the reinforcement can't be fully used, component failure strength depends on the size of the compressive strength of concrete and cross sectional size, and failure also appears brittle. Reasonable amount of reinforcing bars should be in a range of two limits, and so the failure of overreinforced or undereinforced can be avoided.

3.4 Basic calculation principles of load-carrying capacity of normal section for flexural member

Based on the experimental results, the following laws are provided to simplify the calculations.

1) Basic assumptions

(1) Plane section assumption

Under all levels of load, the average strain of cross section keeps a straight line distribution, and the strain of any point on a cross section is proportional to the distance between the point and the neutral axis. This assumption is approximate, but the error resulting from the assumption is small. Therefore, it fully meets the requirements for engineering calculations. Plane section assumption provides the geometrical deformation of flexural capacity of reinforced concrete beam, and can enhance the logical and rational calculation methods. It results in that the formula covers more clear physical meaning.

(2) No consideration of concrete tensile strength

At the cracked sections, most tensile concrete has not worked, and a small part of concrete near the neutral axis still undertakes tensile stress. Tensile stress is small and the arms of internal force are also small. Hence the internal moment is small as well. Then its contribution calculation can be neglected thus simplifying the calculation procedure.

(3) σ-ε curve

①Concrete σ-ε relation curve. The stress-strain curves of concrete have different calculation schemas, and the curve consisting of the parabolic curve and the horizontal line is widely used. Figure 3-18 shows the typical concrete stress-strain curve in CEB-FIP standards and the **CHBC**. The ascending section OA is parabola, and straight line segment AB is horizontal. Its expression is

$$\left. \begin{aligned} \sigma &= \sigma_0 \left[2\frac{\varepsilon}{\varepsilon_0} - \left(\frac{\varepsilon}{\varepsilon_0}\right)^2 \right], \quad \varepsilon \leq \varepsilon_0 \\ \sigma &= \sigma_0, \quad \varepsilon > \varepsilon_0 \end{aligned} \right\} \quad (3\text{-}2)$$

Where σ_0 is the peak stress which is taken as $0.85 f_{ck}$; f_{ck} is the standard cylinder compressive strength of concrete, and 0.85 is the reduction factor. Meanwhile, $\varepsilon_0 = 0.002$ is the strain of the concrete corresponding to the peak stress. The strain of point B $\varepsilon_{cu} = 0.0035$, and ε_{cu} is the concrete strain corresponding to the peak stress.

②Steel reinforcement σ-ε curve. This model uses simplified ideal elastoplastic stress-strain relationship (Figure 3-18). OA section is the elastic phase, and reinforced stress of point A corresponds to yield strength σ_y and the corresponding strain is yielding strain ε_y. The slope of OA line is the elastic modulus E_s. Section AB is the plastic stage, and point B corresponds to the start strain ε_k of the strengthened segment. From Figure 3-18, the stress-strain relations expression for normal steel reinforcement is

$$\left. \begin{aligned} \sigma_s &= \varepsilon_s E_s, \quad 0 \leq \varepsilon_s \leq \varepsilon_y \\ \sigma_s &= \sigma_y, \quad \varepsilon_s > \varepsilon_y \end{aligned} \right\} \quad (3\text{-}3)$$

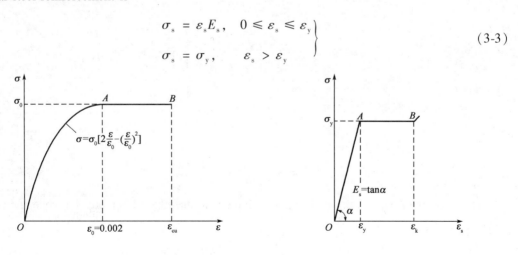

Figure 3-18 σ-ε curves for concrete and steel reinforcement

2) Equivalent rectangular stress block of concrete in compression zone

The premise of calculating flexural capacity M_u of RC beam is to know the concrete compressive stress distribution when structure fails, especially, and to capture the force C and its position of compression concrete (Figure 3-19).

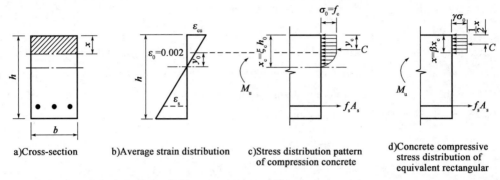

a) Cross-section   b) Average strain distribution   c) Stress distribution pattern of compression concrete   d) Concrete compressive stress distribution of equivalent rectangular

Figure 3-19 Equivalent rectangular stress diagram of compression concrete

When a RC beam damages, concrete compressive stress distribution is similar to the concrete stress-strain curves (in compression), and the model in Figure 3-19 is taken as concrete stress-strain curve model, namely, when $\varepsilon \leqslant \varepsilon_0$, $\sigma = \sigma_0 \left[2 \dfrac{\varepsilon}{\varepsilon_0} - \left(\dfrac{\varepsilon}{\varepsilon_0} \right)^2 \right]$. When $\varepsilon > \varepsilon_0$, $\sigma = \sigma_0$. As shown in Figure 3-19b), the distance between point ε_0 the neutral axis is y_0.

As shown in Figure 3-19c), it is obtained that from the plane section assumption concrete compression zone height $x_c = \xi_c h_0$, and $\varepsilon/\varepsilon_0 = y/y_0$ so $y_0 = \varepsilon_0 \xi_c h_0 / \varepsilon_{cu}$.

As shown in Figure 3-19, the rectangular section as an example is used to compute force C of compressive stress in compression concrete and its location y_c. Concrete stress-strain curve of compression zone includes two segments, and the integral of compressive stress is used to compute force C as

$$C = \int_0^{\xi_c h_0} \sigma(\varepsilon) b \, dy$$

$$= \int_0^{y_0} \sigma_0 \left[\dfrac{2\varepsilon}{\varepsilon_0} - \left(\dfrac{\varepsilon}{\varepsilon_0} \right)^2 \right] b \, dy + \int_{y_0}^{\xi_c h_0} \sigma_0 b \, dy$$

Where $\varepsilon/\varepsilon_0 = y/y_0$, and $y_0 = \dfrac{\varepsilon_0}{\varepsilon_{cu}} \xi_c h_0$.

Therefore, the force C is

$$C = \sigma_0 \xi_c h_0 b \left(1 - \dfrac{1}{3} \dfrac{\varepsilon_0}{\varepsilon_{cu}} \right) \qquad (3\text{-}4)$$

The y_c is the distance from the location of force C to the edge of compression zone and can be calculated using

$$y_c = \xi_c h_0 - \dfrac{\int_0^{\xi_c h_0} \sigma(\varepsilon) b y \, dy}{C}$$

Using integral calculation we have

$$y_c = \xi_c h_0 \left[1 - \frac{\frac{1}{2} - \frac{1}{12}\left(\frac{\varepsilon_0}{\varepsilon_{cu}}\right)^2}{1 - \frac{1}{3}\frac{\varepsilon_0}{\varepsilon_{cu}}} \right] \quad (3\text{-}5)$$

Obviously, using concrete compressive stress-strain $\sigma = \sigma(\varepsilon)$ curve to compute the force C and location y_c is not easy. Simplified method that uses the equivalent rectangular stress block of concrete instead of the actual stress diagram is adopted in calculation.

Using the equivalent rectangular stress block of concrete instead of the actual stress diagram must satisfy the two following conditions:
(1) To keep the position of force C unchangeable;
(2) To keep the size of force C unchangeable.

Introducing two dimensionless parameters β and γ

$$\beta = x/x_c, \quad \gamma = \sigma/\sigma_0$$

From Figure 3-19d)

$$C = \gamma \sigma_0 b x = \gamma \sigma_0 b \beta x_c \quad (3\text{-}6)$$

The acting point of force C is

$$y_c = \frac{x}{2} = \frac{1}{2}\beta x_c \quad (3\text{-}7)$$

Based on the equivalent conditions of Equation (3-4) = Equation (3-6) and Equation (3-5) = Equation (3-7), we have

$$\beta = \frac{\left[1 - \frac{2}{3}\frac{\varepsilon_0}{\varepsilon_{cu}} + \frac{1}{6}\left(\frac{\varepsilon_0}{\varepsilon_{cu}}\right)^2\right]}{1 - \frac{1}{3}\frac{\varepsilon_0}{\varepsilon_{cu}}} \quad (3\text{-}8)$$

and

$$\gamma = \frac{1}{\beta}\left(1 - \frac{1}{3}\frac{\varepsilon_0}{\varepsilon_{cu}}\right) \quad (3\text{-}9)$$

When determining ε_0 and ε_{cu}, the equivalent rectangular stress block in Figure 3-19d) of compression concrete replaces actual stress block in Figure 3-19c). If taking $\varepsilon_0 = 0.002$, concrete ultimate compressive strain $\varepsilon_{cu} = 0.0033$. This value is not based on the CEB-FIP code as $\varepsilon_{cu} = 0.0035$, but from Equations (3-8) and (3-9), we have $\beta = 0.8095$, $\gamma = 0.9608$. The height of equivalent rectangular stress block $x = 0.8095 x_c$, and the equivalent stress value $\gamma \sigma_0 = 0.9608 \sigma_0$. For the ultimate compressive strain ε_{cu} and the corresponding coefficient β of the edge of the compression concrete of flexural member, the **CHBC** specifies its values based on concrete strength levels as listed in Table 3-1. For the equivalent rectangular stress block of compression concrete, combining with experimental data at home and abroad, the **CHBC** takes the equivalent rectangular stress values $\gamma \sigma_0$ as f_{cd}, where f_{cd} is the axial compressive strength design value.

Concrete ultimate compressive strain ε_{cu} and the coefficient value β Table 3-1

Concrete strength	C50 and below	C55	C60	C65	C70	C75	C80
ε_{cu}	0.003 30	0.003 25	0.003 20	0.003 15	0.003 10	0.003 05	0.003 00
β	0.80	0.79	0.78	0.77	0.76	0.75	0.74

3) The relative bound height of compression zone

Bound failure is that the tensile reinforcement reaches the yield strain ε_y and begins to yield and meanwhile the compression concrete zone edge also reaches its ultimate compressive strain ε_{cu} and fails. As shown in Figure 3-20, the height of compression zone is $x_c = \xi_b h_0$. ξ_b is the relative bound height of compression zone.

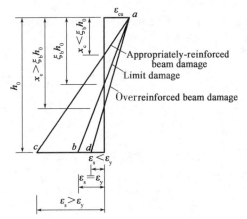

Figure 3-20 Relative bound height of compression zone

The failure of appropriately reinforced concrete flexural members begins the yielding of tensile steel reinforcement and after a period of deformation process, concrete strain reaches ultimate strain ε_{cu}. Then member fails. At this time the tensile strain of tension reinforcement satisfies $\varepsilon_s > \varepsilon_y$, and so strain distribution of appropriately reinforced beam is the line ac in Figure 3-20. At this time the height of compression zone $x_c < \xi_b h_0$.

The failure of overreinforced concrete flexural members begins that concrete strain reaches ultimate strain ε_{cu}. Then member fails. At this time the tensile strain of tension reinforcement is less than yielding strain, and so strain distribution of overreinforced beam is the line ad in Figure 3-20. At this time the height of compression zone $x_c > \xi_b h_0$.

It is seen from Figure 3-20, bound damage is a boundary of overreinforced concrete beam and appropriately reinforced concrete beam. When the actual cross-sectional height of compression zone $x_c > \xi_b h_0$, it is overreinforced beam. When $x_c < \xi_b h_0$, it is appropriately reinforced beam. Therefore, the ξ_b is used as boundary condition, where x_b is the height of concrete compression zone when bound failure occurs, obtained from the plane section assumption.

For the limit height, $x_c = \beta x_b$, the corresponding ξ_b should be

$$\xi_b = \frac{x_c}{h_0} = \frac{\beta x_b}{h_0}$$

By the geometrical relation (plane section assumption)

$$\frac{x_b}{h_0} = \frac{\varepsilon_{cu}}{\varepsilon_{cu} + \varepsilon_y} \qquad (3\text{-}10)$$

Substituting $x_b = \dfrac{\xi_b h_0}{\beta}$ and $\varepsilon_y = \dfrac{f_{sd}}{E_s}$ into Equation (3-10), the bound height of the compression zone based on equivalent rectangular stress block is

$$\xi_b = \frac{\beta}{1 + \dfrac{f_{sd}}{\varepsilon_{cu} E_s}} \qquad (3\text{-}11)$$

Equation (3-11) is a basis of determining the relative height of concrete compression zone ξ_b in the **CHBC**. In Equation (3-11), f_{sd} is the design value of tensile strength of steel rebar. Accordingly, based on the **CHBC**, for the design values of axial compressive strength of concrete, design values of strength and elastic modulus of various steel reinforcement, ξ_b can be obtained from Table 3-2.

Relative bound height of compression zone ξ_b Table 3-2

Steel	Concrete			
	C50 and below	C55, C60	C65, C70	C75, C80
HPB300	0.58	0.56	0.54	—
HRB400, HRBF400, RRB400	0.53	0.51	0.49	—
HRB500	0.49	0.47	0.46	—
Steel strand, steel wire	0.40	0.38	0.36	0.35
Prestressing thread bar	0.40	0.38	0.36	—

Note: For different types of tension steel reinforcement in a flexural member, the ξ_b value should be chosen corresponding to the smaller one of the various steel reinforcements.

4) Minimum reinforcement ratio ρ_{min}

In order to avoid failure mode of underreinforced beam, it is needed to determine the minimum reinforcement ratio ρ_{min}. Minimum reinforcement ratio is the bound of underreinforced beams and appropriately reinforced beams. When the beam reinforcement rate ρ_{min} gradually decreases, the working characteristics of the beam change from reinforced concrete structure to plain concrete structure. Therefore, ρ_{min} is determined by the rule that when reinforced concrete beams fail using minimum reinforcement ratio ρ_{min}, the flexural capacity of normal section M_u is equal to the standard value of cracking moment of the plain concrete beams with same section size and the same material.

Based on the results from the above principles, taking into account temperature changes, shrinkage stress, as well as past design experience, the *General Code for Design of Highway Bridges and Culverts* (*JTG D60—2015*) provides the minimum longitudinal reinforcement ratio $\rho_{min}(\%)$ of flexural reinforced concrete beam (appendix Table 1-9).

For reinforcements on the bending side

$$\rho_{min} = \max\left\{0.2, 45\frac{f_{td}}{f_{sd}}\right\}\%$$

3.5 Singly-reinforced rectangular section for flexural members

1) Basic formulas and application conditions

According to the basic principles of flexural capacity, calculation schematic of a singly reinforced rectangular section for flexural members is given in Figure 3-21.

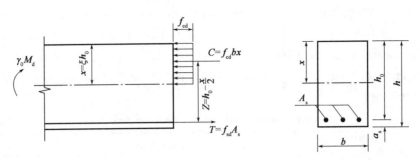

Figure 3-21 Schematic of load-carrying capacity computation of singly reinforced rectangular section for flexural members

Basic principle is that the most disadvantage effects $\gamma_0 M_d$ of basic combination of load on calculated section for a flexural member should not exceed the load-carrying capacity of cross section (resistance) M_u, where it is $\gamma_0 M_d \leq M_u$.

From Figure 3-21 the basic calculation formula (basic equation) of singly reinforced flexural members with rectangulars cross section can be derived.

Using the equilibrium condition of internal forces being of zero, i.e. $T + C = 0$, we have

$$f_{cd} bx = f_{sd} A_s \tag{3-12}$$

From the equilibrium conditions of the sum of bending moment resulting from the point of tension reinforcement force T on a section being zero, the balance equation can be obtained as

$$\gamma_0 M_d \leq M_u = f_{cd} bx \left(h_0 - \frac{x}{2} \right) \tag{3-13}$$

From the equilibrium conditions of the sum of bending moment resulting from the point of compression concrete force C on a section being zero, the balance equation can be obtained as

$$\gamma_0 M_d \leq M_u = f_{sd} A_s \left(h_0 - \frac{x}{2} \right) \tag{3-14}$$

Where M_d is the design value of combinational moment of a cross-section calculation; γ_0 is the structural importance factor; M_u is the flexural capacity of computed section; f_{cd} is the design value of concrete compressive strength; f_{sd} is the design value longitudinal tensile strength of steel reinforcement; A_s is the cross-sectional area of tension steel rebars; x is the computed height of compression zone according to the equivalent rectangular stress diagram; b is the width of cross-section, and h_0 is the effective height of a section.

Two basic equations are Equations (3-12) and (3-13) or (3-14).

Equations (3-12), (3-13) and (3-14) are only suitable for appropriately reinforced beams, while not applied to overreinforced beams and underreinforced beams. Because when the overreinforced beam fails, the actual tensile stress σ_s does not reach the design value of tensile strength, and so the f_{sd} can not be used. Therefore, the formula applying condition is:

(1) In order to avoid overreinforced beams, the height of compression zone x should satisfy

$$x \leq \xi_b h_0 \tag{3-15}$$

Where ξ_b is the relative limit height of compression zone which can be referred to the concrete strength and types of steel reinforcement in Table 3-2.

From Equation (3-12), the calculated depth of compression zone x can be represented as

$$x = \frac{f_{sd}A_s}{f_{cd}b} \qquad (3\text{-}16)$$

So the relative height of compression zone ξ is

$$\xi = \frac{x}{h_0} = \frac{f_{sd}}{f_{cd}} \frac{A_s}{bh_0} = \rho \frac{f_{sd}}{f_{cd}} \qquad (3\text{-}17)$$

From Equation (3-17), ξ not only reflects the reinforcement ratio ρ, but also reflects the strength ratio of materials, and so ξ is also called reinforcement characteristic value, having a more general meaning than the parameter ρ.

When $\xi = \xi_b$, the maximum reinforcement ratio ρ_{max} of appropriately-reinforced beam is

$$\rho_{max} = \xi_b \frac{f_{cd}}{f_{sd}} \qquad (3\text{-}18)$$

Clearly, reinforcement ratio ρ of appropriately-reinforced beam should be met

$$\rho \leqslant \rho_{max} \left(= \xi_b \frac{f_{cd}}{f_{sd}} \right) \qquad (3\text{-}19)$$

Equations (3-19) and (3-15) have the same meaning, whose aim is to prevent the excessive tensile steel reinforcement from forming an overreinforced beam. If Equation (3-19) is satisfied, Equation (3-15) is also met. In the actual calculation, Equation (3-15) is adopted.

(2) In order to avoid underreinforced beams, the reinforcement ratio ρ should satisfy

$$\rho \geqslant \rho_{min} \qquad (3\text{-}20)$$

2) Calculation method

Calculation of normal section of reinforced concrete beams is usually conducted on the controlling section of a component. For equal section of flexural members, the controlling section corresponds to the largest design value of combinational moment. For variable section of flexural members, except the design value of maximum combinational bending moment, controlling sections also include sections that have relatively small sizes but relatively larger design values of combinational bending moment.

Calculation of load-carrying capacity of flexural member mainly includes design of section and checking of section.

(1) Sectional design

①Design items

The design items should include selection of materials, determining sectional sizes, and reinforcement calculation.

Section should be designed to meet the requirement that the bending moment capacity M_u is larger than calculated value $\gamma_0 M_d$. The load-carrying capacity determining the quantity of cross-sectional steel reinforcement is at least equal to the calculated bending moment $\gamma_0 M_d$, and so the basic formula is used for cross-section design where it is generally to take $M_u = \gamma_0 M_d$ for calculation.

②Design steps

Two possible design situations that may arise should be considered.

a. Given: the design values of moment M_d, concrete and steel materials strength f_{cd} and f_{sd}, cross-sectional dimensions b and h.

Unknown: area of steel rebar A_s.

Analysis: namely two basic equations for solving two unknowns x and A_s. The given environmental conditions are used to determine the minimum concrete cover (appendix Table 1-8). γ_0 is based on a given safety level.

Solution:

(a) Assume a_s.

For a beam, generally $a_s = 40$mm (a layer) and $a_s = 65$mm (two layers).

For a slab, generally $a_s = 25$mm or 35mm.

(b) Solve x or ξ using basic equations.

From the basic equation $\gamma_0 M_d = f_{cd} b x \left(h_0 - \dfrac{x}{2} \right)$ (quadratic equation), we have

$$x = h_0 - \sqrt{h_0^2 - \dfrac{2\gamma_0 M_d}{f_{cd} b}} \quad \text{or} \quad \xi = \dfrac{x}{h_0} = 1 - \sqrt{1 - \dfrac{2\gamma_0 M_d}{f_{cd} b h_0^2}}$$

(c) Check $x \leqslant \xi_b h_0$ or $\xi \leqslant \xi_b$.

(d) From the basic equations $f_{cd} b x = f_{sd} A_s$, then

$$A_s = \dfrac{f_{cd} b x}{f_{sd}}$$

(e) Select bar diameter and number.

(f) Check $\rho \geqslant \rho_{min}$, $\rho = \dfrac{A_s}{b h_0}$ (note that A_s is the actual amount of reinforcing steel). If $\rho < \rho_{min}$ taking $\rho = \rho_{min}$, and then $A_s = \rho_{min} b h_0$ (that is construction reinforcement).

(g) Check whether the calculated a_s is equal to the actual a_s or not. If the error is large, re-calculation is needed. If $a_{s,\text{assume}} < a_{s,\text{actual}}$, then $h_{0,\text{assume}} > h_{0,\text{actual}}$. It means that selection is not safe.

(h) Draw reinforcement diagram and check the construction requirements (bar spacing, etc).

b. Given: the design values of moment M_d, concrete and steel materials strength f_{cd} and f_{sd}.

Unknown: b, h, x and A_s.

Analysis: 4 unknowns (x, b, h, A_s).

Solution:

(a) Parameter b is determined by the structural construction requirements (b has little effect on the carrying capacity).

Assume ρ.

For a slab, general economic reinforcement ratio $\rho = 0.3\% \sim 0.8\%$.

For a rectangular beam, $\rho = 0.6\% \sim 1.5\%$.

For a T beam, $\rho = 2\% \sim 3.5\%$.

(b) $\xi = \rho \dfrac{f_{sd}}{f_{cd}}$, $x = \xi h_0$.

(c) From the basic equations $\gamma_0 M_d = f_{cd} b x \left(h_0 - \dfrac{x}{2} \right)$, $h_0 = \sqrt{\dfrac{\gamma_0 M_d}{f_{cd} b \xi (1 - 0.5\xi)}}$. It is needed to estimate the parameter a_s. Selecting $h = h_0 + a_s$, it is needed to be rounded.

(d) Known $b \times h$, it becomes another design situation.

Example

Given: The design bending moment $M_d = 120 \text{kN} \cdot \text{m}$, type I environment and safety level II ($\gamma_0 = 1.0$). Section size is $b \times h = 220\text{mm} \times 500\text{mm}$. Concrete is C30 and steel reinforcement is HRB400.

Unknown: A_s.

Solution:

a. Assume $a_s = 44\text{mm}$ (as it is not concrete cover). Then $h_0 = h - a_s = 500 - 44 = 456(\text{mm})$.

b. From the appendix Table 1-1, for C30 concrete, $f_{cd} = 13.8\text{MPa}$, $f_{td} = 1.39\text{MPa}$.

$$\xi = 1 - \sqrt{1 - \frac{2\gamma_0 M_d}{f_{cd} b h_0^2}}$$

$$= 1 - \sqrt{1 - \frac{2 \times 1.0 \times 12 \times 10^7}{13.8 \times 220 \times 456^2}}$$

$$= 0.38 < \xi_b = 0.53$$

c. $A_s = \dfrac{f_{cd} b h_0 \xi}{f_{sd}} = \dfrac{13.8 \times 220 \times 456 \times 0.38}{330} = 1594.2(\text{mm}^2)$

The case of 2Φ32 is selected, the required minimum beam width (cover thickness on both Side + bar diameter + clear distance).

$b_{min} = 2 \times 30 + 2 \times 35.8 + 35.8 = 167(\text{mm}) < b = 220(\text{mm})$

Satisfying the requirements.

d. Reinforcement ratio.

$$\rho = \frac{A_s}{b h_0} = \frac{1\,608}{220 \times 456} = 1.60\% > \rho_{min}$$

$$\rho_{min} = \text{Max}\left\{45 \frac{f_{td}}{f_{sd}} \times 100\% = 45 \times \frac{1.39}{330} \times 100\% = 0.190\%, 0.2\%\right\} = 0.2\%$$

Actual $a_s = 20 + \dfrac{35.8}{2} = 36(\text{mm}) <$ the assumed value a_s.

e. Figure steel layout (Figure 3-22).

If an engineer modifies the design, the first scenario is to increase h, second one is to increase f_{sd}, and the third one is to increase f_{cd} and b. (Note that the third scenario is not usually adopted).

(2) Section checking

The purpose of the section checking is to check that whether the load-carrying capacity of a designed section meets the requirements or not, meanwhile to check whether it satisfies the construction requirements or not.

①Calculation steps

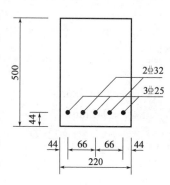

Figure 3-22 (unit: mm)

Given: sectional sizes b and h, area of reinforcing bar A_s, and concrete and steel material strength f_{sd} and f_{cd}.

Unknown: load-carrying capacity M_u.

Analysis: This problem can be solved based on basic equations.

Solution:

a. Check structural construction requirements.

b. a_s is known (no assumptions).

$$a_s = \frac{\sum_{i=1}^{n} f_{sd_i} A_{s_i} a_{s_i}}{\sum_{i=1}^{n} A_{s_i} f_{s_i}}$$

c. Check $\rho \geq \rho_{min}$.

d. From the basic formula $x = \dfrac{f_{sd} A_s}{f_{cd} b}$, and then $\xi = \dfrac{x}{h_0}$.

e. Check condition $\xi \leq \xi_b$.

If $\xi > \xi_b$, the structure is an overreinforced one. When $\xi > \xi_b$, only $\xi = \xi_b$ is taken (more steel reinforcement can not play roles). Then $x = \xi_b h_0$, at this time $M_u = f_{cd} bx \left(h_0 - \dfrac{x}{2} \right) = f_{cd} b \xi_b h_0 \left(h_0 - \dfrac{1}{2} \xi_b h_0 \right)$ (the M_u is the maximum load-carrying capacity of singly reinforced rectangular beams).

When the computed value $M_u < M$, improving concrete strength, changing section size, and using doubly-reinforced can enhance load-carrying capacity.

f. When satisfying $\xi \leq \xi_b$, from the basic formula, we have

$$M_u = f_{cd} bx \left(h_0 - \frac{x}{2} \right)$$

②Example

Given: rectangular section beam with the sizes of $b \times h = 200\text{mm} \times 600\text{mm}$. Concrete strength is C30 and steel strength is HRB400. $A_s = 1250\text{mm}^2$ ($3\Phi20 + 2\Phi14$). As shown in Figure 3-23.

Unknown: M_u.

Solution:

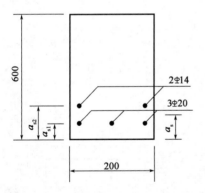

Figure 3-23 A reefangwlar section (unit: mm)

a. Check structural construction requirements.

b. $A_{s1} = 942\text{mm}^2$ ($3\Phi20$), $f_{sd1} = 330\text{MPa}$.

$a_{s1} = 30 + 22.7/2 = 41.35$ (mm) (cover thickness + half of bar diameter)

$A_{s2} = 308\text{mm}^2$ ($2\Phi14$) $f_{sd2} = 330\text{MPa}$

$a_{s2} = 30 + 22.7 + 30 + 16.2/2 = 90.8$ (mm)

c. From appendix Table 1-1, for C30 concrete

$$f_{cd} = 13.8\text{MPa}, f_{td} = 1.39\text{MPa}$$

$$\rho = \frac{A_s}{bh_0} = \frac{1250}{200 \times 546.5} = 1.14\% > \rho_{min}$$

$$\rho_{min} = \text{Max}\left\{45\frac{f_{td}}{f_{sd}} \times 100\% = 45 \times \frac{1.39}{330} \times 100\% = 0.190\%, 0.2\%\right\} = 0.2\%$$

d. $x = \dfrac{f_{sd}A_s}{f_{cd}b} = \dfrac{330 \times 1250}{13.8 \times 200} = 149.5(\text{mm}), \xi = \dfrac{x}{h_0} = 0.274 < \xi_b = 0.53$

e. $M_u = f_{cd}bx(h_0 - \dfrac{x}{2})$

$$= 13.8 \times 200 \times 149.5 \times \left(546.5 - \frac{149.5}{2}\right) = 194653485(\text{N} \cdot \text{mm}) = 194.65(\text{kN} \cdot \text{m})$$

3.6 Doubly-reinforced flexural members with rectangular section

1) Basic concepts

Doubly reinforced section is that steel reinforcements are placed in both tension and compression zones together. That is to say, reinforcing bars in compression zone can assist in concrete in compression zone carrying compression, which makes ξ reduce to $\xi \leq \xi_b$. Hence tension steel stress can reaches its yield strength and so concrete in compression zone does not crush earlier.

The application conditions of doubly reinforced rectangular section are:

(1) Singly reinforced section, when $\xi > \xi_b$, increasing of b, h, f_{cd} is restricted.

(2) In cases of negative of bending moment M, the doubly reinforced section is used.

(3) Construction and anchoring requirements.

(4) The requirements of structural mechanical behavior will result in doubly reinforced section at inner support section.

The disadvantage of doubly reinforced section is not economic from viewpoint of full use of material. The advantage of doubly reinforced section is improving ductile of the section and decreasing deformation of long-term loading.

2) Construction of doubly reinforcement

Construction requirements: closed stirrups must be set to restrict the longitudinal reinforcement compressive deformation (Figure 3-24). Under normal circumstances, the stirrup spacing is not more than 400 mm, and not 15 times greater than the compression bar diameter d'. Stirrup diameter is not less than 8mm or $d'/4$.

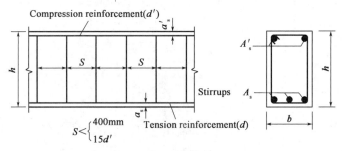

Figure 3-24   Stirrup spacing and form requirements

Mechanical behavior and failure characteristics of doubly reinforced flexural members are basically similar with singly reinforced section. Experimental investigation has shown that as long as it meets $\xi \leqslant \xi_b$, doubly reinforced section has the similar failure characters to singly reinforced section. Therefore, for the establishment of load-carrying capacity formula of doubly reinforced cross-section, the compression zone of concrete can still use the equivalent rectangular stress stock and concrete compressive design strength f_{cd}. The steel stress in compression needs to be determined.

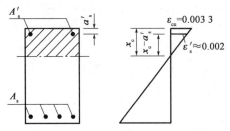

Figure 3-25   Schematic of steel strain of doubly reinforced cross-section

When doubly reinforced beam fails, the steel stress in compression zone depends on its strain. As shown in Figure 3-25, based on the plane section assumption, the compression steel stress is

$$\frac{\varepsilon'_s}{\varepsilon_{cu}} = \frac{x_c - a'_s}{x_c} = 1 - \frac{a'_s}{x_c} = 1 - \frac{0.8 a'_s}{x} \quad (\because x = 0.8 x_c)$$

$$\varepsilon'_s = 0.0033 \left(1 - \frac{0.8 a'_s}{x}\right)$$

(According to Table 3-1, for C50 concrete, $\varepsilon_{cu} = 0.0033$.)

Where x and x_c are respectively the compression height of equivalent rectangular stress stock and compression height based on the plane-section assumption.

When $\dfrac{a'_s}{x} = \dfrac{1}{2}$, namely, $x = 2a'_s$.

$$\varepsilon'_s = 0.0033 \left(1 - \frac{0.8 a'_s}{2a'_s}\right) = 0.00198$$

Based on the **CHBC**, for compression steel strain, when $\varepsilon'_s = 0.002$, for R235 steel rebar $\sigma'_s = \varepsilon_s E'_s = 0.002 \times 2.1 \times 10^5 = 420 \text{MPa} > f'_{sd}(=235\text{MPa})$, and for HRB335, HRB400 and KL400 steel rebar, $\sigma'_s = \varepsilon'_s E'_s = 0.002 \times 2 \times 10^5 = 400 (\text{MPa}) > f'_{sd}(=335\text{MPa or }400\text{MPa})$.

Thus, when $x = 2a'_s$, normal steel rebar can reach the yield strength (not with high-strength steel and not fully used). When $x > 2a'_s$, ε'_s will be greater and steel bar has already yielded. In order to fully play the role of compression steel reinforcement and to determine the yield strength, based on the **CHBC**, when $\sigma'_s = f'_{sd}$,

$$x \geqslant 2a'_s \qquad (3\text{-}21)$$

3) Basic formulas and application conditions

Because the failure mode of doubly-reinforced section is similar with singly-reinforced section, when equations are derived, the assumptions and methods of singly-reinforced section are used for doubly-reinforced section as showed in Figure 3-26.

Basic formula:

$$\sum x = 0 \qquad f_{cd}bx + f'_{sd}A'_s = f_{sd}A_s \qquad (3\text{-}22)$$

$$\sum M_T = 0 \qquad \gamma_0 M_d \leqslant M_u = f_{cd}bx\left(h_0 - \frac{x}{2}\right) + f'_{sd}A'_s(h_0 - a'_s) \qquad (3\text{-}23\text{a})$$

Or $\sum M'_T = 0 \qquad \gamma_0 M_d \leqslant M_u = -f_{cd}bx\left(\frac{x}{2} - a'_s\right) + f_{sd}A_s(h_0 - a'_s) \qquad (3\text{-}23\text{b})$

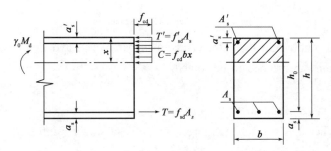

Figure 3-26 Schematic of load-carrying capacity of doubly reinforced flexural members with rectangular section

Where f'_{sd} is the design value of compressive strength of compression steel reinforcement; A'_s is the cross-sectional area of compression steel reinforcement; a'_s is the distance from the locations of compression reinforcements to compression edge of section. Other symbols are the same as those of singly reinforced rectangular section.

The basic application conditions of the basic formula are:

(1) In order to avoid occurrence of overreinforced beam, the compression height x should satisfy

$$x \leqslant \xi_b h_0 \qquad (3\text{-}24)$$

(2) In order to ensure that the compression reinforcement can reach design value f'_{sd} of compressive strength, the following should be satisfied

$$x \geqslant 2a'_s \qquad (3\text{-}25)$$

In the actual design, if the $x < 2a'_s$, it indicates that the compression reinforcement may not reach its design value of compressive strength (more than an unknown basic equation). For the case that concrete cover of compression reinforcement is not large, based on the **CHBC**, $x = 2a'_s$. Assuming the action point of concrete compressive stress stock is coincide with the action point of compression reinforcements (see Figure 3-27), from moment of location points of compression steel reinforcements, the approximate expression of load-carrying capacity of flexural members is

$$M_u = f_{sd}A_s(h_0 - a'_s) \qquad (3\text{-}26)$$

(3) $\rho \geqslant \rho_{min}$ (because tension steel bar can not be increased, so doubly reinforced is used).

4) Calculation methods

(1) Section design

①Solving the tension reinforcement A_s and compression reinforcement A'_s. Two cases was discussed here.

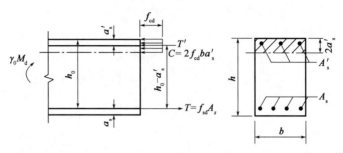

Figure 3-27 When $x < 2a'_s$, schematic of M_u

Analysis: this problem includes three unknown parameters and two basic equations, and so supplement conditions are needed to be solved.

Solutions:

a. Assume a_s and a'_s, and so $h_0 = h - a_s$.

b. Check whether it is a singly reinforced or not (a singly reinforced is not enough and so doubly reinforced section is used).

$$\gamma_0 M_d > M_u = f_{cd} b h_0^2 \xi_b (1 - 0.5\xi_b)$$

If the equations are satisfied, doubly reinforced formulas are used. (Taking $\xi = \xi_b$ into the basic formula of a singly reinforced section, if singly reinforced sections satisfy, doubly reinforced sections are no needed.)

c. Additional conditions. $A_s + A'_s$ is the most economical. (Two basic equations and three unknown parameters add conditions to reduce unknown parameter). Substituting $x = \xi_b h_0$ into the basic equation, and a larger A_s will result in a smaller A'_s. There is full use of tension steel reinforced rather than compression steel.

$$A'_s = \frac{\gamma_0 M_d - f_{cd} b \xi_b h_0^2 (1 - 0.5\xi_b)}{f'_{sd}(h_0 - a'_s)}$$

$$A_s = \frac{\gamma_0 M_d + f_{cd} b x \left(\frac{x}{2} - a'_s\right)}{f_{sd}(h_0 - a'_s)}$$

d. Select bar diameter and number and arrange steel layout.

Check a_s and a'_s, and judge whether these parameters are consistent with the assumption, otherwise it needs re-calculation.

②Solving tensile steel reinforcement A_s.

Given: compression reinforcement A'_s.

Unknown: tensile steel reinforcement A_s.

Analysis: this problem includes two unknown parameters and two basic equations, and so it can be solved.

Solution:
a. Assume a_s, and so $h_0 = h - a_s$.
b. From the basic formula, thus

$$x = h_0 - \sqrt{h_0^2 - \frac{2[\gamma_0 M_d - f'_{sd} A'_s (h_0 - a'_s)]}{f_{cd} b}}$$

Check $x \leqslant \xi_b h_0$. If it is not satisfied, it is needed to modify design or to take the A'_s as an unknown parameter, and then check $x \geqslant 2a'_s$.

c. If $2a'_s \leqslant x \leqslant \xi_b h_0$, the basic formula is

$$A_s = \frac{f_{cd} b x}{f_{sd}} + \frac{f'_{sd} A'_s}{f_{sd}}$$

and select the bar diameter and number of steel reinforcement, arrange layout, and verify a_s.

d. If $x < 2a'_s$, moment calculation is aimed at action point in compression zone, and it is assumed as $x = 2a'_s$, we have

$$A_{s_1} = \frac{\gamma_0 M_d}{f_{sd}(h_0 - a'_s)}$$

e. For singly-reinforced rectangular section, A'_s is not considered in the equation (taking $A'_s = 0$), and so

$$x^d = h_0 - \sqrt{h_0^2 - \frac{2\gamma_0 M_d}{f_{cd} b}}$$

$$A_{s_2} = \frac{f_{cd} b x}{f_{sd}}$$

$A_s = \min\{A_{s_1}, A_{s_2}\}$

(2) Section checking
Given: A_s and A'_s.
Unknown: M_u.
Analysis: this problem can be solved based on two basic equations.
Solution:
① Checking structural construction requirements.
② Calculating height of compression zone, $x = \frac{f_{sd} A_s - f'_{sd} A'_s}{f_{cd} b}$.
③ If $2a'_s \leqslant x \leqslant \xi_b h_0$,

$$M_u = f_{cd} b x \left(h_0 - \frac{x}{2}\right) + A'_s f'_{sd}(h_0 - a'_s)$$

④ If $x < 2a'_s$, form Equation (3-26), so

$$M_{u1} = f_{sd} A_s (h_0 - a'_s)$$

For singly-reinforced rectangular cross-section (taking $A'_s = 0$), we have

$$x = \frac{f_{sd} A_s}{f_{cd} b}$$

$$M_{u2} = f_{cd} b x \left(h_0 - \frac{x}{2}\right)$$

$$M_u = \max\{M_{u_1}, M_{u_2}\}$$

⑤If $x > \xi_b h_0$, the design and the associated maximum load-carrying capacity are modified. $\xi = \xi_b$ and $x = \xi_b h_0$ are taken into the basic formula.

3.7 T-shaped flexural members

1) Introduction

When a rectangular beam fails, the tensile concrete already cracks earlier. At location of cracked section, concrete in tension zones has no contribution to load-carrying capacity flexural members, and so part tensile concrete can be dredged. Steel reinforcements can be placed at the remaining concrete area, and so a reinforced concrete T-shaped beam cross-section is formed. Its load-carrying capacity of T-beam is same as the rectangular beams. But this way can save concrete and reduce the self-weight of beam. Therefore, the reinforced concrete T-beam has a greater span across capacity.

Figure 3-28 shows a typical reinforced concrete T-shaped beam cross-section. Part out of the section is named as the flange plate (the flange), and the width b of the part of the beam web is known as beam rib or beam web. Under positive bending moment, the flange is located in the compression zone of T-shaped beam section, which is called the T-shaped cross-section [Figure 3-28a)]. When the beam is subject to negative bending moment, the flange of upper beam is tensioned and concrete cracks, and then the effective beam cross-section is rib width b and the beam height h of rectangular cross-section [Figure 3-28b)], the bending capacity is calculated according to rectangular section. Therefore, judging whether one section in the calculation is T-shaped cross-section or not, it is not to check its own section shape, but the flange is able to contribute compression. In this sense, I-shaped section, box section, π-shaped section and hollow section are all treated as T-section for calculation of flexural capacity.

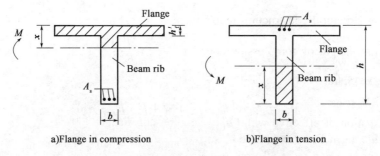

Figure 3-28 T-shaped cross-section in compression zone

Taking a hollow slab as an example with a width of b_f to discuss how to use sectional equivalent method, it is transformed equivalent I-shaped sections.

The principle of hollow deck being converted into an equivalent I-shaped section is bending moment equivalent one that the equivalent cross-section area, moment of inertia and centroid are kept unchanged.

For a hollow deck cross-section, height is set h and hole diameter is D. The distance from centroid of pore area to the upper and lower edge of slab section are y_1 and y_2, respectively. As shown in Figure 3-29a).

The approach of converting a hollow deck section into the equivalent I-shaped section is based on the principle of keeping the area and moment of inertia unchangeable. The hole of hollow plate (diameter D) is conversion into $b_k \times h_k$ rectangular hole, and so

Equivalent area:
$$b_k h_k = \frac{\pi}{4} D^2$$

Equivalent moment of inertia:
$$\frac{1}{12} b_k h_k^3 = \frac{\pi}{64} D^4$$

Solving two equations, so
$$h_k = \frac{\sqrt{3}}{2} D, \quad b_k = \frac{\sqrt{3}}{6} \pi D$$

Then, location of centroid of the hole and width and height of hollow section remain the same conditions, the size of equivalent I-shaped section can be obtained as

The top flange thickness:
$$h_f' = y_1 - \frac{1}{2} h_k = y_1 - \frac{\sqrt{3}}{4} D$$

The lower flange thickness:
$$h_f = y_2 - \frac{1}{2} h_k = y_2 - \frac{\sqrt{3}}{4} D$$

Web thickness:
$$b = b_f - 2b_k = b_f - \frac{\sqrt{3}}{3} \pi D$$

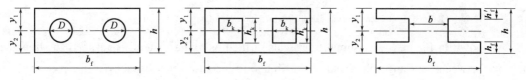

a) Circular hollow section   b) Equivalent rectangular hollow section   c) I-shaped equivalent cross-section

Figure 3-29   Hollow deck cross-section converted to equivalent T-shaped cross section

Figure 3-29c) shows the converted I-shaped cross-section. When the hollow cross-section has a hole with other shapes, it also can be converted into the equivalent I-shaped cross-section according to these principles.

2) Determination of effective width b_f' of compression flange

Through experimental testing and theoretical analysis, when a T-beam is subjected to load, the distribution of vertical compressive stress along flange width is not uniform. The farther away from the rib of the beam will result in a smaller stress. Its distribution depends on relative size of cross-section and the span (length), flange thickness and support conditions, etc.

Flange effective width: in the design and calculation, in order to facilitate the calculation, according to the principle of equivalent force, flange width of beam rib working together is limited

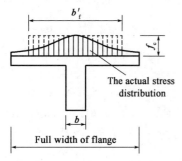

Figure 3-30 T- beam compression flange of the normal stress distribution

to a certain range, which is called the effective width b'_f of compression flange. Within the range of the flange width b'_f, it can be considered to be all involved in the work and assume that the compressive stress is uniformly distributed (Figure 3-30).

Based on the **CHBC**, the minimum effective width b'_f of compression flange of T-shaped beam (within the beam) can be obtained as follows:

(1) For simply supported beam, 1/3 span is used. For positive bending moment region of interior span of continuous beam, 0.2 span is used. For positive bending moment region of side span of continuous beam, 0.27 span is used. For negative bending moment region of interior supports of continuous beam, 0.07 span which is total length of adjacent spans is used.

(2) The average distance between two adjacent beams.

(3) $b + 2b_h + 12h'_f$. When $h_h/b_h < 1/3$, it is taken as $(b + 6h_h + 12h'_f)$, where b, b_h and h'_f shown in Figure 3-31, h_h is the thickness of the supporting rib.

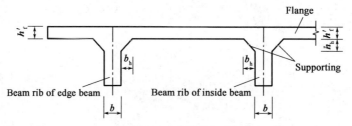

Figure 3-31 Effective width calculation in compression flange of T-shaped cross-section

As shown in Figure 3-31, supporting is also called as axillary stem. Its role is to strengthen the links between the flange plate and beam rib and to increase shear capacity of the root of flange plate.

3) Basic formulas and application conditions

The failure mode of T-section is similar as rectangular section, and so the assumptions and methods of rectangular section are effective for T-section.

T-section can be divided into two categories based on height of compression zone: a. Compression area is in the flange thickness, namely, $x \leq h'_f$ [Figure 3-32a)]. It is the first kind of T-section; b. Compression zone has entered a beam rib, namely $x > h'_f$ [Figure 3-32b)], that is the second kind of T-section.

(1) Basic formulas for flexural capacity of the first kind of T-beam

For the first kind of T-section, the neutral axis is in the compression flange, the depth of compression zone $x \leq h'_f$. Meanwhile, although it is a T-section, the compression zone is rectangle region with a width of b'_f and a height of h. Hence, its flexural capacity is similar to that of rectangular section. When flexural capacity is computed, the section width b of rectangular beam

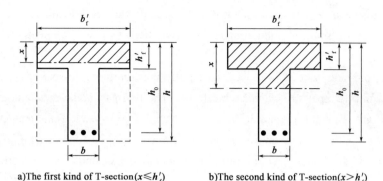

a) The first kind of T-section ($x \leq h'_f$)   b) The second kind of T-section ($x > h'_f$)

Figure 3-32   Two types of T-shaped cross-section

is replaced by the effective width of flange b'_f. From the equilibrium conditions of cross-section (Figure 3-33), the basic formula is

$$f_{cd} b'_f x = f_{sd} A_s \tag{3-27}$$

$$\gamma_0 M_d \leq M_u = f_{cd} b'_f x \left(h_0 - \frac{x}{2} \right) \tag{3-28a}$$

or

$$\gamma_0 M_d \leq M_u = f_{sd} A_s \left(h_0 - \frac{x}{2} \right) \tag{3-28b}$$

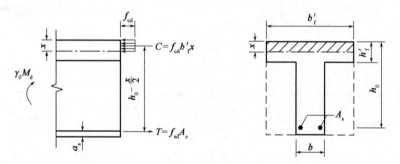

Figure 3-33   Flexural capacity of the first kind of T-section

Basic formulas for conditions:

① $x \leq \xi_b h_0$

For the first kind of T-section, $x = \xi h_0 \leq h'_f$. As the first kind of T-section, x is lower. Thus value ξ is also small, and so this condition is generally satisfied.

② $\rho > \rho_{min}$

$\rho = \dfrac{A_s}{bh_0}$, where b is the width of T-shaped rib section of beam. Minimal reinforcement ratio ρ_{min} is computed based on a condition that flexural strength of cracked section is the same as that of plain concrete. The flexural capacity of plain concrete depends on the strength grade of compression concrete. Load-carrying capacity of plain concrete with T-section is close to that of a rectangular cross-section with a height of h and width b. Therefore, when ρ_{min} of a T-shaped section is calculated, rib width b is approximately taken.

(2) Basic formulas of flexural capacity for the second kind of T-beam

Neutral axis is in beam rib, and compression zone height $x > h'_f$. The compression zone is T-shaped (Figure 3-34).

As the compression zone is T-shaped, generally concrete compressive stress in compression zone is divided into two parts to calculate the force. The first part is the rectangular section with a width of rib b and a height x, and its horizontal force $C_1 = f_{cd} bx$. The other part is also the rectangular section with a width of $b'_f - b$ and a height of h'_f, its force $C_2 = f_{cd} h'_f (b'_f - b)$. As shown in Figure 3-34, from the equilibrium condition of the second T-shaped cross-section, the basic formula is

$$C_1 + C_2 = T \qquad f_{cd} bx + f_{cd} h'_f (b'_f - b) = f_{sd} A_s \tag{3-29}$$

$$\sum M = 0 \qquad \gamma_0 M_d \leq M_u = f_{cd} bx \left(h_0 - \frac{x}{2} \right) + f_{cd} (b'_f - b) h'_f \left(h_0 - \frac{h'_f}{2} \right) \tag{3-30}$$

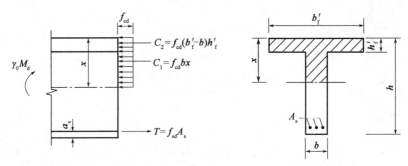

Figure 3-34 The flexural capacity calculation with the second kind of T-section

The application conditions of the basic formulas are:

① $x \leq \xi_b h_0$;

② $\rho \geq \rho_{min}$.

For the second kind of T-section, the reinforcement ratio is higher, in general, and so it can satisfy $\rho \geq \rho_{min}$. Thus, it does not need to be checked.

4) Calculation methods

(1) Cross-sectional design

Given: cross-sectional sizes, material strength and the calculated value of bending moment $\gamma_0 M_d$.

Unknown: sectional area A_s of reinforcing bar.

Calculation steps are:

① Assume a_s, so the effective height $h_0 = h - a_s$.

For the hollow deck cross-section, binding steel skeleton is often used. The work can be conducted according to the equivalent lower flange thickness h_f. In actual cross-sections, arranging one or two layers of steel reinforcement is to assume the a_s value.

For precast or cast-in-site T-beams, welded steel skeleton is often used. Due to the height of multi-layer reinforcement is not more than (0.15 ~ 0.2) h, it can be assumed a_s = 30mm +

$(0.07 \sim 0.1)h$.

②Determine the effective width of flange b'_f.

Based on the **CHBC**, three cases are considered to take the minimum value.

③Determine the type of T-section.

It can be seen from the basic formula that when neutral axis is located at the junction of compression flange and beam rib, $x = h'_f$. For the boundary conditions of two types of T-shaped section, obviously, if satisfied

$$\gamma_0 M_d \leqslant f_{cd} b'_f h'_f \left(h_0 - \frac{h'_f}{2} \right) \tag{3-31}$$

That is to say, calculation value of bending moment $\gamma_0 M_d$ is less than or equal to that generated by compression of concrete of all the flange height together, then $x \leqslant h'_f$, and T-section is the first type, otherwise T-section is the second one.

④For the first kind of T-section.

Taking $b'_f \times h$ as a rectangular cross section, using the basic equations, a height x of compression zone is

$$x = h_0 \xi = h_0 \left(1 - \sqrt{1 - \frac{2\gamma_0 M_d}{f_{cd} b'_f h_0^2}} \right)$$

Then, from Equation (3-27), the area A_s of tensile steel can be solved.

⑤For the second kind of T-section.

From the basic Equation, depth x of compression zone is

$$x = h_0 \xi = h_0 \left[1 - \sqrt{1 - \frac{2\gamma_0 M_d}{f_{cd} b h_0^2} + \frac{2(b'_f - b) h'_f}{b h_0^2} \left(h_0 - \frac{h'_f}{2} \right)} \right]$$

$h'_f < x \leqslant \xi_b h_0$. Substituting the given values and the x value into the basic Equation, the required area A_s of tension reinforcement can be solved.

If it doesn't satisfy the condition of $x \leqslant \xi_b h_0$, the design is needed to be modified.

⑥Select the bar diameter, number and layout, and check a_s.

(2) Section checking

Given: cross-section area, tensile steel reinforcement layout, cross-section dimensions and material strength.

Unknown: cross-sectional flexural capacity.

Calculation steps are:

①Checking whether the reinforcement layout satisfies specifications requirements.

②Determining the effective width b'_f.

③Judging the type of T-shaped cross-section.

Then, if satisfied

$$f_{cd} b'_f h'_f \geqslant f_{sd} A_s \tag{3-32}$$

That is to say, the tension of steel reinforcement $f_{sd} A_s$ is less than or equal to the compressive force ($f_{cd} b'_f h'_f$) provided by concrete compressive stress, then $x \leqslant h'_f$, and so T-shaped cross-section is the first type, otherwise T-shaped cross-section is the second one.

④When it is the first kind of T-section.

From basic Equation (3-27), the depth x of compression zone is obtained. If it satisfies $x \leqslant h'_f$, the given values and the x value are taken into Equation (3-28a) or Eqnation (3-28b) to obtain the flexural capacity M_u. Moreover it must meet $M_u \geqslant \gamma_0 M_d$.

⑤When it is the second kind of T-section.

From Equation (3-29), the depth x of compression zone is obtained. If it satisfies $h'_f < x \leqslant \xi_b h_0$, the given values and the x value are substituted into Equation (3-31) to obtain the flexural capacity M_u. Moreover, it must meet $M_u \geqslant \gamma_0 M_d$.

Example

The prefabricated simply-supported T-beam, section height $h = 1.30$m, the effective width $b'_f = 1.60$m (1.58m prefabricated width), C30 concrete, HRB400 steel rebar. The first kind of environmental conditions. Safety level is 2. The combinational bending moment design value of cross-section $M_d = 3\,000$ kN·m. Performing (welded reinforcement skeleton) calculation of steel reinforcement and section checking.

Solution:

From the appendix, $f_{cd} = 11.5$MPa, $f_{td} = 1.23$MPa, $f_{sd} = 330$MPa, $\xi_b = 0.53$, $\gamma_0 = 1.0$, the calculated bending moment $M = \gamma_0 M_d = 2\,500$kN·m.

In order to facilitate calculation, the Figure 3-35a) for actual T-shaped section is transformed into Figure 3-35b) as the calculation section. $h'_f = \dfrac{100 + 140}{2} = 120\,(\text{mm})$, the other sizes remain same.

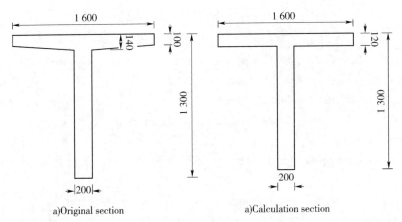

a) Original section a) Calculation section

Figure 3-35 Schematic of T-beam (unit:mm)

1) Section design

(1) Calculating h_0.

Due to the welded steel skeleton, set $a_s = 125$mm, so the section effective height $h_0 = 1\,300 - 125 = 1\,175\,(\text{mm})$.

(2) The effective calculation width b'_f is given, there is no need to be calculated.

(3) Determining T-shaped section type.

$$f_{cd} b'_f h'_f \left(h_0 - \dfrac{h'_f}{2} \right) = 13.8 \times 1\,600 \times 120 \times (1\,175 - 120/2) = 2\,954.3 \times 10^6\,(\text{N} \cdot \text{mm})$$

$= 2\,954.3(\text{kN}\cdot\text{m}) < \gamma_0 M_d(= 3\,000\text{kN}\cdot\text{m})$

It belongs to the second kind of T-shaped section.

(4) Calculating compression zone height.

Using the basic equations, then

$$\begin{cases} f_{cd}bx + f_{cd}h'_f(b'_f - b) = f_{sd}A_s \\ \gamma_0 M_d = f_{cd}bx\left(h_0 - \dfrac{x}{2}\right) + f_{cd}h'_f(b_f - b)\left(h_0 - \dfrac{h'_f}{2}\right) \end{cases}$$

Substituting data into above equations, so the follow equation can be obtained as

$3\,000 \times 10^6 - 13.8 \times 120 \times (1\,600 - 200) \times (1\,175 - 120/2) = 13.8 \times 200 \times (1\,175 - x/2)\cdot x$

After simplifying, then

$$x^2 - 2\,350 + 300\,710 = 0$$

So $x = 136\text{mm} > h'_f(= 120\text{mm})$. It is the second kind of T-beam.

(5) Computing area of steel rebar A_s.

All the known parameters are taken into the basic equations of load-carrying capacity of T-shaped beam, so it can be obtained as

$$A_s = \frac{f_{cd}bx + f_{cd}h'_f(b_f - b)}{f_{sd}} = \frac{13.8 \times [200 \times 136 + 120 \times (1\,600 - 200)]}{330} = 8\,163(\text{mm}^2)$$

Now the type of 10 ϕ 32 is selected as design of steel reinforcement and so sectional area $A_s = 8\,042\text{mm}^2$. The number of layers for reinforcement arrangement is 5, and the layout is shown in Figure 3-36.

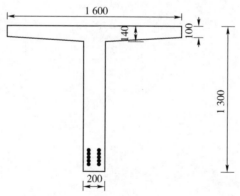

Figure 3-36 Reinforced arrangement (unit:mm)

The concrete cover thickness is 20mm, and specified value in appendix Table 1-8 is 20mm. The horizontal space between steel reinforcements is $S_n = 200 - 2 \times 35 - 2 \times 35.8 = 58(\text{mm}) > 40$ (mm) and $1.25d = 1.25 \times 32 = 40(\text{mm})$. It meets the construction requirements.

2) Section checking

(1) Calculating actual value of a_s

Based on the designed steel reinforcements, the area of the 10 ϕ 32 is $8\,042\text{mm}^2$, and $f_{sd} = 330\text{MPa}$. Using reinforcement arrangement as Figure 3-36, The parameter a_s can be obtained as

$$a_s = 20 + 2.5 \times 35.8 = 110(\text{mm})$$

The actual effective height $h_0 = 1\,300 - 110 = 1\,190(\text{mm})$.

(2) Determining the T-shaped section type

Using basic equations, and then the following equations can be presented as

$$f_{cd}b'_f h'_f = 13.8 \times 1\,600 \times 120 = 2.649 \times 10^6 (\text{N} \cdot \text{mm}) = 2\,649(\text{kN} \cdot \text{m})$$

$$f_{sd}A_s = 8\,042 \times 330 = 2.654 \times 10^6 (\text{N} \cdot \text{mm}) = 2\,654(\text{kN} \cdot \text{m})$$

Because $f_{cd}b'_f h'_f < f_{sd}A_s$, the cross-section is the second kind of T-shaped section.

(3) Calculating compression zone height x

The basic equations are used to compute x, and then

$$x = \frac{f_{sd}A_s - f_{cd}h'_f(b'_f - b)}{f_{cd}b} = \frac{330 \times 8\,042 - 13.8 \times 120 \times (1\,600 - 200)}{13.8 \times 200}$$

$$= 121.5(\text{mm}) > h'_f(=120\text{mm})$$

(4) Flexural load-carrying capacity of normal section

Using the basic equations, the flexural load-carrying capacity of normal section is

$$M_u = f_{cd}bx\left(h_0 - \frac{x}{2}\right) + f_{cd}h'_f(b_f - b)\left(h_0 - \frac{h'_f}{2}\right)$$

$$= 13.8 \times 200 \times 121.5 \times (1\,190 - 121.5/2) +$$
$$13.8 \times 120 \times (1\,600 - 200) \times (1\,190 - 120/2)$$

$$= 2\,998.7 \times 10^6 (\text{N} \cdot \text{mm}) = 2\,998.7(\text{kN} \cdot \text{m}) \approx \gamma_0 M_d(=30\text{kN} \cdot \text{m})$$

Thus, $\rho = \dfrac{A_s}{bh_0} = 8\,042/(200 \times 1\,190) = 3.37\% > \rho_{\min} = 0.2\%$. Therefore section checking meets the requirement of design code.

Chapter 4
Calculation of Inclined Section's Load-carrying Capacity of Flexural Members

4.1 Inclined section's performance characteristics and failure forms of flexural members

For a member under flexural, generally there will be shear V acting in its section besides moment M. What kind of performance characteristics and failure forms will occur in the sections under the combined actions of shear and moment? This is a problem to be solved in this chapter.

Research indicates that: concrete, stirrups, bent-up bars, longitudinal bars can provide the shear resistance of RC structures. In this section, the shear resistance of RC beams without stirrups and bent-up bars (RC beams without web reinforcement) is analyzed firstly, and then to find the influence of the concrete, stirrups, longitudinal bars to the shear resistance.

1) Stress state of simply supported beams without web reinforcement before and after diagonal cracks

Figure 4-1 shows a simply supported beam without web reinforcement under two symmetrical concentrated loads. There is bending moment M at the cross-section only, and CD segment is called pure bending section. There are bending moment M and shear force V at the cross-section in segment AC and segment DB, which called shear-bending section.

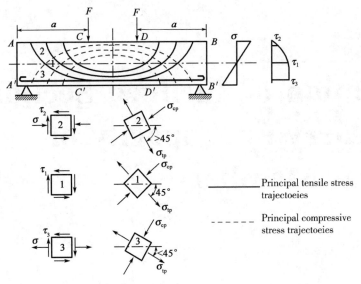

Figure 4-1 The principal stress distribution of beam without web reinforcement

(1) Stress state before diagonal cracks: the stress can be calculated by using the mechanics material stress formula.

In shear-bending section of the beam, two kinds of stress occur (bending and shear), and they can be calculated with the following expressions.

$$\left. \begin{array}{l} \sigma = \dfrac{My_0}{I_0} \\[2ex] \tau = \dfrac{Vs_0}{bI_0} \end{array} \right\} \qquad (4\text{-}1)$$

A typical infinitesimal element in the shear-bending section is chosen for stress analysis. The bending and shear stress are combined into inclined compressive and tensile stresses, which is called principal stress. The principal tensile stress σ_{tp} and the principal compressive stress σ_{cp} can be determined by the following expression

$$\left. \begin{array}{l} \sigma_{tp} = \dfrac{\sigma}{2} + \sqrt{\dfrac{\sigma^2}{4} + \tau^2} \\[2ex] \sigma_{cp} = \dfrac{\sigma}{2} - \sqrt{\dfrac{\sigma^2}{4} + \tau^2} \end{array} \right\} \qquad (4\text{-}2)$$

The direction of principal stress is found, then principal stress trajectories can be obtained, as shown in Figure 4-1.

(2) The reasons for the inclined section form: under the combined action of bending moment M and shear force V, the principal tensile stress and the principal compressive stress arise in the

cross-section. When the principal tensile stress exceeds the tensile strength of concrete, diagonal cracking is produced in the beam, and the cracks are formed skew to the neutral axis of beam, and these cracks are called inclined sectional failure.

(3) The inclined sectional failure: after diagonal cracks are formed, the stress state of cross-section qualitatively changes, and the stress distribution takes place in the beam.

Stress state analysis of the inclined section is shown in Figure 4-2, the shear resistance on inclined section is provided by the following four components:

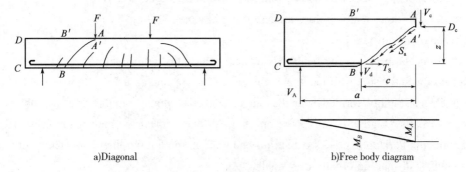

a) Diagonal   b) Free body diagram

Figure 4-2   Stress state of RC beams without web reinforcement after diagonal cracks

①The longitudinal reinforcement tension T_s;

②The compression force D_c and the shear force V_c on the residual section at the top of the diagonal crack;

③The forces due to friction between relative displacement and dislocation on both sides of inclined cracks, and the interlocking force between the aggregates on the two sides of the diagonal crack S_a;

④The dowel force V_d of the longitudinal reinforcing bars crossing the diagonal crack.

(4) After diagonal cracks appear, the stress state in the RC beam changes as following:

①Before cracking, the shear force is resisted by the entire cross-section of the beam. After cracking, the shear force is resisted by the residual cross-section at top of the diagonal crack. The effective cross-sectional area reduces, and the shear stress and compressive stress of the cross-section increases.

②Before cracking, the tension T_s in the longitudinal reinforcing bars is determined by the bending moment at the cross-section. After cracking, the tension T_s at the cross-section is determined by the bending moment of the residual cross-section at top of the diagonal crack, thus the tensile stress in the longitudinal reinforcing bars is increasing.

2) The arch analogy failure mechanism of RC beams without web reinforcement

The stress state of RC beams without web reinforcement is similar to a tied arch structure. The block Ⅰ is thought of as the arch, whereas the longitudinal reinforcement is said to be the tied rod. It is shown in Figure 4-3.

3) Modes of failure mechanism of inclined section of RC simply supported beams without web reinforcement

（1）Shear span ratio $m: m = \dfrac{M}{Vh_0}$, where M is the bending moment and V is shear force of a vertical cross-section in the shear-bending section. For the case of concentrated load the shear span a is the distance between the point of application of the load and the nearest support. $m = \dfrac{M}{Vh_0} = \dfrac{a}{h_0}$ is called narrow shear span ratio.

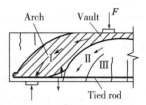

Figure 4-3　Arch analogy

（2）Experimental research shows that RC beams without web reinforcement have three major failure modes of the diagonal section:

①Diagonal tension failure

Production conditions: The beam has a shear span ratio more than $3(m>3)$.

Failure characteristics: When the inclined cracks appear, a main diagonal crack (critical inclined cracks) is rapidly formed, and the diagonal crack extends rapidly upward and splits the beam into two parts [Figure 4-4a)]. The splitting surface is clean without many crush marks, at the same time, the horizontal cracks come into being along the longitudinal bars. The entire process is sudden and rapid. The deformation is small.

Shear capacity: Diagonal tension failure is due to the principal tensile stress exceeding the tensile strength of concrete. The shear capacity of beam is lower, and the ultimate load is equal to or slightly more than the diagonal cracking load.

②Shear compression failure

Production conditions: The beam has a shear span ratio between 1 and 3 ($1 \leqslant m \leqslant 3$).

Failure characteristics: With the increase of the load, several diagonal cracks come into being, and one of them develops to the critical inclined crack [Figure 4-4b)]. After the appearance of critical inclined crack, the beam still works until the concrete at the top of the crack is eventually crushed by the principal compression.

Shear capacity: It is related to the size of the cross-section, the concrete strength and the spacing and strength of stirrups.

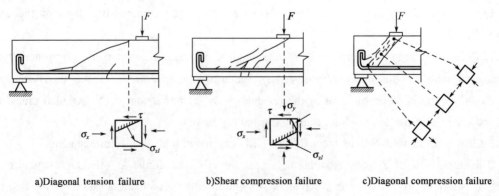

a) Diagonal tension failure　　b) Shear compression failure　　c) Diagonal compression failure

Figure 4-4　Failure modes of the diagonal section

③Diagonal compression failure

Production conditions: The beam has a shear span ratio less than 1 ($m<1$).

Failure characteristics: There is an inclined crack appearing between the support and the concentrated load, and then several parallel inclined cracks come into being which divide the web of beam into several inclined cylinders [Figure 4-4c]. With the increase of the load, the principal compression stress in the concrete near the support exceeds the compressive strength of concrete, and the concrete is crushed by the principal compression, which is similar to the crush of the prism.

Shear capacity: It is depends on the size of cross-section and the concrete strength. The shear capacity is higher.

The order of load-carrying capacity for three failure modes : Diagonal compression failure > Shear compression failure > Diagonal tension failure. They are all brittle failure.

4) Stress state of simply supported beams with web reinforcement

RC beam without web reinforcement is not convenient for engineering application, with a lower shear capacity. In the *CHBC*, some types of web reinforcement are needed in RC beams. The web reinforcement can effectively provide the shear capacity of inclined section.

(1) Effect of web reinforcement

The presence of web reinforcement has no noticeable effect on the general behavior of the member before the diagonal cracks, and it is not the function of web reinforcement to prevent the appearance of diagonal cracks. The stress of web reinforcement is small. After the diagonal cracks, web reinforcement will greatly enhance the capacity of beams, in particular the role of stirrups, which include:

①The stirrup crossing the crack can bear the part of the shear.

②The presence of the stirrup restricts the extension of the diagonal crack and thus retains a deeper residual compression zone. The shear stress and compressive stress are thereby lower. The shear capacity of concrete is increased.

③The stirrup effectively prevents the longitudinal bars from any vertical displacement, and the concrete between the bars will not be torn by the dowel action of the longitudinal bars.

It can be seen that the stirrup to improve the shear capacity of the inclined section is multifaceted and comprehensive.

(2) Failure mode

Similar to the beams without web reinforcement, RC beams with web reinforcement have three major failure modes of the diagonal section: diagonal tension failure, shear compression failure and diagonal compression failure.

4.2 Factors affecting the shear capacity of diagonal section

The experimental study indicates: the major factors that affect the shear capacity of the diagonal section are shear span ratio, strength of concrete, longitudinal steel ratio, stirrup ratio and the strength of stirrup.

1) Shear span ratio

Tests show that the shear capacity of diagonal section decrease with the shear span ratio increasing. For a shear span ratio $m > 3$, the shear capacity of diagonal section is further stabilized, and the influence of the shear span ratio is unnoticeable, as shown in Figure 4-5.

2) The compressive strength of concrete

The shear capacity of a beam increases with the compressive strength of concrete, and the relation is nearly liner. However, due to the fact that the beams with different shear span ratios have different failure modes, the extent of this influence is not the same. Diagonal tension failure affected by the tensile strength of concrete, while diagonal compression failure and shear compression failure affected by the compressive strength of concrete.

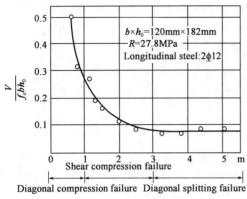

Figure 4-5 The influence of the shear span ratio to the shear capacity

3) Longitudinal steel ratio

Tests show that the shear capacity of diagonal section increases with an increase of longitudinal steel ratio ρ, and the relation is nearly liner.

4) Stirrup ratio and the strength of stirrup

Under otherwise equal conditions, the ultimate shear strength tends to increase with the value of yielding strength and the stirrup ratio increases, and the relation is nearly liner. The stirrup ratio is given by

$$\rho_{sv} = \frac{A_{sv}}{bs_v}$$

Where A_{sv} is total area of stirrups within the stirrup spacing; b is the width of the rectangular section or the web width of T-section or I-section; s_v is spacing of stirrup in the direction of the longitudinal reinforcement.

4.3 The calculation of inclined section's shear capacity of flexural members

1) The basic formula and application conditions of shear capacity

(1) Calculation basis

Shear compression failure.

(2) Calculation diagram (Figure 4-6)

The design block diagram for shear capacity calculation of a beam is showed is Figure 4-6. The shear capacity is composed of three components: shear strength of concrete V_c, shear strength of stirrups V_{sv} and shear strength of bent-up bars V_{sb}. The aggregate interlocking force and the dowel action are neglected.

(3) Basic formula (semi-empirical semi-theoretical)

$$V_u = V_c + V_{sv} + V_{sb}$$
$$V_u = V_{cs} + V_{sb}$$
$$\gamma_0 V_d \leq V_u = \alpha_1 \alpha_2 \alpha_3 \times 0.45 \times 10^{-3} bh_0$$
$$\sqrt{(2 + 0.6p)} \sqrt{f_{cu,k}} \rho_{sv} f_{sv} +$$
$$(0.75 \times 10^{-3}) f_{sd} \sum_{i=1}^{n} A_{sbi} \sin\theta_s \qquad (4\text{-}3)$$

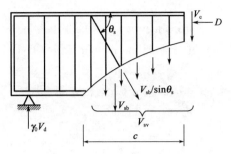

Figure 4-6 The calculation diagram for inclined section's shear capacity

Where V_d is design value of max shear force effect combination (kN); γ_0 is the importance coefficient of bridge structure; α_1 is opposite-sign influence coefficient of bending moment, for simply supported beam and continuous-supported beam near edge support $\alpha_1 = 1.0$, for continuous beam and cantilever beam near intermediate support $\alpha_1 = 0.9$; α_2 is coefficient for the use of prestressed, for reinforced concrete members $\alpha_2 = 1.0$, for prestressed concrete members $\alpha_2 = 1.25$; α_3 is coefficient of the compressed flange, when the cross-section has compressed flange $\alpha_3 = 1.1$; b is the width of the rectangular section or the web width of T-section or I-section (mm); h_0 is effective depth of section (mm); p is longitudinal steel ratio, $p = 100\rho$, $\rho = A_s/(bh_0)$, when $p > 2.5$, take $p = 2.5$; $f_{cu,k}$ is standard value of compressive strength of concrete cube (MPa); ρ_{sv} is stirrup ratio; f_{sv} is design value of stirrups tensile strength (MPa); f_{sd} is design value of tensile strength of the bent-up bars (MPa); A_{sb} is total area of stirrups within a cross-section of the beam (mm²); θ_s is the inclination of the bent-up bars' tangent to the member's axis.

(4) Bounds of shear strength of beams

①Upper limit value of shear strength—this value is dependent on the dimensions of the section and the compressive strength of the concrete. The recommended upper limit values for the ultimate diagonal section resistance help to prevent beams from the diagonal compression failure, when the dimensions of the section is too small and the principal compressive stress is too large. The size of cross-section should meet

$$\gamma_0 V_d \leq 0.51 \times 10^{-3} \sqrt{f_{cu,k}} bh_0 \quad (\text{kN}) \qquad (4\text{-}4)$$

Where V_d is design value of shear force combination in the checking section (kN); $f_{cu,k}$ is standard value of compressive strength of concrete cube (MPa); b is the width of the rectangular section or the web width of T-section or I-section (mm); h_0 is effective depth of section (mm).

②Lower limit value of shear strength—to prevent diagonal splitting failure

According to the **CHBC**, if the inequality is right, it shows the principal tensile stress is too small to bring diagonal splitting failure. The stirrups configure according to detailing requirements.

$$\gamma_0 V_d \leq 0.5 \times 10^{-3} \alpha_2 f_{td} bh_0 \quad (\text{kN}) \qquad (4\text{-}5a)$$

Where f_{td} is design value of concrete tensile strength (kN), $\alpha_2 = 1.0$ for reinforced concrete members.

For the plate structure, the formula (4-5b) is used

$$\gamma_0 V_d \leq 1.25 \times 0.5 \times 10^{-3} \alpha_2 f_{td} bh_0 = 0.625 \times 10^{-3} \alpha_2 f_{td} bh_0 \quad (\text{kN}) \qquad (4\text{-}5b)$$

2) Calculation method

Calculation of shear capacity includes design the shear reinforcement and recheck the shear strength of the inclined section. The following describes the design calculation method with a beam as example.

(1) Primary design of web reinforcement for a simply-supported beam with constant height

Data: $L, b, h, f_{cd}, f_{sd}, f_{sv}, A_s, \gamma_0 V_d$.

Design: The number of web reinforcements.

Design steps:

①Check whether the section size meets the requirements

$$\gamma_0 V_d \leq 0.51 \times 10^{-3} \sqrt{f_{cu,k}} bh_0$$

Otherwise, increase the section size or enhance the concrete strength.

②Check whether web reinforcement must be determined by calculation

$$\gamma_0 V_d \leq 0.5 \times 10^{-3} f_{td} bh_0$$

Otherwise, web reinforcement must be determined by calculation.

③Calculate the value of shear force

In the **CHBC**, the max shear force V is the force in the cross-section that is $h/2$ away from the bearing center. The burden of the concrete and stirrup can not be less than 60%, and the burden of the bent up bars can not be more than 40%.

④Stirrup design

Preset the diameter, number and the space of stirrups are

$$s_v = \frac{\alpha_1^2 \alpha_3^2 0.56 \times 10^{-6}(2 + 0.6p)\sqrt{f_{cu,k}} f_{sv} A_{sv} bh_0^2}{(v')^2} \quad (4\text{-}6)$$

Results s_v are rounded and meet the detailed requirements.

⑤Bent up bars design (Figure 4-7)

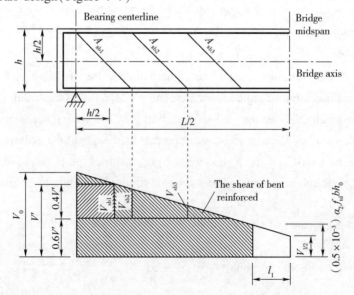

Figure 4-7 The calculation diagram for web reinforcement design

a. The formula of the number of bent up bars is

$$A_{sbi} = \frac{V_{sbi}}{0.75 \times 10^{-3} f_{sd} \sin\theta_s} = \frac{1\,333.33 V_{sbi}}{f_{sd} \sin\theta_s} (\text{mm}^2)$$

b. The value of V_{sbi} is from the **CHBC**.

(a) $V_{sb1} = 0.4 V'$.

(b) The shear force in the bent up point, which burden by reinforcement, is chosen to calculate next each row of bent up bars.

c. The bent up angle of the bent up bars and the position relationship between bars.

(a) Bend angle: 45 degree is typically used, under specific conditions 30 degree can be used. However, it is not more than 60 degree. The arc bending radius of the bent up bar is not less than 20 times the bar diameter (subject to bar axis).

(b) The position relationship between the bending bars: in a simply supported beam, the bending point of bent up bars in the first row (in terms of the support) should be on the center section at support. The bending point of bent up bars in the next row should be on or beyond the before bent up section.

(2) Recheck the inclined section's shear strength

For a designed beam, V_u, $\gamma_0 V_d$ are calculated firstly. If the shear design value $\gamma_0 V_d$ is less than the shear bearing capacity V_u, technically the design is satisfactory and no more calculation is required. However, if the shear design value $\gamma_0 V_d$ is larger than V_u, the design is adequate. There are different shear effects in different inclined sections, and many of inclined sections are chosen for the rechecking of shear strength.

①Position selected for the calculate section (Figure 4-8):

a. Section is half of beam high from the center of support (1-1).

b. Section is at the bent up point of bent-up bars in tensile area, and the anchor on the tensile strength of longitudinal reinforcement is zero (2-2, 3-3, 4-4).

c. Section with the change of number or spacing of stirrup (5-5).

d. Section with the change of ribs width.

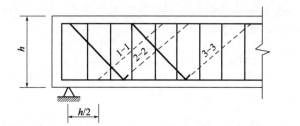 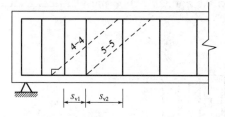

Figure 4-8　The calculation diagram for checking of inclined section's shear capacity

②The projection length c of inclined section: Horizontal projection length of inclined crack is from the top of the crack to the intersection point with the longitudinal reinforcement at the bottom.

$$c = 0.6mh_0 = 0.6\frac{M_d}{V_d} \tag{4-7}$$

Where m is generalized shear span ratio, when $m > 3$, take $m = 3$; V_d is design value of shear force combination in the checking section; M_d is design value of moment combination in the checking section.

③Method for determining the top position of inclined section.

a. Simplified method is usually chosen to determine: the top position of inclined section is on the section where h_0 is away from the bottom position in the direction to midspan.

b. Trial method: calculate the shear span ratio m at the bottom of inclined section. $c = 0.6mh_0$. To trial top position of inclined section using this method.

④The values ($\rho, A_{sb}, \rho_{sv}, p$) obtained by the method above can be substituted in the following formula

$$V_u = \alpha_1\alpha_2\alpha_3 0.45 \times 10^{-3}bh_0\sqrt{(2+0.6p)}\sqrt{f_{cu,k}\rho_{sv}f_{sv}} + 0.75 \times 10^{-3}f_{sd}\Sigma A_{sb}\sin\theta_s \tag{4-8}$$

Shear resistance V_u can be obtained, which is compared with $\gamma_0 V_d$ to recheck the safety of design.

4.4 Inclined section's flexural capacity of flexural members

If the longitudinal bars are maintained over the entire length of the span without being bent up or cut off, the moment of resistance of any diagonal section should be adequate. But the longitudinal bars have to be bent-up or cut-off to enhance shear strength of the beam or to provide the moment of resistance for negative moment in a continuous beam. In these cases, the moment of resistance of any diagonal section may be beyond the bending requirements.

1) Flexural capacity

As the ultimate strength design specifies that design values of max flexural force combination on the normal section must be less than the ultimate moment, as shown in Equation (4-9). The calculation diagram is shown in Figure 4-9.

$$\gamma_0 M_d \leqslant M_u = f_{sd}A_s Z_s + \Sigma f_{sd}A_{sb}Z_{sb} + \Sigma f_{sv}A_{sv}Z_{sv} \tag{4-9}$$

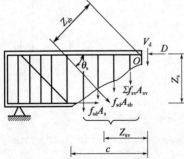

Figure 4-9 The calculation diagram for inclined section's flexural capacity

Where M_d is design value of max flexural force combination (kN); A_s, A_{sv}, A_{sb} are area of longitudinal bars, stirrups and the bent-up bars within a cross-section of the beam; Z_s, Z_{sv}, Z_{sb} are arms of force of reinforcement A_s, A_{sv}, A_{sb} to the center of the compression zone of concrete.

Chapter 4 ▸ Calculation of Inclined Section's Load-carrying Capacity of Flexural Members

2) Bending position of the longitudinal bars

According to the results of analysis, if the bent point extended beyond the full useable point $0.5\ h_0$, the flexural bearing capacity of inclined section can be guaranteed. To guarantee the flexural bearing capacity of inclined section, some construction measures are implemented.

3) Drawing of ultimate moment diagram

Analysis: Taking account the needs of bending and shear, the longitudinal bars have to be bent-up to enhance shear strength of the beam, or cut-off. The ultimate moment diagram and the moment envelope diagram of the beam are compared to determine the bent up point on longitudinal bars.

(1) Several concepts

①Ultimate moment diagram: The connection of the maximum moment of resistance of the normal sections along the member. It is line chart with step.

②Moment envelope diagram: The distribution of design values of moment combination on the normal section.

For example: Moment envelope diagram of a simply supported beam is approximately a parabola, if midpoint of the span is chosen for the abscissa origin, then

$$M_{d,x} = M_{d,l/2}\left(1 - \frac{4x^2}{L^2}\right) \qquad (4\text{-}10)$$

Where $M_{d,x}$ is design value of combination moment in the section x (x away from middle span section); $M_{d,l/2}$ is design value of combination moment in the middle span section; L is calculated span length of simply supported beam.

③Shear envelope diagram: The distribution of design values of shear combination on the normal section. The vertical axis value represents the largest design value of shear on section.

Shear envelope diagram of a simply supported beam is approximately a line.

$$V_{d,x} = V_{d,l/2} + (V_{d,0} - V_{d,l/2})\frac{2x}{L} \qquad (4\text{-}11)$$

Where $V_{d,0}$ is design value of shear combination in center of bearing; $V_{d,l/2}$ is design value of shear combination in the middle span section; L is calculated span length of simply supported beam.

(2) Drawing steps of the ultimate moment diagram

①Resistance moment of each section is proportional to the area of reinforced. The bent up point on longitudinal bars can be determined by the proportion.

$$\frac{M_i}{M_j} \approx \frac{A_{si}}{A_{sj}}$$

②Find the fully developed point and the theoretical cut-off point.

③The bent up point must be extended beyond the fully usable point $0.5\ h_0$.

④Intersection point of bent up bars and axis of beam is the starting point for the next step in the material map.

4.5 Load-carrying capacity checking of the whole beam and the detailing requirements

1) Content and requirements of the load-carrying capacity checking of the whole beam

(1) Ultimate flexural strength requirments: To ensure that each section $M_u \geqslant \gamma'_0 M_d$.

(2) Oblique cross-section flexural strength: Guarantee by the detailing requirements.

(3) Oblique cross-section shear strength. According to the calculation of the shear reinforcement in section 4.3 of this chapter, oblique cross-section strength is guaranteed when the the actual distribution of shear reinforcement calculated meets the requirements of shear calculation. How to guarantee that each section in whole beam is able to meet the requirements for flexural strength after the longitudinal reinforcement bent? It can be known after the comparision of the materials maps and the moment envelope diagram.

2) The detailing requirements

Reinforcement layout should not only meet force requirements but also meet the following detailing requirements.

(1) Anchorage of longitudinal reinforcement at supports

①It should be at least two and more than 1/5 of the lower main reinforcement through the support of a reinforced concrete beam.

②For the main reinforcements in the bottom of beam, the development length which extends beyond the support must satsity: not less than $10d$ (for R235 and with a semi-circular hook), not less than $12.5d$ (for epoxy resin coated steel bar), where d is the diameter of tension reinforcement.

(2) Cut-off and anchorage of the longitudinal reinforcement in the beams

Before a bar is cut-off, it must be extended beyond the cut-off section by a certain distance $(l_a + h_0)$, where l_a is the anchorage length, which is the required bond length to pass the force from reinforcement to concrete. The ***CHBC*** provides the minimum anchorage length of the reinforcement under different loadings, as shown in Table 4-1.

The minimum anchorage length l_a of the reinforcement   Table 4-1

Typs of steel bar	HPB300				HRB400, HRBF400, RRB400			HRB500		
Grades of concrete strength	C25	C30	C35	⩾C40	C30	C35	⩾C40	C30	C35	⩾C40
Compressed steel bar (Straight part)	$45d$	$40d$	$38d$	$35d$	$30d$	$28d$	$25d$	$35d$	$33d$	$30d$

Chapter 4 · Calculation of Inclined Section's Load-carrying Capacity of Flexural Members

continued

Typs of steel bar		HPB300				HRB400,HRBF400,RRB400			HRB500		
Tensile steel bar	Straight part	—	—	—	—	$35d$	$33d$	$30d$	$45d$	$43d$	$40d$
	Curved part	$40d$	$35d$	$33d$	$30d$	$30d$	$28d$	$25d$	$35d$	$33d$	$30d$

Note: ①The variable d denotes the nominal diameter of steel abr.

②For the anchorage length of tensile steel bar with $d_e \leqslant 28$mm and compressed steel bar, the equivalent diameter should be used and each steel bar can be cut at the same destination. For tensile steel bar with the equivalent diameter $d_e \geqslant 28$mm, the anchorage length should be 1.3 times of a single steel bar, and the steel bar should be cut with a ladder shape. From the first anchorage point, the first anchorage length should be 1.3 times of a single steel bar, the second anchorage length should be 2.6 times of a single steel bar, and the third anchorage length should be 3.9 times of a single steel bar length.

③For the steel bar with epoxy coating, the minimum anchorage length of tens le steel bar should increase by 25%.

④When the concrete is disturbed in thd solidification state, the anchorage lengtn should increase by 25%.

⑤When a curved part type of tensile steel bar is used, the anchorage length includes the projected length of the curved part of thde steel bar.

(3) Joints of reinforcement

When the steel bars in the beam need be lengthened, banding lap joints or welded joints and mechanical joints can be used.

①The banding lap joints: l_d is the lap length of banding lap joints for tensile reinforcement. The **CHBC** provides the lap length of banding lap joints for compression reinforcement, which is 0.7 times of the lap length for tensile reinforcement, as shown in Table 4-2. It should not have two connectors between the two points which are 1.3 times of the lap length from the banding joints center on a steel bar. In this section, for the reinforcement which has the banding lap joints, the section area of this kind of reinforcement percentage of the total cross-section area, tension zone should not exceed 25%, and the compression zone should not exceed 50%. When the percentage exceeds the above requirements, the provisions values in Table 4-2 should be multiplied by the following factor: when the sectional area of tensile reinforcement banding joints more than 25% but not more than 50%, multiplied by 1.4, and when more than 50%, multiplied by 1.6. When the sectional area of compression reinforcement banding joints greater than 50%, it should be multiplied by 1.4 (from Table 4-2, the lap length of banding lap joints for compression reinforcement is 0.7 times of the l_d).

The lap length l_d of banding lap joints for tension reinforcement Table 4-2

Types of steel bar	HPB300		HRB400,HRBF400,RRB400	HRB500
Grades of concrete strength	C25	$\geqslant$C30	$\geqslant$C30	$\geqslant$C30
Splicing length	$40d$	$35d$	$45d$	$50d$

Note: ①When the diameter of steel bar $d > 25$mm, the splicing length of tensile steel bar should increase by $5d$. When the diameter of the ribbed steel bar is less than 25mm, the splicing length can be decrease by $5d$.

②When the steel bar is disturbed in construction, the splicing length should increase by $5d$.

③At each condition, the splicing length of tensile steel bar should be greater than 300mm. The splicing length of compressed steel bar should be greater than 200mm.

④For the splicing length of steel bars with epoxy coating, the value of tensile steel bar is in the table.

⑤For the tensile area, the end part of the binding joint of HPB300 steel bar should be curved, but the end part of HRB400, HRB500, HRBF400 and RRB400 should not be curved.

②Welded joints: When using the welded joints, the **CHBC** has the appropriate detailing requirements. For example, arc welding rod with clip, clip bar area should not be less than the cross-sectional area of the welded steel bars. If using double-sided weld, the length of clip should be not less than $5d$; with one side seam, the length of clip should be not less than $10d$ (d is the bar diameter). Another example is the use of stacked arc welding, at end section reinforcement should be pre-bending to one side, so that two contacts steel bars have the same axis. When the two bars are lapping, the double-sided weld length should not less than $5d$; one side seam length is not less than $10d$ (d is the bar diameter). The **CHBC** provides: it should not have two connectors on a steel bar between the two points which 35 times the bar diameter from the joints center and not less than 500mm. In this section, the joint of a reinforcement is not more than one, the section area percentage of jointed reinforcement of the total cross section area should not exceed 50% (tension zone), the compression zone has no such limit.

③Mechanical joints: including the pressed sleeve connector and upsetting straight thread connector, the structural requirements is shown in the **CHBC**.

④The minimum reinforcement ratio of stirrup: $\rho_{sv\,min} = 0.14\%$ (HPB300), $\rho_{sv\,min} = 0.11\%$ (HRB400).

3) Example of the inclined section's shear capacity calculation

The known design data and requirements:

An simply supported reinforced concrete beam with span $L_0 = 19.96\mathrm{m}$. The section of the T-shaped beam is shown in Figure 4-10. The environmental conditions at the bridge is class I. The security level is 2 ($\gamma_0 = 1$).

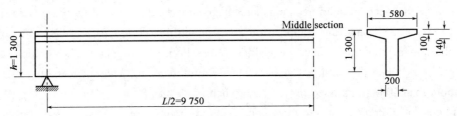

Figure 4-10   Size of the simply supported reinforced concrete beam (unit: mm)

The concrete is of grade C30, the design value of compressive strength of concrete $f_{cd} = 13.8$ MPa, the design value of tensile strength of concrete $f_{td} = 1.39$ MPa. The longitudinal reinforcement is HRB400, the design value of tensile strength of reinforcement $f_{sd} = 330$ MPa. The transverse stirrup is HPB300, with the diameter 8mm, the design value of tensile strength of reinforcement $f_{sd} = 250$ MPa.

On control cross-section, the combination design value of the bending moment and shear is:

In the middle section $M_{d,l/2} = 2\,600\mathrm{kN\cdot m}$, $V_{d,l/2} = 84\mathrm{kN}$

1/4 cross-section $M_{d,l/4} = 1\,900\mathrm{kN\cdot m}$

Support section $M_{d,0} = 0$, $V_{d,0} = 440\mathrm{kN}$

It is required to determine numbers of the longitudinal tensile reinforcement, and design the web reinforcement of the beam.

(1) Calculation of the longitudinal tensile reinforcement in the middle section

①The effective width b'_f of compression flange in T-shaped beam

According to the T-shaped cross-section thickness of compression flange (Figure 4-10), the average thickness $h'_f = \dfrac{140 + 100}{2} = 120 \text{(mm)}$. It can be find

$$b'_{f1} = \frac{1}{3}L = \frac{1}{3} \times 19\,500 = 6\,500 \text{(mm)}$$

$b'_{f2} = 1\,600\text{mm}$ (This example is an assembly T-beam, the average distance of adjacent main beam is 1 600mm, as shown in Figure 4-10, and the width of precast beam flange is 1 580mm.)

$$b'_{f3} = b + 2c + 12h'_f = 200 + 2 \times 0 + 12 \times 120 = 1\,640 \text{(mm)}$$

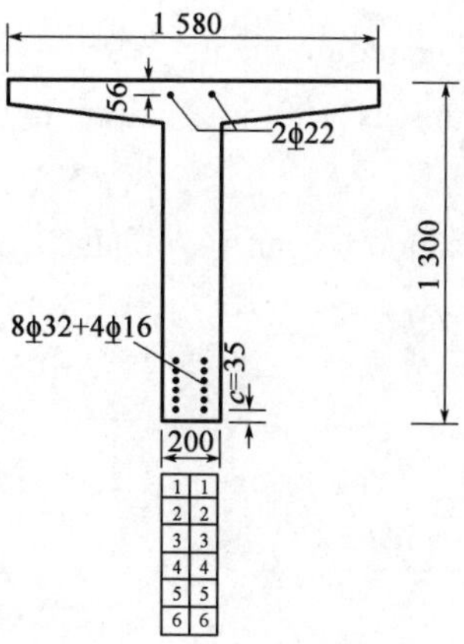

Therefore, the effective compression flange width $b'_f = 1\,600\text{mm}$.

②Calculating the numbers of bars

Reinforcement numbers (in the middle section) and section check calculation (omitted).

In the middle section, the reinforcement is $8 \, \phi 32 + 4 \, \phi 16$, welded steel skeleton layer is 6, and the area of the longitudinal bars $A_s = 7\,238\text{mm}^2$, as shown in Figure 4-11. The effective height of middle section $h_0 = 1\,183\text{mm}$, and flexural capacity $M_u = 2\,695.0\text{kN} \cdot \text{m} > \gamma_0 M_{d,l/2} = 2\,600\text{kN} \cdot \text{m}$.

Figure 4-11 Cross-section of the T-beam (unit: mm)

(2) Design of web reinforcement

①Check of the section size

According to detailing requirements, the bottom steel $2 \, \phi 32$ through the bearing cross-section, and the effective cross-section height at bearing cross-section is

$$h_0 = h - \left(35 + \frac{35.8}{2}\right) = 1\,247 \text{(mm)}$$

$$0.51 \times 10^{-3}\sqrt{f_{cu,k}}\,bh_0 = 0.51 \times 10^{-3} \times \sqrt{30} \times 200 \times 1\,247$$

$$= 96.67 \text{(kN)} > \gamma_0 V_{d,0} \, (= 440\text{kN})$$

Meet the design requirements of section size.

②Check the arrangement of stirrups

The middle section: $0.5 \times 10^{-3} f_{td} b h_0 = 0.5 \times 10^{-3} \times 1.23 \times 200 \times 1\,183 = 145.51 \text{(kN)}$

$$\gamma_0 V_{d,l/2} \, (= 84\text{kN}) < 0.5 \times 10^{-3} f_{td} b h_0$$

Therefore, within the length in the middle of beam span, the stirrups configured according to structural requirements, and stirrups of the remaining sections should be configured by calculation.

③Calculate the shear distribution

The shear envelope is shown in Figure 4-12. The shear force at the bearing cross-section $V_0 = \gamma_0 V_{d,0}$, the shear force at the middle section $V_{l/2} = \gamma_0 V_{d,l/2}$.

The distance from cross-section $V_x = \gamma_0 V_{d,x} = 0.5 \times 10^{-3} f_{td} bh_0 = 64.44\text{kN}$ to the middle section can be obtained from shear envelope by the ratio

$$l_1 = \frac{L}{2} \times \frac{V_x - V_{l/2}}{V_0 - V_{l/2}}$$

$$= 9\,750 \times \frac{64.44 - 84}{440 - 84}$$

$$= 2\,203(\text{mm})$$

Within the length of l_1, the stirrup could be arranged by detailing requirements.

According to the **CHBC**, the maximum space between stirrups is 100mm, within $h/2 = 650$mm of the centerline of supports.

The shear force V' on the section, $h/2$ from the centerline of supports, can be obtained from shear envelope by the ratio

$$V' = \frac{LV_0 - h(V_0 - V_{l/2})}{L}$$

$$= \frac{19\,500 \times 440 - 1\,300 \times (440 - 84)}{19\,500}$$

$$= 416.27(\text{kN})$$

The shear force which shall be borne by concrete and stirrups at least $0.6V' = 249.76\text{kN}$; and should be borne by bent bars (including oblique bar) not more than $0.4V' = 166.51\text{kN}$. Set bent length of bent bars 4 560mm (Figure 4-12).

④Design of stirrups

With a diameter of 8mm double-leg stirrups, the stirrup cross-sectional area is

$$A_{sv} = nA_{sv1} = 2 \times 50.3 = 100.6(\text{mm}^2)$$

In the uniform reinforced concrete beams, the stirrups as far as possible equidistant arrangement. For the more convenient calculation, when the stirrups are designed by Equation (4-13), the p and h_0 can be approximated by average values of bearings cross-section and the middle section, calculated as follows:

In the middle section $\quad p_{l/2} = 3.06 > 2.5, (p_{l/2} = 2.5, h_0 = 1\,183\text{mm})$

Bearing cross-section $\quad p_0 = 0.64, h_0 = 1\,247\text{mm}$

The mean value is, respectively $p = \dfrac{2.5 + 0.64}{2} = 1.57, h_0 = \dfrac{1\,183 + 1\,247}{2} = 1\,215(\text{mm})$

The spacing of stirrups is calculated as

$$s_v = \frac{\alpha_1^2 \alpha_3^2 0.56 \times 10^{-6}(2+0.6p)\sqrt{f_{cu,k}}A_{sv}f_{sv}bh_0^2}{V'^2}$$

$$= \frac{1 \times 1.1^2 \times 0.56 \times 10^{-6}(2+0.6 \times 1.57)\sqrt{30} \times 100.6 \times 250 \times 200 \times 1215^2}{416.27^2}$$

$$= 468(\text{mm})$$

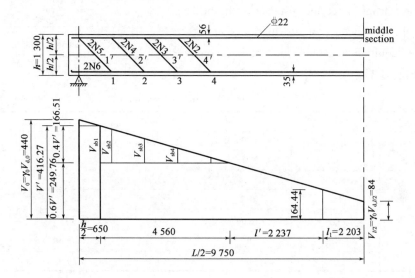

Figure 4-12 Calculation of the shear distribution map (unit: mm, kN)

To determine the spacing of stirrups, the detailing requirements in the **CHBC** should be considered.

If the spacing of stirrups $s_v = 400\text{mm} \leqslant \frac{1}{2}h = 650\text{mm}$ and 400mm, it is to meet the specification requirements.

if double leg stirrups $\phi 8$ are chosen, the stirrup reinforcement ratio is

$$\rho_{sv} = \frac{A_{sv}}{bs_v} = \frac{100.6}{200 \times 400} = 0.12\% < 0.14\% (\text{HPB300})$$

It does not meet the specification requirements. So the stirrups $s_v = 250\text{mm}$ are chosen. The reinforcement ratio $\rho_{sv} = 0.2\% > 0.14\%$, and less than $\frac{1}{2}h = 650\text{mm}$ and 400mm.

From the above calculation, in the 1 300mm range from the bearing center direction to the span length, the spacing of stirrups $s_v = 100\text{mm}$; Within the rest of the beam segment, the uniform spacing of stirrup $s_v = 250\text{mm}$.

⑤Design of bent reinforcement and inclined reinforcement

Assuming skeleton steel type (HRB400) is $\phi 22$, the distance between the reinforced center and the wing panel of the beam compression edge a'_s is 56mm.

The bent angle of bent bars is 45 °. The end of bent bar is welded with erect steel. In order to obtain the distribution shear force of each pair of bent bars, the starting bent point of the bars should be calculated. First of all, the vertical distance Δh_i between the up bent point and the down bent point should be calculated (Figure 4-13).

N1 ~ N5 is chosen to be bent. The Δh_i of cross-section in the starting point, the distance to the bearing center section x_i, the values of distribution shear force V_{sbi}, and the required area of each bent bars A_{sbi} are shown in Table 4-3.

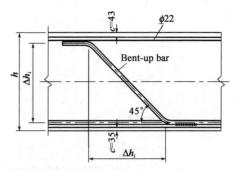

Figure 4-13　The details of bent bars (unit: mm)

The calculation result of bent bars　　　Table 4-3

Bent points	1	2	3	4	5
Δh_i (mm)	1 125	1 090	1 054	1 035	1 017
x_i (mm)	1 125	2 215	3 269	4 304	5 321
V_{sbi} (kN)	166.51	149.17	109.36	70.88	
A_{sbi} (mm²)	952	852	624	405	
Real area of bent bars A_{sbi} (mm²)	1 609 (2ϕ32)	1 609 (2ϕ32)	1 609 (2ϕ32)	402 (2ϕ16)	
x_c' (mm)	564	1 690	2 779	3 841	

Table 4-3 is calculated as follows:

According to the **CHBC**, the end bent point of the first row of bent bars (at the supports) should be at the bearing center section. Δh_1 is calculated as

$$\Delta h_1 = 1\ 300 - [(35 + 35.8 \times 1.5) + (43 + 25.1 + 35.8 \times 0.5)]$$
$$= 1\ 125(\text{mm})$$

The bent angle is 45°, the end bent point of the first row of bent bars (2N5) away from the bearing center 1 125mm.

1' is the intersecting point of bent bars and the beam longitudinal axis, the distance to the bearing center is

$$1125 - \left[\frac{1\ 300}{2} - (35 + 35.8 \times 1.5)\right] = 564(\text{mm})$$

For the second row of bent bars

$$\Delta h_2 = 1\ 300 - [(35 + 35.8 \times 2.5) + (43 + 25.1 + 35.8 \times 0.5)] = 1\ 090(\text{mm})$$

The bent point 2 (2N4) away from the bearing center $1\ 125 + \Delta h_2 = 1\ 125 + 1\ 090 = 2\ 215(\text{mm})$.

V_{sb2} is the distribution shear force of the second row of bent bars, which is calculated by the ratio between available

$$\frac{4\ 560 + 650 - 1\ 125}{4\ 560} = \frac{V_{sb2}}{166.51}$$

Get $\qquad V_{sb2} = 149.17\text{kN}$

Where $0.4V' = 166.51\text{kN}$, $\dfrac{h}{2} = 650\text{mm}$; The segment of bent bars is set as 4 560mm.

The needed cross-section area A_{sb2} is

$$A_{sb2} = \dfrac{1\,333.33 V_{sb2}}{f_{sd}\sin 45°} = \dfrac{1\,333.33 \times 149.17}{280 \times 0.707} = 852(mm^2)$$

The bent point 2' away from the bearing center $1\,125 + \Delta h_2 = 1\,125 + 1\,090 = 2\,215(\text{mm})$.

2' is the intersecting point of the second row of bent bars, the distance to the bearing center is

$$2\,215 - \left[\dfrac{1\,300}{2} - (35 + 35.8 \times 2.5)\right] = 1\,690(\text{mm})$$

The remaining row of bent bars are calculated in the same way as the second row of bent bars.

It can be seen from Table 4-3 that the bent point of the N1 away from the bearing center 5 321mm, and it is more than $4\,560 + h/2 = 4\,560 + 650 = 5\,210(\text{mm})$. It is outside the region that sets the bent-up bars, and not participate in the calculation of bent-up bars. In Figure 4-14, bent-up bars are cut off, but in practical engineering, the bars are not cut off but bent-up to enhance the stiffness of steel frame.

According to the calculated shear force, bent bars arranged preliminary as shown in Figure 4-14.

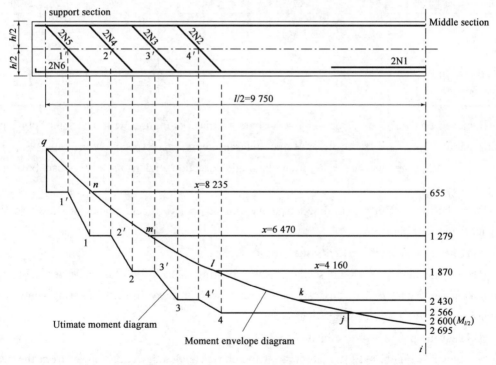

Figure 4-14 Ultimate moment diagram and moment envelope diagram (unit: mm, kN · m)

In order to meet the flexural requirements in the cross-section and inclined section at the same time, the bent position of bent-up bars is determined. The moment in the middle section is $M_{l/2} = \gamma_0 M_{d,l/2} = 2\,200\text{kN} \cdot \text{m}$, and in the bearing cross-section is $M_0 = \gamma_0 M_{d,0} = 0$. The Ultimate moment diagram is obtained as shown in Figure 4-14. In the $\frac{1}{4}l$ section ($x = 4.875\text{m}, l = 19.5\text{m}$), the moment calculation value is

$$M_{l/4} = 2\,600 \times \left(1 - \frac{4 \times 4.875^2}{19.5^2}\right) = 1\,950(\text{kN} \cdot \text{m})$$

Compared with the known values $M_{d,l/4} = 1\,900\text{kN} \cdot \text{m}$, the relative error is 2.6%. It is feasible to use Equation (4-10) to describe the ultimate moment diagram.

After the bending of the steel bars, the corresponding values of flexural capacity M_{ui} is shown in Table 4-4.

Flexural capacity of the bent bar corresponding of cross-section   Table 4-4

Segment of beam	Longitudinal reinforcement	Effective height h_0(mm)	T-shaped cross-section type	Height of compression zone x(mm)	Flexural capacity M_{ui}(kN·m)
bearing center ~ point 1	2 Φ32	1 247	I	24	655.0
point 1 ~ point 2	4 Φ32	1 229	I	48	1 279
point 2 ~ point 3	6 Φ32	1 211	I	72	1 870
point 3 ~ point 4	8 Φ32	1 193	I	96	2 430
point 4 ~ cut off of N1	8 Φ32 + 2 Φ16	1 189	I	102	2 566
cut off of N1 ~ the middle point of span	8 Φ32 + 4 Φ16	1 183	I	108	2 695

Represented the M_{ui} in Table 4-4 by the parallel lines in Figure 4-14. The intersecting points of this parallel lines and moment envelope diagram represent respectively $n, m, \cdots, j$. Using Equation (4-10), the distance x between the middle section and $n, m, \cdots, j$ can be obtained (Figure 4-14).

From the moment envelope diagram, the initial positions of bent points as shown in Figure 4-14 meet the requirements.

(3) Check of inclined section's shear bearing capacity

For the checking of inclined section's shear bearing capacity, it should be according to the requirements in the **CHBC**. In this case, an inclined section away from the bearing center $h/2$ is chosen to check the shear bearing capacity.

①Select the position on inclined section

As shown in Figure 4-12, the abscissa of the chosen inclined section (away from the bearing center $h/2$) $x = 9\,750 - 650 = 9\,100(\text{mm})$. The effective height of this section is $h_0 = 1\,247\text{mm}$. Projection length of the inclined section is taken $c' \approx h_0 = 1\,247\text{mm}$, then the top position A of the inclined section is determined (Figure 4-15), and the abscissa is $x = 9\,100 - 1\,247 = 7\,853(\text{mm})$.

②Check of inclined section's shear bearing capacity

Shear force V_x and the corresponding moment M_x on section A is calculated as follows.

$$V_x = V_{l/2} + (V_0 - V_{l/2})\frac{2x}{L} = 84 + (440 - 84) \times \frac{2 \times 7\,853}{19\,500} = 370.74(\text{kN})$$

$$M_x = M_{l/2}\left(1 - \frac{4x^2}{L^2}\right) = 2\,600 \times \left(1 - \frac{4 \times 7\,853^2}{19\,500^2}\right) = 913.31(\text{kN} \cdot \text{m})$$

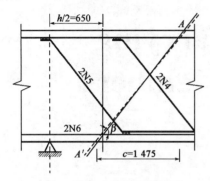

Figure 4-15 Calculation graphic of inclined section's shear bearing capacity (unit: mm)

The effective height on cross-section A is $h_0 = 1\,229\text{mm} = 1.229\text{m}$ (main reinforcement is $4 \phi 32$), the actual generalized shear span ratio m and the projection length of oblique section c are:

$$m = \frac{M_x}{V_x h_0} = \frac{913.3}{370.74 \times 1.229} = 2.0 < 3$$

$$c = 0.6mh_0 = 0.6 \times 2.0 \times 1.229$$
$$= 1.475(\text{m}) > 1.247(\text{m})$$

The chosen section for the check is the inclined section AA' (dashed line in Figure 4-15). The oblique angle the $\beta = \tan^{-1}(h_0/c) = \tan^{-1}(1.229/1.475) \approx 39.8°$.

The average reinforcement ratio p of the tensile longitudinal bars in the oblique section is

$$p = 100\frac{2 \times A_s}{bh_0} = \frac{2 \times 100 \times 1\,608}{200 \times 1\,247} = 1.28 < 2.5$$

Stirrup ratio ($S_v = 250\text{mm}$) ρ_{sv} is:

$$\rho_{sv} = \frac{A_{sv}}{bS_v} = \frac{100.6}{200 \times 250} = 0.201\% > \rho_{\min}(= 0.18\%)$$

There are two kinds of reinforcement intersecting with the inclined section. Bent-up bars are 2N5 ($2\phi32$) and 2N4 ($2\phi32$).

Inclined section's shear bearing capacity of AA' is calculated by Equation (4-3).

$$V_u = \alpha_1\alpha_2\alpha_3 0.45 \times 10^{-3} bh_0\sqrt{(2 + 0.6p)} \sqrt{f_{cu,k}}\rho_{sv}f_{sv} + 0.75 \times 10^{-3}f_{sd}\Sigma A_{sb}\sin\theta_s$$
$$= 1 \times 1 \times 1.1 \times 0.45 \times 10^{-3} \times 200 \times 1\,247$$
$$\times \sqrt{(2 + 0.6 \times 1.28)} \times \sqrt{30} \times 0.002\,01 \times 250 + (0.75 \times 10^{-3}) \times 330 \times 2 \times 1\,608 \times 0.707$$
$$= 316.23 + 633.09$$
$$= 949.32(\text{kN}) > V_x = 370.74(\text{kN})$$

Therefore, inclined section's shear bearing capacity meets the design requirements.

Chapter 5

Calculation of Load-carrying Capacity of Torsion Members

5.1 Overview

For curved bridge, inclined bridge and bridge under the eccentric loads that appearing in civil engineering, the main beam bears the torque. This chapter describes the calculation of torsion members.

1) Common torsion members in engineering

(1) Curved beam (curved bridge), oblique beam (plate);

(2) The beam, plate, box girder bridge under the eccentric loads;

(3) Board of spiral staircase.

In fact, torsion acting singly in structure is rare, mostly bending moment. Shear force and torque affect simultaneously, sometimes axial force affects too. The failure mechanism of bending, shear and torsion interaction in the structure is rather complicated. This chapter discusses the design of torsion component of reinforced concrete structure starting from the pure torsion members.

2) Characteristics of torsion members

In curved girder bridge, even without considering the live load, under the dead load only, there are not only torque T existing in the cross-section of beam (plate), but also the moment M and the shear force V (Figure 5-1).

Due to the effects of torque, bending moment and shear, the cross-section of members will produce a corresponding principal tensile stress. When the principal tensile stress exceeds the tensile strength of concrete, the members would be cracking. Therefore, the appropriate amount of reinforcement (longitudinal reinforcement and stirrups) must be configured to improve the torsion capacity of reinforced concrete members.

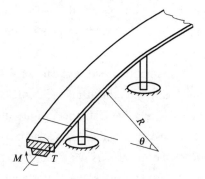

Figure 5-1　Schematic diagram of curved beams

5.2　Failure characteristics and capacity calculation of pure torsion members

1) The loading process experiment of reinforced concrete members under pure torsion

The curve of torque T and torsion reverse angle θ in the whole process of reinforced concrete members damage under torque is shown in Figure 5-2. The mechanical characteristics of damage under torque can be found.

(1) Beginning of loading: Torsion deformation of cross-section is small, and similar to the stress of torsion member that is made of elastic materials.

(2) After the crack appears: Concrete quits working after cracking. As concrete partial unloading, reinforcement stress increases apparently, and horizontal plateau appears in the $T - \theta$ curve. When the reverse angle increases to a certain value, the reinforced strain becomes more stable and forms a new stress state.

(3) Continue to increase load: Deformation increases rapidly, and the number of cracks increases gradually. Crack width increases gradually, then the relationship of $T - \theta$ is generally linear changing.

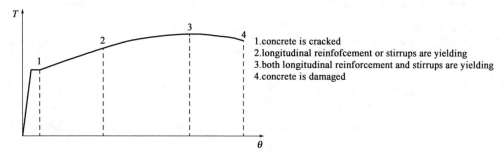

Figure 5-2　T-θ curve of RC torsion members

(4) Load close to the ultimate torque: Among the inclined cracks in the long side of the

member section, one of them develops to critical crack, and part of the stirrups (long limbs) or longitudinal reinforcement intersecting the spatial inclined cracks will be yielded firstly, and have a greater plastic deformation, then $T - \theta$ curves tend to be horizontal.

(5) Load reached ultimate torque: The stirrups (short limbs) and longitudinal reinforcement, intersecting the critical inclined cracks, yields successively and concrete withdraws from the work gradually, and the resistance torque of members decreases gradually. Finally, when the other long side of the member appears plastic hinge or the phenomena that concrete crushed between two cracks, the member is damaged.

Known from the characteristics of pure torsion RC members, the two important measurement indexes of torsional calculation of pure torsion RC members are: cracking torque and damage torque of the members.

2) Cracking torque of pure torsion members with rectangular section

The stress of reinforcement is very small before the torsion RC members cracking. Reinforcement has little effect on the cracking torque, so you can ignore the impact of reinforcement on the cracking torque, and take the member as a pure concrete member to calculate the cracking torque.

The pure torsion member with a rectangular cross-section is shown in Figure 5-3. Under pure torque, the shear stress of homogeneous elastic material of the rectangular section member distributes is shown in Figure 5-4a), where cracks always extend along the direction 45° to the vertical axis of member, and the cracking torque is the torque when the principal tensile stress $\sigma_{tp} = \tau = f_t$.

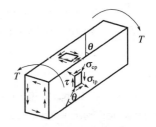

Figure 5-3 Pure torsion member with rectangular section

Assuming member is perfectly plastic material and its section is rectangular. When the stress of some point at the cross-section reaches the yield strength, it only means local materials begin to enter the plastic state. At this time, the member is still able to bear the load. Until all the stress on cross-section reaches the yield strength, the member will reach its ultimate bearing capacity, at this time, the shear stress distribution of the cross-section is shown in Figure 5-4b).

Now calculate the resisting torque according to the shear stress distribution of the perfect plastic material member shown in Figure 5-4b), obtained by the equilibrium condition

$$T = \left\{ 2 \cdot \frac{b}{2} \cdot (h - b) \cdot \frac{b}{4} + 4 \cdot \frac{b}{2} \cdot \frac{b}{2} \cdot \frac{1}{2} \cdot \frac{b}{3} + 2 \cdot \frac{b}{2} \cdot \frac{b}{2} \left[\frac{2}{3} \cdot \frac{b}{2} + \frac{1}{2}(h - b) \right] \right\} \tau_{max}$$

$$= \frac{b^2}{6}(3h - b)\tau_{max} = W_t \tau_{max} \tag{5-1}$$

Where W_t is called the anti-torsion plastic resisting moment of rectangular section

$$W_t = \frac{b^2}{6}(3h - b).$$

For reinforced concrete, the concrete is neither perfect elastic material, nor perfect plastic material, and it is elastic-plastic material between them. If calculated by elastic stress distribution, the cracking torque would be underestimated; while if calculated by fully plastic stress distribution, the cracking torque would be overestimated.

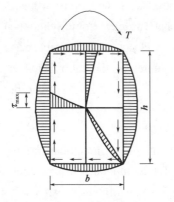
a) Shear stress distribution in elastic state

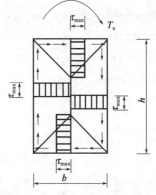
b) Shear stress distribution in plastic state

Figure 5-4   Shear stress distribution of pure torsion member with rectangular section

In summary, the cracking torque of torsion RC member with rectangular section can be calculated approximately by shear stress graph of perfectly plastic material, at the same time through the experiment to be correct. Multiplied by a reduction factor of 0.7, the calculation formula of cracking torque is followed

$$T_{cr} = 0.7 W_t f_{td} \quad (5\text{-}2)$$

Where T_{cr} is cracking torque of pure torsion member with rectangular section; f_{td} is design value of concrete tensile strength; W_t is the anti-torsion plastic resisting moment of rectangular section.

3) Damage characteristics of pure torsion member with rectangular section

(1) Damage type of torsion RC member with rectangular section

①Failure of less reinforcement: When the number of resisting torsion bar is too less, the member fails quickly, because steel bar can not bear the torque that is passed by cracking concrete.

②Failure of proper reinforcement: In condition of normal reinforcement, with the increasing external torque, the torsional stirrups and longitudinal reinforcement reach the yield strength first, and then the main crack extends promptly and finally causes the compression face of concrete crushed.

③Failure of excessive reinforcement: When the number of resisting torsion bar is too high or the strength of concrete is too low, with the increasing external torque, concrete of member is crushed first, resulting in failure of the member. At this time, neither torsional stirrups nor longitudinal reinforcement reaches the yield strength.

④Partial failure of excessive reinforcement: When the number of torsional stirrups or longitudinal reinforcement is too high, only partial longitudinal reinforcement or stirrups yield, while the other part of the torsion bars (stirrups or vertical bars) have not yet reached yield strength. The failure has the characteristic of brittle failure.

(2) Reinforcement strength ratio ζ

The ratio of the number, strength of longitudinal reinforcement and stirrup (short for reinforcement strength ratio, indicated as ζ) has some influence on torsional capacity. When the amount of stirrup is relatively small, the member's torsional capacity is controlled by the stirrup. At this time, increasing in longitudinal reinforcement can not increase the torsional capacity. Conversely, when the amount of longitudinal reinforcement is small, increasing in stirrups will not be fully effective.

If using the volume ratio of reinforcement to represent the ratio between the number of longitudinal reinforcement and stirrup, the strength ratio ζ of reinforcement can be written as

$$\zeta = \frac{f_{sd} A_{st} s_v}{f_{sv} A_{sv1} U_{cor}} \quad (5\text{-}3)$$

Where A_{st}, f_{sd} are the cross-sectional area of all symmetrical arranged longitudinal reinforcement and design value of tensile strength of longitudinal reinforcement; A_{sv1}, f_{sv} are the cross-sectional area of single-limb stirrup and design value of tensile strength of stirrups; s_v is interval of stirrups; U_{cor} is the perimeter of the cross-section of core concrete, and it can be obtained by the length of the epidermis of stirrups.

Strength ratio ζ of reinforcement may change within a certain range, limited to $0.6 \leqslant \zeta \leqslant 1.7$. Design desirable $\zeta = 1.0 \sim 1.2$.

(3) The influence of the amount of reinforcement to failure mode of torsion members

Even in the condition of strength ratio of reinforcement keeps unchanged, the amount of longitudinal reinforcement and stirrups will also affect the failure mode of torsion member. Figure 5-5 shows the relationship between stirrup amount $f_{sv} A_{sv1}/S_v$ and the torsional capacity while $\zeta = 1$. For different strength ratios of reinforcement, the boundaries of rare reinforcement, proper reinforcement and excessive reinforcement are different.

4) Calculation theory of capacity of pure torsion member

There are two calculation theories at present: one is variable-angle spatial truss model, the other is failure theory of skew bending.

(1) Variable-angle spatial truss model

Experimental study and theoretical analysis have shown that when the cracks are fully developed and steel stress is closed to yielding, the core concrete quits working, then the torsion member could be assumed as a box-type section member: a concrete shell with spiral-shaped cracks, longitudinal reinforcement and stirrups constitute the space truss together to resist external torque. Variable-angle spatial truss model calculation theory calculates the resistance of structure by this spatial truss model.

①The basic assumption of variable-angle spatial truss model

a. Concrete only bears pressure, the concrete shell with spiral-shaped cracks compose oblique bar of truss, the obliquity is α; shown as Figure 5-6.

b. Longitudinal reinforcement and stirrups only bear tension. They constitute chord and web member of the truss respectively.

c. Ignore the dowel action of the core concrete and steel bar.

Chapter 5 · Calculation of Load-carrying Capacity of Torsion Members

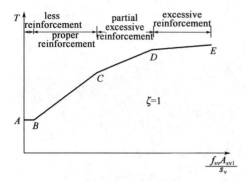

Figure 5-5 The influence of the amount of reinforcement to torsional capacity

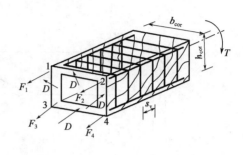

Figure 5-6 Schematic diagram of variable-angle spatial truss model

②The bearing capacity formula

According to stress state of the failure state, the ultimate torsion strength formula can be derived as

$$T_u = 2\frac{A_{sv1}f_{sv}U_{cor}}{s_v}\cot\alpha = 2\sqrt{\zeta}\cdot\frac{A_{sv1}f_{sv}A_{cor}}{s_v} \qquad (5-4)$$

From formula (5-4), the torsional capacity of member is mainly associated with the size of reinforced frame, amount and strength of stirrups, and the parameter ζ characterizing the relative amount of longitudinal reinforcement and stirrups. It is the bearing capacity formula for low reinforcement torsion members.

(2) Failure theory of skew bending

For pure torsion RC members, as cracks appear in the long edge of the section, and oblique section forms, the torque can be decomposed into the moment (torque) perpendicular to the oblique section and the moment (torque) parallel to the oblique section, concrete bending crushing zone occurs on the long edge of oblique section, forms skew bending failure. Failure theory of skew bending is taking the actual failure cross-section as the isolation, deriving the torsional capacity formula associated with the amount of longitudinal reinforcement and stirrups.

①The basic assumptions of calculation theory of skew bending

a. The longitudinal reinforcement and stirrups through the twisted surfaces reach the yield strength as the member failing (Figure 5-7).

b. High compression zone is taken as twice the cover thickness approximately, and assuming that the resultant force of compression zone acting on the centroid of compression zone approximately.

c. Torsional capacity of concrete is negligible, the longitudinal reinforcement and stirrups bear all the torque.

d. The longitudinal torsion reinforcements are arranged around the core of member symmetrically and uniformly, the torsion stirrups are arranged along the axis of member equidistantly, and all are anchored reliably.

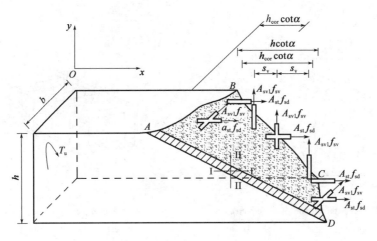

Figure 5-7　Schematic diagram of calculation theory of skew bending

②The bearing capacity formula

Based on the above assumptions, the torque capacity formula (5-5) according to skew bending theory is identical with the calculation formula (5-5) according to variable-angle spatial truss theory.

$$T_u = 2\sqrt{\zeta} \cdot \frac{A_{sv1} f_{sv} A_{cor}}{s_v} \tag{5-5}$$

(3) Characteristics of the two calculation theories

The bearing capacity formulas of torsion members derived from variable-angle spatial truss model or failure theory of skew bending reflect the role of the torsion reinforcement. But experiments show that when low reinforcement, the calculated value tends to be conservative, and when high reinforcement, the calculated value tends to be higher. Because the longitudinal reinforcement and stirrups can not yield simultaneously, it should be avoided in design.

5) Bearing capacity calculation rules of the **CHBC** on the pure torsion member with rectangular section

(1) Bearing capacity formula

Based on the variable-angle spatial truss model, and through the test results of torsion members, the anti-torsion ability takes the lower limit of the experimental data under the premise of ensuring the safety, and the torsion capacity formula of rectangular cross-section members can be obtained as

$$\gamma_0 T_d \leq 0.35\beta_a f_{td} W_t + 1.2\sqrt{\zeta} \frac{f_{sv} A_{sv1} A_{cor}}{s_v} \tag{5-6}$$

ζ is the strength ratio of the longitudinal reinforcement and stirrups, it can be calculated by the formula (5-3); For RC members, the **CHBC** specifies the value due to ζ should be consistent with $0.6 \leq \zeta \leq 1.7$, when $\zeta > 1.7$, taking $\zeta = 1.7$.

(2) Restrictive conditions

①The upper limit of torsional reinforcement—prevent the brittle failure due to excessive reinforcement.

When the amount of torsion reinforcement is excessive, the torsion member perhaps fails before the torsion reinforcements yield, because the concrete is crushed, that is, the failure torque depends on the strength of concrete and section size. Therefore, the **CHBC** prescribes the section size of pure torsion RC member with rectangular cross-section should conform to the requirements as follows

$$\frac{\gamma_0 T_d}{W_t} \leqslant 0.51 \times 10^{-3} \sqrt{f_{cu,k}} \qquad (5\text{-}7)$$

Where T_d is design value of combinational torque (kN·mm); W_t is the anti-torsion plastic resisting moment of rectangular section (mm³); $f_{cu,k}$ is standard value of compressive strength of concrete cube (MPa).

②The lower limit of torsional reinforcement—prevent the brittle failure due to less reinforcement.

When the torsion reinforcements configure too little, the reinforcement will not help the anti-torsion ability of torsion members after cracking, so as to prevent brittle fracture of concrete of pure torsion member when taking low reinforcement configuration, the torque undertaken by pure torsion member should not be less than its cracking torque. The **CHBC** prescribes that torsional capacity of pure torsion RC member can not be calculated if the member meets the requirements as follows, but it must meet the requirements of the minimum reinforcement ratio.

$$\frac{\gamma_0 T_d}{W_t} \leqslant 0.50 \times 10^{-3} f_{td} \ (\text{kN/mm}^2) \qquad (5\text{-}8)$$

Where f_{td} is the design value of concrete tensile strength, and the rest of symbols have the same meaning with Equation (5-7).

The **CHBC** prescribes that the stirrup reinforcement ratio of pure torsion member should meet the formula $\rho_{sv} = \frac{A_{sv}}{s_v b} \geqslant 0.055 \frac{f_{cd}}{f_{sv}}$; the longitudinal reinforcement ratio should meet the formula $\rho_{st} = \frac{A_{st}}{bh} \geqslant 0.08 \frac{f_{cd}}{f_{sd}}$.

5.3 The bearing capacity of rectangular section member under the action of bending, shear and torsion

1) Failure type of bending-shear-torsion member

The stress state of RC members is very complicated under the action of bending moment, shear and torque, and the failure characteristics and bearing capacity of member are related to the external loading conditions and internal factors of members.

(1) External conditions: It is usually represented with the ratio of torsion and bend $\psi \left(\psi = \frac{T}{M} \right)$ and the ratio of torsion and shear $\chi \left(\chi = \frac{T}{Vb} \right)$.

(2) Internal factors: It means the section shape, size, reinforcement and material strength of the member.

When the internal factors of the member keep unchanged, its failure characteristics are only related with the ratio of torsion and bend ψ and the ratio of torsion and shear χ. When the values ψ and χ are the same, the member can result in different types of failure because of the different internal factors (such as the section size) of member.

(1) Bending type failure: Bending-shear-torsion member, in the proper reinforcement and small ratio of torsion and bend ψ condition that bending moment is significantly higher than torque, and the cracks appear in the bottom tension surface of the member; then cracks develop to the two sides of member. The longitudinal reinforcements at the top section of the member bear pressure, and bear the tension caused by the torque. The spiral cracks in three surface form a distorted failure surface. The fourth side is top surface of bending compression. Structures failure appears as that the longitudinal reinforcement and stirrups intersected with the spiral cracks reach the yield strength first, and the concrete on the edge of the section crushes by pressure finally [Figure 5-8a)].

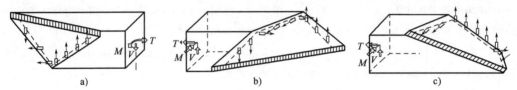

Figure 5-8 Failure type of bending-shear member

(2) Torsion-type failure: If the ratio of torsion and bend ψ and the ratio of torsion and shear χ are larger, that means the torque effect is significant. If the reinforcement at the top of member is less, because of the less moment, resulting in the compressive stress at the top surface of member less, the tension caused by the torque will pull off the stress under bending, and the rest tension is lager. Then the top longitudinal reinforcement reaches the tensile yield strength prior to the bottle longitudinal reinforcement, and the failure surface starts from the top side and develops to the two sides [Figure 5-8b)].

(3) Shear and torsion type failure: If the shear force and torque are larger, and the ratio of torsion and shear χ is larger, the crack appears in the side of member where the direction of stress flow is same as the direction of torque flow, and develops to the top surface and bottom surface. The spiral cracks on the three surfaces form a distorted failure surface. When the member fails, the reinforcements are intersected with the spiral cracks reach the yield strength, the compression zone close to the other side [Figure 5-8c)].

For the member under the action of bending, shear and torsion, besides the above-mentioned three kinds of failure modes, the experiment results show that if the shear force is significant and torque is relatively small, shear failure modes will occur which is very similar to shear-compression. In fact, it is already the shear that plays the controlling role.

2) Calculation method of reinforcement of bending-shear-torsion members

Same as the bearing calculation method of pure torsion members, the bearing capacity calculation method of RC members under the action of bending, shear and torsion is mainly based on the variable angle space truss theory and the oblique bending theory. However, in practice, for the bending-torsion or bending-shear torsion members, it is very complex to calculate by the two theories. Therefore, we need to simplify the calculation method.

At present in China, ***Code for Design of Concrete Structures*** (**GB 50010—2010**) specifies that when the torque withstanded by members is less than 1/4 cracking torque, you can ignore the influence of torque, and calculate as bending-shear member; When the shear withstanded by members is less than 1/2 shearing capacity of members without stirrups, you can ignore the influence of shear, and calculate as bending-torsion member. For the reinforcement calculation of bending-shear-torsion members, you can take the cross-section design method that singly calculate according to bending, shearing and torque, and then superimpose all the corresponding reinforcements.

The ***CHBC*** also takes superposition calculation the simplified cross-section design method. The bending capacity calculation method is as mentioned above. Now, we analyze the torsion and shear capacity calculation problem of shear-torsion members.

(1) The bearing capacity calculation of shear-torsion members

Based on experiments, for shear-torsion members with rectangular section, the shear capacity calculation and torsion capacity calculation in the ***CHBC*** use the following formula.

①Shear capacity of shear-torsion members

$$\gamma_0 V_d \leqslant V_u = 0.5 \times 10^{-4} \alpha_1 \alpha_2 \alpha_3 \frac{10 - 2\beta_t}{20} bh_0 \sqrt{(2+0.6p)} \sqrt{f_{cu,k}} \rho_{sv} f_{sv} \quad (N) \qquad (5\text{-}9)$$

$$\beta_t = \frac{1.5}{1 + 0.5 \dfrac{V_d W_t}{T_d bh_0}} \qquad (5\text{-}10)$$

Where V_d is shear combination design value of shear-torsion members (N); β_t is concrete torsional capacity reduction factor of shear-torsion members, when $\beta_t < 0.5$, taking $\beta_t = 0.5$; when $\beta_t > 1.0$, taking $\beta_t = 1.0$; W_t is the anti-torsion plastic resisting moment of rectangular section, $W_t = \dfrac{b^2}{6}(3h-b)$; Other symbols see the shear capacity formula of oblique section.

②Torsional capacity of shear-torsion members

$$T_u = \beta_t \left(0.35\beta_a f_{td} + 0.05 \frac{N_{po}}{A_o}\right) W_t + 1.2\sqrt{\zeta} \frac{N_{\varphi} f_{sv} A_{sv1} A_{cor}}{s_v} \qquad (5\text{-}11)$$

Significance of β_t in the formula is the same as former; while T_d is the torque combination design value of shear-torsion members, $\dfrac{N_{po}}{A_o}$ is the influence of prestress.

(2) The upper and lower limits of shear and torsion reinforcement

①The upper limit of shear and torsion reinforcement—prevent the failure of excessive reinforcement

When the amount of torsional steel bar is excessive, the concrete will be the first to be crushed and then the member fail. Therefore, the restrictions of cross-section must be provided to prevent this destructive phenomenon.

Under the interaction of bending, shear and torsion, the section size of rectangular section member must meet the conditions

$$\frac{\gamma_0 V_d}{bh_0} + \frac{\gamma_0 T_d}{W_t} \leq 0.51 \times 10^{-3} \sqrt{f_{cu,k}} \ (kN/mm^2) \tag{5-12}$$

Where V_d is the shear combination design value (kN); T_d is the torque combination design value (kN·mm); b is the total width of web of rectangular or box section perpendicular to the bending moment (mm); h_0 is the effective height of rectangular or box section parallel to the bending moment (mm); W_t is the anti-torsion plastic resisting moment of rectangular section (mm³); $f_{cu,k}$ is standard value of compressive strength of concrete cube (MPa).

②The lower limit of shear and torsion reinforcement—prevent the failure of less reinforcement

a. The reinforcement ratio of stirrups of shear and torsion member should meet the condition as follows

$$\rho_{sv} \geq \rho_{sv,min} = (2\beta_t - 1)\left(0.055\frac{f_{cd}}{f_{sv}} - c\right) + c \tag{5-13}$$

β_t in the formula is calculated according to formula (5-10). For the value of in the formula, taking 0.001 8 when the stirrups are made of R235 (Q235) steel bar; taking 0.001 2 when the stirrups are made of HRB335 steel bar.

b. Reinforcement ratio of longitudinal steel bar should meet

$$\rho_{st} \geq \rho_{st,min} = \frac{A_{st,min}}{bh} = 0.08(2\beta_t - 1)\frac{f_{cd}}{f_{sd}} \tag{5-14}$$

Where $A_{st,min}$ is the minimum sectional area of longitudinal reinforcement of pure torsion member; h is length of the long side of rectangular cross-section; b is length of the short side of rectangular cross-section; ρ_{st} is reinforcement ratio of longitudinal torsional steel bar, $\rho_{st} = \frac{A_{st}}{bh}$; A_{st} is all cross-sectional area of longitudinal torsional reinforcement.

c. The rectangular cross-section member bearing the bending, shear and torsion, needn't to calculate the torsional capacity if meeting the condition as formula (5-15), and just need to reinforce according to structural requirements.

$$\frac{\gamma_0 V_d}{bh_0} + \frac{\gamma_0 T_d}{W_t} \leq 0.50 \times 10^{-3} f_{td} \ (kN/mm^2) \tag{5-15}$$

(3) The reinforcement calculation of member under the action of bending moment, shear and torque

For the member under the action of bending moment, shear and torque, its longitudinal steel bars and stirrups should be calculated as following provisions.

①The longitudinal anti-bending steel bars of bending members should calculate the reinforcement area according to the bearing capacity of cross-section, and be configured at the edge of tension zone.

②The longitudinal reinforcements and stirrups are calculated as shear-torsion reinforcement. The longitudinal anti-torsion steel bars calculated by anti-torsion are symmetrically configured along the edge of section, and the space is not more than 300mm, the vertical reinforcement in four corners of the rectangular cross-section must be configured. Stirrups are configured by product of area required by the shearing and torsion calculation.

③The reinforcement ratio of longitudinal steel bar should not less than the sum of minimum reinforcement ratio of longitudinal reinforcements of bending member and shear-torsion member. The reinforcement ratio of stirrups should not be less than minimum reinforcement ratio of stirrups of shear and torsion members.

5.4 Torsion member of T-shaped and I-shaped cross-section

T-shaped and I-shaped cross-section can be seen as a complex cross-section formed by simple rectangle (Figure 5-9), when calculating the cracking torque and ultimate torsion strength, the cross-section can be divided into several rectangular cross-section, and we can allocate the torque T_d by plastic resistance moment of torque in proportion to each rectangular block, in order to achieve the torque withstand by each rectangular block.

(1) For the rectangular block of the ribs

$$T_{wd} = \frac{W_{tw}}{W_t} T_d \qquad (5\text{-}16)$$

(2) For the rectangular block of the top flange

$$T'_{fd} = \frac{W'_{tf}}{W_t} T_d \qquad (5\text{-}17)$$

(3) For the rectangular blocks of the bottom flange

$$T_{fd} = \frac{W_{tf}}{W_t} T_d \qquad (5\text{-}18)$$

Where T_d is the combination design value of torque withstand by member's cross section; T_{wd} is the combination design value of torque withstand by the ribs; T'_{fd}, T_{fd} are the combination design values of torque withstand by the top flange and bottom flange.

The principle of division of each rectangular area is generally agreed that determining the cross-section of rib plate by the total height of section, and then dividing the compression flange and tensile flange (Figure 5-9).

The anti-torsion plastic resisting moment of rectangular section of the part of ribs, compression flange and tensile flange is calculated as follows.

Rib plate:
$$W_{tw} = \frac{b^2}{6}(3h - b) \qquad (5\text{-}19)$$

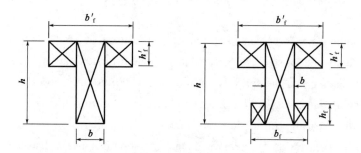

Figure 5-9 Partition diagram of T-shaped and I-shaped cross-section

Compression flange: $$W'_{tf} = \frac{h'^2_f}{2}(b'_f - b) \tag{5-20}$$

Tensile flange: $$W_{tf} = \frac{h^2_f}{2}(b_f - b) \tag{5-21}$$

Where b, h are the sizes of short side and long side of rectangular cross-section; b'_f, h'_f are width and height of compression flange of T-shaped and I-shaped cross-section; b_f, h_f are width and height of tensile flange of I-shaped cross-section.

In the calculation, the width of flange should be consistent with the provisions of $b'_f \leqslant b + 6h'_f$ and $b_f \leqslant b + 6h_f$. Therefore, the anti-torsion plastic resisting moment of T-shaped cross-section is

$$W_t = W_{tw} + W'_{tf} \tag{5-22}$$

The total anti-torsion plastic resisting moment of I-shaped cross-section is

$$W_t = W_{tw} + W'_{tf} + W_{tf} \tag{5-23}$$

The section design of T-shaped cross-section member under the action of the bending moment, shear and torque could be carried out by the following methods:

(1) Calculating the required cross-sectional area of longitudinal reinforcement according to the bending capacity of cross-section of bending member.

(2) Calculating the cross-sectional area of stirrups under shear according to bearing capacity under the action of shear and torsion, and the cross-sectional area of longitudinal reinforcement and stirrups under torsion according to bending capacity.

(3) Superposing the two cross-sectional area of longitudinal steel and stirrups obtained as above, then obtain the last required cross-section area of longitudinal steel, and configure the reinforcement in the appropriate location.

5.5 Torsion member with box section

In bridge engineering, box sections with the advantages of bigger bending rigidity and torsional stiffness, can withstand bending moments of different signs, lightweight structure, are widely used in the continuous girder bridge, curved bridge and the viaduct in the city.

The results of experiment show that the torsional capacity of box girders is similar to solid rectangular beam.

(1) When the ratio of wall thickness of box girder and width of the corresponding direction is: $t_2/b \geqslant 1/4$ or $t_1/h \geqslant 1/4$, its torsional capacity can be calculated as rectangular cross section with the same dimensions (take the empty part of the box as entity), shown in Figure 5-10.

(2) When $1/10 \leqslant t_2/b < 1/4$ or $1/10 \leqslant t_1/h \leqslant 1/4$, the resistance capacity of the member is approximately multiplied by a reduction factor β_a. As a result, the torsional capacity of shear and torsion member with box section is calculated as

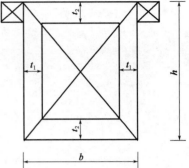

Figure 5-10 Box section member

$$\gamma_0 T_d \leqslant T_u = 0.35\beta_a \beta f_{td} W_t + 1.2\sqrt{\zeta} \frac{f_{sv} A_{sv1} A_{cor}}{S_v} \quad (\text{N} \cdot \text{mm}) \qquad (5\text{-}24)$$

Where β_a is the reduction factor of effective thickness of box section. When $0.1b \leqslant t_2 \leqslant 0.25b$ or $0.1h \leqslant t_1 \leqslant 0.25h$ take small value of $\beta_a = 4\dfrac{t_2}{b}$ and $\beta_a = 4\dfrac{t_1}{h}$. When $t_2 > 0.25b$ or $t_1 > 0.25h$, take $\beta_a = 1.0$.

5.6 Construction requirements

(1) As the external load torque is balanced by the resisting moment of torsional reinforcement, so in the premise of ensuring the necessary protection layer, stirrups and longitudinal reinforcement should be arranged as far as possible around the surface of the component in order to increase the torsional effect. Vertical reinforcement must be arranged in the inside of stirrup, limiting its expansion by stirrups.

(2) According to the requirements of torsional strength, the spacing of longitudinal torsion bars should be less than 300mm, and the amount should be four at least, arranged in four corners of rectangular cross-section of the member, and its diameter should not be less than 8mm. Enough anchorage length should be reserved at the end of longitudinal reinforcement.

(3) If the erect steel and longitudinal crack-resistance reinforcements on the both sides of beam ribs are anchored reliably, they can be taken as torsion bar. In the side of bending steel, large diameter steel can be used to meet the needs of resisting bending moment and torque.

(4) The torsional stirrups must be made to be closed (Figure 5-11), and anchor the stirrups in the concrete core with hooks at the angle of 135°, the anchor length is about equal to 10 times the diameter of stirrups. The maximum stirrup spacing according to the requirements under torsion should not be higher than 1/2 of beam height and not bigger than 400mm, and not be larger than the maximum spacing of shear stirrups. The diameter of stirrup should not be less than 8mm, and not less than 1/4 diameter of the main bar.

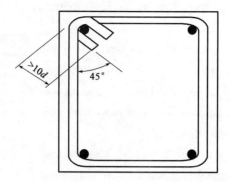

Figure 5-11 Schematic diagram of closed stirrups

(5) For the torsional member which has complex cross-section formed by a number of rectangular cross-section, such as T-shaped, L-shaped and I-shaped cross-section, each torsional reinforcement must be configured to cage-shaped frame, and make the torsion bar in the each rectangular element within the complex cross section interwoven firmly and to form a whole.

Chapter 6
Calculation of Strength of Axially Loaded Members

6.1 Introduction

1) Concept of axially loaded members

When a member is subjected to an axial compressive force, it is called the axially loaded member. Sometimes, it is also called the column. Besides the compressive force, there is no bending moments acting on the section of an axially loaded member.

In actual practice, there are no perfectly axially loaded members, due to non-uniformity of material, tolerable error in construction, and possible eccentricity in loading, etc. However, a discussion of such members provides an excellent starting point for explaining the theory involved in designed real columns with their eccentric loads. If the deviation of action line of a load to the section centroid is very small, the members subjected to such a negligible moment can be designed as the axially loaded members in practice, for example, the web poles in the reinforced concrete truss arch. The calculation theory of axial compression member is simple. It is mainly used to estimate the dimension of section and check the strength.

2) Types of axially loaded reinforced concrete members

According to the contribution and configuration of the stirrups, the axially loaded members can be divided into the following two categories:

(1) The tied-stirrup column, which has a series of closed stirrups, shown in Figure 6-1a);

(2) The spiral-stirrup column, which has a series of spiral stirrups, shown in Figure 6-1b).

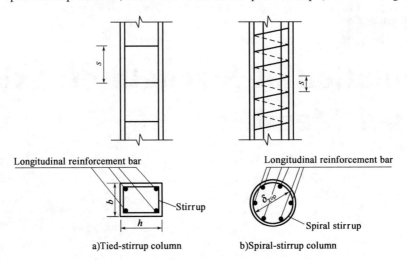

Figure 6-1　Types of axially loaded reinforced concrete member

3) Contribution of stirrup and longitudinal reinforcement

(1) Stirrup: Stirrups constitute the skeleton together with longitudinal bars to provide convenience for construction and avoid local bulking of longitudinal bars. The spiral stirrup also restricts internal concrete and improves strength and ductility of members.

(2) Longitudinal reinforcement: These bars bear the compressive force which can reduce the dimension of section, subject to possible small bending moment and avoid brittle failure of concrete members.

6.2　Calculation on axially loaded members with longitudinal bars and tied stirrups

1) Failure mode

The axially loaded members are classified into the short column and the slender column according to the slenderness ratio of a column.

(1) Short column

①Experimental analysis

As the axial compression force P increases, column A (Figure 6-2) will be compressed. The compressive deformation of concrete and reinforcement bars will be produced.

When the axial compression force P reaches about 90% the ultimate load, the longitudinal cracks appear in the concrete in the middle of column, and then the local cover scales off. The longitudinal reinforcements between stirrups are buckled and extruded to exterior. Then, the concrete is crushed and the failure of the experimental column occurs (Figure 6-3). The failure characteristic of short column displays the material failure, i. e. the concrete is crushed.

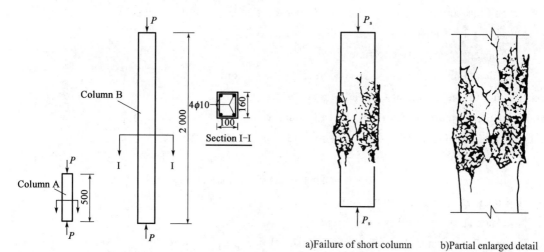

Figure 6-2   Sample of axial compression member (unit: mm)   Figure 6-3   Failure mode of the axial compression short column

The experimental results show the compressive strain of concrete is about 2×10^{-3} and the compressive strength reaches ultimate value when the failure of short column occurs. Meantime, the longitudinal ordinary hot-rolled reinforcement bars would yield at this time.

②Calculation on strength

The failure load P_s is carried by reinforcement and concrete. According to the equilibrium of axial forces, the following relationship can be obtained.

$$P_s = f_c A_c + f'_s A'_s \tag{6-1}$$

(2) Slender column

①Experimental analysis

Under a little compression force P, column B (Figure 6-2) is only subjected to compressive stress. As the compression force increases, the column will deform laterally. The compressive stress on the concave side of slender column is larger than that on the convex side. The lateral deformation increases quickly when the load approaches the failure value, which results in a bulking failure in sudden. The total failure process is described as follows: on the concave side of slender column, the longitudinal cracks appear firstly; then the concrete is crushed; the longitudinal reinforcements are bended to the exterior of column; and the concrete cover is spalled off. On the convex side of slender column, the compressive stress suddenly changes into tensile stress and the horizontal cracks appear (Figure 6-4).

②Calculation on strength

The experimental results show the bulking failure generates in the slender column. The bearing capacity of slender column is less than that of short column with the same section, material property, and reinforcement. The strength of slender column is calculated by the bearing capacity of the corresponding short column to combine a reduction coefficient φ^0 (this coefficient is called the stability coefficient).

$$P_l = \varphi^0 P_s \qquad (6\text{-}2)$$

Where P_s is the axial ultimate compression force of short column; P_l is the bulking compression force of corresponding slender column.

2) Stability coefficient

(1) Concept of stability coefficient

The calculation coefficient φ, considering the reduction of bearing capacity due to the secondary or additional effect generated by the increase of the slenderness ratio, is called the stability coefficient of the axially loaded members. The stability coefficient is the ratio of the critical load capacity P_l of slender column to the ultimate axial compression force P_s of short column. This coefficient indicates the reduction degree of bearing capacity of slender column.

(2) Calculation of stability coefficient

Based on the material mechanics, the critical compression force under all supporting conditions for column can be written as

$$P_l = \frac{\pi^2 EI}{l_0^2} \qquad (6\text{-}3)$$

Figure 6-4 Failure mode of the slender column
a) Failure of slender column b) Partial enlarged detail

Where EI is the flexural rigidity of column; l_0 is the calculating length of column.

Substituting Equations (6-3) and (6-1) into Equation (6-2), as

$$\varphi^0 = \frac{P_l}{P_s} = \frac{\pi^2 EI}{l_0^2 (f_c A + f'_s A'_s)} = \frac{\pi^2 EI}{l_0^2 A (f_c + f'_s \rho')} \qquad (6\text{-}4)$$

Where $\rho' = \dfrac{A'_s}{A}$; A is the cross-section area of column; A'_s is the section area of longitudinal reinforcement.

Considering the cracking of concrete, the rigidity would reduce by 30% ~ 50% at the elastic stage. The EI in Equation (6-4) should be revised and substituted by $\beta_1 E_c I_c$, where β_1 is the reduction coefficient of rigidity. The Equation (6-4) is re-expressed as

$$\varphi^0 = \frac{\pi^2 \beta_1 E_c I_c}{l_0^2 A (f_c + f'_s \rho')} = \frac{\pi^2 \beta_1 E_c}{f_c + f'_s \rho'} \cdot \frac{I_c}{A l_0^2} \qquad (6\text{-}5)$$

Substituting φ^0, f_c, f'_s by φ, f_{cd}, f_{sd} respectively and considering the gyration radius $r = \sqrt{I_c/A}$, the slenderness ratio $\lambda = l_0/r$, Equation (6-5) can be expressed as

$$\varphi = \frac{\pi^2 \beta_1 E_c}{f_{cd} + f'_{sd} \rho'} \cdot \frac{1}{\lambda^2} \qquad (6\text{-}6)$$

(3) Influencing factors for stability coefficient

The stability coefficient φ mainly depends on the slenderness ratio λ when the material and ratio of longitudinal reinforcement have been determined in terms of Equation (6-6). The effect of concrete strength and reinforcement type on φ is slight. Therefore, the stability coefficient φ is only determined by the slenderness ratio in the **CHBC** (Table 1-10).

Chapter 6 · Calculation of Strength of Axially Loaded Members

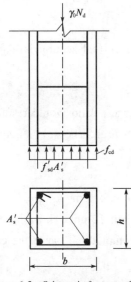

Figure 6-5 Schematic for strength of normal section for tied column

3) Strength of normal section

The **CHBC** defines the calculating equation of the strength of normal section (Figure 6-5) for the axially loaded members with longitudinal reinforcement and tied stirrup as

$$\gamma_0 N_d \leq N_u = 0.9\varphi(f_{cd}A + f'_{sd}A'_s) \tag{6-7}$$

Where N_d is the design value of axial force; φ is the stability coefficient seen in appendix Table 1-10; A is the area of cross-section; A'_s is the total area of longitudinal reinforcement in the section; f_{cd} is the design axial compressive strength of concrete; f'_{sd} is the design compressive strength of longitudinal reinforcement.

When the longitudinal steel ratio is more than 3%, $\rho' = \dfrac{A'_s}{A} > 3\%$, the area of cross-section A in Equation (6-7) should be substituted by the net section area of concrete A_n, where $A_n = A - A'_s$.

The calculation process for the strength of tied column consists of two parts, as follows:

(1) Section design

When the section size, computing length l_0, axial compressive strength of concrete, design compressive strength of reinforcement, and combined axial compressive force of N_d are known, the area of longitudinal bar A'_s needs to be solved by the following steps:

①Computing slenderness ratio and obtaining stability coefficient φ from appendix Table 1-10;

②Computing the area of longitudinal bar A'_s by $N_u = \gamma_0 N_d$ in Equation (6-7), where γ_0 is structural importance coefficient. Then A'_s is obtained from Equation (6-8).

$$A'_s = \frac{1}{f'_{sd}}\left(\frac{\gamma_0 N_d}{0.9\varphi} - f_{cd}A\right) \tag{6-8}$$

③Selecting the longitudinal bar according to the construction requirements and A'_s.

(2) Section strength check

When the section size, computing length l_0, the total area of longitudinal bar A'_s, axial compressive strength of concrete, design compressive strength of reinforcement, and combined axial compressive force of N_d are known, the strength of section N_u needs to be solved.

The checking steps are as follows:

①Checking the construction requirements of longitudinal bars and stirrups;

②Obtaining the stability coefficient φ from appendix Table 1-10 according to the size of section and slenderness ratio;

③Computing the strength of normal section of axially loaded column N_u, the N_u should satisfy $N_u > \gamma_0 N_d$.

4) Construction requirements

(1) Concrete

The C25 ~ C40 concretes are usually used because the strength in normal section is mainly provided by concrete.

(2) Section size

The section size should not be too small because the larger slenderness ratio would result in more reduction of strength, so that the material strength cannot be fully used. The side length of section should be more than 250mm.

(3) Longitudinal bars

①The HPB300, HRB400, HRBF400 and HRB500 longitudinal reinforcements are usually used. The diameter of longitudinal reinforcement should not be less than 12mm. The minimum numbers of longitudinal bars are 4 in each section. The number of the longitudinal bar at the each corner of section is not less than 1.

②The net spacing between longitudinal bars should not be less than 50mm, and not be more than 350mm. For the prefabricated concrete members, in which the concrete is horizontally poured, the minimum net spacing between longitudinal bars satisfies the requirement for flexural members. The requirement for the minimum concrete cover of longitudinal bars can be seen in appendix Table 1-8.

③The longitudinal bar ratio should consider concrete creep and the possible small eccentric bending moment in axially loaded members. The minimum longitudinal bar ratio ρ_{min} (%) is prescribed in the ***CHBC***, which can be seen in appendix Table 1-9. The longitudinal bar ratio can not be more than 5% the gross cross-sectional area of a column, which is usually 1% ~ 2%.

(4) Stirrups

①The stirrups in the tied column must be closed. The diameter of stirrups should not be less than 1/4 diameter of longitudinal bars and 8mm.

②The spacing of stirrups may not be more than 15 times of the diameter of longitudinal bar, the minimum length of side of section (for circular section, 0.8 times diameter), and 400mm.

③Within the overlapping part of longitudinal bars, the spacing of stirrups may not be more than 10 times of the diameter of longitudinal bars and 200mm.

④When the sectional area is more than 3% concrete sectional area, the spacing of stirrups may not be more than 10 times of the diameter of longitudinal bars and 200mm.

6.3 Axially loaded members with longitudinal bars and spiral stirrups

1) Failure feature

For the axially loaded members with longitudinal bars and spiral stirrups, if the spiral stirrups are densely placed, the core will be able to resist an appreciable amount of additional load beyond the design load. The densely distributed loops of the spiral form a cage with the longitudinal bars, confines effectively the concrete. As a result, the transverse deformation of the core is confined and the strength of column can be improved.

The axial compression force-strain curves are shown in Figure 6-6. The curves of tied and spiral columns are approximately the same when the concrete compressive strain ε_c is less than 0.002. When the axial compression force increases continually and the compressive strains of concrete and bar reach 0.003~0.0035, the longitudinal bars yield. And then the concrete cover spalls off. As a result, the effective area of section decreases. In this case, the core concrete is confined by the spirals under the triaxial compression condition. When the actual compressive strength is beyond the axial compressive strength f_c, the strength loss of spalled concrete can be compensated and the axial force re-increases with strain. As the axial force increases and the tensile force of spirals enhances to yield load, the spirals cannot confine the transverse deformation of core concrete. And then the concrete will be crushed and the column fails. When the loads reach the second peak value and the compressive strain may reach more than 0.01.

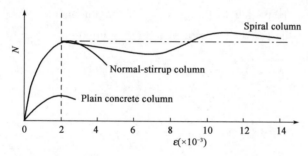

Figure 6-6 Axial force-strain curves for axially loaded column

It can be seen from Figure 6-6 that the ductility of spiral column is very good and its deformation performance is better than that of tied column.

2) Calculation on normal section strength

(1) Calculation equation

The failure feature in normal section for spiral column is that the core concrete is crushed along with yielding of the longitudinal bars. The concrete cover will spall off before failure occurs.

According to equilibrium equation in calculation schematic in Figure 6-7, the strength of spiral column can be expressed as

$$N_u = f_{cc}A_{cor} + f'_s A'_s \tag{6-9}$$

Where f_{cc} is the compressive strength of core concrete under the triaxial compression; A_{cor} is the area of core concrete; A'_s is the area of longitudinal bar.

The constraint effect of spiral on the core concrete can result in the increase of the concrete strength. According to the experimental results of triaxial compression for cylinder concrete, the axial compressive strength of core concrete can be expressed as

$$f_{cc} = f_c + k\sigma_2 \tag{6-10}$$

Where σ_2 is the radial compressive stress on core concrete.

When the spiral column fails, the spiral reaches yield strength and provides the lateral compressive stress σ_2. The spacing of spiral is s. The equilibrium equation of the isolated element along spirals (Figure 6-8) is

$$\sigma_2 d_{cor} s = 2 f_s A_{s01}$$

i. e.

$$\sigma_2 = \frac{2 f_s A_{s01}}{d_{cor} s} \qquad (6\text{-}11)$$

Where A_{s01} is the section area of single spiral stirrup; f_s is the tensile strength of spiral stirrup; s is the spacing of spirals stirrups; d_{cor} is the diameter of core concrete, $d_{cor} = d - 2c$; c is the concrete cover for longitudinal bars.

Based on the principle of the equal volume, the area of the spirals with spacing s is converted to the area A_{s0} of longitudinal bar. The A_{s0} is called as the conversion area of indirect reinforcing bar, and is expressed by

$$\pi d_{cor} A_{s01} = A_{s0} s, \quad A_{s0} = \frac{\pi d_{cor} A_{s01}}{s} \qquad (6\text{-}12)$$

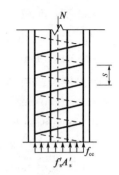

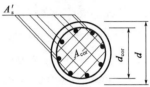

Figure 6-7 Calculation schematic for strength of normal section for spiral column

Substituting Equation (6-12) into Equation (6-11) gives

$$\sigma_2 = \frac{2 f_s A_{s01}}{d_{cor} s} = \frac{2 f_s}{d_{cor} s} \cdot \frac{A_{s0} s}{\pi d_{cor}} = \frac{2 f_s A_{s0}}{\pi (d_{cor})^2} = \frac{f_s A_{s0}}{2 \dfrac{\pi (d_{cor})^2}{4}} = \frac{f_s A_{s0}}{2 A_{cor}}$$

Substituting $\sigma_2 = \dfrac{f_s A_{s0}}{2 A_{cor}}$ into Equation (6-10) gives

$$f_{cc} = f_c + \frac{k f_s A_{s0}}{2 A_{cor}} \qquad (6\text{-}13)$$

Substituting Equation (6-13) into Equation (6-9) and considering the effect of actual indirect reinforcing bar, the strength of normal section for spiral column should satisfy

$$\gamma_0 N_d \leq N_u = 0.9 (f_{cd} A_{cor} + k f_{sd} A_{s0} + f'_{sd} A'_s) \qquad (6\text{-}14)$$

Where k is the influencing coefficient of indirect reinforcing bar. When the concrete strength grade is equal to or lower than C50, $k = 2.0$. When the grade is between C50 and C80, $k = 1.7 \sim 2.0$ is used, and the k takes value according to linear interpolation method.

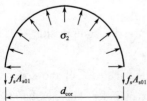

Figure 6-8 Response of spiral column

Chapter 6 · Calculation of Strength of Axially Loaded Members

Note: the spirals stirrups can only improve strength of column, no stability load. Therefore, the strength of slender columns and medium length columns should be computed based on the theory of tied column.

(2) Specific items of calculation equation in the **CHBC**

①The strength of spiral column may not be more than the 50% bearing capacity of tied column [Equation (6-7)], so that the concrete cover of spiral stirrups will not spall off before service load, i.e.

$$0.9(f_{cd}A_{cor} + kf_{sd}A_{s0} + f'_{sd}A'_s) \leq 1.35\varphi(f_{cd}A + f'_{sd}A'_s) \qquad (6\text{-}15)$$

②The contribution of spirals don't need to be considered. The strength of spiral column is computed according to Equation (6-7) when the following items appear:

a. The slenderness ratio $\lambda = \dfrac{l_0}{i} \geq 48$, where i is the minimum gyration radius. For the column with circular section, the slenderness ratio $\lambda = \dfrac{l_0}{d} \geq 12$, where d is the diameter of circular section. This is because the spirals are not effective for the column with large slenderness ratio.

b. The result based on Equation (6-14) is less than value by Equation (6-7). When the cover is larger and the relative core concrete area is smaller, the accuracy for Equation is not high.

c. For $A_{s0} < 0.25A'_s$, the effect or contribution of spiral stirrups is slight.

The design and check of section for spiral column should be conducted according to Equation (6-14).

3) Construction requirements

(1) The longitudinal bars in spiral columns should be uniformly distributed along circle section. The area of longitudinal bars cannot be less than 0.5% area of core concrete. The longitudinal bar ratio is usually from 0.8% to 1.2%.

(2) The area of core concrete may not be less than 2/3 area of member.

(3) The diameter of spiral stirrups may not be less than 1/4 diameter of longitudinal bars and 8mm, and is usually 8~12mm. In order to ensure the contribution of spirals, the spacing between spirals should satisfy the following requirements:

①The S may not be more than 1/5 diameter d_{cor} of core concrete, i.e. $s \leq \dfrac{1}{5}d_{cor}$;

②For the convenience of construction, the S may not be more than 80mm, and not be less than 40mm.

[**Example 6-1**]   Considering an axially loaded column of 40 cm in diameter and 50cm in length. The concrete is C30. The longitudinal reinforcing bar is 8 ⌽ 20 HRB400. The diameter of core concrete d_{cor} is 35cm. The spiral stirrup is ⌽ 20 HPB300 with 4cm spacing. The N_u is required to be solved.

Solution

Check if the column can be calculated according to spiral column theory.

(1) Slenderness ratio

$$\lambda = \frac{l_0}{d} = \frac{0.5 \times 5}{0.4} = 6.25 < 12$$

So it can be calculated based on the equations of spiral column.

(2) Core concrete

$$A_{cor} = \frac{\pi d_{cor}^2}{4} = 96\,163\,\text{mm}^2 > \frac{2}{3}A = \frac{2}{3} \times \frac{\pi}{4} \times 400^2 = 83\,733\,\text{mm}^2$$

(3) Area of longitudinal reinforcement

$$A'_s = 2\,513\,\text{mm}^2 > 0.5\% A_{cor} = 481\,\text{mm}^2$$

(4) Equivalent area of spiral stirrup

$$A_{s0} = \frac{\pi d_{cor} A_{s01}}{s} = \frac{3.14 \times 350 \times 201.1}{40} = 5\,525\,\text{mm}^2 > 0.25 A'_s = 0.25 \times 2\,513 = 628\,\text{mm}^2$$

$$S = 40\,\text{mm} < \frac{1}{5}d_{cor} = \frac{1}{5} \times 350 = 70\,\text{mm},\ \text{and}\ s = 40\,\text{mm} < 80\,\text{mm}\ \text{and}\ \not< 40\,\text{mm}$$

So N_u is obtained as

$$N_u = 0.9(f_{cd}A_{cor} + kf_{sd}A_{s0} + f'_{sd}A'_s)$$
$$= 0.9 \times (13.8 \times 96\,163 + 2 \times 250 \times 5\,525 + 330 \times 2\,513)$$
$$= 4\,918.84 \times 10^3\,\text{N} = 4\,918.84\,\text{kN}$$

Check if the cover shall fall off.

$$N'_u = 0.9\varphi(f_{cd}A + f'_{sd}A'_s)$$
$$= 0.9 \times \left(13.8 \times \frac{\pi}{4} \times 400^2 + 330 \times 2\,513\right) = 2\,274.42\,\text{kN}$$

$$1.5 N'_u = 1.5 \times 2\,274.42 = 3\,411.63\,\text{kN} < N_u$$

So the cover may fall off. The strength of the column is $3\,411.63\,\text{kN}$.

The above example shows that the spiral stirrups can confine the core concrete and improve the bearing capacity. It is similar to the flexural member that the strength of column cannot be enhanced infinitely by only increase of the steel bar ratio. The condition of $N_u < 1.5 N'_u$ is the upper limit for improving the strength of column by spiral stirrups.

Chapter 7
Calculation of Strength of Eccentrically Loaded Members

A member subjected to a compressive force deviating from the centroid of section is named as the eccentrically loaded member as shown in Figure 7-1.

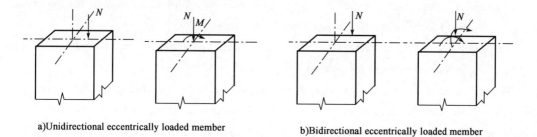

a) Unidirectional eccentrically loaded member   b) Bidirectional eccentrically loaded member

Figure 7-1 Eccentrically loaded member

A member whose section is subjected to the compression forces as well as the bending moments is named as the bending-compression member as shown in Figure 7-2. The calculation principle of strength is the same as that of the eccentrically loaded member.

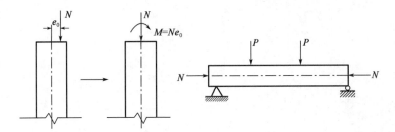

Figure 7-2 Eccentrically loaded members and bending-compression members

The eccentrically loaded members are widely used in bridge engineering, such as piers, towers, and arch rings. Their common section forms include rectangular, circle, I-shape, T-shape, box-section, shown in Figure 7-3. The chapter introduces the calculation on strength of eccentrically loaded members.

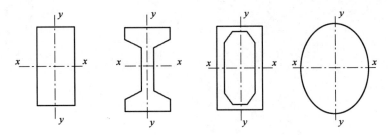

Figure 7-3 Section forms for eccentrically loaded members

7.1 Failure modes and mechanical characteristics

The mechanical characteristics and failure modes of eccentrically loaded members are introduced by the experimental results and phenomena.

1) Failure modes

Two major failure modes occur for eccentrically loaded member under various eccentricities.

(1) Tension failure or compression failure with a large eccentricity

The compression failure with a large eccentricity occurs when the relative eccentricity is rather large and the amount of tensile reinforcing bar is small.

The failure mode which is similar to that of the double-tendon section beam. The one side of the section suffers tensile forces, while the other is subjected to compressive forces. The tensile reinforcing bars will yield first, and then the compressed concrete reaches its ultimate strength and is crushed. The reinforcements in the compression zone also reach strength.

(2) Compression failure or compression failure with small eccentricity

The compression failure with a small eccentricity occurs as follows: the eccentricity is small; the eccentricity is large, and the amount of tensile steel bar is also large; the eccentricity is small, the amount of steel bar on the opposite side of eccentricity is small but the amount of steel bar on the side of eccentricity is excessive.

The failure mode is that the concrete on the side of eccentricity is crushed first, and then the steel bar reaches yield strength. The reinforcement on the opposite side of eccentricity may be in tension or in compression, does not reach yield strength. All types are shown in Figure 7-4.

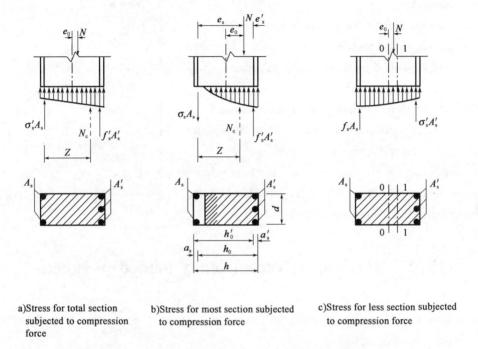

a) Stress for total section subjected to compression force

b) Stress for most section subjected to compression force

c) Stress for less section subjected to compression force

Figure 7-4 Calculation schematics for short column with small eccentricity

2) Boundary of tension failure and compression failure

(1) Definition of balanced failure

When the tensile reinforcing bars reach yield strength, the strain of the compressed concrete reaches the ultimate value and the concrete is crushed at the same time. The failure of the section is called as the balanced failure.

(2) Criterion

The failure mode of the eccentrically loaded column is similar to that of flexural members. Therefore, its judgement criterion is the relative balanced compression height ξ_b for the flexural members.

For $\xi > \xi_b$, the failure mode is the compression failure.

3) Relationship between moment and axial force

According to a large number of experimental results, the relationship between the moment and the axial force is expressed as M-N curve as shown in Figure 7-5.

(1) When the M-N combination in the cross-section falls outside of the curve or on the curve, the member fails. The ab is a quadratic parabola. The cd is an approximate linear quadratic function curve.

(2) $e_0 = \dfrac{M_d}{N_d} = \tan\theta$. The θ is the larger, the θ_0 is the greater. The curve cb corresponds to the compression failure. The curve ab corresponds to the tension failure.

(3) Three characteristic points: Point c (axial compression member), point b (boundary of tension failure and compression failure), and point a (flexural member).

(4) The characteristics of the curve is as follows:

The ab segment (tensile failure section) for large eccentric compression: the increase of axial pressure will improve the flexural capacity.

The bdc segment (compression failure section) for small eccentric compression: the increase in axial pressure will reduce the flexural capacity.

The M-N curve shows moment and axial force restrict each other. Therefore, they should be considered at the same time in calculation on real projects.

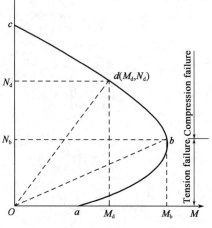

Figure 7-5　M-N curve

7.2　Buckling of eccentrically loaded members

The failure types of the eccentric members are different under various slenderness ratios (short column and long column).

1) Failure types of eccentrically loaded members

(1) Short column

The additional deformation induced by the eccentric compression force is not considered, because the lateral deflection is small. It represents the material failure type as the member loses its carrying capacity when the material reaches its ultimate strength.

(2) Long column

The additional deformation induced by the eccentric compression force can not be ignored, because the lateral deflection is large. The additional moment will increase the speed of failure. The final member failure occurs when its material reaches the ultimate strength.

(3) Slender column

For the column with the large slenderness ratio, when the eccentric compression force reaches the maximum value, the lateral deflection u of columns increases suddenly. During the total process, the strains of steel and the concrete do not reach the ultimate value. Therefore, the eccentrically loaded member will lose its carrying capacity and stability in advance. It is shown by the curve OE in Figure 7-6.

Based on the above result, the slender columns should not be used in real projects. For the long columns, the effect of the additional eccentricity should be also considered.

2) Enhancement factor of eccentricity

For long columns, the effect of additional moments induced by eccentric compression is described by additional deflection (Figure 7-7).

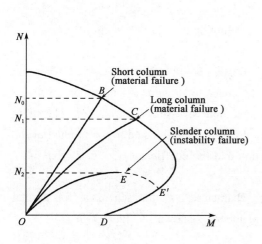

Figure 7-6　Effect of slenderness ratio

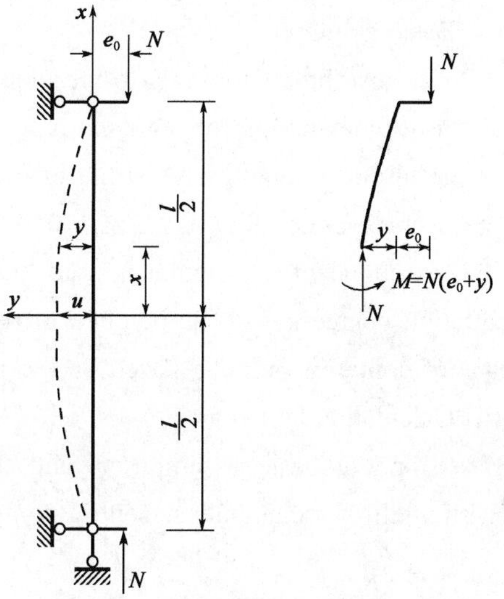

Figure 7-7　Calculation schematic for rectangular eccentric compression

$$M = N(e_0 + u) = N\frac{e_0 + u}{e_0}e_0$$

$$\eta = \frac{e_0 + u}{e_0} = 1 + \frac{u}{e_0}$$

$$M = N \cdot \eta e_0 \tag{7-1}$$

Where η is the enhancement factor.

Based on the **CHBC**, the η are defined as follows

$$\eta = 1 + \frac{1}{1\,300 e_0/h_0}(L_0/h)^2 \zeta_1 \zeta_2 \tag{7-2}$$

$$\zeta_1 = 0.2 + 2.7 \frac{e_0}{h_0} \leqslant 1.0 \tag{7-3a}$$

$$\zeta_2 = 1.15 - 0.01 \frac{l_0}{h} \leqslant 1.0 \tag{7-3b}$$

Where ζ_1 is the influencing coefficient of load eccentricity ratio on the curvature of section; ζ_2 is the influencing coefficient of slenderness ratio on the curvature of section.

Note: the **CHBC** states when $\frac{l_0}{i} > 17.5, \frac{l_0}{h}$ (rectangular section) $> 5, \frac{l_0}{d_1}$ (circular section) > 4.4, the effect of deformation on the eccentricity in the bending plane should be considered. In such a case, ηe_0 will be used to replace e_0.

7.3　Eccentrically loaded rectangular member

1）Strength of normal section

（1）Basic assumption

①Strain distribution submits to the plane section assumption.

②The tensile strength of concrete is neglected.

③The ultimate compressive strain of concrete $\varepsilon_{cu} = 0.0033 \sim 0.003$. $\varepsilon_{cu} = 0.0033$, when the concrete strength is less than C50. $\varepsilon_{cu} = 0.003$, when the concrete strength is larger than C80.

④The equivalent rectangular is used to express the compressive stress distribution. The stress concentration degree is f_{cd}. The height x of rectangular stress diagram is equal to the product of the height x_c of compression zone based on the plane section and the coefficient β, i.e. $x = \beta x_c$.

（2）Calculation equation

Based on the basic assumptions and the failure characteristics, the schematic for normal section strength of rectangular eccentric compression members is shown in Figure 7-8.

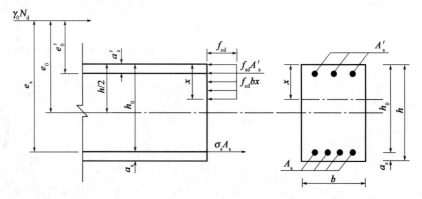

Figure 7-8　Schematic for strength of normal section for rectangular eccentric compression

The equilibrium equation along longitudinal axis can be expressed as

$$\gamma_0 N_d \leqslant N_u = f_{cd}bx + f'_{sd}A'_s - \sigma_s A_s \tag{7-4}$$

The equilibrium equation for A_s can be expressed as

$$\gamma_0 N_d e_s \leqslant f_{cd}bx\left(h_0 - \frac{x}{2}\right) + f'_{sd}A'_s(h_0 - a'_s) \tag{7-5}$$

The equilibrium equation for A'_s can be expressed as

$$\gamma_0 N_d e'_s \leqslant -f_{cd}bx\left(\frac{x}{2} - a'_s\right) + \sigma_s A_s(h_0 - a'_s) \tag{7-6}$$

The equilibrium equation for $\gamma_0 N_d$ can be expressed as

$$f_{cd}bx\left(e_s - h_0 + \frac{x}{2}\right) = \sigma_s A_s e_s - f'_{sd}A'_s e'_s \tag{7-7}$$

Where
$$e_s = \eta e_0 + h/2 - a_s \tag{7-8}$$
$$e'_s = \eta e_0 - h/2 + a'_s \tag{7-9}$$
$$e_0 = M_d/N_d$$

The explanation of these equations are as follows:

①The value of σ_s to A_s:

a. For $\xi = x/h_0 \leqslant \xi_b$, when the members are the compression member with a large eccentricity, $\sigma_s = f_{sd}$.

b. For $\xi = x/h_0 > \xi_b$, when the members are the compression member with a small eccentricity, σ_s should be calculated according to Equation (7-10) and satisfy $f'_{sd} \leqslant \sigma_s \leqslant f_{sd}$.

$$\sigma_s = \varepsilon_{cu} E_s \left(\frac{\beta h_0}{x} - 1 \right) \tag{7-10a}$$

By the empirical fitting formula,

$$\sigma_s = \frac{f_{sd}}{\xi_b - \beta}(\xi_b - \beta) \tag{7-10b}$$

②In order to ensure the compressive steel in large eccentric compression member reaches its design compressive strength just when the member fails, x must satisfy

$$x \geqslant 2a'_s \tag{7-11}$$

If $x < 2a'_s$, take $x = 2a'_s$. The stress distribution of cross section is shown in Figure 7-9. The resultant force positions of compressed concrete and compressed steel bar is coincident.

According to the equilibrium equation for compression reinforcement, it can be expressed as

$$\gamma_0 N_d e'_s \leqslant f_{sd} A_s (h_0 - a'_s) \tag{7-12}$$

③When the eccentricity is very small and the more longitudinal reinforcements A'_s is used near the side of axial force, the gravity center will deviate to the more reinforcements (Figure 7-10). In order to avoid inadequate reinforcements A_s, the following conditions should be satisfied

$$\gamma_0 N_d e' \leqslant f_{cd} bh \left(h'_0 - \frac{h}{2} \right) + f_{sd} A_s (h'_0 - a_s) \tag{7-13}$$

$$\begin{cases} h'_0 = h - a'_s \\ e' = h/2 - e_0 - a'_s \end{cases}$$

2) Calculation of eccentrically loaded rectangular member with asymmetric reinforcement

The asymmetric reinforcement is $A_s \neq A'_s$. The strength should be calculated after the type of member is identified based on the eccentricity.

(1) Section design

When $\eta e_0 \geqslant 0.3h_0$, the cross-section can be assumed to be with large eccentricity. When $\eta e_0 < 0.3h_0$, the cross-section can be assumed to be with small eccentricity.

After that, the eccentric compression members could be designed as follows:

①When $\eta e_0 \geqslant 0.3h_0$, the large eccentric compression member is considered.

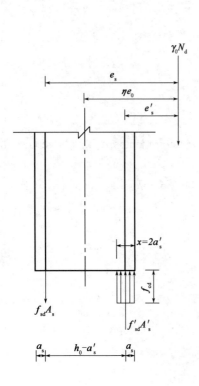
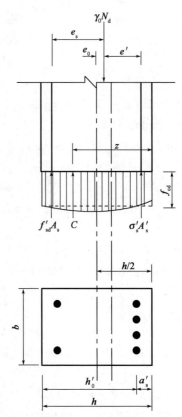

Figure 7-9 Calculation schematic for strength of normal section for compression member with large eccentricity

Figure 7-10 Calculation schematic for strength of normal section for compression member with small eccentricity

a. First case

Given: Section size $b \times h$, calculation length l_0, load effect N_d and M_d, material strength f_{cd}, f_{sd} and f'_{sd}, initial eccentricity e_0.

Finding: The amount of compression and tension reinforcements A'_s and A_s.

Solution:

I. Assuming $a_s, a'_s, h_0 = h - a'_s$, confirmating $\eta e_0 \geq 0.3 h_0$.

II. Calculating η, e_s and e'_s. $\eta = 1.0$, when it's the short column.

III. Drawing the force diagram, listing the basic formula.

IV. Incorprating equation $x = \xi_b h_0$.

V. Soluting A_s.

(a) Taking $\xi = \xi_b$, i. e. $x = \xi_b h_0$, the area of longitudinal bar A'_s in Equation (7-5) can be expressed as

$$A'_s = \frac{\gamma_0 N_d e_s - f_{cd} b h_0^2 \xi_b (1 - 0.5\xi_b)}{f'_{sd}(h_0 - a'_s)} \geq \rho'_{min} bh \tag{7-14}$$

Where ρ'_{min} is the minimum reinforcement ratio on compressive side of section. $\rho'_{min} = 0.2\% = 0.002$.

(b) If $A'_s < \rho'_{min} bh$ or the value of A'_s is negative, the section size is too large. Therefore, the reinforcements A'_s should be based on $A'_s \geq \rho'_{min} bh$. After that, The A_s can be calculated according to given A'_s. Otherwise a new design need to be executed.

(c) If $A'_s \geq \rho'_{min} bh$, substituting A'_s into baisc Equation, and the A_s can be expressed as

$$A_s = \frac{f_{cd} bh_0 \xi_b + f'_{sd} A'_s - \gamma_0 N_d}{f_{sd}} \geq \rho'_{min} bh \qquad (7\text{-}15)$$

Where ρ_{min} is the minimum reinforcement ratio at one side of section and its value is selected from appendix Table 1-9.

b. Second case

Given: Section size $b \times h$, calculation length l_0, load effect N_d and M_d, material strength f_{cd}, f_{sd} and f'_{sd}, initial eccentricity e_0, and the area of compression reinforcement A'_s.

Finding: The area of tension reinforcement A_s.

Solution:

I. Assuming $a'_s, h_0 = h - a'_s$.

II. Calculating η, e_s and e'_s. $\eta = 1.0$, when it's short column. Confirmating $\eta e_0 \geq 0.3 h_0$.

III. Drawing force diagram, listing the basic formula, calculating the depth of compression zone x.

$$x = h_0 - \sqrt{h_0^2 - \frac{2[\gamma_0 N_d e_s - f'_{sd} A'_s (h_0 - a'_s)]}{f_{cd} b}} \qquad (7\text{-}16)$$

(a) Assuming $2a'_s < x \leq \xi_b h_0$

$$A_s = \frac{f_{cd} bx + f'_{sd} A'_s - \gamma_0 N_d}{f_{sd}} \qquad (7\text{-}17)$$

(b) Assuming $x \leq \xi_b h_0$, and $x \leq 2a'_s$, taking $x = 2a'_s$

$$A_s = \frac{\gamma_0 N_d e'_s}{f_{sd}(h_0 - a'_s)} \qquad (7\text{-}18)$$

The A_s should satisfy the minimum reinforcement ratio in the construction requirements.

②When $\eta e_0 < 0.3 h_0$, the small eccentric compression member is considered.

a. First case

Given: Section size $b \times h$, calculation length l_0, load effect N_d and M_d, material strength f_{cd}, f_{sd} and f'_{sd}, initial eccentricity e_0.

Finding: The amount of compression and tension reinforcements A'_s and A_s.

Solution:

I. Assuming $a_s, a'_s, h_0 = h - a_s$.

II. Calculating η, e_s and e'_s. $\eta = 1.0$, when it's short column. Confirmating $\eta e_0 \geq 0.3 h_0$.

III. Drawing force diagram, listing the basic formula.

$A_s = \rho'_{min} bh = 0.002 bh$。

x can be obtained from Equation (7-6) and (7-10).

$$\gamma_0 N_d e'_s = -f_{cd} bx \left(\frac{x}{2} - a'_s\right) + \sigma_s A_s (h_0 - a'_s) \qquad (7\text{-}19)$$

And

$$\sigma_s = \varepsilon_{cu} E_s \left(\frac{\beta h_0}{x} - 1\right)$$

Then a unary cubic equation is expressed as

$$Ax^3 + Bx^2 + Cx + D = 0 \tag{7-20}$$
$$A = -0.5f_{cd}b \tag{7-21a}$$
$$B = f_{cd}ba'_s \tag{7-21b}$$
$$C = \varepsilon_{cu}E_sA_s(a'_s - h_0) - \gamma_0 N_d e'_s \tag{7-21c}$$
$$D = \beta\varepsilon_{cu}E_sA_s(h_0 - a'_s)h_0 \tag{7-21d}$$

and $e'_s = \eta e_0 - h/2 + a'_s$.

The relative compressive depth can be obtained as $\xi = x/h_0$ after the value of x is obtained from Equation (7-20).

Assuming $h/h_0 > \xi > \xi_b$, substituting $\xi = x/h_0$ into Equation (7-10), the stress of reinforcement σ_s can be calculated. Then substituting A_s, σ_s and x into Equation (7-4), the value of required reinforcement area A'_s can be calculated and should satisfy $A'_s \geq \rho'_{min} bh$.

Assuming $\xi \geq h/h_0$ and taking $x = h$, the calculation formula of A'_s is

$$A'_s = \frac{\gamma_0 N_d e_s - f_{cd}bh(h_0 - h/2)}{f'_{sd}(h_0 - a'_s)} \geq \rho'_{min}bh$$

Empirical formula

$$\sigma_s = \frac{f_{sd}}{\xi_b - \beta}(\xi - \beta), \quad -f'_{sd} \leq \sigma_s \leq f_{sd} \tag{7-22}$$

Substituting Equation (7-22) into (7-6), a quadratic equation is expressed as

$$Ax^2 + Bx + C = 0 \tag{7-23}$$

The coefficients can be obtained as

$$A = -0.5f_{cd}bh_0 \tag{7-24a}$$
$$B = \frac{h_0 - a'_s}{\xi_b - \beta}f_{sd}A_s + f_{cd}bh_0 a'_s \tag{7-24b}$$
$$C = -\beta\frac{h_0 - a'_s}{\xi_b - \beta}f_{sd}A_s h_0 - \gamma_0 N_d e'_s h_0 \tag{7-24c}$$

Since an approximate linear relationship lies in between steel stress σ_s and ξ in Equation (7-22), the above method can avoid the difficulty of calculating a cubic equation according to Equation (7-20). This approximate method is suitable for ordinary concrete with strength below C50.

b. Second case

Given: Section size $b \times h$, calculation length l_0, load effect N_d and M_d, material strength f_{cd}, f_{sd} and f'_{sd}, initial eccentricity e_0 and the area of compression reinforcement A'_s.

Finding: The area of tension reinforcement A_s.

Solution:

Assuming a_s, a'_s, then h_0, η, e_s and e'_s can be calculated as the same in front.

By Equation (7-5) the compressive depth x and the relative compressive depth $\xi = x/h_0$ can be obtained. Assuming $h/h_0 > \xi > \xi_b$, the member is a small eccentricity compression member. Substituting ξ into Equation (7-9), the stress σ_s can be calculated. By Equation (7-4) the value of required reinforcement A_s can be calculated.

Assuming $\xi \geq h/h_0$, the entire cross section is compressive. Substituting $\xi = h/h_0$ into

Equation (7-10), the stress σ_s of the steel area A_s can be calculated. Then the area of reinforcement A_{s1} can be calculated from Equation (7-4).

In order to avoid inadequate reinforcements A_s, the following should be satisfied

$$A_{s2} \geq \frac{\gamma_0 N_d e' - f_{cd} bh \left(h'_0 - \frac{h}{2}\right)}{f'_{sd}(h'_0 - a_s)} \tag{7-25}$$

Selecting the large value between A_{s1} and A_{s2}, and the results must satisfy the construction requirement.

(2) Check strength

Calculate the actual strength N_u of the initially designed eccentric compression member, then check if the strength $N_u = \gamma_0 N_d$ is satisfied. The strength should be checked not only in bending plane but also in perpendicular bending plane. The check in perpendicular bending plane is the same as the check of axially loaded member.

Given: Section size $b \times h$, calculation length l_0, load effect N_d and M_d, material strength f_{cd}, f_{sd} and f'_{sd}, initial eccentricity e_0, and the area of compression and tension reinforcement A'_s and A_s.

Finding: N_u.

① Check strength in bending plane

a. The check of large eccentric compression member.

(a) Check the amount of reinforcements to satisfy the construction requirement, calculation η.

(b) Listing the basic Equations (7-4) and (7-5) of large eccentricity compression member, take $\sigma_s = f_{sd}$, the resistance is N_u.

(c) x is calculated as

$$x = (h_0 - e_s) + \sqrt{(h_0 - e_s)^2 + 2 \times \frac{f_{sd} A_s e_s - f'_{sd} A'_s e'_s}{f_{cd} b}} \tag{7-26}$$

i. e.
$$\xi = \frac{x}{h_0}$$

(d) If $2a'_s \leq x \leq \xi_b h_0$, $N_u = f_{cd} bx + f'_{sd} A'_s - f_{sd} A_s$.

(e) If $2a'_s > x$, the strength of section can be calculated by Equation (7-12) as $N_u = \frac{f_{sd} A_s (h_0 - a'_s)}{\gamma_0 e'_s}$.

b. When $\xi > \xi_b$, the member is small eccentricity compression member.

Checking small eccentricity compression member.

(a) Checking reinforcement arrangement to satisfy the construction requirement.

(b) Listing the basic equations for small eccentricity compression member. The resistance is N_u. By Equation (7-7) and (7-10), i. e.

$$f_{cd} bx \left(e_s - h_0 + \frac{x}{2}\right) = \sigma_s A_s e_s - f'_{sd} A'_s e'_s$$

and
$$\sigma_s = \varepsilon_{cu} E_s \left(\frac{\beta h_0}{x} - 1\right)$$

(c) A unary cubic equation about x is expressed as

$$Ax^3 + Bx^2 + Cx + D = 0 \tag{7-27}$$

The coefficients can be expressed as:

$$A = 0.5f_{cd}b \tag{7-28a}$$

$$B = f_{cd}b(e_s - h_0) \tag{7-28b}$$

$$C = \varepsilon_{cu}E_s A_s e_s + f'_{sd}A'_s e'_s \tag{7-28c}$$

$$D = -\beta\varepsilon_{cu}E_s A_s e_s h_0 \tag{7-28d}$$

Where $e'_s = \eta e_0 - h/2 + a'_s$.

If the stress σ_s of reinforced area A_s can be expressed by ξ with linear relationship, i.e. Equation (7-22), a quadratic equation about x can be expressed as

$$Ax^2 + Bx + C = 0 \tag{7-29}$$

The coefficients can be expressed as

$$A = 0.5f_{cd}bh_0 \tag{7-30a}$$

$$B = f_{cd}bh_0(e_s - h_0) - \frac{f_{sd}A_s e_s}{\xi_b - \beta} \tag{7-30b}$$

$$C = \left(\frac{\beta f_{sd}A_s e_s}{\xi_b - \beta} + f'_{sd}A'_s e'_s\right)h_0 \tag{7-30c}$$

By Equation (7-27) or (7-29), the height of compressive zone x and the corresponding stress ξ can be obtained.

(d) When $\xi > \xi_b$, substituting ξ into the basic Equation, the stress σ_s of reinforced area A_s can be calculated. Then, by Equation (7-4), the strength N_u of section can be calculated and checked.

(e) When $\xi > h/h_0$, substituting $\xi = h/h_0$ into the basic Equation, the stress σ_s of reinforced area A_s can be calculated. By Equation (7-4), the strength N_{u1} of section can be calculated. As the entire cross section is under compression, the possibility of damage of the section edge which is far from the longitudinal compression should be considered. Therefore, the strength of section N_{u2} can be calculated by Equation (7-13).

(f) The strength of section N_u is equal to the less value of N_{u1} and N_{u2}. It means that the strength of section could be decided by the less value.

②Check strength in perpendicular bending plane

By the same method as that of checking strength of axially loaded members.

When the ratio of slenderness is $l_0/b > 8$, φ is obtained from appendix Table 1-10 and the strength of section can be expressed as

$$N_{u\perp} = 0.9\varphi[f_{cd}bh + f'_{sd}(A_s + A'_s)]$$

③The strength of the member $N_u = \min\{N_u, N_{u\perp}\}$

3) Construction requirements for eccentrically loaded rectangular members

(1) Section size

$b_{\min} \geq 300\text{mm}$ and $h/b = 1.5 \sim 3$. The length should be a multiple of 50mm in order to construct conveniently. The long side on rectangular section should be in the direction of moment.

(2) Longitudinal reinforcement ratio

The longitudinal reinforcement bars should distribute along the short side on section.

①the minimum longitudinal reinforcement ratio ρ_{min} on the total section is 0.5%; ρ_{min} is not less than 0.6% when the grade of concrete is large than C50.

②the minimum longitudinal reinforcement ratio ρ_{min} on one side section is 0.2%.

Longitudinal reinforcement ratios are usually as follows:

①The compression members with large eccentricity: $\rho = 1\% \sim 3\%$;

②The compression members with small eccentricity: $\rho = 0.5\% \sim 2\%$.

For $h \geqslant 600$mm, the longitudinal constructional steels with 10 ~ 16 mm diameter should be configured along the long side of section. The additional stirrups or complex stirrups could be also used according to actual condition.

[**Example 7-1**] The eccentric compression reinforced concrete members, The section size is $b \times h = 300$mm × 400mm. The calculation length of two directions (bending moment and its perpendicular direction) is $l_0 = 4$m. The design value of axial force $N_d = 232$kN. The design value of bending moment $M_d = 148$kN · m. The C30 concrete is horizontally poured in member. The longitudinal reinforcements is HRB400. The environment condition is I and the safety classes is the first. The reinforcement bars are required to design and check section.

Solution: $f_{cd} = 13.8$MPa, $f_{sd} = f'_{sd} = 330$MPa, $\xi_b = 0.56$, $\gamma_0 = 1.1$.

(1) Section design

The axial force $N = \gamma_0 N_d = 255.2$kN, the bending moment $M = \gamma_0 M_d = 162.8$kN · m, the eccentricity e_0 will be available by

$$e_0 = \frac{M}{N} = \frac{162.8 \times 10^6}{255.2 \times 10^3} = 638 \text{mm}$$

As the slenderness ratio in bending plane is $\frac{l_0}{h} = \frac{4\,000}{400} = 10 > 5$, the modification factor of eccentricity η should be considered. The η can be calculated by Equation (7-2). Assuming $a_s = a'_s = 45$mm, so $h_0 = h - a_s = 400 - 45 = 355$mm.

$$\zeta_1 = 0.2 + 2.7 \frac{e_0}{h_0} = 0.2 + 2.7 \times \frac{638}{355} = 5.05 > 1, \text{ take } \zeta_1 = 1.0$$

$$\zeta_2 = 1.15 - 0.01 \frac{l_0}{h} = 1.15 - 0.010 \times 10 = 1.05 > 1, \text{ take } \zeta_2 = 1.0$$

Therefore, $\eta = 1 + \frac{1}{1\,300 e_0/h_0}\left(\frac{l_0}{h}\right)^2 \zeta_1 \zeta_2 = 1 + \frac{1}{1\,300 \times \frac{638}{355}} \times 10^2 = 1.04$

①Identification for the compression member with large or small eccentricity

As $\eta e_0 = 1.04 \times 638 = 664$mm $> 0.3 h_0$ ($= 0.3 \times 355 = 107$mm), the section can be designed as large eccentricity compression members. $e_s = \eta e_0 + \frac{h}{2} - a_s = 664 + 400/2 - 45 = 819$mm.

②Calculation of area longitudinal bars

According to the above result, A_s and A'_s can be calculated according to the compression member with large eccentricity. Taking $\xi = \xi_b = 0.53$, the value of A'_s can be expressed by Equation (7-14).

$$A'_s = \frac{\gamma_0 N_d e_s - \xi_b(1-0.5\xi_b)f_{cd}bh_0^2}{f'_{sd}(h_0 - a'_s)}$$

$$= \frac{1.1 \times 232 \times 10^3 \times 819 - 0.53 \times (1-0.5 \times 0.53) \times 13.8 \times 300 \times 355^2}{330 \times (355-45)}$$

$$= 56.3(\text{mm}^2) < 0.002 \times 300 \times 400 = 240\text{mm}^2$$

Take $A'_s = 240\text{mm}^2$.

Choosing the type and quantity of compressive bars as 3 ⌀12, the actual area of compression bars is $A'_s = 339\text{mm}^2$, $a'_s = 45\text{mm}$, $\rho' = 0.28\% > 0.2\%$.

The value of height of compressive zone x can be calculated by Equation (7-16) as

$$x = h_0 - \sqrt{h_0^2 - \frac{2[\gamma_0 N_d e_s - f'_{sd}A'_s(h_0 - a'_s)]}{f_{cd}b}}$$

$$= 355 - \sqrt{355^2 - \frac{2 \times [1.1 \times 232 \times 10^3 \times 819 - 330 \times 339 \times (355-45)]}{13.8 \times 300}}$$

$$= 151\text{mm} < \xi_b h_0 = 0.53 \times 355 = 188\text{mm} > 2a'_s = 2 \times 45 = 90\text{mm}$$

Substituting $\sigma_s = f_{sd}$ into Equation (7-17)

$$A_s = \frac{f_{cd}bx + f'_{sd}A'_s - \gamma_0 N_d}{f_{sd}}$$

$$= \frac{13.8 \times 300 \times 151 + 330 \times 339 - 55.2 \times 10^3}{330}$$

$$= 1460\text{mm}^2 > \rho_{min}bh$$

$$= 0.002 \times 300 \times 400 = 240\text{mm}^2$$

Choosing the type and quantity of tensile bars as 2 ⌀20 + 2⌀25, $A_s = 1610\text{mm}^2$, $\rho = 1.34\% > 0.2\%$. $\rho + \rho' = 1.62\% > 0.5\%$.

The designed longitudinal bars on the short side b on section are arranged in a row (Figure 7-11). As the eccentrically loaded member is prefabricated and concrete is horizontally poured, the minimum spacing of longitudinal bars should be 30mm. In the design of section, take $a_s = a'_s = 45\text{mm}$, therefore, the thickness of concrete cover is $45 - \frac{28.4}{2} = 30.4\text{mm}$, and can meet the requirement of criterion.

The minimum width of section

$b_{min} = 2 \times 31 + 3 \times 30 + 2 \times (22.7 + 28.4) = 255\text{mm} < b = 300\text{mm}$

(2) Check section

①Check the strength in perpendicular bending moment plane

The slenderness ratio $l_0/b = 4000/300 = 13 > 8$,

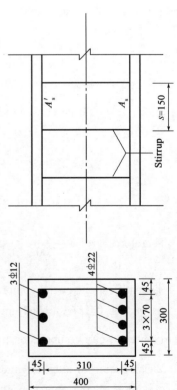

Figure 7-11　Schematic in Example 7-1
(unit:mm)

choosing $\varphi = 0.935$ from appendix Table 1-10, the strength can be expressed as
$$N_{u\perp} = 0.9\varphi[f_{cd}bh + f'_{sd}(A_s + A'_s)]$$
$$= 0.9 \times 0.935 \times [13.8 \times 300 \times 400 + 330 \times (1742 + 339)]$$
$$= 1934.7 \times 10^3 \text{N} = 1934.7(\text{kN}) > \gamma_0 N_d = 255.2\text{kN}$$

The results meet the design requirements.

②Check the strength in bending moment plane

Actual and effective height of section $h_0 = 400 - 45 = 355$mm, calculate $\eta = 1.04$ and $\eta e_0 = 664$mm,
$$e_s = \eta e_0 + \frac{h}{2} - a_s = 664 + \frac{400}{2} - 45 = 819 \text{mm}$$
$$e'_s = \eta e_0 - \frac{h}{2} + a'_s = 664 - \frac{400}{2} + 45 = 509 \text{mm}$$

If the member is a large eccentricity compression member, take $\sigma_s = f_{sd}$, the height of compressive zone x can be calculated by Equation (7-26).

$$x = (h_0 - e_s) + \sqrt{(h_0 - e_s)^2 + 2 \times \frac{f_{sd}A_s e_s - f'_{sd}A'_s e'_s}{f_{cd}b}}$$
$$= (355 - 819) + \sqrt{(355 - 819)^2 + 2 \times \frac{330 \times 1742 \times 819 - 330 \times 339 \times 509}{13.8 \times 300}}$$
$$= 167 \text{mm} \begin{cases} < \xi_b h_0 = 0.56 \times 355 = 199 \text{mm} \\ > 2a'_s = 2 \times 45 = 90 \text{mm} \end{cases}$$

The result shows that the assumption is right.

The strength of section can be calculated by Equation (7-4).
$$N_u = f_{cd}bx + f'_{sd}A'_s - \sigma_s A_s$$
$$= 13.8 \times 300 \times 183 + 330 \times 339 - 330 \times 1742$$
$$= 271.95 \times 10^3 \text{N} = 271.95 \text{kN} > \gamma_0 N_d = 255.2(\text{kN})$$

The results meet the requirement of load-carrying capacity.

4) Calculation on eccentrically loaded rectangular member with symmetric reinforcements

In real structures, when the difference in the amount of two side reinforcements is not large, the symmetric reinforcement method could be used in order to simplify the structure and make the construction easy. For the fabricated eccentric compression member, to ensure construction convenient, the symmetrical reinforcement method is also used.

The symmetrical reinforcement method is that both sides of the section are configured with the same grade and number of steels, i.e. $A_s = A'_s$, $f_{sd} = f'_{sd}$, $a_s = a'_s$. The design principle of the symmetrical reinforcement method is the same as that of the asymmetric reinforcement method.

(1) Section design

①Identification of compression member with large or small eccentricity

The member is assumed to be a large eccentricity compression member, the strength of section can be calculated by Equation (7-4).
$$\gamma_0 N_d = f_{cd}bx$$

Substituting $x = \xi h_0$ into the above equation, as

$$\xi = \frac{\gamma_0 N_d}{f_{cd} b h_0} \tag{7-31}$$

a. For $\xi \leq \xi_b$, the design should refer to the large eccentricity compression member;

b. For $\xi > \xi_b$, the design should refer to the small eccentricity compression member.

②Calculation of large eccentricity compression member ($\xi \leq \xi_b$)

For $2a'_s \leq x \leq \xi_b h_0$, the area of steels can be calculated as

$$A_s = A'_s = \frac{\gamma_0 N_d e_s - f_{cd} b h_0^2 \xi(1 - 0.5\xi)}{f'_{sd}(h_0 - a'_s)} \tag{7-32}$$

Where $e_s = \eta e_0 + \frac{h}{2} - a_s$. For $x < 2a'_s$, the area of A_s can be calculated by Equation (7-18).

③Calculation of small eccentricity compression member ($\xi > \xi_b$)

For the small eccentricity compression member with symmetrical reinforcement as $A_s = A'_s$, even the total section is under compression, the concrete far from the eccentrically-pressed point would never be crushed firstly.

The height of compressive zone x should be calculated first. Based on the code, the relative height of compressive zone ξ should be calculated as

$$\xi = \frac{\gamma_0 N_d - f_{cd} b h_0 \xi_b}{\frac{\gamma_0 N_d e_s - 0.43 f_{cd} b h_0^2}{(\beta - \xi_b)(h_0 - a'_s)} + f_{cd} b h_0} + \xi_b \tag{7-33}$$

After that, the reinforcement area can be obtained by using Equation (7-32).

(2) Check section

The strength both in the perpendicular bending moment plane and in the bending moment plane should be checked. The calculation method is the same as that for asymmetric reinforcement.

[**Example 7-2**] Considering an eccentric compression reinforced concrete member. The size of section is $b \times h = 400\text{mm} \times 500\text{mm}$, the calculation length of two directions (bending moment and its perpendicular direction) is $l_0 = 4\text{m}$. The environment condition is Ⅰ, the safety class is first. The calculation axial force $N = 600\text{kN}$, and the calculation bending moment $M = 360\text{kN} \cdot \text{m}$. The prefabricated members are horizontally poured using C30 concrete and longitudinal reinforcements of HRB400 grade steel. Choosing reinforcements and check the section.

Solution: $f_{cd} = 13.8\text{MPa}, f_{sd} = f'_{sd} = 330\text{MPa}, \xi_b = 0.53$.

(1) Section design

For $N = 600\text{kN}, M = 360\text{kN} \cdot \text{m}$, the eccentricity is

$$e_0 = \frac{M}{N} = \frac{360 \times 10^6}{600 \times 10^3} = 600\text{mm}$$

In the bending moment direction, the slenderness ratio $l_0/h = 4000/500 = 8 > 5$. Set $a_s = a'_s = 45\text{mm}, h_0 = h - a_s = 455\text{mm}$. Based on Equation (7-2), $\eta = 1.034, \eta e_0 = 620\text{mm}$ can be obtained.

①Identification of compression member with large or small eccentricity

The relative height of compressive zone ξ can be calculated by Equation (7-31).

$$\xi = \frac{N}{f_{cd}bh_0} = \frac{600 \times 10^3}{13.8 \times 400 \times 455} = 0.24 < \xi_b = 0.53$$

Thus it can be designed as a large eccentricity compression member.

②Calculation of longitudinal bars area

As $\xi = 0.24$, $h_0 = 455$mm, the height of compressive zone x is
$x = \xi h_0 = 0.24 \times 455 = 109$mm $> 2a'_s = 90$mm, and

$$e_s = \eta e_0 + \frac{h}{2} - a_s = 620 + \frac{500}{2} - 45 = 825 \text{mm}$$

By Equation (7-32), the area of longitudinal bars can be calculated as

$$A_s = A'_s = \frac{\gamma_0 N_d e_s - f_{cd}bh_0^2 \xi(1 - 0.5\xi)}{f_{sd}(h_0 - a'_s)}$$

$$= \frac{600 \times 10^3 \times 825 - 13.8 \times 400 \times 455^2 \times 0.24 \times (1 - 0.5 \times 0.24)}{330 \times (455 - 45)}$$

$$= 1\,875 \text{mm}^2$$

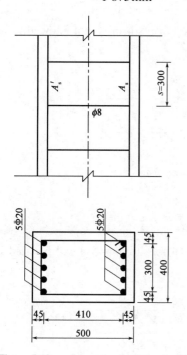

Figure 7-12　Schematic in Example 7-2 (unit: mm)

Choosing the type and quantity of tensile bars on each side as 5⌀20, i.e. $A_s = A'_s = 1\,900$mm² $> 0.002bh$ [$= 0.002 \times 400 \times 500 = 400$(mm²)], the minimum width of each side for the reinforcement bars is $b_{min} = 2 \times 33.65 + 4 \times 50 + 5 \times 25.1 = 392.8$mm $< b = 400$mm. Taking a_s and a'_s for 45mm, the section layout is shown in Figure 7-12.

(2) Check section

①Check strength in perpendicular bending moment plane

As the slenderness ratio $l_0/b = 4\,000/400 = 10$, $\varphi = 0.98$ is selected from appendix Table 1-10, the strength can be obtained by Equation (6-7), $N_{ul} = 2987$kN $> N = 600$kN. It meets the requirements.

②Check strength in bending moment plane

From Figure 7-12, it is obtained that $a_s = a'_s = 45$mm, $A_s = A'_s = 1\,900$mm² and $h_0 = 455$mm. By Equation (7-2), it is obtained that $\eta = 1.034$, $\eta e_0 = 620$mm. $e_s = 825$mm, $e'_s = \eta e_0 - h/2 + a'_s = 620 - 500/2 + 45 = 415$mm.

The member is assumed to be a large eccentricity compression member, $\sigma_s = f_{sd}$, the height of compressive zone x can be calculated by Equation (7-26).

$$x = h_0 - e_s + \sqrt{(h_0 - e_s)^2 + \frac{2f_{sd}A_s(e_s - e'_s)}{f_{cd}b}}$$

$$= 455 - 825 + \sqrt{(455 - 825)^2 + \frac{2 \times 330 \times 1\,900 \times (825 - 415)}{13.8 \times 400}}$$

$$= 110(\text{mm}) \begin{cases} < \xi_b h_0 = 0.53 \times 455 = 241 \text{mm} \\ > 2a'_s = 2 \times 45 = 90 \text{mm} \end{cases}$$

The result shows that the assumption is right.

The section strength is calculated as

$$N_u = f_{cd}bx = 13.8 \times 400 \times 111 = 607.2 \times 10^3 \text{N} = 607.2 \text{kN} > \gamma_0 N_d = 600 \text{kN}$$

It meets the requirements.

7.4 Eccentrically loaded members with I-shaped and T-shaped sections

For eccentrically loaded members with the large size, I-shaped, box-shaped and T-shaped sections are generally used. In order to save concrete and reduce dead load, these section types are widely used in arch ribs, columns, etc. The construction requirements for I-shaped, box-shaped and T-shaped sections are the same as that for rectangular section. The failure forms, calculation methods and principles of I-shaped, box-shaped and T-shaped sections are also the same as that of rectangular section. In order to avoid the concrete cracking, the bevel stirrup is not allowed to be used. The composition of stirrup form is shown in Figure 7-13.

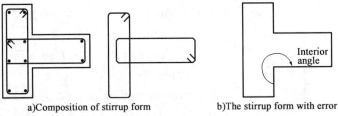

Figure 7-13 The stirrup forms in T-shaped section

1) Strength of normal section

For the I-shaped section, there are three different cases for the height of compression zone before the member reaches its ultimate strength (Figure 7-14). Different equations will be used in different cases.

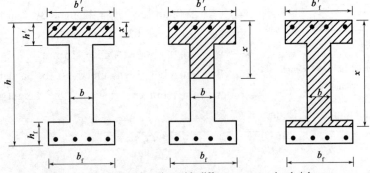

Figure 7-14 I-shaped section with different compressive heights

(1) For $x \leq h'_f$, as the height of compressive zone is located within the compressive wing plate, the strength can be calculated as in a rectangular section with the effective width of wing plate b'_f, effective height h_0 and the height of compressive zone x (Figure 7-15).

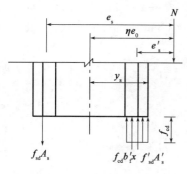

The basic formulas are expressed as

$$\gamma_0 N_d \leqslant N_u = f_{cd} b'_f x + f'_{sd} A'_s - f_{sd} A_s \quad (7\text{-}34)$$

$$\gamma_0 N_d e_s \leqslant f_{cd} b'_f x \left(h_0 - \frac{x}{2} \right) + f'_{sd} A'_s (h_0 - a'_s) \quad (7\text{-}35)$$

$$f_{cd} b'_f x \left(e_s - h_0 + \frac{x}{2} \right) = f_{sd} A_s e_s - f'_{sd} A'_s e'_s \quad (7\text{-}36)$$

Where $e_s = \eta e_0 + h_0 - y_s$, $e'_s = \eta e_0 - y_s + a'_s$. The y_s is the distance between the centroid axis and the edge of compressive zone.

The formulas are only applicable under conditions as

$$x \leqslant \xi_b h_0$$

and

$$2a'_s \leqslant x \leqslant h'_f \quad (7\text{-}37)$$

Where h'_f is the depth of compressive wing plate.

For $x < 2a'_s$, the strength will be calculated by Equation (7-12).

(2) For $h'_f < x \leqslant h - h_f$, the depth of compressive zone is located within the rib (Figure 7-16).

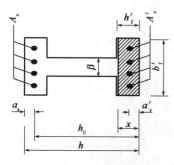

Figure 7-15 Calculation schematic for $x \leqslant h'_f$

Basic formulas are expressed as

$$\gamma_0 N_d \leqslant N_u = f_{cd} [bx + (b'_f - b) h'_f] + f'_{sd} A'_s - \sigma_s A_s \quad (7\text{-}38)$$

$$\gamma_0 N_d e_s \leqslant f_{cd} \left[bx \left(h_0 - \frac{x}{2} \right) + (b'_f - b) h'_f \left(h_0 - \frac{h'_f}{2} \right) \right] + f'_{sd} A'_s (h_0 - a'_s) \quad (7\text{-}39)$$

$$f_{cd} bx \left(e_s - h_0 + \frac{x}{2} \right) + f_{cd} (b'_f - b) h'_f \left(e_s - h_0 + \frac{h'_f}{2} \right) = \sigma_s A_s e_s - f'_s A'_s e'_s \quad (7\text{-}40)$$

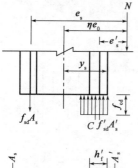

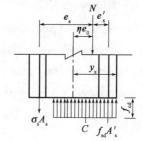

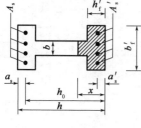

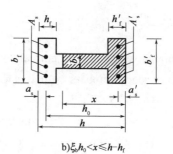

a) $h'_f < x \leqslant \xi_b h_0$ b) $\xi_b h_0 < x \leqslant h - h_f$

Figure 7-16 Calculation schematic for $h'_f < x \leqslant h - h_f$

(3) For $h - h_f < x \leqslant h$, the height of compressive zone is located within the wing plate with little tension or compression zone on I-shaped section. It is a compression member with small eccentricity. Basic formulas are expressed as

$$\gamma_0 N_d \leqslant N_u = f_{cd}[bx + (b'_f - b)h'_f + (b_f - b)(x - h + h_f)] + f'_{sd}A'_s - \sigma_s A_s \quad (7-41)$$

$$Ne_s \leqslant f_{cd}\left[bx\left(h_0 - \frac{x}{2}\right) + (b'_f - b)h'_f\left(h_0 - \frac{h'_f}{2}\right) + \right.$$

$$\left. (b_f - b)(x - h + h_f)\left(h_f - a_s - \frac{x - h + h_f}{2}\right)\right] + f'_s A'_s(h_0 - a'_s) \quad (7-42)$$

$$f_{cd}\left[bx\left(e_s - h_0 + \frac{x}{2}\right) + (b'_f - b)h'_f\left(e_s - h_0 + \frac{h'_f}{2}\right) + (b_f - b)(x - h + h_f)\right.$$

$$\left. \left(e_s + a_s - h_f + \frac{x - h + h_f}{2}\right)\right] = \sigma_s A_s e_s - f'_{sd}A'_s e'_s \quad (7-43)$$

Where $\quad \sigma_s = \varepsilon_{cu} E_s \left(\dfrac{\beta}{\xi} - 1\right)$

The size of section is shown as Figure 7-17.

(4) When $x > h$, the entire section subjects to the compression force, the failure mode is a compression failure with small eccentricity. Taking $x = h$, the basic formulas are

$$N \leqslant N_u = f_{cd}[bh + (b'_f - b)h'_f + (b_f - b)h_f] + f'_{sd}A'_s - \sigma_s A_s \quad (7-44)$$

$$\gamma_0 N_d e_s \leqslant f_{cd}\left[bh\left(h_0 - \frac{h}{2}\right) + (b'_f - b)h'_f\left(h_0 - \frac{h'_f}{2}\right) + (b_f - b)h_f\left(\frac{h_f}{2} - a_s\right)\right] + f'_{sd}A'_s(h_0 - a'_s)$$

$$(7-45)$$

$$f_{cd}\left[bh\left(e_s - h_0 + \frac{h}{2}\right) + (b'_f - b)h'_f\left(e_s - h_0 + \frac{h'_f}{2}\right) + (b_f - b)h_f\left(e_s + a_s - \frac{h_f}{2}\right)\right]$$

$$= \sigma_s A_s e_s - f'_{sd}A'_s e'_s \quad (7-46)$$

For the small eccentricity compression members with $x > h$, the edge concrete is crushed first, which should be avoided. When the concrete is far from the action point of eccentricity compression force, the strength could satisfy as

$$\gamma_0 N_d e'_s \leqslant f_{cd}\left[bh\left(h'_0 - \frac{h}{2}\right) + (b'_f - b)h'_f\left(\frac{h'_f}{2} - a'_s\right)\right] + f_{cd}(b_f - b)$$

$$h_f\left(h'_0 - \frac{h_f}{2}\right) + f'_{sd}A_s(h'_0 - a_s) \quad (7-47)$$

The calculation equations of normal load-carrying capacity of eccentrically loaded members with I-shaped section is shown in Equations (7-34) ~ (7-47).

① For $h_f = 0$ and $b_f = b$, the equations of load-carrying capacity is for eccentrically loaded members with T-shaped section.

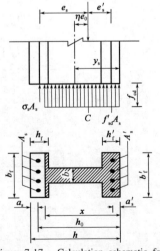

Figure 7-17 Calculation schematic for normal section strength for $h - h_f < x \leqslant h$

②For $h'_f = h_f = 0$ and $b'_f = b_f = b$, the equations is for load-carrying capacity of eccentrically loaded members with rectangular section.

2) Calculation method

In real structures, symmetric reinforcement method is usually used in eccentrically loaded members with I-shape section. The calculation method with I-shape section with symmetric reinforcement is introduced as follows.

Symmetric reinforcement section means both the section and reinforcement distribution are symmetric. The equations of I-shape and box-shape section with symmetric reinforcement is $b'_f = b_f, h'_f = h_f, A'_s = A_s, f'_{sd} = f_{sd}, a_s = a'_s$.

(1) Design of section

For the symmetric reinforcement section, according to Equation(7-38), taking $\sigma_s = f_{sd}$, the relative depth of compressive zone can be expressed as

$$\xi = \frac{\gamma_0 N_d - f_{cd}(b'_f - b)h'_f}{f_{cd}bh_0} \quad (7\text{-}48)$$

When $\xi \leqslant \xi_b$, it is computed based on large eccentricity compression. When $\xi > \xi_b$, it is computed based on small eccentricity compression.

①When $\xi \leqslant \xi_b$

For $h'_f < x \leqslant \xi_b h_0$, the neutral axis is at the location of ribbed panel. Substituting x into Equation(7-39), the area of reinforcement will be

$$A_s = A'_s = \frac{\gamma_0 N_d e_s - f_{cd}\left[bx\left(h_0 - \frac{x}{2}\right) + (b'_f - b)h'_f\left(h_0 - \frac{h'_f}{2}\right)\right]}{f'_{sd}(h_0 - a'_s)} \quad (7\text{-}49)$$

Where $e_s = \eta e_0 + h/2 - a_s$.

For $2a'_s \leqslant x \leqslant h'_f$, the neutral axis locates in the compressive wing plate. Therefore, the depth of compressive zone x should be recomputed as

$$x = \frac{N}{f_{cd}b'_f} \quad (7\text{-}50)$$

The steel area is

$$A_s = A'_s = \frac{\gamma_0 N_d e_s - f_{cd}b'_f x(h_0 - 0.5x)}{f'_{sd}(h_0 - a'_s)} \quad (7\text{-}51)$$

For $x < 2a'_s$, the calculation should be conducted based on the method for the rectangular section, i.e., Equation(7-18).

②When $\xi > \xi_b$

Recomputing the depth of compressive area x, and it is substituting into the corresponding equation to compute the steel area $A_s = A'_s$.

During the course of computing the depth of compressive area x, $\sigma_s = \varepsilon_{cu} E_s \left(\frac{\beta}{\xi} - 1\right)$ is associated with corresponding basic equation. For example, when $h'_f < x \leqslant h - h_f$, associate Equation(7-38) and(7-39). When $h - h_f < x \leqslant h$, associate Equation(7-41) and(7-42). It will result in the solution about cubic equation with x.

The relative depth of compressive area ξ could be designed approximately based on as follows:

a. When $\xi_b h_0 < x \leqslant h - h_f$

$$\xi = \cfrac{\gamma_0 N_d - f_{cd}[(b_f' - b)h_f' + b\xi_b h_0]}{\cfrac{\gamma_0 N_d e_s - f_{cd}\left[(b_f' - b)h_f'\left(h_0 - \dfrac{h_f'}{2}\right) + 0.43 bh_0^2\right]}{(\beta - \xi_b)(h_0 - a_s')} + f_{cd} bh_0} + \xi_b \quad (7\text{-}52)$$

b. When $h - h_f < x \leqslant h$

$$\xi = \cfrac{\gamma_0 N_d + f_{cd}[(b_f - b)(h - 2h_f) - b_f \xi_b h_0]}{\cfrac{Ne_s + f_{cd}[0.5(b_f - b)(h - 2h_f)(h_0 - a_s') - 0.43 b_f h_0^2]}{(\beta - \xi_b)(h_0 - a_s')} + f_{cd} b_f h_0} + \xi_b \quad (7\text{-}53)$$

c. When $x > h$, take $x = h_0$.

(2) Check section

The checking method of section is the same as that of rectangular section with symmetric reinforcement, but the calculation equations are different.

7.5 Eccentrically loaded circular member

A large number of experimental results on the eccentrically loaded circle members show their failure mode is similar to rectangular section. When the e_0 is large ($e_0/r \geqslant 0.86$), many transverse cracks far away the axial force N appear in the tensile zone. Only a part of reinforcements near the tensile edge of concrete reach the yield strength and the members display large longitudinal bending deformation. When the e_0 is little ($e_0/r \leqslant 0.35$), the transverse cracks far away the axial force N appear later with small amount and width. In this situation, the concrete in the compressive zone is crushed and the members fail. Only a part of reinforcements near the compressive edge of concrete reach the yield strength.

The difference between the circle and the rectangular section is the stress of longitudinal reinforcements. The stress in the circle section should be determined by the strain. The circle section cannot be divided into large or small eccentric compression balanced failure completely.

1) Construction requirements

(1) The longitudinal bars should be uniformly distributed along a circle and the number of bars should not be less than 6.

(2) The diameter of longitudinal bars should not be less than 12mm. As the section of the bored pile is large (diameter $D = 800 \sim 1\,500$mm), the diameter of longitudinal bars should not be less than 14mm. The number of bars should not be less than 8, and the net distance of bars should not be less than 50mm. The concrete cover should not be less than $60 \sim 80$mm.

(3) The stirrup diameter should not be less than 8mm, and the stirrup spacing is $200 \sim 400$mm.

2) Basic assumption for calculation of normal section strength

For the eccentric compression member that the steel bars uniformly distributed along the section, the basic assumptions are:

(1) Section deformation satisfies the plane section assumption.

(2) When the member reaches failure state, the ultimate compression strain of concrete at the edge of compressive zone is $\varepsilon_{cu} = 0.0033$.

(3) Equivalent rectangular stress block is used to describe the stress distribution of concrete at the compressive zone.

(4) The tensile force is taken on by steel bars and the concrete in the tensile region is not considered.

(5) Steel bar is assumed as the ideal elastic-plastic material.

For the eccentrically loaded circular members that are uniformly distributed along the section, the calculation formulas of bearing capacity are derived from the equilibrium condition between internal and external forces.

3) Basic equations for strength of normal section

The equilibrium condition along all horizontal forces of the section is

$$N_u = D_c + D_s \tag{7-54}$$

The equilibrium condition on the resultant moment from all forces to the centroidal axis y-y is

$$M_u = M_c + M_s \tag{7-55}$$

Where D_c and D_s are resultant forces of the concrete compression stress and the steel bar stress, respectively; M_c is the moment from resultant force on the concrete stress compressive region to the y axis; and M_s is the moment from the resultant force reinforcement stress to the y axis, respectively.

(1) Resultant stress D_c and resultant moment M_c on the compressive concrete.

The compression zone of the eccentrically loaded circle member is bow. If the radius of the circle section is r, and the central angle corresponding to the compression zone is $2\pi\alpha$ (rad). The area A_c for the concrete at the compressive zone can be expressed as

$$A_c = \alpha\left(1 - \frac{\sin 2\pi\alpha}{2\pi\alpha}\right)A \tag{7-56}$$

Where A is the total area of the section, $A = \pi r^2$.

According to the simplification of the equivalent rectangular stress in the compression zone of the section. It is assumed that the compressive concrete strength in the compression zone is f_c. The D_c and M_c can be expressed as

$$D_c = \alpha f_{cd} A\left(1 - \frac{\sin 2\pi\alpha}{2\pi\alpha}\right) \tag{7-57}$$

$$M_c = \frac{2}{3} f_{cd} A r \frac{\sin^3 \pi\alpha}{\pi} \tag{7-58}$$

(2) Resultant stress D_s and resultant moment M_s.

Generally, a part of steel bars in the section reach the yield strength and other steel bars do not reach the yield strength, i.e. the steel bars near the compression zone or the tension edge may reach the yield strength, while the steel bars close to the neutral axis generally do not reach the yield strength [Figure 7-18c)]. To simplify the calculation, the stress of the steel bars in the tension zone and compression zone are approximately equivalent to the uniform distribution with the strengths of f_s and f'_s. The area ratio of the longitudinal tensile steel bars to all longitudinal steel bars α_t is approximately expressed as

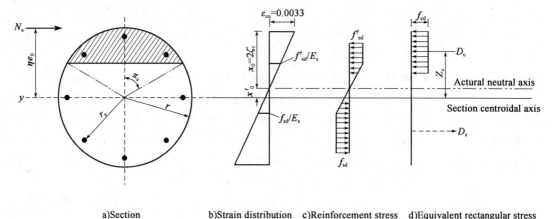

a) Section b) Strain distribution c) Reinforcement stress d) Equivalent rectangular stress distribution of concrete

Figure 7-18 Schematic of eccentrically loaded member of circular section

$$\alpha_t = 1.25 - 2\alpha \geqslant 0 \tag{7-59}$$

A_s represents the total area of the steel bars. Assuming $f_s = f'_s$, the resultant stress D_s and the resultant force will produce a moment M_s to the center of the section, which can be expressed as

$$D_s = (\alpha - \alpha_t) f_{sd} A_s \tag{7-60}$$

$$M_s = f_{sd} A_s r_s \frac{\sin\pi\alpha + \sin\pi\alpha_t}{\pi} \tag{7-61}$$

Substituting the Equations (7-57), (7-58), (7-60), (7-61) into Equations (7-54) and (7-55), respectively, the calculation formulas for normal section bearing capacity of eccentric compression members with circular section can be expressed as

$$N_u = \alpha f_{cd} A \left(1 - \frac{\sin 2\pi\alpha}{2\pi\alpha}\right) + (\alpha - \alpha_t) f_{sd} A_s \tag{7-62}$$

$$N_u e_i = \frac{2}{3} f_{cd} A r \frac{\sin^3 \pi\alpha}{\pi} + f_{sd} A_s r_s \frac{\sin\pi\alpha + \sin\pi\alpha_t}{\pi} \tag{7-63}$$

$$\alpha_t = 1.25 - 2\alpha \geqslant 0$$

$$e_i = \eta e_0$$

Where A is the area of circular cross-section; A_s is the area of all normal longitudinal steel bars; r is the radius of the circular section; r_s is the radius of the circumference of the center of gravity for the longitudinal steel bar; e_0 is the eccentricity of the axial force to the center of gravity of the section; when $\alpha > 0.625, \alpha_t = 0; \eta$ is the enhancement factor of eccentricity of the eccentric compression member.

When calculating the bearing capacity of the normal section of the eccentric compression members with a circular section by hand, it is generally to make assumptions about the value of α and calculate it by the iteration method using the Equations (7-62) and (7-63).

In the engineering applications, the Equation (7-63) is divided by (7-62), i.e.

$$\eta \frac{e_0}{r} = \frac{\frac{2}{3}\frac{\sin^3 \pi\alpha}{\pi} + \rho \frac{f_{sd}}{f_{cd}} \frac{r_s}{r} \frac{\sin\pi\alpha + \sin\pi\alpha_t}{\pi}}{\alpha\left(1 - \frac{\sin 2\pi\alpha}{2\pi\alpha}\right) + (\alpha - \alpha_t)\rho \frac{f_{sd}}{f_{cd}}} \quad (7\text{-}64)$$

And

$$n_u = \alpha\left(1 - \frac{\sin 2\pi\alpha}{2\pi\alpha}\right) + (\alpha - \alpha_t)\rho \frac{f_{sd}}{f_{cd}}$$

Therefore

$$\eta \frac{e_0}{r} = \frac{\frac{2}{3} \cdot \frac{\sin^3 \pi\alpha}{\pi} + \rho \cdot \frac{f_{sd}}{f_{cd}} \cdot \frac{r_s}{r} \cdot \frac{\sin\pi\alpha + \sin\pi\alpha_t}{\pi}}{n_u} \quad (7\text{-}65)$$

Where ρ is longitudinal reinforcement ratio, $\rho = \sum_{i=1}^{n} A_{si}/\pi r^2$; $\sum_{i=1}^{n} A_{si}$ is the total area of the longitudinal steel bars; A_{si} is the area of a single longitudinal steel bar; n is the number of the steel bars.

The bearing capacity of the normal section of the eccentric compression members with a circular section can be obtained according to the Equation (7-62), i.e.

$$N_u = n_u A f_{cd} \quad (7\text{-}66)$$

In the Equation (7-65), the ratio r_s/r can be taken as a representative value. If the values of $\eta e_0/r$ and $\rho f_{sd}/f_{cd}$ are given, the α and n_u can be obtained by the Equation (7-65), and the N_u can also be calculated by the Equation (7-66).

For the eccentric compression member that the steel bars uniformly distributed along the section with a concrete strength grade of C30 ~ C50 and the reinforcement ratio ranges of 0.5% ~ 4%, the *CHBC* provides the calculation table to directly determine the parameters or by interpolation method combining the Equation (7-65) and the corresponding numerical calculations. The eccentric compressive strength of the normal section with a circular section calculated by the table should satisfy the following requirements.

$$\gamma_0 N_d \leq N_u = n_u A f_{cd} \quad (7\text{-}67)$$

Where γ_0 is coefficient for importance of a structure; N_d is the design value of the axial compression member; n_u is the relative load-carrying capacity of the member, as determined from appendix Table 1-11; A is the cross-sectional area of the member; f_{cd} is design compressive strength of concrete.

4) Calculation method

(1) Design of section

The size of section, computing length, strength of material, design axial compression force N, design bending moment M are known. The area of longitudinal reinforcement A_s needs to be solved.

First, calculating the eccentricity e_0. Make sure that whether consider the effect of longitudinal bending on the eccentricity or not. Then, determine α by trial calculation. Calculating the area of steel bars A_s. Finally, select the steel bars and arrange the section.

(2) Check section

Given: the size of section, the actual longitudinal reinforcement area and arrangement, computing length, strength of material, design axial compression force N, and design bending moment M. The strength of section needs to be solved.

Calculating the eccentricity e_0. Make sure that whether consider the effect of longitudinal bending on the eccentricity or not. Calculating the enhancement factor of eccentricity η, and the parameter $\eta e_0/r$ is obtained.

Calculating the $\rho f_{sd}/f_{cd}$ by the actual area of longitudinal steel bars, design strengths of concrete and steel bars.

According to the parameters $\rho f_{sd}/f_{cd}$ and $\eta e_0/r$, the corresponding values n_u can be obtained by Appendix Table. The interpolation is used to obtain the n_u when it cannot directly find the values.

The n_u is substituted into the Equation (7-67) to obtain the bearing capacity N_u, and meet the requirements of Equation (7-67).

Chapter 8
Calculation of Load-carrying capacity of Tensile Members

8.1 Overview

When the section of a member is under the action of an axially tensile force, which coincides with the cross-section centroid axis, the member is an axial tensile member. When there is a moment M acting simultaneously with the axial tensile force, or the tensile force has an eccentricity, the member will be an eccentric tensile member. There are many tensile members in the bridge engineering, such as truss arch, chord in the truss bridge, tie bar of the tied arch bridge and so on.

Reinforced concrete (RC) tensile members need to configure the longitudinal reinforcement and stirrups. The diameter of stirrups should not be less than 6mm, the distance between stirrups is generally 150~200mm.

8.2 Calculation of Load-carrying capacity of axial tensile members

When RC members are under the action of an axial tensile force, the tension will be taken on by both concrete and steel bars before concrete cracks, and the tension will be taken on by the reinforcement alone after concrete cracks.

When the tensile stress of reinforcement reaches the yield strength, the members have reached the ultimate load-bearing capacity. The formula of normal section load-bearing capacity of an axial tension member is shown as follows.

$$\gamma_0 N_d \leq N_u = f_{sd} A_s \tag{8-1}$$

Where N_d is design value of axial tension; N_u is axial tension strength; f_{sd} is design value of reinforcement's tensile strength; A_s is total area of reinforcement within a cross-section.

For an axial tensile member, when the design value of axial tension is known, the design for the reinforcement or the axial tension strength can be obtained by the known design of tensile structures.

The **CHBC** specifies: steel reinforcement ratio of the axial tensile member and the eccentric tensile member should not be less than $45 f_{td}/f_{sd}$, and not less than 0.2; steel reinforcement ratio of the axial tensile member and the small eccentricity tensile member should be calculated using gross section area; steel reinforcement ratio of the large eccentricity tensile member should be calculated using $A_s/(bh_0)$.

8.3 Calculation of strength of eccentric tensile members

According to the longitudinal position of tensile force, the strength calculation of the eccentric tensile member can be divided into two situations: Assuming the section with reinforcement A_s on one side and A'_s on the other side. If the eccentric tension force is acting within the space between A_s and A'_s [$e_0 \leq (h/2 - a_s)$], the member is called a small eccentricity tensile member. If the eccentric tension force is acting outside the space between A_s and A'_s [$e_0 > (h/2 - a_s)$], the member is called a large-eccentricity tensile member.

1) Calculation of strength of small eccentricity tensile members

For a small eccentricity tensile member, cracks occur through the whole section before the failure of the member happens. At the same time, the tension is totally shifted to A'_s and A_s, as shown in Figure 8-1, the tensile strength of concrete is ignored. When the member fails, the stress in reinforcement A'_s and A_s reaches the design value of the tensile strength. According to the equilibrium condition, the basic formula is expressed as

$$\gamma_0 N_d e_s \leq N_u e_s = f_{sd} A_s' (h_0 - a_s') \tag{8-2}$$

$$\gamma_0 N_d e'_s \leqslant N_u e'_s = f_{sd} A_s (h_0 - a_s) \quad (8\text{-}3)$$

e_s and e'_s in Equation (8-2) and Equation (8-3) can be calculated according to

$$e_s = \frac{h}{2} - e_0 - a_s \quad (8\text{-}4)$$

$$e'_s = e_0 + \frac{h}{2} - a'_s \quad (8\text{-}5)$$

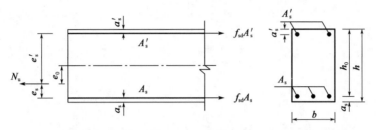

Figure 8-1 A small eccentricity tension section

The meaning of the symbols in Equations (8-2)~(8-3) is shown in Figure 8-1.

The action of eccentricity tension can be divided into axial tension N_d and moment M_d. If the section is to be designed for a combined load (M_d & N_d), then A_s should be designed for the combination of the largest axial tension N_d with the largest moment M_d, and the A'_s should be designed for the combination of the largest axial tension N_d with the smallest moment M_d. It is a very common practice to place the reinforcement symmetrically in an eccentric tension section so that $A_s = A'_s$. In such a case, the reinforcement A'_s will not yield and we may calculate $A_s = A'_s$ from Equation (8-2).

2) Calculation of strength of large eccentricity tensile members

For the normal reinforcement of a rectangular section, a large eccentricity tensile member is similar to a large eccentric compression member. Under the action of N and M, the stress on the section will be a combination of tension and pressure. Close to the eccentric side, the first crack comes out. The concrete far from the axial force remains under pressure. With the increase of load, cracks keep developing while the pressure area is reduced. The failure starts from the yielding of the tension reinforcement followed by the extension of cracks and the crushing of concrete.

The schematic cross-section load-bearing capacity of eccentric rectangular-section tensile members is shown in Figure 8-2. The stress of longitudinal tensile reinforcement reaches the design value of the tensile strength. In the compression zone of concrete, the stress diagram can be simplified to a rectangle, the stress within this rectangle is the design value of concrete compressive strength. According to the equilibrium condition, the basic formula is expressed as

$$\gamma_0 N_d \leqslant N_u = f_{sd} A_s - f'_{sd} A'_s - f_{cd} bx \quad (8\text{-}6)$$

$$\gamma_0 N_d e_s \leqslant N_u e_s = f_{cd} bx \left(h_0 - \frac{x}{2}\right) + f'_{sd} A'_s (h_0 - a'_s) \quad (8\text{-}7)$$

$$f_{sd} A_s e_s - f'_{sd} A'_s e'_s = f_{cd} bx \left(e_s + h_0 - \frac{x}{2}\right) \quad (8\text{-}8)$$

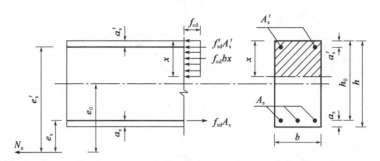

Figure 8-2 Large eccentricity tension section

Where

$$e_s = e_0 - \frac{h}{2} + a_s \tag{8-9}$$

Conditions for the application of the above formulas are as follows

$$2a'_s \leq x \leq \xi_b h_0$$

Where, ξ_b is the limit of the relative height of concrete compression zone. The value of this height is shown in Table 3-2.

When $x < 2a'_s$, the space between the compression steel bars and the neutral axis is small. The stress of concrete can not reach the design value of concrete compressive strength. Then restart the analysis by assuming $x = 2a'_s$, implying that the effect point of the concrete force coincides with the compression of reinforcement, the equation may be expressed by an approximate formula.

$$\gamma_0 N_d e'_s \leq N_u = f_{sd} A_s (h_0 - a'_s) \tag{8-10}$$

If the strength calculated using Equation (8-10) is smaller than the strength unconcerned compressive reinforcement, the act of A'_s will be ignored. When the $b \times h$, N_d, e_0 are known to calculate the area of reinforcement. In order to make full use of the strength, when $x = \xi_b h_0$, the most economic design will be obtained.

$$A'_s = \frac{\gamma_0 N_d e_s - f_{cd} b h_0^2 \xi_b (1 - 0.5\xi_b)}{f'_{sd}(h_0 - a'_s)} \tag{8-11}$$

$$A_s = \frac{\gamma_0 N_d + f'_{sd} A'_s + f_{cd} b h_0 \xi_b}{f_{sd}} \tag{8-12}$$

If the value of A'_s obtained in terms of Equation (8-11) is too small or negative, A'_s is determined according to the minimum reinforcement ratio or construction requirements. Then A_s can be calculated using Equation (8-6) ~ (8-8). Under normal conditions, x is often less than $2a'_s$, and then A_s can be calculated using Equation (8-10).

For a large eccentricity tension member with symmetrically, distributed reinforcement bars, $f_{sd} = f'_{sd}$, $A_s = A'_s$, x can be calculated using Equation (8-6). If x is negative or less than $2a'$, then A_s can be calculated using Equation (8-10).

Chapter 9
Calculation of Stress, Cracking and Deflection of Reinforced Concrete Flexural Members

9.1 Introduction

To ensure the safety, applicability and durability of members, the *CHBC* prescribes that the calculation of reinforced concrete members must be conducted in the two states (load-carrying capacity and serviceability), under the four conditions (short term, long term, accidental events and earthquake). Since the calculation method of load-carrying capacity limit state has been introduced, this chapter will take flexural member as an example to introduce the calculation of serviceability limit state under long-term load effect and the calculation of member stress under short-term load effect.

1) Difference between the two limit states

(1) Calculation of ultimate limit state

Thecalculation of load-carrying capacity of members in various loading states will be discussed. The load-carrying capacity is the primary condition to ensure the structure safety, which determines the material, size, reinforcement and construction of members.

(2) Purpose of checking calculation in serviceability limit state

Reinforced concrete members will be in ultimate limit state because of strength failure or instability, or member deformation and excessive cracking will affect the applicability and durability of members, so members can not meet the normal service requirements. Therefore, besides the calculation of load-carrying capacity, the checking calculation in normal service limit state should be conducted for all reinforced concrete members according to the service condition to meet the normal service requirements.

2) Stress, cracking and deformation calculation of reinforced concrete flexural members

(1) Calculation of concrete and rebar stress in construction stage and service stage.

(2) Deformation in service stage.

(3) Maximum crack width in service stage.

3) Calculation characteristics

(1) Difference of calculation basis: ultimate limit state establishes the calculation schema based on the condition of failure stage (III_a). Generally, the service stage refers to stage II, that is, beam's serving stage with cracks. Construction is usually in stage I and stage II.

(2) Difference of influence degrees: the damage and severity of the consequences (e.g. casualty and economic loss) caused by exceeding the serviceability limit state are relatively less, compared with being caused by ultimate limit state. Therefore, the assurance rate of the reliability should be relaxed.

(3) Difference of calculation contents

①Ultimate limit state: It includes section design and section rechecking. Its calculation determines the design size, material, number of reinforcement and rebar arrangement.

②Construction and service stages: In construction stage, one needs to only check that the concrete and rebar stress cannot exceed the limit value. In normal service stage, hydrochloride crack width and its deformation should be less than the limit value.

(4) Difference of load effect and resistance strength

①Ultimate limit state: Vehicle load should include the impact coefficient. Both the load effect and the resistance strength of structure members adopt the design value considering the partial coefficient. Under the circumstance of various load effects, the design values in various effects should consider the most disadvantageous combination and the combined coefficient in various effects should be used according to the load effect condition combined.

②Serviceability limit state: Vehicle load can exclude the impact coefficient. The load effect should take one or several combinations in frequent combination and quasi-permanent combination.

(5) Difference of calculations of section geometric parameters

Different from the calculation of section geometric parameters of single material construction, transformed section calculation should be adopted, considering the contribution of steel rebars to concrete section.

9.2 Transformed section

1) Definition of transformed section

Transformed section refers to an equivalent section composed of a single material with same tensile and compressive behavior, which is transformed from an actual section, composed of a mixed material of rebar and concrete.

2) Basic assumption in calculation

In the calculation of geometric characteristics of the transformed section, some assumptions are adopted considering that steel rebars and concrete bond to each other perfectly and completely in construction stage and concrete may crack in service stage:

(1) The normal section of beams remains plane after the beam bear load and bends.

(2) Assumption of elasticity: linear distribution of stress in the concrete compression zone.

(3) Ignore the tensile contribution of concrete, and assume the tensile stresses is carried completely by rebars.

The calculation schemais constructed with the above basic assumptions (Figure 9-1).

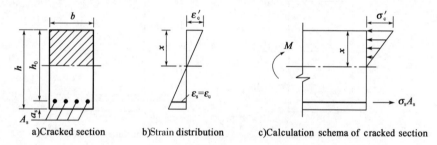

a) Cracked section   b) Strain distribution   c) Calculation schema of cracked section

Figure 9-1   Cracked section of flexural member

3) Transformation principle

The resultant force and the position of action point remain the same as before and after conversion.

$$A_s \sigma_s = A_{sc} \sigma_c$$

$$A_{sc} = A_s \frac{\sigma_s}{\sigma_c} = A_s \frac{E_s \varepsilon_s}{E_c \varepsilon_c} = \alpha_{Es} A_s \frac{\varepsilon_s}{\varepsilon_c} = \alpha_{Es} A_s$$

where A_{sc} is transformed area of steel rebar; $\alpha_{Es} = E_s/E_c$, refers to transformation coefficient of section of reinforced concrete member, which equals the ratio of elastic modulus of steel and concrete.

4) Geometric parameters calculation of transformed section of cracked section

(1) Definition: after reinforced concrete flexural members are stressed and served with cracks, concrete in tensile zone is out of service and the tensile stress carried by concrete is transferred to rebars. The equivalent section that is composed of the concrete area in compression

zone and transformed area of rebars in tensile zone, is called transformed section of cracked section.

(2) Geometric parameters calculation

①Area of transformed section A_0

$$A_0 = bx + \alpha_{Es}A_s \tag{9-1}$$

②Static moment of transformed section to neutral axis S_0

a. Compression zone

$$S_{oc} = \frac{1}{2}bx^2 \tag{9-2}$$

b. Tensile zone

$$S_{ot} = \sigma_{Es}A_s(h_0 - x) \tag{9-3}$$

③Inertial moment of transformed section I_{cr}

$$I_{cr} = \frac{1}{3}bx^3 + \alpha_{Es}A_s(h_0 - x)^2 \tag{9-4}$$

④Calculation of the compression zone height x of transformed section

a. Rectangle section (Figure 9-2):

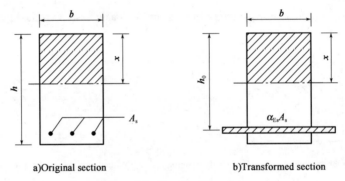

a) Original section b) Transformed section

Figure 9-2 Transformed section

A_s for flexural members, the neutral axis of the cracked section is in the same position as the centroidal axis. It means $S_{oc} = S_{ot}$, and it can be obtained that $\frac{1}{2}bx^2 = \alpha_{Es}A_s(h_0 - x)$. Thus, the depth of compression zone of the transformed section is

$$x = \frac{\alpha_{Es}A_s}{b}\left(\sqrt{1 + \frac{2bh_0}{\alpha_{Es}A_s}} - 1\right) \tag{9-5}$$

b. T-shaped section (Figure 9-3):

When the depth of compression zone $x \leq$ thickness of compressive flange h'_f, it belongs to the first type of T-shaped section. And transformed section geometric characteristics of cracked section can be calculated as if it is a rectangular section with the width of b'_f.

When the depth of compression zone $x > h'_f$, it indicates that the neutral axis is located in the web of T-shaped section, so it belongs to the second type of T-shaped section.

$$x = \sqrt{A^2 + B} - A$$

$$A = \frac{\alpha_{Es}A_s + (b'_f - b)h'_f}{b}, B = \frac{2\alpha_{Es}A_s h_0 + (b'_f - b)(h'_f)^2}{b}$$

The moment of inertia of transformed section of cracked section about its neutral axis I_{cr} is

$$I_{cr} = \frac{b'_f x^3}{3} - \frac{(b'_f - b)(x - h'_f)^3}{3} + \alpha_{Es}A_s(h_0 - x)^2 \qquad (9\text{-}6)$$

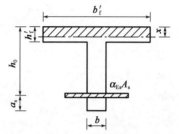
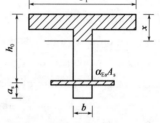

a) The first type of T-shaped section   b) The second type of T-shaped section

Figure 9-3 Calculation schema of transformed T-shaped section in cracked condition

5) Geometric parameters calculation of transformed section of total cross-section (Figure 9-4)

(1) Definition: It refers to the assumption section transformed from the section in which the steel rebar and entire concrete are in service.

(2) Calculation of geometric parameters.

①Area of transformed section

$$A_0 = bh + (b'_f - b)h'_f + (\alpha_{Es} - 1)A_s \qquad (9\text{-}7)$$

②Height of the centroid axis to the top edge

$$x = \frac{\frac{1}{2}bh^2 + \frac{1}{2}(b'_f - b)(h'_f)^2 + (\alpha_{Es} - 1)A_s h_0}{A_0} \qquad (9\text{-}8)$$

③The moment of inertia of transformed section about its centroid axis

$$I_0 = \frac{1}{12}bh^3 + bh\left(\frac{1}{2}h - x\right)^2 + \frac{1}{12}(b'_f - b)(h'_f)^3 + (b'_f - b)h'_f\left(\frac{h'_f}{2} - x\right)^2 +$$

$$(\alpha_{Es} - 1)A_s(h_0 - x)^2 \qquad (9\text{-}9)$$

a) Original section   b) Transformed section

Figure 9-4 Schema of total transformed cross-section

9.3　Checking of stress

For a reinforced concrete flexural member in construction and service stages, the stress calculation will adopt the method in materials mechanics only if the stress in each stage does not exceed the limit value.

1) Stress limit

The **CHBC** prescribes that the stress limit in construction stage can be achieved by conducting internal force combination according to the load that may appear in construction. And the normal section stress of flexural member should meet the following requirements.

(1) the tensile stress of the edge of concrete compression zone

$$\sigma_{cc}^t \leqslant 0.80 f_{ck} \tag{9-10}$$

(2) the tensile stress of rebar

$$\sigma_{si}^t \leqslant 0.75 f_{sk} \tag{9-11}$$

Where f_{ck} is the characteristic value of concrete axial compressive strength; f_{sk} is the characteristic value of tensile strength of regular reinforcement in construction stage; σ_{si}^t is the stress of rebar in layer I in tensile zone calculated in transient condition.

(3) The tensile stress at the neutral axis is

$$\sigma_{tp}^t = \frac{V_k^t}{b z_0} \leqslant f_{tk}' \tag{9-12}$$

Where V_k^t is the shear force caused by the construction load; b is the web width of the rectangle section, the T-shaped section, and the I-shaped section; z_0 is the distance between the pressure resultant and the resultant force from reinforcing steel bars, which is determined according to the triangular distribution of the pressure stress; f_{tk}' is the standard strength for axis tension of concrete.

2) Stress calculation

(1) Stress calculation: after the combination of the loads that appear in construction and service stages, the calculation method in materials mechanics should be adopted and the cross-section geometric value should be paid attention to.

(2) Calculation steps of normal stress of rectangular cross-section beams:

①Calculating the depth of compression zone x;

②Calculating the inertia moment of transformed section of cracked section I_{cr};

③Calculating the stress of cracked section:

Stress of concrete edge of compression zone

$$\sigma_{cc}^t = \frac{M_k^t x}{I_{cr}} \leqslant 0.80 f_{ck} \tag{9-13}$$

Stress of gravity center of tensile steel rebar area

$$\sigma_{si}^t = \alpha_{Es} \frac{M_k^t(h_{oi} - x)}{I_{cr}} \leq 0.75 f_{sk} \qquad (9\text{-}14)$$

Where I_{cr} is inertia moment of transformed section of cracked section; M_k^t is bending moment value formed by the characteristic value of temporary construction load.

(3) The normal stress calculation method of T-section beams is the same as that of rectangular section beams.

9.4 Checking of cracks and crack width of flexural members

Cracking is allowed when the reinforced concrete members are in normal service, but the service performance and durability are degraded if cracks are too wide. Therefore, cracks are allowed, but the crack width must be limited.

1) Reasons of controlling crack width

(1) To meet people's psychological limit, the crack width should not exceed 0.3mm.

(2) To avoid rebar corrosion and maintain the durability of reinforced concrete structure, the crack width should be limited.

2) Reasons of cracks and measures for crack control

(1) Some concrete cracks are induced by load like bending moment, axial force and torsional moment in cross-section. They may be controlled by checking design calculations and construction measures.

(2) Some cracks are produced by imposed deformation and constrained deformation like concrete shrinkage, temperature change, uneven settlement of foundation, etc. They can be controlled by construction measures and construction technology.

(3) The corrosion of rebar can also produce concrete cracks. To avoid these cracks, the enough thickness of concrete cover and sufficient concrete density should be assured to decrease the concrete permeability, and the using amount of early strength agent must be controlled.

3) Calculation theory and method of flexural crack width of flexural members

(1) The first category is theory analysis method based on tests. It establishes a calculation schema of the cracking section through theoretical analysis. Then the calculation formula of crack width is obtained. The uncertain parameters in the formula can be determined by using testing results. The theory includes bond stress slip theory, non-slip theory and combination of both.

①Bond stress slip theory: crack control is determined by the bond performance between rebar and concrete. When cracks appear, there will be slip between the rebar and concrete. Differential deformation leads to the increase of cracks.

②Non-slip theory: the surface crack width is determined by strain gradients from rebar to concrete surface. Crack width increases with the increase of the distance to rebar. The thickness of concrete cover is the main factor that affects crack width.

③Combined theory: considering the effects of concrete cover thickness and the slip between rebar and concrete on crack width.

(2) The second category is mathematical statistics method based on test results. It analyzes the factors that affect crack width according to testing results. And the simple, applicable and reliable crack width calculation formula can be obtained by using mathematical statistics method.

4) Influential factors of crack width

(1) Rebar stress σ_{ss}: It is the main factor. Linear relationship exists between the maximum crack width and σ_{ss}.

(2) Rebar diameter d: The maximum crack width W_{cr} decreases with the decrease of ρ.

(3) Reinforcement ratio ρ: W_{cr} decreases with the increase of ρ; W_{cr} remains constant when ρ is close to a certain value.

(4) Cover thickness c: The experimental results show that a larger c corresponds to a larger W_{cr}.

(5) Rebar shape: The effect of rebar shape can be considered with the introduction of coefficient c_1. Ribbed rebar has better adhesive performance, but round rebar has less adhesive strength and wider crack.

(6) Load effect: To consider the repeated load, the coefficient c_2 was introduced.

(7) Effects of member stress property: Since the crack width differs with the difference of member stress property, the coefficient c_3 is introduced.

5) Calculation formula of maximum crack width

According to the **CHBC**, for reinforced concrete flexural member with rectangular section, T-shaped section and I-shaped section, the maximum crack width can be calculated by

$$W_{cr} = C_1 C_2 C_3 \frac{\sigma_{ss}}{E_s}\left(\frac{c+d}{0.36+1.7\rho_{te}}\right) (\text{mm}) \qquad (9\text{-}15)$$

Where C_1 is surface shape coefficient of rebar, $C_1 = 1.4$ for plain face rebar, $C_1 = 1.0$ for ribbed rebar; C_2 is long-term effect coefficient, $C_2 = 1 + 0.5\dfrac{M_1}{M_s}$, in which M_1 is the designed bending moment (or the designed axial force) calculated by the quasi-permanent combination of actions, and M_s is calculated by the frequent combination of actions; C_3 is coefficient related to member subjected to load conditions, $C_3 = 1.15$ for the flexural member of reinforced concrete plate, $C_3 = 1.0$ for other types of flexural member, $C_3 = 1.1$ for eccentric tensile members, $C_3 = 0.75$ for eccentric compression members with circular section, $C_3 = 0.9$ for other eccentric compressive members, and $C_3 = 1.2$ for axial compressive members; d is the diameter (mm) of longitudinal tensile bars. The transformed diameter d_e is adopted if using different diameters, $d_e = \dfrac{\sum n_i d_i^2}{\sum n_i d_i}$; for reinforced concrete members, n_i is the number of rebar of the ith type of tensile steel; d_i is the nominal diameter of ith type of rebar; when a single steel bar is used, d_i is the nominal diameter; when a bundle of steel bars are used, d_i is the equivalent diameter d_{se}, $d_{se} = \sqrt{n}d$, n is the number of steel bars. For welded steel skeleton, d or d_e is multiplied by a coefficient of 1.3. ρ_{te} is the effective reinforcement ratio of longitudinal tensile bars, for the rectangular, T-shaped and I-shaped flexural members, $\rho_{te} = \dfrac{A_s}{A_{te}}$, A_s is the cross-sectional area of longitudinal reinforcement in the tension zone, A_{te} is the cross-sectional area of the effective tensile concrete, and $2a_s b$ is taken for the reinforced concrete flexural members, a_s is the distance from the center of gravity of tensioned bar to the edge of the tension zone; for the rectangular sections, b is the beam width; for the flange at the tension zone of T-shaped and I-shaped sections, b is the effective flange width in the tension zone. When $\rho_{te} > 0.1, \rho_{te} = 0.1$, when $\rho_{te} < 0.01, \rho_{te} = 0.01$; σ_{ss} is stress of longitudinal rebar. For the flexural reinforced concrete members, $\sigma_{ss} = \dfrac{M_s}{0.87 A_s h_0}$; the σ_{ss} for others members can be found in in the **CHBC**; E_s is elastic modulus of rebars (MPa); h_0 is effective height.

6) Limited value of crack width

The **CHBC** prescribes that the maximum crack width of reinforced concrete flexural members should meet the following requirements when they are under loading conditions:

(1) Class I and class II environment: $[W_{fk}] = 0.2$mm;

(2) Class III and class IV environment: $[W_{fk}] = 0.15$mm.

9.5 Deformation (deflection) checking of flexural members

Excessive deformation affects thefunction and durability of members. Reinforced concrete structure is composed of materials with two different types of property. The method of deflection calculation differs from that of single material. Starting with the deflection calculation method of single elastic material, this chapter considers the characteristic of post-cracking stiffness degradation of reinforced concrete flexural members and calculates the post-cracking stiffness of members to calculate deflection.

1) Reasons of the deformation calculation of reinforced concrete flexural members

(1) Excessive deflection affects their functions: excessive mid-span deflection of simply-supported beams will enlarge the rotation angle of the end of beam. The expansion and contraction joint can be damaged when there is vehicle impacting. Excessive deflection of continuous beam will make bridge surface unsmooth and vehicles passing will cause bumpiness and dashing.

(2) Cracks and crushing between members.

(3) Psychological safety.

(4) Excessive deflection will lead to vibration effect and dynamic effect.

2) Deflection calculation method

As for simply-supported beam of homogeneous elastic solid materials, a general formula of deflection calculation can be written as

$$f = \alpha \frac{ML^2}{B} \quad (9\text{-}16)$$

Where α is deflection coefficient related to load form and constraint; B is flexural stiffness of section. Flexural stiffness $B = EI$ as for homogeneous elastic beams; M is mid-span section bending moment; L is calculation span length.

As for the reinforced concrete members of certain load form and supporting condition, the key of deflection calculation is to determine the stiffness of post-cracking section.

3) Flexural stiffness calculation of reinforced concrete flexural members

(1) Definition of equivalent stiffness.

The reinforcement distribution of all the sections of reinforced concrete flexural members is different and the flexural moment is also different at each section. Sections with small flexural moment may not have flexural cracks and their stiffness is much greater than those with larger flexural moment. Therefore, flexural stiffness is different along the beam span. In order to simplify the calculation, the equivalent stiffness members can be changed into members with equivalent stiffness.

(2) Calculation method of equivalent stiffness (Figure 9-5): according to the principle of members rotation angle equality under the effect of two-end flexural member, the equivalent stiffness B of flexural members can be calculated. It is also the flexural stiffness of equivalent section of cracked members.

(3) The ***CHBC*** suggested calculation formula is

$$B = \frac{B_0}{\left(\dfrac{M_{cr}}{M_s}\right)^2 + \left[1 - \left(\dfrac{M_{cr}}{M_s}\right)^2\right]\dfrac{B_0}{B_{cr}}} \qquad (9\text{-}17)$$

Where B is flexural stiffness of equivalent section of cracked members; B_0 is flexural stiffness of total cross-section, $B_0 = 0.95 E_c I_0$; B_{cr} is flexural stiffness of cracked section, $B_{cr} = E_c I_{cr}$; E_c is the elasticity modulus of concrete; I_0 is transformed section inertial moment of total cross-section; I_{cr} is transformed section inertial moment of cracked section; M_s is flexural moment by frequent combination; M_{cr} is cracking moment, $M_{cr} = \gamma f_{tk} W_0$; f_{tk} is characteristic value of concrete axial tensile strength; γ is concrete plastic influential coefficient of members in tensile area, $\gamma = 2 S_0 / W_0$; S_0 is the area moment about the gravity axis for the area that is above the gravity axis of the total transformed cross-section; W_0 is elastic resistance modulus for the section edge of total transformed cross-section.

4) Calculation requirements of deflection of reinforced concrete flexural members

(1) Checking requirements

The ***CHBC*** prescribes that the long-term deflection value of reinforced concrete flexural members—long-term deflection value due to structure self-weight $\leqslant$ deflection limit value.

Deflection calculation of reinforced concrete flexural members in service stage should consider long-term effect, so the deflection value by frequent combination should be multiplied by deflection long-term magnification coefficient η.

That means

$$f_l = \eta_\theta f_s \qquad (9\text{-}18)$$

Chapter 9 · Calculation of Stress, Cracking and Deflection of Reinforced Concrete Flexural Members

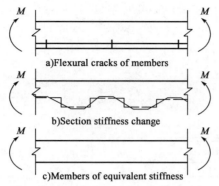

Figure 9-5　Equivalent section of members

Where η_θ is the long-term deflection magnification coefficient. When concrete below C40 is used, $\eta_\theta = 1.60$; when concrete between C40 ~ C80 is used, $\eta_\theta = 1.45$ ~ 1.35. Intermediate strength grade can be determined by linear interpolation.

(2) Deflection value of reinforced concrete flexural members

①Maximum deflection of main girder of beam bridge: $l/600$;

②Cantilever end of main girder of beam bridge: $l_1/300$.

Where l is the calculation span length of flexural members. l_1 is the length of cantilever.

5) Installation of predetermined camber

(1) Definition: it refers to the reversed camber predetermined in construction.

(2) Purpose of installation:

①To eliminate the deformation caused by the long-term load of structure gravity.

②To maintain certain camber of the members without static load effect.

(3) Condition of installation

When long-term deflection caused by load long-term effect of frequent combination does not exceed 1/1600 of calculated span length, the camber is not predetermined; otherwise, predetermined.

(4) Value of predetermined camber

$$\Delta = w_G + \frac{1}{2} w_Q \tag{9-19}$$

Where Δ is the value of predetermined camber; w_G is long-term vertical deflection caused by structure self-weight; w_Q is long-term vertical deflection caused by variable common load.

The attention should be paid that the installation of the camber should follow a smooth curve.

6) Example

A reinforced concrete simply-supported T-beam with a span length $L_0 = 19.96\text{m}$, the calculated span length $L = 19.50\text{m}$. For concrete C30, $f_{ck} = 20.1\text{MPa}$, $f_{tk} = 2.01\text{MPa}$, $E_c = 3.0 \times 10^4\text{MPa}$. The environmental condition is type I, and the safety class is class 1.

The dimension of the main beam section is shown in Figure 9-6a). The main rebar is HRB400. The section area of rebar $A_s = 6836\text{mm}^2$ ($8\Phi32 + 2\Phi16$), $a_s = 111\text{mm}$, $E_s = 2 \times 10^5\text{MPa}$, and $f_{sk} = 335\text{MPa}$.

The lifting point of the simply-supported beam is set at the point away from the edge of beam $a = 400$mm when beams are lifted [Figure 9-6a)]. Flexural moment, $M_{G1} = 505.69$kN · m, is formed in mid-span section because of beam self-weigh.

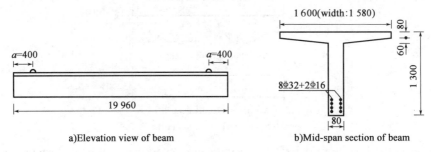

Figure 9-6　Reinaforced concrete simply-supported T-beam

Moment, $M_{Q1} = 596.04$kN · m (vehicle impact coefficient excluded), is formed by vehicle load characteristic value when T-beam mid-span section is in service; moment due to pedestrian load characteristic value $M_{Q2} = 55.30$kN · m; and moment caused by permanent load characteristic value $M_G = 751$kN · m.

Preliminary Checking of reinforced concrete simply-supported T-beam.

Solution: (1) normal stress check in hoisting construction

According to the hoisting point position and main beam deadweight (which can be viewed as uniformly-distributed load) shown in Figure 9-6a), the maximum negative flexural moment is in hoisting point section; the maximum positive flexural moment is in beam mid-span section. And both sections are normal stress checking section. The calculation method introduced in this example is based on normal stress checking of beam mid-span section.

①Calculation of transformed section inertia moment I_{cr} of beam mid-span section

According to the **CHBC**, the effective width of beam compressive flange $b'_f = 1500$mm. The average thickness of compression flange is 110mm. The effective height $h_0 = h - a_s = 1\,300 - 111 = 1\,189$mm.

$$\alpha_{Es} = \frac{E_s}{E_c} = \frac{2 \times 10^5}{3.0 \times 10^4} = 6.667$$

Depth of compression zone is

$$\frac{1}{2} \times 1500 \times x^2 = 6.667 \times 6836 \times (1\,189 - x)$$

As a result　　　　　　　$x = 240.12$mm $> h'_f (= 110$mm$)$

So it belongs to the second kind of T-shaped section.

Depth of compression zone of transformed section x could be obtained by

$$A = \frac{\alpha_{Es} A_s + h'_f (b'_f - b)}{b} = \frac{6.667 \times 6836 + 110 \times (1\,500 - 180)}{180} = 1\,060$$

$$B = \frac{2\alpha_{Es} A_s h_0 + (b'_f - b) h'^2_f}{b} = \frac{2 \times 6.667 \times 6836 \times 1\,189 + (1\,500 - 180) \times 110^2}{180}$$

$= 690\,838$

As a result

$$x = \sqrt{A^2 + B} - A = \sqrt{1\,060^2 + 690\,838} - 1\,060 = 287\text{mm} > h'_f(\ = 110\text{mm})$$

Transformed section inertia moment of cracking section I_{cr} is

$$I_{cr} = \frac{b'_f x^3}{3} - \frac{(b'_f - b)(x - h'_f)^3}{3} + \alpha_{Es} A_s (h_0 - x)^2$$

$$= \frac{1\,500 \times 287^3}{3} - \frac{(1\,500 - 180) \times (287 - 110)^3}{3} + 6.667 \times 6\,836 \times (1\,189 - 287)^2$$

$$= 46\,460.55 \times 10^6 \text{mm}^4$$

②Normal stress check

Hoisting dynamic coefficient is 1.2 (The main beam is overweight in hoisting.). The calculated flexural moment of mid-span section is:

Normal stress of concrete of compression zone edge would be

$$\sigma'_{cc} = \frac{M^t_k x}{I_{cr}} = \frac{606.828 \times 10^6 \times 287}{46\,460.55 \times 10^6}$$

$$= 3.75\text{MPa} < 0.8 f'_{ck} = 0.8 \times 20.1 = 16.08\text{MPa}$$

Rebar stress of gravity center of rebar area would be

$$\sigma^t_s = \alpha_{Es} \frac{M^t_k (h_0 - x)}{I_{cr}} = 6.667 \times \frac{606.828 \times 10^6 \times (1189 - 287)}{46\,460.55 \times 10^6}$$

$$= 78.54\text{MPa} < 0.75 f_{sk} = 0.75 \times 400 = 300\text{MPa}$$

The distance between the gravity center of bottom rebar (2 Φ 32) and the top compression edge is: $h_{01} = 1300 - \left(\frac{35.8}{2} + 35\right) = 1\,247\text{mm}$, and the rebar stress is

$$\sigma_s = \alpha_{Es} \frac{M^t_k}{I_{cr}} (h_{01} - x) = 6.667 \times \frac{606.828 \times 10^6}{46\,460.55 \times 10^6} \times (1\,247 - 287)$$

$$= 83.6\text{MPa} < 0.75 f_{sk} = 300\text{MPa}$$

As check results have shown, both the concrete compressive stress and rebar tensile stress when beam is lifted are smaller than code limited value and the lifting-point position can be determine as Figure 9-6a).

(2) Check of crack width W_{fk}

①Determination of coefficient

a. Coefficient c_1: For ribbed rebar, $c_1 = 1.0$.

b. Flexural calculated value in frequent combination of actions is

$$M_s = M_G + \psi_{11} \times M_{Q1} + \psi_{12} \times M_{Q2}$$

$$= 751 + 0.7 \times 596.04 + 0.4 \times 55.30 = 1\,190.35\text{kN} \cdot \text{m}$$

c. Moment calculated value of quasi-permanent combination of actions is
$$M_1 = M_G + \psi_{21} \times M_{Q1} + \psi_{22} \times M_{Q2}$$
$$= 751 + 0.4 \times 596.04 + 0.4 \times 55.30$$
$$= 1\,011.54 \text{kN} \cdot \text{m}$$
$$c_2 = 1 + 0.5 \frac{M_1}{M_s} = 1 + 0.5 \frac{1\,011.54}{1\,190.35} = 1.42$$

d. Coefficient $c_3 = 1.0$, for non-plate flexural member, $c_3 = 1.0$.

②Calculation of rebar stress σ_{ss}
$$\sigma_{ss} = \frac{M_s}{0.87 h_0 A_s} = \frac{1\,190.35 \times 10^6}{0.87 \times 1\,189 \times 6\,836} = 168 \text{MPa}$$

③Transformed diameter d

Because of the different rebar diameters in tensile zone, d should take the value of transformed diameter d_e. Then it is
$$d = d_e = \frac{8 \times 32^2 + 2 \times 16^2}{8 \times 32 + 2 \times 16} = 30.2 \text{mm}$$

For welded steel reinforcement cage: $d = d_e = 1.3 \times 30.2 = 39.26$mm.

④Effective reinforcement ratio ρ_{te} of longitudinal tensile rebar:

According to the relevant dimensions of the T-beam section, calculating the cross-sectional area of the concrete in the effective tension zone as
$$A_{te} = 2a_s b = 2 \times 111 \times 180 = 39\,960 \text{mm}^2$$

The calculated effective reinforcement ratio ρ_{te} of the longitudinally tensioned steel bars is
$$\rho_{te} = \frac{A_s}{A_{te}} = \frac{6\,836}{39\,960} = 0.171 > 0.1$$
$$\text{So } \rho_{te} = 0.1.$$

⑤Maximum crack width W_{cr}
$$W_{cr} = c_1 c_2 c_3 \frac{\sigma_{ss}}{E_s} \left(\frac{c + d}{0.36 + 1.7 \rho_{te}} \right)$$
$$= 1 \times 1.42 \times 1 \times \frac{168}{2 \times 10^5} \times \left(\frac{35 + 39.26}{0.36 + 1.7 \times 0.1} \right)$$
$$= 0.17 \text{mm} < [W_f] = 0.2 \text{mm}$$

(3) Check of beam mid-span deflection

When calculating the beam deflection, the calculated width of flange should be the full flange after connected with the adjacent beam, that is, $b'_{fl} = 1600$mm and h'_f is 110mm.

①Calculation of T-beam transformed section inertia moment I_{cr} and I_0

For T-beam cracked section
$$\frac{1}{2} \times 1\,600 \times x^2 = 6.667 \times 6\,836 \times (1\,189 - x)$$
$$x = 233.3 \text{mm} > h'_f (= 110 \text{mm})$$

The mid-span section of beam belongs to the second kind of T-shaped section. The height of the compression zone x should be recalculated

$$A = \frac{\alpha_{Es}A_s + h'_f(b'_{fl} - b)}{b}$$

$$= \frac{6.667 \times 6836 + 110 \times (1600 - 180)}{180} = 1120.98$$

$$B = \frac{2\alpha_{Es}A_s h_0 + (b'_{fl} - b)h'^2_f}{b}$$

$$= \frac{2 \times 6.667 \times 6836 \times 1189 + (1600 - 180) \times 110^2}{180} = 697560.0$$

Then

$$x = \sqrt{A^2 + B} - A = \sqrt{1120.98^2 + 697560.0} - 1120.98$$
$$= 277\text{mm} > h'_f(=110\text{mm})$$

Transformed section inertia moment I_{cr} of cracking section is

$$I_{cr} = \frac{1600 \times 277^3}{3} - \frac{(1600 - 180) \times (277 - 110)^3}{3} + 6.667 \times 6836 \times (1189 - 277)^2$$
$$= 47038.14 \times 10^6 \text{mm}^4$$

Transformed section area A_0 of T-beam total cross-section is

$$A_0 = 180 \times 1300 + (1600 - 180) \times 110 + (6.667 - 1) \times 6836 = 428939.6\text{mm}^2$$

Height of the centroid axis to the top edge x is

$$x = \frac{\frac{1}{2} \times 180 \times 1300^2 + \frac{1}{2} \times (1600 - 180) \times 110^2 + (6.667 - 1) \times 6836 \times 1189}{428939.6}$$

$$= 482\text{mm}$$

Calculation of total transformed cross-section inertia moment I_0 is

$$I_0 = \frac{1}{12}bh^3 + bh\left(\frac{h}{2} - x\right)^2 + \frac{1}{12}(b'_{fl} - b)(h'_f)^3$$

$$+ (b'_{fl} - b)h'_f \cdot \left(x - \frac{h'_f}{2}\right)^2 + (\alpha_{ES} - 1)A_s(h_0 - x)^2$$

$$= \frac{1}{12} \times 180 \times 1300^3 + 180 \times 1300 \times \left(\frac{1300}{2} - 482\right)^2 + \frac{1}{12} + (1600 - 180)(110)^3$$

$$+ (1600 - 180) \times 110 \times \left(482 - \frac{110}{2}\right)^2 + (6.667 - 1) \times 6836 \times (1189 - 482)^2$$

$$= 8.76 \times 10^{10} \text{mm}^4$$

②Calculation of flexural stiffness of cracked member

Total cross-section flexural stiffness

$$B_0 = 0.95 E_c I_0 = 0.95 \times 3.0 \times 10^4 \times 8.76 \times 10^{10} = 2.49 \times 10^{15} \text{N} \cdot \text{mm}^2$$

Flexural stiffness of cracking section

$$B_{cr} = E_c I_{cr} = 3.0 \times 10^4 \times 47038.14 \times 10^6 = 1.41 \times 10^{15} \text{N} \cdot \text{mm}^2$$

Elastic resistant modulus of tensile zone edge of transformed section is

$$W_0 = \frac{I_0}{h - x} = \frac{8.76 \times 10^{10}}{1300 - 482} = 1.07 \times 10^8 \text{mm}^3$$

Area moment of transformed section is

$$S_0 = \frac{1}{2}b'_{fl}x^2 - \frac{1}{2}(b'_{fl} - b)(x - h'_f)^2$$

$$= \frac{1}{2} \times 1\,600 \times 482^2 - \frac{1}{2} \times (1\,600 - 180) \times (482 - 110)^2$$

$$= 8.76 \times 10^7 \text{mm}^3$$

Elastic influential coefficient is

$$\gamma = \frac{2S_0}{W_0} = \frac{2 \times 8.76 \times 10^7}{1.07 \times 10^8} = 1.64$$

Cracking moment is

$$M_{cr} = \gamma f_{tk} W_0 = 1.64 \times 2.01 \times 1.07 \times 10^8 = 3.527 \times 10^8 \text{N} \cdot \text{mm} = 352.7 \text{kN} \cdot \text{m}$$

Flexural stiffness of cracked member is

$$B = \frac{B_0}{\left(\frac{M_{cr}}{M_s}\right)^2 + \left[1 - \left(\frac{M_{cr}}{M_s}\right)^2\right]\frac{B_0}{B_{cr}}} = \frac{2.49 \times 10^{15}}{\left(\frac{352.7}{1\,190.35}\right)^2 + \left[1 - \left(\frac{352.7}{1\,190.35}\right)^2\right]\frac{2.49 \times 10^{15}}{1.41 \times 10^{15}}}$$

$$= 1.466 \times 10^{15} \text{N} \cdot \text{mm}^2$$

③ Long-term mid-span deflection of flexural member

Characteristic moment of mid-span section under the frequent combination $M_s = 1190.35$ kN·m. Characteristic moment of mid-span section under structure dead weight $M_G = 751$ kN·m. As for concrete C30, the deflection long-term magnification coefficient $\eta_\theta = 1.60$.

The long-term mid-span deflection of flexural member in service is

$$w_l = \frac{5}{48} \times \frac{M_s L^2}{B} \times \eta_\theta = \frac{5}{48} \times \frac{1\,190.35 \times 10^6 \times (19.5 \times 10^3)^2}{1.466 \times 10^{15}} \times 1.60$$

$$= 51.46 \text{mm}$$

The long-term mid-span deflection of flexural member under structure self-weight is

$$w_G = \frac{5}{48} \times \frac{M_G L^2}{B} \times \eta_\theta = \frac{5}{48} \times \frac{751 \times 10^6 \times (19.5 \times 10^3)^2}{1.466 \times 10^{15}} \times 1.60$$

$$= 32.47 \text{mm}$$

The calculated long-term deflection w_{ll} is

$$w_{ll} = w_l - w_G = 51.46 - 32.47 = 18.99 \text{mm} < \frac{L}{600}\left(= \frac{19.5 \times 10^3}{600} = 33 \text{mm}\right)$$

The requirements in the **CHBC** are fulfilled.

(4) Installation of predetermined camber

Considering the frequency combination and the influence of long-term load effect, the long-term mid-span deflection of beam is

$$w_l = 51.46 \text{mm} > \frac{L}{1\,600} = \frac{19.5 \times 10^3}{1\,600} = 12 \text{mm}$$

The predetermined camber of mid-span section should be built.

According to the codes about predetermined camber in the ***CHBC***, the predetermined camber in mid-span section of beam is

$$\Delta = w_G + \frac{1}{2}w_{ll} = 32.47 + \frac{1}{2} \times 18.99 = 41.97 \text{mm}$$

9.6　Durability of concrete structure

1) Introduction of structure durability

(1) Definition

The durability of concrete structure refers to the ability of maintaining the safety, function and appearance of concrete structure without high expense treatment in condition of natural environment, service and material factor in designed service period.

(2) Main factors affecting concrete structure durability

①Concrete freezing and thawing damage: When concrete in water-saturated state (water content to the limit value of 91.7%) is frozen, the expanded pressure and permeable pressure in capillary porosity will cause internal crack and damage in concrete structure. The repeated action of freezing leads to damage accumulation and will cause structure failure finally.

②Concrete alkali-aggregate reaction: It refers to the chemical reaction between certain active mineral in concrete aggregates and alkali solution in concrete micropore. Alkali-aggregate reaction produces alkali silicate gels and its volume may be increased by three to four times because of water expansion, which will cause concrete spalling, cracks and even destruction.

③Corrosion of corrosive medium: Erosion of certain chemical corrosive medium may cause some concrete components to be dissolved and lost and cracks, pores, loose and break happen. The chemical reaction between certain chemical corrosive medium and some concrete components may generate volume expansion to destruct concrete structure.

④Mechanical wearing: It is usually seen in industrial ground, highway pavement, bridge surface and airport runway, etc.

⑤Concrete carbonization: It means that the chemical reaction between carbon dioxide in atmosphere and the calcium hydroxide in concrete leads to the decrease of pH value in concrete, which destroys the protective cover of rebar in concrete and produces rebar corrosion.

⑥Rebar corrosion: It detaches concrete protective cover and rebar effective area decrease, which causes reduction of load-carrying capacity and structure being damaged.

2) Requirements for the durability design of concrete structures

(1) The design and service life of concrete structures and components of highway bridges and culverts should conform to the regulation according to ***Technical Standard of Highway Engineering*** (***JTG B01***).

(2) The concrete structures and components of highway bridges and culverts should be

determined according to the environment directly contacted by their surfaces, as shown in Table 9-1.

Environmental classification of concrete structures and components of highway bridges and culverts Tabel 9-1

Environmental category	Condition
Class I—General environment	Environment affected only by carbonation of concrete
Class II—Freezing and thawing environment	Environment affected by repeated freezing and thawing
Class III—Offshore or marine chloride environment	Environment affected by chloride salts in the marine environment
Class IV—De-icing salt and other chloride environments	Environment affected by chloride salts such as deicing salts
Class V—Salt crystal environment	Environment affected by sulphate crystal expansion in concrete pores
Class VI—Chemically corrosive environment	Environment eroded by chemicals with strong acidity and alkalinity
Class VII—Abrasion environment	Environment affected by friction, cutting, impact, etc. of wind, water, or inclusions in water

(3) The minimum requirements for the concrete strength grades under various environments should comply with the requirements of Table 9-2.

Environmental classification of concrete structures and components of highway bridges and culverts Tabel 9-2

Component category / Design life	Beam, slab, tower, arch ring, upper culvert		Pier body, lower part of culvert		Bearing platform, basement	
Class I—General environment	100 years	50years, 30years	100 years	50years, 30years	100 years	50years, 30years
Class II—Freezing and thawing environment	C35	C30	C30	C25	C25	C25
Class III—Offshore or marine chloride environment	C40	C35	C35	C30	C30	C25
Class IV—De-icing salt and other chloride environment	C40	C35	C35	C30	C30	C25
Class V—Salt crystal environment	C40	C35	C35	C30	C30	C25
Class VI—Chemically corrosive environment	C40	C35	C35	C30	C30	C25
Class VII—Abrasion environment	C40	C35	C35	C30	C30	C25

(4) The following durability technical measures should be taken for concrete structures and

components of highway bridges and culverts:

①The thickness of the concrete cover should meet the requirements of Schedule Table 1-8.

②The prestressing system in the prestressed concrete structure adopts corresponding multiple protective measures according to the specific conditions.

③Concrete structures with impermeability requirements meet the requirements of relevant standards.

④In the humid environment of severe cold and cold areas, the concrete should meet the anti-freeze requirements, and the concrete anti-freeze level meets the requirements of relevant standards.

⑤The structures of the bridge and culvert are conducive to drainage and ventilation, avoiding condensation of water and gas and accumulation of harmful substances.

Chapter 10
Local Compression

Local compression, which means the load-carrying state of bearing the load with the local area on the surface of member, is constantly visible in bridge members such as the end anchorage zone of the post-tension concrete member, the part of the abutment cap directly bearing the support, and the arch ring compressed by the column on the arch. As shown in Figure 10-1, the transmission and distribution of the stress in the locally compressed area can be acquired by mechanical analysis and has the following characteristics:

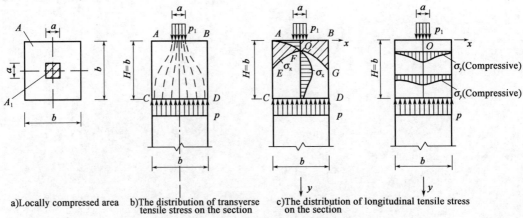

a) Locally compressed area b) The distribution of transverse tensile stress on the section c) The distribution of longitudinal tensile stress on the section

Figure 10-1 Locally compressed area at the end of member

(1) x-direction stress σ_x: In *AOBGFE* section of locally compressed area, σ_x is considered as a compression stress, and tensile stress in other parts. The maximum value of transverse tensile stress is near the center of locally compressed area *ABCD*.

(2) y-direction stress σ_y: In locally compressed area, most of σ_y is the compression stress. The compression near Oy axis is larger than that of other axis. Among all the section the value of stress in the position of point O is the maximum one, which is p_1.

(3) Shear stress τ: Shear stress exists in the sections parallel to y-axis outside of the locally compressed area.

In general, the behavior of the concrete member under local compression has the following characteristics:

(1) The compressive strength of locally compressed structure is higher than those compressed in large area.

(2) There is transverse tensile stress in the middle of locally compressed area (Figure 10-1). The stress can cause cracks of concrete structures, and these cracks are parallel to the direction of pressure in local area.

It is evident from the characteristics of the local compression that the calculation of the crack resistance and the strength should be conducted for the structural member under local compression.

10.1 The mode and mechanism of failure under local compression

1) The failure mode of concrete members under local compression

The failure mode of concrete members under local compression mostly correlates with A_l/A (where, A_l is the concrete area under local compression, A is the section area of a reinforced concrete member) and the position of A_l on the surface of the member. As far as the axial local compression is concerned in which A_l is symmetrically located on the member surface, there are three failure modes as follows.

(1) Cracking induced failure

The characteristic of failure: When the member section area is close to the area of local compression (in general conditions, $A/A_l < 9$) and the load is about 50% ~ 90% of the failure load, the longitudinal cracking is induced first on one side of the member. With the increase of load, the cracks spread, and there are similar cracks occurring on the other sides. Finally, a wedge is formed in the concrete due to punching [Figure 10-2a)], then the member has a splitting failure when concrete is split into pieces.

(2) Failure immediately after cracking

The characteristic of failure: When the section area of member is larger than the local compression area (in general conditions, $9 < A/A_l < 36$), the member fails suddenly as soon as concrete cracks. The cracks develop from top to bottom and their width at the top is wider than that at the bottom, then the zone located outside the local compression area is split into pieces. The local compression area is punched in a wedge shape [Figure 10-2b)].

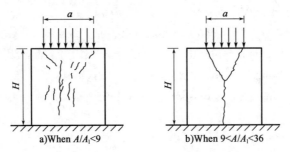

Figure 10-2　Failure modes of concrete member under local compression

(3) Subsidence of local concrete

The characteristic of failure: When the member section area is much larger than the area under local compression (in general conditions, $A/A_1 > 36$), before the whole member fails, the concrete under the local compression subsides first. The concrete surrounding the local compression zone has a shear failure, but the member can still bear the load till the surrounding concrete is split into pieces finally.

2) Analysis of working mechanism of local compression structure

(1) The theory of steel tube confined concrete

Concrete under local compression could be considered to be a concrete core which is confined by pressure from side directions. When the effect of the load increases, the compressed concrete expands, and the surrounding concrete takes effect to prevent the compressed concrete from expanding. Consequently, the concrete bearing the pressure is in a three-dimensional state of stress, which improves the compressive strength of concrete. When the hoop tensile stress born (or carried) by the surrounding concrete reaches the tensile strength of concrete, the member is broken as shown in Figure 10-3.

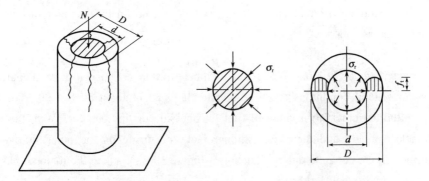

Figure 10-3　The mechanical model of theory of steel tube confined concrete

(2) Shear theory

When a member is subjected to local load, the characteristic of loading in fractural area of the member can be considered to be an arch structure with some tension rods. The concrete below the compressed board, which is seen as the tension rods, is subjected to the tensile force in the transverse direction. While the tensile stress σ_x caused by the local compressed load exceeds the ultimate tensile strength of concrete f_t, part of concrete working as tension rods will crack, but the failure mechanism has not been formed [Figure 10-4b)]. As the load increases continuously, more tensile rods will fail, and cracks will develop. Meanwhile, the internal force will be redistributed. When the ultimate load is reached, the concrete below the compressed board will be punched to wedge blocks due to the effect of shear-compressed load, and a slip plane will be formed. The final failure mode of arch structure is the cleavage of the wedge block as shown in Figure 10-4c).

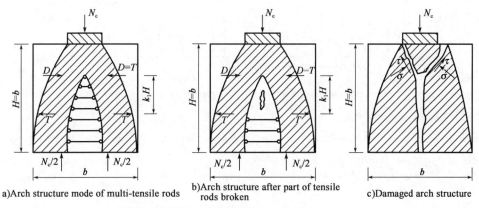

a) Arch structure mode of multi-tensile rods
b) Arch structure after part of tensile rods broken
c) Damaged arch structure

Figure 10-4 Locally compressed mode of shear theory

10.2 Enhancement coefficient of concrete strength for local compression

1) Enhancement coefficient of concrete strength for local compression β

(1) According to the **CHBC**, β can be given as

$$\beta = \sqrt{\frac{A_b}{A_l}} \qquad (10\text{-}1)$$

In which, A_l is the local compressed area when the area increase in the thickness of the steel board along with an angle of 45° is considered. If the forms of the compressed areas are circular, the area of the holes should not be subtracted while there exist holes. A_b is the calculated button area for locally compressed concrete.

(2) Determination of calculated area A_b of locally compressed concrete:

①Determination method: A_b has the same centroid as the actual local compressed zone.

②Generally the way to determine the value can be seen in Figure 10-5.

2) Enhancement coefficient β_{cor} of concrete strength for locally compressed concrete reinforced with indirect steel bars

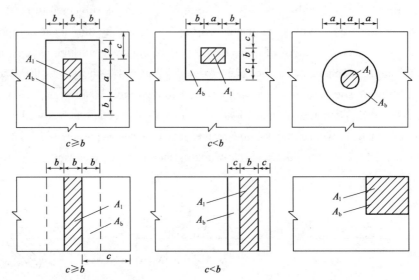

Figure 10-5 Sketch of calculated button area A_b of concrete subjected to locally compressed load

(1) Types of indirect steel bars

Indirect steel bars used in locally compressed structures can be designed in two types: steel grid nets or spiral steels, shown in Figure 10-6.

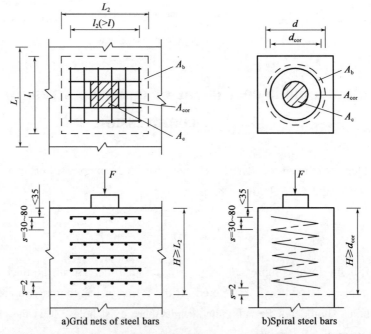

a) Grid nets of steel bars b) Spiral steel bars

Figure 10-6 Arrangement of indirect steel bars in the local compression zone (unit: mm)

(2) Ratio ρ_v of volume of the indirect steels

①It can be defined as the ratio of indirect steel volume and the compression concrete core volume.

②Formula

a. When the indirect steel rebars are designed in the form of steel grid nets, the formula can be given as [Figure 10-6a)]

$$\rho_v = \frac{n_1 A_{s1} l_1 + n_2 A_{s2} l_2}{A_{cor} s} \quad (10\text{-}2)$$

Where s is the distance between steel bar layers; n_1 and A_{s1} are the number of steels in a layer of steel net and the area of one steel along with the direction of l_1, respectively; n_2 and A_{s2} are the number of steels in a layer of steel net and the area of one steel along with the direction of l_2, respectively; A_{cor} is the core concrete area in the range of inner face of grid net of indirect steel rebars, and its gravity center should be coincided with A_1. It must follow the principle of concentricity and symmetry when calculating.

In addition, the difference between the steel rebar areas in the two directions (l_1 and l_2) of nets should be less than 50%. The number of indirect steel rebars subjected to local compression load should not be less than 4.

b. When the indirect steels are designed in the form of spiral steel bars, the formula can be given as [Figure 10-6b)]

$$\rho_v = \frac{4 A_{ss1}}{d_{cor} s} \quad (10\text{-}3)$$

Where A_{ss1} is the area of the section of one spiral steel; d_{cor} is the radius of the center of concrete at the range of inner face of indirect steel bars between the spiral steel bars; s is the distance between spiral steels.

The number of spiral steel bars should not be less than 4.

(3) Formula of enhancement coefficient of concrete strength for local compression concrete reinforced with indirect steel bars β_{cor}

$$\beta_{cor} = \sqrt{\frac{A_{cor}}{A_1}} \geq 1 \quad (10\text{-}4)$$

Where A_{cor} is the area of core concrete at the range of indirect steel bars net or spiral steel bars.

10.3 Calculation of local compression zone

1) Calculation content of local compression zone

The **CHBC** requires that the local compression member should be calculated for its load-carrying capacity and the section size.

2) Load carrying capacity calculation of local compression area

(1) Formula

$$\gamma_0 F_{ld} \leq F_u = 0.9(\eta_s \beta f_{cd} + k \rho_v \beta_{cor} f_{sd}) A_{ln} \quad (10\text{-}5)$$

Where F_{ld} is design value of local pressure on compression zone. For the anchor head of post-tensioned concrete member, it can take 1.2 times the maximum pressure value when tensioned. η_s is correction factor of concrete local pressure, according to Table 10-1. β is enhancement factor of concrete compressive strength. k is influence factor of indirect reinforcement, when the concrete strength is C50 or less, $k = 2.0$; for concrete C50 ~ C80, $k = 2.0$ ~ 1.70, and interpolation method can be used for other values (see Table 10-1). ρ_v is volume reinforcement ratio of indirect reinforcement. β_{cor} is enhancement factor of local compressive capacity with indirect reinforcement used. f_{sd} is design value of tensile strength of indirect reinforcement. A_{ln} is the concrete local compression area excluding holes on the surface (the area according to 45° angle expansion included). That means that A_{ln} is the local compression area minus the area of the holes.

η_s and k of concrete local pressure   Table 10-1

Concrete strength	≤C50	C55	C60	C65	C70	C75	C80
η_s	1.0	0.96	0.92	0.88	0.84	0.80	0.76
k	2.0	1.95	1.90	1.85	1.80	1.75	1.70

(2) Application condition

It is applicable for local compression area with indirect reinforcement, in which $A_{cor} > A_l$ and center of gravity of A_{cor} coincides with that of A_l.

3) Section size of local compression area

$$\gamma_0 F_{ld} \leq 1.3\eta_s \beta f_{cd} A_{ln} \qquad (10\text{-}6)$$

The meanings of symbols are the same as the above.

4) Example

A reinforced concrete spandrel-braced arch bridge, of which the design value of reaction force F_{ld} born by abutment cap is equal to 2 500kN, has a local compression area of 250mm × 300mm (taking into account the rectangular plate thickness and the area increase according to 45° rigid angle), as shown in Figure 10-7. The cubic concrete compressive strength of abutment cap is 30 MPa, thus f_{cd} = 13.8MPa. The strength grade of indirect reinforcement is HPB300, thus f_{sd} = 250MPa. Please calculate the local pressure of this abutment cap.

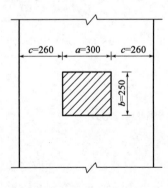

Figure 10-7   Schematic diagram

Solution: As shown in Figure 10-7, we can know: The long edge of rectangular local compression area A_l is a = 300mm, the short edge b = 250mm, the minimum distance between edge of rectangular local compression zone and edge of abutment cap c = 260mm > b, so the bottom area A_b and the length of each side can be obtained as follows

$$L_1 = 250 + 2 \times 250 = 750\text{mm}$$
$$L_2 = 300 + 2 \times 250 = 800\text{mm}$$
$$A_b = 750 \times 800 = 6 \times 10^5 \text{mm}^2$$

Local compression area $A_l = 250 \times 300 = 0.75 \times 10^5 \text{mm}^2$.

According to Figure 10-6, assuming the size of indirectly reinforced grid core concrete: $l_1 =$ 500mm, $l_2 = 600$mm, then $A_{cor} = 500 \times 600 = 3 \times 10^5mm^2 > A_l = 0.75 \times 10^5$mm^2.

So the enhancement factor β of local compressive strength of concrete is

$$\beta = \sqrt{\frac{A_b}{A_l}} = \sqrt{\frac{6 \times 10^5}{0.75 \times 10^5}} = 2.83$$

Because the concrete with the strength grade of C30 is used, from Table 10-1 $\eta_s = 1$ and $\eta_s\beta = 2.83$.

When indirect reinforcement is used, the enhancement factor of local compressive strength of concrete β_{cor} is

$$\beta_{cor} = \sqrt{\frac{A_{cor}}{A_l}} = \sqrt{\frac{3 \times 10^5}{0.75 \times 10^5}} = 2$$

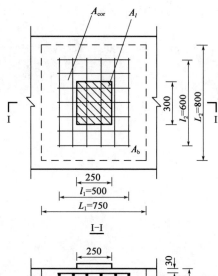

Assuming the number of steel rebars at welded steel grid along the direction l_1 is $n_1 = 6$ (spacing is 100mm), the numbers along the direction l_2 is $n_2 = 7$ (spacing is 100mm), reinforcement diameter is $\phi 6$ and the area is $A_{s1} = A_{s2} = 28.3$mm^2. The setting condition of indirect reinforcement is shown in Figure 10-8.

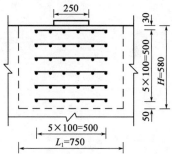

Figure 10-8 Setting condition of indirectly reinforced (unit: mm)

In Figure 10-8, the distance between the top of reinforcement mesh and local compression area is 30mm, spacing of each mesh $s = 100$mm and the set depth: $H = 580$mm, layers $m = 5$, and the reinforcement ratio of indirect reinforcement ratio ρ_v is

$$\rho_v = \frac{n_1 A_{s1} l_1 + n_2 A_{s2} l_2}{A_{cor} s}$$

$$= \frac{6 \times 28.3 \times 500 + 7 \times 28.3 \times 600}{(500 \times 600) \times 100}$$

$$= 0.0068$$

and

$$\frac{n_1 A_{s1}}{n_2 A_{s2}} = \frac{6 \times 28.3}{7 \times 28.3} = 0.86 > 0.5$$

(1) Checking calculation of concrete crack resistance

$1.3\eta_s\beta f_{cd} A_{ln} = 1.3 \times 2.83 \times 11.5 \times 0.75 \times 10^5 = 3173.14$kN $> F_{ld}(= 2500$kN$)$

The size of local compression area satisfies the requirements.

(2) Calculation of the load-carrying capacity of local compression area

$$0.9(\eta_s \beta f_{cd} + R\rho_v \beta_{cor} f_{sd}) A_{ln} = 0.9 \times (2.83 \times 13.8 + 2 \times 0.006\ 8 \times 2 \times 250) \times 0.75 \times 10^5$$
$$= 3\ 095.15 \text{kN} > F_{ld}(= 2\ 500 \text{kN})$$

The carrying capacity of local compression area satisfies the requirements.

PART 2 | Prestressed Concrete Structures

Chapter 11
Basic Concepts and Materials of Prestressed Concrete Structures

11.1 Introduction

1) Disadvantages of Reinforced Concrete Structures

(1) It's allowed that the structure can have cracks, it can't be applied to those in which no cracking are allowed to exist.

(2) The potential functions of high-strength steel wires and the high-grade concrete can not be fully developed.

(3) The use of new materials and technologies are restricted. The dead load is heavy and the span ability is restricted in bridge structure.

2) Basic Concepts

(1) Basic principle

The prestressing force for concrete structures is the inner stress manually introduced beforehand into concrete or reinforced concrete. Furthermore, the value and the distribution can just make the stress from load be offset to a proper degree. The structure with the internal stress applied beforehand to concrete is called the prestressed concrete structure.

Examples of simply-supported beams are shown in Figure 11-1. The stress changed obviously after the simply-supported beam is applied by eccentricity pressure under even load.

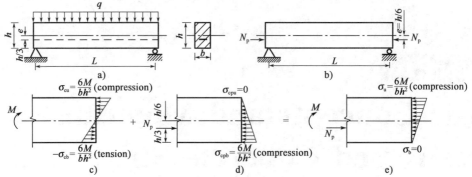

Figure 11-1　Basic principle of prestressed concrete structures

①The stress status under even load q without prestress;

②The stress condition under even load q when the prestress N_y is applied;

③The stress under even load q when the same prestress N_y is applied at the different eccentric position of e_y.

The calculation examples above illustrate two important problems as follows:

①Due to the prestress N_y on the concrete beam, the tensile stress on the lower edge of concrete beam will entirely or partly be offset by prestressing force under even load q so that the cracking of concrete beam will be avoided or delayed.

②In the actual projects, prestressing must be implemented according to the state of stress under load. The prestress N_y depends on both the value of load (or bending moments M) and the position of the application of N_y (i.e. to the eccentric distance e_y). Thus prestressing force (value and position) should be designed.

(2) Classification of reinforced concrete structures

The classification of reinforced concrete structures may usually be carried out according to the degree of the prestressing forces.

①Definition of degree of prestressing

The **CHBC** defines that the degree of prestressing is the ratio of decompression moment M_0 divided by prestressing force to the moment M_s from the external load.

$$\lambda = M_0/M_s \qquad (11\text{-}1)$$

Where M_0 is the decompression moment, i.e. the bending moment when the prestress at the edge of members for anti-cracking has been counteracted to zero; M_s is the moments calculated according to the combination of frequent combination of actions (or load) effects; λ is the degree of prestressing of prestressed concrete members.

②Classification of reinforced concrete members

a. Fully prestressed concrete members: under combination for a short-term effect of an action (load), the tensile stress is forbidden to exist on the tension edge of a cross section, i.e. $\lambda \geq 1$.

b. Partially prestressed concrete members: under the combination for a frequent combination of actions (load), the tensile stress is permitted to appear on the tension edge of a cross section or cracks not exceeding the width in the specification are allowed to exist, i. e. $1 > \lambda > 0$.

c. Non-prestressed concrete members: the concrete members without any prestressing forces, i. e. $\lambda = 0$.

3) Advantages and disadvantages of prestressed concrete structures

(1) Advantages

①Crack resistance and stiffness of members are improved;

②Materials are saved and costs are reduced;

③Quality of structures is safe and reliable;

④Durability of members have been strengthened;

⑤New bridge systems are developed.

(2) Disadvantages

①The process is complicated, and there're strict requirements for the quality;

②There should be some special equipment;

③As to construction with small span and fewer members, the cost is high at the beginning.

11.2 Methods and equipments for prestressing construction

1) Major method for prestressing forces

(1) Pre-tensioning method

The pre-tensioning method is illustrated in Figure 11-2.

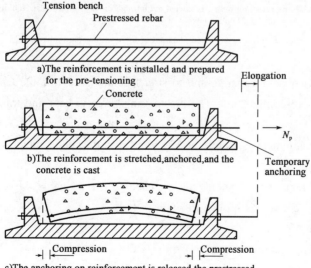

Figure 11-2 Process diagram of the pre-tensioning method

①Definition

The pre-tensioning method is, first of all, to pour concrete and to stretch reinforcement, and then cast concrete members. It relies on the adhesion between reinforcement and concrete to transfer and maintain the prestress forces. As to the reinforcement to be prestressed, steel wires of high strength, steel strands and cold-drawn steel tendons with smaller diameters are usually used.

②Advantages

The process of the pre-tensioning method is simple, the procedure fewer and the efficiency high. It's easy to control the quality and suitable for mass production in factories.

③Disadvantages

It requires special tensioning bench and the investment for infrastructure construction is huge. This method can be adapted to smaller members reinforced in a linear way.

(2) Post-tensioning method

The post-tensioning is illustrated in Figure 11-3.

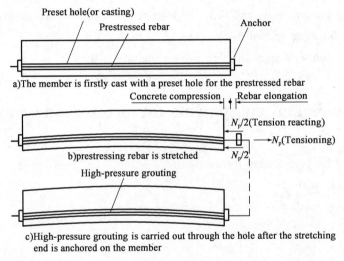

Figure 11-3   Process diagram of the post-tensioning method

①Definition

The method is first pour concrete and to stretch tendons. While the tendons is stretched, the concrete structure is pre-compressed. The prestressing force is kept and transferred by anchoring.

②Advantages

The stretching equipment is simple on site, and it is the major method for prestressing large-scaled concrete members.

The prestressed rebar layout should be matched to a reasonable curve when the moment of load and the shearing change as is seen in Figure 11-4.

③Disadvantages

The construction is relatively more complicated.

2) Anchorage

(1) Concept

The anchorage is the device to anchor rebar abidingly for prestressing when prestressed concrete members are made.

Figure 11-4　Prestressed reinforcement layout

(2) Requirements

The anchorage should have safe and reliable mechanical properties. The losses of prestress would be minimized. The structure should be simple and the anchorage should be convenient for manufacture so that the cost is low with less consumption of steel. The construction equipment should be simple. The tensioning and anchoring would become convenient and quick.

(3) Classification

①The anchorage by friction resistance: for example, anchoring cone, wedges anchorage and JM anchorage.

②The anchorage by compression: for example, head anchorage (BBRV) and twisted steel anchorage.

③The anchorage by cohesive forces: for example, prestressed rebar bonded anchorage for pretensioning and strand-end-patterned anchors for post-tensioning.

(4) Common kinds of anchorage

①Anchoring cone (Figure 11-5)

Operation principles: Steel wires to be prestressed will be put between the bore and the conical plug through the top pressure to the conical plug. After the prestressed reinforcement is released from tensioning jack, steel wires will curve back towards the inner wall of beams and this force makes the conical plug tightened to the interior wall of the bore. In this way the prestressed steel wires will transfer the pretensioning to the bore through the friction resistance.

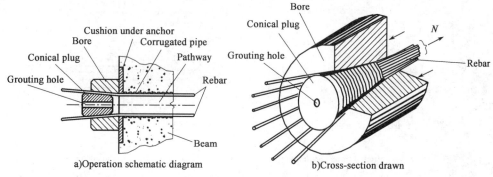

Figure 11-5　Cone anchorage

②Head anchorage (Figure 11-6)

Operation principles: Prestressed wires would be drilled through the sockets one by one at first, and then the end of prestressed wire is upset thickly by special heading machine. When the thickened head directly bears the pressure, the steel wire will be fixed on the anchor ring. On the exterior of the ring threads has been made. After the penetration, the nut would be put on the end, i.e. having rebar fixed on the end of the beam. On the pretensioning end, first of all, the tension rod connecting to the jack will be turned inside the ring for tensioning. After the steel bar or wire is stretched to design requirements by the ring, the nut will be screwed tight on the surface of a member along the thread on the exterior side of the ring. The hoisting jack will be relieved slowly, and the tension rod put out. Thus the steel wire set-back will be transferred to the concrete of a beam through the ring and the plate so as to get it anchored.

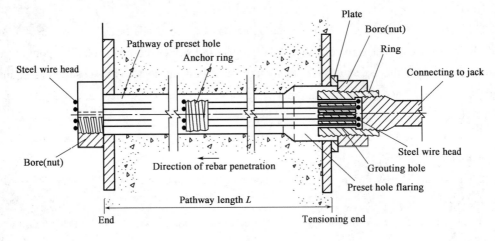

Figure 11-6 Operation diagram of head anchorage

③Screw-thread steel anchorage

Operation principles: With the steel thread on both ends of thick reinforcing bars, the nut is directly tightened for anchorage after steel is tensioned. The set-back of steel will be transferred from nuts through anchoring plates to concrete beams so as to gain the prestressing forces (shown in Figure 11-7).

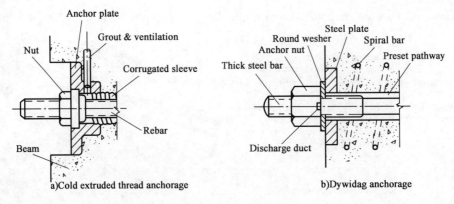

Figure 11-7 Screw-thread steel anchorage

Advantages: the stress from the screw-thread steel anchorage is obvious. The anchorage is reliable, simple structure, and convenient for construction. The anchorage can be tensioned, relaxed or dismantled repeatedly. It can be extended with sleeves conveniently.

④Wedge anchorage

a. Strand wedge anchors

The strand wedge anchors consist of anchor plate and jigs (Figure 11-8). The steel strand would be anchored by jigs after steel is stretched.

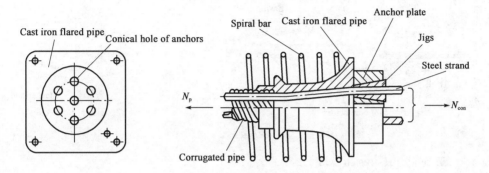

Figure 11-8 Complete wedges anchor schematic diagram

b. Flat wedges anchorage

The flat wedge anchorage is designed for the need to anchor the prestressed reinforcement of the flat and thin cross section of members (such as bridge decks or beams). The operation principle is the same as that for the common wedges anchor system. The difference lies only in the shapes of working anchor plate, steel backing plates and flared pipes under anchors. The corrugated pipes for all of the presetting holes are flat shaped.

⑤Fixed end anchorage

When the tension is carried out on one end, the fixed end anchorage can not only be used to the anchor plate for pretensioning the same end, but also the extruding (Figure 11-9) and bulb anchors adopted (Figure 11-10).

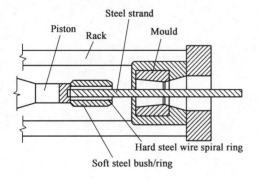
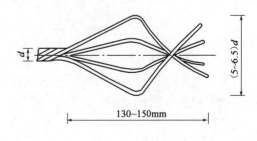

Figure 11-9 Operational principles of pressing-head anchorage

Figure 11-10 Bulb anchorage

⑥Connectors

a. End connectors: The end connectors are used to connect other steel strands, after steel strands are anchored [Figure 11-11a)].

b. Extension connectors: The extension connectors are used directly for two sections of steel strand which need to be extended before pretensioned [Figure 11-11b)].

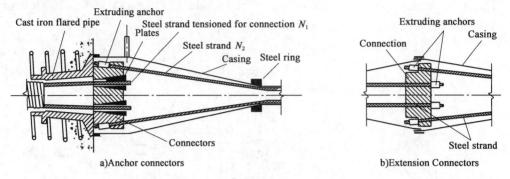

Figure 11-11 Connector construction

3) Jack and other prestressing devices

The jack is used to stretch prestressed steel rebar in actual engineering. Jack can be divided into many types according to stretching force. The center hole single-act jack with larger diameters is illustrated in Figure 11-12.

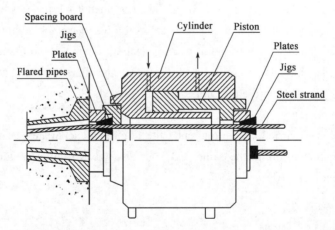

Figure 11-12 Schematic diagram of installing pretensioning jack of wedge anchorage

Other prestressing devices are as follows:

(1) Hole punching machines: The device for reservation of hole of post-tensioned members.
①rubber extraction pipes;
②metal bellows.

(2) Cable threading machines: The device for prestressed steel rebar dragged through the hole.
①hydraulic type usually used in bridge construction;
②electric cable threading machines.

(3) Mud jacking machines of cement grouting: The device for jacking mud into hole after prestressed steel rebar stretched, in bonded prestressed concrete members. It mainly consists of agitating vessel of mud, barrel of storing mud, mortar pump and water supply.

(4) Pre-tensioned beds for pre-stretching construction.

The working beds of anchoring temporary and stretching prestressed steel rebar in the construction of pre-tensioned concrete.

11.3 Materials of prestressed concrete structures

The materials of prestressed concrete structure include the high-strength concrete and prestressed steel rebar. The requirement and characteristics of materials are presented as follows.

1) Concrete

(1) Concrete requirements of prestressed concrete structure

①High strength, sufficient load-carrying capacity;

②Quick hardening and early strength required for quick construction;

③Small shrinkage and creep, large effective prestress and high efficiency.

(2) Strength

The ***CHBC*** states that the concrete strength grade of prestressed concrete structure should be over C40.

(3) Concrete shrinkage and creep

Because of shrinkage and creep of concrete, the prestressed concrete member will shrink. As a result, the stress of tendons will decrease, and loss of prestress will happen.

①Shrinkage

Concrete shrinkage may be calculated by the following formula in the ***CHBC***.

$$\varepsilon_{cs}(t,t_s) = \varepsilon_{cs0} \cdot \beta_s(t - t_s) \qquad (11\text{-}2)$$

Where $\varepsilon_{cs}(t,t_s)$ is the shrinkage stress at the time of age t_s (3d ~ 7d) of shrinkage beginning and age t of calculation; ε_{cs0} is the proportional coefficient of shrinkage, for concrete of C20 ~ C50, the value of this coefficient can be selected as follows in Table 11-1; β_s is the coefficient of shrinkage along with time increasing.

The proportional coefficient of shrinkage ε_{cs0} Table 11-1

40% ≤ RH ≤ 70%	70% ≤ RH ≤ 90%
0.529×10^{-3}	0.310×10^{-3}

②Creep

a. Definition: The plastic deformation along with time increasing in the situation of stress invariability.

b. Main influential factor: Stress, loading time, concrete quality, loading ages, member thickness and work surrounding.

c. Coefficients of creep

It is defined as the proportional coefficient of creep strain and elastic strain.

$$\phi = \varepsilon_c / \varepsilon_e \tag{11-3}$$

Where ε_c is the value of creep strain; ε_e is the value of elastic strain under the stress σ_c.

The creep coefficient may be calculated by the following formula in the **CHBC**.

$$\phi(t, t_0) = \phi_0 \cdot \beta_c(t - t_0) \tag{11-4}$$

Where ϕ_0 is the concrete nominal creep coefficient; $\beta_c(t - t_0)$ is the coefficient of creep along with time increasing after loaded, and the detailed calculation can be found in the **CHBC**.

2) Prestressed reinforcement

(1) Requirements for prestressed reinforcements

①High strength, sufficient load-carrying capacity;

②Superior plastic property and weld-ability, satisfying the requirement of construction;

③Sufficient bond strength with the concrete, offering reliable anchorage;

④Little loss of prestress due to relaxation.

(2) General types of prestressed reinforcement

①Steel wires of high strength strands: The diameter $\phi = 5 \sim 9$mm, round steel wire by stress eliminated, or with screw rib.

②Steel strand: the bundle reinforcement after inertial stress removed by several steel wire of high strength twisted.

③Finished steel rebar: the steel rebar that screw thread rib rolled regularly in the surface, and can be anchored by screw cap.

(3) Other materials

Such prestressed reinforcement as glass fibers, Kevlar fibers and carbon fibers can also be used to improve the plasticity of concrete members.

Chapter 12
Design and Calculation of Prestressed Concrete Flexural Members

12.1 Mechanical phases and calculation characteristics

1) Mechanical phases of prestressed concrete flexural members

What is the mechanical process of the prestressed concrete flexural members when they have to go through such three phases as the monolithic performance, the cracking and the damage? The researches indicate that it can be classified into three phases starting from prestressing to damage.

(1) Construction stage

It also includes two mechanical phases from prestressing to transportation and installation of beams.

①Prestressing stage

The whole process of prestressing starts from prestressing to force transmission and anchoring. As to a simply-supported beam, the member will produce the force going upwards due to the eccentric function of N_p, and thus the first period dead load (gravity load) G_1 will act together with the prestressing force N_p. The stress is shown in Figure 12-1.

a. Key points for designing:

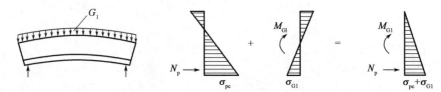

Figure 12-1 Stress distribution of cross-section at the prestressing stage

(a) Control of prestressing (tensioning);

(b) The stress on the upper/lower edge of concrete should satisfy the requirement of the **CHBC**;

(c) The partial compressive capacity in the end of the anchorage zone should meet the requirements and no cracking appears.

b. Characteristics at the phase:

Load: The effective prestressing should be in maximum value and the live load in minimum value, and there will only be the dead load in existence.

The Strength of concrete is relatively low without reaching the design strength completely.

②Transportation and installation

The tensile stress may appear on the top edge of the beam at prestressing stage. In addition, the type of crane beams is different to that of the bridge finished so as to cause the negative moment happening in the suspending part of the beam under its dead load and to increase the tensile stress on the top edge, and this will make the concrete crack. This should not be allowed. Besides if there is the impact from lifting and landing beams, it will increase the effect of stress. The **CHBC** specifies that the dynamic coefficient of 1.2 or 0.85 should be considered for checking the tensile stress of concrete at the fulcrum of members or the upper flange of a cross section of the lifting point.

(2) Service stage

The operation of a bridge will cover the whole process beginning from the bridge opening to traffic for operation, including decompression, appearance of cracking and operation with cracking.

At the service stage, the tensile stress is forbidden in the fully prestressed concrete structure; while the tensile stress is permitted and crack is forbidden in partially prestressed concrete structure of Type A. As to partially prestressed concrete structure of category B, cracks are permitted, but the crack width is forbidden to exceed the value specified.

①Decompression

The member will not only bear the eccentric prestressing of N_p and the first period dead load of G_1, but also bear the second stage dead load G_2 of bridge deck installation, sidewalk and fences added to it and the live load Q caused by vehicles and people. While the normal stress from the controlled cross section of a beam and the live load (shown in Figure 12-2) is equal to that from the eccentric prestressing of N_p, the stress at lower edge should be zero [shown in Figure 12-3b], the prestress would be cancelled.

Chapter 12 ▸ Design and Calculation of Prestressed Concrete Flexural Members

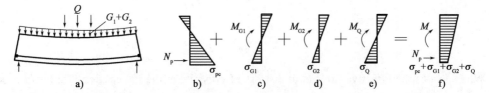

Figure 12-2　Stress distribution of cross section under various functions in operation

②Appearance of cracking

The lower edge of a cross section will bear compressive stress, zero stress, tensile stress and cracking of concrete as seen in Figure 12-3.

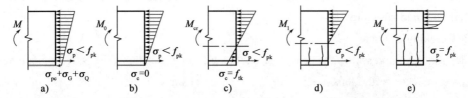

Figure 12-3　Stress of cross section at service and failure stages

a. Key points for design:

(a) The normal stress and major stress of concrete and the stress of prestressed rebar should meet the requirements;

(b) The anti-cracking checking of normal and oblique sections of members;

(c) The deformation checking of routine service of members.

b. Characteristics:

(a) The effective prestressing becomes smaller due to the loss of major prestressing;

(b) The external load will be the largest, including all live loads.

③Operation with cracking

When concrete cracks, the steel stress will gradually increase to the yielding strength with the load mounting, cracks developing and the tensile stress of concrete born by reinforcement.

(3) Failure stage—ultimate load-carrying capacity mode

When the balanced reinforcement beam reaches its limit, the reinforcement bar in tensile areas yields firstly, and then the concrete is crushed in the compressive areas. Thus the member fails [Figure 12-3e)]. The stress condition and the computing method of a cross section will be similar to that for the flexural members of reinforced concrete when it's destroyed.

Special note: the ultimate bearing capacity of the prestressing concrete structures will become exhausted after the material strength comes to an end.

2) Calculation of decompression moment, cracking moment and ultimate moment

(1) Decompression moment

The tensile stress from external load may set off the compressive stress σ_{pc} existing at the lower edge of the controlled cross section. When σ_{pc} reaches to zero, it's called the decompression. At this time, the moment induced by external moment is named as the compressive moment M_0.

$$\sigma_{pc} = \frac{M_0}{I_0} y_b = \frac{M_0}{W_0} \tag{12-1}$$

$$M_0 = \sigma_{pc} \cdot W_0 \tag{12-2}$$

Where σ_{pc} is the effective prestressing stress of concrete at the lower edge of a beam produced by the perpetual prestressing N_p; W_0 is the conversion of the elastic resistance moment of a cross section to the side prestressed.

"Decompression" means that the prestressing is exhausted and it is at the same level with that for the initial condition of reinforced concrete structure without any prestressing. The decompression in the prestressed concrete structures is a condition that doesn't exist in the reinforced concrete structures.

(2) Cracking moment

After the decompression, the tensile strength will appear at the lower edge of concrete until the concrete reaches to the tensile strength f_{tk}, i.e. cracks appear in structures. At this point, the theoretic critical moment is called the cracking moment M_{cr}, and it has one more M_0 than $M_{cr,c}$ of the cracking moment in common concrete structures.

$$M_{cr} = M_0 + M_{cr,c} \tag{12-3}$$

$$M_{cr} = (\sigma_{pc} + \gamma f_{tk}) W_0 \tag{12-4}$$

$$\gamma = 2S_0/W_0 \tag{12-5}$$

Where, $M_{cr,c}$ is the approximate equivalent to the cracking moment of a cross section of a reinforced concrete beam; γ is the plastic influential coefficient of concrete in tensile areas; S_0 is the total cross section conversion of the partial area above (or below) the pivot axis of a cross section to the area moment of gravity axis.

(3) Failure moment

It is defined that the failure moment is the ultimate one when structures reach their limits, the reinforcement yielding and the concrete breaking.

Furthermore, it can be seen that the "essence" of prestressed concrete structures is as follows:

①Due to the prestressing, the concrete will always be elastic under construction and at serve stage. It can be computed in accordance with the method from "material mechanics", i.e. the function of prestressing to turn concrete into the elastic material.

②The fracture moment of the prestressed concrete flexural member is nearly the same to that of the reinforcement concrete beam under the same conditions, i.e. pre-stressing only for balancing load.

③The compressive stress stored in concrete can keep concrete from cracking or having smaller ones. In this way the rebars with high strength will work together with concrete, i.e. prestressing promoting high strength steel bars to work with concrete.

Chapter 12 · Design and Calculation of Prestressed Concrete Flexural Members

12.2 Calculation of ultimate load-carrying capacity

1) Calculation of load-carrying capacity of normal cross sections

(1) Basis for calculation

There is the prestressed rebar A_p and the non-prestressed reinforcement A_s in the tensile area, and the prestressed steel bar A'_s and the non prestressed steel bar A'_s in compressive area. When they fail, the reinforcement in tensile areas all reach the yielding strength of f_{pd} and f_{sd}. The common reinforcement in compressive area will reach to the yielding strength f'_{sd}, while the prestressed one won't and the stress will be σ_{pa}.

(2) Calculation assumptions

①Plane cross sections;

②The concrete tensile resistance not included;

③The concrete rectangle stress diagram in compressive area.

(3) Formula for calculation

The mechanical diagram and the basic formula can be obtained from the basic assumptions. It is shown in Figure 12-4.

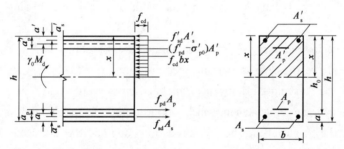

Figure 12-4　Load-carrying capacity for flexural member in rectangular section with prestressed rebar at compression zone of normal cross section

$$f_{sd}A_s + f_{pd}A_p = f_{cd}bx + f'_{sd}A'_s + (f'_{pd} - \sigma'_{p0})A'_p \tag{12-6}$$

$$\gamma_0 M_d \leqslant f_{cd}bx(h_0 - \frac{x}{2}) + f'_{sd}A'_s(h_0 - a'_s) + (f'_{pd} - \sigma'_{p0})A'_p(h_0 - a'_p) \tag{12-7}$$

Where, σ'_{p0} is the effective prestressing stress when the concrete stress equal zero in the position of gravity axis of prestressed tendon.

(4) Application conditions

①The height x in compressive areas should meet the requirement specified by the ***CHBC***

$$x \leqslant \xi_b h_0, \xi_b = \frac{\beta}{1 + \dfrac{0.002}{\varepsilon_{cu}} + \dfrac{f_{pd} - \sigma_{p0}}{\varepsilon_{cu} E_p}} \tag{12-8a}$$

The value of ξ_b is shown in the Table 3-2.

②When the prestressed bar is compressed in compressive area, i.e. $(f'_{pd} - \sigma'_{p0}) > 0$, it should meet the following requirement

$$x \geq 2a' \quad (12\text{-}8b)$$

③When the prestressed bar is subjected to tension in the compressive areas, i.e. $(f'_{pd} - \sigma'_{p0}) < 0$, the following requirement should be met

$$x > 2a'_s \quad (12\text{-}8c)$$

Where a' is the distance from the joint function of reinforcement A'_p and A'_s in compressive area to the nearest edge of a cross section; when the stress of the prestressed reinforcement A'_p is the tensile stress, a' would be replaced by a'_s; a'_p is the distance from the joint function of steel bars (A'_p) to the nearest edge of a cross section.

(5) Design method

①Cross section design: The design of prestressed reinforcement should be carried out in accordance with $A_s = A'_s = A'_p = 0$. A_p should be worked out for the initial computation. The area of prestressed reinforcement can be finally determined in line with the requirement of cracking resistance. It's shown in the computation example.

②Strength checking: The calculation steps are the same to that for the reinforcement concrete structure (the detailed calculation shown in an example).

③Computation of a T-shaped cross-section: It's similar to that mentioned above, the formula of T-shaped or I-shaped cross-section with external prestressing can consult the reguirement in **CHBC**.

2) Calculation of load-carrying capacity of oblique cross-sections

(1) Functions of pre-stressing in oblique sections:

①To improve the carrying capacity of an inclined cross-section;

②To delay the appearance of inclining cracks and to control the development of inclining cracking;

③To increase the height at the shear compression zone and the aggregate interlock forces.

(2) Calculation of shear bearing capacity

The computing formula of the shearing capacity of an inclined cross-section in the **CHBC**.

$$\gamma_0 V_d \leq V_{cs} + V_{pb} + V_{sb} + V_{pb.ex} \quad (12\text{-}9)$$

Where V_d is the maximum shear combination value produced by the function (or load) on the normal cross-section at the compressive end of the inclined cross-section (kN); V_{cs} is the design value of shearing capacity of the concrete and reinforcement within a inclined cross-section (kN); V_{pb} is the design value of shearing capacity of the rebar bent by the prestressing intersected with the inclined cross-section (kN); $V_{pb.ex}$ is the design value of shear capacity of the external prestressing that intersected with the inclined cross-seetion(kN).

$$V_{cs} = \alpha_1 \alpha_2 \alpha_3 0.45 \times 10^{-3} bh_0 \sqrt{(2 + 0.6p)} \sqrt{f_{cu,k}} \rho_{sv} f_{sv} \quad (12\text{-}10)$$

Where α_2 is the coefficient to improve prestressing. As far as the prestressed flexural concrete member is concerned, $\alpha_2 = 1.25$, but if the moment of a cross section by the rebar joint force is in the same direction with that of the external moment, or if the prestressed concrete flexural member with cracks is allowed, $\alpha_2 = 1.0$. p is the reinforcement ratio of the longitudinal tensile steel bars.

$p = 100\rho$, $\rho = (A_p + A_{pb} + A_s)/(bh_0)$; when $p > 2.5$, $p = 2.5$. ρ_{sv} is the reinforcement ratio of stirrup of an inclined cross section, and $\rho_{sv} = A_{sv}/(s_v b)$. In real construction, the vertical prestressed reinforcement has been adopted in composite panels for prestressed concrete box girders. At this time, ρ_{sv} should be replaced by the reinforcement ratio of vertical prestressed reinforcement ρ_{pv}. s_v is the spacing of vertical prestressed reinforcement in an oblique cross-section (mm). f_{sv} is the design value of tensile strength of vertical prestressed reinforcement. A_{sv} is the area of vertical prestressed reinforcement of the same cross-section within an inclined cross-section.

$$V_{pb} = 0.75 \times 10^{-3} f_{pd} \sum A_{pb} \sin\theta_p \quad (12\text{-}11)$$

Where θ_p is the included angle between a horizontal line and a tangent of rebar bent by prestressing (at the normal cross-section of a compressive end in an inclined cross section); A_{pb} is the cross-section area of rebar bent by prestressing on the same bending plane in an inclined cross-section (mm^2). f_{pd} is the design value of tensile strength of prestressed rebar.

Notes: As to the computation of the shear resistance capacity of the flexural member of prestressed concrete, the maximum and the minimum value in a formula to be satisfied should be the same as those for the flexural member of common reinforcement concrete.

(3) Calculation of flexural load-carrying capacity of an inclined cross-section

According to the bending fatigue characteristics of an oblique cross-section, the left part of an oblique cross-section is selected as the separated body (Figure 12-5), and the function point O of an joint force (rotary hinges) of concrete at the compressive zone is the center for moment. Due to $\sum M_O = 0$, the flexural bearing capacity computing formula of flexural members of an oblique cross-section in rectangle, T-shaped and I-shaped cross-sections can be obtained.

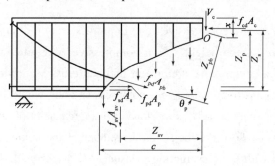

Figure 12-5 Flexural bearing capacity of inclined cross section

$$\gamma_0 M_d \leq f_{sd} A_s Z_s + f_{pd} A_p Z_p + \sum f_{pd} A_{pb} Z_{pb} + \sum f_{sv} A_{sv} Z_{sv} \quad (12\text{-}12)$$

Where M_d is the maximum composite moment design value in a normal cross-section at the compressive end of an oblique cross-section; Z_s and Z_p are the distances between the joint force of common longitudinal tensile bar and that of longitudinal prestressed tensile reinforcement to the center point O of the compressive area, respectively; Z_{pb} is the joint force of reinforcement bent by prestressing on the same bending plane intersected with an inclined cross-section to the center point O at the compressive area; Z_{sv} is the distance from the enforcement joint force of the same plane intersected with an inclined cross-section to the horizontal distance of the compressive edge of an oblique cross-section.

When the flexural bearing capacity of an oblique cross-section is calculated, the most unfavorable place of an inclined cross-section should be chosen in the place where the amount of prestressed reinforcement becomes less, the amount of stirrups and the spacing varies, and the thickness of the composite plate of the cross-section of concrete members differs. But the horizontal projection length c of an oblique cross-section should be calculated according to different oblique angles from the bottom to the top. The worst horizontal projection length of an inclined cross-section should be computed in line with the following formula.

$$\gamma_0 V_d = \sum f_{pd} A_{pb} \sin\theta_p + \sum f_{sv} A_{sv} \quad (12\text{-}13)$$

If α is the included angle between the worst inclined cross-section and the horizontal line, and the horizontal projection length is c, then the reinforcement cross-section area of an inclined cross section should be $\sum A_{sv} = A_{sv} \cdot c/s_v$. When it is substituted into the formula mentioned above the expression of the worst horizontal projection length c can be obtained as follows.

$$c = \frac{\gamma_0 V_d - \sum f_{pd} A_{pb} \sin\theta_p}{f_{sv} \cdot A_{sv}/s_v} \quad (12\text{-}14)$$

Where V_d is the normal cross-section in the compressive edge of an inclined cross-section corresponding to the shear composite design value of the maximum moment composite design value; s_v is the spacing of reinforcement (mm).

The position O of the joint force in a compressive area should be decided in order to compute the length of the force of the arm respectively after the horizontal projection length c is determined. From the mechanical balance condition of an oblique cross-section ($\sum H = 0$), it can be concluded that

$$\sum f_{pd} A_{pb} \cos\theta_p + f_{sd} A_s + f_{pd} A_p = f_{cd} A_c \quad (12\text{-}15)$$

Therefore, the area A_c of a compressive zone in a cross section of concrete can be calculated. Because A_c is the function of the height x in a compressive zone, after the cross section style is decided, the height x of a compressive area of an inclined cross section can be calculated easily. Then the position of the joint force acting in a compressive zone would be determined.

It's very difficult to compute the flexural load-carrying capacity of an oblique cross-section of pre-stressed concrete beams. Just like the flexural member of the common reinforced concrete, it will be usually protected with the construction measures.

12.3 Calculation of tension control stress and loss of prestressing

1) Tension control stress

The tension control stress is σ_{con}

$$\sigma_{con} = \frac{N_{p.con}}{A_p} \quad (12\text{-}16)$$

Where $N_{p.con}$ is the total tensile force before the anchorage of prestressed reinforcement; A_p is the area of prestressed rebar.

In the **CHBC**, it is specified that the control stress σ_{con} on the end of rebar (under the anchorage) should be in accordance with the following points.

For prestressing steel wires and steel strands

$$\sigma_{con} \leq 0.75 f_{pk} \quad (12\text{-}17)$$

For prestressing twisted reinforcement

$$\sigma_{con} \leq 0.85 f_{pk} \quad (12\text{-}18)$$

In which f_{pk} is the tensile strength standard value of prestressed reinforcement.

In real work, the rebar which only needs to be in high stress within a short period of time can be overstretched. But under any circumstances, the maximum tensile control stress of steel wires and steel strands can only exceed $0.05 f_{pk}$.

2) Losses of pre-stressing and computation

The prestress loss would happen in the period from the prestress steel rebar stretching to service stage. It includes 6 types losses in the **CHBC**.

(1) Friction loss σ_{l1}—the stress loss caused by friction between prestressed rebar and the wall of pipes.

The friction loss refers to the phenomenon that when the post tension method is used to stress reinforcement the friction existing between the prestressed rebar and the surrounding wall of pipes causes [shown in Figure 12-6a)], the stress of the prestressed steel bar will reduce gradually with the increase of the distance from the tensioning end. The friction loss consists of two parts as follows:

①The causes for friction loss are the position difference of pipes, the roughness of holes/walls or the roughness of the surface of reinforcement, also called the friction loss impacted by the pipe deviation. The value of friction loss of this type is small.

②As far as the prestressed rebar for tension curve is concerned, due to the curvature of pipe, the friction loss may be caused by the radial pressure between the prestressed steel bars and pipes. This friction loss is also called the winding course having impact on the friction loss. The value can be large as seen in Figure 12-6.

The frictional loss σ_{l1} can be calculated as follows.

$$\sigma_{l1} = \frac{N_{con} - N_x}{A_p} = \sigma_{con}[1 - e^{-(\mu\theta + kx)}] \quad (12\text{-}19)$$

Where σ_{con} is the tension control stress under anchorage, $\sigma_{con} = \frac{N_{con}}{A_p}$. N_{con} is the tension control force under anchorage. A_p is the cross section area of prestressed rebar. θ is the sum of the included angle [seen in Figure 12-6a)] of the curve of the pipe plane between the tensile ends and the computed cross section, i.e. the wrap angle of a curve. In line with the absolute value it should be added up with the radian as the unit. If the duct is a 3D spatial curve duct that bent in a vertical plane and a horizontal plane at the same time, then θ may be calculated by the following formula

$$\theta = \sqrt{\theta_H^2 + \theta_V^2} \quad (12\text{-}20)$$

θ_H, θ_V are the angles of bend in a horizontal plane or an angle of bend in a vertical plane in the same section of a pipe, respectively. x is the projection length of the pipe length from the tensile end to the computed cross section on the longitudinal axis or the length of 3D spatial curve pipe (m). k is the impacting coefficient of partial deviation of the length/meter of a pipe to friction, and it can be adopted in accordance with appendix Table 2-6. μ is the coefficient of friction between the tendon and the duct. It can be adopted according to appendix Table 2-6.

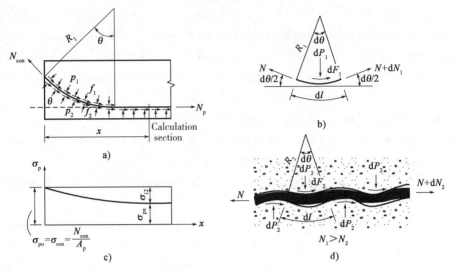

Figure 12-6 Simplified diagram of calculating the frictional loss of the prestressed rebar by the friction of pipes

In order to reduce the friction loss, the following measures should be utilized generally:

①Stretching on both ends to decrease the value θ and the length x of a pipe;

②The over-stretching method can be adopted. As to the post-tensioned reinforcement, the tension technology should be applied as follows.

For steel strand:

0→initial stress($0.1\sigma_{con} \sim 0.15\sigma_{con}$)→$1.05\sigma_{con}$(load holding 2min)→$\sigma_{con}$(stretching)

For steel wire:

0→initial stress($0.1\sigma_{con} \sim 0.15\sigma_{con}$)→$1.05\sigma_{con}$(load holding 2min)→0→$\sigma_{con}$(stretching)

Note: As to the common wedge anchorage, it's not proper to adopt the over stretching technology.

(2) Loss of anchorage σ_{l2}—the stress loss because of the deformed anchorage, the reinforcement shrinkage and the joint compression.

When tendons are stretched and anchored, the anchorage device is deformed, gaps are pressed closer and the inward shrinkage of reinforcement happens. In addition, the joint of assembled members would be pressed tightly and becomes deformed so as to cause stress loss σ_{l2} after the anchorage is placed

$$\sigma_{l2} = \frac{\sum \Delta l}{l} E_P \quad (12\text{-}21)$$

Where $\sum \Delta l$ is the sum of the anchorage deformation at the tensile end, the steel shrinkage and the joint compression (mm). It can be determined in accordance with tests. When there are no materials, what is listed in appendix Table 2-7 will be utilized. l is the distance between the tensile end and the anchorage end. E_p is the elastic modulus of prestressed tendons.

The **CHBC** specifies that the computation of prestressed concrete members by the post tensioning should include the loss of prestressing caused by the friction resistance from anchorage deformation and rebar shrinkage.

The methods of reducing σ_{l2} are as follows:
①Over stretching;
②Application of anchorage with smaller $\sum \Delta l$, especially to shorter members.

(3) Temperature difference σ_{l3}—the stress loss by the temperature difference between reinforcement and abutments.

When the steam-heating or other heating methods are employed to cure concrete by post tensioning, the newly cast concrete hasn't hardened yet, and the rebar is heated and expanded. The abutment is fixed steadily, i. e. the length of rebar keeps stable. The stress within the reinforcement will decrease with the increase of temperature so as to cause the stress loss σ_{l3}.

If the temperature for both between the tendon and the platform is supposed to be t_1 in tendon stretching, the highest temperature will be t_2 when concrete is heated for curing. The tendon hasn't bonded with concrete at this time. The free deformation Δl_t of the tendon in concrete will develop freely as the temperature increases from t_1 to t_2

$$\Delta l_t = \alpha \cdot (t_2 - t_1) \cdot l \qquad (12\text{-}22)$$

Where α is the coefficient of tendon expansion, generally $\alpha = 1 \times 10^{-5}$; l is the effective edge of reinforcement; t_1 is the site temperature (C°) for tensioning rebar; t_2 is the highest temperature (C°) of tensioned rebar for heating and curing concrete.

When the heating and the curing stop, the concrete has connected together with the tendons. They will expand and shrink together as temperature changes. Since the stress lost due to the curing at higher temperature can not be resumed, the stress loss by temperature difference σ_{l3} can be calculated as follows.

$$\sigma_{l3} = \frac{\Delta l_t}{l} \cdot E_p = \alpha(t_2 - t_1) \cdot E_p \qquad (12\text{-}23)$$

Assuming the modulus of elasticity of the prestressed tendons $E_p = 2 \times 10^5 \text{MPa}$, then the following can be achieved.

$$\sigma_{l3} = 2(t_2 - t_1) \qquad (12\text{-}24)$$

In order to reduce the loss from temperature differences, the method of raising temperature for a second time will be usually applied for curing, i. e. the first time from the atmospheric temperature t_1 to t_2' increased for concrete curing. The initial temperature increase will be usually controlled within 20℃. After the concrete reaches to a certain strength (for example, 7.5 ~ 10MPa) and is able to prevent reinforcement from moving in concrete freely, the temperature will be raised to t_2' for curing. At the time, the concrete and the tendon would deform together and this won't produce the loss of stress due to the second rising of temperature. Therefore the computation of temperature differences will only be $t_2' - t_1$.

If the member and platforms are heated and deform together, the loss of stress σ_{l3} should not be considered.

(4) Loss σ_{l4} of stress by the elastic compression of concrete

When the concrete was compressed by prestressing and deforms, there would appear the compression strain $\varepsilon_p = \varepsilon_c$ so as to cause the loss of stress, i.e. the loss of stress σ_{l4} of concrete elastic compression.

① Pre-tensioned members

When the tensioning is relaxed, the loss due to elastic shrinkage of concrete can be written as follows

$$\sigma_{l4} = \varepsilon_p \cdot E_p = \varepsilon_c \cdot E_p = \frac{\sigma_{pc}}{E_c} \cdot E_p = \alpha_{Ep} \cdot \sigma_{pc} \qquad (12\text{-}25)$$

Where α_{Ep} is the ratio of modulus E_p of the elasticity of the prestressed tendon to the elastic modulus E_c of concrete. σ_{pc} is the prestressing stress by prestressing force N_{p0} at the gravity of the reinforcement in a cross section for the pre-tensioned member. It can be calculated as follows: $\sigma_{pc} = \frac{N_{p0}}{A_0} + \frac{N_{p0} e_p^2}{I_0}$. N_{p0} is the effective prestressing force of all steel bars (the prestressing loss at the relevant stage has been deducted). A_0, I_0 are the conversed cross section area and the conversed cross section moment of inertia of a full cross section of members, respectively. e_p is the distance between the gravity of prestressed tendon and the gravity axis of a cross section conversed.

② Post-tensioned members

The post-tensioned member will be usually stretched and anchored in turn. When the latter group is post-tensioned and the deformation of the concrete elastic shrinkage will appear so as to cause the stress loss from the prestressed reinforcement of the former group tensioned and anchored. It is called the stress loss of reinforcement tensioned in groups and can be expressed with σ_{l4}.

The **CHBC** specifies that the loss σ_{l4} can be calculated as follows

$$\sigma_{l4} = \alpha_{Ep} \sum \Delta \sigma_{pc} \qquad (12\text{-}26)$$

Where α_{Ep} is the ratio of the elastic modulus of the prestressed reinforcement concrete to the elastic modulus of concrete; $\sum \Delta \sigma_{pc}$ is the sum of the normal stress of concrete from reinforcement post-tensioned in groups at the gravity of rebar pretensioned where the cross-section should be computed.

Due to the relative location of the reinforcement of the post-tensioned member is different (there're such three different points as the reinforcement location, the tension force each time and the cross-section computed), the calculation of $\sum \Delta \sigma_{pc}$ will vary. If the detailed calculation is required, it'll be complicate. In order to simplify the calculation, the following similar calculation will be carried out:

a. The cross-section controlled by stress will be considered to be the even cross-section of the whole beam for computation and the rest part of the cross-section won't be taken into consideration. As to that of the simply-supported beam, $l/4$ cross-section will be used for it.

b. Assuming that all prestressed rebars in the same cross-section (for instance $l/4$ cross-section) are all located at the joint force in a concentration way, and if the tension for prestressed reinforcement in all groups is equal, it will be equal to the even value of the tension of rebar in all groups. When the tension for the rebar in each group is obtained, the normal stress of the concrete at the gravity of the reinforcement post-tensioned in the previous group will be $\Delta\sigma_{pc}$, i. e.

$$\Delta\sigma_{pc} = \frac{N_p}{m}\left(\frac{1}{A_n} + \frac{e_{pn} \cdot y_i}{I_n}\right) \quad (12\text{-}27)$$

Where N_p is the joint force of all prestressed reinforcement (after the loss σ_{l1} and σ_{l2} of stress are deducted at the relevant stage); m is the total batch figure of pre-stretched tendons; e_{pn} is the distance between the joint force N_p of the prestressed stress of the prestressed rebar and the gravity axis of a net cross-section; y_i is the distance from the gravity of the rebar pre-tensioned in the previous group (i. e. the gravity of the all prestressed reinforcement assumed) to the gravity axis of a net concrete cross-section, therefore $y_i \approx e_{pn}$; A_n and I_n are the net cross-section area and the inertia moment of a net cross-section of concrete beams, respectively.

The sum of the normal stress $\Delta\sigma_{pc}$ of concrete from each batch of reinforcement tensioned will be the joint force N_p of all batch tendons (in m batches). The normal stress of concrete at the acting point (or the gravity of all reinforcement) will be as follows.

$$\Sigma\Delta\sigma_{pc} = m\Delta\sigma_{pc} = \sigma_{pc}$$

Or it can be expressed in this way.

$$\Delta\sigma_{pc} = \frac{\sigma_{pc}}{m} \quad (12\text{-}28)$$

c. If the further assumption is that the total even value of the stress loss of concrete elastic shrinkage at the gravity of all prestressed rebar on the same cross section ($l/4$ cross section) is the stress loss caused by the concrete elastic compression from all batches of rebar, the stress loss $\sigma_{l4(i)}$ of batch i can be calculated as follows.

$$\sigma_{l4(i)} = (m - i) \cdot \alpha_{Ep}\Delta\sigma_{pc} \quad (12\text{-}29)$$

The average value of the elastic compression loss of the tendons from each batch in a cross section can be written as

$$\sigma_{l4} = \frac{\sigma_{l4(1)} + \sigma_{l4(m)}}{2} = \frac{m-1}{2} \cdot \alpha_{Ep}\Delta\sigma_{pc} \quad (12\text{-}30)$$

If the number of various batch of tendon is equal, the even loss of stress of each batch for tensioning will be the following by Formula (12-28) substituted into Formula (12-30).

$$\sigma_{l4} = \frac{m-1}{2m} \cdot \alpha_{Ep}\sigma_{pc} \quad (12\text{-}31)$$

Where σ_{pc} is the concrete normal stress from tensioning all prestressed reinforcement at the gravity of all rebar in a cross section computed.

The measures to reduce σ_{l4} include over-stretching and raising the concrete strength grade.

(5) Loss of stress due to relaxation of tendon σ_{l5}

Assuming the length of tendon is fixed without any changes, the stress within reinforcement will decrease as time goes on and the relaxation of stress appears. Figure 12-7 expresses the typical relaxation curve of tendon prestressed.

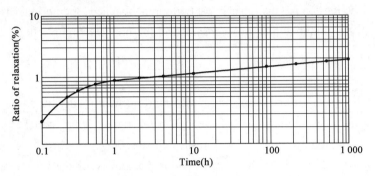

Figure 12-7　Typical relaxation curve of tendon prestressed

The final value of stress from reinforcement relaxation will be calculated in accordance with the following requirements:

For the fine rolling twisted reinforcement:

As to reinforcement stretched once

$$\sigma_{l5} = 0.05\sigma_{con} \tag{12-32}$$

As for rebar over-stretched

$$\sigma_{l5} = 0.035\sigma_{con} \tag{12-33}$$

For prestressed steel wires and strand

$$\sigma_{l5} = \psi \cdot \zeta \cdot \left(0.52\frac{\sigma_{pe}}{f_{pk}} - 0.26\right) \cdot \sigma_{pe} \tag{12-34}$$

Where, ψ is the coefficient of tensioning, when the rebar is tensioned once, $\psi = 1.0$, and if the rebar is overstretched, $\psi = 0.9$; ζ is the coefficient of relaxation of reinforcement, for Class I relaxation (normal relaxation), $\zeta = 1.0$, and for Class II relaxation (low relaxation), $\zeta = 0.3$; σ_{pe} is the stress of reinforcement when the anchorage transmits forces; for member post-tensioned, $\sigma_{pe} = \sigma_{con} - \sigma_{l1} - \sigma_{l2} - \sigma_{l4}$; while for the member pre-tensioned, $\sigma_{pe} = \sigma_{con} - \sigma_{l2}$.

The **CHBC** requires that when $\frac{\sigma_{pe}}{f_{pk}} \leq 0.5$, the loss value of relaxation is zero if carbon steel wires and steel strand are used.

The method of decreasing σ_{l5} includes over-stretching or slow relaxation of tendons prestressed.

(6) Loss of stress due to shrinkage and creep of concrete σ_{l6}

The shrinkage and the creep of concrete will both make the length of members shorter so as to cause the loss of stress.

①The stress loss σ_{l6} of tendons prestressed in the tensioning zone can be written

$$\sigma_{l6}(t) = \frac{0.9[E_p\varepsilon_{cs}(t,t_0) + \alpha_{Ep}\sigma_{pc}\phi(t,t_0)]}{1 + 15\rho\rho_{ps}} \quad (12\text{-}35)$$

Where $\sigma_{l6}(t)$ is the prestressing loss due to shrinkage and creep of concrete at the gravity of a cross section of all longitudinal reinforcement in the tensile zone. σ_{pc} is the prestressing at the gravity of a cross section of all longitudinal reinforcement in the tensile zone (the prestressing loss at relative stages deducted) and the normal stress of concrete from the self-weight of members (MPa); for simply-supported beams, the even value of a mid-span cross section and 1/4 cross section will be used for computing various cross sections for a whole beam; when $\sigma_{pc} \leq 0.5 f''_{cu}$, f''_{cu} will be the pressure strength of the cubic concrete as the prestressed rebar transmits forces and is anchored. E_p is the modulus of elasticity of the tendons prestressed. α_{Ep} is the ratio of the modulus of elasticity of tendons prestressed to that of concrete. ρ is the reinforcement ratio of all longitudinal tendons of members in tension zones; for pre-tensioned members, $\rho = (A_p + A_s)/A_0$, and as for post-tensioned member, $\rho = (A_p + A_s)/A_n$; of which A_p, A_s are the cross section areas of the tendon prestressed and non-prestressed reinforcement in tension zones respectively; A_0 and A_n are the conversion cross section area and the net cross section, respectively. $\rho_{ps} = 1 + \frac{e_{ps}^2}{i^2}$. i is the radius of gyration, and $i^2 = \frac{I}{A}$; as far as pre-tensioned members are concerned, assuming $I = I_0$, $A = A_0$, for post-tensioned members, given that $I = I_n$, $A = A_n$, in which I_0 and I_n are the conversion cross section of moment of inertial and the moment of inertia of net cross section, respectively. e_{ps} is the distance from the cross section gravity of prestressed reinforcement and non-prestressed rebar in the tensioning zone for members to the gravity axis of the cross section of members, $e_{ps} = (A_p e_p + A_s e_s)/(A_p + A_s)$. e_p is the distance from the cross section gravity of prestressed reinforcement in the tensioning zone for members to the gravity axis of the cross section of members. e_s is the distance from the cross section gravity of non-prestressed rebar in the tensioning zone for members to the gravity axis of the cross section of members. $\varepsilon_{cs}(t,t_0)$ is the shrinkage strain of concrete at the computed age t, when the age of the transmitting of forces and anchorage of prestressed reinforcement is assumed as t_0. $\phi(t,t_0)$ is the coefficient of creep at the computed age t, when the load age is assumed as t_0.

When the member with prestressed reinforcement A'_p and non-prestressed reinforcement A'_s in compression zones is considered, $A'_p = A'_s = 0$ might be the stress loss of prestressed tendon in tensile zones. The computation will be in an approximate way.

②As far as the member with prestressed reinforcement A'_p and non-prestressed reinforcement A'_s in compression zones is concerned, the loss of prestressing of the prestressed rebar of members in the compressive zone because of the concrete shrinkage and the creep should be as follows.

$$\sigma'_{l6} = \frac{0.9[E_p\varepsilon_{cs}(t,t_0) + \alpha_{Ep}\sigma'_{pc}\phi(t,t_0)]}{1 + 15\rho'\rho'_{ps}} \quad (12\text{-}36)$$

Where $\sigma'_{l6}(t)$ is the loss of prestressing of members because of the concrete shrinkage and the creep at the gravity of a cross section of all longitudinal reinforcement in the tensile zone. σ'_{pc} is the prestressing at the gravity of a cross section of all longitudinal reinforcement in the tensile zone (the prestressing loss at relative stages deducted) and the normal stress of concrete from the self-weight of members (MPa), $\sigma'_{pc} \leq 0.5 f'_{cu}$. When σ'_{pc} is the tensile stress, the value should be zero. ρ' is the reinforcement ratio of all longitudinal tendons in the compression zone of members. As to members pre-tensioned, $\rho = (A'_p + A'_s)/A_0$. As for post-tensioned members, $\rho = (A'_p + A'_s)/A_n$, in which A'_p, A'_s represent the areas of cross-sections of prestressed reinforcement and non-prestressed rebar in compression zone respectively; $\rho'_{ps} = 1 + \dfrac{e'^2_{ps}}{i^2}$. e'_{ps} is the distance from the gravity of a cross-section of prestressed rebar and non-prestressed steel bar of member in compressive zones to the gravity axis of the cross-section of members, $e'_{ps} = (A'_p e'_p + A'_s e'_s)/(A'_p + A'_s)$. e'_p is the distance from the gravity of a cross-section of prestressed steel tendon of member in compressive zones to the gravity of a cross-section of members. e'_s is the distance from the gravity of a cross-section of the non-prestressed longitudinal reinforcement of members in compressive zones to the gravity of a cross-section of members.

It should be noted that there exists the mutual impact among the concrete shrinkage and the loss of creeping stress and the loss of relaxation stress of rebar. Nowadays, the method of separately computing the single factor needs to be improved.

The measures for reducing σ_{l6} include utilization of high strength concrete and the improvement of the quality of construction and maintenance.

3) Calculation of the effective prestressing of reinforcement

The effective prestressing σ_{pe} of tendon is the residual stress value after the stress loss σ_l is deducted from the controlled stress σ_{con} of prestressed rebar. The details are shown in Table 12-1.

Combination of the loss of prestressing at various stages　　　　Table 12-1

Combination of the loss of prestressing	Member pre-tensioned	Member post-tensioned
Loss at transmission & anchorage (batch I) σ_{lI}	$\sigma_{l2} + \sigma_{l3} + \sigma_{l4} + 0.5\sigma_{l5}$	$\sigma_{l1} + \sigma_{l2} + \sigma_{l4}$
Loss after transmission & anchorage (Batch II) σ_{lII}	$0.5\sigma_{l5} + \sigma_{l6}$	$\sigma_{l5} + \sigma_{l6}$

The effective prestressing σ_{pe} of tendon prestressed can be expressed as follows.

At the pre-stressing stage, the effective prestressing in prestressed reinforcement will be

$$\sigma_{pe} = \sigma_{pI} = \sigma_{con} - \sigma_{lI} \tag{12-37}$$

At the service stage, the effective prestressing, i.e. the permanent prestressing in prestressed reinforcement will be

$$\sigma_{pe} = \sigma_{pII} = \sigma_{con} - (\sigma_{lI} + \sigma_{lII}) \tag{12-38}$$

12.4 Calculation and checking of stress for the flexural member of prestressed concrete

1) Introduction

(1) Target of stress computation (checking): To meet the requirement specified by the specification for the stress of concrete during construction and at service stage.

(2) Contents of stress checking: The major compressive stress of concrete, the tensile stress of steel bar and the normal stress of concrete.

Notes:

(1) As far as the simply supported structure of prestressed concrete is concerned, only the major stress from the prestressing should be calculated.

(2) As to the super stabilized structure of continuous prestressed concrete beams, the minor effect from prestressing should be computed as well.

2) Stress checking for short-term stage

(1) Contents of checking: The stress from the normal and oblique cross sections caused by prestressing, the self-weight of members and the construction loads under manufacture during transportation and installation.

(2) Principles of calculation: The computation of stress for temporary conditions will belong to that of strength for members at the elastic stage, and can be calculated by material mechanics.

(3) Calculation of cross-sections: As to the simply supported beams, the most unfavorable cross-section will always be near supporting points, especially the cross-section of a simply supported beam with prestressed reinforcement concrete assembled in a linear way, in which the tensile stress at the upper edge of a supporting point. This stress will always be the control stress computed.

(4) Formula of calculation

①Normal stress under prestressing

The normal stress during prestressed stage can be calculated by the material mechanics. As shown in Figure 12-8, the effect M_{G1} from the eccentric prestressing N_p and the first permanent load (self-weight) of a beam G_1 will be mainly born.

σ_{ct}^t and σ_{cc}^t are the total normal stress of concrete on the top and bottom edges of cross sections at the prestressing stage, respectively, and they can be written as follows.

Pre-tensioned members

$$\left.\begin{array}{l} \sigma'_{ct} = \dfrac{N_{p0}}{A_0} - \dfrac{N_{p0}e_{p0}}{W_{0u}} + \dfrac{M_{G1}}{W_{0u}} \\[2ex] \sigma'_{cc} = \dfrac{N_{p0}}{A_0} + \dfrac{N_{p0}e_{p0}}{W_{0b}} - \dfrac{M_{G1}}{W_{0b}} \end{array}\right\} \qquad (12\text{-}39)$$

Post-tensioned members

$$\left.\begin{array}{l}\sigma'_{ct} = \dfrac{N_p}{A_n} - \dfrac{N_p e_{pm}}{W_{nu}} + \dfrac{M_{G1}}{W_{nu}} \\[2mm] \sigma'_{cc} = \dfrac{N_p}{A_n} + \dfrac{N_p e_{pn}}{W_{nb}} - \dfrac{M_{G1}}{W_{nb}}\end{array}\right\} \quad (12\text{-}40)$$

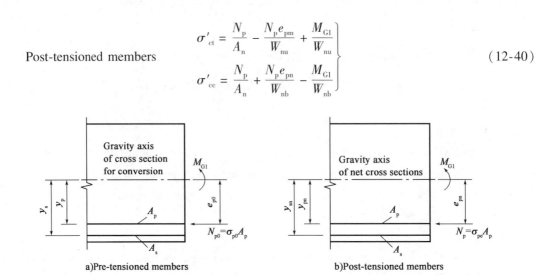

Figure 12-8 Joint force and the eccentric moment of the prestressed and non-prestressed rebar at the prestressing stage

Where σ'_{p0} is the stress of prestressed tendon, equal to zero when the concrete normal stress at the joint force point of the prestressed reinforcement in the tensile zone. $\sigma'_{p0} = \sigma'_{con} - \sigma'_{l1} + \sigma'_{l4}$, in which σ'_{l4} is the loss of prestressing of the prestressed rebar in the tension zone due to elastic relaxation of concrete, σ'_{l1} is the loss of prestressing when the prestressed steel bar transmits and is anchored in the compressive zone. e_{p0} is the eccentric moment of the joint force of the prestressed rebar to the cross-section gravity of the whole cross-section conversion for members. N_p is the joint force of the prestressed steel bar of members post-tensioned, and it can be calculated as follows

$$N_p = \sigma_{pe} A_p \quad (12\text{-}41)$$

As to the member with the prestressed reinforcement for spatial curves, the A_p in the above formula should be $A_p = A_p + A_{pb}\cos\theta_p$, in which A_{pb} is the cross-section area of the tendon bent, and θ_p is the included angle between the tangent of prestressed tendon which bends and the axis line of a member in a cross-section computed; A_p is the area of a cross-section of the tendon prestressed in tensile zones. σ_{pe} is the effective prestressing of the tendon prestressed in tension zones, $\sigma_{pe} = \sigma_{con} - \sigma_{II}$, σ_{II} is the prestressing loss of the prestressed tendon in tension zone during transmitting forces and anchoring. e_{pn} is the eccentric moment of the joint force of the prestressed tendon to the gravity of the net cross-section of members. A_n is the area of the net cross-section. W_{0u} and W_{0b} are the resistance moments of the cross-section at the upper and lower edges of the cross-section for the conversion of the whole cross-section of members, respectively. W_{nu} and W_{nb} are the resistance moments of the cross-section at the upper and lower edges of the net cross-section of members, respectively.

②Calculation of normal stress during the transportation and the hoisting stages

The method of stress calculation at these stages is the same as those for the prestressing application stage. The only attention should be that the prestressing force has decreased at these stages. When the moment from the first permanent load is computed, the change of the calculation diagram should be taken into consideration, including the dynamic coefficient.

③Limit stress of concrete at the construction stage

In the **CHBC**, the normal stress of concrete at this stage obtained must meet the following requirements:

a. Compressive stress of concrete σ^t_{cc}

$$\sigma^t_{cc} \leqslant 0.70 f'_{ck} \qquad (12\text{-}42)$$

In this formula, f'_{ck} is the standard value of the compressive strength of concrete at the center of the axis at various stages of manufacture, transportation and hoisting.

b. Concrete tension stress σ_{pc}:

When $\sigma^t_{ct} \leqslant 0.70 f'_{tk}$, the ratio of reinforcement assembly in the pretensioning zones should not be less than 0.2% for the longitudinal non-prestressing rebar;

When $\sigma^t_{ct} = 1.15 f'_{tk}$, the ratio of reinforcement assembly in the pretensioning zones should not be less than 0.4% for the longitudinal non-prestressing rebar;

When $0.70 f'_{tk} < \sigma^t_{ct} < 1.15 f'_{tk}$, the ratio of the longitudinal non-prestressed reinforcement in the prestressing zone should be decided by the linear interpolation of the two examples mentioned above. The tensile stress σ^t_{ct} should not exceed $1.15 f'_{tk}$.

3) Sustainability of operation

(1) Contents of checking: The normal stress of concrete, the principal stress and the stress of prestressed tendons.

(2) Features of calculation: The prestress loss is finished; the effective pre-stress σ'_{pe} is the minimum and is calculated by the standard value; the vehicle load should be included; the impact factor must be considered; the distribution coefficient of load is 1.0.

(3) Principles of calculation: As to the computation of sustainable stress and the complete prestressed concrete and the partially prestressed concrete structures of category A, it is still the computation of members at the elastic stage. The material mechanics method will be used for the computation.

(4) Controlled cross-sections: As for the most unfavorable cross-section, as far as the simply supported beams with cross-sections of rebar assembled in a straight line way are concerned, the mid-span cross-section will be regarded as the most unfavorable one. But as for those with cross-sections of reinforcement arranged in a fold line way or those with variable cross-sections, the mid-span, $l/4$ cross-section, $l/8$ cross-section of the supporting point and the cross-section where it varies will be used for computation according to the bending condition of the prestressed rebar and the change of the cross-section of concrete.

(5) Calculation formula

①Calculation of the normal stress: The normal stress of prestressed concrete members with common reinforcement can be written as follows (Figure 12-9).

a. Pre-tensioned members

The normal compressive stress of concrete at the upper edge of a cross-section of a member from standard values and prestressing can be written as

$$\sigma_{cu} = \sigma_{pt} + \sigma_{kc} = \left(\frac{N_{p0}}{A_0} - \frac{N_{p0} \cdot e_{p0}}{W_{0u}}\right) + \frac{M_{G1}}{W_{0u}} + \frac{M_{G2}}{W_{0u}} + \frac{M_Q}{W_{0u}} \quad (12\text{-}43)$$

The maximum tensile stress of prestressed tendons is calculated as follows.

$$\sigma_{pmax} = \sigma_{pe} + \alpha_{EP}\left(\frac{M_{G1}}{I_0} + \frac{M_{G2}}{I_0} + \frac{M_Q}{I_0}\right) \cdot y_{p0} \quad (12\text{-}44)$$

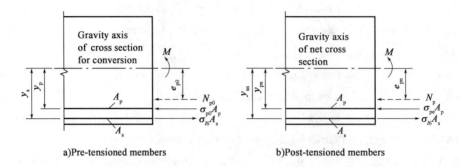

Figure 12-9   Joint forces and eccentric moment of prestressed tendon and non-prestressed reinforcement at service stage

Where σ_{kc} is the normal compressive stress of concrete by standard values. σ_{pe} is the permanent prestressing of prestressed reinforcement, i. e. $\sigma_{pe} = \sigma_{con} - \sigma_{II} - \sigma_{III} = \sigma_{con} - \sigma_l$. N_{p0} is the joint force of the prestressed rebar and the non-prestressed reinforcement at service stage [see Figure 12-9a)], it can be calculated as follows

$$N_{p0} = \sigma_{p0}A_p - \sigma_{l6}A_s \quad (12\text{-}45)$$

σ_{p0} is the stress of the prestressed rebar at the place of joint forces of the prestressed rebar in the tensile zone when the normal stress of concrete is zero; $\sigma_{p0} = \sigma_{con} - \sigma_l + \sigma_{l4}$, in which σ_{l4} is the loss of stress of the prestressed rebar caused by the elastic shrinkage of concrete in the tension zone, σ_l is the loss of stress.

σ_{l6} is the loss of prestressing due to the shrinkage and creep of concrete in the tension zone.

e_{p0} is the distance from the point of the joint force by the prestressed and non-prestressed reinforcement to the gravity axis of the cross-section of members for conversion, e_{p0} should be computed as follows

$$e_{p0} = \frac{\sigma_{p0}A_p y_p - \sigma_{l6}A_s y_s}{\sigma_{p0}A_p - \sigma_{l6}A_s} \quad (12\text{-}46)$$

A_s is the area of a cross-section of the non-prestressed rebar in the tension zone.

y_s is the distance between the gravity of the non-prestressed reinforcement and the gravity of the cross-section for conversion.

W_{0u} is the resistance of moment of the conversed cross-section of concrete members to the upper edge of the cross-section.

α_{EP} is the ratio of modulus of elasticity between the prestressed reinforcement and the concrete.

M_{G2} is the moment standard value by the second dead loads of deck pavement, side walk and fences.

M_Q is the most unfavorable moment in a cross-section calculated with the standard value composite of the variable loads; the coefficient of impact will take into consideration for vehicle loads.

b. Post-tensioned members

The normal compressive concrete stress σ_{cu} from the characteristic value of an action (or load) and prestressing on the upper edge of post-tensioned members can be written as follows.

$$\sigma_{cu} = \sigma_{pt} + \sigma_{kc} = \left(\frac{N_p}{A_n} - \frac{N_p \cdot e_{pn}}{W_{nu}}\right) + \frac{M_{G1}}{W_{nu}} + \frac{M_{G2}}{W_{0u}} + \frac{M_Q}{W_{0u}} \tag{12-47}$$

The maximum tensile stress of tendon prestressed is expressed in the following way.

$$\sigma_{pmax} = \sigma_{pe} + \alpha_{EP} \frac{M_{G2} + M_Q}{I_0} \cdot y_{0p} \tag{12-48}$$

Where N_p is the joint force of the prestressed and non-prestressed reinforcement, N_p can be calculated as follows.

$$N_p = \sigma_{pe} A_p - \sigma_{l6} A_s \tag{12-49}$$

σ_{pe} is the effective prestressing of the prestressed tendon in a tensile zone, $\sigma_{pe} = \sigma_{con} - \sigma_l$.

W_{nu} is the resistance moment of the net cross-section of members to the upper edge of a cross-section.

e_{pn} is the point of the joint force of the prestressed and the non-prestressed reinforcement to the gravity axis of the net cross-section of members, it can be written

$$e_{pn} = \frac{\sigma_{pe} A_p y_{pn} - \sigma_{l6} A_s y_{sn}}{\sigma_{pe} A_p - \sigma_{l6} A_s} \tag{12-50}$$

y_{sn} is the distance between the gravity of the non-prestressed reinforcement and the gravity of the net cross-section.

y_{0p} is the distance between the gravity of the prestressed reinforcement and the gravity of the cross-section for conversion.

If the tendon A'_p is placed in the cross-section of a compression zone, the action of A'_p should be considered in the formula.

②Calculation of the principle stress of concrete

The principal compressive stress σ_{cp} and the principal tensile stress σ_{tp} of concrete from the function of the characteristic value of an action (or loads) and prestressing on the flexural member of prestressed concrete can be calculated according to the following formula, i. e.

$$\sigma_{tp} \atop \sigma_{cp} = \frac{\sigma_{cx} + \sigma_{cy}}{2} \mp \sqrt{\left(\frac{\sigma_{cx} - \sigma_{cy}}{2}\right)^2 + \tau^2} \tag{12-51}$$

In this formula, σ_{cx} is the concrete normal stress at the principal stress point to be computed from the characteristic value of an action and the prestressing.

Pre-tensioned members $\quad \sigma_{cx} = \dfrac{N_{p0}}{A_0} - \dfrac{N_{p0} e_{p0}}{I_0} y_0 + \dfrac{M_{G1} + M_{G2} + M_Q}{I_0} y_0 \tag{12-52}$

Post-tensioned members $\quad \sigma_{cx} = \dfrac{N_p}{A_n} - \dfrac{N_p e_{pn}}{I_n} y_n + \dfrac{M_{G1}}{I_n} y_n + \dfrac{M_{G2} + M_Q}{I_0} y_0 \tag{12-53}$

Where y_0 and y_n are the distance from the major stress point computed to the gravity axis of a cross-section for conversion and the net cross-section, respectively.

I_0 and I_n are the inertia moment of conversed cross-sections and the net cross-section, respectively.

σ_{cy} is the vertical compressive stress of concrete by the prestressing from the prestressed vertical tendons calculated as follows

$$\sigma_{cy} = 0.6 \frac{n\sigma'_{pe} A_{pv}}{b \cdot s_v} \tag{12-54}$$

n is the number of vertical bar in the same cross-section.

σ'_{pe} is the effective stress after the total prestressing loss, deducted from the vertical prestressed reinforcement.

A_{pv} is the area of a cross-section of an individual vertical prestressed steel bar.

s_v is the spacing of the vertical prestressed tendon.

τ is the shear stress of concrete, computed in accordance with that from the shear of the standard composite value of an action; while the shear stress from the standard composite value of an action at any point of a cross-section of a beam with high prestressing force can be calculated as follows.

Pre-tensioned members: $\tau = \dfrac{V_{G1} S_0}{b I_0} + \dfrac{(V_{G2} + V_Q) S_0}{b I_0}$ \hfill (12-55)

Post-tensioned members: $\tau = \dfrac{V_{G1} S_n}{b I_n} + \dfrac{(V_{G2} + V_Q) S_0}{b I_0} - \dfrac{\sum \sigma''_{pe} A_{pb} \sin\theta_p S_n}{b I_n}$ \hfill (12-56)

Where V_{G1} and V_{G2} are the standard value of shear by the first and the second dead load, respectively.

V_Q is the standard value composition of shear by variable action (or load); as to simply supported beams, it can be calculated as follows

$$V_Q = V_{Q1} + V_{Q2} \tag{12-57}$$

V_{Q1} and V_{Q2} are effect of vehicle loads (coefficient of impact included) and the standard value of shear by vehicles and human activities, respectively.

S_0 and S_n are the area of cross section conversion above (or below) the principle stress to be computed to the area moment between the area of the gravity axis of cross-sections and the net cross-section, respectively.

θ_p is the calculation of the tangent of the prestressed reinforcement bent of a cross-section and the included angle of the longitudinal axis line of members (shown in Figure 12-10).

b is the width of the composite panel of members at the principle point to be calculated.

σ''_{pe} is the effective stress after the total loss of prestressing has been deducted from the longitudinal tendons bent.

A_{pb} is the cross-section area of the prestressed rebar bent in the same plane of a cross-section.

Notes: a. The compressive stress will be the positive and the tensile one the negative in the above formula.

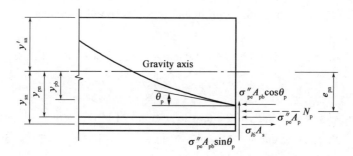

Figure 12-10 Calculation of shear

b. As to such extra stable structure as continuous beams, the minor effect of prestressing and temperature should be taken into consideration.

c. As far as the prestressed concrete continuous beam with variable height is concerned, when the shear stress by functions is computed, the additional shear stress from the moment of axial force in a cross-section should be calculated.

③Value of the stress limit of reinforcement and concrete under permanent conditions

When the flexural member of complete prestressed concrete or partially prestressed concrete of category A is considered, the **CHBC** specifies the following:

a. The maximum compressive stress of the concrete in a normal cross-section at service stage should satisfy the following

$$\sigma_{kc} + \sigma_{pc} \leqslant 0.5 f_{ck} \quad (12\text{-}58)$$

Where σ_{kc} is the normal compressive stress of concrete by the characteristic value of an action; σ_{pt} is the normal tensile stress of concrete by prestressing; f_{ck} is the standard value of compressive strength of concrete at the center of an axis.

b. The maximum limited value of tensile stress of the prestressed tendon at service stage:

The **CHBC** specifies the maximum tensile stress for reinforcement should be as follows.

Internal steel wires and strands: $\quad \sigma_{pe} + \sigma_p \leqslant 0.65 f_{pk} \quad (12\text{-}59\text{a})$

Prestressing twisted steel bar: $\quad \sigma_{pe} + \sigma_p \leqslant 0.75 f_{pk} \quad (12\text{-}59\text{b})$

Where σ_{pe} is the effective after the total loss of prestressing has been deducted; σ_p is the increase of the prestressed rebar from actions; f_{pk} is the standard value of the tensile strength of prestressed tendons.

c. The limit value of the principal stress of concrete at service stage:

The principal compressive stress of concrete should meet the following requirement

$$\sigma_{cp} \leqslant 0.6 f_{ck} \quad (12\text{-}60)$$

In the formula, f_{ck} is the standard value of the compressive strength of concrete at the center of an axis.

The calculated principal tensile stress of concrete σ_{tp} is the complement to the calculation of the shearing resistance of an oblique cross section of a member, and then the rebar can be arranged as follows:

a. In the zone of $\sigma_{tp} \leq 0.5 f_{tk}$, the reinforcement can be only assembled in accordance with the structural requirement;

b. In the zone of $\sigma_{tp} > 0.5 f_{tk}$, the rebar spacing s_v can be calculated as follows

$$s_v = \frac{f_{sk} A_{sv}}{\sigma_{tp} b} \tag{12-61}$$

Where f_{sk} is the standard value of the tensile strength of reinforcement; f_{tk} is the standard value of the tensile strength of concrete at the center of an axis; A_{sv} is the total area of cross sections of the reinforcement in the same cross section; b is the width of a rectangle cross section or the width of T-shaped or I-shaped composite panels.

If the number of reinforcement computed by the above formula is fewer than that required by the computation of load-carrying capacity of shear resistance of an included cross section, the reinforcement of members will be arranged according to the requirement of the load-carrying capacity of the shear resistance.

12.5 Checking of anti-cracking

1) Introduction

Brief introduction: Since the cracking in the completely prestressed concrete and the partially pre-stressed structure of category A are not allowed, the **CHBC** requires that the checking of anti-cracking for normal and inclined cross sections be carried out for the completely prestressed concrete and the partially pre-stressed structure of category A. While for the partially prestressed structure of category B, the **CHBC** allows that the flexural crack won't exceed the value fixed. But it is permitted for an inclined cross section to crack. Therefore, the checking of the anti-cracking of an inclined cross section must be carried out.

Principles: The checking of the anti-cracking of the normal cross section of structures will examine whether the tensile stress of concrete of the normal cross section exceeds the limit specified or not and sets it as the standard and expressed, while the checking of the anti-cracking of the inclined cross section will test whether major tensile stress of concrete members exceeds the limit required or not and be expressed in this way. Therefore, it is only necessary to compute the normal and the major stress.

2) Requirement for anti-cracking checking of the normal section

(1) The completely prestressed member under the combination of the frequent combination of actions should meet the following requirements.

Pre-cast members
$$\sigma_{st} \leq 0.85 \sigma_{pc} \tag{12-62}$$

Casting block by block or the longitudinal blocking of the joint of mortar

$$\sigma_{st} \leq 0.80 \sigma_{pc} \tag{12-63}$$

(2) The partially prestressed member of category A under the combination of the frequent combination of actions should meet the following requirement.

Chapter 12 • Design and Calculation of Prestressed Concrete Flexural Members

$$\sigma_{st} - \sigma_{pc} \leq 0.7 f_{tk} \quad (12\text{-}64)$$

The partially prestressed member of category A under the combination of the quasi-permanent of actions should meet the following requirement.

$$\sigma_{lt} - \sigma_{pc} \leq 0 \quad (12\text{-}65)$$

①The checking of the normal tensile stress σ_{st} of concrete of the edge for the anti-cracking of members from the short term effect of an action will be as follows.

Pre-tension members:
$$\sigma_{st} = \frac{M_s}{W} = \frac{M_{G1} + M_{G2} + M_{Qs}}{W_0} \quad (12\text{-}66)$$

Post-tension members:
$$\sigma_{st} = \frac{M_s}{W} = \frac{M_{G1}}{W_n} + \frac{M_{G2} + M_{Qs}}{W_0} \quad (12\text{-}67)$$

Where σ_{st} is the checking of the normal tensile stress of concrete of the edge for the anti-cracking in accordance with the frequent combination of actions. M_s is the flexural moment computation according to the frequent combination of actions. M_{Qs} is the flexural moment of variable loads in accordance with the frequent combination of an actions. As to the simply supported beam, the computation is as follows.

$$M_{Qs} = \psi_{11} M_{Q1} + \psi_{12} M_{Q2} = 0.7 M_{Q1} + 1.0 M_{Q2} \quad (12\text{-}68)$$

ψ_{11} and ψ_{12} are the effect of vehicle load in accordance with the frequent combination of actions and the coefficient of the frequent value of an action from the effect of the human load, respectively. M_{Q1} and M_{Q2} are the standard value of the flexural moment from the vehicle load effect (regardless of impact) and the effect of human load, respectively. W_0 and W_n are the anti-cracking checking of the elastic resistance moment at the edge to the conversed cross section and the net cross section of members, respectively.

For the prestressed concrete continuous beam and continuous structures, such indirect action as temperature difference, shrinkage and creeping of concrete should be considered besides direct loads, dead loads and vehicles on the beam.

②Calculation of the normal stress of concrete at the lower edge under the quasi-permanent combination of actions.

The long-term effect combination of an action should include that of the standard value for a permanent action and that of the quasi-permanent value of a variable action. The calculation of the normal stress of concrete at edges with the long-term effect combination of an action is basically the same as those with the short-term effect combination of an action. It can be calculated as follows.

Pre-tension members:
$$\sigma_{lt} = \frac{M_l}{W} = \frac{M_{G1} + M_{G2} + M_{Ql}}{W_0} \quad (12\text{-}69)$$

Post-tension members:
$$\sigma_{lt} = \frac{M_l}{W} = \frac{M_{G1}}{W_n} + \frac{M_{G2} + M_{Ql}}{W_0} \quad (12\text{-}70)$$

Where σ_{lt} is the computation of the normal tensile stress of concrete at the edge for the check of the anti cracking of members in accordance with the quasi-permanent combination of an action. M_l is the flexural moment in line with the quasi-permanent combination of an action. M_{Ql} is the flexural moment of a variable action in accordance with the quasi-permanent combination of an action. If such direct action as vehicles and human beings on the flexural moment from the load of members is considered, it can be calculated as follows

$$M_{Q1} = \psi_{21}M_{Q1} + \psi_{22}M_{Q2} = 0.4M_{Q1} + 0.5M_{Q2} \quad (12\text{-}71)$$

M_{Q1} and M_{Q2} are the standard value of the flexural moment from vehicle action (regardless of impact) and human action effects, respectively. ψ_{21} and ψ_{22} are the quasi-permanent coefficient of the vehicle load and the human being effects in line with the long-term effect combination of an action, respectively.

3) Requirements for the checking of the anti-cracking of an inclined cross section

(1) If the completely prestressed member is considered, the following should be met in accordance with the frequent combination of actions.

Pre-cast members: $\quad \sigma_{tp} \leq 0.6f_{tk} \quad (12\text{-}72)$

Cast in place members: $\quad \sigma_{tp} \leq 0.4f_{tk} \quad (12\text{-}73)$

(2) When the partially prestressed member of category A and B are concerned, the following should be met in accordance with the frequent combination of actions.

Pre-cast members: $\quad \sigma_{tp} \leq 0.7f_{tk} \quad (12\text{-}74)$

Cast in place members: $\quad \sigma_{tp} \leq 0.5f_{tk} \quad (12\text{-}75)$

In this formula f_{tk} is the standard value of the concrete tensile strength of the center of an axis. σ_{tp} can be calculated as follows

$$\sigma_{tp} = \frac{\sigma_{cx} + \sigma_{cy}}{2} - \sqrt{\left(\frac{\sigma_{cx} - \sigma_{cy}}{2}\right)^2 + \tau^2} \quad (12\text{-}76)$$

In this formula the computation for the normal stress σ_{cx}, σ_{cy} and the shear stress τ to be calculated is shown in Formula (12-51) ~ (12-56). When the shear stress τ is computed, the shear V_Q will be chosen from the shear value V_{Qs} from the frequent combination of variable actions.

4) Selection of the cross section of anti-cracking checking of an inclined cross section

The cross section with larger shear and flexural moments should be chosen along the direction of the length of a member, or the cross section with a sudden change externally; along the height of a cross section the heart shaped cross section for conversion and the changed width of a cross section should be selected. As to the pre-tensioned member, the actual number of the pre-stressing within the range of the length of transmission l_{cr} should be considered.

12.6 Calculation of deformation and setting of pre-camber

1) The total deflection of the flexural member of prestressed concrete in accordance with the frequent combination of load w_s

The total deflection will be calculated as follows.

$$w_s = -\delta_{pe} + w_{Ms} \quad (12\text{-}77)$$

Where δ_{pe} is the camber from the permanent prestressing force N_{pe}; w_{Ms} is the deflection value due to the value of flexural moment in line with the frequent combination of an action.

(1) The camber δ_{pe} due to the permanent prestressing force N_{pe} can be calculated by unit-force method and expressed as follows

$$\delta_{pe} = \int_0^l \frac{M_{pe} \cdot \overline{M_x}}{B_0} dx \qquad (12\text{-}78)$$

Where, M_{pe} is the value of the flexural moment due to the permanent prestressing force (the joint force of the permanent prestressing force) at the any cross section x; $\overline{M_x}$ is the flexural moment value at any cross section x when the mid-span acts on a unit force; B_0 is the flexural rigidity to be computed in accordance with the value of the real compression stage.

(2) The value w_{Ms} of deflection from the value of flexural moment in line with the frequent combination of actions.

According to the requirements of the **CHBC**, the deflection of the simply supported beam of uniform height and the cantilever beam of completely prestressed members and partially prestressed concrete member of category A will be expressed as follows

$$w_{Ms} = \frac{\alpha M_s l^2}{0.95 E_c I_0} \qquad (12\text{-}79)$$

Where l is the computed span of a beam; α is the coefficient of deflection, relating to the shape of the flexural rectangle diagram and the constraining condition for supporting (seen in Table 12-2); M_s is the flexural moment computed in line with the frequent combination of actions (or load); I_0 is the inertia moment of a cross section conversed for the total cross section of members.

Maximum flexural moment M_{max} of beams and coefficients α for deflection at the mid-span (or on the cantilever end)　　　　　　　　　　　Table 12-2

Diagram of loads	Diagram of flexural moment and the max. ones M_{max}	Coefficient of deflection, α
(uniform load q over span l)	$\dfrac{ql^2}{8}$	$\dfrac{5}{48}$
(partial uniform load q over βl)	$\dfrac{\beta^2(2-\beta)^2 ql^2}{8}$	$\beta \leqslant \dfrac{1}{2} : \dfrac{3-\beta}{12(2-\beta)^2};$ $\beta > \dfrac{1}{2} : \dfrac{4\beta^4 - 10\beta^3 + 9\beta^2 - 2\beta + 0.25}{12\beta^2(\beta-2)^2}$
(triangular load q)	$\dfrac{ql^2}{15.625}$	$\dfrac{5}{48}$
(point load F at βl, simply supported)	$F\beta(1-\beta l)$	$\beta \geqslant \dfrac{1}{2} : \dfrac{4\beta^2 - 8\beta + 1}{-48\beta}$
(point load F at βl, cantilever)	$F\beta l$	$\dfrac{\beta(3-\beta)}{6}$
(uniform load q over βl, cantilever)	$\dfrac{q\beta^2 l^2}{2}$	$\dfrac{\beta(4-\beta)}{12}$

2) Frequent combination of load with the consideration of the deflection value w_l of the long-term effect influence

The deflection of the flexural member of prestressed concrete will increase due to the creeping of concrete as time goes on. The increase will be realized through the coefficient η_θ of the long term increase of deflection in the **CHBC**. The detailed computation is as follows

$$w_l = -\eta_{\theta,pe} \cdot \delta_{pe} + \eta_{\theta,Ms} \cdot w_{Ms}$$
$$= -\eta_{\theta,pe} \cdot \delta_{pe} + \eta_{\theta,Ms} \cdot (w_{G1} + w_{G2} + w_{Qs}) \qquad (12\text{-}80)$$

Where w_l is the deflection value of the long-term effects of loads; $\eta_{\theta,pe}$ is the coefficient of the increase of the long term effects considered in the computation of the camber value of prestressing; when the counter vault of the prestressing at the service period is computed, the total loss of prestressing should be deducted from the prestressing of the prestressed rebar, $\eta_{\theta,pe} = 2$; $\eta_{\theta,Ms}$ is the coefficient of the deflection increase of the frequent combination of load considering long-term effects of an action; the value of this factor can be obtained from Table 12-3.

The coefficient of the deflection increase of frequent combination of load with the consideration of long-term effects of an action   Table 12-3

Grade of concrete strength	< C40	C40	C45	C50	C55	C60	C65	C70	C75	C80
$\eta_{\theta,Ms}$	1.60	1.45	1.44	1.43	1.41	1.40	1.39	1.38	1.36	1.35

The maximum vertical deflection value permitted for the flexural member of prestressed concrete with the frequent combination of loads and with the consideration of the influence of the long term effects of an action will be the same as that for the reinforced concrete beams.

3) Setting of pre-camber

In the **CHBC**, it requires that the pre-camber may not be implemented when the long term counter-vault value of the flexural member of the prestressed concrete from the prestressing stress is larger than the long term deflection computed in line with the frequent combination of loads, while when the long term counter vault value of the prestressed stress is smaller than that of the long term deflection calculated with the short term effect combination of load, the pre-camber should be built. The difference between the deflection of this load and the long term counter-vault value of the prestressing force will be used as the pre-camber value Δ, i.e.

$$\Delta = \eta_{\theta,Ms} w_{Ms} - \eta_{\theta,pe} \delta_{pe} \qquad (12\text{-}81)$$

When the pre-camber is set, a smooth curve will be drawn in line with the max. pre-camber value and along the direction of a bridge.

12.7 Calculation of the end in the anchorage zone

1) Calculation of the partial load-carrying in the anchorage zone of post-tensioned members

The huge prestressing force F_{ld} of the post-tensioned member is transmitted to concrete through anchorage and a small tie plate placed under it. After a part of concrete is compressed, the computation for compressive strength and the checking tensile resistance for this part should be carried out. The requirements for the member should be met.

(1) Strength calculation of the partial load-carrying capacity with the anchorage

The compressive resistance capacity of the end of the spiral reinforcement should meet the following requirement

$$\gamma_0 F_{ld} \leq 0.9(\eta_s \beta f_{cd} + k\rho_{st}\beta_{cor}f_{sd})A_{ln} \qquad (12-82)$$

If this requirement is not satisfied, then the following measures can be taken:

①To enlarge the partial area which is under compression.

②To adjust the place for anchorage.

③To enhance the grade of concrete strength.

(2) Tensile resistance checking when a part of concrete is under compression

In order to prevent the appearance of cracks in the part where concrete is under compression in the direction of the length of members, the checking of tensile resistance in the partial compression zone in which indirect reinforcement is placed should be calculated according to the following:

$$\gamma_0 T_{(.),d} \leq f_{sd} \cdot A_s \qquad (12-83)$$

(3) Requirement for structures

①The tie plates of no less than 16 mm thickness in the anchorage should be placed or the flared anchorage cushion panel be used.

②The length of spiral rebar coils should be less than that of the flared pipe.

③The closed stirrup is applied to resist the horizontal bursting force in the anchorage zone, the distance should not more than 120mm.

④The anti-cracking steel bar is placed in the end section to resist the surface bursting force.

⑤The plane size of the end of a girder will be decided by the size of anchorage, the spacing among anchorages and the requirement of a jack for tensioning.

2) Transmission length and the anchorage length of prestressed tendons of pre-tensioned members

(1) Transmission length l_{tr} and anchorage length l_a

The transmission and the retaining of prestressing will be completed through the cohesive force of steel bars and the concrete by the pre-tensioned member. The stress and strain of the prestressed reinforcement exposing to the open air will be zero. Due to the cohesive function on a certain length of the reinforcement, the stress of the prestressed rebar will be the effective prestressed stress σ_{pe}, and this length of rebar is called the transmission one l_{tr} of the prestressed steel bar. The mechanical condition of reinforcement within the range of transmission length is very complicated as the stress changes. In the **CHBC**, it is described the stress will approximately change in accordance with the linear law as shown in Figure 12-11. Similarly, when the member reaches its the ultimate bending capacity, the stress of prestressed tendon will increase from zero at the end to the designed strength f_{pd}, and this length is called the anchorage length l_a. The

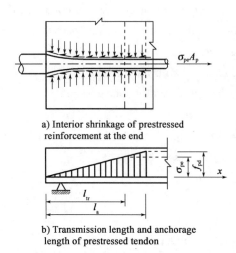

a) Interior shrinkage of prestressed reinforcement at the end

b) Transmission length and anchorage length of prestressed tendon

Figure 12-11 Anchorage of prestressed tendon by pre-tensioning

anchorage length should ensure that tendons will reach to its yielding strength without being pulled out.

The ***CHBC*** specifies the value for the transmission length l_{tr} and the anchorage length l_a of the prestressed reinforcement after many factors have been taken into consideration as seen in appendix Table 2-8. In the meantime it's assumed that the stress of the prestressed reinforcement within the range of transmission and anchorage length ($0 \sim \sigma_{pe}$ or f_{pd}) may be computed in accordance with the linear interpolation [Figure 12-11b)]. Therefore, the stress of the prestressed reinforcement σ_{pe} should be obtained with the linear interpolation and in accordance with the place at which the inclined cross section is located when the load-carrying capacity of a inclined cross section is computed within the range of the anchorage length l_a at the end. The actual stress of the prestressed rebar should also be achieved with the linear interpolation and in line with place where the checking of a cross section is carried out when the anti-cracking computation is calculated within the range of transmission length l_{tr} of prestressing at the end.

(2) Other requirements

The partial strengthening measures should be adopted at the end of a member pre-tensioned. The common measures for the concrete around the end of the prestressed reinforcement are the following:

When a single steel wire is applied, no less than 150 mm of spiral rebar should be placed at the end; while when multiple steel wires are used, three to five steel meshes should be placed at the spacing of $10d$ (d is the diameter of the prestressed rebar) at the end.

12.8 Design of simply-supported prestressed concrete beams

1) Contents of design and calculation

The post-tensioned simply supported beam will be taken as an example to describe the contents of design and computation as follows:

(1) According to the design requirement, the type and the size of a cross section of members should be chosen, or the initial evaluation of the maximum sizes of cross sections of moment should be carried out in line with the flexural requirement of the cross section with the direct consideration of the maximum cross section of the moment.

(2) The maximum design moment and shear of a controlled cross section.

(3) The estimate of the quantity of prestressed reinforcement and proper arrangement of rebar.

(4) The computation of the geometric characteristics of the cross section of a main girder.

(5) The calculation of the normal and inclined bearing capacity.

(6) The determination of the tensile controlled stress of reinforcement and the estimate of the loss of stress of various members and the computation of the relevant effective stress of different stages.

(7) The stress checking of members during the temporary and permanent conditions.

(8) The anti-cracking checking of the normal and inclined cross section.

(9) The computation of the deformation of the major beam.

(10) The computation of the partial bearing capacity of anchorage and design of the anchoring zone.

The key point of design is how to decide the sizes, to estimate the quantity of tendons and to arrange rebar properly. Why? It's because the design itself will involve assumptions first and the computation is done until the requirements of the specification are met. Otherwise, redo the design and repeat the computation until the requirements are satisfied.

2) Efficiency index of a cross-section

There is the relevant relationship between the load effect and the cross-section. The larger the load is, the bigger the cross-section should be. But the load includes the dead (self weight) and live ones. The size of a cross-section will directly have an impact on the total load effect.

The efficient index ρ of the flexural cross-section shall be used to check the rationality

$$\rho = \frac{k_u + k_b}{h}$$

If ρ is larger, the higher the flexural efficiency will become. ρ is usually within 0.4 ~ 0.55. k_u is the upper core point and k_b is the lower core point.

3) Common types of cross-sections

Some common types of cross-sections in bridge construction are shown in Figure 12-12, including hollow slab, T-beam, I-beam and box-girder.

4) Estimation of quantity of prestressed tendons

The estimation of the area of a cross-section of prestress reinforcement should be considered in three aspects.

(1) Estimation of the area of a cross-section of prestressed reinforcement

The estimation of the quantity of tendons should be made according to the anti-cracking requirement of the normal cross-section of members.

The checking of the anti-cracking of the normal cross-section for the completely prestressed concrete beam will be carried out in accordance with the frequent combination of actions. The normal tensile stress obtained by such computations should meet the requirement for the critical value as follows.

$$\frac{M_s}{W} - 0.85 N_{pe} \left(\frac{1}{A} + \frac{e_p}{W} \right) \leq 0 \qquad (12\text{-}84)$$

If a little change is made to the formula mentioned above, the effective prestressing can be achieved for the need of the anti-cracking checking and the satisfaction of the completely prestressed concrete beam to the frequent combination of actions, as

$$N_{pe} \geqslant \frac{\dfrac{M_s}{W}}{0.85\left(\dfrac{1}{A} + \dfrac{e_p}{W}\right)} \qquad (12\text{-}85)$$

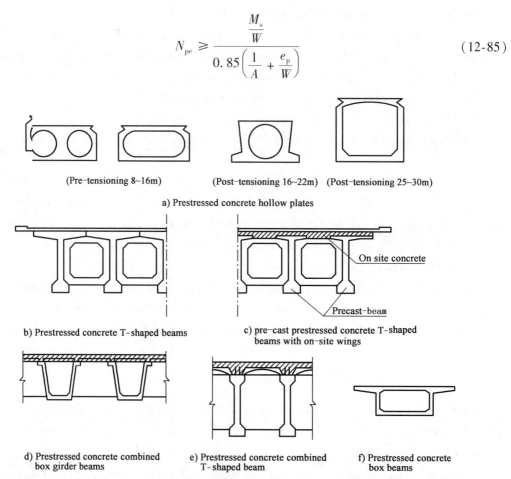

Figure 12-12 Common types of cross-sections of prestressed concrete beams

Where M_s is the moment value computed in accordance with the frequent combination of actions; N_{pe} is the joint force of the permanent stress of the prestressed reinforcement at the service stage; A is the area of concrete of the total cross-section; W is the elastic resistance moment at the edge of the whole cross-section of members for anti-cracking checking; e_p is the distance between the function point of the prestressed rebar to the gravity axis of a cross-section.

For the partially prestressed concrete member of category A, a similar equation can be obtained as follows

$$N_{pe} \geqslant \frac{M_s/W - 0.7f_{tk}}{\dfrac{1}{A} + \dfrac{e_p}{W}} \qquad (12\text{-}86)$$

When N_{pe} is obtained, the proper tensile controlled stress σ_{con} will be determined, and the relevant loss σ_l of stress deducted (for the post-tensioned member of high strength steel wire or steel strand, σ_l will be about $0.2\sigma_{con}$), and the total area of the prestressed reinforcement will be estimated as

$$A_p = \frac{N_{pe}}{(1 - 0.2)\sigma_{con}}.$$

After A_p has been determined, the prestressed steel strand n_1 will be computed in line with the area A_{pl} of a bunch of prestressed tendons. In the formula, A_{pl} represents the cross-section area of a bunch of prestressed tendons.

$$n_1 = \frac{A_p}{A_{pl}} \qquad (12\text{-}87)$$

(2) Estimate of the quantity of non-prestressed reinforcement in accordance with the requirement of the critical condition of the bearing capacity of members

After the quantity of the prestressed reinforcement is determined, the non-prestressed rebar of the rectangular cross-section beam and the T-shaped cross-section beam will be determined in line with the critical condition of the bearing capacity of the normal cross-section.

(3) Requirement of the minimum ratio of reinforcement

The quantity of reinforcement estimated with the above method must also meet the requirement of the minimum reinforcement ratio. The **CHBC** specifies that the minimum ratio of rebar of the flexural member of the prestressed concrete should meet the following conditions

$$\frac{M_u}{M_{cr}} \geqslant 1.0 \qquad (12\text{-}88)$$

Where M_u is the design value of the flexural bearing capacity of the normal cross-section of the flexural member, computed in line with the right part of Formula (3-29) or Formula (3-21); M_{cr} is the cracking moment value in a normal cross section of flexural members; calculated by the following formula

$$M_{cr} = (\sigma_{pc} + \gamma f_{tk}) W_0 \qquad (12\text{-}89)$$

Where σ_{pc} is the pre-compressive stress of concrete at the anti-cracking edge of members when the joint force of the prestressed reinforcement and the common rebar N_{p0} of the whole loss of prestressed stress; W_0 is the elastic resistance moment of the anti cracking of a conversed cross-section; γ is the calculation parameter and in line with $\gamma = \frac{2S_0}{W_0}$ for computation, of which S_0 is the area above the gravity axis of the conversed cross-section of the whole cross-section to the area moment of the gravity axis.

5) Arrangement of prestressed tendon

(1) Core of a cross-section

The core of the rectangular and T-shaped cross-sections is shown in Figure 12-13. When the prestressing acts within the core, there'll be no tensile stress in the cross-section. The arrangement of rebar will be placed in this way.

(2) Center of gravity limit of reinforcement

According to the principle that there'll be no tensile stress at the upper and lower edges of concrete of the cross-section of completely prestressed concrete member, the critical value N_p of

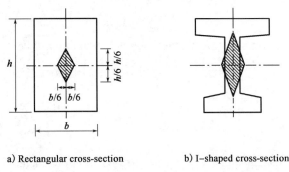

a) Rectangular cross-section   b) I-shaped cross-section

Figure 12-13 Core zone of cross sections

the eccentric moment of various cross-sections should be determined. The critical value lines E_1 and E_2 of two e_p are shown in Figure 12-14. As long as the location of the action point N_p (approximate as the center of gravity of a cross-section) is within the range of E_1 and E_2, it can be ensured that there'll be no stress to appear at the upper and lower edges of concrete under the minimum external load and the most unfavorable load. Therefore, the critical limit of the center of gravity of reinforcement is called the center of gravity limit of reinforcement (strand) when the prestressed reinforcement is placed around the two curves of E_1 and E_2.

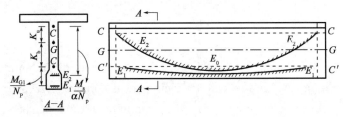

Figure 12-14 Limit of the center of gravity of reinforcement of completely prestressed concrete simply supported beams

At the pre-stressing stage, the condition under which there should be no appearance of tensile stress at the upper edge of a beam should be the following

$$\sigma_{ct} = \frac{N_{pI}}{A} - \frac{N_{pI} e_{pI}}{W_u} + \frac{M_{GI}}{W_u} \geq 0 \quad (12\text{-}90)$$

From the above it can be obtained

$$e_{pI} \leq E_1 = K_b + \frac{M_{GI}}{N_{pI}} \quad (12\text{-}91)$$

Where e_{pI} is the eccentric moment of prestressing force; if the joint force is located under the center of gravity of axis of a cross-section, e_{pI} will be the positive value; otherwise the negative value; K_b is the lower core point under the cross-section of concrete, $K_b = \dfrac{W_u}{A}$; W_u is the elastic resistant moment of the whole cross-section of members to the upper edge of a cross-section; N_{pI} is the joint force of the prestressing at the stage of the transmission of forces and the anchorage.

Similarly with the function of the flexural moment computed in accordance with the frequent combination of actions, the eccentric moment e_{p2} of the prestressing joint force can be obtained as follows

Chapter 12 · Design and Calculation of Prestressed Concrete Flexural Members

$$e_{p2} \geqslant E_2 = \frac{M_s}{\alpha N_{pI}} - k_u \qquad (12\text{-}92)$$

Where M_s is the flexural moment value calculated with the frequent combination of actions; α is the ratio value of the permanent prestressing N_{pe} at service stage and the effective prestressing N_{pI} at the stage of transmitting forces and anchoring, and the approximate value of $\alpha = 0.8$ may be selected; k_u is the upper core moment of the cross section of concrete section, $k_u = \frac{W_b}{A}$; W_b is the elastic resistant moment of the whole cross section to the lower edge of a cross section.

It can be seen that e_{p1} and e_{p2} have the similar change rules to those for moment M_{GI} and moment M_s respectively, and all can be regarded as the para-curve, which will change along the span. The area between the critical value of E_2 and E_1 will be the range to place steel bars. It can be known from this that the condition followed for computing of the position of the center of gravity of rebar (e_p) should be as follows

$$\frac{M_s}{\alpha N_{pI}} - k_u \leqslant e_p \leqslant k_b + \frac{M_{GI}}{N_{pI}} \qquad (12\text{-}93)$$

For the partially prestressed concrete member with tensile stress or cracks, when different limit values of the tensile stress (including the nominal tensile stress) of concrete at the upper and lower edges of members are worked out, the limit of the center of gravity of reinforcement can be determined.

(3) Principles of the arrangement of prestressed tendons

①As far as the arrangement of the prestressed rebar is concerned, the center line should be designed without exceeding the range of the limit of the center of gravity of reinforcement. In the meantime, the prestressed steel bar with the edge of members gradually bending will lead to the preshearing. This will be very beneficial to setting off the shearing from the larger external shearing near the support point.

②The angle of prestressed reinforcement bent should match the change of shearing. The preshearing V_p from the prestressed rebar bent should set off part of the shear combination design value V_d of an action (or load). The rest of the shear after the setting off is called the minus surplus shear and will be the basis for the arrangement of the antishearing reinforcement.

③The arrangement of the prestressed reinforcement should be in line with the requirement of structures.

(4) Determination of the bending point of the prestressed rebar

The bending point of tendon should be determined by considering the load-carrying capacity against the shear and moment.

①From the viewpoint of the shearing, the bending point will start from the cross section of $\gamma_0 V_d \geqslant V_{cs}$ theoretically so as to provide a part of the preshearing V_p to resist the shear from an action. Thus, the position of the bending point will usually be located at 1/3 or 1/4 of a span.

②From the viewpoint of the bending, after the bending of the prestressed rebar, the center of gravity line will move upwards to make the eccentric distance e_p smaller, i. e. the moment M_p of the prestressing will be smaller. The requirement of the normal cross section flexural bearing capacity should be noted after the bending of the prestressed rebar.

③It is necessary to satisfy requirement of the flexural bearing capacity of an inclined cross section for the bending point of the prestressed reinforcement, i. e. it should be ensured that the flexural bearing capacity of an inclined cross section after the bending of the prestressed rebar won't be less that of the flexural bearing capacity of the normal cross section located on top of the end of the inclined cross section.

(5) Angle of the prestressed rebar bent

In theory, the best design of the tendon bent will be to control the angle θ_p for bending of the prestressed rebar under the condition of $N_{pd}\sin\theta_p = V_{G1} + V_{G2} + \dfrac{V_Q}{2}$. But as to a beam with larger dead load (due to larger span), if the value θ_p calculated by this formula is obviously too large, only the condition where a larger value θ_p is allowed will be chosen. As to the section of a beam near the support point, the quantity of the prestressed rebar bent should be as much as possible when the condition of the flexural bearing capacity can be satisfied. But in order to reduce the loss of friction stress from the prestressing of the curve of the prestressed rebar, the angle θ_p for bending should be no larger than 20°. As to the anchored rebar with the bent part exposing the top of a beam, the angle θ_p will always over 20°, usually between 25° and 30°. If the prestressed tendon with a larger angle θ_p is taken into consideration, it should be noted that the measures to reduce the coefficient of friction should be adopted to reduce the loss of the friction stress.

(6) Shapes of curves of the prestressed rebar bent

There are three kinds of curve of the prestressed rebar bent, circular arc, the para-curve and the catenary curve. The circular arc will be mainly used in the construction of bridges and roads. The **CHBC** specifies that the radius of curvature of the curve shaped for the prestressed steel bar of the prestressed member and post-tensioned members should satisfy the following requirements:

①Steel wires and strands: If the diameter is $d \leqslant 5$mm, the diameter shouldn't be less than 4m; if $d > 5$mm, it should be no smaller than 6m.

②The prestressing twisted reinforcement: If the diameter is $d \leqslant 25$mm, the diameter should be no smaller than 12m; if $d > 25$mm, it shouldn't be smaller than 15m.

③As to the prestressed reinforcement for special usages (semi circle prestressed steel bar for hooping of the cable-stayed bridges with a radius about 1.5 meters), there should be no restricting requirement for it due to the special measures applied.

(7) Structural requirements for the arrangement of the prestressed reinforcement

①Post-tensioned members

The post-tensioned member with curve shaped prestressed steel is shown in Figure 12-15. The smallest thickness for concrete protection course of ducts inside and outside curve plane can be calculated as

A. Within the curve plane

Chapter 12 ▸ Design and Calculation of Prestressed Concrete Flexural Members

$$c_{in} \geq \frac{P_d}{0.266r \sqrt{f'_{cu}}} - \frac{d_s}{2} \quad (12\text{-}94)$$

Where, c_{in} is the smallest thickness for concrete protection course within the curve plane; d_s is the peripheral diameter of ducts (mm); P_d is the tensioning design value of prestressed steel wire (N), the tensioning should be timed the factor of 1.2 after the friction at the anchorage loop, the steel bar shrinkage and the friction loss of the ducts in the calculated section are deducted; r is the radium of ducts and the calculation shall follow the **CHBC** Article 9.3.15; f'_{cu} is the cube compression strength (MPa) of concrete with 150mm side length.

When the calculated value of thickness for concrete protection course by Equation (12-94) is beyond that of the linear ducts protecting thickness specified above, the minimum protective thickness of the linear ducts should be applied. However the stirrups should be settled within the bend plane in the curve section of ducts. The area of cross section of a single stirrup is expressed as

$$A_{sv1} \geq \frac{P_d s_v}{2 r f_{sv}} \quad (12\text{-}95)$$

Where, A_{sv1} is the area of cross section of a single stirrup (mm^2); s_v is the spacing of the stirrups (mm), and f_{sv} is designed value for tension strength of stirrup (MPa).

B. Outside the curve plane

$$c_{out} \geq \frac{P_d}{0.266 \pi r \sqrt{f'_{cu}}} - \frac{d_s}{2} \quad (12\text{-}96)$$

Where, c_{out} is the smallest thickness for concrete protection course of ducts outside the curve plane.

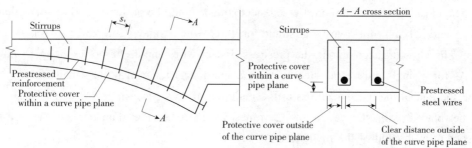

Figure 12-15 Schematic diagram of the curve duct protective cover of prestressed steel

If the calculated values of c_{out} is less than that of the straight ducts under each environment, the thickness for concrete protection course of straight ducts under corresponding environment should be applied.

The layout of the post-tensioning ducts should follow these specifications below:

a. The horizontal clear distance among the straight ducts should not less than 40mm and 0.6 times the diameter of ducts. Two ducts could be overlapped vertically for metal or plastic corrugated ducts and iron duct.

b. The clear distance c_{in} is defined the smallest distance among the neighboring ducts within the plane of curve shaped prestressed steel ducts (shown in Figure 12-16), and it can be calculated as Equation (12-94). Where, P_d and r are defined separately the tensioning design value and radium of one prestressed steel with a larger curve between two neighboring ducts. The smallest clear distance of the straight ducts should be applied when the calculated value by the Equation (12-94) is less than the clear distance of the relevant linear duct.

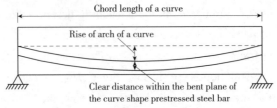

Figure 12-16   The clear distance within the bent plane of the curve shaped prestressed steel

The clear distance c_{out} is defined the smallest distance among the neighboring ducts outside the plane of curve shaped prestressed steel ducts, and it is expressed as Equation (12-96).

c. The cross section area of the inner diameter should not be less than two times area of the prestressed steel.

d. The preset duct should be arched in the meantime when pre-camber settled by calculation.

②Pre-tensioned members

a. The steel strand, the spiral ribbed steel wires or the indented steel wires should be used for the prestressed concrete member post-tensioned. When the plain steel wire is used for prestressed reinforcement, proper measures should be taken to ensure that the steel bar in concrete can be anchored in a reliable way so as to prevent the shortage of adhesion between steel wires and concrete, because the insufficient adhesion will cause the steel wire to loose and result in loss of prestressing force.

b. The clear distance of the prestressed steel strand should be no less than 1.5 times that of its diameter, and 20 mm for two-wire strand or three-wire strand, and 25 mm for seven-wire tendons. The clear distance among the prestressed steel wires should be no less than 15 mm.

c. As for a single prestressed steel wire, the spiral steel rebar of no less than 150 mm should be set at the end of the wire; while as to the multiple prestressed rebar, three or five steel meshes should be placed with the range of 10 times as large as the diameter of the prestressed reinforcement at the end of the member.

d. The minimum concrete cover thickness of common and linear steel bar (the distance from the peripheral edge of steel bars or pipes to the concrete surface) should be no less than the nominal diameter of steel wires, and in compliance with the requirement of appendix Table 1-8.

(8) Arrangement of the non-prestressed reinforcement

①Stirrup bars

The **CHBC** specifies that the arrangement of stirrup bars should be in accordance with the following:

a. The stirrup bars in diameter of no less than 10 mm and 12 mm should be arranged for the T-shaped, the I-shaped cross section beams of prestressed concrete and within the composite panel of the box girder beam, and the ribbed rebar should be used with the spacing no larger than 200 mm; the closed stirrup bars should be adopted within the range of no less than 1 times the beam height from the center the support platform with the spacing of no larger than 120 mm.

b. In the "horse's hoof" part which is under the cross section of the T-shaped and the I-shaped beams, the closed stirrup bars in diameter of no less than 8 mm should be set up and the spacing should be no larger than 200 mm. In addition, the positioned rebar in diameter of no less than 12 mm should be arranged within the "horse's hoof" area.

②Longitudinal and horizontal secondary steel bars

There will be the flange at the edge and the "horse's hoof" at the lower edge of a cross section of the T-shaped prestressed concrete beam. Their horizontal sizes of a beam shall be larger than the thickness of the composite panel. As to the prestressed concrete beam, the steel meshes in smaller diameter should be used for this kind of reinforcement. The rebar should be placed closely to the stirrup bars and arranged on both sides of the composite panel so as to enhance the adhesion with concrete and to reduce the spacing of cracks and width.

③Steel bars strengthened partially

As to the part where a certain part will undergo larger compression, the strengthened steel bar should be placed, such as the closed stirrup bars in the "horse's hoof" and that in the anchorage zone at the edge of a beam. Besides, the steel mesh should be arranged at the base support of a beam.

④Installation and positioning of steel wires

The erection of steel is used for the support of stirrup bars, and it is usually round the steel bar in diameter ranging from 12 mm to 20 mm; the steel bar positioned refers to those for a hole puncher of fixing the reserved hole, usually the grid-shaped.

(9) Protection of anchorage

For the anchorage device embedded into the body of a beam, after the pre-stressing is finished, the connection of the structural reinforcement to the body of a beam in the surrounding should be arranged, and then the concrete should be cast to cover the anchorage. The concrete strength grade for closing anchorage should be no less than 80% of the strength grade of the concrete of the member and no lower than C30.

12.9 Example for calculation of the hollow-core deck of prestressed concrete

1) Materials of design

(1) Span: Standard span $l_k = 13.00$m, and calculated span $l = 12.6$m.

(2) Deck width: $2.5m + 4 \times 3.75m + 2.5m = 20.0m$.

(3) Designed load: Highway class Ⅱ load; pedestrain load: 3.0kN/m².

(4) Materials.

Prestressed steel bars:

The 1×7 steel strand is utilized, the normal diameter is 2.7mm and the nominal area is 98.7mm², with $f_{pk} = 1860$MPa, $f_{pd} = 1260$MPa, and $E_p = 1.95 \times 10^5$MPa.

Steel rebar:

HRB400 is used, $f_{sk} = 400\text{MPa}$, $f_{sd} = 330\text{MPa}$. HPB 300 is used, $f_{sk} = 300\text{MPa}$ and $f_{sd} = 250\text{MPa}$.

Concrete:

The C50 concrete is used for the hollow-core deck with $f_{ck} = 32.4\text{MPa}$, $f_{cd} = 22.4\text{MPa}$, $f_{tk} = 2.65\text{MPa}$ and $f_{tk} = 1.83\text{MPa}$; The C30 fine aggregate concrete is used for hinged joints; the C30 asphalt concrete is used for bridge decking with the thickness of 12cm; the C20 concrete is used for railing and sidewalks.

(5) Design requirements: According to the **CHBC**, the design will be in line with that for prestressed concrete members of category A. The beam is designed as category A.

(6) Construction method: The pre-tensioned method will be carried out.

2) Sizes of hollow-core slabs

The 20 pieces precast hollow-core-core slabs of the C40 prestressed concrete will be made for decking the full width of a bridge, 99cm in width, 62cm in height and 12.96m full length for each deck. The full deck length of the cross-section is shown in Figure 12-17. The cross-section and the structural size of each hollow-core-core deck are shown in Figure 12-18.

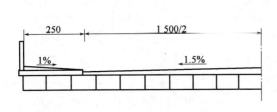
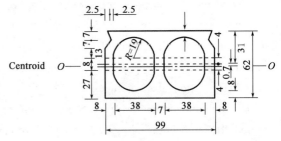

Figure 12-17 Cross section of a bridge (unit: cm)

Figure 12-18 Cross-section and sizes of a hollow-core slab (unit: cm)

3) Calculation of geometrical characteristics of the gross cross section of the hollow-core deck

(1) Area A of a gross cross-section

$$A = 99 \times 62 - 2 \times 38 \times 8 - 4 \times \frac{\pi \times 19^2}{2} - 2 \times \left(\frac{1}{2} \times 7 \times 2.5 + 7 \times 2.5 + \frac{1}{2} \times 7 \times 5\right) = 3\,174.3\text{cm}^2$$

(2) Position of the center of gravity of a gross cross-section

The static moment from 1/2 of the height of a slab to a gross cross section is as follows

$$S = 2 \times \left[\frac{1}{2} \times 2.5 \times 7 \times \left(24 + \frac{7}{3}\right) + 7 \times 2.5 \times \left(24 + \frac{7}{2}\right) + \frac{1}{2} \times 7 \times 5 \times \left(24 - \frac{7}{3}\right)\right]$$

$$= 2\,181.7\text{cm}^3$$

The areas of a hinged joint will be calculated as follows

$$A_{\text{hinged joint}} = 2 \times \left(\frac{1}{2} \times 2.5 \times 7 + 2.5 \times 7 + \frac{1}{2} \times 5 \times 7\right) = 87.5\text{cm}^2$$

The distance from the center of gravity of a gross cross-section to 1/2 of the height of a slab will be computed as the following

$$d = \frac{S}{A} = \frac{2\,181.7}{3\,174.3} = 0.687 \text{cm} \approx 0.7 \text{cm} = 7\text{mm (moving downwards)}$$

The distance from the center of a hinged joint to 1/2 of the height of the deck will be the following

$$d_{\text{hinged joint}} = \frac{2\,181.7}{87.5} = 24.9\text{cm}$$

(3) Distance from the gross section of a hollow-core deck to the moment of inertia of the center of gravity

From Figure 12-19, it's assumed that the area A of the semi circle of each hole excavated is computed as follows

$$A = \frac{1}{8}\pi d^2 = \frac{1}{8}\pi \times 38^2 = 567.1 \text{cm}^2$$

The gravity axis of a semi circle is calculated as follows

$$y = \frac{4d}{6\pi} = \frac{4 \times 38}{6 \times \pi} = 8.06 \text{cm} = 80.6 \text{mm}$$

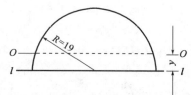

The moment of inertia I itself from the O-O gravity axis to the semi circle is computed in the following way

Figure 12-19 a semi circle excavated (unit: cm)

$$I' = 0.006\,86 d^4 = 0.006\,86 \times 38^4 = 14\,304 \text{cm}^4$$

The moment of inertia I from the gross cross section of a hollow-core slab to the gravity axis is as follows

$$I = \frac{99 \times 62^3}{12} + 99 \times 62 \times 0.7^2 - 2 \times \left(\frac{38 \times 8^3}{12} + 38 \times 8 \times 0.7^2\right) - 4 \times 143\,04 - 2 \times 567.1 \times$$

$$[(8.06 + 4 + 0.7)^2 + (8.06 + 4 - 0.7)^2] - 87.5 \times (24.9 + 0.7)^2 = 1\,520\,077.25 \text{cm}^4$$

$$= 1.520\,1 \times 10^{10} \text{mm}^4$$

(The moment of inertia from the hinged joint to the gravity axis itself is ignored.)

The torsion stiffness of the cross section of hollow-core decks can be simplified into a cross section of a single box for approximate computation as shown in Figure 12-20.

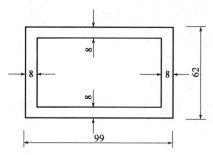

Figure 12-20 Simplified diagram of the cross section of a hollow-core slab for I_T computation (unit: cm)

$$I_T = \frac{4b^2h^2}{\frac{2h}{t_1} + \frac{2b}{t_2}} = \frac{4 \times (99-8)^2 \times (62-8)^2}{\frac{2 \times (62-8)}{8} + \frac{2 \times (99-8)}{8}}$$

$$= 2.664\,5 \times 10^6 \text{cm}^4$$

$$= 2.664\,5 \times 10^{10} \text{mm}^4$$

4) Combination of effects of actions

The basic combination for the design of the critical condition of bearing capacity is the following

$$\gamma_0 S_{ud} = \gamma_0(1.2 S_{Gk} + 1.4 S_{Q1k} + 0.8 \times 1.4 S_{Qjk})$$

Where γ_0 is the important coefficient of

structures, $\gamma_0 = 0.9$ for this bridge design; S_{ud} is the combined design of the effect; S_{Gk} is the standard value of the effect of a permanent action; S_{Qlk} is the standard value of the vehicular action (including the vehicle impact); S_{Qjk} is the standard value of the crowd load.

As far as the design of the normal service of the limit state of bearing capacity is concerned, the design should be in line with different requirement and two kinds of combination utilized as follows:

(1) The calculation for the frequent of actions

$$S_{sd} = S_{Gk} + 0.7 \times S'_{Qlk} + 1.0 \times S_{Qjk}$$

Where S_{sd} is the designed value of the frequent of actions; S_{Gk} is the designed value of the quasi-permanent combination of actions; S'_{Qlk} is the standard value of the vehicle effect, regardless of the impact factor; S_{Qjk} is the standard value of the effect of crowd load.

(2) The calculation for the long term effect of an action

$$S_{ld} = S_{Gk} + 0.4 \times S'_{Qlk} + 0.4 \times S_{Qjk}$$

The **CHBC** specifies that the standard combination of the effect of an action should be used for the cross section stress computation at the elastic stage for structural members. The expression of the combination of effects should be as follows

$$S = S_{Gk} + S_{Qlk} + S_{Qjk}$$

Where S is the design for the standard value of the effect combination; S_{Gk}, S_{Qlk}, S_{Qjk} are the standard value of the permanent effect of an action of the vehicle effect and the pedestrain load effect, respectively.

According to the effect of an action computed and various combinations of expressions of the **CHBC**, each combination of effects can be achieved. All kinds of computation have been collected as shown in Table 12-4.

Calculation of the combined effect of an action of a hollow-core deck   Table 12-4

Number	Kinds of load			$M(kN \cdot m)$		$V(kN)$		
				Mid-span	1/4	Mid-span	1/4	fulcrum
Standard of the effect of an action	Effects of a permanent action		g_I	157.49	118.12	0	25.00	50.00
			g_{II}	65.17	48.88	0	10.34	20.69
			$g = g_I + g_{II}$ (S_{Gk})	222.66	167.00	0	35.34	70.69
	Effects of a variable action	Lane load	Regardless of impact S'_{Qlk}	131.32	98.41	21.52	34.16	108.36
			$\times (1+\mu) S_{Qlk}$	172.69	129.42	28.31	44.92	142.51
			Pedestrain load S_{Qjk}	13.40	10.05	1.06	2.39	3.19
Critical condition of bearing capacity	Basic combination	1.2 S_{Gk}	(1)	267.19	200.40	0	42.41	84.83
		1.4 S_{Qlk}	(2)	241.77	181.19	39.63	62.89	199.51
		0.8×1.4 S_{Qjk}	(3)	15.01	11.26	1.19	2.68	3.57
		$S_{ud} = (1)+(2)+(3)$		523.97	392.85	40.82	107.98	287.91

Continued

Number	Kinds of load			M(kN·m)		V(kN)		
				Mid-span	1/4	Mid-span	1/4	fulcrum
Critical condition of normal service	Combination of the short term effect of an action S_{sd}	S_{Gk}	(4)	222.66	167.00	0	35.34	70.69
		$0.7\,S'_{Q1k}$	(5)	91.92	68.89	15.06	23.91	75.85
		S_{Qjk}	(6)	13.40	10.05	1.06	2.39	3.19
		$S_{sd} = (4)+(5)+(6)$		327.98	245.94	16.12	61.64	149.73
Critical condition of normal service	Combination of the long term effect of an action S_{ld}	S_{Gk}	(7)	222.66	167.00	0	35.34	70.69
		$0.4\,S'_{Q1k}$	(8)	52.53	39.36	8.61	13.66	43.34
		$0.4\,S_{Qjk}$	(9)	5.36	4.02	0.42	0.96	1.28
		$S_{ld} = (7)+(8)+(9)$		280.55	210.38	9.03	49.96	115.31
Calculation of the stress of the cross section at the elastic stage	Combination of the effect of the standard value S	S_{Gk}	(10)	222.66	167.00	0	35.34	70.69
		S_{Q1k}	(11)	172.69	129.42	28.31	44.92	142.51
		S_{Qjk}	(12)	13.40	10.05	1.06	2.39	3.19
		$S = (10)+(11)+(12)$		408.75	306.47	29.37	82.65	216.39

5) Quantity evaluation and arrangement of the prestressed reinforcement

(1) Quantity evaluation of prestressed steel bars

The prestressed concrete hollow-core deck pre-tensioned will be used. The design should meet the requirement for control conditions specified in the **CHBC** for different design conditions, such as bearing capacity, cracking resistance, crack width, deformation and stress. The quantity of prestressed steel bars should be determined first of all according to the cracking resistance of the normal cross section or the width of cracks in critical conditions of normal service of the structure under normal conditions when the design of prestressed concrete bridges are carried out, and the quantity can be decided in line with the requirement of the critical condition of the bearing capacity of members. What has been discussed in this example is the design of the partially prestressed members of category A. First of all, the effective prestressing N_{pe} should be determined according to the cracking resistance of the normal cross-section.

In accordance with article 6.3.1 of the **CHBC**, the cracking resistance of the normal cross-section of the prestressed concrete members of category A should be the normal tensile stress for controlling concrete. The following requirement should be met.

Under the combination of the short term effect of an action, the requirement of $\sigma_{st} - \sigma_{pc} \leq 0.70 f_{tk}$ should be satisfied.

Where σ_{st} is the normal tensile stress of marginal concrete for the checking of the cracking resistance of members under the function of the combination M_{sd} of the short term effect of an action; σ_{pc} is the effective prestressing of the marginal concrete for checking of the cracking resistance of members.

When the preliminary design is carried out, the approximate computation of σ_{st} and σ_{pc} can be done according to the following formula

$$\sigma_{st} = \frac{M_{sd}}{W}$$

$$\sigma_{pc} = \frac{N_{pc}}{A} + \frac{N_{pc}e_p}{W}$$

Where A and W are the resistant moments of elasticity for the areas of the gross cross section of members and the tensioned edges of the gross cross section respectively; e_p is the eccentric moment from the center of gravity of the prestressed steel bars to the gravity axis of the gross cross section. As to $e_p = y - a_p$, first of all, a_p can be assumed.

When $\sigma_{st} - \sigma_{pc} \leq 0.70 f_{tk}$, the effective prestressing required by the normal cross section of the partially prestressed member of category A shall follows

$$N_{pc} = \frac{\frac{M_{sd}}{W} - 0.70 f_{tk}}{\frac{1}{A} + \frac{e_p}{W}}$$

Where f_{tk} is the standard value of the tensile strength of concrete.

The C50 concrete is used for the prestressed hollow-core slab with $f_{tk} = 2.65$, $M_{sd} = 327.98$ kN·m $= 327.98 \times 10^6$ N·mm to be obtained from Table 12-4. The area conversed for the gross cross section of a hollow-core slab is $A = 3\,174.3 \text{cm}^2 = 3\,174.3 \times 10^2 \text{mm}^2$.

$$W = \frac{I}{y_{lower}} = \frac{1\,520.1 \times 10^3}{31 - 0.70} = 50.17 \times 10^3 \text{cm}^3 = 50.17 \times 10^6 \text{mm}^3$$

Assuming $a_p = 4\text{cm}$, $e_p = y_{lower} - a_p = 31 - 0.7 - 3 = 26.3\text{cm} = 263\text{mm}$.

The computation is as follows

$$N_{pe} = \frac{\frac{327.98 \times 10^6}{50.17 \times 10^6} - 0.7 \times 2.65}{\frac{1}{3\,174.3 \times 10^2} + \frac{263}{50.17 \times 10^6}} = 557\,642.2 \text{ N}$$

The areas A_p of the cross section of the prestressed steel bars is

$$A_p = \frac{N_{pe}}{\sigma_{con} - \sum \sigma_l}$$

Where σ_{con} is the controlled tensile stress of the prestressed reinforcement bars; $\sum \sigma_l$ is the total loss of prestressing force estimated as 20% of the controlled tensile stress.

The seven-wire strands (1×7) is used as prestressed steel bars in this example. Its diameter is 12.7mm and the nominal area of a cross section 98.7mm^2. $f_{pk} = 1\,860$MPa, $f_{pd} = 1\,260$MPa and $E_p = 1.95 \times 10^5$MPa.

According to $\sigma_{con} \leq 0.75 f_{pk}$ and $\sigma_{con} = 0.70 f_{pk}$ specified in the **CHBC**, if 20% of the controlled tensile stress is approximately estimated as the total prestressing loss the total prestressed loss, the following can be achieved

$$A_p = \frac{N_{pe}}{\sigma_{con} - \sum \sigma_l} = \frac{N_{pe}}{\sigma_{con} - 0.2\sigma_{con}} = \frac{557\,642.2}{0.8 \times 0.70 \times 1\,860} = 535.37\text{mm}^2$$

If the seven-wire strand (1×7) is used, i.e. $\phi^s 12.7$, the nominal area of individual tension

Chapter 12 · Design and Calculation of Prestressed Concrete Flexural Members

will be 98.7mm^2, and $A_p = 7 \times 98.7 = 690.9\text{mm}^2$ will meet the requirement.

(2) Arrangement of prestressed steel wires

The seven-wire steel strand (1×7) for a prestressed hollow-core slab will be placed at the lower edge of the slab, $a_p = 40\text{mm}$. The reinforcement will be arranged in a linear way along the length of the span of a hollow-core deck, i.e. $a_p = 40\text{mm}$ will remain unchanged if the arrangement along the length of the span is designed as shown in Figure 12-21. The steel bar arrangement should meet the structural requirement of the **CHBC**. The clear distance of the steel strand should be no less than 25 mm, and the length of the end of spiral steel bars no less than 150 mm.

(3) Evaluation and the layout of common steel bars

When the amount of the prestressed steel bar is decided, the quantity of the traditional rebar can be obtained from the requirement of the critical condition for the bearing capacity of the normal cross section. The arrangement of the prestressed reinforcement bars in the compressive zone will be out of consideration temporarily and so is the case for the influence from the traditional reinforcement. The cross section of a hollow-core slab can be converted into and considered to be the equivalent I-shaped cross section.

From the following

$$b_k h_k = \frac{\pi}{4} \times 38^2 + 8 \times 38 = 1\,438.115\text{cm}^2$$

It can be obtained

$$b_k = \frac{1\,438.115\text{cm}^2}{h_k}$$

$$\frac{1}{12} b_k h_k^3 = \frac{38 \times 8^3}{12} + 2 \times 0.006\,86 \times 38^4 + 2 \times 567.1 \times (8.06 + 4)^2 = 195\,191.53\text{cm}^4$$

And if $b_k = \dfrac{1\,438.114\text{cm}^2}{h_k}$ is put into $\dfrac{1}{12} b_k h_k^3 = 195\,191.53\text{cm}^4$, the following can be achieved: $h_k = 40.4\text{cm}$ and $b_k = \dfrac{1\,438.114\text{cm}^2}{40.4} = 35.6\text{cm}$.

The thickness h'_f of the upper flange slab of the equivalent I-shaped cross section will be as follows

$$h'_f = y_{\text{upper}} - \frac{h_k}{2} = 31 - \frac{40.4}{2} = 10.8\text{cm}$$

The thickness h_f of the lower flange slab of the equivalent I-shaped cross section will be computed

$$h_f = y_{\text{lower}} - \frac{h_k}{2} = 31 - \frac{40.4}{2} = 10.8\text{cm}$$

The thickness of the ribbed slab of the equivalent I-shaped cross section will be calculated as the following

$$b = b'_f - 2b_k = 99 - 2 \times 35.6 = 27.8\text{cm}$$

The size of the equivalent I-shaped cross section is shown in Figure 12-22.

When the traditional reinforcement is evaluated, first it can be assumed that $x \leqslant h'_f$, the height x of the compressive zone will be achieved from the following formula, if $h_0 = h - a_{ps} = 62 - 4 = 58\text{cm} = 580\text{mm}$.

$$\gamma_0 M_{ud} \leqslant f_{cd} b'_f x \left(h_0 - \frac{x}{2} \right)$$

From the **CHBC**, $\gamma_0 = 0.9$, C50 and $f_{cd} = 22.4$ MPa can be obtained. From Table 12-4, $M_{ud} = 523.97\text{kN} \cdot \text{m} = 523.97 \times 10^6 \text{N} \cdot \text{mm}$, $b'_f = 990\text{mm}$ for the mid span. These figures will be computed with the above formula, and the following achieved

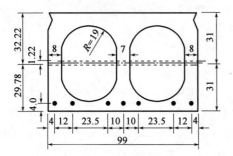 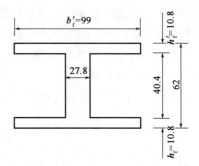

Figure 12-21 Layout of prestressed steel bars in a cross section of the mid span of a hollow-core slab (unit: cm)

Figure 12-22 Equivalent I-shaped cross section for the conversion of a hollow-core slab (unit: cm)

$$0.9 \times 523.97 \times 10^6 \leqslant 22.4 \times 990 \times x \times \left(580 - \frac{x}{2} \right)$$

When simplified, the following can be obtained

$$x^2 - 1160x + 42.53 \times 10^3 \leqslant 0$$

Then, the below result can be achieved

$$x = 37.8\text{mm} < h'_f = 108\text{mm}, \text{ and } x < \xi_b h_0 = 0.4 h_0 = 232\text{mm}$$

The above result demonstrates that the neutral axis is located within the flange panel, thus, the area A_s of the common reinforcement can be computed with the following formula

$$A_s = \frac{f_{cd} b'_f x - f_{pd} A_p}{f_s d} = \frac{22.4 \times 990 \times 37.8 - 1260 \times 690.9}{330} < 0$$

If HRB400 is used for common steel wires, then $f_{sd} = 330\text{MPa}$ and $E_s = 2 \times 10^5 \text{MPa}$. According to the **CHBC**, $A_s \geqslant 0.003 bh_0 = 0.003 \times 278 \times 580 = 483.72\text{mm}^2$.

If 5 ⌽ 12 is adopted for use as the common reinforcement, $A_s = 5 \times \frac{\pi (12)^2}{4} = 565.5\text{mm}^2 > 483.72\text{mm}^2$.

The reinforcement of 5 ⌽ 12 will be placed in one row below the lower flange of a hollow-core slab (the tensioned edge of a cross section), 40mm from the lower flange to the center of gravity of reinforcement, i.e. $a_s = 40\text{mm}$, in a linear way along the span length of a hollow-core slab.

6) Calculation of the geometrical property for the conversion of cross-sections

From the calculation discussed above, the geometrical property of the gross cross-section for a hollow-core slab has been computed. The area of a gross cross-section is $A = 317\,430\,\text{mm}^2$, the distance from the gravity axis of a gross cross-section to $\frac{1}{2}$ of the deck height is $d = 7\,\text{mm}$ (downwards). The moment of inertial from the gross cross-section to the gravity axis is $I = 15\,201 \times 10^6\,\text{mm}^4$.

(1) The area conversion of a cross-section A_0 is as follows

$$A_0 = A + (\alpha_{Ep} - 1)A_p + (\alpha_{Es} - 1)A_s$$

$$\alpha_{Ep} = \frac{E_p}{E_c} = \frac{1.95 \times 10^5}{3.45 \times 10^4} = 5.65, A_p = 690.9\,\text{mm}^2$$

$$\alpha_{Es} = \frac{E_s}{E_c} = \frac{2 \times 10^5}{3.25 \times 10^4} = 5.79, A_s = 565.5\,\text{mm}^2$$

$$A = 317\,430\,\text{mm}^2$$

A_0 is calculated as follows

$$A_0 = 317\,430 + (5.65 - 1) \times 690.9 + (5.79 - 1) \times 565.5 = 323\,797\,\text{mm}^2$$

(2) Position for the center of gravity for cross section conversion

The static moment from the cross-section to the center of gravity of a gross cross-section for all reinforcement conversion will be the following

$$\begin{aligned}S_{01} &= (\alpha_{Ep} - 1)A_p \times (310 - 7 - 40) + (\alpha_{Es} - 1)A_s \times (310 - 7 - 40)\\ &= (5.65 - 1) \times 690.9 \times 263 + (5.79 - 1) \times 565.5 \times 263\\ &= 1\,557\,336.1\,\text{mm}^3\end{aligned}$$

The distance between the center of gravity of a conversion cross-section and the center of a gross cross-section of a hollow-core slab is

$$d_{01} = \frac{S_{01}}{A_0} = 4.8\,\text{mm}(\text{move down})$$

The distance from the center of gravity of a conversion cross-section to the lower flange of a cross-section of a hollow-core slab is

$$y_{01l} = 310 - 7 - 4.8 = 298.2\,\text{mm}$$

The distance from the center of gravity of a conversion cross-section to the center of prestressed reinforcement is

$$y_{01u} = 310 + 7 + 4.8 = 321.8\,\text{mm}$$

The distance from the center of gravity of a conversion cross section to the center of common reinforcement is

$$e_{01p} = 298.2 - 40 = 258.2\,\text{mm}$$

(3) Moment of inertial I_0 of conversion cross-sections

$$\begin{aligned}I_0 &= I + Ad_{01}^2 + (\alpha_{Ep} - 1)A_p e_{01p}^2 + (\alpha_{Es} - 1)A_s e_{01s}^2\\ &= 15\,201 \times 10^6 + 317\,430 \times 4.8^2 + (5.65 - 1) \times 690.9 \times 258.2^2 + (5.79 - 1) \times 565.5 \times 258.2^2\\ &= 1.5633 \times 10^{10}\,\text{mm}^4\end{aligned}$$

(4) Elastic resistance moment of conversion cross-sections

Lower flange: $W_{0ll} = \dfrac{I_0}{y_{0ll}} = \dfrac{1.5633 \times 10^{10}}{298.2} = 52.45 \times 10^6 \text{mm}^3$

Upper flange: $W_{0lu} = \dfrac{I_0}{y_{0lu}} = \dfrac{1.5633 \times 10^{10}}{321.8} = 48.57 \text{mm}^3$

7) Calculation of the critical condition of the load-carrying capacity for permanent load

(1) Calculation of the flexural capacity of the normal cross section of the mid-span cross-section

The structural size and reinforcement arrangement of the mid-span cross section are seen in Figure 12-21. The distance from the point of the joint action of the prestressed steel strand to the base of a cross-section is $a_p = 40$mm. The distance from the base of the cross-section of the common reinforcement is $a_s = 40$mm. The distance between the joint force of the prestressed and common reinforcement to the base of a cross-section is the following

$$a_{ps} = \dfrac{f_{sd}A_s a_s + f_{pd}A_p a_p}{f_{sd}A_s + f_{pd}A_p} = \dfrac{330 \times 565.5 \times 40 + 1260 \times 690.9 \times 40}{330 \times 565.5 + 1260 \times 690.9} = 40 \text{mm}$$

$$h_0 = h - a_{ps} = 620 - 40 = 580 \text{mm}$$

The calculation with the equivalent I-shaped cross-section to be converted can be seen in Figure 12-22. The thickness of the upper flange is $h'_f = 108$mm, the width of the upper flange $b'_f = 990$mm, and the rib width $b = 278$mm. First, the type of a cross-section will be determined in accordance by the following formula

$$f_{pd}A_p + f_{sd}A_s \leqslant f_{cd}b'_f h'_f$$

$$f_{pd}A_p + f_{sd}A_s = 1260 \times 690.9 + 330 \times 565.5 = 10\,577\,149\text{N}$$

$$\leqslant f_{cd}b'_f h_f = 22.4 \times 990 \times 108 = 2\,395\,008\text{N}$$

Therefore, it belongs to a T-shaped one of the kind of I. The flexural bearing capacity will be computed in line with the width of $b_f = 990$mm of a rectangular cross-section. The height x of the compressive zone of the concrete will be worked out from $\sum x = 0$.

From $\qquad f_{pd}A_p + f_{sd}A_s = f_{cd}b'_f x$

the following result is obtained

$$x = \dfrac{f_{pd}A_p + f_{sd}A_s}{f_{cd}b'_f} = \dfrac{1260 \times 690.9 + 330 \times 565.5}{22.4 \times 990} = 47.7 \text{mm}$$

$$< \xi_b h_0 = 0.4 \times 580 = 232 \text{mm}$$

$$< h'_f = 108 \text{mm}$$

The flexural bearing capacity M_{ud} of the cross-section of the mid span will be calculated with $x = 46.4$mm to be substituted in the following formula

$$M_{ud} = f_{cd}b'_f x \left(h_0 - \dfrac{x}{2} \right) = 22.4 \times 990 \times 46.4 \times \left(580 - \dfrac{46.4}{2} \right)$$

$$= 572.92 \times 10^6 \text{N} \cdot \text{mm}$$

$$= 572.92 \text{kN} \cdot \text{m} > \gamma_0 M_d = 0.9 \times 523.97 = 471.57 \text{kN} \cdot \text{m}$$

The results show that the flexural bearing capacity of the mid-span cross section satisfies the requirements.

(2) Calculation of shear load-carrying capacity of an oblique cross-section

①Checking of the upper and lower limits of the shear strength of a cross-section

The cross-section at $h/2$ from the fulcrum will be used for calculating the shear resistance capacity. The structural size and the reinforcement assembly will be seen in Figure 12-23. According to article 5.2.9 of the **CHBC**, the checking of the upper and lower limits of the shear strength of a cross-section should be carried out first.

$$\gamma_0 V_d \leqslant 0.51 \times 10^{-3} \sqrt{f_{cu,k}} bh_0 (\text{kN})$$

Where V_d is the composite design value of shearing for a certain part in a cross-section where it should be checked. From Table 12-4, the shearing at the fulcrum and that of the cross-section of the mid span can be obtained. The shearing V_d of the cross-section at $h/2 = 310$ mm from the fulcrum will be achieved with the interpolation method

$$V_d = 287.91 - \frac{310 \times (287.91 - 40.82)}{6300} = 275.75 \text{kN}$$

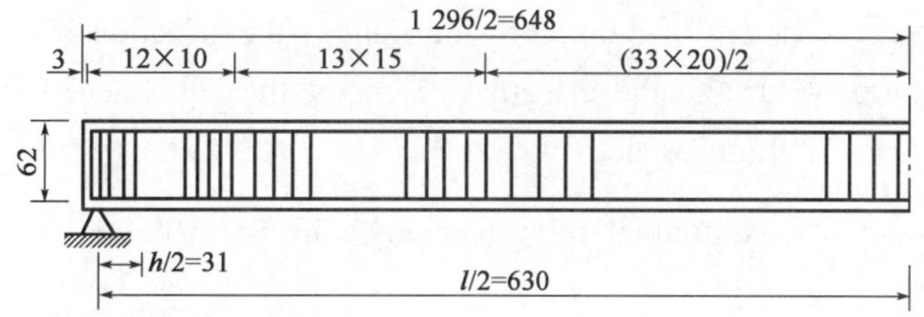

Figure 12-23 The stirrup arrangement along the span length of hollow-core slabs

h_0 is the effective height of a cross-section. Since the prestressed reinforcement and the traditional steel wire are all arranged in a linear way in this example, the effective height h_0 will be the same as the mid-span cross-section, i.e. $h_0 = 580$mm.

$f_{cu,k}$ is the compressive strength of the 150mm side length of a concrete cube and the C50 hollow-core slab as follows: 50MPa, $f_{td} = 1.83$MPa.

b is the width of the composite panel of an equivalent I-shaped cross-section with $b = 278$mm.

Therefore, the below calculations are carried out

$$\gamma_0 V_d = 0.9 \times 275.5 = 248.18 \text{kN}$$

$$\gamma_0 V_d \leqslant 0.51 \times 10^{-3} \sqrt{50} \times 278 \times 580 = 581.47 \text{kN}$$

The result shows that the size of a hollow-core cross-section meets the requirements.

In line with article 5.2.9 in the **CHBC**:

$1.25 \times 0.5 \times 10^{-3} \times \alpha_2 f_{td} bh_0 = 1.25 \times 0.5 \times 10^{-3} \times 1.0 \times 1.83 \times 278 \times 580 = 184.42 \text{kN}$

Where, $\alpha_2 = 1.25$, based on article 5.2.9 in the **CHBC**. The coefficient can be improved by multiplying 1.25 for flat slab flexural members.

Due to $\gamma_0 V_d = 0.9 \times 275.75 = 248.18 \text{kN} > 1.25 \times 0.5 \times 10^{-3} \times \alpha_2 f_{td} bh_0 = 184.42 \text{kN}$, with the contrast of the combined design value of the controlled shearing of various cross-sections along the length of a span in Table 12-4, the shear stirrup bars should be placed in some zones from 1/4 to the pivot point according to the computation requirement. The stirrup bars should be arranged in accordance with the structural demands in other sections.

For the convenience of structure design and construction, the bent reinforcements are not considered in this example for the prestressed concrete hollow-core slab. The shear computed will be all born by concrete and stirrup bars. The shear capacity of an inclined cross-section should be computed as follows

$$\gamma_0 V_d \leqslant V_{cs}$$

$$V_{cs} = \alpha_1 \alpha_2 \alpha_3 \times 0.45 \times 10^{-3} bh_0 \sqrt{(2+0.6P)} \sqrt{f_{cu,k}} \rho_{sv} f_{sv}$$

In the formula, every coefficient will be decided in accordance with article 5.2.7 in the **CHBC**.

α_1 is the influential coefficients of bearing opposite sign moment; for simply supported beams, $\alpha_1 = 1.0$.

α_2 is the coefficient for improving the prestressing force. As the partially prestressed member of Category A has been chosen in this example, for the safety sake, the value chosen is $\alpha_2 = 1.0$.

α_3 is the influence coefficient of a compression flange; the selection is $\alpha_3 = 1.1$.

b and h_0 are the rib width and the effective height of the equivalent I-shaped cross section with $b = 278$mm and $h_0 = 580$mm respectively.

P is the rebar ratio of longitudinal reinforcement with $P = 100\rho = 100 \times \dfrac{690.9 + 565.5}{278 \times 580} = 0.78$.

ρ_{sv} is the assembly ratio of stirrup bars with $\rho_{sv} = \dfrac{A_{sv}}{b s_v}$; As $\phi 10$ double-leg stirrup bars are used, $A_{sv} = 2 \times \dfrac{\pi \times 10^2}{4} = 157.08 \text{mm}^2$, and the spacing computation for stirrup bars will be the following

$$s_v = \dfrac{\alpha_1^2 \alpha_2^2 \alpha_3^2 \times 0.2 \times 10^{-6}(2+0.6P)\sqrt{f_{cu,k}} f_{sv} A_{sv} bh_0^2}{(\gamma_0 V_d)^2}$$

$$= \dfrac{1.0^2 \times 1.0^2 \times 1.1^2 \times 0.2 \times 10^{-6}(2+0.6 \times 0.78)\sqrt{50} \times 330 \times 157.8 \times 278 \times 580^2}{(0.9 \times 275.75)^2}$$

$$= 336.5 \text{mm}$$

$f_{cu,k} = 50$MPa.

If HRB400 is used as the stirrup bar, $f_{sv} = 330$MPa.

When the spacing of stirrup bars is $s_v = 150$mm, and in accordance with the requirement of the **CHBC**, the spacing of stirrup bars will be 100mm within the range of being twice as high as the height of a beam from the center of a fulcrum towards the direction of the mid span.

The ratio of stirrup bars will be

$$\rho_{sv} = \dfrac{A_{sv}}{b_{sv}} = \dfrac{157.08}{278 \times 150} = 0.0038 = 0.38\% > \rho_{svmin} = 0.12\%$$

As for as the composite design value $\gamma_0 V_d \leqslant 1.25 \times 0.5 \times 10^{-3} \times \alpha_2 f_{td} bh_0 = 184.42$kN for certain sections of a beam is concerned, the arrangement of reinforcement can be carried out only according to the structural requirement. If the $\phi 10$ double-leg stirrup bars are used, the ratio ρ_{sv}

Chapter 12 · Design and Calculation of Prestressed Concrete Flexural Members

will be ρ_{svmin}. The spacing for structural stirrup bars obtained from this will be

$$s'_v = \frac{A_{sv}}{b\rho_{\text{svmin}}} = \frac{157.08}{278 \times 0.0012} = 470.9 \text{mm}.$$

The spacing will be assumed as $s'_v = 200$mm.

After comparison and comprehensive consideration, the stirrup arrangement along the span length of hollow-core slabs is shown in Figure 12-23.

②Calculation of the shear load-carrying capacity of an inclined cross section

According to Figure 12-23, the following three locations will be chosen for the calculation of the shear resistance capacity of an inclined cross section of a hollow-core slab.

a. The cross section is $h/2 = 310$mm away from the center of a support, $x = 5990$mm.

b. The cross section is $x = 3300$mm away from the center of the mid span (the location where the spacing of stirrup bars changes).

c. The cross section is $x = 3300 + 13 \times 150 = 5250$mm away from the mid span (the location where the spacing of stirrup bars changes).

The shear combination value of a cross section will be calculated by the interpolation method and the design value of the mid span and the fulcrum in Table 12-4. The result are listed in Table 2-5.

Computation of the shear composite design values of various cross sections Table 12-5

Location x(mm) of cross sections	Fulcrum $x = 6300$	$x = 5990$	$x = 5250$	$x = 3300$	Mid-span $x = 0$
Shear composite design value V_d(kN)	287.91	275.75	246.73	170.25	40.82

a. The cross section is $h/2 = 310$mm away from the center of the support, i.e. $x = 5990$mm

Since the prestressed reinforcement and the common steel bars of hollow-core slabs are assembled in a linear way, the effective height of a cross section will be approximately the same as that for the mid span, i.e. $h_0 = 580$mm. The rib width of the equivalent I-shaped cross section is $b = 278$mm. With the rebar bent not taken into consideration, therefore, the shear strength of an inclined cross section will be computed as follows

$$V_{cs} = \alpha_1 \alpha_2 \alpha_3 \times 0.45 \times 10^{-3} bh_0 \sqrt{(2 + 0.6P)\sqrt{f_{cu,k}} \rho_{sv} f_{sv}}$$

Where, $\alpha_1 = 1.0, \alpha_2 = 1.0, \alpha_3 = 1.1, b = 278$mm, $h_0 = 580$mm,

$$P = 100\rho = 100 \times \frac{690.9 + 565.5}{278 \times 580} = 0.78$$

Assuming $s_v = 100$mm, $2 \oplus 10$ and $A_{sv} = 157.08$mm², the following can be obtained

$$\rho_{sv} = \frac{A_{sv}}{bs_v} = \frac{157.08}{278 \times 100} = 0.0565 = 0.565\% > \rho_{\text{svmin}} = 0.12\%$$

$$f_{cu,k} = 50 \text{MPa}, f_{sv} = 330 \text{MPa}$$

The following can be achieved after the substitution

$$V_{cs} = 1.0 \times 1.0 \times 1.1 \times 0.45 \times 10^{-3} \times 278 \times 580 \times \sqrt{(2 + 0.6 \times 0.78) \times \sqrt{50} \times 0.00565 \times 330}$$
$$= 455.28 \text{kN}$$

$$\gamma_0 V_d = 0.9 \times 275.75 = 248.18\text{kN} < V_{cs} = 455.28\text{kN}$$

The result shows that the shear strength can meet the requirement.

b. The cross section is $x = 3\,300\text{mm}$ away from the mid-span

If the spacing of stirrups is $s_v = 200\text{mm}$ and $V_d = 170.25\text{kN}$, the following can be calculated

$$\rho_{sv} = \frac{A_{sv}}{bs_v} = \frac{157.08}{278 \times 200} = 0.283\% > \rho_{svmin} = 0.11\%$$

The shear strength of an inclined cross section will be worked out as follows

$$V_{cs} = 1.0 \times 1.0 \times 1.1 \times 0.45 \times 10^{-3} \times 278 \times 580 \times \sqrt{(2 + 0.6 \times 0.78) \times \sqrt{50} \times 0.0028 \times 330}$$
$$= 322.21\text{kN}$$

$$\gamma_0 V_d = 0.9 \times 170.25 = 153.23\text{kN} < V_{cs} = 322.21\text{kN}$$

The result shows that the shear strength of a cross section can satisfy the demand.

c. The cross section is $x = 5\,250$ mm away from the mid-span

If the spacing of stirrups is $s_v = 150\text{mm}$ with $V_d = 246.73\text{kN}$, the following can be obtained

$$\rho_{sv} = \frac{A_{sv}}{bs_v} = \frac{157.08}{278 \times 150} = 0.00377 = 0.377\% > \rho_{svmin} = 0.11\%$$

The shear strength of an inclined cross section will be calculated as follows

$$V_{cs} = 1.0 \times 1.0 \times 1.1 \times 0.45 \times 10^{-3} \times 278 \times 580 \times \sqrt{(2 + 0.6 \times 0.78) \times \sqrt{50} \times 0.00377 \times 330}$$
$$= 371.90\text{kN}$$

$$\gamma_0 V_d = 0.9 \times 246.73 = 222.06\text{kN} < V_{cs} = 371.90\text{kN}$$

The calculation shows that the shear carrying capacity of an inclined cross section can meet the requirements.

8) Calculation of the loss of prestressing

The 1×7 steel strand of 12.7mm in diameter is used as prestressed rebar in this example. $E_p = 1.95 \times 10^5 \text{MPa}, f_{pk} = 1\,860\text{MPa}$, and the controlled stress is $\sigma_{con} = 0.7f_{pk} = 0.7 \times 1\,860 = 1\,302\text{MPa}$.

(1) Stress loss σ_{l2} from anchorage deformation and shrinkage

The length of the tension bench will be the effective length for prestressed steel strand. Assuming that the bench length is $L = 50\text{m}$, the steel strand will be tensioned at one end and the wedge anchorage utilized. When the top pressure appears and $\Delta l = 4\text{mm}$, the stress loss will be as follows

$$\sigma_{l2} = \frac{\sum \Delta l}{L} E_p = \frac{4}{50 \times 10^3} \times 1.95 \times 10^5 = 15.6\text{MPa}$$

(2) Temperature loss σ_{l3} due to heating curing

The heating curing will be used for the prestressed concrete hollow-core slab pretensioned to reduce the prestressing loss caused by temperature. The slab will be cured section by section. If the maximum temperature difference between the controlled prestressed steel strand and the bench is $\Delta t = t_2 - t_1 = 15°C$, the loss can be computed as follows

$$\sigma_{l3} = 2\Delta t = 2 \times 15 = 30 \text{MPa}$$

(3) The prestressed loss caused by the stress relaxation of the prestressed steel strand

$$\sigma_{l5} = \psi\xi\left(0.52\frac{\sigma_{pe}}{f_{pk}} - 0.26\right)\sigma_{pe}$$

Where ψ is the coefficient of tension. When the tension is carried out once, $\psi = 1.0$. ξ is the coefficient of the relaxation of the prestressed steel strand, and the low relaxation $\xi = 0.3$. f_{pk} is the standard values of the tensile strength of steel strand and $f_{pk} = 1\,860 \text{MPa}$. σ_{pe} is the stress of the reinforcement transferring forces and the anchoring according to the CHBC as far as the member pretensioned is concerned: $\sigma_{pe} = \sigma_{con} - \sigma_{l2} = 1\,302 - 15.6 = 1\,286.4$ MPa.

Thus, the result of σ_{l5} can be calculated as follows

$$\sigma_{l5} = 1.0 \times 0.3 \times \left(0.52 \times \frac{1\,286.4}{1\,860} - 0.26\right) \times 1\,286.4 = 38.45 \text{MPa}$$

(4) Prestressing loss σ_{l4} from concrete elastic shrinkage

As to the pretensioned units, $\sigma_{l4} = \alpha_{Ep}\sigma_{pe}$.

Where α_{Ep} is ratio of the elastic modulus of prestressed strand and that of concrete, $\alpha_{Ep} = \frac{1.95 \times 10^5}{3.45 \times 10^4} = 5.65$; σ_{pe} is the normal concrete stress (MPa) caused by the total prestressing of steel strand at the center of gravity of a cross section computed.

The value will be as follows

$$\sigma_{pe} = \frac{N_{p0}}{A_0} + \frac{N_{p0}e_{p0}}{I_0}y_0$$

$$N_{p0} = \sigma_{p0}A_p - \sigma_{l6}A_s$$

$$\sigma_{p0} = \sigma_{con} - \sigma'_l$$

Where σ'_l is The total prestressing loss when the prestressed reinforcement transfers force and is anchored; according to article 6.2.8 of the **CHBC**, the stress loss is $\sigma_l = \sigma_{l2} + \sigma_{l3} + 0.5\sigma_{l5}$.

The total stress loss will be calculated as

$$\sigma_{p0} = \sigma_{con} - (\sigma_{l2} + \sigma_{l3} + 0.5\sigma_{l5})$$
$$= 1\,302 - 15.6 - 30.0 - 0.5 \times 38.45 = 1\,237.18 \text{MPa}$$

$$N_{p0} = \sigma_{p0}A_p - \sigma_{l6}A_s = 1\,237.18 \times 690.9 - 0 = 854.77 \times 10^3 \text{N}$$

Since the area of the conversion cross section of the hollow-core slab computed above is $A_0 = 323351 \text{mm}^2$, and $I_0 = 1.5633 \times 10^{10} \text{mm}^4$, $e_{p0} = 258.2 \text{mm}$ and $y_0 = 257.8 \text{mm}$. The following can be calculated

$$\sigma_{pe} = \frac{854.77 \times 10^3}{323\,351.4} + \frac{854.77 \times 10^3 \times 258.2}{1.5633 \times 10^{10}} \times 258.2 = 6.27 \text{MPa}$$

$$\sigma_{l4} = \alpha_{Ep}\sigma_{pe} = 5.65 \times 6.27 = 35.42 \text{MPa}$$

(5) The stress loss σ_{l6} caused by concrete shrinkage and creeping

$$\sigma_{l6} = \frac{0.9[E_p\varepsilon_{cs}(t,t_0) + \alpha_{Ep}\sigma_{pc}\phi(t,t_0)]}{1 + 15\rho\rho_{ps}}$$

Where ρ is the ratio of reinforcement of the total longitudinal reinforcement in the tensile zone

for members
$$\rho = \frac{A_p + A_s}{A_0} = \frac{690.9 + 565.5}{323\,797} = 0.003\,88.$$

$$\rho_{ps} = 1 + \frac{e_{ps}^2}{i^2}$$

e_{ps} is the distance from the center of gravity of all longitudinal reinforcement of the cross section of members to the center of gravity of members, $e_{ps} = 297.8 - 40 = 257.8\,\text{mm}$; i is the radius of gyration of the cross section of members, $i^2 = \frac{I_0}{A_0} = \frac{1.5633 \times 10^{10}}{323\,351.4} = 48\,340.0\,\text{mm}^2$. σ_{pc} is the normal stress of concrete caused by prestressing force (the prestressing loss from the relevant stage deducted) and the dead weight of structures at the center of gravity of all longitudinal reinforcement of the cross section of members in compressive zones is computed as follows

$$\sigma_{pc} = \frac{N_{p0}}{A_0} + \frac{N_{p0}\lambda_{p0}}{I_0} y_0$$

N_{p0} is the prestressing force from the prestressed reinforcement with the force being transferred and rebar anchored

$$N_{p0} = \sigma_{p0}A_p - \sigma_{l6}A_s = [\sigma_{con} - (\sigma_{l2} + \sigma_{l3} + \sigma_{l4} + 0.5\sigma_{l5})]A_p - 0$$
$$= [1\,302 - (15.6 + 30.0 + 35.42 + 0.5 \times 38.45)] \times 690.9$$
$$= 830\,292.5\,\text{N}$$

$$\lambda_{p0} = \frac{\sigma_{p0}A_p y_p - \sigma_{l6}A_s y_s}{N_{p0}} = \frac{830\,292.5 \times 258.2}{830\,292.5}$$
$$= 258.2\,\text{mm}\,(y_p = y_s = 258.2\,\text{mm})$$

y_0 is the distance from the center of gravity of all longitudinal reinforcement to the center of gravity of cross section in compressive zones, and the value is $y_0 = \lambda_{p0} = 258.2\,\text{mm}$ computed above.

$\varepsilon_{cs}(t, t_0)$ is the age t_0 of the prestressed rebar transferring forces and being anchored, and the computed age will be the shrinking strain of concrete at a t time.

$\phi(t, t_0)$ is the loading age is t_0 as the coefficient of creeping at a t time.

$$\sigma_{pc} = \frac{N_{p0}}{A_0} + \frac{N_{p0}\lambda_{p0}}{I_0} y_0 = \frac{830\,292.5}{323\,351.4} + \frac{830\,292.5 \times 258.2}{1.5633 \times 10^{10}} \times 258.2 = 6.11\,\text{MPa}$$

$$\rho_{ps} = 1 + \frac{l_{ps}^2}{i^2} = 1 + \frac{258.2^2}{48\,340} = 2.379, E_p = 1.95 \times 10^5\,\text{MPa}, \alpha_{Ep} = 5.65$$

With consideration of the influence of dead load and due to the longer sustainable time for shrinkage and creeping, the permanent action has been adopted for all components. The total permanent action has been adopted. The moment M_{Gk} of the total permanent action of a cross section of the mid span of a hollow-core slab can be obtained from Table 12-4 as $M_{Gk} = 222.66\,\text{kN}\cdot\text{m}$. The tensile stress of the dead load at the center of gravity of all reinforcement will be the following.

A cross-section of a mid-span

$$\sigma_t = \frac{M_{Gk}}{I_0} y_0 = \frac{222.66 \times 10^6}{1.5633 \times 10^{10}} \times 258.2 = 3.67 \text{MPa}$$

1/4 of a cross-section

$$\sigma_t = \frac{167.00 \times 10^6}{1.5633 \times 10^{10}} \times 258.2 = 2.75 \text{MPa}$$

The cross-section of a fulcrum: $\sigma_t = 0$.

The compressive stress at the center of gravity of all longitudinal reinforcement is as follows

Mid-span cross-section: $\sigma_{pc} = 6.11 - 3.67 = 2.44 \text{MPa}$.

1/4 of a cross-section: $\sigma_{pc} = 6.11 - 2.75 = 3.36 \text{MPa}$.

Cross-section of a fulcrum: $\sigma_{pc} = 6.11 \text{MPa}$.

According to article 6.2.7 of the **CHBC**, σ_{pc} shouldn't be 0.5 times larger than the compressive strength f'_{cu} of the cubic concrete at the time of transferring forces and anchoring rebar. If the concrete reaches C30 when the force is transferred and the anchorage is installed, $f'_{cu} = 30 \text{MPa}$ and $0.5 f'_{cu} = 0.5 \times 30 = 15 \text{MPa}$. Therefore, the compressive stress at the center of gravity of all reinforcement of the mid span, 1/4 of the cross section and at the cross section of a fulcrum are 2.44 MPa, 3.36 MPa and 6.11 MPa, respectively, and they are less than $0.5 f'_{cu} = 0.5 \times 30 = 15 \text{MPa}$. Therefore, the results meet the requirement.

Assuming that the transferring and anchoring age is 7 days, and the age computed is the concrete ultimate value t_u and the relative moisture of the atmosphere in which a bridge is located is 75%, according to the computation above, the area of the gross cross section of a hollow-core slab is $A = 3174.3 \times 10^2 \text{mm}^2$, and the peripheral length of a hollow-core slab coming into contact with atmosphere is u.

$$u = 2 \times 990 + 2 \times 620 + 2\pi \times 380 + 4 \times 80 = 5927.6 \text{mm}$$

The theoretical thickness

$$h = \frac{2A}{u} = \frac{2 \times 3174.3 \times 10^2}{5927.6} = 107.1 \text{mm}$$

According to Table 6.2.7 of the **CHBC**, the following can be obtained by the linear interpolation method

$$\varepsilon_{cs}(t, t_0) = 0.000297, \phi(t, t_0) = 2.308$$

With the substitution of the values into the formula, the following can be achieved

For a mid-span

$$\sigma_{l6}(t) = \frac{0.9 \times (1.95 \times 10^5 \times 0.000297 + 5.65 \times 2.44 \times 2.308)}{1 + 15 \times 0.00388 \times 2.379} = 70.94 \text{MPa}$$

For 1/4 of a cross-section

$$\sigma_{l6}(t) = \frac{0.9 \times (1.95 \times 10^5 \times 0.000297 + 5.65 \times 3.36 \times 2.308)}{1 + 15 \times 0.00388 \times 2.379} = 80.42 \text{MPa}$$

For the cross-section of a fulcrum

$$\sigma_{l6}(t) = \frac{0.9 \times (1.95 \times 10^5 \times 0.000297 + 5.65 \times 6.11 \times 2.308)}{1 + 15 \times 0.00388 \times 2.379} = 108.77 \text{MPa}$$

(6) Prestressing loss combination

The loss $\sigma_{l,1}$ by the time of transferring and anchoring for the first time

$$\sigma_{l,1} = \sigma_{l2} + \sigma_{l3} + \sigma_{l4} + \frac{1}{2}\sigma_{l5} = 15.6 + 30 + 35.42 + \frac{1}{2} \times 38.45 = 100.25 \text{MPa}$$

The total prestressing loss σ_l after transferring and anchoring

$$\sigma_l = \sigma_{l2} + \sigma_{l3} + \sigma_{l4} + \sigma_{l5} + \sigma_{l6} = 15.6 + 30 + 35.42 + 38.45 + 70.94 = 190.41 \text{MPa}$$

1/4 of a cross-section

$$\sigma_l = 15.6 + 30 + 35.42 + 38.45 + 80.42 = 199.89 \text{MPa}$$

Cross-section of a fulcrum

$$\sigma_l = 15.6 + 30 + 35.42 + 38.45 + 108.77 = 228.24 \text{MPa}$$

The effective stress of each cross-section: $\sigma_{pe} = \sigma_{con} - \sigma_l$.

The cross-section at a mid-span: $\sigma_{pe} = 1\,302 - 190.41 = 1\,111.59 \text{MPa}$.

1/4 of a cross-section: $\sigma_{pe} = 1\,302 - 199.89 = 1\,102.11 \text{MPa}$.

The cross-section of a fulcrum: $\sigma_{pe} = 1\,302 - 228.24 = 1\,073.76 \text{MPa}$.

9) Critical condition of normal service

(1) Checking of cracking resistance of a normal cross-section

The calculation of crack resistance of a normal cross-section is to check the tensile stress of concrete on a cross-section at the mid span of a unit, and it should satisfy the requirement of article 6.3 of the **CHBC**. As to the partially prestressed member of category A, the following two requirements should be met.

Under the short-term effect combination of an action, $\sigma_{st} - \sigma_{pc} \leq 0.7 f_{tk}$.

Under the long-term effect of loading: $\sigma_{lt} - \sigma_{pc} \leq 0$, i.e. the tensile stress won't appear.

In this formula, σ_{st} is the normal tensile stress of edge concrete of a hollow-core slab to be checked for anti-cracking under the short term effect combination of an action. From Table 12-4, the moment of a cross-section at the mid span of a hollow-core slab is $M_{sd} = 327.98 \text{kN} \cdot \text{m} = 327.98 \times 10^6 \text{N} \cdot \text{mm}$. According to the elastic resistance moment $W_{011} = 52.495 \times 10^6 \text{mm}^3$ of the lower flange of a cross-section computed above, the substitution is carried out and the following obtained

$$\sigma_{st} = \frac{M_{sd}}{W_{011}} = \frac{327.98 \times 10^6}{52.459 \times 10^6} = 6.25 \text{MPa}$$

Where, σ_{pc} is the prestressing deducted from the total prestressing stress. The prestressing force from the edge of a member for anti-cracking checking is computed as follows

$$\sigma_{pc} = \frac{N_{p0}}{A_0} = \frac{N_{p0} e_{p0}}{I_0} y_0$$

$$\sigma_{p0} = \sigma_{con} - \sigma_l + \sigma_{l4} = 1302 - 190.4 + 35.42 = 1147.01 \text{MPa}$$

$$N_{p0} = \sigma_{p0} A_p - \sigma_{l6} A_s = 1\,147.01 \times 690.9 - 70.94 \times 565.5 = 752\,352.64 \text{N}$$

$$e_{p0} = \frac{\sigma_{p0} A_p y_p - \sigma_{l6} A_s y_s}{N_{p0}} = 258.2 \text{mm}$$

The prestressing stress σ_{pc} from the lower flange of a cross-section of the mid span of a hollow-core slab is

$$\sigma_{pc} = \frac{N_{p0}}{A_0} + \frac{N_{p0}e_{p0}}{I_0}y_0 = \frac{752\,352.64}{3\,233\,351.4} + \frac{752\,352.64 \times 258.2}{1.563\,3 \times 10^{10}} \times 298.2 = 6.04\,\text{MPa}$$

From Table 12-4, the cross-section at the mid span is $M_{ld} = 280.55\,\text{kN} \cdot \text{m} = 280.55 \times 10^6$ N · mm. And so is $W_{0ld} = 52.45 \times 10^6 \text{N} \cdot \text{mm}^3$. With the substitution into Formula σ_{lt}, the following can be achieved

$$\sigma_{lt} = \frac{M_{ld}}{W_{0ld}} = \frac{280.55 \times 10^6}{520\,495 \times 10^6} = 5.34\,\text{MPa}$$

Hence

$$\sigma_{st} - \sigma_{pc} = 6.25 - 6.11 = 0.14\,\text{MPa} < 0.7f_{tk} = 0.7 \times 2.65 = 1.855\,\text{MPa}$$
$$\sigma_{lt} - \sigma_{pc} = 5.34 - 6.11 = -0.77\,\text{MPa} < 0$$

This meets the specifications in the **CHBC** for members of category A.

The temperature difference stress should be based on appendix B in the **CHBC**. The thickness for bridge decking in the example is 100mm. According to article 4.3.10 in the **CHBC**, $T_1 = 14\,°C$, $T_2 = 5.5\,°C$. The vertical temperature gradient is seen in Figure 12-24. The height of a hollow-core deck is 620mm and larger than 400mm, therefore, $a = 300$mm.

For simply-supported slab bridges, the temperature difference stress is as follows

$$N_t = \sum A_y t_y a_c E_c$$
$$M_t^0 = \sum A_y t_y a_c E_c e_y$$

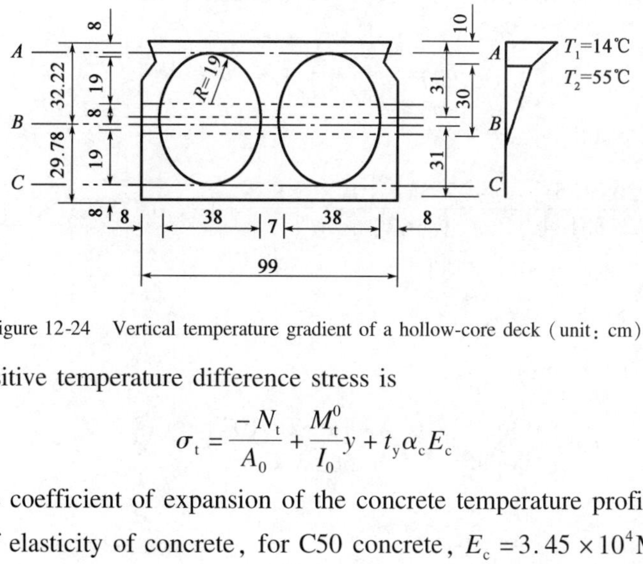

Figure 12-24 Vertical temperature gradient of a hollow-core deck (unit: cm)

The normal positive temperature difference stress is

$$\sigma_t = \frac{-N_t}{A_0} + \frac{M_t^0}{I_0}y + t_y\alpha_c E_c$$

Where α_c is the coefficient of expansion of the concrete temperature profile, $\alpha_c = 0.000\,01$; E_c is the modulus of elasticity of concrete, for C50 concrete, $E_c = 3.45 \times 10^4\,\text{MPa}$; A_y is unit area within a cross section; t_y is the average value of the temperature gradient with the unit area A_y, all value shall be position in the calculation; y is the distance from stress point calculated to the gravity axis of a cross section converted, the value above the gravity axis should be a positive one and that below it negative; A_0 and I_0 are the area and the moment of inertia of a cross section converted respectively; e_y is the distance from the center of gravity of the unit area a_y to the gravity axis of a cross section conversed, the value above the gravity axis should be positive one and that below it negative.

The computation of A_y, t_y and e_y is listed in Table 12-6.

Values of parameters A_y, t_y and e_y Table 12-6

No.	Unit area A_y (mm^2)	Temperature t_y (℃)	The distance e_y from the center of gravity of the unit area A_y to the gravity axis of a cross section conversed
1	$80 \times 990 = 79\,200$	$\dfrac{14+7.2}{2} = 10.6$	$e_y = 322.2 - \dfrac{80 \times (14 + 2 \times 7.2)}{3 \times (14 + 7.2)} = 286.5$
2	$(2 \times 80 + 70) \times 20 = 4\,600$	$\dfrac{7.2+5.5}{2} = 6.35$	$e_y = 322.2 - \dfrac{80 \times (7.2 + 2 \times 5.5)}{3 \times (7.2 + 5.5)} = 232.6$
3	$(2 \times 80 + 70) \times 300 = 69\,000$	$\dfrac{5.5}{2} = 2.75$	$e_y = 322.2 - 80 - 20 - \dfrac{1}{3} \times 300 = 122.2$

Thus, it is obtained that

$$N_t = \sum A_y t_y \alpha_c E_c = (79\,200 \times 10.6 + 4\,600 \times 6.35 + 69\,000 \times 2.75) \times 0.000\,01 \times 3.45 \times 10^4$$
$$= 365\,175.6 \text{ N}$$

$$M_t^0 = \sum A_y t_y \alpha_c E_c e_y = -(79\,200 \times 10.6 \times 286.5 + 4\,600 \times 6.35 \times 232.6 + 690\,00 \times 2.75 \times 122.2) \times 0.000\,01 \times 3.45 \times 10^4$$
$$= -93.92 \times 10^6$$

The positive temperature difference stress is calculated as follows

$$\sigma_t = \dfrac{-N_t}{A_0} + \dfrac{M_t^0}{I_0} y + t_y \alpha_c E_c$$

At the top of a beam

$$\sigma_t = \dfrac{-365\,175.6}{323\,351.4} + \dfrac{-93.32 \times 10^6 \times 321.8}{1.563\,3 \times 10^{10}} + 14 \times 0.000\,01 \times 3.45 \times 10^4$$
$$= -1.13 - 1.92 + 4.83$$
$$= 1.78 \text{ MPa}$$

At the bottom of a beam

$$\sigma_t = \dfrac{-365\,175.6}{323\,351.4} + \dfrac{-93.32 \times 10^6}{1.563\,3 \times 10^{10}} \times (-298.2) + 0$$
$$= -1.13 + 1.78$$
$$= 0.65 \text{ MPa}$$

At the center of gravity of prestressed reinforcement

$$\sigma'_t = \dfrac{-365\,175.6}{323\,351.4} + \dfrac{-93.92 \times 10^6}{1.563\,3 \times 10^{10}} \times (-258.2)$$
$$= -1.13 + 1.54$$
$$= 0.41 \text{ MPa}$$

At the center of gravity of common reinforcement

$$\sigma'_t = \frac{-365\,175.6}{323\,351.4} + \frac{-87.913\,9 \times 10^6}{1.563\,3 \times 10^{10}} \times (-258.2)$$
$$= -1.13 + 1.54$$
$$= 0.41 \text{MPa}$$

The temperature difference stress of prestressed steel bars

$$\sigma_t = \alpha_{Ep}\sigma'_t = 5.65 \times 0.41 = 2.32 \text{MPa}$$

The temperature difference stress of traditional steel bars

$$\sigma_t = \alpha_{Ep}\sigma'_t = 5.79 \times 0.41 = 2.37 \text{MPa}$$

The reverse temperature difference stress is calculated as follows:

According to article 4.2.10 in the **CHBC**, the reverse temperature difference should be computed with the positive temperature difference multiplied by -0.5, and the reverse temperature difference stress will be obtained as follows.

The reverse temperature difference stress of the top of a beam

$$\sigma_t = 1.78 \times (-0.5) = -0.89 \text{MPa}$$

The stress for the bottom of a beam

$$\sigma_t = 0.65 \times (-0.5) = -0.32 \text{MPa}$$

The reverse temperature difference stress of the prestressed steel strand is

$$\sigma_t = 2.32 \times (-0.5) = -1.16 \text{MPa}$$

The reverse temperature difference stress of common steel strands is

$$\sigma_t = 2.39 \times (-0.5) = -1.20 \text{MPa}$$

The positive value computed above represents the compressive pressure, while the negative value means the tensile stress.

It is assumed that the coefficient of the frequent value of temperature difference is 0.8 and the temperature difference stress is taken into consideration with the combination of the short term effect of an action, thus, the total tensile stress of the bottom of a beam will be computed as follows

$$\sigma_{st} = 6.25 + 0.8 \times 0.31 = 6.5 \text{MPa}$$

And $\sigma_{st} - \sigma_{pc} = 6.50 - 6.11 = 0.39 \text{MPa} < 0.7f_{tk} = 0.7 \times 2.65 = 1.855 \text{MPa}$.

This meets the conditions of the prestressed concrete member of category A.

The total tensile stress at the bottom of a beam under the combination of the long-term effect of an action will be calculated as follows

$$\sigma_{lt} = 5.34 + 0.8 \times 0.32 = 5.59 \text{MPa}$$

And $\sigma_{lt} - \sigma_{pc} = 5.54 - 6.11 = -0.77 \text{MPa} < 0$, which is in conformity with the condition of prestressed concrete of category A.

The computation above shows that the cracking resistance of the normal cross-section can meet the requirement with the temperature difference stress taken into consideration under the combination of the short term and long term effect of an action.

(2) Checking of the cracking resistance of an inclined cross-section

The checking of the cracking resistance of an inclined cross-section of partially prestressed members of category A is done for the control of the main tensile stress. The combinations of a short term effect of an action and the temperature difference will be taken into consideration. The computation of temperature difference stress for the cracking resistance of a normal cross-section in Table 12-6 and Figure 12-24 can be used for evaluating the temperature difference effect of an action. The cross-section of a fulcrum will be selected for computing the main tensile stress at fiber A-A (top of hollow holes of the cross-section of a fulcrum), fiber B-B (the gravity axis of the cross-section of a hollow-core slab converted) and fiber C-C (bottom of hollow holes). The following condition should be met for the partially prestressed members of category A

$$\sigma_{tp} \leq 0.7 f_{tk}$$

Where f_{tk} is the standard value of concrete tensile strength, and for C40 concrete, f_{tk} = 2.4MPa; σ_{tp} is the principle tensile stress induced by the combination of the short-term effect of an action and the prestressing with the consideration of the function of temperature differences.

From Table 12-6 and Figure 12-24, the temperature difference stress will be calculated first.

①Positive temperature difference stress

Fiber A-A

$$\sigma_t = \frac{-N_t}{A_0} + \frac{M_t^0}{I_0} y + t_y \alpha_c E_c$$

$$= \frac{-365\,175.6}{323\,351.4} + \frac{-93.32 \times 10^6}{1.563\,3 \times 10^{10}} \times (321.8 - 80) + 7.2 \times 0.000\,01 \times 3.45 \times 10^4$$

$$= -1.13 + (-1.43) + 2.48$$

$$= -0.08 \text{MPa}$$

Fiber B-B

$$\sigma_t = \frac{-365\,175.6}{323\,351.4} + \frac{-93.32 \times 10^6}{1.563\,3 \times 10^{10}} \times 0 + 1.42 \times 0.000\,01 \times 3.45 \times 10^4$$

$$= -1.13 + 0.49$$

$$= -0.64 \text{MPa}$$

Fiber C-C

$$\sigma_t = \frac{-365\,175.6}{32\,351.4} + \frac{-93.32 \times 10^6}{1.563\,3 \times 10^{10}} \times [-(297.8 - 80)] + 0$$

$$= -1.13 + 1.30$$

$$= 0.17 \text{MPa}$$

Chapter 12 · Design and Calculation of Prestressed Concrete Flexural Members

②Reverse temperature difference stress

The reverse temperature difference stress is obtained using the positive temperature difference stress times a factor of -0.5 as follows

Fiber $A\text{-}A:\sigma_t = (-0.08) \times (-0.5) = 0.04$ MPa.
Fiber $B\text{-}B:\sigma_t = (-0.6) \times (-0.5) = 0.3$ MPa.
Fiber $C\text{-}C:\sigma_t = (-0.16) \times (-0.5) = 0.08$ MPa.

The positive figure from the computation above represents the compressive stress, while the negative one represents the tensile stress.

③Principle tensile stress σ_{tp}

a. Fiber A-A (top of hollow core holes)

$$\sigma_{tp} = \frac{\sigma_{cx}}{2} - \sqrt{\left(\frac{\sigma_{cx}}{2}\right) + \tau^2}$$

$$\tau = \frac{V_d S_{01A}}{bI_0}$$

Where V_d is the shear design value of the short term effect combination of an action of the cross-section of a fulcrum, and from Table 12-4, $V_d = 149.73$ kN $= 149.73 \times 10^3$ N; b is the total width of the composite panel of a cross-section of the main tensile stress to be $b = 70 + 2 \times 80 = 230$ mm; I_0 is the flexural moment of inertia at the location of the principle tensile stress to be $I_0 = 1.5633 \times 10^{10}$ mm^4; S_{01A} is the static moment from the cross-section above fiber $A\text{-}A$ of a hollow core holes slab, $S_{01A} = 990 \times 80 \times (321.8 - 80/2) = 22.32 \times 10^{10}$ mm^3.

Hence

$$\tau = \frac{V_d S_{01A}}{bI_0} = \frac{149.73 \times 10^3 \times 22.35 \times 10^6}{230 \times 1.5633 \times 10^{10}} = 0.93 \text{ MPa}$$

$$\sigma_{sx} = \sigma_{pc} + \frac{M_s y_0}{I_0} + \Psi_{ij}\sigma_t$$

Where

$N_{p0} = \sigma_{p0} A_p - \sigma_{l6} A_s$
$\sigma_{p0} = \sigma_{con} - \sigma_l + \sigma_{l4} = 1302 - 228.2 + 35.42 = 1038.38$ MPa
$N_{p0} = \sigma_{p0} A_p - \sigma_{l6} A_s = 1038.38 \times 690.9 - 108.77 \times 565.5 = 655907.3$ N

$$e_{p0} = \frac{\sigma_{p0} A_p y_p - \sigma_{l6} A_s y_s}{N_{p0}} = \frac{1038.38 \times 690.9 \times 258.2 - 108.77 \times 565.5 \times 258.2}{655907.3} = 258.2 \text{ mm}$$

$$\sigma_{pc} = \frac{N_{p0}}{A_0} - \frac{N_{p0} e_{p0}}{I_0} y_0 = -0.64 \text{ MPa}$$

y_0 is the distance from fiber $A\text{-}A$ to the gravity axis, and $y_0 = 322.2 - 80 = 242.2$ mm.

Therefore, $\sigma_{sx} = \sigma_{pc} + \frac{M_s y_0}{I_0} + \Psi_{ij}\sigma_t = -0.64 + 0 + 08 \times (-0.08) = -0.70$ MPa (the positive temperature difference is considered).

Where M_s is the moment induced by vertical loading, and at the support, $M_s = 0$; Ψ_{ij} is the coefficient of the frequency value of temperature difference, and $\Psi_{ij} = 0.8$.

If the reverse temperature difference effect is considered, $\sigma_{cx} = -0.64 + 0.8 \times 0.04 = -0.61 \text{MPa}$.

So, the principle tensile stress: $\sigma_{tp} = \dfrac{\sigma_{cx}}{2} - \sqrt{\left(\dfrac{\sigma_{cx}}{2}\right)^2 + \tau^2} = -1.34 \text{MPa}$ (the positive temperature difference is considered).

If the anti-temperature difference stress is considered, the following can be achieved

$$\sigma_{tp} = \dfrac{-0.61}{2} - \sqrt{\left(\dfrac{-0.61}{2}\right)^2 + (0.93)^2} = -1.28 \text{MPa}$$

The negative values indicate the tensile stress in the formula.

The prefabricated member of the prestressed concrete member of category A under the short term effect combination of an action should meet the following: $\sigma_{tp} \leq 0.7 f_{tk} = 0.7 \times 2.65 = 1.855 \text{MPa}$.

For fiber A-A: $\sigma_{tp} = 1.34 \text{MPa} < 1.68 \text{MPa}$ (the positive temperature difference influence is considered) or $\sigma_{tp} = -1.28 \text{MPa}$ (the negative temperature difference is considered). The results satisfy requirements.

b. Fiber B-B (the center of gravity of the cross-section conversed of a hollow-core slab)

The details can be seen in Figure 12-24.

$$\tau = \dfrac{V_d S_{01B}}{bI_0}$$

Where S_{01B} is the static moment of the cross-section above fiber B-B to the gravity axis.

$S_{01B} = 990 \times 322.2 \times \dfrac{322.2}{2} - 2 \times \dfrac{\pi \times 380}{8} \times (322.2 - 80 - 190 + 80.6) - 2 \times 380 \times (322.2 - 80 - 90) \times \dfrac{322.2 - 80 - 190}{2} = 35.29 \times 10^6 \text{mm}^3$ (hinged joints are not deducted)

$$\tau = \dfrac{V_d S_{01B}}{bI_0} = \dfrac{149.73 \times 10^3 \times 35.29 \times 10^6}{230 \times 1.5633 \times 10^{10}} = 1.47 \text{MPa}$$

$\sigma_{cx} = \sigma_{pc} + \dfrac{M_s y_0}{I_0} + \Psi_{ij} \sigma_t$ (y_0 is the distance from the gravity axis to fiber B-B, $y_0 = 0$)

$$\sigma_{pc} = \dfrac{655\,907.3}{323\,351.4} - \dfrac{655\,907.3 \times 258.2}{1.5633 \times 10^{10}} \times 0 = 2.03 \text{MPa}$$

Similarly, $M_s = 0$, $\Psi_{ij} = 0.8$.

$$\sigma_{cx} = 2.03 + 0.8 \times (-0.6) = 1.55 \text{MPa}$$
$$\sigma_{cx} = 2.03 + 0.8 \times 0.3 = 2.27 \text{MPa}$$

$\sigma_{tp} = \dfrac{\sigma_{cx}}{2} - \sqrt{\left(\dfrac{\sigma_{cx}}{2}\right)^2 + \tau^2} = -0.88 \text{MPa}$ (the positive temperature difference stress is considered)

$\sigma_{tp} = \dfrac{\sigma_{cx}}{2} - \sqrt{\left(\dfrac{\sigma_{cx}}{2}\right)^2 + \tau^2} = -0.68 \text{MPa}$ (the negative temperature difference stress is considered)

Fiber B-B:

$\sigma_{tp} = -0.88\text{MPa}$ (the positive temperature difference stress is considered)

$\sigma_{tp} = -0.68\text{MPa}$ (the negative temperature difference stress is considered)

The negative values are the tensile stress and should be less than $0.7f_{tp} = 0.7 \times 2.65 = 1.855$ (MPa). This is in conformity with the requirement of the cracking resistance of an inclined cross section of partially prestressed member of category A in the ***CHBC***.

c. Fiber C-C (Bottom of hollow cores)

$$\sigma_{tp} = \frac{\sigma_{cx}}{2} - \sqrt{\left(\frac{\sigma_{cx}}{2}\right)^2 + \tau^2}$$

$$\tau = \frac{V_d S_{01C}}{bI_0}$$

Where S_{01C} is the static moment of the cross section below fiber C-C to the gravity axis of hollow-core slabs.

$S_{01C} = 990 \times 80 \times (298.2 - 80/2) + (5.65 - 1) \times 690.9 \times 258.2 + (5.79 - 1) \times 565.5 \times 258.2$

$\quad = 21.98 \times 10^6 \text{mm}^3$

$\tau = V_d S_{01C}/(bI_0) = 149.73 \times 10^3 \times 21.98 \times 10^6/(230 \times 1.5633 \times 10^{10}) = 0.92\text{MPa}$

$\sigma_{pc} = N_{p0}/A_0 + (N_{p0}e_{p0}/I_0) \times y_0$

$\quad = 655\,907.3/323\,351.4 + 655\,907.3 \times 258.2 \times 218.2/(1.5633 \times 10^{10})$

$\quad = 2.03 + 2.36 = 4.39\text{MPa}$

[y_0 is the distance from fiber C-C to the gravity axis, $y_0 = 297.8 - 80 = 217.8\text{mm}$]

$\sigma_{cx} = \sigma_{pc} + (M_s y_0)/I_0 + \psi_{ij}\sigma_t = 4.39 + 0 + 0.8 \times 0.17 = 4.53\text{MPa}$ (the positive temperature difference stress is considered)

$\sigma_{cx} = 4.39 + 0 + 0.8 \times (-0.08) = 4.33\text{MPa}$ (the negative temperature difference stress is considered)

$$\sigma_{tp} = \sigma_{cx}/2 - \sqrt{\left(\frac{\sigma_{cx}}{2}\right)^2 + \tau^2}$$

$\quad = 4.53/2 - \sqrt{\left(\frac{4.53}{2}\right)^2 + 0.92^2} = -0.18\text{MPa}$ (the positive temperature difference stress is considered)

$\sigma_{tp} = 4.33/2 - \sqrt{\left(\frac{4.33}{2}\right)^2 + 0.92^2} = -0.19\text{MPa}$ (the negative temperature difference stress is considered)

The negative values represent the tensile stress.

The main tensile stress of fiber C-C, $\sigma_{tp} = 0.18\text{MPa} < 0.7f_{tp} = 0.7 \times 2.65 = 1.855\text{MPa}$ and $\sigma_{tp} = 0.19\text{MPa} < 0.7f_{tp} = 0.7 \times 2.65 = 1.855\text{MPa}$.

The results calculated above indicate the computed stress meets the requirements of the cracking resistance of an inclined cross section of partially prestressed members of category A in the ***CHBC***.

10) Calculation of deformation

(1) Deflection computation of the normal service stage

The deflection at the service stage should be computed in accordance with the short term effect combination of an action with the consideration η_θ of the long term increasing coefficient η_θ of deflection. As to the C50 concrete, $\eta_\theta = 1.60$. As for the partially prestressed member of category A, when the deflection at the service stage is computed, the flexural strength will be $B_0 = 0.95 E_c I_0$. The value B_0 will be determined according to the size of a cross section of a mid span and the reinforcement assembly

$$B_0 = 0.95 E_c I_0 = 0.95 \times 3.45 \times 10^4 \times 1.5633 \times 10^{10} = 5.124 \times 10^{14} \text{mm}^2$$

The deflection under the short-term effect of an action may be simplified into the equivalent effect of load evenly distributed for computation

$$f_s = \frac{5}{48} \frac{l^2 M_s}{B_0} = \frac{5 \times 12\,600^2 \times 327.98 \times 10^6}{48 \times 5.124 \times 10^{14}} = 10.5 \text{mm}$$

The deflection induced by the dead weight can be computed according to the equivalent effect of load evenly distributed

$$f_G = \frac{5}{48} \frac{l^2 M_{Gk}}{B_0} = \frac{5 \times 12\,600^2 \times 222.66 \times 10^6}{48 \times 5.124 \times 10^{14}} = 7.2 \text{mm}$$

M_s and M_{Gk} can be obtained from Table 12-4.

After the deflection from dead load is eliminated and the long term influence factor η_θ is taken into consideration, the defection under the normal service stage is computed as follows

$$f_l = \eta_\theta (f_s - f_g) = 1.6 \times (10.5 - 7.2) = 5.28 (\text{mm}) < l/600 = 12\,600/600 = 21 \text{mm}$$

The results indicate the deflection values meet requirements in the **CHBC**.

(2) Calculation of the inverted camber from prestressing and design of pre-camber

① Calculation of the inverted camber from prestressing

The inverted camber happens when the prestressed steel strand of a hollow-core slab is relaxed. It is assumed that the concrete strength of a hollow-core slab reaches to C30. The computation of the inverted camber from prestressing will be carried out in accordance with the size of a cross section and the reinforcement ratio with the consideration of the long term increasing coefficient $\eta_\theta = 2.0$ of the inverted camber. First of all, the flexural strength will be worked out: $B'_0 = 0.95 E'_c I'_0$.

When the prestressed steel strand is released, it's assumed that the concrete strength of a hollow-core slab reaches to C30, and $E'_c = 3.0 \times 10^4$ MPa.

The result will be the following

$$\alpha'_{Ep} = \frac{E_p}{E'_c} = \frac{1.95 \times 10^5}{3.0 \times 10^4} = 6.5, a_p = 690.9 \text{ mm}^2$$

$$\alpha'_{Es} = \frac{E_s}{E'_c} = \frac{2.0 \times 10^5}{3.0 \times 10^4} = 6.5, a_s = 565.5 \text{mm}^2$$

The area of cross-section conversion is computed as follows

$$A'_0 = 317\,430 + (6.5-1) \times 690.9 + (6.7-1) \times 565.5 = 324\,453 \text{mm}^2$$

The static moment of the cross-section conversion of all reinforcement to the center of gravity of gross cross-section is worked out

$$S'_{01} = (\alpha'_{Ep} - 1) \times a_p \times (310 - 7 - 40) + (\alpha'_{Es} - 1) \times a_s \times (310 - 7 - 40)$$
$$= (6.5 - 1) \times 690.9 \times 263 + (6.7 - 1) \times 565.5 \times 263$$
$$= 1\,847\,128 \text{mm}^2$$

The distance from the center of gravity of cross section conversion to the center of gravity of gross cross-section will be computed as follows

$$d'_{01} = \frac{S'_{01}}{A'_0} = \frac{1\,847\,128}{324\,453} = 5.7 \text{mm (moving downwards)}$$

Hence, the distance from the center of gravity of cross-section conversion to the lower flange of hollow-core slabs will be worked out

$$y'_{0ll} = 310 - 7 - 5.7 = 297.3 \text{mm}$$

The distance from the center of gravity of cross-section conversion to the upper flange of hollow-core slabs will be worked out

$$y'_{0lu} = 310 + 7 + 5.7 = 322.7 \text{mm}$$

The distance from the prestressed steel strand to the center of gravity of the cross-section conversed will be worked out

$$e'_{01p} = 297.3 - 40 = 257.3 \text{mm}$$

The distance from the common steel strand to the center of gravity of the cross-section conversed will be worked out

$$e'_{01s} = 297.3 - 40 = 257.3 \text{mm}$$

The moment of inertial for conversion cross-sections

$$I'_0 = 15\,201 \times 10^6 + 317\,430 \times 5.7^2 + (6.5 - 1) \times 690.9 \times 257.8^2 +$$
$$(6.7 - 1) \times 565.5 \times 257.8^2$$
$$= 1.5676 \times 10^{10} \text{mm}^4$$

The elastic resistance moment of transformed cross-section is

The lower flange: $W'_{0ll} = I'_0 / y'_{0ll}, = 1.5676 \times 10^{10}/297.3 = 52.495 \times 10^6 \text{mm}^3$.

The upper flange: $W'_{0lu} = I'_0 / y'_{0lu}, = 1.5676 \times 10^{10}/322.7 = 48.5776 \times 10^6 \text{mm}^3$.

The geometric parameters of transformed cross-sections of hollow-core decks are listed in Table 12-7.

The prestressing force after the prestressing loss is deducted will be calculated according to the computation above: $N_{p0} = 752\,352.64 \text{N}$.

$$M_{p0} = 752\,352.64 \times 257.3 = 193.58 \times 10^6 \text{N} \cdot \text{mm}$$

Therefore, the following can be obtained with the reverse camber of the mid span from the prestressing, and after the result is multiplied by the long term increasing coefficient $\eta_\theta = 2.0$, the computation will be as follows

$$f_p = 2.0 \times \frac{5l^2 M_{p0}}{48 \times 0.95 E'_c I'_c}$$

$$= 2.0 \times \frac{5 \times 193.586 \times 10^6 \times 12\,600^2}{48 \times 0.95 \times 3.0 \times 10^4 \times 1.567\,6 \times 10^{10}}$$

$$= 14.33\,\text{mm}$$

Geometrical properties of transformed cross-sections of hollow-core decks Table 12-7

Items	Symbol	Unit	C30, $\alpha'_{Ep} = 6.5$	C40, $\alpha'_{Ep} = 5.65$
Area of conversion cross-sections	A'_0	mm²	324453	323351.4
Distance from the center of gravity of conversion cross section to the lower flange of the cross-section	y'_{01l}	mm	297.3	298.2
Distance from the center of gravity of conversion cross-section to the upper flange of the cross-section	y'_{01u}	mm	322.7	298.2
Distance from the prestressed steel strand to the gravity axis of the cross-section	e'_{01u}	mm	257.3	258.2
Distance from the common reinforcement to the gravity axis of cross-sections	e'_{01s}	mm	257.3	258.2
Inertia of moment of the conversion cross-section	I'_0	mm⁴	$1.567\,6 \times 10^{10}$	$1.563\,3 \times 10^{10}$
Elastic resistance moment of conversion cross-section	W'_{01l}	mm³	52.495×10^6	52.450×10^6
Elastic resistance moment of conversion cross-sections	W'_{01u}	mm³	$48.577\,6 \times 10^6$	$48.57^0 \times 10^6$

②Design of reverse camber

According to article 6.5.5 in the **CHBC**, when the long-term reverse camber from the prestressing force f_p is less than the long term deflection f_{sl} resulted from the computation of the short term effect combination of an action, the reverse camber should be set up, and the value should be chosen from the difference between the deflection of this load and the long term reverse camber of prestressing.

In this example, $f_p = 14.3\,\text{mm} < f_{sl} = 1.6 \times 10.5 = 16.8\,\text{mm}$. This shows that the structure rigidity is not enough. Measurements should be taken to improve the rigidity to satisfy the requirement. There's another measurement that the bridge decking layer when it is under constructed might serve as a way to increase the rigidity of the structure. The computation based on experiences might be a solution to be considered.

11) Checking of the permanent stress

The normal concrete compressive stress σ_{kc} of a cross section at the serve stage, the tensile stress σ_p of the prestressed reinforcement and the main compressive stress σ_{cp} of an inclined cross section to be computed should be included in the checking of the permanent stress. The standard value of an action should be adopted for computation, regardless of the coefficient of subentries. The factor of impact should be considered in the vehicle load, and so is the temperature difference stress.

(1) Normal compressive stress at mid-span σ_{kc}

The checking of the normal concrete compressive stress σ_{kc} at mid-span

$$\sigma_p = \sigma_{con} - \sigma_l = 1\,302 - 190.41 = 1\,111.59 \text{MPa}$$

The effective prestressing force of a cross-section at the mid-span

$$N_p = \sigma_p A_p = 1111.59 \times 690.9 = 767\,997.5 \text{N}$$

From Table 12-4, the effect combination of standard values can be obtained as $M_s = 408.75$ kN·m $= 408.75 \times 10^6$ N·mm.

Hence

$$\sigma_{kc} = \frac{A_p}{A_p} - \frac{N_p e_p}{W_{01u}} + \frac{M_s}{W_{01u}} + \sigma_t$$

$$= \frac{767\,997.5}{323\,351.4} - \frac{767\,997.5}{48.57 \times 10^6} \times 258.2 + \frac{408.75 \times 10^6}{48.7 \times 10^6} + 1.78$$

$$= 8.50(\text{MPa}) < 0.5 f_{ck} = 0.5 \times 32.4 = 16.2 \text{MPa}$$

(2) The checking of the tensile stress σ_p of the prestressed steel strand of a cross section at mid-span

$$\sigma_p = \sigma_{pe} + \sigma_{ep}\sigma_{kt} \leq 0.65 f_{pk}$$

Where σ_{kt} is the normal concrete stress at the center of gravity of the prestressed steel strand computed in accordance with the standard value of the effect of load: $\sigma_{kt} = 408.75 \times 106 \times 257.8 / (1.5633 \times 10^{10}) = 6.75 \text{MPa}$.

The effective prestressing

$$\sigma_{pe} = \sigma_{con} - \sigma_l = 1\,302 - 190.41 = 1\,111.59 \text{MPa}$$

If the temperature difference stress is taken into consideration, the tensile stress in the prestressed steel strand will be computed as follows

$$\sigma_p = \sigma_{pe} + \sigma_{ep}\sigma_{kt} + \sigma_t$$

$$= 1\,111.59 + 5.65 \times 6.75 + 1.16$$

$$= 1\,150.88(\text{MPa}) < 0.65 f_{pk} = 0.65 \times 1\,860 = 1\,209 \text{MPa}$$

(3) Checking of the principle stress of an oblique cross-section

The main compressive stress σ_{cp} and the tensile stress σ_{tp} of fiber A-A (the top of hollow core holes), fiber B-B (the central axis of hollow-core slabs) and fiber C-C (the bottom of hollow core holes) of the cross-section of fulcrums under the effect combination of the standard value and under the function of prestressing should be selected for the computation of the main stress of an inclined cross-section. The requirement of $\sigma_{cp} \leqslant 0.6 f_{ck} = 0.6 \times 26.8 = 16.08 \text{MPa}$ should be met.

$$\sigma_{cp} = \sigma_{cxk}/2 + \sqrt{\left(\frac{\sigma_{cxk}}{2}\right)^2 + \tau_k^2}$$

$$\sigma_{tp} = \sigma_{cxk}/2 - \sqrt{\left(\frac{\sigma_{cxk}}{2}\right)^2 + \tau_k^2}$$

$$\sigma_{cxk} = \sigma_{pc} + \frac{M_k y_0}{I_0} + \sigma_t$$

$$\tau_k = \frac{V_d S_{01}}{b I_0}$$

①Fiber A-A (the top of a hollow core hole)

$$\tau_k = \frac{V_d S_{01A}}{b I_0} = \frac{216.39 \times 10^3 \times 22.32 \times 10^6}{230 \times 1.5633 \times 10^{10}} = 1.35 \text{MPa}$$

Where V_d is the design value of the effect combination of the standard value of a cross-section of a fulcrum. From Table 12-4, the following is obtained: $V_d = 216.391 \text{kN} = 216.391 \times 10^3 \text{N}$; B is the width of the composite panel, $b = 230 \text{mm}$; S_{01A} is the static moment of the cross-section above fiber A-A to the gravity axis of a hollow core slab.

$$S_{01A} = 22.32 \times 10^6 \text{mm}^3$$

$$\sigma_{cxk} = \sigma_{pc} + \frac{M_k y_0}{I_0} + \sigma_t = -0.64 + 0 + (-0.08) = -0.72 \text{MPa}$$

Where σ_{pc} is the positive stress from the prestressing at fiber A-A, $\sigma_{pc} = -0.64$ MPa; M_k is the moment of the cross-section from the vertical loading and the cross-section of a fulcrum is $M_k = 0$; σ_t is the positive temperature difference stress at fiber A-A, $\sigma_t = -0.08 \text{MPa}$, the reverse temperature difference stress $\sigma_t = 0.04$ MPa and so on.

Hence, the principle stress at fiber A-A (the positive temperature difference stress is included) is computed as follows

$$\sigma_{cp} = \frac{-0.27}{2} + \sqrt{\left(\frac{-0.72}{2}\right)^2 + 1.35^2} = 1.04 \text{MPa}$$

$$\sigma_{tp} = \frac{-0.27}{2} - \sqrt{\left(\frac{-0.72}{2}\right)^2 + 1.35^2} = -1.76 \text{MPa}$$

When the reverse temperature difference stress is considered

$$\sigma_{cxk} = -0.64 + 0 + 0.04 = -0.60 \text{MPa}$$

Therefore,

$$\sigma_{cp} = (-0.60/2) + \sqrt{\left(\frac{-0.6}{2}\right)^2 + 1.35^2} = 1.08 \text{MPa}$$

$$\sigma_{tp} = (-0.60/2) - \sqrt{\left(\frac{-0.6}{2}\right)^2 + 1.35^2} = -1.68 \text{MPa}$$

The main stress limit of the C50 concrete is $0.6 f_{ck} = 0.6 \times 32.4 = 19.44 \text{MPa}$.

$$\sigma_{cpmax} = 1.08 \text{ MPa} < 19.44 \text{ MPa}$$

This meets the requirement of the **CHBC**.

②Fiber B-B

$$\tau_k = \frac{V_d S_{01B}}{b I_0} = \frac{216.39 \times 10^3 \times 35.29 \times 10^6}{230 \times 1.5633 \times 10^{10}} = 2.12 \text{MPa}$$

Where S_{01B} is the static moment of the cross-section above fiber B-B to the gravity axis of a hollow core slab, and $S_{01B} = 35.29 \times 10^6 \text{mm}^3$.

From the computation mentioned above, $\sigma_{pc} = 2.03$ MPa and $\sigma_t = -0.60 \text{MPa}$ (the positive temperature difference is included), or $\sigma_t = -0.3 \text{MPa}$ (the reverse temperature difference is included).

Therefore,

$$\sigma_{cxk} = \sigma_{pc} + \frac{M_k y_0}{I_0} + \sigma_t = 2.03 + 0 + (-0.6) = 1.43 \text{MPa} \text{ (the positive temperature difference is considered)}$$

$$\sigma_{cxk} = \sigma_{pc} + \frac{M_k y_0}{I_0} + \sigma_t = 2.03 + 0 + 0.3 = 2.33 \text{MPa} \text{ (the negative temperature difference is considered)}$$

The principle stress at fiber B-B (the positive temperature difference is considered) is

$$\sigma_{cp} = 1.43/2 + \sqrt{\left(\frac{1.43}{2}\right)^2 + 2.12^2} = 2.95 \text{MPa}$$

$$\sigma_{tp} = 1.43/2 - \sqrt{\left(\frac{1.43}{2}\right)^2 + 2.12^2} = -1.52 \text{MPa}$$

If the negative temperature difference stress is considered

$$\sigma_{cp} = 2.33/2 + \sqrt{\left(\frac{2.33}{2}\right)^2 + 2.12^2} = 3.58 \text{MPa}$$

$$\sigma_{tp} = 2.33/2 - \sqrt{\left(\frac{2.33}{2}\right)^2 + 2.12^2} = -1.25 \text{MPa}$$

The limit value of the concrete principle compressive stress is 16.08 MPa > 3.68 MPa, and in accordance with the requirements of the **CHBC**.

③Fiber C-C

$$\tau_k = \frac{V_d S_{01C}}{b I_0} = \frac{216.39 \times 10^3 \times 21.98 \times 10^6}{230 \times 1.5633 \times 10^{10}} = 1.33 \text{MPa}$$

Where S_{01C} is the static moment of the cross section above fiber *C-C* to the gravity axis of a hollow core slab, and $S_{01C} = 22.06 \times 10^6 \text{mm}^3$.

Similarly, from the computation mentioned above, this can be obtained: $\sigma_{pc} = 4.39 \text{MPa}$, $\sigma_t = 0.18 \text{MPa}$ (the positive temperature difference stress is considered), and $\sigma_t = -0.19 \text{MPa}$ (the negative temperature difference stress is considered).

Hence

$$\sigma_{cxk} = \sigma_{pc} + \frac{M_k y_0}{I_0} + \sigma_t = 4.39 + 0 + 0.18 = 4.57 \text{MPa} \text{ (the positive temperature difference}$$

stress is considered)

$$\sigma_{cxk} = \sigma_{pc} + \frac{M_k y_0}{I_0} + \sigma_t = 4.39 + 0 + (-0.19) = 4.2 \text{MPa} \text{ (the negative temperature difference}$$

stress is considered)

Therefore, the main stress at fiber *C-C* (the positive temperature difference stress is considered) will be the following

$$\sigma_{cp} = 4.57/2 + \sqrt{\left(\frac{4.57}{2}\right)^2 + 1.33^2} = 4.93 \text{MPa}$$

$$\sigma_{tp} = 4.57/2 - \sqrt{\left(\frac{4.57}{2}\right)^2 + 1.33^2} = -0.36 \text{MPa}$$

If the negative temperature difference stress is considered,

$$\sigma_{cp} = 4.2/2 + \sqrt{\left(\frac{4.2}{2}\right)^2 + 1.33^2} = 4.59 \text{MPa}$$

$$\sigma_{tp} = 4.2/2 - \sqrt{\left(\frac{4.2}{2}\right)^2 + 1.33^2} = -0.39 \text{MPa}$$

The concrete principle compressive stress is $\sigma_{cp} = 4.93 \text{MPa} < 16.08 \text{MPa}$, it is in conformity with the requirements in the CHBC.

The calculated results indicate the normal stress of a normal cross section, the tensile stress of prestressed tendons and the principle compressive stress of an oblique cross section in service have met the requirements of the specification.

The maximum principle tensile stress at fiber *A-A* mentioned above is 1.76MPa. According to article 7.1.6 of the CHBC, at the section of $\sigma_{tp} \leq 0.5 f_{tk} = 0.5 \times 2.65 = 1.33 \text{MPa}$, the stirrup bars should be arranged for the purpose of structure.

If it is located at the section of $\sigma_{tp} > 0.5 f_{tk} = 1.33 \text{MPa}$, the spacing s_v of the stirrup bars should be set up in line with the following equation

$$s_v = \frac{f_{sk} A_{sv}}{\sigma_{tp} b}$$

Where f_{sk} is the standard value of the tensile strength of stirrup bars, and HRB335 is used as stirrup bars with $f_{sk} = 335\text{MPa}$; A_{sv} is the total area of a cross section of the stirrup bar within the same cross section, and the stirrup bar is the double-leg one of $2\phi 10$, $A_{sv} = 157.08\text{mm}^2$; b is the width of the composite panel, and $b = 230\text{mm}$.

Therefore, the computation of the spacing of the stirrup bar is as follows

$$s_v = \frac{f_{sk}A_{sv}}{\sigma_{tp}b} = \frac{335 \times 157.08}{1.76 \times 230} = 130\text{mm}$$

$s_v = 100\text{mm}$ will be used as the spacing of stirrup bars.

Now the stirrup ratio is the following

$$\rho_{sv} = \frac{A_{sv}}{s_v b} = \frac{157.08}{100 \times 230} = 0.0068 = 0.68\%$$

In line with article 9.3.13 of the **CHBC**, as far as HRB 335 is concerned, ρ_{sv} will be no less than 12% and satisfy the requirement. The spacing of the stirrup bars near the fulcrum will be 100 mm. Other cross sections will be widened to a proper degree and there'll be the need to determine the size to be widened according to the computation. The layout for stirrup bars can be seen in Figure 12-23. Not only can it meet the requirement of the shear resistance of an inclined cross section, but also it can satisfy the computing demand for the main tensile stress. The spacing of the stirrup bars can fulfill the requirement that the spacing won't be larger than half of the height of the slab, i.e. $h/2 = 310$ mm, and no larger than 400 mm for structures.

12) Checking of temporary stress

For the calculation of the temporary stage of prestressed concrete flexural members, the cross sectional stress induced by prestressing force (the relevant stress loss deducted), dead weight of members and other construction loads should be calculated and checked so as to meet requirement of the **CHBC** when members are fabricated, transported and assembled. Hence, the compressive stress at the bottom and the tensile stress at the top of a precast hollow-core deck should be calculated as the prestressed steel strand is releasing in this example.

It is assumed that the prestressed steel strand should be released when the concrete strength of precast hollow core slabs reaches to C30. At this time, the hollow core is under the initial prestressing and the dead load of hollow core slabs, and the normal stress of the top (upper flange) and the bottom (lower flange) of a hollow core slab should be calculated.

For concrete C30, $E'_c = 3 \times 10^4\text{MPa}$, $f'_{tk} = 20.1\text{MPa}$, $f'_{tk} = 2.01\text{MPa}$, $E_p = 1.95 \times 10^5\text{MPa}$, $\alpha'_{Ep} = \frac{E_p}{E'_c} = \frac{1.95 \times 10^5}{3.0 \times 10^4} = 6.5$, $\alpha'_{Es} = \frac{2.0 \times 10^5}{3.0 \times 10^4} = 6.7$. The geometrical properties of the cross section hollow-core decks are showed in Table 13-7.

When the prestressing strand is relaxed, such three cross sections of the normal stress of the cross section of hollow core slabs as the mid span, 1/4 of a cross section and the fulcrum should be calculated as follows:

(1) Cross section at the mid span:

①The normal stress of concrete induced by prestressing force (according to article 6.5.1 of the *CHBC*):

Compressive stress of deck bottom
Tensile stress of deck top
$$\genfrac{}{}{0pt}{}{\sigma_{\text{lower}}}{\sigma_{\text{upper}}} = \frac{N_{p0}}{A_0} \pm \frac{N_{p0}}{I_0} e_{p0} \times \genfrac{}{}{0pt}{}{y_{0\text{ll}}}{y_{0\text{lu}}}$$

Where, N_{p0} is the joint force of the pretensioned prestressed reinforcement and the common steel strand, and the value is as follows

$$N_{p0} = \sigma_{p0} A_p - \sigma_{l6} A_s$$
$$\sigma_{p0} = \sigma_{\text{con}} - \sigma_l + \sigma_{l4}$$

Where σ_l is the prestressing loss when prestressing strands are relaxed. In line with article 6.2.8 of the *CHBC*:

As to pretensioned members

$$\sigma_l = \sigma_{l1} = \sigma_{l2} + \sigma_{l3} + \sigma_{l4} + 0.5\sigma_{l5}$$

Hence

$$\sigma_{p0} = \sigma_{\text{con}} - \sigma_{l1} + \sigma_{l4} = \sigma_{\text{con}} - (\sigma_{l2} + \sigma_{l3} + \sigma_{l4} + 0.5\sigma_{l5}) + \sigma_{l4}$$
$$= 1\,302 - 15.6 - 30 - 0.5 \times 38.45$$
$$= 1\,237.18\,\text{MPa}$$

$$N_{p0} = \sigma_{p0} A_p - \sigma_{l6} A_s = 1\,237.18 \times 690.9 - 70.94 \times 38.45 = 814\,651.09\,\text{N}$$

$$e_{p0} = \frac{\sigma_{p0} A_p y_p - \sigma_{l6} A_s y_s}{N_{p0}}$$
$$= \frac{1\,237.18 \times 690.9 \times 257.3 - 70.94 \times 565.5 \times 257.3}{814\,651.09} = 257.3\,\text{mm}$$

Lower flange stress
Upper flange stress
$$\genfrac{}{}{0pt}{}{\sigma_{\text{lower}}}{\sigma_{\text{upper}}} = \frac{N_{p0}}{A_0} \pm \frac{N_{p0} e_{p0}}{I_0} \times \genfrac{}{}{0pt}{}{y_{0\text{ll}}}{y_{0\text{lu}}}$$

$$= \frac{813\,949.9}{324\,453} \pm \frac{813\,949.9 \times 257.3}{1.567\,6 \times 10^{10}} \times \genfrac{}{}{0pt}{}{297.3}{322.7}$$

$$= 2.51 \pm \genfrac{}{}{0pt}{}{3.97}{4.31}$$

$$= \genfrac{}{}{0pt}{}{6.48}{-1.80}\,\text{MPa}$$

②Stress from the upper and lower flanges induced by the dead load of decks

According to Table 12-4, the moment of the dead load of the cross section of slabs at the mid span of hollow core slabs is $M_{G1} = 157.49\,\text{kN} \cdot \text{m} = 157.49 \times 10^9\,\text{N} \cdot \text{mm}$.

The normal sectional stress of the dead weight of decks is:

Lower flange stress
Upper flange stress
$$\genfrac{}{}{0pt}{}{\sigma_{\text{lower}}}{\sigma_{\text{upper}}} = \frac{M_{G1}}{I_0} \times \genfrac{}{}{0pt}{}{y_{0\text{ll}}}{y_{0\text{lu}}} = \frac{157.49 \times 10^6}{1.567\,6 \times 10^{10}} \times \genfrac{}{}{0pt}{}{-297.3}{322.7} = \genfrac{}{}{0pt}{}{-2.99}{3.24}\,\text{MPa}$$

The normal stress of the lower and upper flanges of hollow-core decks induced by prestressing forces and the dead weight as the prestressed strand is relaxed

Stress of the lower flange: $\sigma_{lower} = 6.48 - 2.99 = 3.49 \text{MPa}$

Stress of the upper flange: $\sigma_{upper} = -1.8 + 3.24 = 1.44 \text{MPa}$

The stress at the upper and lower flanges of the cross sectional is the compressive stress and less than $0.7f'_{ck} = 0.7 \times 20.1 = 14.07 \text{MPa}$ to be in conformity with the requirement of the **CHBC**.

(2) $l/4$ cross-section

$$\sigma_{p0} = \sigma_{con} - \sigma_{l1} + \sigma_{l4} = \sigma_{con} - (\sigma_{l2} + \sigma_{l3} + \sigma_{l4} + 0.5\sigma_{l5}) + \sigma_{l4}$$

$$= \sigma_{con} - \sigma_{l2} - \sigma_{l3} - 0.5\sigma_{l5}$$

$$= 1\,302 - 15.6 - 30 - 0.5 \times 38.45$$

$$= 1\,237.18 \text{MPa}$$

$$N_{p0} = \sigma_{p0}A_p - \sigma_{l6}A_s = 1\,237.1 \times 690.9 - 80.42 \times 565.5 = 809\,234.88 \text{ N}$$

$$e_{p0} = \frac{\sigma_{p0}A_p y_p - \sigma_{l6}A_s y_s}{N_{p0}} = 257.3 \text{mm}$$

Lower flange stress σ_{lower}
Upper flange stress σ_{upper} $= \dfrac{N_{p0}}{A_0} \pm \dfrac{N_{p0}e_{p0}}{I_0} \times \dfrac{y_{0ll}}{y_{0lu}}$

$$= \frac{808\,255.3}{324\,453} + \frac{808\,255.3 \times 257.3}{1.567\,6 \times 10^{10}} \times \frac{297.3}{322.7}$$

$$= 2.49 \pm \frac{3.94}{4.28}$$

$$= \frac{6.43}{-1.79} \text{MPa}$$

If from Table 12-4 the bending moment at 1/4 of a cross-section induced by the dead load is $M_{G1} = 118.12 \text{kN} \cdot \text{m} = 118.12 \times 10^6 \text{N} \cdot \text{mm}$, the stress of the lower and upper flanges at 1/4 of a cross-section induced by the dead weight is

Lower flange stress σ_{lower}
Upper flange stress σ_{upper} $= \dfrac{M_{G1}}{I_0} \times \dfrac{y_{0ll}}{y_{0lu}} = \dfrac{118.12 \times 10^6}{1.567\,6 \times 10^{10}} \times \dfrac{-297.3}{322.7} = \dfrac{-2.24}{2.43} \text{MPa}$

The normal stress of the lower and upper flanges of hollow-core decks induced by prestressing forces and the dead weight as the prestressed strand is relaxed.

Lower flange stress is $\sigma_{lower} = 6.43 - 2.24 = 4.19 \text{MPa}$.

Upper flange stress is $\sigma_{upper} = -1.79 + 2.43 = 0.64 \text{MPa}$.

The stress at the upper and lower flanges of the cross-sectional is the compressive stress and less than $0.7f'_{ck} = 0.7 \times 20.1 = 14.07 \text{MPa}$ to be in conformity with the requirement of the **CHBC**.

(3) Fulcrum of a cross-section

The normal stress of the upper and lower flange of a cross-section of a fulcrum induced by prestressing forces is

Lower flange stress σ_{lower}
Upper flange stress σ_{upper} $= \dfrac{N_{p0}}{A_0} \pm \dfrac{N_{p0}e_{p0}}{I_0} \times \dfrac{y_{0ll}}{y_{0lu}}$

$$\sigma_{p0} = \sigma_{con} - \sigma_{l1} + \sigma_{l4} = \sigma_{con} - (\sigma_{l2} + \sigma_{l3} + \sigma_{l4} + 0.5\sigma_{l5}) + \sigma_{l4}$$
$$= \sigma_{con} - \sigma_{l2} - \sigma_{l3} - 0.5\sigma_{l5}$$
$$= 1\,302 - 15.6 - 30 - 0.5 \times 38.45$$
$$= 1\,237.18 \text{MPa}$$

$$N_{p0} = \sigma_{p0}A_p - \sigma_{l6}A_s = 1\,237.18 \times 690.9 - 108.77 \times 565.5 = 793\,247.35 \text{N}$$

$$e_{p0} = \frac{\sigma_{p0}A_p y_p - \sigma_{l6}A_s y_s}{N_{p0}} = 257.3 \text{mm}$$

Lower flange stress σ_{lower}
Upper flange stress σ_{upper}
$$= \frac{N_{p0}}{A_0} \pm \frac{N_{p0}e_{p0}}{I_0} \times \frac{y_{0l1}}{y_{0lu}}$$

$$= \frac{791\,228.1}{324\,453} \pm \frac{791\,228.1 \times 257.3}{1.5676 \times 10^{10}} \times \frac{297.3}{322.7}$$

$$= 2.44 \pm \genfrac{}{}{0pt}{}{3.86}{4.19} = \genfrac{}{}{0pt}{}{6.30}{-1.75} \text{MPa}$$

The moment of the cross-section of a fulcrum induced by the dead weight of decks is 0 and so the normal stress of the mid span of the cross-section of a fulcrum is:

The stress of the lower flange is $\sigma_{lower} = 6.30 \text{MPa}$.

The stress of the upper flange is $\sigma_{upper} = -1.75 \text{MPa}$.

The stress at the lower flange: $\sigma_{lower} = 6.30 \text{MPa} < 0.7f'_{ck} = 0.7 \times 20.1 = 14.07 \text{MPa}$.

The computation result for the stress of the upper and lower flanges as the prestressed steel strand is released at cross-sections of the mid span, 1/4 of a cross-section and a fulcrum is shown in Table 12-8.

Normal stresses of cross-sections for temporary conditions of hollow-core decks

Table 12-8

Cross-sections Stress		At the mid span		At $l/4$ cross-section		At a fulcrum	
		σ_{lower}	σ_{upper}	σ_{lower}	σ_{upper}	σ_{lower}	σ_{upper}
Types of actions	Prestressing forces	-1.8	6.48	-1.79	6.43	-1.75	6.3
	Dead weight	3.24	-2.99	2.43	-2.24	0	0
Total stress(MPa)		1.44	3.49	0.64	4.19	-1.57	6.30
Limit values of compressive stress $0.7f'_{ck}$ = 14.07MPa		14.07	14.07	14.07	14.07		14.07

Note: the negative values are the tensile stress and the positive values are the compressive stress. All the compressive stress meets the requirements in the ***CHBC***.

From the calculations mentioned above, the tensile stress of the upper flange of a cross-section of the fulcrum is the following when the prestressed strand is relaxed

$$\sigma_{upper} = 1.75 \text{MPa} \quad \begin{matrix} > 0.7f'_{tk} = 0.7 \times 2.01 = 1.407 \text{MPa} \\ < 1.15f'_{tk} = 1.15 \times 2.01 = 2.312 \text{MPa} \end{matrix}$$

According to article 7.2.8 in the ***CHBC***, the longitudinal reinforcement should be arranged at the pretension area (the upper flange of a cross-section), and arranged according to the following principles:

When $\sigma_{upper} \leqslant 0.7f'_{tk}$, the reinforcement ratio should be no less than 0.2% at the pretensioning area.

When $\sigma_{upper} = 1.15f'_{tk}$, the reinforcement ratio should be no less than 0.4% at the pretensioning area.

When $0.7f'_{tk} < \sigma_{upper} < 1.15f'_{tk}$, the reinforcement ratio at the pretension area should be obtained according to the interpolation method of both the two calculations mentioned above.

The reinforcement ratio mentioned above is $\dfrac{A'_s}{A}$. A'_s is the area of the cross-section of the common reinforcement at the pretension area, and A is the area of a gross cross-section and $A = 317\,430(\text{mm}^2)$.

Furthermore, from above computation, the result will be obtained with the interpolation method, i.e. $\sigma_{upper} = 1.75\text{MPa}$, and the longitudinal reinforcement ratio is 0.00276, then $A'_s = 0.00276 \times 317\,430 = 876.1\text{mm}^2$.

When the ribbed reinforcement is utilized for the longitudinal reinforcement at the pretensioning zone with its diameter no larger than 14 mm and the HRB 400 8 ⌽12 rebar is used, then $A'_s = 8 \times \dfrac{\pi \times 12^2}{4} = 904.8\text{mm}^2$. This value is larger than 876.1mm², meaning the requirements are satisfied. The reinforcement should be placed at the flange of a cross section of a fulcrum of a hollow core slab as shown in Figure 12-25.

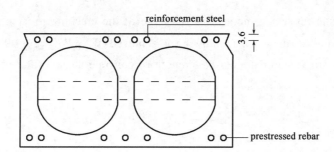

Figure 12-25 Reinforcement arrangement of a cross-section of a fulcrum of hollow-core decks (unit: cm)

In order to present the upper flange tensile stress from the fulcrum of a cross-section from becoming excessive, the method of reducing the prestressing compressive stress of the fulcrum of a cross-section will be used, i.e. sleeves to be placed near the fulcrum to isolate the prestressed steel strand from the part of concrete from transmitting forces. If only five-wire strand near the cross-section of a fulcrum transmits the prestressing and the other two-wire strands are separated, the tensile stress of the upper flange of a hollow core slab will be reduced to $\sigma_{up} = 1.75 \times \dfrac{5}{7} = 1.25\text{MPa} < 0.7f'_{tk} = 1.407\text{MPa}$. according to the requirement of the ***CHBC*** no less than 0.2%

rebar ratio at the pretension zone should be designed for the longitudinal common reinforcement with the value of 0.2% and $0.002 \times 317\,430 = 634.9 \text{mm}^2$, the $6 \phi 12$ reinforcement should be used and $A'_s = 6 \times \dfrac{\pi \times 12^2}{4} = 678.6 \text{mm}^2$.

13) Checking of the minimum reinforcement ratio

According to article 9.1.12 of the **CHBC**, the minimum reinforcement ratio of prestressed concrete flexural member should satisfy the following requirement

$$\frac{M_{ud}}{M_{cr}} \geq 1.0$$

Where M_{ud} is the design value of the bearing capacity of the normal cross-section of flexural members. According to the value computed, $M_{ud} = 567.86 \text{kN} \cdot \text{m}$.

M_{cr} is the cracking moment of the normal cross-section of flexural members to be computed with the formula as $M_{cr} = (\sigma_{pc} + rf_{tk})W_0$, in which $r = \dfrac{2S_0}{W_0}$.

σ_{pc} is the prestressing stress of concrete from the edge of the cracking resistance of members caused by the joint force N_{p0} of the total prestressing reinforcement and the common rebar to be deducted, and $\sigma_{pc} = 6.01 \text{MPa}$ to be achieved. S_0 is the static moment above the gravity axis of a cross section conversed to the gravity center with the following value

$$S_0 = 990 \times 322.2 \times \frac{321.8}{2} - 2 \times 380 \times \frac{(321.8 - 80 - 192)^2}{2} - 2 \times \frac{\pi \times 380^2}{8} \times$$

$$[80.6 + (322.2 - 80 - 190)] = 35\,290\,870 \text{mm}^2$$

W_0 is the elastic resistance moment of the edge of the cracking resistance of a cross-section conversed, and from the above, the following is achieved: $W_0 = W_{0l} = 52.495 \times 10^6 \text{mm}^3$.

f_{tk} is the standard value of the tensile strength of concrete at the center of an axis, and as for C50, $f_{tk} = 2.65 \text{MPa}$.

$$r = \frac{2S_0}{W_0} = \frac{2 \times 35\,290\,870}{52.495 \times 10^6} = 1.345$$

M_{cr} is obtained as follows

$$M_{cr} = (\sigma_{pc} + rf_{tk})W_0 = (6.01 + 1.345 \times 2.65) \times 52.495 \times 10^6$$

$$= 502.60 \times 10^6 \text{N} \cdot \text{mm} = 502.60 \text{kN} \cdot \text{m}$$

$\dfrac{M_{ud}}{M_{cr}} = \dfrac{567.86}{502.60} = 1.13 > 1.0$, the requirements in the **CHBC** are satisfied.

In line with article 9.1.12 in the **CHBC**, the area of a cross-section of the common tensioned rebar of the partially prestressed flexural member should be no less than $0.003bh_0$ The area of the common tensioned rebar in this example is as follows

$$A_s = 565.5\text{mm}^2 > 0.003bh_0 \, (0.003bh_0 = 0.003 \times 178 \times 580 = 483.7\text{mm}^2)$$

In the formula, b is the rib width of an equivalent I-shaped cross-section of a hollow core slab with $b = 278$mm. The computation result meets the requirement of the ***CHBC***.

Chapter 13
Partially Prestressed Concrete Flexural Members

13.1 Concepts and characteristics of PPC members

Although prestressed members have the advantages of good crack resistance, fatigue resistance and leakage resistance, there are also some shortages while applied in engineering. Large camber causes bigger changes for the actual depth of deck pavement and does harm to the bridge deck and traffic driving comfort. Too large pre-tension force will lead to irreclaimable cracks along the prestressing bars because the transverse tensile strain of concrete under the anchorage is beyond the ultimate tensile strain.

The partially prestressed concrete (PPC) structure is used to overcome the shortage of PC structures, as a kind of structure between PC and RC with proper prestressing force imposed and regular reinforcement allocated to assure the load-carrying capacity.

PPC structures not only make full use of prestressed steel but also exploit regular steel to improve the service performance. At the same time, cracking is allowed in service period. These advantages extend the applications of PC structure and accelerate the development of design idea, allowing designers to select the prestress degree according to the operating requirements of structures.

13.2 Classification and mechanic characteristic of PPC structure

1) Classification of PPC

In the *CHBC*, PPC structures are classified into two types according to prestress degree λ: PC and PPC structures. Moreover, PPC structures are divided into category A and category B.

(1) Category A

Small tension stress is allowable in normal section in service period, but cracking in service period is not permitted. The design of category A structures is the same as that of PC members.

(2) Category B

Finite cracks in normal section in service period are permitted.

2) Moment-deflection curve of PPC

Based on test results, the moment-deflection curves under different prestress degrees are drawn in Figure 13-1.

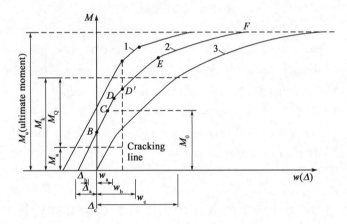

Figure 13-1 Moment-deflection curve

From the figure above, we can see that the load-bearing characteristics of PPC beam are as follows:

(1) Linear stage

When the load is small, the force-bearing characteristics of PPC beams (curve 2) is similar to PC beams (curve 1). The cambers of PPC beams under dead load and prestressing force are less than PC beams.

(2) Zero-deflection stage

With the increase of load, the bending moment increases to point B, which means the deflection under load equals to the camber and the whole deflection is zero.

(3) Zero-stress stage

As the load increases to point C in curve 2, the efficient compressive stress caused by prestressing force counteracts the tension stress by loads. Then the stress at concrete tension edge is zero. The corresponding moment is called eliminating moment.

(4) Crack-appearing stage

When the load is increased beyond point D, the tension stress at concrete edge is close to the tensile strength. At point D', cracks occur. The corresponding moment is called cracking moment. Then $M_{cr} = M_{pcr} - M_0$.

(5) Failure stage

From point D' cracks develop, the stiffness reduces and the deflection increases rapidly. At point E, steel bars yield. From point E to point F, cracks develop and stiffness reduces further, and deflection increases faster. At point F, members are in the state of the ultimate bearing capacity.

3) Feasible implementation method of PPC members

(1) All adopted are high-strength steel bars, with some of them being stretched to maximum allowable stress and others acting as regular bars. By this way anchorage device and pulling work are economized.

(2) All prestressed steel bars are stretched to lower stress.

(3) High strength steel bars and regular bars are both arranged. Prestressed bars can balance some loads, improve the safety against cracking, reduce the deflection and provide part or most of the bearing capacity. Regular steels act to scatter cracks, increase the capacity and the ductility, and reinforce the structures that are hard to arrange prestressed steel.

For category B of PPC the last method mentioned above is commonly used. Because the members consist of both prestressed and regular steels, they have the advantages and avoid weaknesses of these two steel materials.

13.3 The design calculation for category B of PPC

There are not only many similarities in designing and calculating between PPC and PC, but also differences in some aspects:

(1) Similarities: long-term capacity for cross section and inclined section, properties under partial loading, and estimating prestressing loss.

(2) Differences: calculating normal stress, crack width, deflection, fatigue, designing the reinforcement and construction requirement under service stage.

1) Calculating normal stress under service period

Category B PPC flexural members will have cracked in service period. The computation becomes more complex because the neutral axis position and the geometric features are decided by the dimension and position of the prestressed force. From the moment-deflection curve, we can see that the cracked beams work elastically. Elastic analysis method is still used for stress calculation of cracked beams.

The stress state of cracked PPC beams is similar to that under load of large eccentricity. The stress of cracked RC beams in large eccentricity state can be calculated by elastic analysis. It should be noted that when the external force is zero the stress of PC members with a load of large eccentricity is zero (that is called "zero stress" state).

The moment M of category B PPC flexural members could be substituted by equivalent eccentric compression, but the stress is not the same. Because of the presence of prestressing force, even without any loads there is axial stress on the concrete cross-section.

Seeing that the design formula for PC members under load of large eccentricity is based on "zero stress" state, removing the compression stress caused by prestressing, we can calculate the stress of concrete and bars by exploiting calculation method for large-eccentricity compression members.

To calculate the stress of crack PC member, it is reasonable to transfer flexural member under moment M_k and compression force N_p to eccentric compression member with lever arm e_{0N}. For post-stressed redundant structures such as continuous bridges, the moment M_k mentioned above should take the secondary moment M_{p2} created by prestressing force into account.

Figure 13-2 shows that for post-tensioned T-section beam of category B ($M > M_{cr}$), and the stress state can be decomposed into stages as follows:

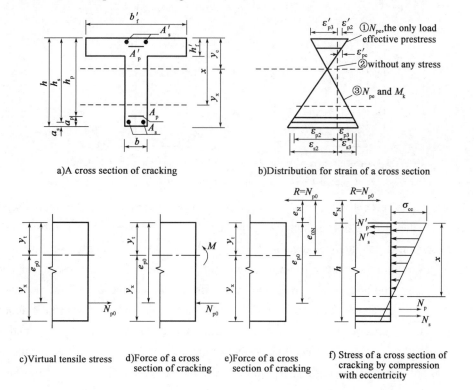

Figure 13-2 Equivalent course for compression member with large eccentricity

(1) Effective prestress N_{pe}

The strain under the only load of effective prestress N_{pe} is shown as curve ① in Figure 13-2b), where the tension stress in prestressed steel does not take the self-weight of the beam into account.

$$\sigma_{pe} = \sigma_{con} - \sigma_l \quad (13\text{-}1)$$

$$\sigma'_{pe} = \sigma'_{con} - \sigma'_l \quad (13\text{-}2)$$

Because of shrinkage and creep of concrete, internal force comes into being within regular bars towards the opposite direction of prestressed bars, reducing the effective prestress at tension block. To simplify the calculation method, in the follows the prestress loss due to shrinkage and creep of concrete is used approximately as regular bar stress (Figure 13-3). The relation of compressive prestress σ_{pc} and σ'_{pc} under composite force N_p of prestressed and regular bars at prestressed bars gravity center between upper and lower fibers is

$$\sigma_{pc} = \frac{N_p}{A_n} + \frac{N_p e_{pn}}{I_n} y_{pn} \quad (13\text{-}3a)$$

$$\sigma'_{pc} = \frac{N_p}{A_n} - \frac{N_p e_{pn}}{I_n} y'_{pn} \quad (13\text{-}3b)$$

$$N_p = \sigma_{pe} A_p + \sigma'_{pe} A'_p - \sigma_{l6} A_s - \sigma'_{l6} A'_s \quad (13\text{-}4)$$

Where, σ_{pe} and σ'_{pe} can be calculated by Equations (13-1) and (13-2), and e_{pn} by equation (13-5).

$$e_{pn} = \frac{\sigma_{pe} A_p y_{pn} - \sigma'_{pe} A'_p y'_{pn} - \sigma_{l6} A_s y_{sn} + \sigma'_{l6} A'_s y'_{sn}}{\sigma_{pe} A_p + \sigma'_{pe} A'_p - \sigma_{l6} A_s - \sigma'_{l6} A'_s} \quad (13\text{-}5)$$

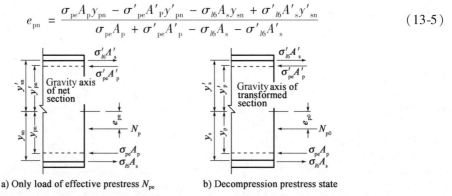

a) Only load of effective prestress N_{pe} b) Decompression prestress state

Figure 13-3 Joint force and the eccentric moment of the prestressed and non-prestressed rebar

Where e_{pn} is the eccentricity from the action point of N_p to the mass axis of the net section; A_n is the area of the net section summarized by the area of concrete deducted the pipes and transformed section of longitudinal regular steels; I_n is the net rotational inertia. In post-tensioning method, before grouting the ducts the prestressed force is applied to the net section of concrete. Therefore, the area, the rotational inertia, the mass axis and the eccentricity of the net section are used in calculation. y_{pn} and y'_{pn} are respectively the distance from the mass axis to the action point of resultant force of prestressed steels in tensile region and compressive region; y_{sn} and y'_{sn} are respectively the distance from the mass axis to the action point of resultant force of regular steels in tensile region and compressive region; σ_{l6} and σ'_{l6} are respectively the loss of prestress in tensile and compressive region due to the shrinkage and creep of concrete.

(2) Decompression prestress state

This is a "virtual" stage for calculating "zero action". Under "virtual" action, the whole section is without any stress [shown as curve ② in Figure 13-2b)].

To reach this state, a tension force N_{p0} (also called as virtual force) should be applied to the section to eliminate the pre-stress.

In decompression state, the concrete strain at the action point of prestressed steel in tensile region and compressive region changes from ε_{pc} or ε'_{pc} to zero, the incremental stress of prestressed steel in tensile region and compressive region is ($-\varepsilon_{p2}$) and ($-\varepsilon'_{p2}$), whose absolute values are equal to the concrete stress of mass axis ε_{pc} and ε'_{pc} respectively. The tension stress increase of prestressed steel at tension and compression block is

$$\sigma_{p2} = E_p \cdot (-\varepsilon_{p2}) = E_p \varepsilon_{pc} = \alpha_{Ep} \sigma_{pc} \tag{13-6}$$

$$\sigma'_{p2} = E_p \cdot (-\varepsilon'_{p2}) = E_p \varepsilon'_{pc} = \alpha_{Ep} \sigma'_{pc} \tag{13-7}$$

Where, α_{Ep} is the elastic modular ratio of prestressed steel to concrete.

The steel at tension and compression block is supposed to be the same type, and $\alpha_{Ep} = E_p / E_c$.

Without any compression, the stress σ_{p0} and σ'_{p0} respectively of prestressed steel at compression and tension blocks are calculated as follows

$$\sigma_{p0} = \sigma_{pe} + \sigma_{p2} = \sigma_{con} - \sigma_l + \alpha_{Ep} \sigma_{pc} \tag{13-8}$$

$$\sigma'_{p0} = \sigma'_{pe} + \sigma'_{p2} = \sigma'_{con} - \sigma'_l + \alpha_{Ep} \sigma'_{pc} \tag{13-9}$$

In decompression state, the strain of regular steel ε_{s1} and ε'_{s1} vanishes because of effective prestress N_{pe}. But there is still the strain of regular steel ε_{s2} and ε'_{s2} due to the creep and shrinkage of concrete, so the compressive stress of regular steel exists in tensile and compressive regions, which is approximately the loss value of prestress due to the creep and shrinkage of concrete. Compressive force N_{s2} and N'_{s2} are written as follows

$$N_{s2} = -\sigma_{l6} A_s \tag{13-10}$$

$$N'_{s2} = -\sigma'_{l6} A'_s \tag{13-11}$$

Without any compression, resultant force of prestressed and regular steel is written as follows

$$N_{p0} = \sigma_{p0} A_p - \sigma_{l6} A_s + \sigma'_{p0} A'_p - \sigma'_{l6} A'_s \tag{13-12}$$

The distance from action spot to compression edge [Figure 13-2c)] is

$$e_{p0} = \frac{\sigma_{p0} A_p y_{pn} - \sigma'_{p0} A'_p y'_{pn} - \sigma_{l6} A_s y_{sn} + \sigma'_{l6} A'_s y'_{sn}}{\sigma_{p0} A_p - \sigma_{l6} A_s + \sigma'_{p0} A'_p - \sigma'_{l6} A'_s} \tag{13-13}$$

For pre-tensioned members

$$\sigma_{p0} = \sigma_{con} - \sigma_l + \sigma_{l4} \tag{13-14}$$

$$\sigma'_{p0} = \sigma'_{con} - \sigma'_l + \sigma'_{l4} \tag{13-15}$$

(3) Method to handle virtual state

The virtual tension force (N_{p0}) is set for calculation, so to eliminate the influence a force equal in magnitude and opposite in direction should be applied at the resultant force spot.

Then, the moment M_k and eccentric force N_{p0} acting at the crack section may be replaced by an equivalent force R, and the magnitude and lever arm e_N of which can be resolved (see Figure 13-4).

Supposed that $R = N_{p0}$, according to moment balance

$$M_k = R(h_{ps} + e_N) = N_{p0}(h_{ps} + e_N)$$

$$e_N = \frac{M_k}{N_{p0}} - h_{ps} \tag{13-16}$$

So category B PPC flexural member under resultant force N_p and moment M_k is converted to a PC eccentric compression member loaded by compression $R = N_{p0}$ whose lever arm is e_N (see Figure 13-4).

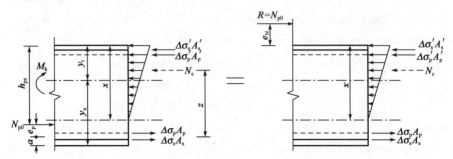

Figure 13-4 Eccentric compressive member of equivalent reinforcement concrete

(4) Using the method for PC large-eccentricity case to calculate the effective depth of crack section

The assumptions made for PPC category B cracked section flexural moments are as follows:
①Plane section before bending remains plane after bending;
②The compression concrete stress may be simplified as triangle distribution;
③The tensile strength in the concrete may be neglected.

Supposing the neutral axis of cracked section lie in the rib (see Figure 13-5), according to moment balance, the equation to solve compression height x.

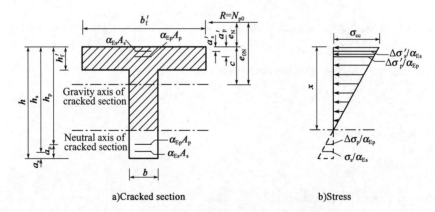

a) Cracked section b) Stress

Figure 13-5 Cracked section and stress

$$Ax^3 + Bx^2 + Cx + D = 0 \tag{13-17}$$

$$A = b \tag{13-18}$$

$$B = 3be_N \tag{13-19}$$

$$C = 3b_0 h'_f (2e_N + h'_f) + 6\alpha_{Ep}(A_p g_p + A_p g'_p) + 6\alpha_{Es}(A_s g_s + A'_s g'_s) \tag{13-20}$$

Chapter 13 ▸ Partially Prestressed Concrete Flexural Members

$$D = -b_0 h'^2_f (3e_N + 2h'_f) - 6\alpha_{Ep}(A_p h_p g_p + A'_p a'_p g'_p) - 6\alpha_{Es}(A_s h_s g_s + A'_s a'_s g'_s)$$

(13-21)

Where b is the width of the web of T-shaped and I-shaped section or the width of rectangular section; e_N is the distance from the action point of N_{p0} to compressive edge, positive and negative sign separately stands for N_{p0} located outside or in the section; b_0 is the width difference of the compressive flange and the web of T-shaped or I-shaped sections, $b_0 = b'_f - b$; h'_f is the thickness of the compressive flange of T-shaped or I-shaped sections; h_p, h_s are respectively the distance from the compressive edge to the gravity center of prestressed steel and regular steel in tension region; g_p, g_s are respectively the distance from the action point of N_{p0} to the gravity center of prestressed steel and regular steel in tension region, $g_p = h_p + e_N$, $g_s = h_s + e_N$; g'_p, g'_s are respectively the distance from the action point of N_{p0} to the gravity center of prestressed steel and regular steel in compressive region, $g'_p = a'_p + e_N$, $g'_s = a'_s + e_N$; a'_p, a'_s are respectively the distance from the compressive edge to the gravity center of prestressed steel and regular steel in compressive region.

It should be noted while calculating the cracked compression height x:

①The stress of regular tension steel should comply with the standard.

②When tension stress exists in prestressed steel, that is $\alpha_{Es}\sigma_{cc} - \sigma'_{p0} < 0$, in Formulas (13-20) and (13-21), the items containing A'_p should be marked with the positive sign other than negative sign, where σ_{cc} is the compressive stress of concrete at the action point of the resultant force in compressive region.

③Without any prestressed or regular steel, A'_p or A'_s is set as zero.

Formula (13-17) is applicable to category B PPC T-section and I-section. As to rectangular beam, h'_f should be set to zero.

(5) Compression stress of cracked PC section

According to the **CHBC**, the equations for compression stress of category B PPC flexural moment are as follows

$$\sigma_{cc} = \frac{N_{p0}}{A_{cr}} + \frac{N_{p0} e_{0N} c}{I_{cr}} \leq 0.5 f_{ck} \quad (13-22)$$

$$e_{0N} = e_N + c \quad (13-23)$$

$$e_N = \frac{M_k}{N_{p0}} - h_{ps} \quad (13-24)$$

$$h_{ps} = \frac{\sigma_{p0} A_p h_p - \sigma_{l6} A_s h_s + \sigma'_{p0} A'_p a'_p - \sigma'_{l6} A'_s a'_s}{N_{p0}} \quad (13-25)$$

Where N_{p0} is the resultant force of prestressed and regular steel while normal prestressed force of concrete is zero, pre-tensioned and post-tensioned members can be calculated by Formula (13-22); $\sigma_{p0}, \sigma'_{p0}$ are the stress of prestressed steel while normal prestressed force of concrete is zero, pre-tensioned members can be calculated by Formulas (13-24) or (13-25) and post-tensioned members by Formulas (13-8) and (13-9); c is the gravity center of transformed cracked section from compressive edge; e_{0N} is the gravity center of transformed cracked section from the action point of N_{p0}; e_N is the distance from the action point of N_{p0} to the bending edge, and the positive and negative sign separately stands for when N_{p0} lies outside or in the section; h_{ps} is the action point of resultant force of prestressed and regular steels from compressive edge; h_p, a'_p are the action point of resultant force of prestressed steels respectively in tension and compress region from compressive edge; h_s, a'_s are the action point of composite force of regular steels respectively in tension and compress region from compressive edge; A_{cr} is transferred area of the cracked section; I_{cr} is transferred rotational inertia of the cracked section.

(6) The prestressed steel stress of cracked section

The stress increment of prestressed steel of cracked section

$$\Delta\sigma_p = \alpha_{Ep} \left[\frac{N_{p0}}{A_{cr}} - \frac{N_{p0} e_{0N}(h_p - c)}{I_{cr}} \right] \quad (13\text{-}26)$$

The total stress of prestressed tension steel of cracked section is $\sigma_p = \sigma_{p0} + \Delta\sigma_p$, Where, σ_{p0} is the prestressed steel stress when the concrete axial stress reaches zero at the steel resultant force spot of tension block.

In service stage, the total stress of prestressed steel in a cracked section should follow the relationship as follows.

For steel wire and steel strand: $\sigma_{p0} + \Delta\sigma_p \leq 0.65 f_{pk}$ (13-27)

For rolled threaded bar: $\sigma_{p0} + \Delta\sigma_p \leq 0.80 f_{pk}$ (13-28)

The stress of regular steel at PC tension block in service stage is too small to be taken into account.

2) Calculation of crack width

Category B PPC flexible members allow cracks in regular service stage. Therefore, crack width control and the calculation are important.

(1) Calculation of crack width

For category B PPC members, the equation to calculate the crack depth in the **CHBC** is written as follows

$$W_{tk} = C_1 C_2 C_3 \frac{\sigma_{ss}}{E_s} \frac{C+d}{0.36 + 1.7\rho_{te}} \quad (\text{mm}) \quad \rho_{te} = \frac{A_s}{A_{te}} \quad (13\text{-}29)$$

Where, A_{te} is the effective area of cross-section of tensile concrete. ρ_{te} is effective ratio of longitudial tensile reinforcement, ρ_{te} is ratio of reinforcement, $\rho_{te} = \frac{A_s}{A_{te}}$, $\rho = 0.01$ when $\rho > 0.01$ and $\rho = 0.01$ when $\rho = 0.01$; C_1 is surface configuration factor of steel, for plain steel $C_1 = 1.4$, for ribbed steel $C_1 = 1.0$; C_2 is the factor affected by long-term effects, $C_2 = 1 + 0.5 \frac{N_l}{N_s}$; N_l and N_s respectively are the calculated internal force (moment or axial force) under long-term and short-term combined effect; C_3 is the factor about mechanics properties of member, for RC plate type flexural members $C_3 = 1.15$, and others $C_3 = 1.0$; d is the dimension of longitudinal tension reinforcement (mm). If different dimensions are used, d will be replaced by transferred dimension d_e, $d_e = \dfrac{\sum n_i d_i^2}{\sum n_i d_i}$, in the **CHBC**, for mixed reinforcement PC members, because prestressed steel is made up of steel tendons or steel strand, where d_i is the nominal diameter of regular steel or equivalent diameter of steel tendon or strand d_{pe}, $d_{pe} = \sqrt{n}\,d$, where n stands for the number of the steel tendons or strands, d is the nominal diameter of simple tendon or strand. σ_{ss} is the tension stress of longitudinal steel in cracked section under short-term combined effect considering long-term effect, calculated by the following formula

$$\sigma_{ss} = \dfrac{M_s - N_{p0}(z - e_p)}{(A_p + A_s)z} \quad (13\text{-}30)$$

$$z = \left[0.87 - 0.12(1 - \gamma'_f)\left(\dfrac{h_0}{e}\right)^2\right] h_0 \quad (13\text{-}31)$$

$$\gamma'_f = \dfrac{(b'_f - b) h'_f}{b h_0} \quad (13\text{-}32)$$

$$e = e_p + \dfrac{M_s}{N_{p0}} \quad (13\text{-}33)$$

Where M_s is the moment calculated under short-term combined effect; N_{p0} is the resultant force of prestressed and regular steels when normal stress of concrete is zero, calculated by Equation (13-12); z is the distance from resultant force of longitudinal prestressed and regular steels in tension region to compression region (Figure 13-4); γ'_f is the ratio of effective compress area of flange to ribbed plate; b'_f and h'_f are the width and height of compression flange, if $h'_f > 0.2 h_0$, $h'_f = 0.2 h_0$; e_p is the distance from the action spot of resultant force N_{p0} of longitudinal prestressed and regular steels to that in tension region when normal stress of concrete is zero (Figure 13-4).

It should be noted that for category B PC hyper-static flexural members the secondary moment M_{p2} caused by prestressed force N_p should be taken into consideration while σ_{ss} is calculated, so Equation (13-30) and Equation (13-33) can be converted as follows

$$\sigma_{ss} = \dfrac{M_s \pm M_{p2} - N_{p0}(z - e_p)}{(A_p + A_s)z} \quad (13\text{-}34)$$

$$e = e_p + \dfrac{M_s \pm M_{p2}}{N_{p0}} \quad (13\text{-}35)$$

Where, positive and negative sign separately stands for the same and reverse direction of M_{p2} and M_s.

(2) Control of crack width

In the **CHBC**, the maximum crack width of category B PPC is specified as follows:

For PC members with prestressing rolled threaded bars, 0.20mm under class Ⅰ and Ⅱ conditions, and 0.15mm under class Ⅲ and Ⅳ conditions.

For steel wire and steel strand, 0.10mm under class Ⅰ, Ⅱ, Ⅲ and Ⅳ conditions and no crack under V conditions.

3) Calculation of deflection

In the **CHBC**, to calculate the deflection of category B PPC flexural members, we should consider short and long term effect combination. The principle is the same as that of PC flexural members. The bending stiffness of category B PPC flexural members is adopted by the section moment M_s under combination for short-term action effects. Its flexural rigid piecewise interpolation:

Under cracking moment M_{cr}: $\qquad B_0 = 0.95 E_c I_0 \qquad$ (13-36)

Under moment $(M_s - M_{cr})$: $\qquad B_{cr} = E_c I_{cr} \qquad$ (13-37)

Where I_0 and I_{cr} are respectively the transferred rotational inertia of the total section or cracked section.

4) Calculation of fatigue

The steel rebars in high stress region of PPC flexural members often have fatigue breakdown. Without cracks, the members will not have fatigue failure. Thus, the purpose of fatigue check for members is to check the steel stress in tension region of the perpendicular section and control the stirrups stress of the diagonal section.

The fatigue check of stress range ability of tension steel is done by $\Delta\sigma_m = \sigma_{max}^p - \sigma_{min}^p$ and the allowable value $[\Delta\sigma_p]$ is determined by tests or by Table 13-1 if test data is not enough.

The allowable value of stress rangeability of steel   Table 13-1

kinds of steel	Plain surface steel bars	Regular deformed bars	Plain surface prestressed steel wire	Steel strands	High-strength bars
$[\Delta\sigma_p]$ (MPa)	250	150	200	200	80

13.4 Design of PPC flexural member crack allowed

1) Distribution of steel rebars by method of prestress degree λ

λ stands for the prestress degree: $\lambda = M_0 / M_s$.

Where, M_0 is the decompression moment when the compression pre-stress at lower limb reaches zero for flexural members and M_0 can be calculated as follows

$$M_0 = \sigma_{pc} W_0$$

Besides $\qquad \sigma_{pc} = \dfrac{N_{pe}}{A}\left(1 + \dfrac{e_p \cdot y_x}{i^2}\right)$

So

$$N_{pe} = \frac{\lambda M_s}{W_0} \cdot \frac{A}{1 + \dfrac{e_p \cdot y_x}{i^2}} \qquad (13\text{-}38)$$

Chapter 13 · Partially Prestressed Concrete Flexural Members

$$A_p = \frac{N_{pe}}{\sigma_{con} - \sigma_l} \tag{13-39}$$

Where σ_{con} is the control stress; σ_l is the combination of loss of prestress, to estimate the loss we can assume 20% ~ 30% of the control stress for pre-tensioned members and 15% ~ 25% of the control stress except for frictional loss for post-tensioned ones.

After the area of prestressed steels A_p is determined by Equation (13-39), the area A_s of regular steels can be calculated by the bearing capacity of normal section of flexural members. For example, for rectangular beams with single prestressed steel A_p and reinforcement A_s in tension region, sized with width b and height h, the following equations can be derived from basic mechanical equilibrium equation

$$f_{cd}bx = f_{pd}A_p + f_{sd}A_s \tag{13-40}$$

$$\gamma_0 M_d = f_{pd}A_p\left(h - a_p - \frac{x}{2}\right) + f_{sd}A_s\left(h - a_s - \frac{x}{2}\right) \tag{13-41}$$

Where M_d is the designed value of combined moment.

From Equations (13-40) and (13-41) the compress height x and the area of regular steels A_s can be calculated and x should obey the rule of $x \leq \xi_b h_0$.

Calculation process (taking rectangular section flexural beam as an example) is generally as follows:

(1) Calculate the geometrical characteristics of concrete gross section such as A, I, W and y_x.

(2) Assuming the action spot of the composite prestress force we can get the eccentricity; assuming the action spot of the composite force of prestressed and regular steel we can get the effective depth.

(3) Prestress degree λ is generally selected in the range of 0.6 to 0.8.

(4) The area of prestressed steel is calculated by Equation (13-39), and the area of corresponding regular steel is obtained by Equations (13-40) and (13-41).

(5) Arranging proper prestressed and regular steel, check whether the section can meet the ultimate load-carrying capacity.

2) The prestressed steel design by method of nominal tension stress

According to PPC control of crack width under service stage, this method aims to calculate the area of prestressed and regular steel at controlled section.

Steps can be expressed as: cracked members are treated as non-cracked; the maximum tension stress at tension edge is calculated by theory of mechanics of materials. Caused by cracking, the calculated tension stress is certainly greater than the tensile strength of concrete. Thus this tension stress is nominal stress that must be greater than the bending tensile strength because of cracked section. Based on large quantity of tests nominal tension stress is related to different limited crack width.

Under service load, by material mechanics method, the maximum tensile stress at the edge σ_{st} of uniform non-cracked section plus effective prestressing σ_{pc} is the total tensile stress that is nominal tensile stress. This nominal tensile stress is limited as

$$\sigma_{st} - \sigma_{pc} \leqslant [\sigma_{ct}] \tag{13-42}$$

Where, $[\sigma_{ct}]$ is the allowable nominal tensile stress of concrete, calculated by Equations (13-44) and (13-45).

Steps to estimate the steel area by method of nominal tension stress:

(1) Calculating the geometrical properties of concrete.

(2) Calculating σ_{st},

$$\sigma_{st} = \frac{M_s}{W} \tag{13-43}$$

Where, M_s is the moment caused by short-term effect of combined loads; W is the elastic modulus at the tensile edge which could be calculated by gross section.

(3) Determining allowable nominal tensile stress of concrete $[\sigma_{ct}]$: Calculating the allowable crack width $[\sigma_{ct}]$ according to the operating requirement and condition by equation as:

Post-tensioned method $\qquad [\sigma_{ct}] = \beta[\sigma'_{ct}] + 4\rho \leqslant f_{cu,k}/4 \tag{13-44}$

Pre-tensioned method $\qquad [\sigma_{ct}] = \beta[\sigma'_{ct}] + 3\rho \leqslant f_{cu,k}/4 \tag{13-45}$

Where, $[\sigma'_{ct}]$ is the basic nominal tensile stress of concrete that is related to prestress mode, concrete strength and crack width, can be selected from Table 13-2; β is the coefficient of correction of height and can be got from Table 13-3; ρ is the ratio of reinforcement of regular steels in tensile region of flexural members $[\rho = A_s/(bh_0)]$, for 1% of ρ, allowable stress of pre-tensioned members $[\sigma'_{ct}]$ will increase by 3MPa, and post-tensioned by 4MPa; $f_{cu,k}$ is the strength grade of concrete.

(4) Calculating the effective prestress N_{pe} and corresponding area A_p of prestressed steels

From Equation (13-42), the essential effective compression stress σ_{pc} of concrete tensile edge is as follows

Basic allowable tensile stress of concrete $[\sigma'_{ct}]$ (MPa)   Table 13-2

Member sort	Allowable crack width (mm)	Strength grade of concrete		
		C30	C40	≥C50
Pre-tensioned members	0.1	—	4.6	5.5
	0.15	—	5.3	6.2
	0.20	—	6.0	6.9
Post-tensioned members	0.1	3.2	4.1	5.0
	0.15	3.5	4.6	5.6
	0.20	2.8	5.1	6.2

Note: Only applicable to concrete strength grades under C60.

Coefficient of correction of height for allowable nominal tensile stress of concrete   Table 13-3

The height of members (mm)	≤200	400	600	800	≥1 000
Coefficient of correction	1.1	1.0	0.9	0.8	0.7

$$\sigma_{pc} \geq \sigma_{st} - [\sigma_{ct}] \quad (13\text{-}46)$$

$$\sigma_{pc} = N_{pe}\left(\frac{1}{A} + \frac{e_p}{W}\right) \quad (13\text{-}47)$$

From Equations (13-46) and (13-47), expressing σ_{st} by Equation (13-43), it can be get

$$N_{pe} \geq \frac{\dfrac{M_s}{W} - [\sigma_{ct}]}{\dfrac{1}{A} + \dfrac{e_p}{W}} \quad (13\text{-}48)$$

The essential area of prestressed steels is

$$A_p = \frac{N_{pe}}{\sigma_{con} - \sum \sigma_l} \quad (13\text{-}49)$$

Where e_p is the eccentricity from prestressed steel to the mass axis of un-cracked section, $e_p = y - a_p$, y and a_p are respectively the mass axis and the distance from prestressed steel to tensile edge, and a_p can be assumed first; A is the area of members and gross section can be used in calculation; $\sum \sigma_l$ is total prestressing loss. When estimating the total loss for pre-tensioned members, $(0.2 \sim 0.3)\sigma_{con}$ can be used, While for post-tensioned members, $(0.25 \sim 0.35)\sigma_{con}$ can be used.

(5) Calculating the area of regular steels A_s according to the normal bearing capacity of flexural members.

(6) According to the strength calculation of normal section check the compressive height x whether $x \leq \xi_b h_0$ to avoid over-reinforced damage.

3) Construction requirement

Considering the characteristics of PPC flexural members, in the **CHBC** and **the Design and Suggestion for PPC Members** issued by China Civil Engineering Society (CCES), the construction requirements are proposed as follows:

(1) Mixed reinforcement is suggested to be used in PPC beams. Ribbed regular steel with smaller diameter could be arranged near the tension edge with closer bar spacing.

(2) In **the Design and Suggestion for PPC Members**, the amount of regular steels in mixed-reinforced flexural members can be determined by the prestress degree and the rule as:

①If high prestress degree ($\lambda > 0.7$) is used, to guarantee the safety and ductility, small diameter and close bar spacing should be used. Minimum percentage of reinforcement $\rho_s = A_s/A_{he} = (0.2 \sim 0.3)\%$ is used to calculate the area of regular steels, where A_s is the area of regular steel and A_{he} is the area of concrete in tensile region.

②If middle prestress degree ($0.4 \leq \lambda \leq 0.7$) is used, because of the increase of regular steels, the diameter of steels, especially the outer steels, should be oversized.

③If low prestress degree ($\lambda < 0.4$) is used, the amount of regular steels is greater than that of prestressed steels, so the mechanical performance is near that of RC members and we can arrange regular steels according to the construction requirement.

(3) HRB335 and HRB400 hot ribbed steels are to be properly used.

(4) Percentage of reinforcement.

①Minimal percentage of reinforcement

In ***the Design and Suggestion for PPC Members***, minimum percentage of reinforcement of PPC flexural members should obey that

$$M_u/M_{cr} > 1.25 \tag{13-50}$$

Where M_u is the normal bending bearing capacity of flexural members; M_{cr} is the cracking moment.

PPC flexural members that meet the challenge should have enough emergency capacity to avoid being damaged soon after cracking.

②Maximal percentage of reinforcement should meet the compression height $x \leq \xi_b h_0$.

Chapter 14
Calculation of the Flexural Members of Unbonded Prestressed Concrete

    The unbonded prestressed concrete beam refers to the post-tensioned concrete beam with the main reinforcement composed of the unbonded prestressed steel bars, while the unbonded prestressed rebar refers to a single steel wire or multi-wire prestressed steel tendons of high strength, steel strands or thick steel bars with the special anti-corrosion grease coating for the full length and with the external coating so as to make them have no connection to the concrete around it and the corresponding longitudinal sliding happen when they are tensioned.

    The construction method similar to the common reinforced concrete member can be used for the unbonded prestressed concrete member. The layout is the same as that for common reinforcement. When the concrete is placed, the tensioning and anchoring for prestressed rebar will be carried out after concrete reaches its strength specified. In this way, it will save procedures for preembedding of pipes, penetration of wires and pumping of cement slurry so that the need for engineering equipment is reduced, the process is simplified, and the construction period is shortened. Therefore, the comprehensive economic benefit is improved.

14.1 Flexural strength of unbonded prestressed concrete members

1) Concepts and classifications

Unbonded prestressed concrete beams can be classified into two types, unbonded prestressed concrete beam and unbonded partially prestressed concrete beams. As to the former, the unbonded prestressed rebar is used for all flexural reinforcement, while, as for the latter, the mixed rebar beam with unbonded prestressed reinforcement and the proper amount of bonded non-prestressed steel bars are adopted for the flexural rebar.

2) Mechanical flexural properties and failure characteristics

(1) Comparisons of the mechanical properties and failure characteristics between unbonded and bonded prestressed concrete beams

①Development of cracks and modes of failure

Figure 14-1 shows the whole cracking development process of two different beams under loading. The test results show that the brittle failure of unbonded prestressed concrete beams is obvious under testing load as the dramatic increase of the width and height, when the cracking of unbonded prestressed concrete beam happens and results in concrete crushing.

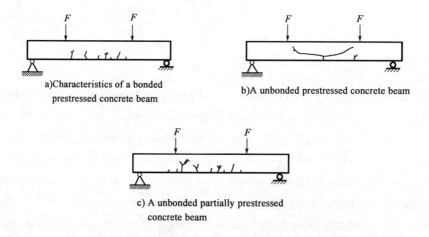

Figure 14-1 Cracking of bonded and unbonded concrete beams

②Comparison of the deflection curves of loads and the mid span

It's seen from Figure 14-2 that there are three types of the load curvature for bonded prestressed concrete beams, while as to unbonded prestressed concrete beams, not only will there appear no straight lines at the third phase, but also no obvious ones at the second phase.

③Comparison between the rebar stress and the change of load on a cross section of maximum moment of beams

Chapter 14 · Calculation of the Flexural Members of Unbonded Prestressed Concrete

It can be clearly seen from Figure 14-3 that the increase of load from the unbonded prestressed reinforcement will always be less than that of the bonded prestressed reinforcement, and with the increasing of load, the difference will become greater. At the cross section of the maximum moment of beams, the stress increment is less than that of the bonded rebar. From the diagram, it can be known that from the initial loading to failure, the stress from unbonded prestressed steel bars is less than that of the bonded rebar.

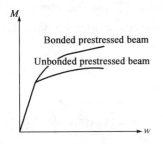

Figure 14-2 Impact of bonded force on the deflection of a beam

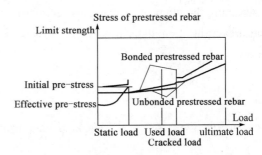

Figure 14-3 Impact of bonded force on the stress of prestressed rebar

Before the concrete members crack, the stress increment induced by unbonded rebar under load can be calculated through the compatibility condition of longitudinal deformation, i. e. the total extension of the unbonded steel rebar should be equal to that of the concrete around the total length of rebar.

Assuming the moment M at the arbitrary cross section, then, the strain of an arbitrary cross section can be obtained as

$$\varepsilon_c = \frac{\sigma_c}{E_c} = \frac{M}{E_c I_c} y \qquad (14\text{-}1)$$

The total extension of the concrete for members along the full length of the unbonded reinforcement will be computed at this time as follows

$$\Delta = \int \varepsilon_c dx = \int \frac{M}{E_c I_c} y dx \qquad (14\text{-}2)$$

If the length of the unbonded steel rebar is l, the stress increment of the unbonded steel rebar is determined by the following

$$\frac{\Delta}{l} = \int \frac{M}{E_c I_c l} y dx \qquad (14\text{-}3)$$

The corresponding stress increment of the unbonded steel rebar is as follows

$$\Delta \sigma_p = E_p \frac{\Delta}{l} = \frac{E_p}{E_c l} \int \frac{M}{I_c} y dx \qquad (14\text{-}4a)$$

Due to $\alpha_{Ep} = \dfrac{E_p}{E_c}$, the following can be achieved

$$\Delta \sigma_p = \frac{\alpha_{Ep}}{l} \int \frac{M}{I_c} y dx \qquad (14\text{-}4b)$$

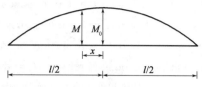

Figure 14-4 Bonded prestressed concrete of a simply-supported beam

To compare the stress increment of the unbonded steel rebar under the function of the moment of the unbonded reinforcement, the simply-supported beam of a single span with the rectangle cross section assembled in a linear unbonded way is set up as an example under the evenly distributed load, and shown in Figure 14-4.

Assuming the moment of the cross section in the mid-span is M_0 and the eccentric moment of the prestressed in the mid-span is e, the cross section at the position x from the mid-span (m) is as follows

$$M = M_0\left[1 - 4\left(\frac{x}{l}\right)^2\right] \tag{14-5}$$

From Equation (14-5), the stress of unbonded steel rebar is computed as follows

$$\Delta\sigma_p = \frac{\alpha_{Ep}}{lI_c}\int_{-l/2}^{l/2} M_0\left[1 - 4\left(\frac{x}{l}\right)^2\right]e\,dx = \frac{2}{3}\frac{\alpha_{Ep}M_0 e}{I_c} \tag{14-6}$$

In this formula, $\frac{\alpha_{Ep}M_0 e}{I_c}$ is the mid-span stress increment of the bonded steel rebar. Therefore, under the same moment for the unbonded and the bonded steel rebar, the stress increment of the bonded rebar is larger than that of the unbonded rebar. For example, if the rebar is placed in a linear way, the stress increment of the unbonded rebar is two third of that for the bonded rebar.

The stress increment at the cross section of the unbonded reinforcement placed in a parabolic curve way can be computed by the same method, and it's the eight fifteenth of the stress increment of the bonded reinforcement.

From the contents discussed above, the flexural strength of the unbonded steel rebar is lower than that of the bonded steel rebar; and under the function of load, there'll be a few cracks and they will develop fast. The mode of failure is obviously the brittle failure. To avoid these disadvantages, the partially prestressed unbonded rebar are usually used. The calculation of the bonded prestressed concretes is introduced in following section.

(2) Comparison of the mechanical properties and the failure characteristics between unbonded partially prestressed concrete beams and bonded prestressed concrete beams

According to the research results of many research institutes and universities, the following main points have been summarized.

①The moment and curvature (seen in Figure 14-5) of unbonded partially prestressed concrete beams will display a three-straight-line shape just as that of bonded partially prestressed concrete beams.

②Due to the constraint from the common steel rebar, the number and space are very close to those of the common concrete beam with the same reinforcement [shown in Figure 14-1c)].

③Under normal conditions, a partially prestressed unbonded concrete beam won't bend and fail until the common steel wire yields, cracks extend upwards till the edge of the area under pressure and the concrete strain reaches to the limit.

④ Although there still exist the characteristics of the equal stress of unbonded partially prestressed reinforcement for the full length (regardless of friction) and of the limit stress not exceeding yielding strength $\sigma_{0.2}$, the value of limit stress will be far larger than that of unbonded beams.

⑤ The stress increment $\Delta\sigma_y$ of the unbonded partially prestressed reinforcement is closely related to the comprehensive rebar assembly index of a beam when a beam fails. Their relationship can be expressed as follows

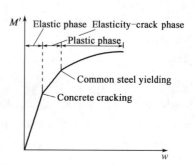

Figure 14-5 Moment-curvature curves of an unbonded partially prestressed concrete beam

$$\beta_0 = \beta_p + \beta_s = \frac{A_p \sigma_{pe}}{bh_p f_{cd}} + \frac{A_s f_{sd}}{bh_s f_{cd}} \quad (14\text{-}7)$$

Where A_p and σ_{pe} are the cross section area and the effective prestressing of the unbonded prestressed steel bar, respectively; A_s and f_{sd} are the cross section area and the design value of the tensile strength of the bonded prestressed steel bar, respectively; b is the width of a beam; h_p is the distance from the center of gravity of a cross section to the edge of a cross section under pressure; f_{cd} is the design value of tensile strength of concrete.

The test results show that the linear relationship between $\Delta\sigma_p$ and β_0 is good, and $\Delta\sigma_p$ will increase when β_0 decreases. In fact, β_0 can almost reflect the neutral axis and the rotation capacity of a beam as the normal section is destroyed, while the ultimate stress increment of the unbonded steel rebar is closely related to the position and rotation capacity of the neutral axis of a beam.

⑥ As to unbonded partially prestressed concrete beam, the span-height ratio has no obvious impact on $\Delta\sigma_p$ under three point load. Under one point load at the middle span, the span-height ratio will have impact on $\Delta\sigma_p$. When the span-height ratio is different and the value of β_0 close to $\Delta\sigma_p$ will decrease while the span-height ratio increases. If the span-height ratio is the same or quite close to β_0, $\Delta\sigma_p$ under one point concentrated load will be less than the value of $\Delta\sigma_p$ under three point load.

14.2 Calculation of flexural members of unbonded partially prestressed concrete

The ultimate carrying capacity, the stress calculation at construction and the service stages, as well as the calculation of deformation and the widest width of cracks (members of category B) will all be carried out in the same way as those for bonded prestressed concrete beams. These computing methods can be found in the relevant chapters of the **CHBC** or other design specifications. The calculating method for unbonded prestressed rebar will be introduced here.

1) Ultimate stress σ_{pu} of unbonded prestressed steel rebar

The ultimate stress σ_{pu} is a key factor for calculating of the flexural strength of unbonded prestressed steel rebar. But lots of factors will affect the ultimate stress σ_{pu}, such as the effective stress of unbonded reinforcement, the comprehensive reinforcement assembly index β_0, the span-height ratio and the loading condition. Therefore, the ultimate stress σ_{pu} must be obtained and analyzed from the test when the steel rebar bends and fails.

The common formula expression of σ_{pu} is as follows

$$\sigma_{pu} = \sigma_{pe} + \Delta\sigma_p \tag{14-8}$$

Where σ_{pe} is the effective stress of unbonded prestressed steel rebar; $\Delta\sigma_p$ is the stress increment of unbonded prestressed steel rebar under ultimate loading.

In line with the **Technology Code for Unbonded Prestressed Concrete Structures**, the carbon steel wire and steel strand will be used to make flexural members of unbonded prestressed steel rebar, and the designed ultimate stress under the ultimate carrying condition can be calculated as follows.

If the span-height ratio of a member is ≤35

$$\sigma_{pu} = \frac{1}{\gamma_s}[\sigma_{pe} + (500 - 770\beta_0)] \tag{14-9a}$$

If the span-height ratio of a member is >35

$$\sigma_{pu} = \frac{1}{\gamma_s}[\sigma_{pe} + (250 - 380\beta_0)] \tag{14-9b}$$

Where β_0 is the comprehensive reinforcement assembly index, $\beta_0 \leq 0.45$; γ_s is the coefficient of materials, $\gamma_s = 1.2$.

In the meantime, the value of σ_{pu} should be less than that of the designed tensile strength value of unbonded prestressed steel rebar, and should be more than that of the effective stress σ_{pe} of unbonded prestressed reinforcement.

The **Suggestions for Unbonded Partially Prestressed Concrete Structure** developed by the China Association of Civil Engineering suggests that the equation for calculating the ultimate stress Equation (14-8) for the flexural member of unbonded prestressed reinforcement should be adopted, while the stress increment $\Delta\sigma_p$ of unbonded prestressed rebar should be selected in Table 14-1 when the unbonded prestressed reinforcement is made of steel wires or steel strand of high strength.

Stress increment $\Delta\sigma_p$ (MPa) of unbonded prestressed steel rebar under ultimate load

Table 14-1

Indexes of reinforcement $\beta_p + \beta_s$	$\frac{l}{h_p}$		Indexes of reinforcement assembly $\beta_p + \beta_s$	$\frac{l}{h_p}$	
	10	20		10	20
0.05	500	500	0.20	350	300
0.10	500	500	0.25	250	200
0.15	450	400			

Note: l is the total length of a beam and h_p is the distance from the center of gravity of a cross-section of unbonded prestressed reinforcement to the compressive edge of concrete. When $\beta_0 = \beta_p + \beta_s$, it can be calculated in line with Equation (14-7).

As $\sigma_{pu} = \sigma_{0.2}$ is obtained by Formula (14-8), the **Suggestions for Unbonded Partially Prestressed Concrete Structures** specifies $\sigma_{pu} > \sigma_{0.2}$, i. e. the designed tensile strength should be obtained and it requires that the value of σ_{pe} in the equation be no less than $0.6\sigma_{0.2}$.

2) Calculation of the friction loss of unbonded prestressed steel rebar

Before calculating the effective stress σ_{pe}, the friction loss σ_l of the pre-stress must be estimated. The estimation of the prestressing loss of unbonded prestressed reinforcement might be computed in accordance with the relevant calculation methods for members of prestressed concrete with the post tensioning method in Chapter 12.

When estimating the prestressing loss caused by friction, it's very important to make good use of the coefficients k and μ of friction between the prestressed steel bars and the wall of pores. The coefficient of friction between the unbonded prestressed reinforcement and the concrete around it will change greatly when different coating, packing materials for external layers and process of production, and the different shapes of cross sections are involved.

The coefficients of friction of unbonded reinforcement suggested by the **Technique Code for Unbonded Prestressed Concrete Structures** in China are given in Table 14-2.

Coefficients of friction of unbonded prestressed steel bars   Table 14-2

Unbonded prestressed steel	k	μ	Unbonded prestressed steel	k	μ
$7\phi^P 5$ round wires	0.003 5	0.10	$\phi^S 15.2$ strands	0.004 0	0.12

3) Carrying capacity of the normal cross section for a unbonded partially prestressed concrete beam

The calculating method for the unbonded partially prestressed concrete beam is similar to that of the bonded prestressed concrete beams. But as to the unbonded steel bars, the ultimate stress σ_{pu} will be adopted instead of the designed tensile strength f_{sp}.

Take a cross section of a rectangular beam for example, where the unbonded prestressed steel bar A_p and the non-prestressed reinforcement A_s are placed at the compressive area (shown in Figure 14-6). Based on the free body balance and references of the **CHBC**, the carrying capacity of the normal section will be obtained as follows

$$\sigma_{pu} A_p + f_{sd} A_s = f_{cd} bx \tag{14-10}$$

$$\gamma_0 M_d \leqslant M_u = f_{cd} bx \left(h_0 - \frac{x}{2} \right) \tag{14-11}$$

The effective height of a cross section, h_0, in Formula (14-11) is

$$h_0 = h - a$$

While a can be calculated in the following way

$$a = \frac{\sigma_{pu} A_p a_p + f_{sd} A_s a_s}{\sigma_{pu} A_p + f_{sd} A_s} \tag{14-12}$$

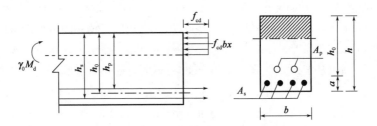

Figure 14-6　Carrying capacity of unbonded partially prestressed concrete

Where, a_p and a_s are the distance from the center of gravity to the compression edge of unbonded steel bars and non-prestressed reinforcement, respectively.

The following conditions should be satisfied if the depth of compression x at the compression area is obtained as follows

$$x \leqslant \xi_b h_0$$

But the ***Suggestion of Unbonded Partially Prestressed Concrete Structure*** developed by the China Association of Civil Engineering suggests the following conditions in consideration of the requirement of improving the ductility of the flexural members of partially prestressed concrete at a proper degree:

(1) For the common members, $x \leqslant 0.4 h_p$;

(2) For the members of high ductility, $x \leqslant 0.3 h_p$.

The meaning and calculation of h_p can be seen in Table 14-1.

14.3　Design of a cross-section of a flexural member of unbonded partially prestressed concrete member

When the structure of partially prestressed concrete is designed and calculated to satisfy performance requirements, the proper prestressing must be determined as far as the design is concerned, so as to decide the number of steel wires and to monitor the tension stress, or to obtain the ratio of the prestressed and non-prestressed reinforcement assembly for mixed rebar assembly, i.e. the degree of prestressing λ introduced in Chapters 11 and 13. With the utilization of the degree of prestressing, the reinforcement estimate of unbonded partially prestressed concrete beams will carried out.

Besides the method of the degree of prestressing, the method of partial prestressing ratio is also used in estimating the non-cohesive partially prestressed concrete beams.

The partial prestressing ratio can be expressed by the ratio of the ultimate resistance moment from prestressed steels $(M_u)_p$ to the flexural moment from the total steels $(M_u)_{p+s}$ under stress (for mixed rebar assembly). When the partially prestressing ratio is expressed by the characteristics of prestressed reinforcement and non-prestressed rebar, it may be expressed as follows

Chapter 14 · Calculation of the Flexural Members of Unbonded Prestressed Concrete

$$PPR = \frac{A_p \sigma_{pu}}{A_p \sigma_{pu} + A_s f_{sd}} \tag{14-13}$$

Where A_p and σ_{pu} are the areas of a cross section and the ultimate tensional stress of prestressed steel bar, respectively; A_s and f_{sd} are the areas and the tension strength design of non-prestressed steel wires, respectively.

The rectangular cross section is used to introduce the method of PPR for estimating reinforcement of unbonded partially prestressed concrete beams.

1) Calculation in accordance with the carrying capacity of a normal cross section

Due to the following, it can be seen from Figure 14-6 that

$$(A_p \sigma_{pu} + A_s f_{sd}) h_0 = A_p \sigma_{pu} h_p + A_s f_{sd} h_s$$

$$h_0 = \frac{A_p \sigma_{pu} h_p + A_s f_{sd} h_s}{A_p \sigma_{pu} + A_s f_{sd}}$$

Therefore,
$$h_0 = (PPR) h_p + [1 - (PPR)] h_s \tag{14-14}$$

From Equation (14-16), it is assumed that $M_u = \gamma_0 M_d$, and the following can be computed

$$\gamma_0 M_d = f_{cd} b h_0^2 \xi (1 - 0.5\xi)$$

The result is
$$\xi = 1 - \sqrt{1 - \frac{2\gamma_0 M_d}{b h_0^2 f_{cd}}} \tag{14-15}$$

From Equations (14-10) and (14-13), the following can be obtained

$$f_{cd} b x = \sigma_{pu} A_p + f_{sd} A_s = \frac{\sigma_{pu} A_p}{PPR}$$

Substituted into Equation (14-16) and supposing $M_u = \gamma_0 M_d$, the result can be achieved

$$\gamma_0 M_d = \frac{A_p \sigma_{pu}}{PPR} h_0 (1 - 0.5\xi)$$

From Equation (14-11) the following can be resolved

$$A_p = \frac{(PPR) \gamma_0 M_d}{(1 - 0.5\xi) h_0 \sigma_{pu}} \tag{14-16}$$

$$A_s = [1 - PPR] \frac{A_p \sigma_{pu}}{(PPR) f_{sd}} \tag{14-17}$$

The above calculation can be obtained by adopting the estimate method of the partial prestressing ratio A_p and A_s. The detailed calculation steps include:

(1) Assuming and optimizing PPR from 0.7 to 0.95.

(2) Assuming the distance between the center of the gravity of the prestressed steel bars and non-prestressed steel bars to the lower edge of the cross section of a beam is a_p and a_s respectively, $h_p = h - a_p$ and $h_s = h - a_s$ are computed. h_0 can be obtain according to Equation (14-14).

(3) ξ can be worked out by Equation (14-15). The value of ξ is checked to see if it satisfies the condition of $\xi \leq \xi_b$. If not, PPR should be adjusted or the sizes of a cross section rectified, and the calculation will be carried out again.

(4) Assuming σ_{pu} is the ultimate stress of unbonded reinforcement, according to the experiences gained from designing of bridges, $\sigma_{pu} = (0.6 \sim 0.8)f_{pd}$ is suggested.

(5) A_p and A_s can be worked out from Equations (14-16) and (14-17).

2) Estimation in line with construction and phases requirements

At this time, the estimate area A_p of unbonded prestressed reinforcement will mainly be carried out. As to members of category A of partially prestressed concrete, it can be obtained with the reference to the method introduced in Chapter 13; while as for members of category B, the nominal tensioning stress can be used for the estimating of A_p when there is a requirement for estimation during construction.

In consideration of the above two aspects for choosing proper A_p and A_s, the following requirement should be satisfied in the meantime

$$\frac{A_s f_{sd}}{A_s f_{sd} + A_p \sigma_{pu}} > 0.25$$

14.4 Structure of flexural members of unbonded partially prestressed concrete

As far as the structure of flexural members of unbonded partially prestressed concrete is concerned, the unbonded characteristics are described on the basis of structural requirements for the bonded partially prestressed concrete in the following paragraphs.

(1) The carbon steel wires, steel strand and heat treated reinforcement should be used as much as possible, and correspondingly the grade of concrete strength should not be lower than C40. Hot-rolled reinforcement of HRB335 and HRB400 should be used for non-prestressed steel bars, and the diameter of the steel wires should range from 12mm to 14mm but not exceeding 20mm.

(2) If flexural members of mixed reinforcement assembly are adopted, the placement of non-prestressed rebar should be close to the compression edge of a cross section, and there should be enough thickness of concrete for protection of rebars that can satisfy the requirement of the specification. While, the unbonded prestressed steel bar should be arranged above the non-prestressed reinforcement, i.e. $h_p < h_s$, so as to increase the thickness of concrete protecting layer of unbonded prestressed steel bars. Furthermore, once cracks appear, the non-prestressed reinforcement will control the width of cracking.

(3) The minimum number of non-prestressed steel rebar should satisfy $\dfrac{A_s f_{sd}}{A_s f_{sd} + A_p \sigma_{pu}} > 0.25$ while f_{sd} will be no larger than 400MPa.

(4) In order to ensure the durability of unbonded prestressed concrete beams, first of all, there'll be the need to ensure that the unbonded prestressed reinforcement won't corrode during construction. Therefore, the anti-corrosion coating on the surface of the prestressed steel bars for

Chapter 14 · Calculation of the Flexural Members of Unbonded Prestressed Concrete

the full length should be used. Such corrosion protection as asphalt coating, stone matrix asphalt, butter and wax will be used. Nowadays, special oils have been adopted for the treatment of the unbonded prestressed reinforcement by manufacturers because of their good chemical stability of no erosion to external layer, good waterproof and lubrication performance. The coating layer should be protected in a special way, such as plastic casing and waterproof plastic paper bags. There should be certain space between the protecting layer and the prestressed reinforcement so as to let the prestressed rebar slide in a relative way.

(5) The test and acceptance of the anchorage of unbonded prestressed steel bars must be in accordance with the procedure specified in the *Anchorage, Grips and Couplers* (GB/T 14370—2007). They won't be used until they pass the test. The reliable long term anti-corrosion measures should be adopted for anchorage in structures, and the method of post casting and sealing concrete will be used.

[**Example 14-1**] The area of the rectangular cross section of a unbonded partially prestressed concrete beam is 300mm × 600mm (see in Figure 14-7), the steel strand for non-prestressed reinforcement is $4\phi^s15.2$, and the area is $A_p = 560\text{mm}^2$; HRB335 is used for the non-prestressed steel bars, and the area of $4\phi16$ is $A_s = 804\text{mm}^2$; and the concrete is C40. The effective prestressed stress of unbonded prestressed steel bars is $\sigma_{pe} = 720\text{MPa}$. The computed span of a beam is 8m, and the total length is 8.5m. The structure is placed in the normal atmospheric environment, and the importance of structures is grade II. The moment under the mid-span cross section of a beam is that from the dead load of a beam $M_{G1} = 3\text{kN} \cdot \text{m}$. The dead load from phase II is $M_{G2} = 34\text{kN} \cdot \text{m}$, and the moment induced by the vehicle load is $M_{Q1} = 144.4\text{kN} \cdot \text{m}$.

Compute the carrying capacity of a normal cross section of the beam.

The calculation is as follows.

In line with the rule in the **CHBC**, the mid-span cross section of a beam will be calculated in accordance with the basic combination value of the function effect value when the ultimate condition of carrying capacity is designed

$$M_d = 1.2 \times (36 + 34) + 1.4 \times 144.4 = 286.2\text{kN} \cdot \text{m}$$

The coefficient for the grade of structural importance is $\gamma_0 = 1.0$.

(1) Ultimate tension stress of the unbonded prestressed stress σ_{pu}.

σ_{pu} can be calculated by the method from the **Suggestion of Unbonded Partially Prestressed Concrete Structure**.

Where, $h = 600, b = 300, a_p = 82$ and $a_s = 40$.

From the cross section shown in Figure 14-7, it's seen that $a_p = 82\text{mm}$ and $a_s = 40\text{mm}$. The following can be obtained: $h_p = h - a_p = 518\text{mm}$ and $h_s = h - a_s = 560\text{mm}$, and then the following achieved

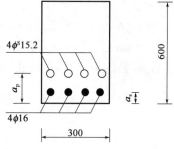

Figure 14-7 Cross section of the beam (unit: mm)

$$\frac{l}{h_p} = \frac{8\,500}{518} = 16.4$$

The comprehensive reinforcement assembly index β_0 will be calculated by Formula 14-2, i.e.

$$\beta_0 = \beta_p + \beta_s$$

$$= \frac{A_p \sigma_{pe}}{b h_p f_{cd}} + \frac{A_s f_{sd}}{b h_s f_{cd}}$$

$$= \frac{560 \times 720}{300 \times 518 \times 18.4} + \frac{804 \times 280}{300 \times 560 \times 18.4}$$

$$= 0.214$$

Due to $\frac{l}{h_p} = 16.4$ and $\beta_0 = 0.214$, the stress increment is $\Delta\sigma_p = 290\text{MPa}$ from Table 14-1.

From Equation (14-3), the ultimate tension stress σ_{pu} of the unbonded prestressed steel bar can be worked out as follows

$$\sigma_{pu} = \sigma_{pe} + \Delta\sigma_p$$

$$= 720 + 290$$

$$= 1\,010(\text{MPa}) < \sigma_{0.2} = f_{pd} = 1\,260\text{MPa}$$

And
$$\frac{A_s f_{sd}}{A_s f_{sd} + A_p \sigma_{pu}} = \frac{280 \times 804}{280 \times 804 + 1\,010 \times 560}$$

$$= 0.285 > 0.25$$

It satisfies the requirement of the minimum content for non-prestressed reinforcement.

(2) Calculation of the compression height x of a cross section

From Equation (14-5), the following can be calculated

$$x = \frac{\sigma_{pu} A_p + f_{sd} A_s}{f_{cd} b}$$

$$= \frac{1\,010 \times 560 + 280 \times 804}{18.4 \times 300}$$

$$= 143\text{mm} < 0.3 h_p = 0.3 \times 518 = 155\text{mm}$$

(in accordance with the requirement for ductility specified by the **Suggestion of Unbonded Partially Prestressed Concrete Structure**)

$$< \xi_b h_0 = 0.4 \times 512 = 208\text{mm}$$

It will meet the requirement specified by the **CHBC**.

(3) Checking of the normal carrying capacity of a cross section

From Equation (14-7), the following can be obtained

$$a = \frac{\sigma_{pu} A_p a_p + f_{sd} A_s a_s}{\sigma_{pu} A_p + f_{sd} A_s}$$

$$= \frac{1\,010 \times 560 \times 82 + 280 \times 804 \times 40}{1\,010 \times 560 + 280 \times 804} = 70\text{mm}$$

and the effective height of a cross section is $h_0 = h - a = 600 - 70 = 530\text{mm}$.

Chapter 14 · Calculation of the Flexural Members of Unbonded Prestressed Concrete

From Equation(14-6), the flexural capacity of a mid-span cross section of a beam will be computed

$$M_u = f_{cd} bx \left(h_0 - \frac{x}{2} \right)$$
$$= 18.4 \times 300 \times 143 \times \left(530 - \frac{143}{2} \right)$$
$$= 361.92 \text{kN} \cdot \text{m} > \gamma_0 M_d = 286.2 \text{kN} \cdot \text{m}$$

The results satisfy the design requirement.

PART 3 | Masonry structures

Chapter 15
Basic Concepts and Materials of Masonry Structures

15.1 Basic concepts of masonry structures

Masonry structures are built according to a certain rule with bricks, stones and concrete precast blocks.

Masonry structures $\begin{cases} \text{brick masonry structure} & \boxed{\text{built with brick, stone and together with little stone and mortar}} \\ \text{concrete structure} & \boxed{\text{built with precast concrete blocks or integral cast concrete}} \end{cases}$

1) Mechanical characteristics and engineering applications of masonry structures

Masonry structures are strong in compressive strength but much weaker in tensile strength. These structures are generally used in foundations of bridges, culverts, arch bridges and retaining walls.

2) Advantages and disadvantages of masonry structures

(1) Advantages

①The use of materials such as bricks and stones can improve the thermal performance of a building, giving increased comfort both in hot summer and in cold winter. These materials may be ideal for passive solar applications.

②Bricks typically require no painting and so can provide a structure with reduced life-cycle costs. Sealing appropriately will reduce potential spalling due to frost damage. Non-decorative concrete block generally is painted or stuccoed if exposed.

③The appearance, especially when well crafted, can impart an impression of solidity and permanence.

④Masonry is very heat resistant and thus provides good fire protection.

⑤Masonry walls are more resistant to projectiles, such as debris from hurricanes or tornadoes than walls of wood or other softer, less dense materials.

(2) Disadvantages

①Extreme weather causes degradation of masonry wall surfaces due to frost damage. This type of damage is common with certain types of bricks, though rare with concrete blocks. If non-concrete (clay-based) brick is to be used, the care should be taken to select bricks suitable for the climate in question.

②Masonry tends to be heavy and must be built upon a strong foundation (usually reinforced concrete) to avoid settling and cracking.

③If expansive soils (such as adobe clay) are present, the foundation needs to be quite elaborate and the services of a qualified structural engineer may be required, particularly in earthquake prone regions.

15.2　Materials of the masonry

The common materials of bridges and culverts are stone, concrete, and mortar.

1) Stone masonry

(1) Concept

Stone masonry is the craft of shaping rough pieces of rock into accurate geometrical shapes, mostly simple, but some of considerable complexity, and then arranging the resulting stones, often together with mortar to form structures.

(2) Advantages

Stone masonry has good compressive strength and anti-freezing property.

(3) Engineering applications

Stone masonry is generally used in the bridge foundation, pier and abutment, retaining wall and so on. The stone used in the bridge or culvert is usually strong, uncracked, and resistant to weathering.

(4) Types of stone

Stone is classified, by the size, shape, and mining method, into rubble, block stone, fine dressed stone, half fine dressed stone and coarse stone (Table 15-1).

Chapter 15 · Basic Concepts and Materials of Masonry Structures

Technical requirements for stone materials Table 15-1

Classification	Mining method	Requirements
Rubble	Irregular stone block obtained by explosion	a. Of various shapes; b. Usually more than 150mm in height; c. Exclusive of the cobble stone and thin section stone
Block stone	Stone block obtained by the explosion with the stone bedding	a. Approximately rectangle and smooth in both sides; b. 200~300mm in height, width is 1.0~1.5 times of the height, and length is 1.5~3.0 times of the height; c. No need to refine
Fine dressed stone	Refined block stone	a. The shape should be hexahedron, and its surface roughness is usually less than 10mm; b. The height belongs to 200~300mm, width is 1.0~1.5 times of height; length is the 2.5~4.0 times of the height
semi-fine stone	Similar to fine dressed stone	The requirement is similar to those for fine stone, the surface roughness is generally less than 15mm
Coarse stone	Similar to fine dressed stone	The requirement is similar to those for fine stone, the surface roughness is generally less than 20mm

(5) Strength

The strength grades of the stone used in the bridge structure can be classified into MU30, MU40, MU50, MU60, MU80, MU100 and MU120. MU stands for the strength grade, where the figure is the compressive strength of the $70 \times 70 \times 70 mm^3$ cubic stone, and the unit is MPa.

2) Concrete

(1) Strength: The strength grades of the concrete are classified into C15, C20, C25, C30, C35 and C40.

(2) Types: The concrete can be classified into precast concrete block and integral in-situ casting concrete.

①Precast concrete block

Characteristics: The precast concrete block is convenient for constructing and thus resulting in shortened construction period.

The size and shape of the precast concrete block are regular so as to make the structures look better in appearance.

②Integral in-situ casting concrete

Characteristics:

a. The disadvantage of this concrete is the large shrinkage deformation, and thus, it is easy to produce shrinkage and temperature cracks;

b. Massive using leads to waste wood and its construction period is very long.

③Smaller stone concrete

It is composed of cement, coarse aggregate smaller than 20mm, fine aggregate and water.

This concrete has high compressive strength, and thus the amount of cement and sand in the concrete can be reduced largely.

3) Mortar

(1) Definition: It is typically a mixture made of sand, binder such as cement or lime, and water.

(2) Function: It is used to bind construction blocks together, and usually fills the gaps between the blocks. The block materials may be stone, brick, cinder blocks, etc.

(3) Types

①Non plastic materials mortar: It is composed of cement, sand and water that are mixed in a certain ratio; the non-plastic materials mortar has high strength property.

②Plastic materials mortar: It is made of cement, sand, plaster and water that are mixed in a certain ratio.

③Lime mortar: It is made of sand, plaster and water in a certain ratio; the lime mortar has low strength property.

(4) Strength: The strength grades of the mortar include M5, M7.5, M10, M15 and M20.

Where, M stands for the strength grade, the figure is the compressive strength of the cubic stone with side length of 70.7mm, and the unit is MPa.

(5) Basic requirements: The mortar should have strength, plasticity and water retention properties as follows:

①Enough strength, durability and bonding strength between blocks;

②Good plasticity convenient for constructing and helpful to improve the strength of the masonry structure;

③Enough water retention capacity.

4) Types of masonry structures

Based on the different materials, the masonry structures are classified as follows:

(1) Rubble masonry: The main building material of this masonry is the rubble, and the small stone is used as filling material.

(2) Block stone masonry: The main building material of this masonry is the block stone, such a masonry structure is built following the rule of tooth seams.

(3) Fine stone masonry: The main building material of this masonry is the fine stone, and the structure is made by the building method of controlling the surface roughness and the width of the joints.

(4) Semi-fine stone masonry: The main building material of masonry is the semi-fine stone, and the structure is built by the building method of controlling the surface roughness and the width of the joints.

(5) Coarse stone masonry: The main building material of this masonry is the precast concrete block, and the structure is built by the building method of controlling the surface roughness and the width of the joints.

15.3 The strength and displacement of the masonry

1) Compressive strength of masonry

(1) Mode of compression failure

Three stages are obtained from the beginning of axial loading to its failure:

①The first stage is called the integral service stage: From the beginning of loading to the beginning of cracking. The value of pressure is 50% ~ 70% of the ultimate load.

②The second stage is called the service stage with cracking: With the pressure increasing, the cracks in masonry keep developing, and the local cracks connect with each other to form the continuous cracks.

③The third stage is called the failure stage: With the load increasing, the development of cracks increases quickly. The masonry is divided into several zones by the several continuous cracks, and finally, the masonry is completely damaged.

(2) The stress stage of the compressed masonry

The compressed masonry has an important character that the component materials will crack first, because the masonry is compressed in a complex stress state.

(3) Factors influencing the compressive strength of the masonry

①The materials strength: The strength of the masonry depends on the building materials, because the component materials will crack first when the masonry is loaded.

②The shape and sizes: More regular the shape of the blocks is, higher the corresponding strength will be, and the strength of the masonry increases as its height increases.

③Mechanical properties of the mortar: The strength grade of mortar, the plasticity of mortar, the elastic modulus of mortar will all affect the compressive strength of the masonry structure.

④Thickness of joint formed in laying: Its strength decreases as thickness increases, the reasonable thickness should be 10 ~ 12mm.

⑤Construction quality: The strength of the masonry is also related to the conditions of the uniformity and compactness of the construction, and as shown in Appendix Table 3-4 to Appendix Table 3-6.

2) The tensile, flexural and shear strengths of the masonry

The test results show that tensile, flexural and shear failures of masonry generally occur at the faying surface of the mortar and blocks. Therefore, the strength of the masonry depends on the strength of mortar.

(1) Axial tensile failure

The characters of the failure are as follows: The first failure mode is that the failure occurs along with the tooth seams of a shape looking like the tooth [Figure 15-1a)]; the second failure mode is that the failure occurs along with vertical seams and is the block failure [Figure 15-1b)]; the third failure mode is that the failure occurs along with straight joints [Figure 15-1c)].

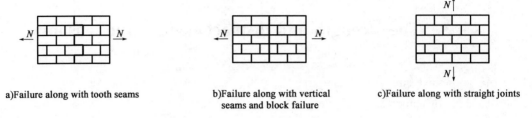

a) Failure along with tooth seams b) Failure along with vertical seams and block failure c) Failure along with straight joints

Figure 15-1 Failure of axial tensile masonry

(2) Flexural tensile failure

The characters of the failure are as follows: The first mode is that the failure occurs along with straight joints [Figure 15-2a)]; the second failure mode is that the failure occurs along with vertical seams and block failure [Figure 15-2b)].

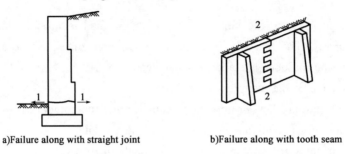

a) Failure along with straight joint b) Failure along with tooth seam

Figure 15-2 Failure of bending tensile masonry

(3) Shear failure

The characters of the failure are as follows: The first failure mode is that the failure occurs along with straight joint [Figure 15-3a)]; the second mode is that the failure occurs along with straight joints and tooth seams [Figure 15-3b)].

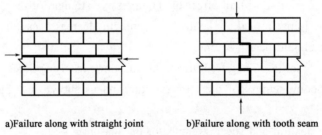

a) Failure along with straight joint b) Failure along with tooth seam

Figure 15-3 Failure of shear masonry

3) Other properties of masonry

(1) Elastic modulus

Elastic modulus: When the upper bound equals 40% ~ 50% of the compressive strength, the secant modulus can be seen as the elastic modulus.

The *Code for Design of Masonry Highway Bridges and Culverts* (*JTG D61—2005*) suggests that the elastic modulus of the masonry structures depends on the strength grade mortar, the elastic modulus and the compressive strength of the component materials.

(2) Linear expansion coefficient, shrinkage displacement and friction coefficient

①Linear expansion coefficient: In the statically indeterminate structures, the linear expansion coefficients are listed in Table 15-2.

Linear expansion coefficients of masonry structures　　Table 15-2

Masonry type	Linear expansion coefficient($10^{-6}/°C$)
Concrete	10
Precast concrete	9
Rubble, block stone, fine dressed stone, semi-fine stone, and coarse stone	8

②Shrinkage displacement: It can be determined by the shrinkage test, for the concrete precast block, the shrinkage displacement is 0.2mm/m at 28d age.

③Friction coefficient: It varies with the types of materials, and can be obtained from Table 15-3.

Friction coefficients μ_f of masonry　　Table 15-3

Material type	The situation of friction surfaces	
	Dry	Moist
Masonry slip with the masonry or concrete	0.70	0.60
Wood slip with masonry	0.60	0.50
Steel slip with masonry	0.45	0.35
Masonry slip with sand and stone	0.60	0.50
Masonry slip with silt	0.55	0.40
Masonry slip with cohesive soil	0.50	0.30

Chapter 16
Calculation of Load-carrying Capacity of Masonry Structures

16.1 Principles of calculation

Masonry structures need calculation of load-carrying capacity. The corresponding construction measurements are needed to meet the requirements of the serviceability limit state.

(1) The *Code for Design of Masonry Highway Bridges and Culverts* (*JTG D61—2005*) specifies that the limit state design method of masonry structures is based on the probability theory and method of considering the partial safety factor.

(2) In the limit state design method of masonry structures, the criterion is

$$\gamma_0 S \leqslant R(f_d, a_d) \tag{16-1}$$

Where γ_0 is the factor for importance of a bridge structure; for the first grade, $\gamma_0 = 1.1$; the second grade, $\gamma_0 = 1.0$; the third grade, $\gamma_0 = 0.9$; S is design value of combination for loading effect; $R(\cdot)$ is the function of the design load-carrying capacity; f_d is design value of materials capacity; a_d is design value of geometry factor, it equals the standard value a_k.

16.2 Load-carrying capacity of the compressed member

1) Experimental investigation on short column masonry structures

(1) The characters of the compressed short column masonry structure

Based on the test results, the characters of the forced state are as follows.

①For the member compressed by the axial pressure, the cross-section is produced by the symmetrical compressive stress [Figure 16-1a)]. When the member fails, the maximum compressive force on the normal section is the load-carrying capacity.

②For the member loaded with the eccentric compression, the cross-section has the unsymmetrical compressive stress. The compressive stress distribution changes with eccentricity e. When the eccentricity is small, the whole section is in compression. The stress figure is curve distribution, and the side carrying more compression stress will be failure at firstly [Figure 16-1b)].

③When the eccentricity e increases, the other side far away from the compression would be tensioned, and the side near the pressures will fail first, the other side will not fail [Figure 16-1c)].

④When the eccentricity e increases, the tensioned side would be cracked. As the cracked section cannot continue working, the compressed section will be decreased [Figure 16-1d)].

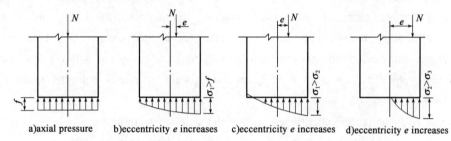

a) axial pressure   b) eccentricity e increases   c) eccentricity e increases   d) eccentricity e increases

Figure 16-1 The stress changes of the cross section of compressed member

(2) Eccentricity factor of the longitudinal force

Based on lots of studies and the equations in text-books of materials mechanics, the **Code for Design of Masonry Highway Bridges and Culverts** defines the eccentricity factor of the longitudinal force as follows

$$\alpha_x = \frac{1-(e_x/x)^m}{1+(e_x/i_y)^2} \quad (16\text{-}2)$$

$$\alpha_y = \frac{1-(e_y/y)^m}{1+(e_y/i_x)^2} \quad (16\text{-}3)$$

Where α_x, α_y are the eccentricity of axial force in x and y direction. x, y are the distance from the gravity of the cross-section to the eccentricity in x and y directions (Figure 16-2).

$$e_x = M_{yd}/N_d, e_y = M_{xd}/N_d$$

Where M_{yd} and M_{xd} are the design value for bending moment in x and y directions respectively;

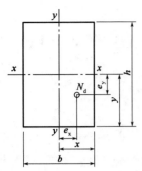

Figure 16-2 Eccentric compression of the masonry member

N_d is the design value for axial force (Figure 16-2).

m is shape factor of the cross-section; for the circular section, $m = 2.5$; for the T-section and U-section, $m = 3.5$; for the box section including the curve and circular section used in the ends of rectangular pier, $m = 8.0$. i_x, i_y are the radius of gyration in the bending plane, $i_x = \sqrt{I_x/A}$ and $i_y = \sqrt{I_y/A}$.

Equations (16-2) and (16-3) are obtained from both theory and experience studies, and they both satisfy the boundary conditions of axial and eccentric compression. If $e_x = 0$ and $e_y = 0$, and then $\alpha_x = 1$ and $\alpha_y = 1$, as the member is compressed member, the eccentricity can not affect the load-carrying capacity. If either $e_x \neq 0$ or $e_y \neq 0$, then the member is forced by the eccentric compression from one direction. If both $e_x \neq 0$ and $e_y \neq 0$, the member is forced by eccentric compressions from two directions.

2) Load-carrying capacity of long column masonry structures

(1) Load-carrying capacity of long column masonry structures

①The long column is forced by the axial pressure, due to the asymmetry characters of multiple materials and any other factors, thus the axial pressure cannot easily be loaded at the center of cross-section, and then the eccentricity is easily produced.

②The eccentricity is composed by two parts, which are eccentricity e and u respectively, where e is induced by the force, u is induced by the foundation, and therefore, the eccentricity is $e + u$.

③In the masonry member, the mortar decreases the integrity of macrocosmic capacity, the bending along with the longitudinal direction is easily to occur compared with reinforced concrete, the occurrence probability of such bending of the masonry increases with the slenderness ratio (Figure 16-3).

Because of those reasons mentioned above, the slenderness ratio λ can substantially affect the load-carrying capacity for the long column masonry concrete. Therefore, the load-carrying capacity of the long column masonry concrete is certainly up to the slenderness ratio λ, and mortar capacity and so on.

(2) Calculation of the longitudinal bending factor for the eccentric compressed member

The *Code for Design of Masonry Highway Bridges and Culverts* specifies the longitudinal bending factor for the eccentric compressed member as follows.

x direction: $\phi_x = \dfrac{1}{1 + \alpha\lambda_x(\lambda_x - 3)[1 + 1.33(e_x/i_y)^2]}$ (16-4)

y direction: $\phi_y = \dfrac{1}{1 + \alpha\lambda_y(\lambda_y - 3)[1 + 1.33(e_y/i_y)^2]}$ (16-5)

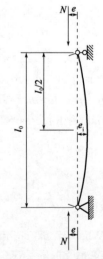

Figure 16-3 Eccentricity of the masonry member

Chapter 16 · Calculation of Load-carrying Capacity of Masonry Structures

3) The load-carrying capacity of the compressed member

(1) Equation of calculating load-carrying capacity

Based on the study of compressed short and long column, a factor φ can be used to consider the longitudinal bending and axial eccentric of the compressed member. The ***Code for Design of Masonry Highway Bridges and Culverts*** suggests the equation for calculating the load-carrying capacity in the range of the limit value for the compressive eccentricity as

$$\gamma_0 N_d \leqslant N_u = \varphi A f_{cd} \tag{16-6}$$

Where N_d is axial design value; A is the member area; φ is influence factor of considering eccentricity e and slenderness ratio λ.

(2) Influence factor φ

Influence factor φ of the eccentric compressed member is a factor that considers the effects of both eccentricity e and slenderness ratio λ.

The ***Code for Design of Masonry Highway Bridges and Culverts*** suggests that the influence factor φ can be expressed as

$$\varphi = \frac{1}{\dfrac{1}{\varphi_x} + \dfrac{1}{\varphi_y} - 1} \tag{16-7}$$

$$\varphi_x = \alpha_x \phi_x = \frac{1 - \left(\dfrac{e_x}{x}\right)^m}{1 + \left(\dfrac{e_x}{i_y}\right)^2} \cdot \frac{1}{1 + \alpha \lambda_x (\lambda_x - 3)[1 + 1.33(e_x/i_y)^2]} \tag{16-8}$$

$$\varphi_y = \alpha_y \phi_y = \frac{1 - \left(\dfrac{e_y}{y}\right)^m}{1 + \left(\dfrac{e_y}{i_x}\right)^2} \cdot \frac{1}{1 + \alpha \lambda_y (\lambda_y - 3)[1 + 1.33(e_y/i_x)^2]} \tag{16-9}$$

Where, φ_x and φ_y are the x direction and y direction influence factor, respectively.

4) Load-carrying capacity calculation of the compressed concrete member

The ***Code for Design of Masonry Highway Bridges and Culverts*** suggests that the compressed member's eccentricity e can be obtained from Table 16-1, the figure of its normal stress is assumed as the rectangular, and its stress is the design value of compressed concrete. The load-carrying capacity is defined as

$$\gamma_0 N_d \leqslant N_u = \varphi A_c f_{cd} \tag{16-10}$$

Where N_d is design value of the axial force; φ is bending factor of the axial compressed member, it can be obtained in Table 16-2; f_{cd} is the design capacity value of compressed concrete, it can be obtained in appendix Table 3-3; A_c is the compressive area of concrete.

Limit value e of eccentricity for the compressive member Table 16-1

Action combination	Limit value e of eccentricity
Basic combination	$\leqslant 0.6s$
Occasional combination	$\leqslant 0.7s$

Note: s is the distance of the gravity axial for section/calculated section to the section edge of the eccentric direction.

313

Bending factor of the compressed member　　　　　　Table 16-2

l_0/b	<4	4	6	8	10	12	14	16	18	20	22	24	26	28	30
l_0/i	<14	14	21	28	35	42	49	56	63	70	76	83	90	97	104
φ	1.00	0.98	0.96	0.91	0.86	0.82	0.77	0.72	0.68	0.63	0.59	0.55	0.51	0.47	0.44

When calculating the compressed area A_c, based on the theory of eccentricity e, the **Code for Design of Masonry Highway Bridges and Culverts** suggests that the distance of the center of gravity to the axle of gravity is $e_c = e$, the compressed area A_c is then obtained from the center of gravity.

(1) Uniaxial eccentric compression

Height of compression zone h_c can be obtained as (Figure 16-4)

$$h_c = h - 2e \qquad (16-11)$$

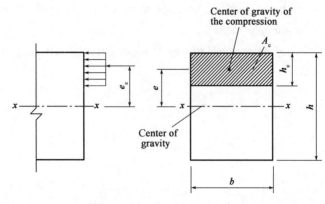

Figure 16-4　Eccentric compression of the concrete member (uniaxial eccentric compression)

The compressed capacity of the rectangle section can be calculated as

$$\gamma_0 N_d \leqslant N_u = \varphi f_{cd} b(h - 2e) \qquad (16-12)$$

Where e_c is the distance of an action point of the normal stress to the center of gravity; e is the eccentricity of axial force; b is the width of the rectangle section; h is the height of the rectangle section.

(2) Biaxial eccentric compression

The height and width of compression zone can be obtained as (Figure 16-5)

$$h_c = h - 2e_y \qquad (16-13)$$
$$b_c = b - 2e_x \qquad (16-14)$$

Eccentric compression of the rectangle can be obtained as

$$\gamma_0 N_d \leqslant N_u = \varphi f_{cd}(h - 2e_y)(b - 2e_x) \qquad (16-15)$$

Where φ is the bending factor of the axial compressed member obtained from Table 16-2; e_{cy} is the distance of action point for normal stress to the center of gravity of section in y direction; e_{cx} is the distance of action point for normal stress to the center of gravity of section in x direction; e_x, e_y are the eccentricity of x and y directions, respectively.

5) Eccentricity checking

The tests show that the load-carrying capacity decreases as the eccentricity increases. To keep the stability of the section, eccentricity e should be limited.

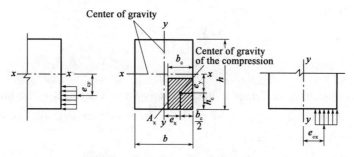

Figure 16-5 Biaxial eccentric compression of the concrete member

Based on the ***Code for Design of Masonry Highway Bridges and Culverts***, the eccentricity e should follow the specifications in Table 16-1. Otherwise, the eccentricity e should be calculated as

Uniaxial eccentric compression:

$$\gamma_0 N_d \leq N_u = \varphi \frac{Af_{tmd}}{\frac{Ae}{W} - 1} \tag{16-16}$$

Biaxial eccentric compression:

$$\gamma_0 N_d \leq N_u = \varphi \frac{Af_{tmd}}{\frac{Ae_x}{W_y} + \frac{Ae_y}{W_x} - 1} \tag{16-17}$$

16.3 Sectional local compression, bended, and shear calculations of the load-carrying capacity

1) Sectional local compressive calculation
(1) Concept
The partial section is forced by the axial force.
(2) Failure modes
Three modes of failure are generalized as follows:
①The failure induced by the longitudinal cracking development;
②The splitting failure;
③Local masonry failure of contacting the load-carrying.
(3) Compressive characters
①The load-carrying capacity of local position loaded directly by the pressure increases with the pressure action, this maybe because of the casing hoop effects.
②The phenomena of local crushing may be present in the real structures.
(4) Checking local compressed capacity
For local compressive masonry structures, to protect the distributed stress not lager than the

design value, the ***Code for Design of Masonry Highway Bridges and Culverts*** suggests the local compressed capacity equation as follows

$$\gamma_0 N_d \leq 0.9 \beta A_1 f_{cd} \quad (16\text{-}18)$$

$$\beta = \sqrt{\frac{A_b}{A_1}} \quad (16\text{-}19)$$

Where N_d is designed axial forces at the local compressed area; β is the increasing factor by the local pressure; A_1 is local compressed area; A_b is the bottom area of local compressed position; f_{cd} is the axial tension design value of concrete.

2) Checking of load-carrying capacity of the bended member

When the bended member is loaded, the tensioned failure may occur along with the straight and tooth cracks. The tensioned edge of the bended normal load-carrying capacity will be less than the corresponding design value.

Considering the construction safety factor and the importance factor, the ***Code for Design of Masonry Highway Bridges and Culverts*** suggests as

$$\gamma_0 M_d \leq W f_{tmd} \quad (16\text{-}20)$$

Where M_d is designed bending moment value; W is elastic resistance moment of the sectional tension edge; f_{tmd} is designed bended tension value of the sectional tension edge.

3) Load-carrying capacity of the shear member

The test results show that the load-carrying capacity of the shear member equals the sum of the shear capacity of the masonry and friction induced by the loading. Therefore, the ***Code for Design of Masonry Highway Bridges and Culverts*** suggests that the load-carrying capacity of the shear member can be calculated as

$$\gamma_0 V_d \leq V_u = A f_{vd} + \frac{1}{1.4} \mu_f N_k \quad (16\text{-}21)$$

Where V_d is the designed shear value; A is the section area by the shear; f_{vd} is the designed shear capacity of concrete or masonry; μ_f is friction factor, $\mu_f = 0.7$; N_k is the standard value of the pressure of shear section.

Appendix

Standard value and design value of concrete strength Table 1-1

Strength kind		Symbol	Grade of concrete strength												
			C20	C25	C30	C35	C40	C45	C50	C55	C60	C65	C70	C75	C80
Standard value	Axially compressed	f_{ck}	13.4	16.7	20.1	23.4	26.8	29.6	32.4	35.5	38.5	41.5	44.5	47.4	50.2
	Axially tensioned	f_{tk}	1.54	1.78	2.01	2.20	2.40	2.51	2.65	2.74	2.85	2.93	3.0	3.05	3.10
Design value	Axially compressed	f_{cd}	9.2	11.5	13.8	16.1	18.4	20.5	22.4	24.4	26.5	28.5	30.5	32.4	34.6
	Axially tensioned	f_{td}	1.06	1.23	1.39	1.52	1.65	1.74	1.83	1.89	1.96	2.02	2.07	2.10	2.14

Note: when in-site reinforced concrete is axially compressed and axially tensioned, if length or diameter is lower than 300 mm, the design value of concrete should multiply 0.8. when the quality of member is assured, it is allowable not to conform.

Elastic modulus of concrete ($\times 10^4$ MPa)　　　　　Table 1-2

Grade of concrete strength	C20	C25	C30	C35	C40	C45	C50	C55	C60	C65	C70	C75	C80
E_c	2.55	2.80	3.00	3.15	3.25	3.35	3.45	3.55	3.60	3.65	3.70	3.75	3.80

Notes: 1. The shear modulus of concrete G_c is adopted based on the $0.4\times$ values in Table.
　　　2. For high strength concrete, when gas additive and sand ratio is used and no field data, the E_c of C50~C80 should discount of 0.95.

Standard value and design value of reinforcement (MPa)　　　　　Table 1-3

Reinforcement	Diameter d (mm)	Symbol	Standard value of tensile value f_{sk}	Design value of tensile strength f_{sd}	Design value of compression strength f'_{sd}
HPB235	8~20	ϕ	235	195	195
HRB335	6~50	ϕ	335	280	280
HRB400	6~50	ϕ	400	330	330
KL400	8~40	ϕR	400	330	330

Notes: 1. d is reinforcement diameter in national standard.
　　　2. When reinforced concrete is axially tensioned or small eccentrically tensioned, tensile strength is larger than 330MPa, the strength is still 330MPa.
　　　3. There are various reinforcement in a reinforced concrete beam, each has its own strength.

Elastic modulus of reinforcement ($\times 10^5$ MPa)　　　　　Table 1-4

Reinforcement	Elastic modulus E_s	Reinforcement	Elastic modulus E_s
HPB235	2.1	HRB335, HRB400, KL400	2.0

Calculation parameters for load-carrying capacity of reinforced concrete member in flexure
Table 1-5

ξ	A_0	ζ_0	ξ	A_0	ζ_0
0.01	0.010	0.995	0.16	0.147	0.920
0.02	0.020	0.990	0.17	0.155	0.915
0.03	0.030	0.985	0.18	0.164	0.910
0.04	0.039	0.980	0.19	0.172	0.905
0.05	0.048	0.975	0.20	0.180	0.900
0.06	0.058	0.970	0.21	0.188	0.895
0.07	0.067	0.965	0.22	0.196	0.890
0.08	0.077	0.960	0.23	0.203	0.885
0.09	0.085	0.955	0.24	0.211	0.880
0.10	0.095	0.950	0.25	0.219	0.875
0.11	0.104	0.945	0.26	0.226	0.870
0.12	0.113	0.940	0.27	0.234	0.865
0.13	0.121	0.935	0.28	0.241	0.860
0.14	0.130	0.930	0.29	0.248	0.855
0.15	0.139	0.925	0.30	0.255	0.850

Continued

ξ	A_0	ζ_0	ξ	A_0	ζ_0
0.31	0.262	0.845	0.49	0.370	0.755
0.32	0.269	0.840	0.50	0.375	0.750
0.33	0.275	0.835	0.51	0.380	0.745
0.34	0.282	0.830	0.52	0.385	0.740
0.35	0.289	0.825	0.53	0.390	0.735
0.36	0.295	0.820	0.54	0.394	0.730
0.37	0.301	0.815	0.55	0.399	0.725
0.38	0.309	0.810	0.56	0.403	0.720
0.39	0.314	0.805	0.57	0.408	0.715
0.40	0.320	0.800	0.58	0.412	0.710
0.41	0.326	0.795	0.59	0.416	0.705
0.42	0.332	0.790	0.60	0.420	0.700
0.43	0.337	0.785	0.61	0.424	0.695
0.44	0.343	0.780	0.62	0.428	0.690
0.45	0.349	0.775	0.63	0.432	0.685
0.46	0.354	0.770	0.64	0.435	0.680
0.47	0.359	0.765	0.65	0.439	0.675
0.48	0.365	0.760			

Area of cross-section and weight of reinforcement Table 1-6

Diameter (mm)	Area (mm^2)									Weight (kg/m)	Ribbed reinforcement	
	1	2	3	4	5	6	7	8	9		Diameter (mm)	Outer diameter (mm)
6	28.3	57	85	113	141	170	198	226	254	0.222	6	7.0
8	50.3	101	151	201	251	302	352	402	452	0.395	8	9.3
10	78.5	157	236	314	393	471	550	628	707	0.617	10	11.6
12	113.1	226	339	452	566	679	792	905	1 018	0.888	12	13.9
14	153.9	308	462	616	770	924	1 078	1 232	1 385	1.21	14	16.2
16	201.1	402	603	804	1 005	1 206	1 407	1 608	1 810	1.58	16	18.4
18	254.5	509	763	1 018	1 272	1 527	1 781	2 036	2 290	2.00	18	20.5
20	314.2	628	942	1 256	1 570	1 884	2 200	2 513	2 827	2.47	20	22.7
22	380.1	760	1 140	1 520	1 900	2 281	2 661	3 041	3 421	2.98	22	25.1
25	490.9	982	1 473	1 964	2 454	2 945	3 436	3 927	4 418	3.85	25	28.4
28	615.8	1 232	1 847	2 463	3 079	3 695	4 310	4 926	5 542	4.83	28	31.6
32	804.2	1 608	2 413	3 217	4 021	4 826	5 630	6 434	7 238	6.31	32	35.8

Area of reinforcement in unit width of 1m for a constant reinforcement spacing　　Table 1-7

Reinforcement spacing (mm)	Steel diameter (mm)								
	6	8	10	12	14	16	18	20	22	24
70	404	718	1 122	1 616	2 199	2 873	3 636	4 487	5 430	6 463
75	377	670	1 047	1 508	2 052	2 681	3 393	4 188	5 081	6 032
80	353	628	982	1 414	1 924	2 514	3 181	3 926	4 751	5 655
85	333	591	924	1 331	1 811	2 366	2 994	3 695	4 472	5 322
90	314	559	873	1 257	1 711	2 234	2 828	3 490	4 223	5 027
95	298	529	827	1 190	1 620	2 117	2 679	3 306	4 001	4 762
100	283	503	785	1 131	1 539	2 011	2 545	3 141	3 801	4 524
105	269	479	748	1 077	1 466	1 915	2 424	2 991	3 620	4 309
110	257	457	714	1 028	1 399	1 828	2 314	2 855	3 455	4 113
115	246	437	683	984	1 339	1 749	2 213	2 731	3 305	3 934
120	236	419	654	942	1 283	1 676	2 121	2 617	3 167	3 770
125	226	402	628	905	1 232	1 609	2 036	2 513	3 041	3 619
130	217	387	604	870	1 184	1 547	1 958	2 416	2 924	3 480
135	209	372	582	838	1 140	1 490	1 885	2 327	2 816	3 351
140	202	359	561	808	1 100	1 436	1 818	2 244	2 715	3 231
145	195	347	542	780	1 062	1 387	1 755	2 166	2 621	3 120
150	189	335	524	754	1 026	1 341	1 697	2 084	2 534	3 016
155	182	324	507	730	993	1 297	1 642	2 027	2 452	2 919
160	177	314	491	707	962	1 257	1 590	1 964	2 376	2 828
165	171	305	476	685	933	1 219	1 542	1 904	2 304	2 741
170	166	296	462	665	905	1 183	1 497	1 848	2 236	2 661
175	162	287	449	646	876	1 149	1 454	1 795	2 172	2 585
180	157	279	436	628	855	1 117	1 414	1 746	2 112	2 513
185	153	272	425	611	832	1 087	1 376	1 694	2 035	2 445
190	149	265	413	595	810	1 058	1 339	1 654	2 001	2 381
195	145	258	403	580	789	1 031	1 305	1 611	1 949	2 320
200	141	251	393	565	769	1 005	1 272	1 572	1 901	2 262

Minimum cover thickness for reinforcement and prestressing strands (mm)　　Table 1-8

No.	Element type	Environmental conditions		
		I	II	III, IV
1	Foundation pier: (1) pier bottom with cushion and templates; (2) pier bottom without cushion and templates	40 60	50 75	60 85
2	Abutment, culvert, beam, deck, arch, architectures	30	40	45
3	Sidewalk member, railing	20	25	30
4	stirrups	20	25	30

Appendix

Continued

No.	Element type	Environmental conditions		
		I	II	III, IV
5	Curb, medium strip, fence	30	40	45
6	Reinforcement for shrinkage, temperature, distribution and anti-crack	15	20	25

Notes: 1. For epoxy coating reinforcement, using I.
2. For anchorage systems of posttensioned prestressed concrete, the minimum concrete cover for I, II, and III (IV) environmental conditions, is 40mm, 45mm and 50mm.
3. For pretensioned prestressed concrete, the reinforcements at ends should be protected.
4. I-type means non-cold or cold atmosphere condition, no corrosive water and soil.
   II-type means cold region under atmospheric conditions, no corrosive water and soil. But using dicing salts and marine environment.
   III-type means seawater environment.
   IV-type means man-made or natural corrosive environments.

The minimum reinforcement ratio for reinforced concrete member (%)   Table 1-9

Structural type		Minimum reinforcement ratio
Member in compression	All main rebars	0.5
	Side rebars	0.2
Side rebars for flexural, eccentrically tensioned and axially-tensioned members		Max $[0.2$ and $45f_{td}/f_{sd}]$
Torsional members		$0.08f_{cd}/f_{sv}$ (for torsion), $0.08(2\beta_t - 1)f_{cd}/f_{sv}$ (for torsion and shear)

Notes: 1. For minimum reinforcement ratio of compressed member, when concrete strength is C50 and above, should not lower than 0.6.
2. For minimum reinforcement ratio should not lower than 0.2 when compressive reinforcement is needed at compression zone for large-eccentrically tensioned members.
3. For axially-loaded, eccentrically-loaded members, when reinforcement ratio and side reinforcement is computed, the area of gross section is used. For axially-tensioned, small eccentrically-tensioned members, when reinforcement ratio and side reinforcement is computed, the area of gross section is used. For flexural and large eccentrically-tensioned members, when reinforcement ratio and side reinforcement is computed, $100 A_s/(bh_0)$ is adopted, where A_s is the area of reinforcement, b is the web width and h_0 is the effective height.
4. When reinforcement is arranged around the section, side tension reinforcement or side compressive reinforcement means the longitudinal reinforcement along a side carrying load.
5. For torsional members, the minimum ratio is $A_{st,min}/(bh)$, $A_{st,min}$ is the minimum area of torsional members, h is the length of long side for unit, b is the length of short side and f_{sv} is the tensile strength of stirrup.

Stability factor φ of axially-loaded member   Table 1-10

l_0/b	≤8	10	12	14	16	18	20	22	24	26	28
l_0/d	≤7	8.5	10.5	12	14	15.5	17	19	21	22.5	24
l_0/r	≤28	35	42	48	55	62	69	76	83	90	97
φ	1.0	0.98	0.95	0.92	0.87	0.81	0.75	0.70	0.65	0.60	0.56
l_0/b	30	32	34	36	38	40	42	44	46	48	50
l_0/d	26	28	29.5	31	33	34.5	36.5	38	40	41.5	43
l_0/r	104	111	118	125	132	139	146	153	160	167	174
φ	0.52	0.48	0.44	0.40	0.36	0.32	0.29	0.26	0.23	0.21	0.19

Notes: 1. l_0 is computed length, b is the short side of rectangular section; d is the diameter of cyclic section, r is minimum radius of section.
2. For two-ends fixed, the l_0 is $0.5\ l$; For one-end fixed and one-end free, the l_0 is $0.7\ l$; For two-ends joint without moving, the l_0 is l; and for one-end fixed and one-end free, the l_0 is $2\ l$.

Factors for load-carrying capacity of eccentrically-loaded cyclic section member Table 1-11

$\eta \dfrac{e_0}{r}$	$\rho f_{sd}/f_{cd}$										
	0.06	0.09	0.12	0.15	0.18	0.21	0.24	0.27	0.3	0.4	0.5
0.01	1.048 7	1.078 3	1.107 9	1.137 5	1.167 1	1.196 8	1.226 4	1.256 1	1.285 7	1.384 6	1.483 5

Standard value of tensile strength of prestressing steels Table 2-1

Steel type		Symbol	d(mm)	f_{pk}
Prestressing strands	1×2 two strands	ϕ^S	8.0,10.0	1 470,1 570,1 720,1 860
			12.0	1 470,1 570,1 720
	1×3 three strands		8.6,10.8	1 470,1 570,1 720,1 860
			12.9	1 470,1 570,1 720
	1×7 severn strands		9.5,11.1,12.7	1 860
			15.2	1 720,1 860
Non stress steel wires	Round wires	ϕ^P	4,5	1 470,1 570,1 670,1 770
			6	1 570,1 670
	Screw rib wires	ϕ^H	7,8,9	1 470,1 570
	Nicked wires	ϕ^I	5,7	1 470,1 570
Finishing thread steel wires		JL	40	540
			18,25,32	540,785,930

Note: d is diameter of steel strands, wires, and finishing thread steel wires in national standard.

Design values of tensile and compressive strength of prestressing steels Table 2-2

Steel types	f_{pk}	f_{pd}	f'_{pd}
Steel strands 1×2 1×3 1×7	1 470	1 000	390
	1 570	1 070	
	1 720	1 170	
	1 860	1 260	
Non-stress wires, screw rib wires	1 470	1 000	410
	1 570	1 070	
	1 670	1 140	
	1 770	1 200	
Nicked wires	1 470	1 000	410
	1 570	1 070	
Finishing thread wires	540	450	400
	785	650	
	930	770	

Elastic modulus of prestressing steels($\times 10^5$ MPa) Table 2-3

Steels types	E_p	Steels types	E_p
Finishing thread steel	2.0	Steel strands	1.95
Non-stress wires, screw rib wire, nicked wires	2.05		

Appendix

Nominal diameter, sectional area and theoretical weight for prestressing strands   Table 2-4

Steel types	Diameter(mm)	Area(mm^2)	Theoretical weight per 1 000m (kg)
1×2	8	25.3	199
	10	39.5	310
	12	56.9	447
1×3	8.6	37.4	199
	10.8	59.3	465
	12.9	85.4	671
1×7 standard type	9.5	54.8	432
	11.1	74.2	580
	12.7	98.7	774
	15.2	139	1 101
1×7 pull type	12.7	112	890
	15.2	165	1 295

Nominal diameter, sectional area and theoretical weight for prestressing wires   Table 2-5

Nominal diameter(mm)	Nominal area(mm^2)	Theoretical reference weight (kg/m)
4.0	12.57	0.099
5.0	19.63	0.154
6.0	28.27	0.222
7.0	38.48	0.302
8.0	50.26	0.394
9.0	63.62	0.499

Factor k and μ   Table 2-6

| Duct type | k | μ | |
		Steel strands	Finishing thread steel
Embedded metal bellows	0.001 5	0.20~0.25	0.50
Embedded plastic bellows	0.001 5	0.14~0.17	—
Embedded steel bellows	0.003 0	0.35	0.40
Embedded steel tube	0.001 0	0.25	—
Pumping centre type	0.001 5	0.55	0.60

Anchorage deformation, steel retraction and joint compression   Table 2-7

Anchorage and joint types		Δl
Conical anchorage		6
Clip type anchorage	With top compression	4
	Without top compression	6
Nut slot		1
Heading anchorage		1
plate slot		1
cement sand joint		1
epoxy mortar joint		1

Prestress transmission length l_{tr} and anchorage length l_a Table 2-8

Items	Steel types	Concrete strength	l_{tr} (mm)	l_a (mm)
1	Steel strands $1\times2, 1\times3$ $\sigma_{pe}=1\,000\text{MPa}$ $f_{pd}=1\,170\text{MPa}$	C30	75d	—
		C35	68d	—
		C40	63d	115d
		C45	60d	110d
		C50	57d	105d
		C55	55d	100d
		C60	55d	95d
		≥C65	55d	90d
	strands 1×7 $\sigma_{pe}=1\,000\text{MPa}$ $f_{pd}=1\,260\text{MPa}$	C30	80d	—
		C35	73d	—
		C40	67d	130d
		C45	64d	125d
		C50	60d	120d
		C55	58d	115d
		C60	58d	110d
		≥C65	58d	105d
2	Thread rib steel $\sigma_{pe}=1\,000\text{MPa}$ $f_{pd}=1\,200\text{MPa}$	C30	70d	—
		C35	64d	—
		C40	58d	95d
		C45	56d	90d
		C50	53d	85d
		C55	51d	83d
		C60	51d	80d
		≥C65	51d	80d
3	Nicked wires $\sigma_{pe}=1\,000\text{MPa}$ $f_{pd}=1\,070\text{MPa}$	C30	89d	—
		C35	81d	—
		C40	75d	125d
		C45	71d	115d
		C50	68d	110d
		C55	65d	105d
		C60	65d	103d
		≥C65	65d	100d

Notes: 1. l_{tr} is based on effective prestress σ_{pe} given in the Table; l_a is based on tensile strength f_{pd} given in the Table.
    2. Prestress transition length is based on concrete cubic compression strength f_{cu}, and f_{cu} falls in a range and then interpolation is used.
    3. When sudden relaxation construction method is used, the start point of anchorage length and transition length is a distance from end of member of $0.25\,l_{tr}$.
    4. When the f_{pd} and σ_{pe} are different from the Table and the l_{tr} and l_a change proportionally.

Appendix

Design value of strength of stone (MPa)   Table 3-1

Strength type \ Stone grade	MU120	MU100	MU80	MU60	MU50	MU40	MU30
Compressive strength f_{cd}	31.78	26.49	21.19	15.89	13.24	10.59	7.95
Flexural tensile strength f_{tmd}	2.18	1.82	1.45	1.09	0.91	0.73	0.55

Conversion factor of stone strength grade   Table 3-2

Speciment size (mm)	200	150	100	70	50
Conversion factor	1.43	1.28	1.14	1.00	0.86

Design value of strength of conrete (MPa)   Table 3-3

Strength \ Concrete strength grade	C40	C35	C30	C25	C20	C15
Compressive strength f_{cd}	15.64	13.69	11.73	9.78	7.82	5.87
Flexural tensile strength f_{tmd}	1.24	1.14	1.04	0.92	0.80	0.66
Shear strength f_{vd}	2.48	2.28	2.09	1.85	1.59	1.32

Design value of compressive strength of prefabricated mortar for concrete f_{cd} (MPa)

Table 3-4

Concrete strength grade	Strength grade of mortar					strength of mortar
	M20	M15	M10	M7.5	M5	0
C40	8.25	7.04	5.84	5.24	4.64	2.06
C35	7.71	6.59	5.47	4.90	4.34	1.93
C30	7.14	6.10	5.06	4.54	4.02	1.79
C25	6.52	5.57	4.62	4.14	3.67	1.63
C20	5.83	4.98	4.13	3.70	3.28	1.46
C15	5.05	4.31	3.58	3.21	2.84	1.26

Design value of compressive strength of mortar for block stone f_{cd} (MPa)   Table 3-5

Stone strength grade	Strength grade of mortar					Strength of mortar
	M20	M15	M10	M7.5	M5	0
MU120	8.42	7.19	5.96	5.35	4.73	2.10
MU100	7.68	6.56	5.44	4.88	4.32	1.92
MU80	6.87	5.87	4.87	4.37	3.86	1.72
MU60	5.95	5.08	4.22	3.78	3.35	1.49
MU50	5.43	4.64	3.85	3.45	3.05	1.36
MU40	4.86	4.15	3.44	3.09	2.73	1.21
MU30	4.21	3.59	2.98	2.67	2.37	1.05

Notes: ①For the all stone masonry the factor should be multiplied as: fine stone masonry 1.5; half fine stone masonry 1.3; coarse masonry 1.2.

②For the dry block stone, the strength of mortar should be zero.

Design value of compressive strength of mortar for brick stone f_{cd} (MPa)　　Table 3-6

Masonry type	Mortar strength grade					Mortar strength
	M20	M15	M10	M7.5	M5	0
MU120	1.97	1.68	1.39	1.25	1.11	0.33
MU100	1.80	1.54	1.27	1.14	1.01	0.30
MU80	1.61	1.37	1.14	1.02	0.90	0.27
MU60	1.39	1.19	0.99	0.88	0.78	0.23
MU50	1.27	1.09	0.90	0.81	0.71	0.21
MU40	1.14	0.97	0.81	0.72	0.64	0.19
MU30	0.98	0.84	0.70	0.63	0.55	0.16

Note: Design value of the compressive strength of mortar should be zero for dry brick stone masonry.

Design value of axial tensile strength, flexural tensile and shear strength (MPa)

Table 3-7

Strength	Damaged Condition	masonry	Mortar strength grade				
			M20	M15	M10	M7.5	M5
Axial tensile f_{td}	Tooth-shaped joint	Regular block masonry	0.104	0.090	0.073	0.063	0.052
		Brick masonry	0.096	0.083	0.068	0.059	0.048
Flexural tensile f_{tmd}	Tooth-shaped joint	Regular block masonry	0.122	0.105	0.086	0.074	0.061
		Brick masonry	0.145	0.125	0.102	0.089	0.072
	Through joint	Regular block masonry	0.084	0.073	0.059	0.051	0.042
Shear strength f_{vd}	—	Regular block masonry	0.104	0.090	0.073	0.063	0.052
		Brick masonry	0.241	0.208	0.170	0.147	0.120

Notes: ①Age of masonry is 28d.

②Regular block masonry includes block stone masonry, corase stone masonry, fine stone and half fine stone masonry, and prefabricated concrete masonry.

③When the regular block masonry be sheared along tooth-shaped joint, the damage is sheared with the block and mortar.

Design value of axial compressive strength of small stone concrete block masonry f_{cd} (MPa)

Table 3-8

Strength	Strength grade of small concrete masonry					
	C40	C35	C30	C25	C20	C15
MU120	13.86	12.69	11.49	10.25	8.95	7.59
MU100	12.65	11.59	10.49	9.35	8.17	6.93
MU80	11.32	10.36	9.38	8.37	7.31	6.19

Appendix

Continued

Strength	Strength grade of small concrete masonry					
	C40	C35	C30	C25	C20	C15
MU60	9.80	9.98	8.12	7.24	6.33	5.36
MU50	8.95	8.19	7.42	6.61	5.78	4.90
MU40	—	—	6.63	5.92	5.17	4.38
MU30	—	—	—	—	4.48	3.79

Notes: When the block is the corase stone masonry, the axial compressive strength should be multiplied 1.2; When the block is fine stone and half fine stone, the axial compressive strength should be multiplied 1.4.

Design value of axial compressive strength of small stone concrete brick masonry f_{cd} (MPa)

Table 3-9

Strength	Strength grade of small stone concrete masonry			
	C30	C25	C20	C15
MU120	6.94	6.51	5.99	5.36
MU100	5.30	5.00	4.63	4.17
MU80	3.94	3.74	3.49	3.17
MU60	3.23	3.09	2.91	2.67
MU50	2.88	2.77	2.62	2.43
MU40	2.50	2.42	2.31	2.16
MU30	—	—	1.95	1.85

Design value of axial tensile, flexural tensile, and shear strength for small concrete block and brick masonry (MPa)

Table 3-10

Strength	Damaged condition	masonry	Strength grade of small concrete masonry					
			C40	C35	C30	C25	C20	C15
Axial tensile f_{td}	Tooth-shaped joint	Block	0.285	0.267	0.247	0.226	0.202	0.175
		Brick	0.425	0.398	0.368	0.336	0.301	0.260
Flexural tensile f_{tmd}	Tooth-shaped joint	Block	0.335	0.313	0.290	0.265	0.237	0.205
		Brick	0.493	0.461	0.427	0.387	0.349	0.300
	Through joint	Block	0.232	0.217	0.201	0.183	0.164	0.142
Shear strength f_{vd}	—	Block	0.285	0.267	0.247	0.226	0.202	0.175
		Brick	0.425	0.398	0.368	0.336	0.301	0.260

Notes: For the other regular block, the strength should multiplied as: corase stone masonry is 0.7; fine and half fine stone masonry is 0.35.

Elastic modulus E_m of all masonry (MPa)　　　　Table 3-11

Masonry	Mortar strength grade				
	M20	M15	M10	M7.5	M5
Prefabricated concrete masonry	$1\ 700f_{cd}$	$1\ 700f_{cd}$	$1\ 700f_{cd}$	$1\ 600f_{cd}$	$1\ 500f_{cd}$
Coarse, block and brick masonry	7 300	7 300	7 300	5 650	4 000
Fine stone and half fine stone	22 000	22 000	22 000	17 000	12 000
Small conorete masonry	$2\ 100f_{cd}$				

Notes: f_{cd} is the design value of the axial compressive strength of masonry.

References

[1] Ye Jianshu. Principle of Structural Design[M]. 2nd ed. Beijing: China Communications Press, 2005.

[2] Liu Lixin. Basic Principles of Concrete Structures[M]. Wu han: Wuhan University of Science and Technology Press, 2004.

[3] Ministry of Transport of P. R China. JTG D60—2015 General code for design of highway bridge and culverts[S]. Beijing: China Communications Press, 2004.

[4] Ministry of Transport of P. R China. JTG 3362—2018 Code for design of reinforced concrete and prestressed concrete highway bridges and culverts[S]. Beijing: China Communications Press, 2018.

[5] Ministry of Transport of P. R China. JTG D61—2005 Code for design of masonry highway bridges and culverts[S]. Beijing: China Communications Press, 2004.

[6] GB 50153—2008 Uniform standard for engineering structural reliability[S]. Beijing: China Architecture & Building Press, 2008.

[7] GB/T 50283—1999 Uniform standard for highway engineering structural reliability[S]. Beijing: China Planning Press, 1999.

[8] CEB-FIP Specification.

[9] McComac Jack C, Nelson James K. Design of reinforced concrete[M]. John Wiley & Sons, Inc, 2006.

目 录

总论 ·· 335
  0.1 概述 ·· 335
  0.2 基本概念 ·· 336
  0.3 本课程的特点与学习本课程的指导意见 ·· 336

第一部分 钢筋混凝土结构

第1章 钢筋混凝土结构的基本概念及材料的物理力学性能 ························ 341
  1.1 钢筋混凝土结构的基本概念 ··· 341
  1.2 混凝土 ·· 343
  1.3 钢筋 ··· 350
  1.4 钢筋与混凝土的黏结 ·· 355
  思考练习题 ·· 356

第2章 结构按极限状态法设计计算的原则 ··· 358
  2.1 概述 ··· 358
  2.2 概率极限状态设计法的基本概念 ·· 360
  2.3 我国《公桥规》的计算原则 ·· 362
  2.4 材料强度的取值 ·· 364
  2.5 作用、作用的代表值和作用效应组合 ·· 365
  思考练习题 ·· 368

第3章 受弯构件正截面承载力计算 ·· 370
  3.1 受弯构件基本概念 ··· 370
  3.2 受弯构件的截面形式与构造 ·· 371
  3.3 受弯构件正截面受力全过程和破坏形态 ·· 376
  3.4 受弯构件正截面承载能力计算的基本原则 ······································· 381
  3.5 单筋矩形截面受弯构件强度计算 ·· 385
  3.6 双筋矩形截面受弯构件强度计算 ·· 390
  3.7 T形截面受弯构件计算 ··· 394
  思考练习题 ·· 401

第4章 受弯构件斜截面承载力计算 ·· 404
  4.1 受弯构件斜截面的受力特点和破坏形态 ·· 404
  4.2 影响受弯构件斜截面抗剪能力的主要因素 ······································· 407

4.3 受弯构件斜截面抗剪承载力的计算 …………………………… 408
  4.4 受弯构件的斜截面抗弯承载力 …………………………… 411
  4.5 全梁承载能力校核与构造要求 …………………………… 412
  　思考练习题 …………………………………………………… 421

第5章　受扭构件承载力计算 ……………………………………… 423
  5.1 概述 ………………………………………………………… 423
  5.2 纯扭构件的破坏特征和承载力计算 ……………………… 424
  5.3 在弯、剪、扭共同作用下矩形截面构件的承载力计算 … 428
  5.4 T形和I形截面受扭构件 …………………………………… 431
  5.5 箱形截面受扭构件 ………………………………………… 432
  5.6 构造要求 …………………………………………………… 433
  　思考练习题 …………………………………………………… 433

第6章　轴心受压构件的正截面承载力计算 …………………… 435
  6.1 概述 ………………………………………………………… 435
  6.2 配有纵向钢筋和普通箍筋的轴心受压构件计算 ………… 436
  6.3 配有纵向钢筋和螺旋箍筋的轴心受压构件 ……………… 439
  　思考练习题 …………………………………………………… 442

第7章　偏心受压构件的正截面承载力计算 …………………… 444
  7.1 偏心受压构件正截面受力特点和破坏形态 ……………… 445
  7.2 偏心受压构件的纵向弯曲 ………………………………… 447
  7.3 矩形截面偏心受压构件 …………………………………… 448
  7.4 工字形和T形截面偏心受压构件 ………………………… 459
  7.5 圆形截面偏心受压构件 …………………………………… 463
  　思考练习题 …………………………………………………… 466

第8章　受拉构件的承载力计算 …………………………………… 469
  8.1 概述 ………………………………………………………… 469
  8.2 轴心受拉构件的承载力计算 ……………………………… 469
  8.3 偏心受拉构件的承载力计算 ……………………………… 470
  　思考练习题 …………………………………………………… 472

第9章　钢筋混凝土受弯构件的应力、裂缝和变形计算 ……… 473
  9.1 概述 ………………………………………………………… 473
  9.2 换算截面 …………………………………………………… 474
  9.3 应力验算 …………………………………………………… 477
  9.4 受弯构件的裂缝和裂缝宽度验算 ………………………… 478
  9.5 受弯构件的变形(挠度)验算 ……………………………… 479
  9.6 混凝土结构的耐久性 ……………………………………… 486

思考练习题……487

第10章 局部承压 489
10.1 局部承压的破坏形态和破坏机理……490
10.2 混凝土局部承压强度提高系数……491
10.3 局部承压区的计算……493
思考练习题……495

第二部分 预应力混凝土结构

第11章 预应力混凝土结构的基本概念及其组成材料 499
11.1 概述……499
11.2 预加应力的方法与设备……501
11.3 预应力混凝土结构的组成材料……506
思考练习题……507

第12章 预应力混凝土受弯构件的设计与计算 508
12.1 预应力混凝土受弯构件的受力阶段与计算特点……508
12.2 预应力混凝土受弯构件承载力计算……511
12.3 张拉控制应力与预应力损失的计算……514
12.4 预应力混凝土受弯构件的应力计算与验算……519
12.5 抗裂验算……524
12.6 变形计算与预拱度设置……526
12.7 端部锚固区计算……527
12.8 预应力混凝土简支梁设计……529
12.9 预应力混凝土空心板计算示例……536
思考练习题……565

第13章 部分预应力混凝土受弯构件 568
13.1 部分预应力混凝土结构的概念与特点……568
13.2 部分预应力混凝土结构分类与受力特性……568
13.3 B类部分预应力混凝土受弯构件的计算……570
13.4 允许开裂的部分预应力混凝土受弯构件的设计……577
思考练习题……580

第14章 无黏结预应力混凝土受弯构件计算 581
14.1 无黏结预应力混凝土受弯构件的受力性能……581
14.2 无黏结部分预应力混凝土受弯构件的计算……584
14.3 无黏结部分预应力混凝土受弯构件的截面设计……586
14.4 无黏结部分预应力混凝土受弯构件的构造……588
思考练习题……590

第三部分 砌 工 结 构

第15章 砌工结构的基本概念与材料 ··· 593
- 15.1 砌工结构的基本概念 ··· 593
- 15.2 材料种类 ·· 594
- 15.3 砌体的强度与变形 ··· 596
- 思考练习题 ·· 598

第16章 砌工结构构件的承载力计算 ·· 599
- 16.1 计算原则 ·· 599
- 16.2 受压构件的承载力计算 ··· 599
- 16.3 局部承压以及受弯、受剪构件的承载力计算 ································ 604
- 思考练习题 ·· 605

附表 ··· 606

参考文献 ··· 617

总 论

0.1 概 述

人类早期主要利用土、石、木等天然材料从事建筑活动。

土木工程在人类历史上经历了三次大飞跃：

第一次飞跃：公元前 11 世纪~公元前 3 世纪，发明了砖、瓦等以承压为主的人工材料，形成了木、砖(石)结构，例如，万里长城、金字塔、阿房宫。

第二次飞跃：17 世纪的生铁、19 世纪的熟铁、19 世纪中叶的优质钢材(人工制造的耐拉、压材料)出现，钢结构兴起，例如，出现了高 300m 的埃菲尔铁塔、主跨 521m 的福斯铁路桥等钢结构。

第三次飞跃：19~20 世纪，钢筋混凝土结构出现。钢铁和水泥产生后，直至 20 世纪初，解决了钢筋混凝土结构计算理论以及预应力混凝土结构的应力损失、锚固等问题，配筋混凝土结构有了蓬勃发展，并在土木工程中占据了主要地位。

结构设计原理是一门重要的专业技术基础课，它是在学习材料力学、建筑材料等先修课程的基础上，结合桥梁工程中实际构件的工作特点来研究结构构件设计的一门学科。因此，它是一门理论性和实践性都很强的学科。

0.2 基本概念

1) 构件

构件是指承受荷载而起骨架作用的部分。按其受力特点,分为受弯构件、受压构件、受拉构件和受扭构件等典型的基本构件。

2) 结构

结构是指由若干构件连接而成的承受作用的空间或平面体系,是构造物的承重骨架部分。例如,桥梁的桥跨、墩(台)及基础组成了桥的承重体系,它们就被称为桥梁结构。

3) 结构分类

(1)按结构的建筑材料可分为混凝土结构、砌体结构、钢结构和木结构。其中混凝土结构包括素混凝土结构、钢筋混凝土结构、预应力混凝土结构和钢-混组合结构。

(2)按结构的用途可分为桥梁结构、建筑结构、地下结构、水工结构和特种结构。对于桥梁结构,根据主要承重结构受力的不同,可将其分为梁式桥(受弯为主)、拱式桥(受压为主)、刚架桥(梁柱组合)、悬索桥(受拉为主)和斜拉桥(组合受力)。

4) 钢筋混凝土结构

钢筋混凝土结构由钢筋和混凝土两种性能不同的材料组成一体,是能发挥不同材料各自特点的一种结构,具有就地取材、造价低、耐久性和耐火性好等特点。广泛用于中小跨径梁、板、墩台、拱和塔结构中。

5) 预应力混凝土结构

预应力混凝土结构是预先人为地给混凝土引入一个应力,以改善混凝土在使用期间的应力状态的混凝土结构。它能够充分利用高强材料,具有跨越能力大、抗裂性能好等特点。用于各种跨径规模的桥梁中。

6) 砌体结构

砌体结构是将砖石等块材按照一定砌筑规则建筑而成的结构。其优点为材料来源广、抗压能力大、耐久和施工简便。该结构主要适用于以受压为主的构件(墩台、护坡)。

7) 钢结构

钢结构的优点为自重轻、抗压强度高、材质均匀。钢结构主要适用于大跨径桥梁上部结构和轻钢住宅等。

0.3 本课程的特点与学习本课程的指导意见

1) 本课程的主要任务

(1)学习钢筋混凝土和预应力混凝土结构的基本概念和基本理论。

(2)学习各种工程材料设计计算原则。

(3)掌握钢筋混凝土结构、预应力混凝土结构、砖石结构、钢结构在各种受力状态下的工作性能、破坏形态、计算原理、设计和构造要求等问题,通过课程设计,掌握梁的设计方法,为工程设计打下基础。

2）本课程的特点

（1）本课程为专业过渡课程。由基础课过渡到专业课,研究问题的方法既有继承性,又有创造性,要注意其中的本质区别,学习新的研究方法。

（2）钢筋混凝土构件的设计方法是建立在试验基础上的,构件的设计方法对试验的依赖性很强,应注意各理论分析的适用范围和应用条件。

（3）设计方案具有多样性的特点,结构设计是一个综合性的问题,需要遵循"安全、适用、经济、合理"的原则。在结构设计中,当给定荷载时,同一构件的设计答案往往不是唯一的,需进行综合分析比较,才能做出合理的选择。

（4）学习本课程应特别注意熟悉分析过程,掌握计算步骤。

（5）学会运用规范。与本课程有关的规范有：

①《公路桥涵设计通用规范》（JTG D60—2015）。

②《公路钢筋混凝土及预应力混凝土桥涵设计规范》（JTG 3362—2018）（以下简称《公桥规》）。

③《公路圬工桥涵设计规范》（JTG D61—2005）。

④《工程结构可靠性设计统一标准》（GB 50153—2008）。

⑤《公路工程结构可靠度设计统一标准》（GB/T 50283—1999）。

⑥《CEB-FIP Model Code 2010》（以下简称 CEB-FIP）。

在学习中要逐步熟悉运用规范。

3）本课程要解决的问题

本课程要解决的问题,归纳起来有：

（1）选择结构的材料类型。

（2）选择截面形式。

（3）拟定截面尺寸。

（4）配筋设计计算。

（5）进行各项验算（强度条件、刚度、稳定性、抗裂性）。

4）本课程的学习方法

学好本课程,应从以下几个方面下功夫：

（1）认真学好课堂内容,把握内容体系的内在联系和细微差别。

（2）注意理论联系实际,重视试验、经验和规范的规定。

（3）深入掌握材料性能,因为结构的许多特性都是由材料性能决定的。

（4）根据结构设计方案的多样性特点,尽可能寻找最优的设计方案。

（5）熟悉、运用规范,重视构造要求。

PART 1 | 第一部分

钢筋混凝土结构

第1章
钢筋混凝土结构的基本概念及材料的物理力学性能

1.1 钢筋混凝土结构的基本概念

1)钢筋混凝土结构的定义

混凝土是一种非匀质材料,其抗压强度高,抗拉强度很低;而钢筋是一种拉压强度都很高的材料。将钢筋和混凝土结合在一起使用,混凝土承受压力,钢筋承受拉力,能充分发挥各自的优势,就形成了钢筋混凝土结构。所以,钢筋混凝土结构是由配置受力的普通钢筋或钢筋骨架的混凝土制成的结构。

2)实例

下面以一个实例来说明素混凝土构件与钢筋混凝土构件的受力状况:

某跨度为4m,跨中作用集中荷载的矩形截面试验梁,梁截面尺寸为200mm×300mm,混凝土为C25。在跨中施加集中荷载,如图1-1所示。

(1)试验结果

①图1-1a)中,素混凝土梁极限荷载 $P=8kN$,由混凝土抗拉强度控制,破坏形态为脆性破坏;

②图1-1b)中,钢筋混凝土梁极限荷载 $P=36kN$,它是由于钢筋受拉屈服混凝土压碎而破坏。

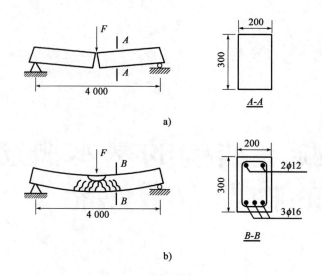

图 1-1 素混凝土梁和钢筋混凝土梁(尺寸单位：mm)

(2) 试验结论

钢筋和混凝土结合,大大提高了结构的承载力并使结构的受力性能得到改善。

3) 钢筋和混凝土可共同工作的原因

钢筋和混凝土之所以能够共同工作,研究发现有以下三个方面原因：

(1) 混凝土和钢筋之间有着良好的黏结力,使两者能可靠地结合成一个整体,在荷载作用下能够很好地共同变形,完成其结构功能。

(2) 钢筋的温度膨胀系数 $\alpha_{st} = 1.2 \times 10^{-5}$,混凝土的温度膨胀系数 $\alpha_{ct} = (1.0 \sim 1.5) \times 10^{-5}$。二者具有相近的温度线膨胀系数,不会由于温度变化产生较大的温度应力和相对变形而破坏黏结力。

(3) 在保护层厚度足够的前提下,呈碱性的混凝土可以保护钢筋不易锈蚀,保证了钢筋与混凝土的共同作用。

4) 钢筋混凝土结构的主要优缺点

(1) 优点

①就地取材,比较经济；

②耐久性和耐火性好；

③可模性好,便于做成各种各样的结构形式；

④现浇或装配式结构的整体性好、刚度大。

(2) 缺点

①自重大；

②抗裂性差；

③施工受气候条件影响较大；

④隔热和隔声性能较差。

由于钢筋混凝土结构存在一些不足,在工程中应注意扬长避短,比如,可以使用轻质集料来减轻结构自重,从而提高其跨越能力。

1.2 混 凝 土

1) 概述

混凝土是由水泥、石、砂、水按一定的配合比制成的不同等级的建筑材料。钢筋混凝土结构对混凝土品质的要求为:具有高强度、工作性(和易、泌水)、耐久性和经济性等。混凝土的性能特点表现为:①抗压强度高,抗拉强度低;②影响强度的因素很多,如水泥等级、集料性质、混凝土龄期、制作方法、养护条件和试验方法等;③混凝土的抗压强度是有条件的——混凝土的抗压强度受横向变形、尺寸效应等因素的影响大。

2) 混凝土的强度

混凝土的基本强度指标有三个:立方体抗压强度f_{cu}、轴心抗压强度f_c和抗拉强度f_t。影响混凝土强度的因素很多,包括:材料的性质、混凝土配合比、养护环境、施工方法、试件的形状与尺寸、试验方法、加载条件和试件的受力性质等,因此,《公桥规》对各种强度的测试作出了规定。

(1) 混凝土立方体抗压强度(f_{cu})

混凝土立方体抗压强度是基本强度指标,以边长为150mm的立方体试件,按照标准方法制作养护28d(在温度20℃±3℃的条件下,置于相对湿度不小于90%的潮湿空气中),用标准试验方法(试件表面不涂润滑剂,全截面受压,加载速度0.15~0.25MPa/s)测得的抗压强度(以MPa为单位)作为混凝土的立方体抗压强度,用符号f_{cu}表示。影响立方体强度的主要因素为试件尺寸和试验方法,其尺寸效应关系为:

$$\left. \begin{array}{l} f_{cu}(150) = 0.95 f_{cu}(100) \\ f_{cu}(150) = 1.05 f_{cu}(200) \end{array} \right\} \tag{1-1}$$

混凝土强度从C25~C80共分为12个等级,中间以5MPa进级。通常C50以下为普通强度混凝土,C50及其以上为高强度混凝土。根据《公桥规》规定:钢筋混凝土构件不应低于C25,当采用HRB400、RRB400级钢筋配筋时,不应低于C30;对于预应力混凝土构件,不应低于C40。

值得一提的是,国外常采用混凝土圆柱体试件来确定混凝土轴心抗压强度。例如,美国、日本和欧洲混凝土协会、CEB-FIP采用直径6英寸(152mm)、高12英寸(305mm)的圆柱体标准试件的抗压强度作为轴心抗压强度指标,记作f'_c。对C60以下的混凝土,圆柱体抗压强度f'_c和立方体抗压强度标准值$f_{cu,k}$之间的关系可按下式换算。

$$f'_c = 0.79 f_{cu,k} \tag{1-2}$$

当$f_{cu,k}$超过60MPa后,随着抗压强度提高,f'_c与$f_{cu,k}$的比值(公式中的系数)提高。CEB-FIP给出的具体值为:对C60的混凝土,比值为0.833;对C70的混凝土,比值为0.857;对C80混凝土,比值为0.875。

(2) 轴心抗压强度(f_c)

轴心抗压强度是以150mm×150mm×300mm棱柱体为标准试件测得的抗压强度。其试件制作、养护和加载试验方法同立方体试件。轴心抗压强度标准值f_{ck}与立方体抗压强度标准值$f_{cu,k}$有一定的关系,试验测试统计结果得到其关系为:

$$f_{ck} = 0.88 \alpha f_{cu,k} \tag{1-3}$$

其中:对于 C50 以下混凝土,$a = 0.76$;对于 C55 ~ C80 混凝土,$a = 0.78 \sim 0.82$。

(3)轴心抗拉强度(f_t)

混凝土的抗拉强度比抗压强度小得多,为抗压强度的 $1/18 \sim 1/8$。主要的测试方法有:

①轴心对拉直接测试方法

如图 1-2 所示,试件为 $100\mathrm{mm} \times 100\mathrm{mm} \times 500\mathrm{mm}$ 的柱体,通过钢筋试件施加拉力,破坏时试件中部产生横向裂缝,破坏截面上的平均拉应力即为轴心抗拉强度 f_t。

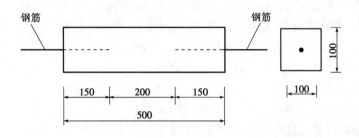

图 1-2　混凝土抗拉强度试验试件(尺寸单位:mm)

②劈裂试验的间接测试方法

如图 1-3 所示,采用 $d = 150\mathrm{mm}$ 圆柱体作为标准试件进行混凝土抗拉强度测定,按照规定的试验方法操作,则可测得混凝土的劈裂抗拉强度 f_{ts}。

$$f_{ts} = \frac{2F}{\pi A} = 0.637 \frac{F}{A} \tag{1-4}$$

式中,f_{ts} 为混凝土劈裂抗拉强度(MPa);F 为劈裂破坏荷载(N);A 为试件劈裂面面积(mm^2)。

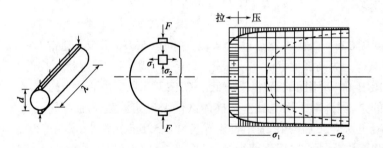

图 1-3　劈裂试验

需要说明的是,采用上述试验方法测得的混凝土劈裂抗拉强度值换算成轴心抗拉强度时,应乘以换算系数 0.9,即 $f_t = 0.9 f_{ts}$。

轴心抗拉强度标准值 f_{tk} 与立方体抗压强度标准值 $f_{cu,k}$ 的换算关系为:

$$f_{tk} = 0.88 \times 0.395 f_{cu,k}^{0.55} (1 - 1.645 \delta_f)^{0.45} \tag{1-5}$$

式中,δ_f 为立方体抗压强度变异系数。

(4)复合应力状态下混凝土强度

①双向正应力作用

如图 1-4 所示,当 σ_1、σ_2(压-压)作用时,混凝土强度增加;当 σ_1、σ_2(拉-压)作用时,混凝土强度降低;当 σ_1、σ_2(拉-拉)作用时,混凝土强度基本不变。

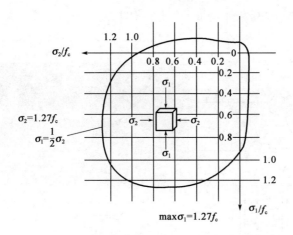

图 1-4　双向应力状态下混凝土强度变化曲线

② 正应力和剪应力作用

如图 1-5 所示,当 $\sigma/f_c < (0.5 \sim 0.7)$ 时,混凝土抗剪强度随压应力的增大而增大;当 $\sigma/f_c > (0.5 \sim 0.7)$ 时,混凝土抗剪强度随压应力的增大而减小。

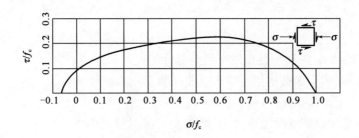

图 1-5　法向应力与剪应力组合时的强度曲线

当压应力在 $0.6f_c$ 左右时,抗剪强度达到最大。工程中对于压剪构件,应控制二者之间的比例。

③ 三轴受压

如图 1-6 所示,当混凝土三向受压时,其抗压强度为:

$$f'_{cc} = f'_c + k'\sigma_2 \tag{1-6}$$

三向受压混凝土的强度大大高于单向受压强度,因此较多应用在钢管混凝土和密配螺旋箍筋等构件(图 1-7)。

3) 混凝土的变形

(1) 混凝土变形性能的特点

混凝土变形主要有受力变形和体积变形。影响混凝土变形的因素主要有:加载方式、荷载作用时间、温度、湿度、试验的尺寸、形状、混凝土强度等。受力变形包括:单调短期加载下的变形、长期

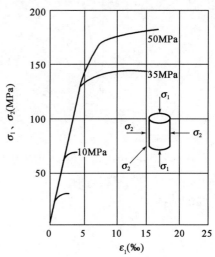

图 1-6　三向受压状态下混凝土强度

荷载作用下的变形、多次重复荷载作用下的变形。体积变形主要是指收缩变形和温度变化引起的变形。

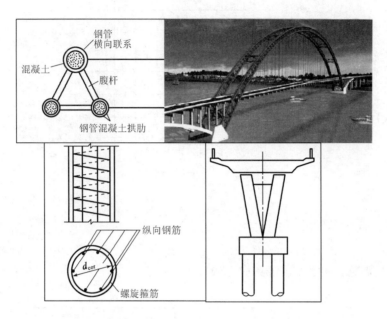

图 1-7　工程应用——钢管混凝土、密配螺旋箍筋

(2) 混凝土在单调、短期加载作用下的变形性能

混凝土在单调、短期加载作用下变形性能试验的试验条件为：①单轴受压；②等应变速度加载；③在试件旁附设高弹性元件与试件一同受压。

单轴受压试验得到的混凝土应力-应变曲线(σ-ε 曲线)，如图 1-8 所示。

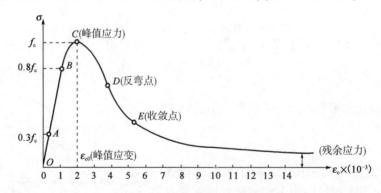

图 1-8　混凝土受压时应力-应变曲线

根据图 1-8，可以得到混凝土应力-应变曲线的三个特征值分别为：①轴心抗压强度 f_c；②峰值点应变 ε_{c0}；③混凝土极限压应变 ε_{cu}，根据《公桥规》：当混凝土强度等级为 C50 及以下时，取 $\varepsilon_{cu}=0.0033$。

混凝土的受力过程分为三个阶段，分别为：OA 弹性阶段，$\sigma<0.3f_c$；AB 弹塑性阶段，或称为裂缝稳定阶段，$\sigma=0.3f_c \sim 0.8f_c$；$BC$ 裂缝不稳定阶段，$\sigma=0.8f_c \sim 1.0f_c$。

不同强度混凝土的应力-应变曲线有明显差异，影响混凝土轴心受压应力-应变曲线的主要因素为：

①混凝土强度。如图1-9所示,低强度混凝土受压时,曲线的下降段比较长,延性比高强度混凝土好。

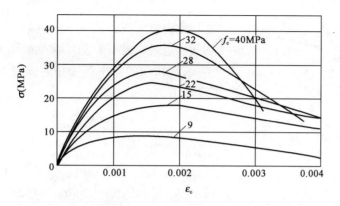

图1-9 强度等级不同的混凝土的应力-应变曲线

②应变速率。应变速率小,峰值应力f_c降低,ε_{c0}增大,下降段曲线坡度显著地减小。

③侧向约束。如图1-10所示,试件的上下表面与试验机承压板之间存在摩阻力,破坏时,形成两个对顶叠置的截头方锥体如图1-10b)所示,测得强度较高。若试件的上下表面与试验机承压板之间涂抹润滑剂,其破坏形态如图1-10c)所示,测得的强度较低且无法测得下降段。

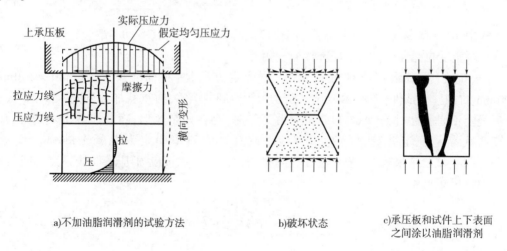

a)不加油脂润滑剂的试验方法　　b)破坏状态　　c)承压板和试件上下表面之间涂以油脂润滑剂

图1-10 混凝土强度的测试

④测试技术和试验条件。应采用等应变加载,如果采用等应力加载,则很难测得下降段曲线。试验机的刚度对下降段的影响很大,如果试验机的刚度不足,无法测出应力-应变曲线的下降段;应变测量的标距对其有影响,应变量测的标距越大,曲线坡度越大;标距越小,坡度越小。

(3)混凝土弹性模型和变形模量

图1-11所示为混凝土三种弹性模量示意图。

①原点弹性模量

原点弹性模量是在混凝土受压应力-应变曲线图的原点作切线,该切线的斜率即为原点弹

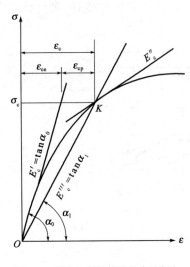

图 1-11 混凝土弹性模量的表示方法

性模量,可以表示为:

$$E'_c = \frac{\sigma}{\varepsilon_{ce}} = \tan\alpha_0 \tag{1-7}$$

②切线模量

切线模量是在混凝土应力-应变曲线上某一应力 σ_c 处作一切线,该切线的斜率即为相应于应力 σ_c 时的切线模量,可以表示为:

$$E''_c = \frac{d\sigma}{d\varepsilon} \tag{1-8}$$

③割线模量

连接混凝土应力-应变曲线的原点 O 及曲线上某一点 K 作割线,K 点混凝土应力为 $\sigma_c(=0.5f_c)$,则该割线(OK)的斜率即变形模量,也称割线模量或弹塑性模量,可用下式计算:

$$E'''_c = \tan\alpha_1 = \frac{\sigma_c}{\varepsilon_c} \tag{1-9}$$

混凝土的剪切模量可用下式表示:

$$G_c = \frac{E_c}{2(1+\mu_c)} \tag{1-10}$$

式中,μ_c 为混凝土横向变形系数(泊松比),当 $\mu_c = 0.2$ 时,$G_c = 0.4E_c$。

混凝土受拉弹性模量与受压弹性模量相等。

《公桥规》推荐混凝土弹性模量 E_c 的确定方法是:对标准尺寸 150mm × 150mm × 300mm 的棱柱体试件,先加载至 $\sigma_c = 0.5f_c$,然后卸载至零,再重复加载、卸载 5~10 次。由于混凝土不是弹性材料,每次卸载至应力为零时,存在残余变形,随着加载次数增加,应力-应变曲线渐趋稳定并基本上趋于直线。该直线的斜率即定为混凝土的弹性模量,见式(1-11)。试验结果表明:按上述方法测得的弹性模量比按应力-应变曲线原点切线斜率确定的弹性模量要略低一些。

$$E_c = \frac{10^5}{2.2 + \frac{34.74}{f_{cu,k}}} \tag{1-11}$$

(4)混凝土单轴向受压应力-应变曲线的数学模型

如图 1-12 所示,国内外广泛采用的描述混凝土单轴向受压应力-应变曲线的数学模型是:上升段为二次抛物线,下降段为斜直线,可表示为:

$$\left.\begin{array}{l}\text{上升段}:\sigma = f_c\left[\dfrac{2\varepsilon}{\varepsilon_0} - \left(\dfrac{\varepsilon}{\varepsilon_0}\right)^2\right], 0 \leq \varepsilon \leq \varepsilon_0 \\ \text{下降段}:\sigma = f_c\left(1.0 - 1.15\dfrac{\varepsilon - \varepsilon_0}{\varepsilon_u - \varepsilon_0}\right), \varepsilon_0 \leq \varepsilon \leq \varepsilon_u\end{array}\right\} \tag{1-12}$$

式中,混凝土峰值应变 $\varepsilon_0 = 0.002$,极限压应变 $\varepsilon_{cu} = 0.0038$。

我国规范建议采用的模型形式较简单,如图 1-13 所示,上升段采用二次抛物线,下降段采用水平直线。该模型可表示为:

$$\left.\begin{aligned}\text{上升段}: \sigma_c &= f_c\left[1-\left(1-\frac{\varepsilon_{cu}}{\varepsilon_0}\right)^n\right], \varepsilon \leqslant \varepsilon_0\\ \text{水平段}: \sigma_c &= f_c, \varepsilon_0 \leqslant \varepsilon \leqslant \varepsilon_{cu}\end{aligned}\right\} \qquad (1\text{-}13)$$

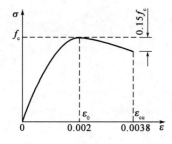
图 1-12 混凝土应力-应变曲线

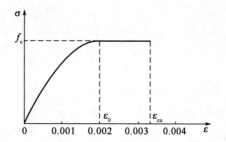
图 1-13 我国规范建议采用的混凝土应力-应变曲线

式中,参数 n、ε_0 和 ε_{cu} 的取值如下:

$$n = 2 - \frac{1}{60}(f_{cu,k} - 50) \leqslant 2.0 \qquad (1\text{-}14)$$

$$\varepsilon_0 = 0.002 + 0.5(f_{cu,k} - 50) \times 10^{-5} \leqslant 0.002 \qquad (1\text{-}15)$$

$$\varepsilon_{cu} = 0.0033 - (f_{cu,k} - 50) \times 10^{-5} \leqslant 0.0033 \qquad (1\text{-}16)$$

(5)混凝土在荷载长期作用下的变形 ε_{cc}——徐变

在荷载的长期作用下,混凝土的变形将随时间推移而增加,即在应力不变的情况下,混凝土的应变随时间延长继续增长,这种现象被称为徐变。试验得到的混凝土应变-时间关系曲线,如图 1-14 所示。

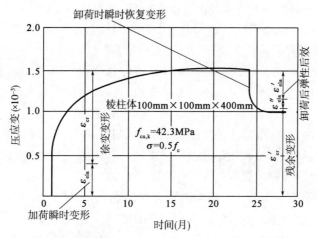

图 1-14 混凝土的徐变曲线

混凝土徐变是在荷载长期作用下,混凝土凝胶体中的水分逐渐压出,水泥石逐渐发生黏性流动,微细空隙逐渐闭合,结晶体内部逐渐滑动,微细裂缝逐渐发生等各种因素的综合结果。

影响混凝土徐变的因素很多,主要有:a. 内在因素,包括混凝土组成、龄期等,龄期越早,徐变越大,因此,工程实际中应避免过早地施加预应力;b. 环境条件,是指养护和使用时的温度、湿度,受力后环境温度越高,湿度越低,徐变就越大;c. 应力条件,当压应力 σ 小于 $0.5f_c$ 时,徐变与应力呈线性关系;当压应力 σ 为 $(0.5\sim0.8)f_c$ 时,徐变的增长较应力的增长快,这种情况称为非线性徐变;当压应力 $\sigma>0.8f_c$ 时,混凝土的非线性徐变往往是不收敛的。需要说明的是:当应力过大时,徐变急剧增加,会导致混凝土破坏,如在预应力混凝土结构中,给混凝土施加过高的预应力是危险的。

混凝土徐变对结构的影响是比较大的,主要体现在:a. 使构件的变形增加;b. 静定结构会使截面中产生应力重分布;c. 超静定结构引起赘余力;d. 在预应力混凝土结构中产生预应力损失。

(6)体积变形——混凝土收缩

在凝结和硬化的物理化学过程中,混凝土体积随时间推移而减小的现象称为收缩。混凝土收缩的主要原因包括以下两个方面:a. 硬化初期水泥石在凝固过程中产生体积变化(化学性收缩,本身的体积收缩);b. 后期主要由于混凝土内自由水分蒸发而引起干缩(物理收缩,失水干燥)。

影响混凝土收缩的主要因素有:a. 混凝土的组成和配比;b. 构件的养护条件、使用环境的温度和湿度,以及凡是影响混凝土中水分保持的因素;c. 构件的体表比,比值越小,收缩越大。

混凝土收缩对结构的影响主要表现为:a. 构件未受荷之前可能产生裂缝;b. 预应力构件中引起预应力损失;c. 超静定结构产生次内力。

1.3 钢 筋

1)基本概念

钢筋是一种具有较高的强度、良好的塑性和可焊性,与混凝土有较好的黏结性能的材料。钢筋的品种和级别比较多。

钢筋按化学成分分类,可分为碳素钢和普通低合金钢。碳素钢包括低碳钢(含碳量在 0.25% 以下)、中碳钢(含碳量为 0.25%~0.6%)和高碳钢(含碳量大于 0.6%)。普通低合金钢是指在钢筋中加入少量低合金元素,以提高其强度和塑性。

钢筋按加工方法分类,可分为热轧钢筋、热处理钢筋和冷加工钢筋。其中冷加工钢筋包括冷拉钢筋、冷轧带肋钢筋和冷轧扭钢筋。但是由于冷拉钢筋延性较差,目前较少使用,当工程中使用时,应遵守专门规程的规定。

热轧钢筋按外形特征分类,可分为热轧光圆钢筋(HPB300)和热轧带肋钢筋(HRB400,HRBF400,RRB400,HRB500),见图 1-15,热轧钢筋的种类和符号如表 1-1 所示。一般对于普

a) 光面钢筋

b) 螺纹钢筋

c) 人字纹钢筋

d) 月牙纹钢筋

图 1-15 热轧钢筋

通钢筋可选取 HPB300、HRB400、HRBF400、HRB500 热轧钢筋,而预应力钢筋应选用钢绞线和钢丝(图1-16),中小型构件或竖、横向钢筋也可选用精轧螺纹钢筋。

热轧钢筋的种类和符号(单位:mm)　　　　　表1-1

钢筋种类	符　号	钢筋种类	符　号
HPB300($d=6\sim22$)	ϕ	HRBF400($d=6\sim50$)	ϕ^F
HRB500($d=6\sim50$)	Φ	HRB400($d=6\sim50$)	Φ

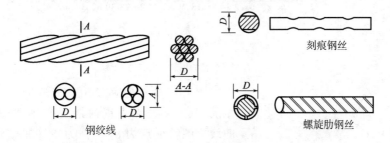

图1-16　预应力钢筋

钢筋按力学性能不同分类,可分为软钢:有明显屈服台阶的钢筋(热轧钢筋、冷拉钢筋);硬钢:无明显屈服台阶的钢筋(高强碳素钢丝、钢绞线)。

2)钢筋的力学性能

如图1-17所示的软钢应力-应变曲线有明显流服。其受力过程分为四个阶段:a. oa 段,弹性阶段;b. $abcd$ 段,屈服阶段;c. fd 段,强化阶段;d. de 段,颈缩阶段。

钢筋的三个应力特征值分别为:比例极限 σ_a、屈服强度 σ_b 和极限强度 σ_d。钢筋的主要物理力学指标有:

(1)两个强度指标。

一是屈服强度,它是钢筋混凝土结构设计计算中强度取值的主要依据,因为钢筋应力达到屈服极限后,荷载不增加,应变继续增大,使得混凝土裂缝开展过宽,变形过大,结构不能正常使用,一般采用屈服的下限值作为屈服强度。

二是极限抗拉强度,即材料的实际破坏强度,用于衡量钢筋屈服后的抗拉能力,不能作为计算依据。钢筋混凝土结构的屈服强度和极限抗拉强度的比值为屈强比,可以表示结构的潜力,屈强比小则结构的可靠性高,但太小则钢材利用率太低。在实际检验钢筋的质量时,仍要检验它的极限抗拉强度并满足检验标准的要求,特别在抗震结构中,由于构件要进入大变形状况工作,钢筋可能进入强化阶段。

$$\frac{屈服强度}{极限抗拉强度} = 屈强比 \tag{1-17}$$

(2)两个塑性指标。

一是伸长率,又称延伸率,如图1-17中 e 点所对应的横坐标,用标注为10倍(δ_{10})或5倍(δ_5)钢筋直径表示为:

$$\delta = \frac{l_2 - l_1}{l_1} \tag{1-18}$$

式中,l_1、l_2 分别为钢筋在标距范围的原长度和变形后的长度。

二是冷弯性能,指钢材在冷加工过程和使用时不开裂、弯断或脆断的性能。

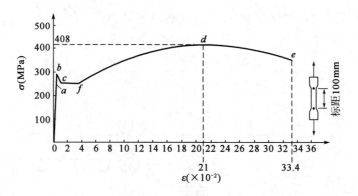

图 1-17 有明显流幅的钢筋应力-应变曲线

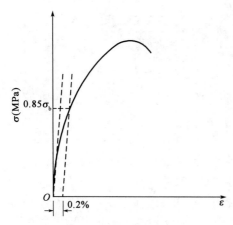

图 1-18 没有明显流幅的钢筋应力-应变曲线

图 1-18 所示为应力-应变关系曲线无明显流幅的钢筋,其力学特点表现为强度高、塑性差和脆性大。曲线上的特征点为 $\sigma_{0.2}$,称为条件屈服强度,一般为极限强度的 85%。

3) 钢筋应力-应变关系的数学模型

常用的钢筋应力-应变曲线模型有以下几种:

(1) 理想完全弹塑性的双直线模型

双直线模型适用于流幅较长的低强度钢材,将钢筋的应力-应变曲线简化为如图 1-19a)所示的两段直线,其数学表达式如下:

$$\left.\begin{array}{l} \sigma_s = E_s \varepsilon_s \left(E_s = \dfrac{f_y}{\varepsilon_y} \right), \varepsilon_s < \varepsilon_y \\ \sigma_s = f_y, \varepsilon_y < \varepsilon_s < \varepsilon_{s,h} \end{array}\right\} \quad (1\text{-}19)$$

(2) 完全弹塑性加硬化的三折线模型

如图 1-19b)所示,三折线模型适用于流幅较短的软钢,可以描述钢筋屈服后立即发生应变硬化(应力强化)的钢材,正确地估计屈服应变后的应力,其数学表达形式如下:

当 $\varepsilon_s \leqslant \varepsilon_y, \varepsilon_y \leqslant \varepsilon_s \leqslant \varepsilon_{s,h}$ 时,表达式同式(1-19);

当 $\varepsilon_{s,h} \leqslant \varepsilon_s \leqslant \varepsilon_{s,u}$ 时,表达式为:

$$f_s = f_y + (\varepsilon_s - \varepsilon_{s,h}) \tan\theta' \quad (1\text{-}20)$$

式中,$\tan\theta' = E_s' = 0.01 E_s$。

(3) 弹塑性的双斜线模型

如图 1-19c)所示,双斜线模型可以描述没有明显流幅的高强钢筋或钢丝的应力-应变曲线,其数学表达形式如下:

$$\left.\begin{array}{l} \sigma_s = E_s \varepsilon_s \left(E_s = \dfrac{f_y}{\varepsilon_y} \right), \varepsilon_s < \varepsilon_y \\ \sigma_s = f_y + (\varepsilon_s - \varepsilon_y) \tan\theta'', \varepsilon_y \leqslant \varepsilon_s \leqslant \varepsilon_{s,u} \end{array}\right\} \quad (1\text{-}21)$$

式中,$\tan\theta'' = E_s'' = \dfrac{f_{s,u} - f_y}{\varepsilon_{s,u} - \varepsilon_y}$。

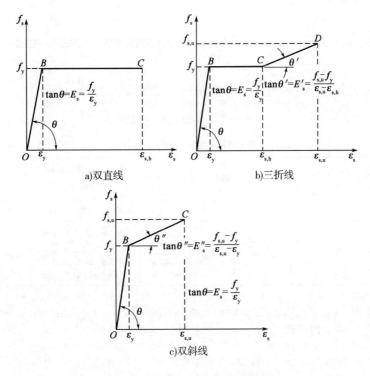

图 1-19　钢筋应力-应变曲线的数学模型

4) 钢筋的接头、弯钩和弯折

如图 1-20 和图 1-21 所示,钢筋的接头有三种:焊接接头、机械连接接头和绑扎接头。其中焊接接头包括对焊和电弧搭接焊。机械连接接头包括套筒挤压接头和镦粗直螺纹接头。绑扎接头主要是指绑扎接头的钢筋直径一般不宜大于 28mm;受压构件中的受压钢筋直径不大于 32mm;钢筋的搭接长度规定为绑扎接头的受力钢筋的截面积:受拉区,不超过 25%;受压区不超过 50%。

根据《公桥规》,纵向受拉钢筋绑扎接头的搭接长度的具体规定见表 1-2,纵向受压区钢筋绑扎接头的搭接长度应取受拉钢筋绑扎搭接长度的 0.7 倍。

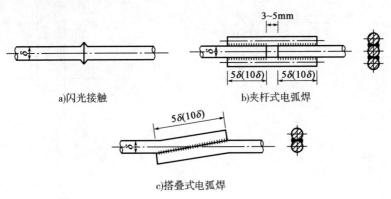

图 1-20　普通钢筋的焊接接头
注:括号内数字为单面焊缝。

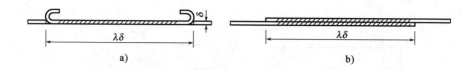

图 1-21　钢筋的绑扎搭接接头

钢筋最小锚固长度 l_a（mm） 表 1-2

钢筋种类		HPB300				HRB400, HRBF400, RRB400			HRB500		
混凝土强度等级		C25	C30	C35	≥C40	C30	C35	≥C40	C30	C35	≥C40
受压钢筋（直端）		45d	40d	38d	35d	30d	28d	25d	35d	33d	30d
受拉钢筋	直端	—	—	—	—	35d	33d	30d	45d	43d	40d
	弯钩端	40d	35d	33d	30d	30d	28d	25d	35d	33d	30d

注：1. d 为钢筋公称直径（mm）。
2. 对于受压束筋和等代直径 $d_e \leqslant 28$mm 的受拉束筋的锚固长度，应以等代直径按表值确定，束筋的各单根钢筋可在同一锚固终点截断；对于等代直径 $d_e > 28$mm 的受拉束筋，束筋内各单根钢筋，应自锚固点开始，以表内规定的单根钢筋的锚固长度 1.3 倍，呈阶梯形延伸后截断，即自锚固起点开始，第一根延伸 1.3 倍单根钢筋的锚固长度，第二根延伸 2.6 倍单根钢筋的锚固长度，第三根延伸 3.9 倍单根钢筋的锚固长度。
3. 采用环氧树脂涂层钢筋时，受拉钢筋最小锚固长度应增加 25%。
4. 当混凝土在凝固过程中易受扰动时，锚固长度应增加 25%。
5. 当受拉钢筋末端采用弯钩时，锚固长度为包括弯钩在内的投影长度。

如图 1-22 所示，钢筋的弯钩和弯折需要满足：受拉的光面钢筋需在端头设半圆弯钩，带肋钢筋设直角形弯钩，弯折处钢筋内侧弯曲直径 D 不得小于 $20d$。

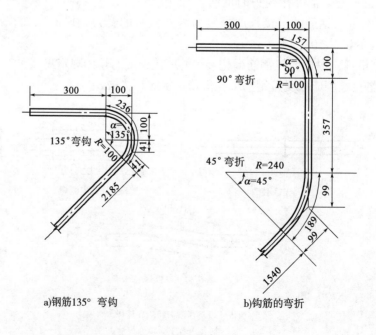

a) 钢筋 135° 弯钩　　　b) 钩筋的弯折

图 1-22　钢筋的弯钩与弯折示意图（尺寸单位：mm）

5)混凝土结构对钢筋的要求

高强度是构件承载力的重要保证;为了使钢筋在断裂前有足够的变形,要求钢材有一定的塑性。钢筋的伸长率和冷弯性能是钢筋塑性的主要指标。可焊性是评定钢筋焊接后的接头性能的指标;可焊性要好,即要求在一定的工艺条件下钢筋焊接后不产生裂纹及过大的变形;对钢筋的耐火性也有要求,热轧钢筋的耐火性能最好,冷轧钢筋其次,预应力钢筋最差;为了保证钢筋与混凝土共同工作,要求钢筋与混凝土之间必须有足够的黏结力。

1.4 钢筋与混凝土的黏结

黏结力是钢筋和混凝土界面之间的一种相互作用力,黏结应力是沿钢筋与混凝土接触面所产生的分布剪应力。通过这种作用来传递钢筋和混凝土之间的应力,协调变形。黏结应力使钢筋应力沿长度变化;反之,没有应力的变化就不存在黏结应力。黏结强度是黏结失效(钢筋被拔出或混凝土被劈裂)时的最大平均黏结应力。

钢筋与混凝土之间黏结的作用为:a. 抵抗钢筋滑动,是保证两种材料共同工作的基础;b. 对钢筋混凝土进行有限元分析,必须要有黏结-滑移关系;c. 黏结作用的退化,对结构的疲劳和抗震有很大影响。

黏结力由三部分组成:化学胶着力,即钢筋和混凝土表面的化学吸附作用;摩擦力,由混凝土收缩紧握钢筋而产生;机械咬合力,钢筋表面凹凸不平和混凝土之间产生机械咬合力。

以拉拔试验为例对黏结应力进行研究,如图 1-23 所示,在加载端钢筋的应力为:

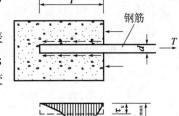

图 1-23 拉拔试验

$$\sigma_s = \frac{F}{A_s}, \varepsilon_s = \frac{\sigma_s}{E_s} \tag{1-22}$$

钢筋在试件端的应力为:

$$\sigma_s = 0, \varepsilon_s = 0 \tag{1-23}$$

黏结的传力过程是根据应变差产生黏结应力 τ,将钢筋拉力逐步向混凝土传递,钢筋的应力减小,而混凝土的应力增加,到钢筋尾部处 $\varepsilon_c = \varepsilon_s, \tau = 0$。

拔出试验统计结果如图 1-24 所示,可以看出黏结应力分布呈曲线形。光圆钢筋和带肋钢筋的黏结应力分布图形不同,带肋钢筋与混凝土的黏结强度比光圆钢筋高得多,统计结果是:光圆钢筋的平均黏结应力 $\bar{\tau}=1.5\sim3.5$MPa,带肋钢筋的平均黏结应力 $\bar{\tau}=2.5\sim6.0$MPa。

对于光面钢筋,黏结力由胶着力、摩擦力和机械咬合力组成。在没有产生滑移前,只有化学吸附作用,一旦混凝土和钢筋发生滑移,则由摩擦力和钢筋表面粗糙不平产生的机械咬合力提供黏结。破坏形态为剪切破坏(钢筋从混凝土中被拔出),破坏面是钢筋和混凝土之间的接触面。避免发生黏结失效的措施为端部做弯钩和有足够的锚固长度(受压时,可不做弯钩),其锚固长度见规范规定。

对于带肋钢筋,黏结力由胶着力、摩擦力和机械咬合力组成。在没有产生滑移前,只有化学吸附作用,一旦混凝土和钢筋发生滑移,则由摩擦力和机械咬合力(主要)提供黏结。带肋钢筋黏结的破坏形态有两种,分别为剪切型黏结破坏(保护层厚度较厚或有环向箍筋约束)和劈裂型黏结破坏(保护层厚度较小或未配环向箍筋)。避免黏结失效的措施是要有足够的保护层厚度、满足锚固长度要求和设置弯钩等。

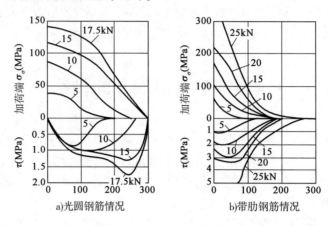

图 1-24　钢筋的黏结应力分布图

研究表明,影响钢筋与混凝土黏结强度的主要因素有混凝土的强度等级、浇筑混凝土时钢筋所处的位置、钢筋之间的净距、混凝土保护层厚度和钢筋的表面形状等。为了保证钢筋和混凝土之间有足够的黏结力,除了满足钢筋锚固长度的要求外,还要使钢筋净距、保护层厚度达到规范要求。

思考练习题

1. 钢筋和混凝土为什么能有效地结合在一起而共同工作?
2. 混凝土的主要力学指标有哪些? 最基本的强度指标是什么?
3. 混凝土强度等级是怎样确定的? 不同强度等级的混凝土有什么用途? 《公桥规》中混凝土强度等级是如何划分的?
4. 试绘出混凝土棱柱体试件在一次短期加载时典型的应力-应变曲线形状,并说明主要特征值。
5. 混凝土弹性模量是怎样确定的? 什么是弹性特征系数? 其数值范围是多少?
6. 试绘出钢筋的应力-应变曲线形状并标出特征值,钢筋的强度指标和塑性指标有哪些?
7. 为什么取屈服强度作为钢筋混凝土构件计算的强度限值?
8. 钢筋混凝土结构对钢筋有哪些基本要求?
9. 锚固长度的确定原则是什么?
10. 钢筋与混凝土之间的黏结力由哪几部分组成? 在工程上采取哪些措施提高钢筋与混凝土之间的黏结力?
11. 混凝土轴心受压的应力-应变曲线有何特点? 影响混凝土轴心受压应力-应变曲线的

因素有哪几个？

12. 混凝土的变形如何分类？混凝土的徐变和收缩变形都是随时间延长而增长的变形，两者有何不同之处？

13. 什么是钢筋和混凝土之间的黏结应力和黏结强度？为保证钢筋和混凝土之间有足够的黏结力，要采取哪些措施？

14. 什么是混凝土的徐变？影响徐变的因素有哪些？什么是线性徐变？什么是非线性徐变？

15. 简述影响黏结强度的因素，各种因素如何影响黏结强度？

第2章
结构按极限状态法设计计算的原则

2.1 概 述

结构设计的目的是设计满足功能要求的结构,也就是把外界作用对结构的效应与结构本身的抵抗力加以比较,以达到结构设计安全、耐久、环保、经济和美观的目的。为了保证结构设计的要求,首先要在设计理论和方法上给予保证。

结构设计的发展历史悠久,至今300余年里,结构设计经历了各种演变,从结构设计理论上来说,由弹性理论发展到以概率论为基础的极限状态设计理论;从设计方法上来说,由定值设计法演变到概率设计法。

目前,结构设计计算的理论和方法有:容许应力法、破损阶段设计法、多系数极限状态设计法和基于可靠性理论的概率极限状态法。

容许应力法是指构件在外界作用下,某截面的最大应力达到或超过材料的容许应力时,构件即失效(破坏),因此设计要满足:

$$\sigma \leqslant [\sigma] = \frac{材料强度}{安全系数} \tag{2-1}$$

对于钢筋混凝土构件：

钢筋 $$\sigma_s \leqslant [\sigma_s] = \frac{f_s}{k_s} \tag{2-2a}$$

混凝土 $$\sigma_c \leqslant [\sigma_c] = \frac{f_c}{k_c} \tag{2-2b}$$

式中，k_c、k_s 分别为混凝土和钢筋的安全系数；$[\sigma_c]$、$[\sigma_s]$ 分别为混凝土和钢筋的容许应力。

以图 2-1 所示的钢筋混凝土构件为例分析，首先做出如下假设：弹性假定，钢筋和混凝土均为弹性材料；平截面假定，变形前的截面变形后保持不变；忽略混凝土的抗拉能力。

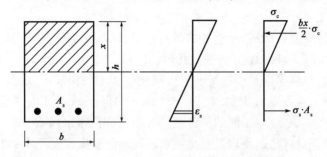

图 2-1　钢筋混凝土构件容许应力计算简图

容许应力法的设计特点为：a. 安全系数 k 是个大于 1 的数字，k 越大，结构的安全度就越高，导致材料的用量就越多；b. 没有考虑结构功能的多样性要求，如结构一方面要考虑承载能力，另一方面要考虑正常使用的要求；c. 安全系数主要凭借经验确定，缺乏严格的科学依据。

容许应力法的适用情况为：当结构是非杆件结构(如大体积坝体、空间薄壳结构等)时，因规范没有给出明确的计算公式，则基于弹性力学的容许应力方法仍是较实用的分析方法。

破损阶段设计法发展于 20 世纪 30 年代，指构件在外界作用下，某截面的内力达到极限内力时，构件即失效(破坏)。以受弯构件为例，其计算表达式为：

$$M \leqslant \frac{M_u}{k} \tag{2-3}$$

式中，M 为截面的计算弯矩；M_u 为截面所能承受的极限弯矩；k 为安全系数。

以如图 2-2 所示的钢筋混凝土构件为例，计算时进行如下假设：a. 以构件破坏阶段为计算依据；b. 不考虑混凝土的抗拉力；c. 计算时受压区混凝土应力分布为矩形应力图。

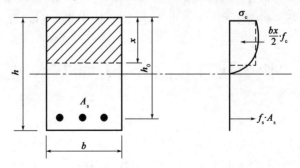

图 2-2　钢筋混凝土构件破损阶段计算简图

该设计方法的特点为:a.考虑了材料塑性和强度的充分发挥,极限荷载可以直接由试验验证,构件的总安全度较为明确;b.安全系数依赖经验确定,且是一个定值;c.没有考虑结构功能的多样性要求。

该方法由于采用了极限平衡的理论,对荷载作用下结构的应力分布及位移变化,无法做出适当的预计。

多系数极限状态设计法是指构件的极限状态不仅包括承载力的极限状态,而且包括挠度(变形)及裂缝宽度的正常使用极限状态,它包含了结构安全性和适用性概念。

对于承载能力极限状态,针对荷载、材料的不同变异性,不再采用单一系数,而是采用多系数表达,如承载能力极限多系数状态的表达式为:

$$M(\sum n_i q_{ik}) \leq mM_u(k_s f_s, k_c f_c, a \cdots) \tag{2-4}$$

式中,q_{ik} 为标准荷载或效应;n_i 为相应的超载系数;f_s、f_c 为钢筋及混凝土的强度;k_s、k_c 为相应的系数;m 为工作条件系数;a 为截面几何特性;其余符号意义同上。

需要说明的是:材料强度是根据统计后按照95%以上的保证率,取其下限分位值;荷载值也尽可能根据各种荷载的统计资料,按照95%以上的保证率,取其下限分位值;材料强度系数、荷载系数仍按经验确定,对不同的荷载变异大小,取用不同的系数。

多系数极限状态设计法的特点是安全系数的选取已经从纯凭经验变为部分采用概率统计值,但是,该设计方法的本质依然是一种半经验半概率的方法。

《公桥规》采用了多项式、单系数表达的极限状态设计法。

基于可靠性理论的概率极限状态设计法是在20世纪40年代美国学者A. M. Freadentbal提出的结构可靠性理论基础上发展起来的。到了20世纪60~70年代,结构可靠性理论有了很大的发展,70年代以来,结构可靠性理论在国际土木工程领域逐步进入了实用阶段,并逐步发展起来。

我国从20世纪70年代中期开始研究,至80年代后期就在建筑结构领域率先应用,先后制定了一系列国家标准,比如:《工程结构可靠度设计统一标准》(GB 50153—92)(现已被GB 50153—2008 替代)和《公路工程结构可靠度设计统一标准》(GB/T 50283—1999)。

结构概率设计的方法按发展进程划分为三个水准,分别为:a.水准Ⅰ,半概率设计法,只对影响结构可靠度的某些参数用数理统计进行分析,并与经验相结合,然后引入某些经验系数,该法对结构的可靠度还不能作出定量的估计;b.水准Ⅱ,近似概率设计法,运用概率论和数理统计理论,对工程结构、构件或截面设计的可靠概率作出较为近似的相对估计,分析中忽略或简化了变量随时间的关系,非线性极限状态方程线性化;c.水准Ⅲ,全概率设计法,该设计方法正在研究中。我国《公桥规》采用的是水准Ⅱ,即近似概率极限状态设计法。

2.2 概率极限状态设计法的基本概念

1)功能要求

按照概率极限状态设计法,通常结构的功能表现在:a.安全性,也称为强度要求,结构应能承受在正常施工和正常使用期间可能出现的各种荷载、外加变形、约束变形等的作用;b.适用性,结构在正常使用条件下应具有良好的工作性能;c.耐久性,结构在正常使用和正常维护条件下,在规定的时间内,应具有完成预定功能的能力;d.稳定性,在偶然荷载作用下或偶然事件发生后,结构仍能保持整体性,不发生倒塌。

2)结构的可靠性

结构的可靠性是指结构在规定时间、规定条件下,完成预定功能的能力,由安全性(强度和稳定性)、适用性和耐久性组成。对结构的可靠性进行概率描述可以得到结构可靠度。

规定时间是指设计基准期,可按《建筑结构可靠度设计统一标准》(GB 50068—2001)确定,一般桥梁结构的设计基准期为100年。结构的设计使用年限虽与结构使用寿命有联系,但不等同。当结构的使用年限超过设计使用年限后,并不意味着结构就要报废,但其可靠度将逐渐降低。规定条件为正常设计、施工和使用条件。

3)结构的极限状态

当整个结构或结构的一部分超过某一特定状态而不能满足设计规定的功能要求时,则此特定状态称为该功能的极限状态。国际上将极限状态分为承载能力极限状态、正常使用极限状态和破坏-安全状态。承载能力极限状态对应于结构或构件达到最大承载能力或不适于继续承载的变形,具体表现为以下几个方面:a.整个结构或结构的一部分作为刚体失去平衡;b.结构构件或连接处因超过材料强度而破坏;c.结构转变成机动体系;d.结构或构件丧失稳定;e.变形过大,不能继续承载和使用。而正常使用极限状态主要对应于结构或构件达到正常使用或耐久性能的某项规定限值,具体表现为以下几个方面:a.由于外观变形影响正常使用;b.由于耐久性能的局部损坏影响正常使用;c.由于震动影响正常使用;d.由于其他特定状态影响正常使用。破坏-安全状态是指偶然事件造成局部损坏后,其余部分不至于发生连续倒塌的状态。在《公桥规》中,破坏-安全极限状态归到承载能力极限状态中。

4)结构失效和可靠指标

在可靠度分析中,结构的极限状态用功能函数描绘。如当有 n 个随机变量($X_1, X_2, \cdots, X_n$)会影响结构的可靠度时,结构的功能函数表示为 Z,它可由荷载效应 S 和结构抗力 R 所组成:

$$Z = Z(X_1, X_2, \cdots, X_n) = g(R, S) = R - S = \begin{cases} > 0, \text{结构处于可靠状态} \\ = 0, \text{结构处于极限状态} \\ < 0, \text{结构失效或破坏} \end{cases} \quad (2\text{-}5)$$

式中,R 为结构抗力,指结构构件承受内力和变形的能力;S 为荷载效应(或作用效应),指荷载或作用对结构产生的效应。

作用是使结构产生内力、变形、应力和应变的所有原因,分为直接作用和间接作用。直接作用为施加在结构上的荷载(集中荷载和分布荷载),如自重、汽车荷载等;间接作用为引起结构外加变形和约束变形的原因。外加变形使结构被强制地产生变形,例如,基础不均匀沉降、地震等等。约束变形为结构材料发生收缩或膨胀等变化,受到结构的支座或节点的约束而使结构间接产生的变形,如混凝土收缩、钢材焊接、大气温度变化引起的变形等。

失效是指结构或结构的一部分不能满足设计所规定的某一功能要求,即达到或超过了承载能力极限状态或正常使用极限状态中的某一限值。失效概率为结构处于失效状态下的概率,可用下列公式表达:

结构的失效概率 $\qquad p_f = p(Z < 0) \qquad (2\text{-}6)$

结构的可靠概率 $\qquad p_r = 1 - p_f \qquad (2\text{-}7)$

可靠指标 β 是用以度量结构可靠度的指标，可以表示为：

$$\beta = \frac{m_Z}{\sigma_Z} = \frac{m_R - m_S}{\sqrt{\sigma_R^2 + \sigma_S^2}} \tag{2-8}$$

式中，m_Z 为功能函数 Z 的平均值；σ_Z 为功能函数 Z 的均方差；m_R、m_S 分别为抗力和荷载效应的平均值；σ_R、σ_S 分别为抗力和荷载的均方值。

当 R 与 S 服从正态分布时，如图 2-3 所示，可以推导出：

$$P_f = \Phi(-\beta) \tag{2-9}$$

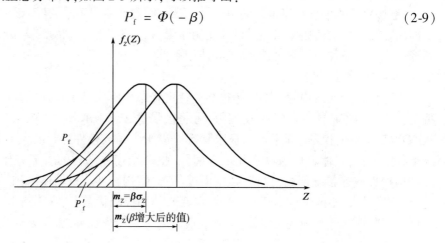

图 2-3　可靠指标 β 与平均值 m_Z 关系图

P_f 与 β 之间可以建立一一对应的数量关系，P_f 越大，β 越小，结构越不可靠。

目标可靠指标是用作结构设计依据的可靠指标，它主要是采用"校准法"并结合工程经验和经济优化原则加以确定的。校准法是根据各基本变量的统计参数和概率分布类型，运用可靠度的计算方法，揭示以往《公路钢筋混凝土及预应力混凝土桥涵设计规范》隐含的可靠度，以此作为确定目标可靠指标的依据。这种方法在总体上承认了以往《公路钢筋混凝土及预应力混凝土桥涵设计规范》的设计经验和可靠度水平，同时也考虑了渊源于客观实际的调查统计分析资料。

根据《公路工程结构可靠度设计统一标准》(GB/T 50283—1999) 规定，公路桥梁结构的目标可靠度如表 2-1 所示。

公路桥梁结构的目标可靠度　　　　　　　　表 2-1

构件的破坏类型	结构安全等级		
	一级	二级	三级
延性破坏	4.7	4.2	3.7
脆性破坏	5.2	4.7	4.2

2.3　我国《公桥规》的计算原则

1) 基本概念

我国《公桥规》规定的结构设计的四种状况为持久状况、短暂状况、偶然状况和地震状况。

持久状况是指桥涵建成后承受自重、车辆荷载等作用持续时间很长的状况,需要进行承载能力极限状态和正常使用极限状态的设计。短暂状况是指桥涵施工过程中承受临时性荷载的状况,该状况对应的是桥梁的施工阶段,一般只进行承载能力极限状态设计。偶然状况为在桥涵使用过程中偶然出现的状况,只进行承载能力极限状态设计。地震状况为特殊的偶然状况,只进行承载力极限状态设计。

根据桥涵结构破坏所产生的后果的严重程度,分为一级、二级和三级三个安全等级,具体见表2-2。

公路桥涵结构的安全等级 表2-2

安全等级	破坏后果	桥涵类型	结构重要性系数 γ_0
一级	很严重	(1)各等级公路上的特大桥、大桥、中桥; (2)高速公路、一级公路、二级公路、国防公路及城市附近交通繁忙公路上的小桥	1.1
二级	严重	(1)三、四级公路上的小桥; (2)高速公路、一级公路、二级公路、国防公路及城市附近交通繁忙公路上的涵洞	1.0
三级	不严重	三、四级公路上的涵洞	0.9

2)极限状态设计表达

承载能力极限状态设计是以塑性理论为基础建立表达式,其设计原则为作用效应最不利组合(基本组合)设计值必须小于或等于结构抗力的设计值。

承载能力极限状态的基本表达式为:

$$\gamma_0 S_d \leqslant R = R(f_d, a_d) \tag{2-10}$$

式中,γ_0 为结构重要性系数,按结构设计安全等级采用,对应于设计安全等级一级、二级和三级分别为1.1、1.0和0.9;f_d 为材料强度设计值;a_d 为几何参数设计值,当无可靠数据时,可采用几何参数标准值 a_k,即设计文件规定值。

持久状况正常使用极限状态的设计原则是以结构弹性理论或弹塑性理论为基础,采用作用(或荷载)的短期效应组合、长期效应组合或短期效应组合并考虑长期效应组合的影响,对构件的应力、裂缝宽度和挠度进行验算,并使各项计算值不超过《公桥规》规定的各项限值。该极限状态设计表达式为:

$$S \leqslant c_1 \tag{2-11}$$

式中,S 为正常使用极限状态的作用(或荷载)效应组合设计值;c_1 为结构构件达到正常使用要求所规定的限值,如变形、裂缝宽度和截面抗裂的应力限值。

持久、短暂状态的应力设计比较复杂,针对不同的构件有不同的设计原则:对钢筋混凝土和预应力混凝土受力构件,按短暂状况设计时,计算其在制作、运输及安装等施工阶段由自重、施工荷载产生的应力,不应超过限值。按持久状况设计预应力混凝土受弯构件时,应计算其使用阶段的应力不超过限值。总体来说:应力不超过限值。持久、短暂极限状态设计表达式为:

$$S \leqslant c_2 \tag{2-12}$$

式中,S 为作用(或荷载)标准值(其中汽车荷载应考虑冲击系数)产生的效应,当有组合时不考虑荷载组合系数;c_2 为结构的功能限值。

2.4 材料强度的取值

1) 材料强度指标的取值原则

材料强度的标准值是材料强度的一种特征值,是由标准试件按标准试验方法,经数理统计以概率分布的 0.05 分位值确定的强度值。

取值原则是在符合规定质量的材料强度实测值的总体中,材料的强度标准值应具有不小于95%的保证率。其基本表达式为:

$$f_k = f_m(1 - 1.645\delta_f) \tag{2-13}$$

式中,f_m 为材料强度平均值;δ_f 为材料强度变异系数。

材料强度的设计值为材料强度标准值除以材料性能分项系数,见下式:

$$f_s = \frac{f_k}{\gamma} \tag{2-14}$$

上式中参数取值方法对不同材料有所不同。例如,对于混凝土材料,$\gamma_c = 1.45$;对于钢筋,$\gamma_s = 1.2$;对于受拉高强钢丝、钢绞线,$\gamma_s = 1.47$。

2) 混凝土强度标准值和设计值

混凝土立方体抗压强度标准值 $f_{cu,k}$ 是边长为 150mm 的立方体试件按照标准方法制作和养护,经 28d 龄期,用标准试验方法测得的具有 95% 保证率的抗压强度。根据《公桥规》,受力构件的混凝土强度等级应按下列规定采用:a. 钢筋混凝土构件不应低于 C25;用 HRB400、RRB400 级钢筋配筋时,不应低于 C30;b. 预应力混凝土构件不应低于 C40。

在假设混凝土轴心抗压强度 f_c 的变异系数与立方体抗压强度 f_{cu} 的变异系数相同的条件下,混凝土轴心抗压强度标准值 f_{ck} 可由下式确定:

$$f_{ck} = f_{c,m}(1 - 1.645\delta_f) = 0.88\alpha_{c1}\alpha_{c2}f_{cu,m}(1 - 1.645\delta_f) = 0.88\alpha_{c1}\alpha_{c2}f_{cu,k} \tag{2-15}$$

式中,$f_{c,m}$、$f_{cu,m}$ 分别为混凝土轴心抗压强度平均值和立方体抗压强度平均值;α_{c1} 为混凝土轴心抗压强度与立方体抗压强度的比值,对于 C50 及以下取 0.76,对于 C80 取 0.82,其间按线性插入;α_{c2} 为混凝土脆性折减系数,对于 C40 及以下取 $\alpha_{c2} = 1.0$,对于 C80 取 $\alpha_{c2} = 0.87$,其间按线性插入。

抗拉强度标准值 f_{tk} 基于混凝土轴心抗拉强度 f_t 的变异系数与立方体抗压强度 f_{cu} 的变异系数相同的假设,其表达式为:

$$f_{tk} = 0.348\alpha_{c2}f_{cu,k}^{0.55}(1 - 1.645\delta_f)^{0.45} \tag{2-16}$$

式中,f_{tk}、$f_{cu,k}$ 分别为混凝土轴心抗拉强度标准值和立方体抗压强度标准值。

根据《公桥规》的规定,取混凝土轴心抗压强度标准值和轴心抗拉强度的材料性能分项系数为1.45,用标准值除以分项系数即可得到混凝土轴心抗压强度设计值。

3) 钢筋强度标准值和设计值

(1) 标准值

国家标准中规定的钢筋屈服强度标准值,即为钢筋出厂检验的废品限值,其保证率不小于95%。

对于有明显流幅的热轧钢筋,钢筋的抗拉强度标准值 f_{sk} 采用国家标准中规定的屈服强度标准值。对于无明显流幅的钢筋,如钢丝、钢绞线等,根据国家标准中规定的极限抗拉强度值确定,其保证率不小于95%,取 $0.85\sigma_b$(σ_b 为国家标准中规定的极限抗拉强度)作为设计取用的条件屈服强度(相应于残余应变为0.2%时的钢筋应力)。

《公桥规》对热轧钢筋和精轧螺纹钢筋的材料性能分项系数取1.20,对钢绞线、钢丝等的材料性能分项系数取1.47。

(2)设计值

钢筋抗拉强度的设计值通过将钢筋的强度标准值除以相应的材料性能分项系数1.20或1.47获得。钢筋抗压强度设计值按 $f'_{sd} = \varepsilon'_s E'_s$ 或 $f_{pd} = \varepsilon'_p E'_p$ 确定。E_s 和 E_p 分别为热轧钢筋和钢绞线等的弹性模量;ε'_s 和 ε'_p 为相应钢筋种类的受压应变,取 $\varepsilon'_s(\varepsilon'_p)$ 等于0.002,且 f'_{sd}(或 f'_{pd})不得大于相应钢筋的抗拉强度设计值。

2.5 作用、作用的代表值和作用效应组合

作用是使结构产生内力、变形、应力、应变的所有原因。作用分为:永久作用、可变作用、偶然作用和地震作用。永久作用是在结构使用期内,其量值不随时间变化,或其变化与平均值相比可忽略不计的作用;可变作用是在结构使用期内,其量值随时间变化,且其变化值与平均值相比不可忽略的作用;偶然作用是在结构使用期间出现的概率小,一旦出现其值很大且持续时间很短的作用。

1)作用的代表值

公路桥涵设计时,对不同的作用效应按下列规定采用不同的代表值:

(1)永久作用的代表值为其标准值。永久作用标准值可根据统计、计算,并结合工程经验综合分析确定。

(2)可变作用的代表值包括标准值、组合值、频遇值和准永久值。组合值为使组合后的作用效应的超越概率与该作用单独出现时期标准值作用效应的超越概率趋于一致的作用值;或组合后使结构具有规定可靠指标的作用值。它可通过组合值系数 ψ_c 对作用标准值的折减来表示。频遇值为在设计基准期内被超越的总时间占设计基准期的比率较小的作用值;或被超越的频率限制在规定频率内的作用值。它可通过频遇值系数 ψ_f 对作用标准值的折减来表示。准永久值为在设计基准期内被超越的总时间占设计基准期的比率较大的作用值。它可通过准永久值系数 ψ_q 对作用标准值的折减来表示。

(3)偶然作用取其设计值作为代表值,可根据历史记载、现场观测和试验,并结合工程经验综合分析确定,也可根据有关标准的专门规定确定。

(4)地震作用的代表值为其标准值。地震作用的标准值应根据现行《公路工程抗震规范》(JTG B02—2013)的规定确定。

(5)作用的设计值应为作用的标准值或组合值乘以相应的作用分项系数。分项系数取值见《公路桥涵设计通用规范》(JTG D60—2015)(以下简称《桥通规》)。

公路桥涵结构设计应考虑结构上可能同时出现的作用,按承载能力极限状态、正常使用极限状态进行作用组合,均应按下列原则取其最不利组合效应进行设计:

(1)只有在结构上同时出现的作用,才进行组合。当结构或结构构件需做不同受力方向

的验算时,则应以不同方向上最不利的作用组合效应进行计算。

(2) 当可变作用的出现对结构或结构构件产生有利影响时,该作用不应参与组合。实际不可能同时出现的作用或同时参与组合概率很小的作用,按《桥通规》中"表4.1.4"规定不考虑其参与组合。

(3) 施工阶段的组合效应,应按计算需要及结构所处条件而定,结构上的施工人员和施工机具设备均应作为可变作用加以考虑。对于组合式桥梁,当把底梁作为施工支撑时,作用组合效应宜分两个阶段计算,底梁受荷为第一个阶段,组合梁受荷为第二个阶段。

(4) 多个偶然作用不同时参与组合。

(5) 地震作用不与偶然作用同时参与组合。

2) 作用效应组合

公路桥涵结构按承载能力极限状态设计时,对持久设计状况和短暂设计状况应采用作用的基本组合,对偶然设计状况应采用作用的偶然组合,对地震设计状况应采用作用的地震组合。

(1) 极限状态作用效应

①基本组合。对于承载能力极限状态,《桥通规》规定采用的基本组合为:

$$S_{ud} = \gamma_0 S(\sum_{i=1}^{m} \gamma_{Gi} G_{ik}, \gamma_{Q1} \gamma_{L1} Q_{1k}, \psi_c \sum_{j=2}^{n} \gamma_{Lj} \gamma_{Qj} Q_{jk}) \tag{2-17}$$

或

$$S_{ud} = \gamma_0 S(\sum_{i=1}^{m} G_{id}, Q_{1d}, \sum_{j=2}^{n} Q_{jd}) \tag{2-18}$$

式中,γ_0 为桥梁结构的重要性系数,按结构设计安全等级采用,对于公路桥梁,安全等级一级、二级和三级,分别为1.1、1.0和0.9;γ_{Gi} 为第 i 个永久作用效应的分项系数,当永久作用效应(结构重力和预应力作用)对结构承载力不利时,$\gamma_G = 1.2$,对结构的承载能力有利时,其分项系数 γ_G 的取值为1.0,其他永久作用效应的分项系数详见《桥通规》;γ_{Q1} 为汽车荷载(含汽车冲击力、离心力)的分项系数,采用车道荷载计算时取 $\gamma_{Q1} = 1.4$,采用车辆荷载计算时取 $\gamma_{Q1} = 1.8$;当某个可变作用在组合中的效应值超过汽车荷载效应时,则该作用取代汽车荷载,其分项系数取 $\gamma_{Q1} = 1.4$;对于专为承受某作用而设置的结构或装置,设计时该作用的分项系数取为 $\gamma_{Q1} = 1.4$。Q_{1k}、Q_{1d} 为汽车荷载(含汽车冲击力、离心力)的标准值和设计值;γ_{Qj} 为在作用组合中除汽车荷载(含汽车冲击力、离心力)、风荷载外的其他第 j 个可变作用的分项系数,一般取 $\gamma_{Qj} = 1.4$,风荷载的分项系数取 $\gamma_{Qj} = 1.1$;Q_{jk}、Q_{jd} 为在作用组合中除汽车荷载(含汽车冲击力、离心力)外的其他第 j 个可变作用的标准值和设计值;ψ_c 为在作用组合中除汽车荷载(含汽车冲击力、离心力)外的其他可变作用的组合值系数,取 $\psi_c = 0.75$;γ_{Lj} 为第 j 个可变作用的结构设计使用年限荷载调整系数,一般取 $\gamma_{Lj} = 1.0$。

当作用与作用效应可按线性关系考虑时,作用基本组合的效应设计值 S_{ud} 可通过作用效应代数相加计算。

②偶然组合。偶然组合为永久作用标准值与可变作用某种代表值、一种偶然作用设计值的组合。与偶然作用同时出现的可变作用,可根据观测资料和工程经验取用频遇值或准永久值。

作用偶然组合的效应设计值可按下式计算:

$$S_{ad} = S(\sum_{i=1}^{m} G_{ik}, A_d, (\psi_{f1} 或 \psi_{q1}) Q_{1k}, \sum_{j=2}^{n} \psi_{qj} Q_{jk}) \tag{2-19}$$

式中，S_{ad}为承载能力极限状态下作用偶然组合的效应设计值；A_d为偶然作用的设计值；ψ_{f1}为汽车荷载(含汽车冲击力、离心力)的频遇值系数，取$\psi_{f1}=0.7$，当某个可变作用在组合中的效应值超过汽车荷载效应时，则该作用取代汽车荷载，人群荷载$\psi_{f1}=1.0$，风荷载$\psi_{f1}=0.75$，温度梯度作用$\psi_{f1}=0.8$，其他作用$\psi_{f1}=1$；$\psi_{f1}Q_{1k}$为汽车荷载的频遇值；ψ_{q1}、ψ_{qj}为第1个和第j个可变作用的准永久值系数，汽车荷载(含汽车冲击力、离心力)$\psi_q=0.4$，人群荷载$\psi_q=0.4$，风荷载$\psi_q=0.75$，温度梯度作用$\psi_q=0.8$，其他作用$\psi_q=1.0$；$\psi_{q1}Q_{1k}$、$\psi_{qj}Q_{jk}$分别为第1个和第j个可变作用的准永久值。

③地震组合。针对桥梁工程，承载能力极限状态下地震作用偶然组合时的承载能力为：

$$\gamma_0 \left(\sum_{i=1}^{m} \gamma_{Gi} S_{Gik} + \sum_{j=1}^{n} S_{Qjk} + Q_e \right) \leq R(\gamma_f, f_k, \gamma_a, \alpha_k) \qquad (2-20)$$

式中，γ_0为结构重要性系数；S_{Gik}为第i个永久作用效应；S_{Qjk}为可能与地震作用同时作用的第j个可变作用的一定量级的效应；Q_e为地震作用效应；γ_{Gi}为永久作用分项系数，具体取值见现行《公路圬工桥涵设计规范》(JTG D61—2005)和《公桥规》的有关规定计算；γ_f为结构材料、岩土性能的分项系数；γ_a为结构或构件几何参数的分项系数；f_k为材料、岩土性能的标准值；α_k为几何参数的标准值。

(2)正常使用极限状态组合

正常使用极限状态应根据不同的设计要求，采用作用的频遇组合或准永久组合，其效应组合表达式为：

①频遇组合效应：

$$S_{fd} = S\left(\sum_{i=1}^{m} G_{ik}, \psi_{f1} Q_{1k}, \sum_{j=2}^{n} \psi_{qj} Q_{jk} \right) \qquad (2-21)$$

式中，S_{fd}为作用频遇组合的效应设计值；G_{ik}为第i个永久作用的标准值；ψ_{f1}为汽车荷载(不计汽车冲击力)频遇值系数，取0.7。当作用与作用效应可按线性关系考虑时，作用基本组合的效应设计值S_{fd}可通过作用效应代数相加计算。

②准永久组合效应：

$$S_{qd} = S\left(\sum_{i=1}^{m} G_{ik}, \sum_{j=1}^{n} \psi_{qj} Q_{jk} \right) \qquad (2-22)$$

式中，S_{qd}为作用准永久组合的效应设计值；ψ_{qj}为汽车荷载(不计汽车冲击力)准永久值系数，取0.4。

当作用与作用效应可按线性关系考虑时，作用基本组合的效应设计值S_{qd}可通过作用效应代数相加计算。

3)算例

钢筋混凝土简支梁桥主梁在结构重力、汽车荷载和人群荷载作用下，分别得到在主梁的1/4跨径处截面的弯矩标准值为：结构重力产生的弯矩$M_{Gk}=512\text{kN}\cdot\text{m}$；汽车荷载弯矩(按车道荷载计算)$M_{Q1k}=425\text{kN}\cdot\text{m}$，冲击系数$1+\mu=1.19$；人群荷载弯矩$M_{Q2k}=38.5\text{kN}\cdot\text{m}$，结构的重要性系数取1.0。通过已知条件，进行设计时的作用效应组合计算。

解：(1)承载能力极限状态设计时作用效应组合

①基本组合：

已知：$\gamma_0=1.0, \gamma_{G1}=1.2, \gamma_{Q1}=1.4, \gamma_{Q2}=1.4, \psi_c=0.75, \gamma_{L1}=\gamma_{L2}=1.4$。

基本组合设计值为(按作用效应代数相加计算)：

$$M_{ud} = \gamma_0 (\sum_{i=1}^{m}\gamma_{Gi}M_{Gik} + \gamma_{Q1}\gamma_{L1}M_{Q1k} + \psi_c \sum_{j=2}^{n}\gamma_{Lj}\gamma_{Qj}M_{Qjk})$$
$$= 1.0 \times (1.2 \times 512 + 1.4 \times 1.0 \times 425 + 0.75 \times 1.0 \times 1.4 \times 38.5)$$
$$= 1249.825(kN \cdot m)$$

②偶然组合：
$$M_{ud} = \sum_{i=1}^{m}M_{Gik} + \psi_{f1}M_{Q1k} + \sum_{j=2}^{n}\psi_{qj}M_{Qjk}$$
$$= 512 + 0.7 \times 425 + 1.0 \times 38.5$$
$$= 848(kN \cdot m)$$

(2) 正常使用极限状态设计时作用效应组合

①作用频遇组合效应为：
$$M_{Q1k} = 425/(1+\mu) = 357.1, \psi_{f1} = 0.7, \psi_{q2} = 1.0$$

作用频遇组合效应组合设计值为：
$$M_{fd} = \sum_{i=1}^{m}M_{Gik} + \psi_{f1}M_{Q1k} + \sum_{j=2}^{n}\psi_{qj}M_{Qjk}$$
$$= 512 + 0.7 \times 357.1 + 1.0 \times 38.5$$
$$= 800.5(kN \cdot m)$$

②准永久组合效应为：
$$M_{Q1k} = 425/(1+\mu) = 357.1, \psi_{q1} = \psi_{q2} = 0.4$$

准永久组合效应组合设计值为：
$$M_{qd} = \sum_{i=1}^{m}G_{ik} + \sum_{j=1}^{n}\psi_{qj}Q_{jk}$$
$$= 512 + 0.4 \times 357.1 + 0.4 \times 38.5$$
$$= 693.34(kN \cdot m)$$

思考练习题

1. 可靠度、永久作用、可变作用、偶然作用、混凝土结构的耐久性、作用、直接作用、间接作用、抗力的基本概念是什么？
2. 材料强度有哪些代表值？《公桥规》对强度标准值的取值有什么规定？
3. 什么是荷载标准值、频遇值、准永久值，它们之间有什么联系？
4. 《公桥规》规定正常使用极限状态设计时应考虑哪些效应组合？
5. 结构承载能力极限状态和正常使用极限状态设计计算的原则是什么？
6. 试说明材料分项系数和荷载分项系数的物理意义。
7. 规范规定汽车荷载怎样分类？车道荷载和车辆荷载怎样构成？
8. 试述用"容许应力法""破坏阶段法"和"极限状态法"计算钢筋混凝土结构的主要区别有哪些。
9. 何谓结构的极限状态？结构的极限状态可分为哪几类？

10. 混凝土的标准强度、设计强度与平均强度、标准差之间有何关系?
11. 何谓安全系数、构件工作条件系数? 它们如何取值?
12. 桥梁结构的功能包括哪几个方面的内容? 何谓结构的可靠性?
13. 结构的设计基准期和使用寿命有何区别?
14. 我国《公桥规》规定结构设计有哪三种状况?
15. 什么是材料强度的标准值和设计值? 两者的关系是什么? 钢筋和混凝土材料分项系数是什么? 为什么不同的材料会采用不同的分项系数?
16. 钢筋与混凝土能共同工作原因之一在于其间的黏结力,黏结力主要由哪几部分组成?
17. 有一批短柱,其抗力 R 呈正态分布($\mu_R = 3\,560\text{kN}, \sigma_R = 377\text{kN}$),用于桥梁结构,其作用效应 S 也呈正态分布($\mu_S = 1\,960\text{kN}, \sigma_S = 316\text{kN}$),求结构可靠指标 β。

第3章
受弯构件正截面承载力计算

3.1 受弯构件基本概念

受弯构件是指截面上通常有弯矩和剪力共同作用而轴力可忽略不计的构件(图3-1)。钢筋混凝土梁和板是土木工程中典型的受弯构件,在桥梁工程中应用很广泛,还有桥梁中的人行道板、行车道板等(图3-2),也都为受弯构件。

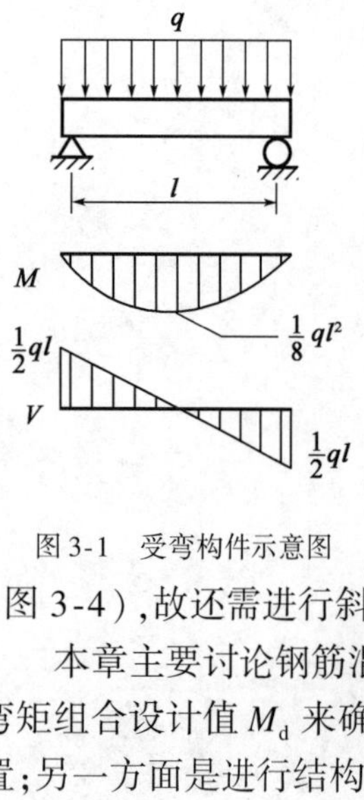

在荷载作用下,受弯构件的截面将承受弯矩 M 和剪力 V 的作用。因此,设计受弯构件时,一般应满足下列两方面要求:

(1)由于弯矩 M 的作用,构件可能沿某个正截面(与梁的纵轴线或板的中面正交的面)发生破坏(图3-3),故需要进行正截面承载力计算。

(2)由于弯矩 M 和剪力 V 的共同作用,构件可能沿剪压区段内的某个斜截面(与梁的纵轴线或板的中面斜交的面)发生破坏

图3-1 受弯构件示意图

(图3-4),故还需进行斜截面承载力计算。

本章主要讨论钢筋混凝土梁和板的正截面承载力计算,目的一方面是进行结构设计,根据弯矩组合设计值 M_d 来确定钢筋混凝土梁和板截面上纵向受力钢筋的面积,并进行钢筋的布置;另一方面是进行结构验算。

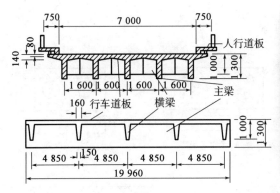
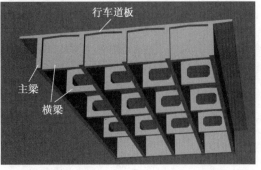

图 3-2 钢筋混凝土 T 形梁桥中的受弯构件示意图(尺寸单位:mm)

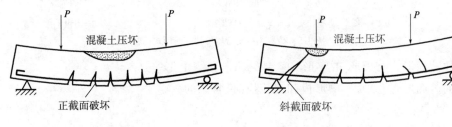

图 3-3 正截面破坏形式　　　　　　图 3-4 斜截面破坏形式

3.2 受弯构件的截面形式与构造

1)构造要求

构造要求是根据规范要求或经验总结,对构件尺寸、材料强度、等级、品种、钢筋数量、布置位置、间距、直径、连接等做出的限制性规定。

结构涉及多方面问题,有些问题非主要但也不容忽视,有些问题还未弄清,用计算理论无法解决和定量确定。设计计算公式不可能反映所有问题,根据工程实践经验和科研成果,考虑施工可能性以及技术经济要求,将设计计算理论和公式中未反映的问题、未定量确定的方面总结为构造要求。主要体现在以下几个方面:

(1)弥补理论上的不足和不确定因素的影响。
(2)考虑施工要求,如板最小厚度要求。
(3)工程实践经验总结,如分布钢筋的间距。
(4)其他技术经济要求,如材料用量。

构造要求对结构设计的作用为:

(1)为初拟构件尺寸提供参考(如梁 $h/l = 1/18 \sim 1/10, h/b = 2 \sim 4$)。
(2)与计算相辅相成。
(3)反映实际工程设计的特点。

2) 截面形式和尺寸

（1）截面形式

钢筋混凝土梁（板）的截面形式如图 3-5 所示。

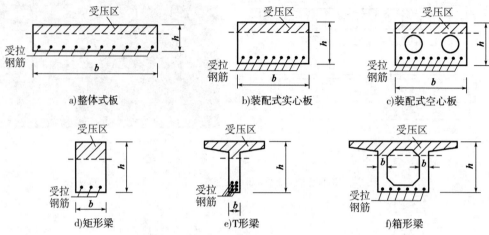

图 3-5　桥梁结构受弯构件的截面形式

①梁的截面形式有矩形、T 字形、工字形、箱形和倒 T 形等。

②板的截面形式有平板、槽形板和多孔板（最常用的）。

（2）尺寸要求

①板的尺寸要求

a. 整体现浇板，截面宽度较大[图 3-5a]，通常取单位宽度（如以 1m 为计算单位）的矩形截面进行计算。

b. 预制板，板宽度一般控制在 1~1.5m，以便规模生产，满足运输和吊装要求。如果施工条件好，不仅能采用矩形实心板[图 3-5b]，还能采用截面形状较复杂的矩形空心板[图 3-5c]，以减轻自重。

c. 板的厚度 h 由其控制截面上的最大弯矩和板的刚度要求决定，并应满足构造要求。为了保证施工质量及耐久性要求，《公桥规》规定了各种板的最小厚度：人行道板不宜小于 80mm（现浇整体）和 60mm（预制）；空心板的顶板和底板厚度均不宜小于 80mm。

②梁的尺寸要求

a. 现浇矩形截面梁的宽度 b 常取 120mm、150mm、180mm、200mm、220mm 和 250mm，其后按 50mm 一级增加（当梁高 $h \leqslant 800$mm 时）或按 100mm 一级增加（当梁高 $h > 800$mm 时）。矩形截面梁的高宽比 h/b 一般可取 2.0~2.5。

b. 预制的 T 形截面梁，其截面高度 h 与跨径 l 之比（称高跨比）一般为 $h/l = 1/16 \sim 1/11$，跨径较大时取用偏小比值。梁肋宽度 b 常取为 150~180mm，根据梁内主筋布置及抗剪要求而定。

c. T 形截面梁翼缘悬臂端部厚度不应小于 100mm，梁肋处翼缘厚度不宜小于梁高 h 的 1/10。

3) 受弯构件的钢筋构造

（1）基本概念

①单筋截面、双筋截面

钢筋混凝土梁(板)正截面承受弯矩作用时,以中和轴为界,一部分受压,另一部分受拉,只在梁(板)的受拉区配置纵向受拉钢筋,此种构件称为单筋截面;如果同时在截面受压区和受拉区配置受力钢筋,则此种构件称为双筋截面。

②配筋率 ρ

截面上配置钢筋的多少,通常用配筋率来衡量,配筋率是指所配置的钢筋截面面积与混凝土截面有效面积的比值。对于矩形截面和T形截面,其受拉钢筋的配筋率 $\rho(\%)$ 表示为:

$$\rho = \frac{A_s}{bh_0} \tag{3-1}$$

式中,A_s 为截面纵向受拉钢筋全部截面积;b 为矩形截面宽度或T形截面梁肋宽度;h_0 为截面的有效高度(图3-6),$h_0 = h - a_s$,其中,h 为截面高度,a_s 为纵向受拉钢筋全部截面的重心至受拉边缘的距离。

③保护层

图3-6中的 c 被称为混凝土保护层厚度,它是指钢筋边缘至构件截面表面之间的最短距离。设置保护层是为了保护钢筋不直接受到大气的侵蚀、氯盐和其他环境因素作用,也是为了保证钢筋和混凝土有良好的黏结,它是钢筋混凝土结构耐久性设计的重要指标。

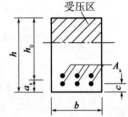

图3-6 配筋率 ρ 的计算图

(2)板的钢筋

①板按受力特点分类

a. 悬臂板。

b. 四边支承板:根据力的传递不同分单向板和双向板。

单向板:l_2(长)/l_1(短) $\geqslant 2$,荷载主要沿短边单方向传递,如图3-7所示的桥面板,需在短边单向配受力主钢筋。

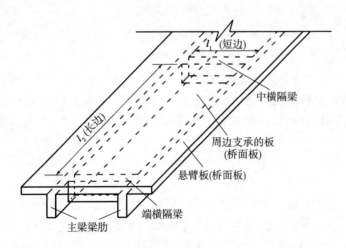

图3-7 周边支承桥面板与悬臂桥面板示意图

双向板:$l_2/l_1 < 2$,荷载沿两个方向传递,板的双向都受力,成为双向板,需要在双向配受力主钢筋。

c. 两边支承:单向板。

②板内钢筋种类及作用分类

a. 主钢筋(纵向受力筋)。

单向板内主钢筋沿板的短边方向布置在板的受拉区,钢筋数量由计算确定,并满足构造要求。受力主钢筋的直径不宜小于10mm(行车道板)或8mm(人行道板)。近梁肋处的板内主钢筋,可在沿板高中心纵轴线的($1/6 \sim 1/4$)计算跨径处按$30° \sim 45°$弯起,但通过支承而不弯起的主钢筋,每米板宽内不应少于3根,且截面面积不少于主钢筋截面积的1/4。

在简支板的跨中和连续板的支点处,板内主钢筋间距应不大于200mm。

行车道板受力钢筋的最小混凝土保护层厚度c(图3-8)应不小于钢筋的公称直径,且应满足规范规定的最小厚度要求。

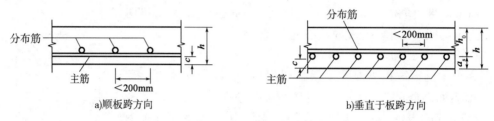

图3-8 单向板内的钢筋

b. 分布钢筋。

分布钢筋是在主筋上按一定间距设置的连接用钢筋,属于构造钢筋,其数量不需要计算,而是按照设计规范规定选择的。分布钢筋的作用是使主钢筋受力更均匀,能固定主筋,且抵抗温度应力和混凝土收缩应力。

分布钢筋应放置在受力钢筋的上侧(图3-8)。《公桥规》规定,行车道板内分布钢筋直径不小于8mm,其间距应不大于200mm,截面面积不宜小于板截面面积的0.1%。在所有主钢筋的弯折处,均应设置分布钢筋。人行道板内分布钢筋直径不应小于6mm,其间距不应大于200mm。

c. 对于周边支承的双向板,板的两个方向同时承受弯矩,所以两个方向均应设置主钢筋。

d. 预制板广泛用于装配式板桥中。板桥的行车道板是由数块预制板利用各板间企口缝填入混凝土拼连而成的。从结构受力性能上分析,在荷载作用下,它并不是双向受力的整体宽板,而是一系列单向受力的窄板式的梁,板与板之间企口缝内的混凝土(称为混凝土铰)借铰缝传递剪力而共同受力,也称预制板为梁式板(或板梁)。因此,预制板的钢筋布置要求与矩形截面梁相似。

(3)梁的钢筋

①钢筋成形方式

梁内钢筋骨架的形式有绑扎钢筋骨架(图3-9)和焊接钢筋骨架(图3-10)两种。绑扎骨架是将纵向钢筋与横向钢筋通过绑扎而成的空间钢筋骨架。焊接骨架是先将纵向受拉钢筋(主钢筋)、弯起钢筋或斜筋和架立钢筋焊接成平面骨架,然后用箍筋将数片焊接的平面骨架组成空间骨架。

②钢筋种类及构造布置

梁内的钢筋有纵向受拉钢筋(主钢筋)、弯起钢筋或斜钢筋、箍筋、架立钢筋和水平纵向钢筋等。

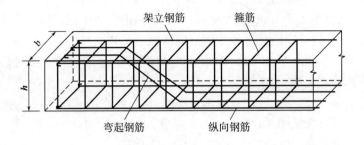

图 3-9　绑扎钢筋骨架示意图

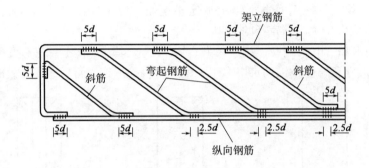

图 3-10　焊接钢筋骨架示意图

a. 主钢筋分受拉主钢筋和受压主钢筋,协助混凝土受力,以提高梁的抗弯能力。数量由正截面承载力计算确定,并应满足构造要求。其直径一般为 12～32mm,且≤40mm。

排列规则为:简支梁的主钢筋尽量排成一层,减少主钢筋的层数(可以增大力臂节约钢筋);采用绑扎骨架时,主钢筋不宜多于 3 层;直径较粗的钢筋布在底层;布置两层或两层以上时,上下层钢筋应当对齐。排列总原则为由下至上,下粗上细,对称布置。

钢筋的最小混凝土保护层厚度应不小于钢筋的公称直径,且应符合规范要求。例如,当桥梁处于 I 类环境条件时,钢筋混凝土梁内主钢筋(钢筋公称直径为 d)与梁底面的混凝土保护层厚度、布置距梁侧面最近的主钢筋与梁侧面的混凝土保护层 c(图 3-11)应不小于钢筋的公

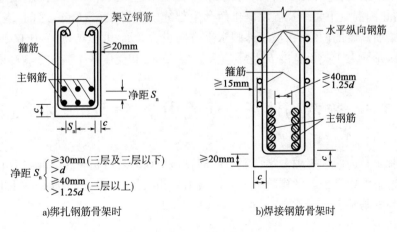

图 3-11　梁主钢筋净距和混凝土保护层

称直径 d 和 30mm。当受拉区主筋的混凝土保护层厚度大于 50mm 时，应在保护层内设置直径不小于 6mm 且间距不大于 100mm 的钢筋网，钢筋网的混凝土保护层厚度不宜小于 25mm。

绑扎钢筋骨架中，钢筋净距是指各主钢筋的横向净距和层与层之间的竖向净距，当钢筋为三层及三层以下时，不应小于 30mm，并不小于钢筋直径；当钢筋为三层以上时，不应小于 40mm，并不小于钢筋直径的 1.25 倍。

焊接钢筋骨架中，多层主钢筋在竖向不留空隙，用焊缝连接，钢筋层数一般不宜超过 6 层。焊接钢筋骨架的净距要求见图 3-11。

伸入支承处的主钢筋，其根数不应少于 2 根，其面积不少于 20% 受拉主钢筋面积。

b. 箍筋及其数量由斜截面承载力计算确定，并应满足构造要求，在梁内是必须设置的。箍筋与纵筋、架立钢筋等形成钢筋骨架，并固定主钢筋的位置，用来提高梁的抗剪能力。箍筋直径 $d \geqslant 8$mm 且应大于主钢筋直径的 1/4。

箍筋形式有开口、闭口、四肢、双肢、单肢（图 3-12）。其中，单肢一般不采用；双肢一般采用单箍双肢；四肢在所箍受拉钢筋每层多于 5 根或所箍受压钢筋每层多于 3 根时采用。

a) 开口式双肢箍筋　　b) 封闭式双肢箍筋　　c) 封闭式四肢箍筋

图 3-12　箍筋的形式

c. 斜筋（弯起钢筋），如图 3-10 所示，设置方式及数量均由斜截面承载力计算确定，并满足构造要求。梁内弯起钢筋是由主钢筋按规定的部位和角度弯至梁上部后，并满足锚固要求的钢筋；斜钢筋是专门设置的斜向钢筋。斜筋的弯起角一般为 45°。

d. 架立钢筋为构造钢筋，按构造要求布置，其作用是固定箍筋并使主钢筋和箍筋能绑扎成骨架。它的直径通常是在 10~14mm 之间。

e. 纵向水平钢筋也是构造钢筋，一般按构造要求布置。其作用是抵抗温度应力与混凝土收缩应力，防止因混凝土收缩及温度变化而产生裂缝。直径在 6~8mm 之间，当梁高时，纵向水平钢筋沿梁肋高度的两侧、在箍筋外侧水平方向设置。面积为 $(0.001 \sim 0.002)bh$。布置间距为在受拉区不应大于腹板宽度 b，且不应大于 200mm，在受压区不应大于 300mm。

3.3　受弯构件正截面受力全过程和破坏形态

钢筋混凝土是由两种物理力学性能不同的材料组成的复合材料，又是非均质、非弹性的材料，受力后，按材料力学公式计算的结果与试验结果相差甚远，因此，钢筋混凝土的计算方法必须建立在试验的基础上。

1) 试验研究

试验是为了研究梁在荷载作用下正截面受力和变形的变化规律，以如图 3-13a) 所示的跨长为 1.8m 的钢筋混凝土简支梁作为试验梁。

(1) 试验目的

试验是为了研究钢筋混凝土梁的受力破坏过程,梁在极限荷载作用下正截面受力和变形特点,以建立正截面强度计算公式。

(2) 试验简介

试验梁为矩形截面简支梁,尺寸为 $b \times h = 100mm \times 160mm$,配有 2ϕ10 钢筋。试验梁混凝土棱柱体抗压强度实测值 $f_c = 20.2MPa$,纵向受力钢筋抗拉强度实测值 $f_s = 395MPa$。采用油压千斤顶施加两个集中荷载 F,其弯矩图和剪力图如图 3-13b)、c)所示。在梁 CD 段,剪力为零(忽略梁自重),而弯矩为常数,称为"纯弯曲"段,它是试验研究的主要对象。集中力 F 大小用测力传感器测读;挠度用百分表测量,测点设置在试验梁跨中的 E 点;混凝土应变用标距为 200mm 的手持应变仪测读,沿梁跨中截面段的高度方向上布置测点 a、b、c、d、e。在试验全过程要测读荷载施加力值、挠度和应变的数据,观测裂缝。

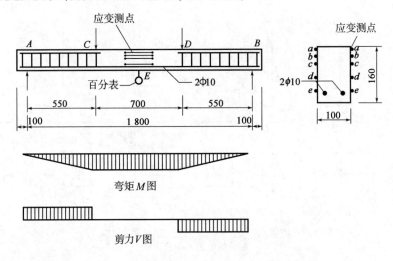

图 3-13 试验梁布置示意图(尺寸单位:mm)

2) 试验梁受力特点

(1) 荷载-挠度曲线

由试验梁的 $F\text{-}w$ 曲线(图 3-14)可以看到,$F\text{-}w$ 曲线上有两个明显的转折点,从而把梁的受力和变形全过程分为整体工作阶段、带裂缝工作阶段和破坏阶段三个阶段。

①阶段Ⅰ:无裂缝出现的整体工作阶段。

②阶段Ⅱ:出现裂缝工作阶段。

③阶段Ⅲ:失效破坏阶段,裂缝迅速开展,钢筋达到屈服强度。

三个特征点:第Ⅰ阶段末(用Ⅰ$_a$表示),裂缝即将出现;第Ⅱ阶段末(用Ⅱ$_a$表示),纵向受力钢筋屈服;第Ⅲ阶段末(用Ⅲ$_a$表示),梁受压区混凝土被压碎,整个梁截面破坏。

(2) 应力-应变曲线

图 3-15 为试验梁在各级荷载下,截面的混凝土应变的实测平均值及相应于各工作阶段截面上的正应力分布图。由图 3-15a)可见,随着荷载的增加,应变值也不断增加,但应变图基本仍是上下两个对顶的三角形。同时还可以看到,随着荷载的增加,中和轴逐渐上升。

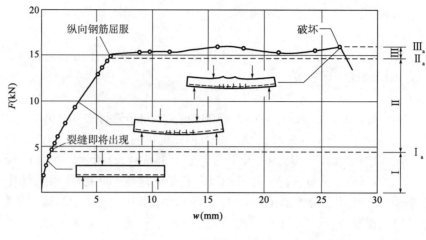

图 3-14 试验梁的荷载-挠度(F-w)图

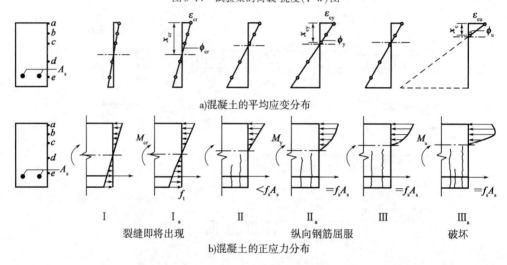

图 3-15 梁正截面各阶段的应力-应变图和应力图

图 3-15b)的应力图是根据图 3-15a)的各测点(a、b、c、d、e 测点)的实测应变值以及图 3-16 中材料的应力-应变图,沿截面从上到下逐个测点推求出来的。

(3)各阶段的受力变形特点

梁三个阶段的受力和变形具有以下特点:

第 I 阶段:梁混凝土全截面工作,混凝土的压应力和拉应力基本上都呈三角形分布。纵向钢筋承受拉应力。混凝土处于弹性工作阶段,应力与应变成正比。

第 I 阶段末:混凝土受压区的应力基本上仍呈三角形分布。但由于受拉区混凝土塑性变形的发展,拉应变增长较快,根据混凝土受拉时的应力-应变曲线[图 3-16c)],拉区混凝土的应力图形为曲线形。这时,受拉边缘混凝土的拉应变临近极限拉应变,拉应力达到混凝土抗拉强度,表示裂缝即将出现,梁截面上作用的弯矩用 M_{cr} 表示。

第 II 阶段:荷载作用弯矩到达 M_{cr} 后,在梁混凝土抗拉强度最弱截面上出现了第一批裂缝。这时,在有裂缝的截面上,拉区混凝土退出工作,把它原承担的拉力转给了钢筋,发生了明显的应力重分布。钢筋的拉应力随荷载的增加而增加,混凝土的压应力不再呈三角形分布,而形成微曲的曲线形,中和轴位置向上移动。

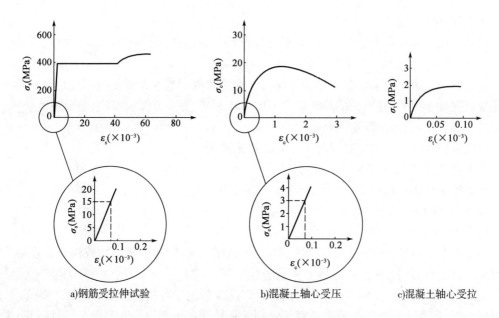

a) 钢筋受拉伸试验　　b) 混凝土轴心受压　　c) 混凝土轴心受拉

图 3-16　试验梁材料的应力-应变图

第Ⅱ阶段末:钢筋应力达到其屈服强度,第Ⅱ阶段结束。

第Ⅲ阶段:在这个阶段里,钢筋屈服后拉应变增加很快,但钢筋的拉应力一般仍维持在屈服强度不变(对具有明显流幅的钢筋)。这时,裂缝急剧开展,中和轴继续上升,混凝土受压区不断缩小,压应力也不断增大,压应力图呈明显的丰满曲线形。

第Ⅲ阶段末:这时,截面受压边缘的混凝土压应变达到其极限压应变值,压应力图呈明显曲线形,并且最大压应力已不在上边缘而是在距上边缘稍下处,这都是混凝土受压时的应力-应变图所决定的。在第Ⅲ阶段末,压区混凝土的抗压强度耗尽,在临界裂缝两侧的一定区段内,压区混凝土出现纵向水平裂缝,随即混凝土被压碎、梁破坏,在这个阶段,钢筋的拉应力仍维持在屈服强度。

(4) 适筋梁正截面的受力特点分析

① 与匀质弹性体梁比较

匀质弹性体梁:σ 与 M 成正比($\sigma = M/I$),中性轴位置、应力图形状不变(直线分布),只有量的变化。

钢筋混凝土梁:随 M 增加,σ 的大小改变,中性轴位置、应力图形状改变;大部分阶段带裂缝工作,应力 σ、挠度 w(刚度 EI 减小)与 M 不成正比。

② 弯曲后仍维持平面变形

钢筋混凝土梁变形后,截面的应变沿梁高度基本保持线性,这成为平截面假定的重要依据。

破坏时,混凝土极限压应变 $\varepsilon_{cu} = 0.0025 \sim 0.0045$,《公桥规》取 0.003。

③ 中性轴随荷载增加不断上升

这是在应力图形变化和开裂的情况下,保持截面静力平衡的结果。在第Ⅱ阶段,中和轴变化较小,主要靠增加 σ_s 来抵抗外弯矩,内力偶臂基本不变。在第Ⅲ阶段,σ_s 不变,为保持截面静力平衡,内力偶臂加大,中和轴上升。

④塑性破坏特征

适筋梁破坏为钢筋先屈服,然后混凝土被压碎。破坏前,钢筋经历较大的塑性伸长,裂缝充分发展,挠度急剧增加,有明显的破坏特征。

上述特点反映了钢筋混凝土结构材料力学性能的两个基本方面:一方面,混凝土的抗拉强度比抗压强度小很多,在不大的拉伸变形下即出现裂缝;另一方面,混凝土是弹塑性材料,当应力超过一定限度时,将出现塑性变形。

从试验研究结果得到钢筋混凝土梁的计算依据为:

① I_a 作为受弯构件抗裂度计算的依据。

② Ⅱ 作为使用阶段的变形和裂缝开展计算时的依据。

③ $Ⅲ_a$ 作为极限状态的承载力计算的依据。

3) 受弯构件正截面破坏形态

钢筋混凝土受弯构件有两种破坏性质:一种是塑性破坏,指的是结构或构件在破坏前有明显变形或其他征兆[图3-17a)];另一种是脆性破坏,指的是结构或构件在破坏前无明显变形或其他征兆[图3-17b)、c)]。根据试验研究,钢筋混凝土受弯构件的破坏性质与配筋率 ρ、钢筋强度等级、混凝土强度等级有关。对于常用的热轧钢筋和普通强度混凝土,破坏形态主要受到配筋率 ρ 的影响。因此,按照钢筋混凝土受弯构件的配筋情况及发生破坏时的性质,得到正截面破坏的三种形态(图3-17)。

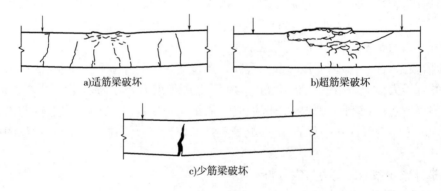

图3-17 梁的破坏形态

(1) 适筋梁的塑性破坏

梁的受拉区钢筋首先达到屈服强度,其应力保持不变,而应变显著地增大,直到受压区边缘混凝土的应变达到极限压应变时,受压区出现纵向水平裂缝,梁随混凝土压碎而破坏。这种梁破坏前,梁的裂缝急剧开展,挠度较大,梁截面产生较大的塑性变形,因而有明显的破坏预兆,属于塑性破坏。

受弯构件的截面曲率 ϕ 是一项综合表达构件的刚度、变形能力的指标。钢筋混凝土梁截面曲率的表达式是 $\phi = \varepsilon_c / \xi_i h_0$ (图3-15)。式中,ε_c 为截面边缘混凝土应变;h_0 为截面有效高度;ξ_i 为相对受压区高度,而受压区高度 $x_i = \xi_i h_0$。图3-15 中,ϕ_y 为钢筋屈服时截面曲率,ϕ_u 为梁破坏时的极限曲率,由于 ε_c 的急剧增大和 $\xi_i h_0$ 的迅速变小,使得 ϕ_u 比 ϕ_y 大很多,即 $\phi_u - \phi_y$ 较大,说明构件刚度降低、变形增大,却表现了较好的耐受变形的能力,即延性。延性是承受地震及冲击荷载作用时构件的一项重要受力特性。

(2) 超筋梁的脆性破坏

当梁截面配筋率 ρ 增大时，钢筋应力增加缓慢，压区混凝土应力有较快的增长，ρ 越大，则纵向钢筋屈服时的弯矩 M_y 越趋近梁破坏时的弯矩 M_u，这意味着第Ⅲ阶段缩短。当 ρ 增大到使 $M_y = M_u$ 时，受拉钢筋屈服与压区混凝土压碎几乎同时发生，这种破坏称为平衡破坏或界限破坏，相应的 ρ 值被称为最大配筋率 ρ_{max}。

当实际配筋率 $\rho > \rho_{max}$ 时，梁破坏时压区混凝土被压坏，而受拉区钢筋应力尚未达到屈服强度。破坏前梁的挠度及截面曲率没有明显的转折点，受拉区的裂缝开展不宽，延伸不高，破坏是突然的，没有明显预兆，属于脆性破坏，称为超筋梁破坏。

超筋梁的破坏是由于压区混凝土抗压强度耗尽，而钢筋的抗拉强度没有得到充分发挥，因此，超筋梁破坏时的弯矩 M_u 与钢筋强度无关，仅取决于混凝土的抗压强度。

(3) 少筋梁的脆性破坏

当梁的配筋率 ρ 很小时，梁受拉区混凝土开裂后，钢筋应力趋近于屈服强度，即开裂弯矩 M_{cr} 趋近于受拉区钢筋屈服时的弯矩 M_y，这意味着第Ⅱ阶段缩短，当 ρ 减少到使 $M_{cr} = M_y$ 时，裂缝一旦出现，钢筋应力立即达到屈服强度，这时的配筋率称为最小配筋率 ρ_{min}。梁中实际配筋率 $\rho < \rho_{min}$ 时，梁受拉区混凝土一开裂，受拉钢筋到达屈服，并迅速经历整个流幅而进入强化阶段，梁仅出现一条集中裂缝，不仅宽度较大，而且沿梁高延伸，此时受压区混凝土还未压坏，而裂缝宽度已很宽，挠度过大，钢筋甚至被拉断。由于破坏很突然，故属于脆性破坏。

少筋梁的抗弯承载力取决于混凝土的抗拉强度，在桥梁工程中不允许采用。

综上所述，受弯构件正截面的破坏特征随配筋多少而变化，其规律是：a. 配筋太少时，构件的破坏强度取决于混凝土的抗拉强度及截面大小，破坏呈脆性；b. 配筋过多时，配筋不能充分发挥作用，构件破坏强度取决于混凝土的抗压强度及截面大小，破坏亦呈脆性。合理的配筋量应在这两个限度之间，可避免发生超筋或少筋的破坏情况。

3.4 受弯构件正截面承载能力计算的基本原则

根据试验结果，为方便计算，提出以下原则。

1) 基本假定

(1) 平截面假定

在各级荷载作用下，截面上的平均应变保持为直线分布，即截面任意点的应变与该点到中和轴的距离成正比。这一假定是近似的，但由此而引起的误差不大，完全能符合工程计算要求。平截面假定为钢筋混凝土受弯构件正截面承载力计算提供了变形协调的几何关系，加强了计算方法的逻辑性和条理性，使计算公式具有更明确的物理意义。

(2) 不考虑混凝土的抗拉强度

在裂缝截面处，受拉区混凝土已大部分退出工作，但在靠近中和轴附近，仍有一部分混凝土承担着拉应力。由于其拉应力较小，且内力偶臂也不大，因此，所承担的内力矩是不大的，故在计算中可忽略不计，从而也简化了计算。

(3) 材料的 $\sigma\text{-}\varepsilon$ 曲线

① 混凝土的 $\sigma\text{-}\varepsilon$ 曲线，有多种不同的计算图式，较常用的是由一条二次抛物线及水平线组

成的曲线。图 3-18 是 CEB-FIP 采用的典型混凝土应力-应变曲线。曲线的上升段 OA 为二次抛物线，直线段 AB 为水平线，其表达式为：

$$\left.\begin{array}{ll}\sigma = \sigma_0 \left[2\dfrac{\varepsilon}{\varepsilon_0} - \left(\dfrac{\varepsilon}{\varepsilon_0}\right)^2\right], & \varepsilon \leqslant \varepsilon_0 \\ \sigma = \sigma_0, & \varepsilon > \varepsilon_0\end{array}\right\} \tag{3-2}$$

式中，σ_0 为峰值应力，CEP-FIP 取 $\sigma_0 = 0.85 f_{ck}$，f_{ck} 为混凝土标准圆柱体抗压强度，0.85 为折减系数；同时，取 $\varepsilon_0 = 0.002$。B 点的应变 $\varepsilon_{cu} = 0.0035$，$\varepsilon_{cu}$ 为混凝土极限压应变。

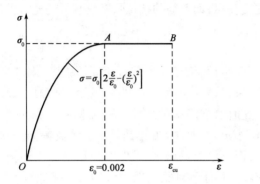
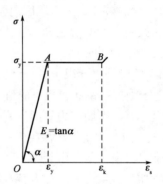

图 3-18　混凝土、钢筋的 σ-ε 曲线

②钢筋的 σ-ε 曲线，多采用理想弹塑性应力-应变关系（图 3-18）。对于有明显屈服台阶的钢筋，OA 为弹性阶段，A 点对应的应力为钢筋屈服强度 σ_y，相应的应变为屈服应变 ε_y，OA 的斜率为弹性模量 E_s。AB 为塑性阶段，B 点对应的应变为强化段开始的应变 ε_k，由图 3-18 可得到普通钢筋的应力-应变关系表达式为：

$$\left.\begin{array}{l}\sigma_s = \varepsilon_s E_s, 0 \leqslant \varepsilon_s \leqslant \varepsilon_y \\ \sigma_s = \sigma_y, \varepsilon_s > \varepsilon_y\end{array}\right\} \tag{3-3}$$

2）压区混凝土等效矩形应力图

钢筋混凝土受弯构件正截面承载力 M_u 的计算前提是要知道破坏时混凝土压应力的分布图形，特别是压区混凝土的压应力合力 C 及其作用位置 y_c，见图 3-19。

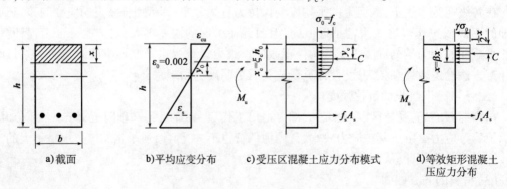

图 3-19　受压区混凝土等效矩形应力图

钢筋混凝土梁正截面破坏时,混凝土压应力的分布图形与混凝土的应力-应变曲线(受压时)是相似的,现取图3-19所示的混凝土应力-应变曲线模式图,即当$\varepsilon \leqslant \varepsilon_0$时,$\sigma = \sigma_0 \left[2\dfrac{\varepsilon}{\varepsilon_0} - \left(\dfrac{\varepsilon}{\varepsilon_0} \right)^2 \right]$;当$\varepsilon > \varepsilon_0$时,$\sigma = \sigma_0$,而$\varepsilon = \varepsilon_0$的点距中和轴的距离为$y_0$,如图3-19b)所示。

如图3-19c)所示,由平截面假定可得到混凝土受压区高度$x_c = \xi_c h_0$,同样得到$\varepsilon / \varepsilon_0 = y / y_0$及$y_0 = \varepsilon_0 \xi_c h_0 / \varepsilon_{cu}$。

现以如图3-19所示的矩形截面,来推导破坏时压区混凝土的压应力合力C及其合力作用位置y_c的表达式。压区混凝土的应力-应变曲线为两段,须分段积分才能得到压应力合力C:

$$C = \int_0^{\xi_c h_0} \sigma(\varepsilon) b \mathrm{d}y = \int_0^{y_0} \sigma_0 \left[\dfrac{2\varepsilon}{\varepsilon_0} - \left(\dfrac{\varepsilon}{\varepsilon_0} \right)^2 \right] b \mathrm{d}y + \int_{y_0}^{\xi_c h_0} \sigma_0 b \mathrm{d}y$$

注意到$\dfrac{\varepsilon}{\varepsilon_0} = \dfrac{y}{y_0}$及$y_0 = \dfrac{\varepsilon_0}{\varepsilon_{cu}} \xi_c h_0$,积分后可得到:

$$C = \sigma_0 \xi_c h_0 b \left(1 - \dfrac{1}{3} \dfrac{\varepsilon_0}{\varepsilon_{cu}} \right) \tag{3-4}$$

混凝土压应力合力C的作用点至受压边缘的距离y_c,可由下式计算:

$$y_c = \xi_c h_0 - \dfrac{\int_0^{\xi_c h_0} \sigma(\varepsilon) b y \mathrm{d}y}{C}$$

将式中的积分计算后,可得到:

$$y_c = \xi_c h_0 \left[1 - \dfrac{\dfrac{1}{2} - \dfrac{1}{12} \left(\dfrac{\varepsilon_0}{\varepsilon_{cu}} \right)^2}{1 - \dfrac{1}{3} \dfrac{\varepsilon_0}{\varepsilon_{cu}}} \right] \tag{3-5}$$

显然,用混凝土受压时的应力-应变曲线$\sigma = \sigma(\varepsilon)$来求应力合力$C$和合力作用点$y_c$是比较麻烦的。采用简化方法,用等效矩形应力图代替混凝土实际应力图。

用等效矩形应力图代替混凝土实际应力图必须满足以下两个等代条件:
(1)保持C的作用点位置不变。
(2)保持C的大小不变。

引入两个无纲量参数β、γ,得:

$$\beta = \dfrac{x}{x_c}, \gamma = \dfrac{\sigma}{\sigma_0}$$

由图3-19d)可得:

$$C = \gamma \sigma_0 b x = \gamma \sigma_0 b \beta x_c \tag{3-6}$$

合力C作用点为:

$$y_c = \dfrac{x}{2} = \dfrac{1}{2} \beta x_c \tag{3-7}$$

按等代条件,可得:

$$\beta = \dfrac{1 - \dfrac{2}{3} \dfrac{\varepsilon_0}{\varepsilon_{cu}} + \dfrac{1}{6} \left(\dfrac{\varepsilon_0}{\varepsilon_{cu}} \right)^2}{1 - \dfrac{1}{3} \dfrac{\varepsilon_0}{\varepsilon_{cu}}} \tag{3-8}$$

$$\gamma = \frac{1}{\beta}\left(1 - \frac{1}{3}\frac{\varepsilon_0}{\varepsilon_{cu}}\right) \tag{3-9}$$

当确定 ε_0、ε_{cu} 值后,即可将图 3-19c)的压区混凝土实际压应力分布图,换成图 3-19d)的等效矩形压应力分布图形。若取 $\varepsilon_0 = 0.002$,混凝土极限压应变 $\varepsilon_{cu} = 0.0033$,而不是按 CEB-FIP 那样取 $\varepsilon_{cu} = 0.0035$。由式(3-8)和式(3-9)可得到 $\beta = 0.8095$,$\gamma = 0.9608$,即等效矩形压应力图形高度 $x = 0.8095x_c$,等效压应力值为 $\gamma\sigma_0 = 0.9608\sigma_0$。

对于受弯构件截面受压区边缘混凝土的极限压应变 ε_{cu} 和相应的系数 β,《公桥规》按混凝土强度级别来分别取值,详见表 3-1。基于上述受压区混凝土应力计算图形采用等效矩形图形的分析,结合国内外试验资料,《公桥规》对所取用的混凝土受压区等效矩形应力值取 $\gamma\sigma_0 = f_{cd}$,f_{cd} 为混凝土的轴心抗压强度设计值。

混凝土极限压应变 ε_{cu} 与系数 β 值　　　　表 3-1

混凝土强度等级	C50 以下	C55	C60	C65	C70	C75	C80
ε_{cu}	0.0033	0.00325	0.0032	0.00315	0.0031	0.00305	0.003
β	0.8	0.79	0.78	0.77	0.76	0.75	0.74

3)相对界限受压区高度 ξ_b

界限破坏为当受拉区钢筋达到屈服应变 ε_y 而开始屈服时,受压区混凝土边缘也同时达到其极限压应变 ε_{cu} 而破坏。根据图 3-20,此时的受压区高度 $x_c = \xi_b h_0$,ξ_b 被称为相对界限受压区高度。

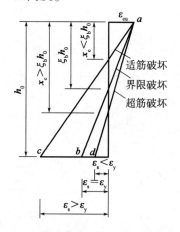

图 3-20　相对界限受压区高度

适筋截面受弯构件破坏始于受拉区钢筋屈服,经历一段变形过程后压区边缘混凝土达到极限压应变 ε_{cu} 后才破坏,而这时受拉区钢筋的拉应变 $\varepsilon_s > \varepsilon_y$,由此可得到适筋截面破坏时的应变分布如图 3-20 中的 ac 直线。此时受压区高度 $x_c < \xi_b h_0$。

超筋截面受弯构件破坏是压区边缘混凝土先达到极限压应变 ε_{cu} 破坏,这时受拉区钢筋的拉应变 $\varepsilon_s < \varepsilon_y$,由此可得到超筋截面破坏时的应变分布如图 3-20 中的 ad 直线,此时受压区高度 $x_c > \xi_b h_0$。

由图 3-20 可以看到,界限破坏是适筋截面和超筋截面的鲜明界线;当截面实际受压区高度 $x_c > \xi_b h_0$ 时,为超筋梁截面;当 $x_c < \xi_b h_0$ 时,为适筋梁截面。因此,一般用 ξ_b 来作为界限条件,x_b 为按平截面假定得到的界限破坏时受压区混凝土高度。

对于界限高度 $x_c = \beta x_b$,相应的 ξ_b 应为:

$$\xi_b = \frac{x_c}{h_0} = \frac{\beta x_b}{h_0}$$

由平截面假定和几何关系得到:

$$\frac{x_b}{h_0} = \frac{\varepsilon_{cu}}{\varepsilon_{cu} + \varepsilon_y} \tag{3-10}$$

将 $x_b = \frac{\xi_b h_0}{\beta}$、$\varepsilon_y = \frac{f_{sd}}{E_s}$ 代入上式并整理得到按等效矩形应力分布图形的受压区相对界限高度:

$$\xi_b = \frac{\beta}{1 + \dfrac{f_{sd}}{\varepsilon_{cu} E_s}} \qquad (3\text{-}11)$$

式(3-11)为《公桥规》确定混凝土受压区高度 ξ_b 的依据,其中 f_{sd} 为受拉钢筋的抗拉强度设计值。据此,按混凝土轴心抗压强度设计值、不同钢筋的强度设计值和弹性模量值可得到《公桥规》规定的 ξ_b 值(表3-2)。

相对界限受压区高度 ξ_b 表3-2

钢筋种类	混凝土强度等级			
	C50 及以下	C55、C60	C65、C70	C75、C80
HPB300	0.58	0.56	0.54	—
HRB400、HRBF400、RRB400	0.53	0.51	0.49	—
HRB500	0.49	0.47	0.46	—
钢绞线、钢丝	0.40	0.38	0.36	0.35
预应力螺纹钢筋	0.40	0.38	0.36	—

注:截面受拉区内配置不同种类钢筋的受弯构件,其 ξ_b 值应选用相应于各种钢筋的较小值。

4)最小配筋率 ρ_{min}

为了避免少筋梁破坏,必须确定钢筋混凝土受弯构件的最小配筋率 ρ_{min}。最小配筋率是少筋梁与适筋梁的界限。当梁的配筋率由 ρ_{min} 逐渐减少时,梁的工作特性也从钢筋混凝土结构逐渐向素混凝土结构过渡,所以,ρ_{min} 可按采用最小配筋率 ρ_{min} 的钢筋混凝土梁在破坏时,正截面承载力 M_u 等于同样截面尺寸、同样材料的素混凝土梁正截面开裂弯矩标准值的原则确定。

由上述原则的计算结果,同时考虑到温度变化、混凝土收缩应力的影响以及过去的设计经验,《公桥规》规定了受弯构件纵向受力钢筋的最小配筋率 ρ_{min}(%),详见附表1-9。

3.5 单筋矩形截面受弯构件强度计算

1)基本公式及适用条件

根据受弯构件正截面承载力计算的基本原则,可以得到单筋矩形截面受弯构件承载力计算简图,如图3-21 所示。

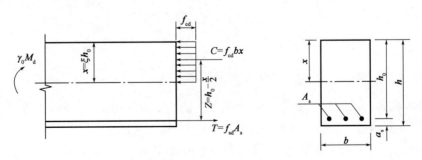

图3-21 单筋矩形截面受弯构件正截面承载力计算图式

基本计算原则为在受弯构件计算截面上的最不利荷载基本组合效应计算值 $\gamma_0 M_d$ 不应超过截面的承载能力(抗力) M_u,即 $\gamma_0 M_d \leq M_u$。

由图3-21可以写出单筋矩形截面受弯构件正截面计算的基本公式(或称基本方程)。

由截面上水平方向内力之和为零的平衡条件,即 $T + C = 0$,可得到:

$$f_{cd} bx = f_{sd} A_s \tag{3-12}$$

由截面上对受拉钢筋合力 T 作用点的力矩之和等于零的平衡条件,可得到:

$$\gamma_0 M_d \leq M_u = f_{cd} bx \left(h_0 - \frac{x}{2} \right) \tag{3-13}$$

由对压区混凝土合力 C 作用点取力矩之和为零的平衡条件,可得到:

$$\gamma_0 M_d \leq M_u = f_{sd} A_s \left(h_0 - \frac{x}{2} \right) \tag{3-14}$$

式中,M_d 为计算截面上的弯矩组合设计值;γ_0 为结构的重要性系数;M_u 为计算截面的抗弯承载力;f_{cd} 为混凝土轴心抗压强度设计值;f_{sd} 为纵向受拉钢筋抗拉强度设计值;A_s 为纵向受拉钢筋的截面面积;x 为按等效矩形应力图计算的受压区高度;b 为截面宽度;h_0 为截面有效高度。

两个基本方程为式(3-12)、式(3-13)或者式(3-14)。

式(3-12)、式(3-13)和式(3-14)仅适用于适筋梁,不适用于超筋梁和少筋梁。因为超筋梁破坏时钢筋的实际拉应力 σ_s 并未到达抗拉强度设计值,故不能按 f_{sd} 来考虑。因此,公式的适用条件为:

(1)为防止出现超筋梁情况,计算受压区高度 x 应满足:

$$x \leq \xi_b h_0 \tag{3-15}$$

式中的相对界限受压区高度 ξ_b,可根据混凝土强度级别和钢筋种类,由表3-2查得。

由式(3-12)可以得到计算受压区高度 x 为:

$$x = \frac{f_{sd} A_s}{f_{cd} b} \tag{3-16}$$

则相对受压区高度 ξ 为:

$$\xi = \frac{x}{h_0} = \frac{f_{sd}}{f_{cd}} \frac{A_s}{bh_0} = \rho \frac{f_{sd}}{f_{cd}} \tag{3-17}$$

由式(3-16)可见,ξ 不仅反映了配筋率 ρ,而且反映了材料强度比值的影响,故 ξ 又被称为配筋特征值,它是一个比 ρ 更有一般性的参数。

当 $\xi = \xi_b$ 时,可得到适筋梁的最大配筋率 ρ_{max} 为:

$$\rho_{max} = \xi_b \frac{f_{cd}}{f_{sd}} \tag{3-18}$$

显然,适筋梁的配筋率 ρ 应满足:

$$\rho \leq \rho_{max} \left(= \xi_b \frac{f_{cd}}{f_{sd}} \right) \tag{3-19}$$

式(3-19)和式(3-15)具有相同意义,目的都是防止受拉区钢筋过多形成超筋梁,满足其中一式,另一式必然满足。在实际计算中,多采用式(3-15)。

(2)为防止出现少筋梁的情况,计算的配筋率 ρ 应当满足:

$$\rho \geq \rho_{min} \tag{3-20}$$

2) 计算方法

钢筋混凝土受弯构件的正截面计算,一般仅需对构件的控制截面进行。控制截面是指弯矩组合设计值最大的截面。

受弯构件正截面承载力计算分为截面设计和截面复核两类计算问题。

(1) 截面设计

①设计内容:选材料、确定截面尺寸、配筋计算。

截面设计应满足承载力 $M_u \geq$ 弯矩计算值 $\gamma_0 M_d$ 的条件,即确定钢筋数量后的截面承载力至少要等于弯矩计算值 $\gamma_0 M_d$,所以在利用基本公式进行截面设计时,一般取 $M_u = \gamma_0 M_d$ 来计算。

②设计步骤(可能出现的两种设计情况)。

a. 已知弯矩设计值 M_d,混凝土和钢筋材料级别(f_{cd}、f_{sd}),截面尺寸 b、h,求钢筋面积 A_s。

分析:此问题为两个基本方程求解两个未知数 x、A_s,根据给定的环境条件确定最小混凝土保护层厚度(附表1-8),根据给定的安全等级确定 γ_0。

解:(a) 假定 a_s。

对于梁,一般 $a_s = 40$ mm(一排),$a_s = 65$ mm(两排)。

对于板,一般 $a_s = 25$ mm 或 35 mm。

所以:$h_0 = h - a_s$。

(b) 列出基本公式,求 x 或 ξ。

由基本方程 $\gamma_0 M_d = f_{cd} b x (h_0 - x/2)$(一元二次方程)可得:

$$x = h_0 - \sqrt{h_0^2 - \frac{2\gamma_0 M_d}{f_{cd} b}} \text{ 或 } \xi = \frac{x}{h_0} = 1 - \sqrt{1 - \frac{2\gamma_0 M_d}{f_{cd} b h_0^2}}$$

(c) 检查 $x \leq \xi_b h_0$ 或 $\xi \leq \xi_b$ 条件。

(d) 由基本方程 $f_{cd} b x = f_{sd} A_s$,可得:

$$A_s = \frac{f_{cd} b x}{f_{sd}}$$

(e) 选择钢筋直径、根数。

(f) 校核 $\rho \geq \rho_{min}$,其中 $\rho = A_s / bh_0$(注意 A_s 为实际配筋量)。

如果 $\rho < \rho_{min}$,则取 $\rho = \rho_{min}$,$A_s = \rho_{min} b h_0$(构造取筋)。

(g) 检查假定 a_s 是否等于实际 a_s,如误差大,重新计算(如果 $a_{s假} < a_{s实}$,则 $h_{0假} > h_{0实}$,偏不安全)。

(h) 绘配筋简图,并检查构造要求(钢筋净距等)。

b. 已知弯矩设计值 M_d,混凝土和钢筋材料级别(f_{cd}、f_{sd}),求 b、h、x 及钢筋面积 A_s。

分析:此问题有4个未知数(x、b、h、A_s),必须先确定或假定其中两个。

解:(a) b 由构造定(b 对承载力影响小)。

假定 ρ。一般经济配筋率 ρ;对于板,0.3%~0.8%;对于矩形梁,0.6%~1.5%;对于T形梁,2%~3.5%。

(b) $\xi = \rho f_{sd}/f_{cd}$,$x = \xi h_0$。

(c) 由基本方程 $\gamma_0 M_d = f_{cd} b x (h_0 - x/2)$ 得:

$$h_0 = \sqrt{\frac{\gamma_0 M_d}{f_{cd} b \xi(1-0.5\xi)}}$$

估计 a_s,选定 $h = h_0 + a_s$ 后取整数。

(d) 已知 $b \times h$,则变成了第①种设计情况。

例:已知弯矩设计值 $M_d = 120\text{kN} \cdot \text{m}$,Ⅰ类环境,安全等级二级($\gamma_0 = 1.0$),截面 $b \times h = 220\text{mm} \times 500\text{mm}$,C30 混凝土,HPB400 钢筋,求 A_s。

解:(a) 假定 $a_s = 44\text{mm}$(注意不要把 a_s 与保护层混淆)。

$$h_0 = h - a_s = 500 - 44 = 456(\text{mm})$$

(b) 由附表 1-1,查得 C30,$f_{cd} = 13.8\text{MPa}$,$f_{td} = 1.39\text{MPa}$。

$$\xi = 1 - \sqrt{1 - \frac{2\gamma_0 M_d}{f_{cd} b h_0^2}}$$

$$= 1 - \sqrt{1 - \frac{2 \times 0.9 \times 12 \times 10^7}{9.2 \times 220 \times 456^2}}$$

$$= 0.38 < \xi_b = 0.53$$

[ξ_b 值见表 3-2;写出计算过程,便于复核]。

(c) 由附表 1-3,查得 $f_{sd} = 330\text{MPa}$。

$$A_s = \frac{f_{cd} b h_0 \xi}{f_{sd}} = \frac{13.8 \times 220 \times 456 \times 0.38}{195} = 1594.2(\text{mm}^2)$$

由附表 1-6 可得钢筋选择的多方案,如选用 2 ⌀32, $A_s = 1608\text{mm}^2 >$ 计算 $A_s = 1430.86\text{mm}^2$,满足要求。2 ⌀28 + 1 ⌀25, $A_s = 1723\text{mm}^2 >$ 计算 $A_s = 1430.86\text{mm}^2$,也满足要求(注意不要比计算值超出过多,不经济,可能超筋,最好在 5% 以内。钢筋直径类型不宜太多)。

选用 3 ⌀25,则所需最小梁宽(两侧保护层 + 钢筋直径 + 钢筋净距):

$$b_{min} = 2 \times 30 + 2 \times 35.8 + 35.8 = 167(\text{mm}) < b = 220\text{mm}$$

满足要求。

(d) 配筋率:

$$\rho = \frac{A_s}{bh_0} = \frac{1608}{220 \times 456} = 1.60\% > \rho_{min}$$

$$\rho_{min} = \max\left\{45\% \frac{f_{td}}{f_{sd}} = 45\% \times \frac{1.39}{330} = 0.190\%, 0.2\%\right\} = 0.2\%$$

实际 $a_s = 20 + 35.8/2 = 36(\text{mm}) <$ 假定 a_s。

(e) 绘钢筋布置图(图 3-22)。如果修改设计,第一选择增加 h,第二选择增加 f_{sd},第三选择增加 f_{cd}、b(第三种选择不常用)。

(2) 截面复核

截面复核目的为对已经设计好的截面检查其承载力是否满足要求,同时检查是否满足构造要求。

① 计算步骤

已知截面尺寸 b、h,钢筋面积 A_s,混凝土和钢筋材料级别 f_{sd}、f_{cd},求截面承载力 M_u。

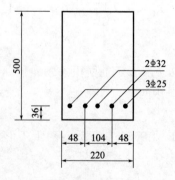

图 3-22 钢筋布置图(尺寸单位:mm)

分析：此问题可以根据基本公式直接求解。

解：a. 检查构造要求。

b. 根据实际情况计算 a_s：

$$a_s = \frac{\sum_{i=1}^{n} f_{sdi} A_{si} a_{si}}{\sum_{i=1}^{n} A_{si} f_{si}}$$

c. 检查 $\rho \geqslant \rho_{\min}$ 是否满足要求。

d. 由基本公式 $x = f_{sd} A_s / (f_{cd} b)$，求得 $\xi = x/h_0$。

e. 检查 $\xi \leqslant \xi_b$ 的条件。

如果 $\xi > \xi_b$，属于超筋，实际上当 $\xi > \xi_b$ 时，只能取 $\xi = \xi_b$（多配钢筋不能发挥作用），则 $x = \xi_b h_0$，此时 $M_u = f_{cd} bx(h_0 - x/2) = f_{cd} b \xi_b h_0 (h_0 - 1/2 \xi_b h_0)$（此即单筋矩形截面适筋梁的最大承载能力）。

当求得的 $M_u < M$ 时，可采取提高混凝土级别、修改截面尺寸，或改为双筋截面等措施。

f. 满足 $\xi \leqslant \xi_b$ 时，则由基本公式计算 M_u：

$$M_u = f_{cd} bx \left(h_0 - \frac{x}{2} \right)$$

② 算例

已知矩形截面梁 $b \times h = 200\text{mm} \times 600\text{mm}$，采用 C30 混凝土和 HRB400 级钢筋，$A_s = 1\,250\text{mm}^2$（$3 \oplus 20 + 2 \oplus 14$），见图 3-23，求 M_u。

解：a. 经检查满足构造要求。

b. $A_{s1} = 942\text{mm}^2$（$3 \oplus 20$），$f_{sd1} = 330\text{MPa}$；

$a_{s1} = 30 + 22.7/2 = 41.35(\text{mm})$（梁底保护层 + 钢筋外径的一半）；

$A_{s2} = 308\text{mm}^2$（$2 \oplus 14$），$f_{sd2} = 330\text{MPa}$；

$a_{s2} = 30 + 22.7 + 30 + 16.2/2 = 90.8(\text{mm})$；

所以 $a_s = \dfrac{942 \times 330 \times 41.35 + 308 \times 330 \times 90.8}{942 \times 330 + 308 \times 330}$

$= 53.53(\text{mm})$。

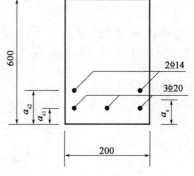

图 3-23　（尺寸单位：mm）

$h_0 = h - a_s = 600 - 53.53 = 546.5(\text{mm})$。

c. 由附表 1-1，查得 C30：$f_{cd} = 13.8\text{MPa}$，$f_{td} = 1.39\text{MPa}$，得：

$\rho = \dfrac{A_s}{bh_0} = \dfrac{1\,250}{200 \times 546.5} = 1.14\% > \rho_{\min}$；

$\rho_{\min} = \max\left\{ 45 \dfrac{f_{td}}{f_{sd}} = 45 \times \dfrac{1.39}{330} = 0.190\%, 0.2\% \right\} = 0.2\%$。

d. $x = \dfrac{f_{sd} A_s}{f_{cd} b} = \dfrac{330 \times 1\,250}{13.8 \times 200} = 149.5(\text{mm})$；

$\xi = \dfrac{x}{h_0} = 0.247 < \xi_b = 0.53$。

e. $M_u = f_{cd} bx \left(h_0 - \dfrac{x}{2} \right)$

$$= 13.8 \times 200 \times 149.5 \times \left(546.5 - \frac{149.5}{2}\right)$$
$$= 194\,653\,485(\text{N} \cdot \text{mm}) = 194.65(\text{kN} \cdot \text{m})。$$

3.6 双筋矩形截面受弯构件强度计算

1) 基本概念

双筋截面是在受拉和受压区都配置受力钢筋的截面,即在截面受压区配置钢筋来协助混凝土承担压力,破坏时受拉区钢筋应力可达到屈服强度,而受压区混凝土不致过早压碎。

双筋矩形截面的使用条件：

(1) 采用单筋截面设计出现 $\xi > \xi_b$ 时,而 b、h、f_{cd} 的改善受到限制；

(2) 截面承受弯矩异号 M,则必须采用双筋截面；

(3) 构造和锚固的需要；

(4) 结构本身受力图式的原因,例如连续梁的内支点处截面,将会产生事实上的双筋截面。

该截面形式的优点是能提高截面的延性并可减少长期荷载作用下受弯构件的变形；缺点是不十分经济。

2) 双筋截面的构造要求

必须设置封闭式箍筋,如图 3-24 所示,以约束受压钢筋的纵向压屈变形,一般情况下,箍筋的间距不大于 400mm,并不大于受压钢筋直径 d' 的 15 倍；箍筋直径不小于 8mm 或 $d'/4$。

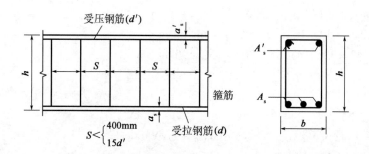

图 3-24　箍筋间距及形式要求

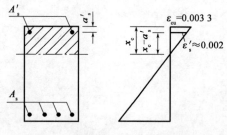

图 3-25　双筋截面受压钢筋应变计算分析图

3) 双筋截面梁的受力特点

双筋截面受弯构件的受力特点和破坏特征基本上与单筋截面相似,试验研究表明,只要满足 $\xi \leq \xi_b$,双筋截面仍具有适筋破坏特征。因此,在建立双筋截面承载力的计算公式时,受压区混凝土仍可采用等效矩形应力图形和混凝土抗压设计强度 f_{cd},而受压钢筋的应力尚待确定。

双筋截面梁破坏时,受压钢筋的应力取决于它的应变 ε'_s。如图 3-25 所示,由平截面假定,受压钢筋应力:

$$\frac{\varepsilon'_s}{\varepsilon_{cu}} = \frac{x_c - a'_s}{x_c} = 1 - \frac{a'_s}{x_c} = 1 - \frac{0.8a'_s}{x} \quad (因为 x = 0.8x_c)$$

$$\varepsilon'_s = 0.0033\left(1 - \frac{0.8a'_s}{x}\right)$$

式中,x 和 x_c 分别为等效矩形应力图形的计算受压区高度和实际受压高度。

当 $\dfrac{a'_s}{x} = \dfrac{1}{2}$,即 $x = 2a'_s$ 时,可得到:

$$\varepsilon'_s = 0.0033\left(1 - \frac{0.8a'_s}{2a'_s}\right) = 0.00198$$

《公桥规》规定,当取受压钢筋应变 $\varepsilon'_s = 0.002$,对 HPB300 级钢筋,$\sigma'_s = \varepsilon'_s E'_s = 0.002 \times 2.1 \times 10^5 = 420(\text{MPa}) > f'_{sd}(=250\text{MPa})$;对 HRBF400、HRB400 和 RRB400 级钢筋,$\sigma'_s = \varepsilon'_s E'_s = 0.002 \times 2 \times 10^5 = 400(\text{MPa}) > f'_{sd}(=330\text{MPa})$。

由此可见,当 $x = 2a'_s$ 时,普通钢筋均能达到屈服强度。当 $x > 2a'_s$ 时,ε'_s 将更大,钢筋亦早已受压屈服。为了充分发挥受压钢筋的作用并确定保其达到屈服强度,《公桥规》规定取 $\sigma'_s = f'_{sd}$ 时必须满足:

$$x \geqslant 2a'_s \tag{3-21}$$

4) 基本计算公式及适用条件

由于双筋截面的受力破坏特征与单筋截面梁类似,在推导计算公式时,采用了与单筋截面相同的计算假定和方法。由计算假定得到了图 3-26 的计算图式。

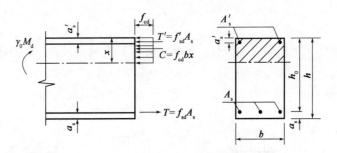

图 3-26 双筋矩形截面的正截面承载力计算图式

根据计算图式,得到基本计算公式:

$$\sum x = 0 \qquad f_{cd}bx + f'_{sd}A'_s = f_{sd}A_s \tag{3-22}$$

$$\sum M_T = 0 \qquad \gamma_0 M_d \leqslant M_u = f_{cd}bx\left(h_0 - \frac{x}{2}\right) + f'_{sd}A'_s(h_0 - a'_s) \tag{3-23a}$$

或

$$\sum M'_T = 0 \qquad \gamma_0 M_d \leqslant M_u = -f_{cd}bx\left(\frac{x}{2} - a'_s\right) + f_{sd}A_s(h_0 - a'_s) \tag{3-23b}$$

式中,f'_{sd} 为受压区钢筋的抗压强度设计值;A'_s 为受压区钢筋的截面面积;a'_s 为受压区钢筋合力点至截面受压边缘的距离;其他符号与单筋矩形截面相同。

基本公式的适用条件为:

(1) 为了防止出现超筋梁情况,计算受压区高度 x 应满足:
$$x \leqslant \xi_b h_0 \tag{3-24}$$
(2) 为了保证受压钢筋达到抗压强度设计值 f'_{sd},应满足:
$$x \geqslant 2a'_s \tag{3-25}$$

在实际设计中,若求得 $x < 2a'_s$,则表明受压钢筋可能达不到其抗压强度设计值。对于受压钢筋保护层混凝土厚度不大的情况,《公桥规》规定这时可取 $x = 2a'_s$,即假设混凝土压应力合力作用点与受压区钢筋合力作用点相重合(图 3-27),对受压钢筋合力作用点取矩,可得到正截面抗弯承载力的近似表达式为:
$$M_u = f_{sd} A_s (h_0 - a'_s) \tag{3-26}$$

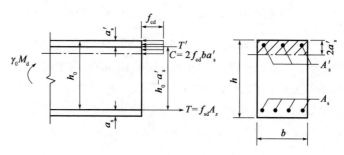

图 3-27 $x < 2a'_s$ 时 M_u 的计算图式

(3) $\rho \geqslant \rho_{min}$ 一般能满足。

5) 计算方法

(1) 截面设计(针对两种情况介绍)

①已知弯矩组合设计值 M_d,材料等级 f_{cd}、f_{sd},截面尺寸 b、h,求受拉钢筋 A_s 和受压钢筋 A'_s。

分析:此问题有三个未知数(A_s, A'_s, x),基本方程只有两个,需补充条件才能计算。

解:a. 假设 a_s、a'_s,求得 $h_0 = h - a_s$。

b. 验算是否需采用单筋(由于按单筋会超筋,所以才用双筋)。

当满足 $\gamma_0 M_d > M_u = f_{cd} b h_0^2 \xi_b (1 - 0.5\xi_b)$,则按双筋计算。

c. 补充条件,使 $A_s + A'_s$ 为最经济。

通常将 $x = \xi_b h_0$ 代入基本方程:

$$A'_s = \frac{\gamma_0 M_d - f_{cd} b \xi_b h_0^2 (1 - 0.5\xi_b)}{f'_{sd}(h_0 - a'_s)}$$

$$A_s = \frac{\gamma_0 M_d + f_{cd} b x \left(\frac{x}{2} - a'_s\right)}{f_{sd}(h_0 - a'_s)}$$

d. 选择钢筋直径、根数并布置。

校核 a_s、a'_s 是否与假定一致,否则重算。

②已知弯矩组合设计值 M_d,材料等级 f_{cd}、f_{sd},截面尺寸 b、h,受压钢筋 A'_s,求受拉钢筋 A_s。

分析:此问题只有两个未知数(A_s, x),利用基本方程可以求解。

解:a. a'_s 已知,假设 a_s,求得 $h_0 = h - a_s$。

b. 由基本公式可得:

$$x = h_0 - \sqrt{h_0^2 - \frac{2[\gamma_0 M_d - f'_{sd} A'_s (h_0 - a'_s)]}{f_{cd} b}}$$

校核 $x \leq \xi_b h_0$, 如不满足, 修改设计或按 A'_s 为未知计算; 校核 $x \geq 2a'_s$。

c. 若 $2a'_s \leq x \leq \xi_b h_0$, 则由基本公式得:

$$A_s = \frac{f_{cd} b x}{f_{sd}} + \frac{f'_{sd} A'_s}{f_{sd}}$$

选择钢筋直径、根数并布置, 校核 a_s。

d. 若 $x < 2a'_s$, 对受压区合力作用点取矩, 并假设 $x = 2a'_s$, 有:

$$A_{s_1} = \frac{\gamma_0 M_d}{f_{sd}(h_0 - a'_s)}$$

e. 与不计 A'_s (取 $A'_s = 0$) 的单筋矩形截面进行比较:

$$x^d = h_0 - \sqrt{h_0^2 - \frac{2\gamma_0 M_d}{f_{cd} b}}$$

$$A_{s_2} = \frac{f_{cd} b x}{f_{sd}}$$

比较后取小值, $A_s = \min\{A_{s_1}, A_{s_2}\}$。

(2) 截面复核

已知材料等级 f_{cd}、f_{sd}, 截面尺寸 b、h, 钢筋面积 A_s 和 A'_s, 求 M_u。

分析: 此问题可以根据基本公式直接求解。

解: ① 检查构造要求。
② 计算受压区高度:

$$x = \frac{f_{sd} A_s - f'_{sd} A'_s}{f_{cd} b}$$

③ 若 $2a'_s \leq x \leq \xi_b h_0$,

$$M_u = f_{cd} b x \left(h_0 - \frac{x}{2}\right) + A'_s f'_{sd} (h_0 - a'_s)$$

④ 若 $x < 2a'_s$, 则由式(3-26):

$$M_{u1} = f_{sd} A_s (h_0 - a'_s)$$

再与不计 A'_s (取 $A'_s = 0$) 的单筋矩形截面进行比较:

$$x = \frac{f_{sd} A_s}{f_{cd} b}$$

$$M_{u2} = f_{cd} b x \left(h_0 - \frac{x}{2}\right)$$

$$M_u = \max\{M_{u_1}, M_{u_2}\}$$

⑤ 若 $x > \xi_b h_0$,
修改设计或求此种情况下的最大承载力, 即取 $\xi = \xi_b$、$x = \xi_b h_0$ 代入基本公式。

3.7　T形截面受弯构件计算

1) 概述

矩形截面梁在破坏时,受拉区混凝土早已开裂。在开裂截面处,受拉区的混凝土对截面的抗弯承载力已不起作用,因此可将受拉区混凝土挖去一部分,将受拉钢筋集中布置在剩余拉区混凝土内,形成了钢筋混凝土 T 形梁的截面,其承载能力与原矩形截面梁相同,但节省了混凝土和减轻了梁自重。因此,钢筋混凝土 T 形梁具有更大的跨越能力。

典型的钢筋混凝土 T 形梁截面见图 3-28,截面伸出部分称为翼缘板(简称翼板),其宽度为 b 的部分称为梁肋或梁腹。在正弯矩作用下,翼板位于受压区的 T 形梁截面,称为 T 形截面[图 3-28a)];当受负弯矩作用时,位于梁上部的翼板受拉后混凝土开裂,这时梁的有效截面是肋宽 b、梁高 h 的矩形截面[图 3-28b)],其抗弯承载力则应按矩形截面来计算。因此,判断一个截面在计算时是否属于 T 形截面,不是看截面本身形状,而是要看其翼缘板是否能参与抗压作用。从这个意义上来讲,工字形、箱形截面、π 形截面以及空心板截面,在正截面抗弯承载力计算中均可按 T 形截面来处理。

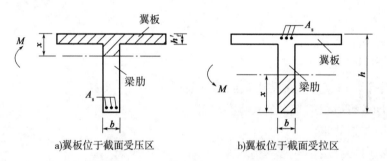

a)翼板位于截面受压区　　　　　b)翼板位于截面受拉区

图 3-28　T 形截面的受压区位置

下面以板宽为 b_f 的空心板截面为例,通过截面换算,将其等效成工字形截面,计算中即可按 T 形截面处理。空心板截面换算成等效的工字形截面的换算原则是抗弯等效的原则即保持截面面积、惯性矩和形心位置不变。

设空心板截面高度为 h,圆孔直径为 D,孔洞面积形心轴距板截面上、下边缘距离分别为 y_1 和 y_2,如图 3-29a)所示。

a)圆孔空心板截面　　　b)等效矩形空心板截面　　　c)等效工字形截面

图 3-29　空心截面换算成等效工字形截面

将空心板截面换算成等效的工字形截面的方法,是先根据面积、惯性矩不变的原则,将空心板的圆孔(直径为 D)换算成 $b_k \times h_k$ 的矩形孔,可按下列各式计算:

按面积相等
$$b_k h_k = \frac{\pi}{4} D^2$$

按惯性矩相等
$$\frac{1}{12} b_k h_k^3 = \frac{\pi}{64} D^4$$

联立求解上述两式,可得到:

$$h_k = \frac{\sqrt{3}}{2} D, \quad b_k = \frac{\sqrt{3}}{6} \pi D$$

然后,在圆孔的形心位置和空心板截面宽度、高度都保持不变的条件下,可一步得到等效工字形截面尺寸:

上翼板厚度
$$h_f' = y_1 - \frac{1}{2} h_k = y_1 - \frac{\sqrt{3}}{4} D$$

下翼板厚度
$$h_f = y_2 - \frac{1}{2} h_k = y_2 - \frac{\sqrt{3}}{4} D$$

腹板厚度
$$b = b_f - 2 b_k = b_f - \frac{\sqrt{3}}{3} \pi D$$

换算工字形截面见图 3-29c)。当空心板截面孔洞为其他形状时,均可按上述原则换算成相应的等效工字形截面。

2)受压翼缘有效宽度 b_f' 的确定

通过试验和分析得知,T 形截面梁承受荷载作用产生弯曲变形时,在翼板宽度方向上纵向压应力的分布是不均匀的。离梁肋愈远,压应力愈小,其分布规律主要取决于截面与跨径(长度)的相对尺寸、翼板厚度、支承条件等。

在设计计算中,为了便于计算,根据等效受力原则,把与梁肋共同工作的翼板宽度限制在一定的范围内,称为受压翼板的有效宽度 b_f'。在 b_f' 宽度范围内的翼板可以认为是全部参与工作,并假定其压应力是均匀分布的,如图 3-30 所示。

《公桥规》规定,T 形截面梁的受压翼板有效宽度 b_f' 考虑下列三中情况后采取最小值:

(1)对于简支梁取计算跨径的 1/3,对于连续梁各中间跨正弯矩区段,取该跨计算跨径的 0.2 倍;边跨正弯矩区段,取该跨计算跨径的 0.27 倍;各跨支点负弯矩区段,则取该支点相邻两跨计算跨径之和的 0.07 倍。

(2)相邻两梁的平均间距 b_f。

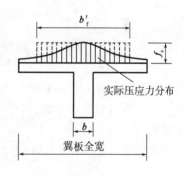

图 3-30 T 形梁受压翼板的正应力分布

(3)$b + 2b_h + 12 h_f'$。当 $\dfrac{h_h}{b_h} < \dfrac{1}{3}$ 时,取 $(b + 6 h_h + 12 h_f')$。此处,b、b_h 和 h_f' 见图 3-31,h_h 为承托根部厚度。

如图 3-31 所示承托,又称梗腋,它是为增强翼板与梁肋之间联系的构造措施,并可增强翼板根部的抗剪能力。

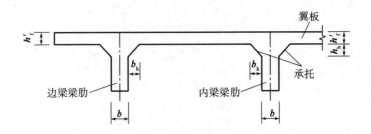

图 3-31　T形截面受压翼板有效宽度计算示意图

3) 基本公式及适用条件

T形截面的受力破坏形式与矩形截面类似,计算时采用与矩形截面相同的假定和方法。

T形截面按受压区高度的不同可分为两类:受压区高度在翼板厚度内,即 $x \leqslant h'_f$ [图3-32a)],为第一类T形截面;受压区已进入梁肋,即 $x > h'_f$ [图3-32b)],为第二类T形截面。

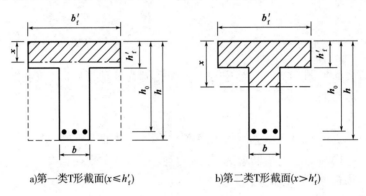

图 3-32　两类 T 形截面

(1) 第一类 T 形截面梁正截面抗弯承载力计算基本公式

第一类 T 形截面,中和轴在受压翼板内,受压区高度 $x \leqslant h'_f$。此时,截面虽为 T 形,但实际是受压区形状为宽 b'_f 和高 h 的矩形,而受拉区截面形状与截面抗弯承载力无关,故按以 b'_f 为宽度的矩形截面进行抗弯承载力计算。计算时只需将单筋矩形截面公式中梁宽 b 以翼板有效宽度 b'_f 置换即可。

由截面平衡条件(图3-33)可得到基本计算公式:

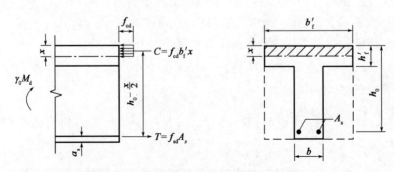

图 3-33　第一类 T 形截面抗弯承载力计算图式

$$f_{cd}b'_f x = f_{sd}A_s \qquad (3\text{-}27)$$

$$\gamma_0 M_d \leqslant M_u = f_{cd}b'_f x\left(h_0 - \frac{x}{2}\right) \qquad (3\text{-}28a)$$

或

$$\gamma_0 M_d \leqslant M_u = f_{sd}A_s\left(h_0 - \frac{x}{2}\right) \qquad (3\text{-}28b)$$

基本公式适用条件为：

a. $x \leqslant \xi_b h_0$（控制超筋梁）

由于第一类 T 形截面的 x 较小，因而 ξ 值也小，所以很容易满足这个条件。

b. $\rho > \rho_{min}$（控制少筋梁）

这里的 $\rho = \dfrac{A_s}{bh_0}$，b 为 T 形截面的梁肋宽度。最小配筋率 ρ_{min} 是根据开裂后梁截面的抗弯强度等于同样截面的素混凝土梁抗弯承载力这一条件得出的，而素混凝土梁的抗弯承载力主要取决于受拉区混凝土的强度等级，素混凝土 T 形截面梁的抗弯承载力与高度为 h、宽度为 b 的矩形截面素混凝土梁的抗弯承载力相接近，因此，在验算 T 形截面的 ρ_{min} 值时，近似地取梁肋宽 b 来计算。

(2) 第二类 T 形截面梁正截面抗弯承载力计算基本公式

中和轴在梁肋部位，受压区高度 $x > h'_f$，受压区为 T 形，如图 3-34 所示。

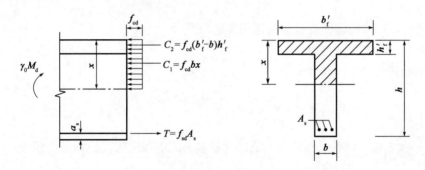

图 3-34 第二类 T 形截面抗弯承载力计算图式

由于受压区为 T 形，故一般将受压区混凝土压应力的合力分为两部分求得：一部分是肋宽为 b、高度为 x 的矩形，其合力 $C_1 = f_{cd}bx$；另一部分是宽度为 $b'_f - b$、高度为 h'_f 的矩形，其合力 $C_2 = f_{cd}h'_f(b'_f - b)$。由图 3-34 的截面平衡条件可得到第二类 T 形截面的基本计算公式为：

$$C_1 + C_2 = T \qquad f_{cd}bx + f_{cd}h'_f(b'_f - b) = f_{sd}A_s \qquad (3\text{-}29)$$

$$\sum M = 0 \qquad \gamma_0 M_d \leqslant M_u = f_{cd}bx\left(h_0 - \frac{x}{2}\right) + f_{cd}(b'_f - b)h'_f\left(h_0 - \frac{h'_f}{2}\right) \qquad (3\text{-}30)$$

基本公式适用条件为：a. $x \leqslant \xi_b h_0$；b. $\rho \geqslant \rho_{min}$。

第二类 T 形截面的配筋率较高，一般情况下均能满足 $\rho \geqslant \rho_{min}$ 的要求。

4) 计算方法

(1) 截面设计

已知截面尺寸、材料强度级别、弯矩计算值 $\gamma_0 M_d$，求受拉钢筋截面面积 A_s。

① 假设 a_s，得到有效高度 $h_0 = h - a_s$。

对于空心板等截面，往往采用绑扎钢筋骨架，因此可根据等效工字形截面下翼板厚度 h_f，在实际截面中布置一层或布置两层钢筋来假设 a_s 值。对于预制或现浇 T 形梁，往往采用多层焊接钢筋骨架，由于多层钢筋的叠高一般不超过 $(0.15 \sim 0.2)h$，故可假设 $a_s = 30\text{mm} + (0.07 \sim 0.1)h$。

②确定翼缘的有效宽度 b'_f。

根据《公桥规》规定，考虑三种情况，取最小值。

③判定 T 形截面类型。

由基本公式可见，当中和轴恰好位于受压翼板与梁肋交界处，即 $x = h'_f$ 为两类 T 形截面的界限情况。显然，若满足：

$$\gamma_0 M_d \leq f_{cd} b'_f h'_f \left(h_0 - \frac{h'_f}{2} \right) \tag{3-31}$$

即弯矩计算值 $\gamma_0 M_d$ 小于或等于由全部翼板高度 h'_f 作为受压混凝土产生的力矩，则 $x \leq h'_f$，属于第一类 T 形截面，否则属于第二类 T 形截面。

④当为第一类 T 形截面时，

按 $b'_f \times h$ 的矩形截面计算，由基本公式求得受压区高度 x：

$$x = h_0 \xi = h_0 \left(1 - \sqrt{1 - \frac{2\gamma_0 M_d}{f_{cd} b'_f h_0^2}} \right)$$

再由式(3-27)求所需的受拉钢筋面积 A_s。

⑤当为第二类 T 形截面时，

由基本公式求受压区高度 x：

$$x = \xi h_0 = h_0 \left[1 - \sqrt{1 - \frac{2\gamma_0 M_d}{f_{cd} b h_0^2} + \frac{2(b'_f - b) h'_f}{b h_0^2} \left(h_0 - \frac{h'_f}{2} \right)} \right]$$

并满足 $h'_f < x \leq \xi_b h_0$。将各已知值及 x 值代入基本公式求得所需受拉钢筋面积 A_s。

不满足 $x \leq \xi_b h_0$，则修改设计。

⑥选择钢筋直径、根数并布置，校核 a_s。

(2) 截面复核

已知受拉钢筋截面面积及钢筋布置、截面尺寸和材料强度级别，要求复核截面的抗弯承载力。

计算步骤：

①检查钢筋布置是否符合规范要求。

②确定翼缘的有效宽度 b'_f。

③判定 T 形截面的类型。

这时，若满足：

$$f_{cd} b'_f h'_f \geq f_{sd} A_s \tag{3-32}$$

即钢筋所承受的拉力 $f_{sd} A_s$ 小于或等于由全部受压翼板高度 h'_f 内混凝土提供的压应力合力 $f_{cd} b'_f h'_f$，则 $x \leq h'_f$，属于第一类 T 形截面，否则属于第二类 T 形截面。

④当为第一类 T 形截面时，

由基本公式求得受压区高度 x，满足 $x \leq h'_f$。将各已知值及 x 值代入基本公式，求得正截面抗弯承载力,满足 $M_u \geq \gamma_0 M_d$,表示强度符合要求。

⑤当为第二类 T 形截面时,

由基本公式(3-29)求受压区高度 x,满足 $h'_f < x \leq \xi_b h_0$ 时。将各已知值及 x 值代入式(3-30)即可求得正截面抗弯承载力,满足 $M_u \geq \gamma_0 M_d$,表示强度达到要求。

例:预制钢筋混凝土简支 T 梁截面高度 $h = 1.30\text{m}$,翼板有效宽度 $b'_f = 1.60\text{m}$(预制宽度 1.58 m),C30 混凝土,HRB400 级钢筋。I 类环境条件,安全等级为二级。跨中截面弯矩组合设计值 $M_d = 3\,000 \text{kN} \cdot \text{m}$。试进行钢筋(焊接钢筋骨架)计算及截面复核。

解:由附表查得 $f_{cd} = 13.8\text{MPa}$,$f_{td} = 1.39\text{MPa}$,$f_{sd} = 330\text{MPa}$,$\xi_b = 0.53$,$\gamma_0 = 1.0$,弯矩计算值 $M = \gamma_0 M_d = 3\,000 \text{kN} \cdot \text{m}$。

为了便于计算,将图 3-35a)的实际 T 形截面换成图 3-35b)所示的计算截面,$h'_f = \frac{100 + 140}{2} = 120(\text{mm})$,其余尺寸不变。

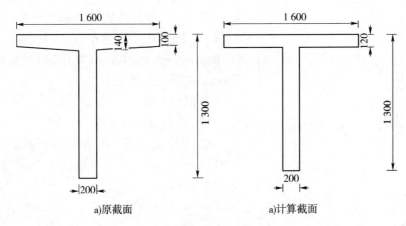

图 3-35 计算示意图(尺寸单位:mm)

1)截面设计

(1)计算截面有效高度 h_0

因采用焊接钢筋骨架,设 $a_s = 125\text{mm}$,则截面有效高度 $h_0 = 1\,300 - 125 = 1\,175(\text{mm})$。

(2)因翼缘有效计算宽度 b'_f 已给出,不再计算。

(3)判定 T 形截面类型

$$f_{cd} b'_f h'_f \left(h_0 - \frac{h'_f}{2}\right) = 13.8 \times 1\,600 \times 120 \times (1\,175 - 120/2) = 2\,954.3 \times 10^6 (\text{N} \cdot \text{mm})$$

$$= 2\,954(\text{kN} \cdot \text{m}) < \gamma_0 M_d (= 3\,000\text{kN} \cdot \text{m})$$

故属于第二类 T 形截面。

(4)求受压区高度

建立基本公式:

$$\begin{cases} f_{cd} b x + f_{cd} h'_f (b'_f - b) = f_{sd} A_s \\ \gamma_0 M_d = f_{cd} b x \left(h_0 - \frac{x}{2}\right) + f_{cd} h'_f (b'_f - b)\left(h_0 - \frac{h'_f}{2}\right) \end{cases}$$

求解可得到:

$$3\,000 \times 10^6 - 13.8 \times 120 \times (1\,600 - 200) \times (1\,175 - 120/2) = 13.8 \times 200 \times (1\,175 - x/2)$$

整理后可得 $x^2 - 2\,350x + 300\,710 = 0$。

$x = 136\,\text{mm} > h'_f (= 120\,\text{mm})$，为第二类 T 形截面。

(5) 求受拉钢筋面积 A_s

将各已知值代入基本公式，可得到：

$$A_s = \frac{f_{cd}bx + f_{cd}h'_f(b'_f - b)}{f_{sd}}$$

$$= \frac{13.8 \times [200 \times 136 + 120 \times (1\,600 - 200)]}{330}$$

$$= 8\,163\,(\text{mm}^2)$$

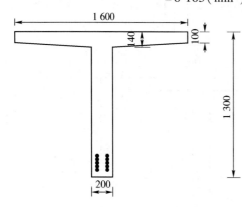

图 3-36　钢筋布置图(尺寸单位:mm)

现选择钢筋为 10 ⌀ 32，截面积 $A_s = 8\,042\,\text{mm}^2$。钢筋叠高层数为 5 层，布置如图 3-36 所示。

混凝土保护层厚度取 20mm（附表 1-8 中规定的 20mm），钢筋间横向净距 $S_n = 200 - 2 \times 35 - 2 \times 35.8 = 58\,(\text{mm}) > 40\,\text{mm}$ 及 $1.25d = 1.25 \times 32 = 40\,(\text{mm})$，故满足构造要求。

2) 截面复核

(1) 计算实际的 a_s 值

已设计的受拉钢筋中，10 ⌀ 32 的面积为 $8\,042\,\text{mm}^2$，$f_{sd} = 330\,\text{MPa}$。由图 3-36 钢筋布置图可求得 a_s，即：

$$a_s = 20 + 2.5 \times 35.8 = 110\,(\text{mm})$$

则实际有效高度：$h_0 = 1\,300 - 110 = 1\,190\,(\text{mm})$。

(2) 判定 T 形截面类型

由基本公式计算：

$$f_{cd}b'_f h'_f = 13.8 \times 1\,600 \times 120 = 2.649 \times 10^6\,(\text{N}\cdot\text{mm}) = 2\,649\,(\text{kN}\cdot\text{m})$$

$$f_{sd}A_s = 8\,042 \times 330 = 2.654 \times 10^6\,(\text{N}\cdot\text{mm}) = 2\,654\,(\text{kN}\cdot\text{m})$$

由于 $f_{cd}b'_f h'_f < f_{sd}A_s$，故为第二类 T 形截面。

(3) 求受压区高度 x

由基本公式求得 x，即：

$$x = \frac{f_{sd}A_s - f_{cd}h'_f(b'_f - b)}{f_{cd}b} = \frac{330 \times 8\,042 - 13.8 \times 120 \times (1\,600 - 200)}{13.8 \times 200}$$

$$= 121.5\,(\text{mm}) > h'_f (= 120\,\text{mm})$$

(4) 正截面抗弯承载力验算

由基本公式求得正截面抗弯承载力 M_u 为：

$$M_u = f_{cd}bx\left(h_0 - \frac{x}{2}\right) + f_{cd}h'_f(b_f - b)\left(h_0 - \frac{h'_f}{2}\right)$$

$$= 13.8 \times 200 \times 121.5 \times (1\,190 - 139/2) +$$

$$\quad 13.8 \times 120 \times (1\,600 - 200) \times (1\,190 - 120/2)$$

$$= 2\,998.7 \times 10^6 (\mathrm{N \cdot mm}) = 2\,998.7(\mathrm{kN \cdot m}) \approx \gamma_0 M_\mathrm{d}(=3\,000\mathrm{kN \cdot m})$$

又 $\rho = \dfrac{A_\mathrm{s}}{bh_0} = 8\,042/(200 \times 1\,190) = 3.37\% > \rho_{\min} = 0.2\%$，故截面复核满足要求。

思考练习题

1. 什么叫钢筋混凝土少筋梁、适筋梁和超筋梁？各自有什么样的破坏形态？在实际工程中为什么应避免少筋梁和超筋梁？

2. 钢筋混凝土适筋梁正截面受力全过程可划分为几个阶段？各阶段受力的主要特点是什么？

3. 为什么钢筋要有足够的混凝土保护层厚度？钢筋的最小混凝土保护层厚度的选择应考虑哪些因素？

4. 什么叫受弯构件纵向受拉钢筋的配筋率？配筋率的表达式中，h_0 的含义是什么？

5. 钢筋混凝土受弯构件正截面承载力计算有哪些基本假定？

6. 什么叫作钢筋混凝土受弯构件的截面相对受压区高度 ξ 和相对界限受压区高度 ξ_b？ξ_b 在正截面承载力计算中起什么作用？ξ_b 的取值与哪些因素有关？

7. 什么叫受弯构件纵向受拉钢筋配筋率？在配筋率的表达式中，h_0 的含义是什么？

8. 钢筋混凝土双筋截面梁正截面承载力计算公式的适用条件是什么？试说明原因。

9. 什么叫作 T 形梁受压翼板的有效宽度？《公桥规》对 T 形梁的受压翼板有效宽度取值有何规定？

10. 在截面设计时，如何判别两类 T 形截面？在截面复核时又如何判别？

11. 某桥梁工程中，有一矩形截面钢筋混凝土梁，弯矩组合设计值 $M_\mathrm{d} = 122\mathrm{kN \cdot m}$，结构重要性系数 $\gamma_0 = 1.0$；拟采用 C25 混凝土，HRB400 钢筋。设计人员有三种意见，其截面 $b \times h$ 分别为 180mm × 400mm、400mm × 1 200mm、200mm × 500mm，按单筋矩形截面计算，试判断哪种截面合适，说明原因，并计算所需受拉钢筋面积。

12. 求单筋矩形截面在下列情况的 x、ξ、A_s，并加以分析，绘出钢筋布置图。承受弯矩组合设计值 $M_\mathrm{d} = 128\mathrm{kN \cdot m}$，结构重要性系数 $\gamma_0 = 1.0$。

习题 12 表

情况	截面尺寸 $b \times h$ (mm × mm)	混凝土强度等级	钢筋种类	x (mm)	$\xi = \dfrac{x}{h_0}$	计算钢筋面积 A_s (mm²)	实际配筋根数、直径、面积 (mm²)
1	200 × 500	C20	HPB300				
2	200 × 500	C30	HPB300				
3	200 × 500	C20	HRB400				
4	300 × 500	C20	HPB300				
5	200 × 600	C20	HPB300				

13. 有一矩形截面简支板桥,板厚 $h=350\text{mm}$,每米板宽承受跨中结构自重弯矩 $M_{Gk}=43.7\text{kN}\cdot\text{m}$,汽车荷载弯矩 $M_{Qk}=111.5\text{kN}\cdot\text{m}$,结构重要性系数 $\gamma_0=1.0$;拟采用 C20 混凝土,HRB335 钢筋。试计算所需受拉钢筋面积 A_s,并绘钢筋布置图。

14. 某楼面简支梁,承受均布荷载 42.8 kN/m(包括自重),计算跨长 5m,计算下表中 5 种情况的用钢量。根据上述计算结果,从下表中可得出什么结论?该结论在设计上有何实际意义?

习题 14 表

情况	梁高 (mm)	梁宽 (mm)	混凝土强度等级	钢 筋 种 类	钢筋面积 A_s (mm²)
1	500	200	C25	HPB235	
2	500	200	C35	HPB235	
3	500	200	C25	HRB335	
4	500	300	C25	HPB235	
5	600	200	C25	HPB235	

15. 双筋矩形截面 $b\times h=200\text{mm}\times450\text{mm}$,承受弯矩 $M_d=168\text{kN}\cdot\text{m}$,结构重要性系数 $\gamma_0=1.0$;采用 C20 混凝土,HRB335 钢筋。

(1)求 A_s、A_s'。

(2)若受压钢筋 $A_s'=763\text{mm}^2$(3Φ18),求受拉钢筋 A_s。

(3)若受压钢筋 $A_s'=982\text{mm}^2$(2Φ20),求受拉钢筋 A_s。

16. 双筋矩形截面 $b\times h=200\text{mm}\times500\text{mm}$,C25 混凝土,HPB300 钢筋,受拉钢筋为 3Φ20($A_s=942\text{mm}^2$),受压钢筋为 3Φ16($A_s'=603\text{mm}^2$);承受弯矩设计值 $M_d=85\text{kN}\cdot\text{m}$,结构重要性系数 $\gamma_0=1.0$。试验算该截面是否安全。

17. 钢筋混凝土矩形截面梁的截面尺寸 $b\times h=200\text{mm}\times500\text{mm}$,采用 C25 混凝土和 HRB400 级钢筋;I 类环境条件,安全等级为一级;最大弯矩组合设计值 $M_d=190\text{kN}\cdot\text{m}$,试按双筋截面求所需的钢筋截面面积,并进行截面布置。

18. 已知简支 T 梁截面尺寸如习题 18 图所示,主梁计算跨径 $l_0=12\,600\text{mm}$,相邻两主梁中心距为 2 100mm,C30 混凝土,HRB335 钢筋,该截面弯矩为:永久作用 $M_{Gk}=492\text{kN}\cdot\text{m}$,汽车荷载 $M_{Qk}=401\text{kN}\cdot\text{m}$,结构重要性系数 $\gamma_0=1.1$。求受拉钢筋面积并复核截面承载力,绘出钢筋布置图。

19. 已知计算跨径 $l_0=15\,000\text{mm}$ 的 Π 形截面梁,C30 混凝土,HRB400 钢筋,梁截面尺寸如习题 19 图所示。该截面弯矩为:结构自重 $M_{Gk}=542\text{kN}\cdot\text{m}$,汽车荷载 $M_{Qk}=514\text{kN}\cdot\text{m}$,结构重要性系数 $\gamma_0=1.0$。求受拉钢筋面积并复核截面承载力,绘出钢筋布置图。

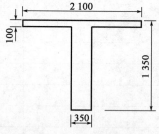

习题 18 图 (尺寸单位:mm)

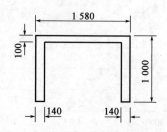

习题 19 图 (尺寸单位:mm)

20. 某钢筋混凝土矩形梁跨度及受到的荷载如习题 20 图所示,$b \times h = 250\text{mm} \times 600\text{mm}$,混凝土采用 C35,纵筋用 HRB400 级($f_y = 360\text{MPa}$),箍筋用 HPB235 级($f_y = 210\text{MPa}$),全梁长配置箍筋 $\phi 6@200$,试按承载力计算该梁所需正截面配筋,并进行承载力校核。

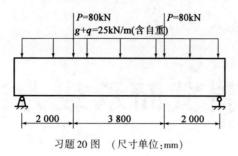

习题 20 图　(尺寸单位:mm)

第4章
受弯构件斜截面承载力计算

4.1 受弯构件斜截面的受力特点和破坏形态

受弯构件在荷载作用下,截面上除作用有弯矩 M 外,往往同时作用有剪力 V。在弯矩 M 和剪力 V 共同作用的区段内,构件受力状况及破坏形态怎样?如何设计相应的钢筋,这是本章要解决的问题。

研究表明:混凝土、箍筋、斜筋、纵筋都能够提供抗剪能力,本节首先从分析无箍筋和斜筋(简称无腹筋梁)梁的抗剪能力入手,通过对比逐步获得混凝土、箍筋、斜筋的抗剪力。

1)无腹筋简支梁斜裂缝出现前后的受力状态

图 4-1 为一无腹筋简支梁,作用有两个对称的集中荷载。CD 段仅有弯矩,称为纯弯段;AC 段和 DB 段内的截面上既有弯矩 M 又有剪力 V,称为剪弯段。下面分析梁受力后的截面应力状态。

(1)斜裂缝产生前的应力状态:因为无裂缝出现,采用材料力学公式计算应力。

在弯剪区段,由于 M 和 V 的存在产生正应力和剪应力:

$$\left. \begin{array}{l} \sigma = \dfrac{M y_0}{I_0} \\[4pt] \tau = \dfrac{V s_0}{b I_0} \end{array} \right\} \tag{4-1}$$

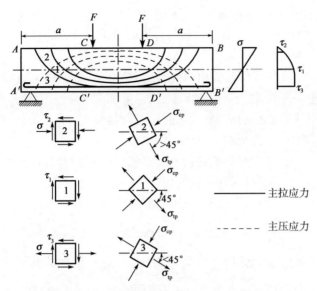

图 4-1 无腹筋梁的主应力分布

将弯剪区段的典型微元进行应力分析,可以求得主拉应力和主压应力:

$$\left.\begin{array}{l}\sigma_{tp} = \dfrac{\sigma}{2} + \sqrt{\dfrac{\sigma^2}{4} + \tau^2} \\ \sigma_{cp} = \dfrac{\sigma}{2} - \sqrt{\dfrac{\sigma^2}{4} + \tau^2}\end{array}\right\} \qquad (4\text{-}2)$$

进一步求得主应力方向和主应力轨迹线,梁内剪弯区段的主应力轨迹线如图 4-1 所示。

(2)斜截面形成:在弯矩 M 和剪力 V 的共同作用下,截面上分别产生正应力和剪应力,形成主拉应力和主压应力,当主拉应力 $\sigma_{tp} > f_t$ 时,即可能产生斜裂缝。

(3)斜截面破坏原因:斜裂缝形成后,截面的应力状态发生了质变,梁内发生了应力重分布。斜截面上的受力状态如图 4-2 所示,斜截面上的抵抗力有:

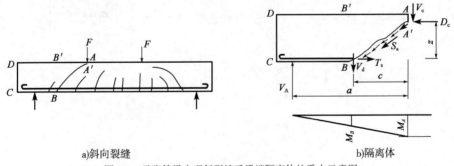

a)斜向裂缝　　b)隔离体

图 4-2 无腹筋梁出现斜裂缝后梁端隔离体的受力示意图

①纵向钢筋的拉力 T_s。

②斜截面上端混凝土剪压面上的压力 D_c 和剪力 V_c。

③斜裂缝两侧混凝土发生相对位移和错动时产生的摩擦力以及集料凹凸不平时相互间的咬合力 S_a。

④由于斜裂缝两边有相对的上下错动,纵向钢筋会承担一定的剪力,即纵筋的销栓力 V_d。

(4) 在斜裂缝出现后,梁内的应力状态有如下变化:

①对于无腹筋梁,斜裂缝出现后,剪力由顶部的剪压面承担,剪压面积减小,剪应力和压应力明显增大。

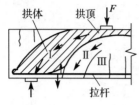

图 4-3 无腹梁沿斜截面受力的拱机理示意图

②斜裂缝出现前,任意截面纵筋的拉应力由该截面处的弯矩决定。斜裂缝出现后,纵筋需要承担斜裂缝顶端截面处弯矩所产生的应力,纵筋的拉应力会增大。

2) 无腹筋梁受力的拱机理

无腹筋梁的受力状态类似一个设拉杆的拱结构:块体 I 相当于受压的拱,纵筋相当于拉杆,如图 4-3 所示。

3) 无腹筋简支梁斜截面的破坏形态

(1) 剪跨比的定义:用 $m = \dfrac{M}{Vh_0}$ 来表示,此处 M 和 V 分别为剪弯区段中某个竖直截面的弯矩和剪力,h_0 为截面有效高度。M 也称为"广义剪跨比"。对于集中荷载作用下的简支梁:$m = \dfrac{M}{Vh_0} = \dfrac{a}{h_0}$,其中 a 为集中力作用点至离简支梁最近的支座之间的距离,称为"剪跨",称 $m = \dfrac{a}{h_0}$ 为"狭义剪跨比"。

(2) 试验研究表明,无腹筋简支梁斜截面的破坏形态主要有三种:

①斜拉破坏

产生条件:一般发生在剪跨比较大($m > 3$)的梁段。

破坏特征:当斜裂缝一出现,很快便形成一条主要斜裂缝(临界斜裂缝),并迅速延伸至荷载作用点,使梁斜向被拉断成两部分,见图 4-4a)。破坏面较整齐,无压碎痕迹,同时,伴随纵向钢筋往往产生水平撕裂裂缝。破坏发生突然,变形很小,无明显预兆。

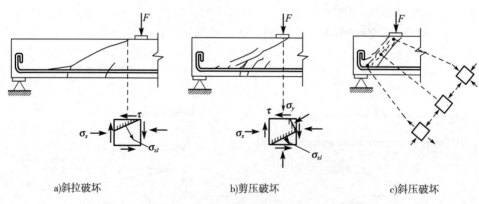

a) 斜拉破坏   b) 剪压破坏   c) 斜压破坏

图 4-4 斜截面破坏形态

抗剪能力:斜拉破坏主要是由于主拉应力超过混凝土的抗拉强度,因此梁的受剪承载力很低,破坏荷载等于或略高于主要斜裂缝出现时的荷载。

②剪压破坏

产生条件:一般发生在剪跨比适中($1 \leq m \leq 3$)的梁段。

破坏特征:梁在剪弯区段内出现斜裂缝,随着荷载的增大,陆续出现几条斜裂缝,其中一条发展成为临界斜裂缝,见图 4-4b)。临界斜裂缝出现后,梁还能继续承受荷载,直到斜裂缝顶端的混凝土在正应力和剪应力的共同作用下被压碎而破坏,这种破坏称为剪压破坏。

抗剪能力:主要与截面大小、混凝土强度、箍筋强度和数量有关。

③斜压破坏

产生条件:在剪跨比较小($m<1$)的梁段。

破坏特征:在加载点和支座之间出现一条斜裂缝,然后出现若干条大体平行的斜裂缝。梁腹被分割成若干个倾斜的小柱体,见图4-4c)。随着荷载增大,梁腹发生类似混凝土棱柱体被压坏的情况,即破坏时斜裂缝多而密,但没有主裂缝,故称为斜压破坏。

抗剪能力:主要取决于构件截面尺寸和混凝土抗压强度,抗剪承载力较高。

三种破坏形态下梁的抗剪承载力大小顺序为:斜压 > 剪压 > 斜拉,均属于脆性破坏。

4) 有腹筋简支梁的受力状态

无腹筋梁斜截面抗剪承载力很低,不便于工程应用。故《公桥规》规定,一般的梁内都需设置腹筋以提高梁的斜截面抗剪承载力。

(1) 腹筋的作用

斜裂缝出现前,腹筋承受的应力很小,腹筋对阻止和推迟斜裂缝出现的作用也很小。在斜裂缝出现后,腹筋将大大提高梁斜截面的抗剪承载力,特别是箍筋的作用明显,主要表现在:

① 与斜裂缝相交的箍筋直接参与抗剪,承担部分剪力。

② 箍筋抑制斜裂缝开展宽度,使斜裂缝顶端剪压面的面积不至于迅速减小,提高了混凝土的抗剪能力。

③ 箍筋限制了纵向钢筋的竖向位移,阻止混凝土沿纵向钢筋的撕裂,保证了纵向钢筋的销栓作用。

可见,箍筋对提高斜截面受剪承载力的作用是多方面的和综合性的。

(2) 破坏形态

有腹筋梁的破坏形态与无腹筋梁相同,也是斜拉破坏、剪压破坏、斜压破坏三种破坏形态。

4.2 影响受弯构件斜截面抗剪能力的主要因素

试验研究表明,影响梁斜截面抗剪承载力的主要因素是剪跨比、混凝土强度、纵向受拉钢筋配筋率和箍筋数量及强度。

1) 剪跨比

试验表明:剪跨比(m)越大,抗剪强度越低,当 $m>3$ 时,斜截面抗剪承载力趋于稳定。剪跨比 m 与梁的抗剪承载力的关系见图4-5。

2) 混凝土抗压强度

梁的抗剪力随混凝土抗压强度的提高而提高,其影响大致按线性规律变化。但是,由于在不同剪跨比下梁的破坏形态不同,所以,强度影响的程度亦不相同。斜拉破坏、受混凝土抗拉强度影响,斜压、剪压破坏受混凝土抗压强度影响。

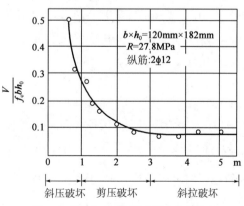

图4-5 剪跨比 m 对梁抗剪承载力的影响

3）纵向钢筋配筋率

试验表明，梁的抗剪力随纵向钢筋配筋率 ρ 的提高而增大，大致呈线性关系。

4）配箍率和箍筋强度

试验表明：当其他条件相同时，梁的抗剪承载力与配箍率和箍筋强度的乘积大致呈线性关系。

配箍率：
$$\rho_{sv} = \frac{A_{sv}}{bs_v}$$

式中，A_{sv} 为斜截面内配置在沿梁长度方向一个箍筋间距 s_v 范围内的箍筋各肢总截面积；b 为截面宽度，对 T 形或 I 形截面梁，取 b 为肋宽；s_v 为沿梁长度方向箍筋的间距。

4.3 受弯构件斜截面抗剪承载力的计算

钢筋混凝土梁沿斜截面的主要破坏形态有斜压破坏、斜拉破坏和剪压破坏等。在设计时，对于斜压和斜拉破坏，一般是采用截面限制条件和一定的构造措施予以避免。对于常见的剪压破坏形态，梁的斜截面抗剪力变化幅度较大，必须进行斜截面抗剪承载力的计算。《公桥规》的基本公式就是针对这种破坏形态的受力特征而建立的。

1）斜截面抗剪承载力计算的基本公式及适用条件

（1）计算依据：剪压破坏。

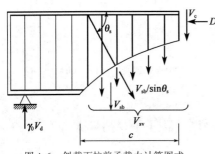

图 4-6 斜截面抗剪承载力计算图式

（2）计算图式（图 4-6）：以剪压破坏时的破坏面为脱离体，梁的抗剪力主要考虑混凝土的抗力 V_c、箍筋的抗力 V_{sv} 和斜筋的抗力 V_{sb}，不计咬合力和销栓力。

（3）基本公式：《公桥规》根据国内外有关的试验资料，采用了下列公式：

$$V_u = V_c + V_{sv} + V_{sb}$$

也可写成
$$V_u = V_{cs} + V_{sb}$$

$$\gamma_0 V_d \leqslant V_u = \alpha_1 \alpha_2 \alpha_3 (0.45 \times 10^{-3}) bh_0 \sqrt{(2+0.6P)\sqrt{f_{cu,k}} \rho_{sv} f_{sv}} + (0.75 \times 10^{-3}) f_{sd} \sum_{i=1}^{n} A_{sbi} \sin\theta_s \qquad (4-3)$$

式中，V_d 为斜截面受压端正截面上由作用（或荷载）效应所产生的最大剪力组合设计值（kN）；γ_0 为桥梁结构的重要性系数；α_1 为异号弯矩影响系数，计算简支梁和连续梁近边支点梁段的抗剪承载力时，$\alpha_1 = 1.0$，计算连续梁和悬臂梁近中间支点梁段的抗剪承载力时，$\alpha_1 = 0.9$；α_2 为预应力提高系数，对钢筋混凝土受弯构件，$\alpha_2 = 1$，对预应力钢筋混凝土受弯构件，$\alpha_2 = 1.25$；α_3 为受压翼缘的影响系数，对矩形截面，取 $\theta_3 = 1.0$，对具有受压翼缘的截面，取 $\alpha_3 = 1.1$；b 为斜截面受压区顶端截面处矩形截面宽度（mm），或 T 形和 I 形截面腹板宽度（mm）；h_0 为斜截面受压端正截面上的有效高度，自纵向受拉钢筋合力点到受压边缘的距离（mm）；P 为斜截面内纵向受拉钢筋的配筋率，$P = 100\rho$，$\rho = \dfrac{A_p + A_s}{bh_0}$，当 $P > 2.5$ 时，取 $P = 2.5$；$f_{cu,k}$ 为混凝土立方体抗压强度标准值（MPa）；ρ_{sv} 为箍筋配筋率；f_{sv} 为箍筋抗拉强度设计值（MPa）；f_{sd} 为弯起钢筋的抗拉强度设计值(MPa);A_{sbi}为斜截面内在某一个弯起钢筋平面内的弯起钢筋截面面积(mm^2);θ_s为弯起钢筋的切线与构件水平纵向轴线的夹角。

(4)适用条件:

①上限值——控制截面最小尺寸

为了防止因梁的尺寸太小主压应力过大而出现斜压破坏,截面尺寸应满足:

$$\gamma_0 V_d \leq 0.51 \times 10^{-3} \sqrt{f_{cu,k}} b h_0 \quad (kN) \tag{4-4}$$

式中,V_d为验算截面处由作用(或荷载)产生的剪力组合设计值(kN);$f_{cu,k}$为边长150mm的混凝土立方体抗压强度标准值(MPa);b为相应于剪力组合设计值处矩形截面的宽度(mm),或T形和I形截面腹板宽度(mm);h_0为相应于剪力组合设计值处截面的有效高度(mm)。

②下限值——按构造要求配置箍筋,防止出现斜拉破坏

《公桥规》规定,若符合式(4-5a),说明主拉应力较小,截面就不会出现斜拉破坏,梁内仅按构造要求配置箍筋:

$$\gamma_0 V_d \leq 0.5 \times 10^{-3} \alpha_2 f_{td} b h_0 \quad (kN) \tag{4-5a}$$

式中,f_{td}为混凝土轴心抗拉强度设计值(MPa);对钢筋混凝土结构,$\alpha_2 = 1.0$。

对于板,可采用式(4-5b)进行计算:

$$\gamma_0 V_d \leq 1.25 \times 0.5 \times 10^{-3} \alpha_2 f_{td} b h_0 = 0.625 \times 10^{-3} \alpha_2 f_{td} b h_0 \quad (kN) \tag{4-5b}$$

2)计算方法

梁的抗剪承载力计算包括抗剪钢筋的设计和斜截面抗剪强度复核,下面以等高度简支梁为例来介绍设计计算方法。

(1)等高度简支梁腹筋的初步设计

已知:梁的跨径,截面尺寸,钢筋数量及强度,剪力效应,即 L、b、h、f_{cd}、f_{sd}、f_{sv}、A_s、$\gamma_0 V_d$。

求:腹筋的数量。

设计步骤:

①验算截面尺寸是否满足要求

$$\gamma_0 V_d \leq 0.51 \times 10^{-3} \sqrt{f_{cu,k}} b h_0$$

否则加大截面尺寸或提高混凝土强度等级。

②确定是否需按计算配箍筋

当 $\gamma_0 V_d \leq 0.5 \times 10^{-3} f_{td} b h_0$ 时,仅需按构造要求配置箍筋。

③计算剪力值的确定与分配

对剪力包络图进行剪力分配。《公桥规》规定:最大剪力取用距支座中心 $h/2$(梁高一半)处截面的数值(记为 V'),其中混凝土和箍筋共同承担不少于60% V',弯起钢筋承担不超过40% V'。

④箍筋设计

假设箍筋直径和肢数,箍筋间距为:

$$s_v = \frac{\alpha_1^2 \alpha_3^2 0.56 \times 10^{-6} (2 + 0.6 p) \sqrt{f_{cu,k}} f_{sv} A_{sv} b h_0^2}{(V')^2} \tag{4-6}$$

s_v 取整并同时满足构造要求。

⑤斜筋设计

斜筋承担剪力如图 4-7 所示。

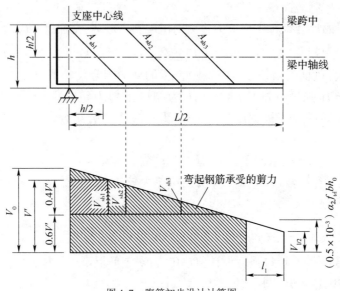

图 4-7 腹筋初步设计计算图

a. 斜筋数量的计算公式：

$$A_{sbi} = \frac{V_{sbi}}{0.75 \times 10^{-3} f_{sd} \sin\theta_s} = \frac{1\,333.33 V_{sbi}}{f_{sd} \sin\theta_s} \quad (\text{mm}^2)$$

b. 关于 V_{sbi} 的取值方法，《公桥规》规定：

（a）$V_{sb1} = 0.4V'$；

（b）计算以后每排弯起钢筋时，取用前一排弯起钢筋弯起点处由弯起钢筋承担的那部分剪力。

c. 起弯角及弯筋之间的位置关系：

（a）弯起角：一般采用 45°，特殊情况下可取 30°或不大于 60°，圆弧弯折，以钢筋轴线为准，圆弧半径不宜小于 20 倍钢筋直径；

（b）弯筋之间的位置关系：简支梁第一排(对支座而言)弯起钢筋的末端弯折点应位于支座中心截面处，以后各排钢筋的弯折末端点应落在或超过前一排弯起钢筋弯起点截面。

(2)斜截面抗剪强度的复核

对已经设计好的梁验算斜截面的强度，基本思路是，求出斜截面上梁的实际抗剪承载力 V_u，并与设计剪力效应 $\gamma_0 V_d$ 比较，是否安全。由于不同的斜截面剪力效应都不同，因此需要验算多个斜截面。

①计算截面位置的选取：

a. 距支座中心(梁高一半)处的截面(1-1)。

b. 受拉区弯起钢筋弯起点处的截面，以及锚于受拉区的纵向钢筋开始不受力处的截面(2-2,3-3,4-4)。

c. 箍筋数量或间距有改变处的截面,如图 4-8 中截面 5-5。

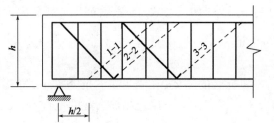 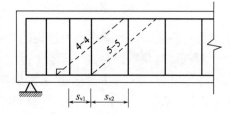

图 4-8 斜截面抗剪承载力的复核截面位置示意图

d. 梁的肋板宽度改变处的截面。

②斜截面投影长度 c:从纵向钢筋与斜裂缝底端相交点至斜裂缝顶端距离的水平投影长度。

$$c = 0.6mh_0 = 0.6\frac{M_d}{V_d} \tag{4-7}$$

式中,m 为斜截面顶端正截面处的广义剪跨比,当 $m>3$ 时,取 $m=3$;V_d 为通过斜截面顶端正截面的剪力组合设计值;M_d 为相应于上述最大剪力组合设计值的弯矩组合设计值。

③斜截面顶端位置的确定方法

a. 通常用简化方法来确定。以验算斜截面底端位置向跨中方向取距离为 h_0 的截面,认为验算斜截面顶端就在此正截面上。

b. 以验算斜截面底端的剪跨比 m 来计算 c 值,$c=0.6mh_0$,由此来推断顶端位置。

由斜截面投影长度确定与斜截面相交的纵向受拉钢筋配筋率 ρ、弯起钢筋数量 A_{sb} 和箍筋的配箍率 ρ_{sv}(h_0、b 取顶端位置处),计算斜截面范围内梁的抗剪承载力。

④将上述各值及与斜裂缝相交的箍筋和弯起钢筋数量代入公式:

$$V_u = \alpha_1\alpha_2\alpha_3 0.45\times10^{-3}bh_0\sqrt{(2+0.6p)\sqrt{f_{cu,k}}\rho_{sv}f_{sv}} + 0.75\times10^{-3}f_{sd}\sum A_{sb}\sin\theta_s \tag{4-8}$$

即可求得斜截面抗剪承载力 V_u,V_u 与设计效应 $\gamma_0 V_d$ 比较进行复核。

4.4 受弯构件的斜截面抗弯承载力

梁内纵筋如果不切断、不弯起,必然满足斜截面的抗弯要求。但是当纵筋被切断或弯起后,斜截面抗弯能力有可能不足。

1)抗弯能力计算

受弯构件斜截面的抗弯强度的计算是要求斜截面内梁的抗弯承载力不小于设计弯矩值,见式(4-9)。计算图式见图 4-9。

$$\gamma_0 M_d \leq M_u = f_{sd}A_s Z_s + \sum f_{sd}A_{sb}Z_{sb} + \sum f_{sv}A_{sv}Z_{sv} \tag{4-9}$$

式中,M_d 为斜截面受压顶端正截面的最大弯矩组合设计值;A_s、A_{sv}、A_{sb} 分别为与斜截面相交的纵向受拉钢筋、箍筋与弯起钢筋的截面面积;Z_s、Z_{sv}、Z_{sb} 分别为钢筋 A_s、A_{sv} 和 A_{sb} 的合力点对混凝土受压区中心点 O 的力臂。

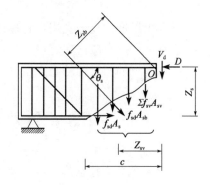

图 4-9 斜截面抗弯承载力计算图式

2) 纵向钢筋弯起位置

经计算发现,纵向钢筋的弯起点在其充分利用点外不小于 $\frac{h_0}{2}$ 处起弯,就能保证斜截面抗弯承载力。斜截面抗弯承载力通过构造措施予以保证。

3) 抵抗弯矩图的绘制

原因分析:纵向钢筋的布置通常是兼顾抗弯和抗剪的需要,在满足抗弯的条件下,将其弯起或截断。通过绘制抵抗弯矩图(材料图)与弯矩包络图比较来确定纵筋弯起的位置。

(1) 几个基本概念

①抵抗弯矩图(又称材料图):各截面实际的纵向受拉钢筋所能承受的弯矩值连线,是一种台阶折线图。

②弯矩包络图:沿梁长度各截面上弯矩组合设计值的连线图,即截面上实际布筋所需的抗弯值的连线图。

例如:简支梁的弯矩包络图近似为一条二次抛物线,若以梁跨中截面处为横坐标原点,则:

$$M_{d,x} = M_{d,\frac{L}{2}}\left(1 - \frac{4x^2}{L^2}\right) \tag{4-10}$$

式中,$M_{d,x}$ 为距跨中截面为 x 处截面上的弯矩组合设计值;$M_{d,\frac{L}{2}}$ 为跨中截面处的弯矩组合设计值;L 为简支梁的计算跨径。

③剪力包络图:沿梁长各截面上剪力组合设计值 V_d 的连线图,其纵坐标表示该截面上作用的最大设计剪力。

对于简支梁的剪力包络图可用直线方程来描述:

$$V_{d,x} = V_{d,\frac{L}{2}} + (V_{d,0} - V_{d,\frac{L}{2}})\frac{2x}{L} \tag{4-11}$$

式中,$V_{d,0}$ 为支座中心处截面的剪力组合设计值;$V_{d,\frac{L}{2}}$ 为简支梁跨中截面的剪力组合设计值;L 为简支梁的计算跨径。

(2) 抵抗弯矩图的绘制步骤

①根据每个截面的抵抗弯矩与截面的钢筋面积成正比的特点,按比例确定起弯点截面的抵抗弯矩值。

$$\frac{M_i}{M_j} \approx \frac{A_{si}}{A_{sj}}$$

②找出纵筋的充分利用点和理论截断点。

③钢筋弯起点应该在超过充分用点 $0.5h_0$ 的地方再弯起。

④弯起钢筋与梁轴的交点是材料图中下一个台阶的起点。

4.5 全梁承载能力校核与构造要求

1) 全梁承载力校核的内容与要求

全梁承载力校核包括正截面抗弯强度、斜截面抗弯强度和抗剪强度校核。

(1) 正截面抗弯强度:保证每个截面 $M_u \geq \gamma_0 M_d$。

(2) 斜截面抗弯强度:由构造要求保证。

(3) 斜截面抗剪强度:由于本章 4.3 节计算了抗剪钢筋,实际布筋如果能满足抗剪计算要求,斜截面强度有保障。纵筋弯起后全梁各个截面能否满足正截面抗弯强度要求,只有绘出材料图和弯矩包络图,进行对比才能知道。

2) 有关的构造要求

钢筋布置除了满足受力要求外,还需要满足下列构造要求:

(1) 纵向钢筋在支座处的锚固

①在钢筋混凝土梁的支点处,应至少有两根并不少于 1/5 的下层受拉主钢筋通过。

②底层两侧之间不向上弯曲的受拉主筋,伸出支点截面以外的长度,对 HPB300 钢筋应不小于 $10d$(并带半圆弯钩),对环氧树脂涂层钢筋应不小于 $12.5d$,d 为受拉主筋直径。

(2) 纵向钢筋在梁跨间的截断与锚固

必须将钢筋从理论切断点外伸一定的长度$(l_a + h_0)$再截断,l_a 称为钢筋的锚固长度(受力钢筋通过混凝土与钢筋黏结将所受的力传递给混凝土所需的长度)。《公桥规》规定了不同受力情况下钢筋的最小锚固长度,见表 4-1。

普通钢筋最小锚固长度　　　　表 4-1

钢筋种类		HPB300				HRB400、HRBF400、RRB400			HRB500		
混凝土强度等级		C25	C30	C35	≥C40	C30	C35	≥C40	C30	C35	≥C40
受压钢筋(直端)		$45d$	$40d$	$38d$	$35d$	$30d$	$28d$	$25d$	$35d$	$33d$	$30d$
受拉钢筋	直端	—	—	—	—	$35d$	$33d$	$30d$	$45d$	$43d$	$40d$
	弯钩端	$40d$	$35d$	$33d$	$30d$	$30d$	$28d$	$25d$	$35d$	$33d$	$30d$

注:1. d 为钢筋公称直径(mm)。

2. 对于受压束筋和等代直径 $d_e \leq 28mm$ 的受拉束筋的锚固长度,应以等代直径按表值确定,束筋的各单根钢筋可在同一锚固终点截断;对于等代直径 $d_e > 28mm$ 的受拉束筋,束筋内各单根钢筋,应自锚固起点开始,以表内规定的单根钢筋的锚固长度的 1.3 倍,呈阶梯形逐根延伸后截断,即自锚固起点开始,第一根延伸 1.3 倍单根钢筋的锚固长度,第二根延伸 2.6 倍单根钢筋的锚固长度,第三根延伸 3.9 倍单根钢筋的锚固长度。

3. 采用环氧树脂涂层钢筋时,受拉钢筋最小锚固长度应增加 25%。

4. 当混凝土在凝固过程中易受扰动时,锚固长度应增加 25%。

5. 当受拉钢筋末端采用弯钩时,锚固长度包括弯钩在内的投影长度。

(3) 钢筋的接头

当梁内钢筋需要接长时,可以采用绑扎搭接接头或焊接接头和机械接头

①绑扎搭接接头:对于受拉钢筋的绑扎接头的搭接长度 l_d,《公桥规》规定见表 4-2;受压区钢筋绑扎接头的搭接长度,应取受拉钢筋绑扎搭接长度的 0.7 倍。在任一绑扎接头中心至搭接长度 1.3 倍的长度区段内,同一根钢筋不得有两个接头;在该区段内有绑扎接头的受力钢筋截面面积占受力钢筋总截面面积的百分数,受拉区不宜超过 25%,受压区不宜超过 50%。当百分数超过上述规定时,应按表 4-2 的规定值乘以下列系数:当受拉钢筋绑扎接头截面面积大于 25%,但不大于 50% 时,乘以 1.4,当大于 50% 时,乘以 1.6;受压钢筋绑扎截面面积大于 50% 时,应乘以 1.4(表 4-2 受压钢筋绑扎接头长度,仍为受拉钢筋绑扎接头长度的 0.7 倍)。

受拉钢筋绑扎接头搭接长度 l_d   表 4-2

钢筋种类	HPB300		HRB400、HRBF400、RRB400	HRB500
混凝土强度等级	C25	≥C30	≥C30	≥C30
搭接长度(mm)	40d	35d	45d	50d

注:1. 当带肋钢筋直径 d 大于 25mm 时,其受拉钢筋的搭接长度应按表值增加 5d 采用;当带肋钢筋直径小于 25mm 时,搭接长度可按表值减少 5d 采用。
2. 当混凝土在凝固过程中受力钢筋易受扰动时,其搭接长度应增加 5d。
3. 在任何情况下,受拉钢筋的搭接长度不应小于 300mm;受压钢筋的搭接长度不应小于 200mm。
4. 环氧树脂涂层钢筋的绑扎接头搭接长度,受拉钢筋按表值的 1.5 倍采用。
5. 受拉区段内,HPB300 钢筋绑扎接头的末端应做成弯钩,HRB400、HRB500、HRBF400 和 RRB400 钢筋的末端可不做成弯钩。

② 焊接接头:当采用焊接接头时,《公桥规》也有相应的构造要求。例如采用夹杆式电弧焊接时,夹杆的总截面积应不小于被焊钢筋的截面积。夹杆长度,若用双面焊缝时应不小于 5d;用单面焊缝时应不小于 10d(d 为钢筋直径)。又例如采用搭叠式电弧焊时,钢筋端段应预先折向一侧,使两根接触的钢筋轴线一致。搭接时,双面焊缝的长度不小于 5d;单面焊缝的长度不小于 10d(d 为钢筋直径)。

《公桥规》还规定,在任一接头中心至搭接长度 1.3 倍长度的区段内,同一根钢筋不得有两个接头,在该区段内有接头的受力钢筋截面积占受力钢筋总截面面积的百分数不宜超过 50%(受拉区钢筋),受压区钢筋的焊接接头无此限值。

③ 机械接头:包括套筒挤压接头和镦粗直螺纹接头,构造规定见《公桥规》。

④ 箍筋最小配箍率: $\rho_{svmin} = 0.14\%$ (HPB300), $\rho_{svmin} > 0.11\%$ (HRB400)。

3) 斜截面抗剪承载力计算示例

钢筋混凝土简支梁全长 $L_0 = 19.96$m,计算跨径 $L = 19.50$m。T 形截面梁的尺寸如图 4-10 所示,桥梁处于 I 类环境条件,安全等级为二级,$\gamma_0 = 1$。

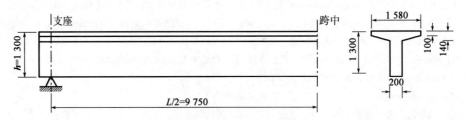

图 4-10 钢筋混凝土简支梁尺寸(尺寸单位:mm)

梁体采用 C30 混凝土,轴心抗压强度设计值 $f_{cd} = 13.8$MPa,轴心抗拉强度设计值 $f_{td} = 1.39$MPa。主筋采用 HRB400 钢筋,抗拉强度设计值 $f_{sd} = 330$MPa;箍筋采用 HPB300 钢筋,直径 8mm,抗拉强度设计值 $f_{sd} = 250$MPa。

简支梁控制截面的弯矩组合设计值和剪力组合设计值为:

跨中截面 $M_{d,l/2} = 2600$ kN·m, $V_{d,l/2} = 84$kN

1/4 跨截面 $M_{d,l/4} = 1900$ kN·m

支点截面 $M_{d,0} = 0, V_{d,0} = 440$kN

要求确定纵向受拉钢筋数量和进行抗剪钢筋设计。

(1)跨中截面的纵向受拉钢筋计算

①T形截面梁受压翼板的有效宽度 b'_f

由图4-10所示的T形截面受压翼板的厚度尺寸,可得平均厚度。则可得到:

$$b'_{f1} = \frac{1}{3}L = \frac{1}{3} \times 19\,500 = 6\,500\,(\text{mm})$$

$b'_{f2} = 1\,600\,\text{mm}$(本算例为装配式T梁,相邻两主梁的平均间距为1 600mm,图4-10所示1 580mm为预制梁翼板宽度)。

$$b'_{f3} = b + 2c + 12h'_f = 200 + 2 \times 0 + 12 \times 120 = 1\,640\,(\text{mm})$$

故取受压翼板的有效宽度 $b'_f = 1\,600\,\text{mm}$。

②钢筋数量计算

钢筋数量(跨中截面)计算及截面复核略。

跨中截面主筋为 8Φ32 + 4Φ16,焊接骨架的钢筋层数为6层,纵向钢筋面积 $A_s = 7\,238\,\text{mm}^2$,布置如图4-11所示。截面有效高度 $h_0 = 1\,183\,\text{mm}$,抗弯承载力 $M_u = 2\,695.0\,\text{kN} \cdot \text{m} > \gamma_0 M_{d,l/2} = 2\,600\,\text{kN} \cdot \text{m}$。

(2)抗剪钢筋设计

①截面尺寸检查

根据构造要求,梁最底层钢筋 2Φ32 通过支座截面,支点截面有效高度为 $h_0 = h - \left(35 + \dfrac{35.8}{2}\right) = 1\,247\,\text{mm}$。

$$0.51 \times 10^{-3}\sqrt{f_{cu,k}}bh_0 = 0.51 \times 10^{-3} \times \sqrt{30} \times 200 \times 1\,247$$
$$= 696.67\,(\text{kN}) > \gamma_0 V_{d,0}\,(=440\,\text{kN})$$

截面尺寸符合设计要求。

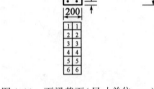

图4-11 下梁截面(尺寸单位:mm)

②检查是否需要根据计算配置箍筋

跨中段截面: $0.5 \times 10^{-3} f_{td} b h_0 = 0.5 \times 10^{-3} \times 1.39 \times 200 \times 1\,183 = 164.44\,(\text{kN})$

因 $\gamma_0 V_{d,l/2}\,(=84\,\text{kN}) < 0.5 \times 10^{-3} f_{td} b h_0$

故可在梁跨中的某长度范围内按构造配置箍筋,其余区段应按计算配置箍筋。

③计算剪力图分配

在如图4-12所示的剪力包络图中,支点处剪力计算值 $V_0 = \gamma_0 V_{d,0}$,跨中处剪力计算值 $V_{l/2} = \gamma_0 V_{d,l/2}$。

$V_x = \gamma_0 V_{d,x} = 0.5 \times 10^{-3} f_{td} b h_0 = 164.44\,\text{kN}$ 的截面距跨中截面的距离可由剪力包络图按比例求得,为:

$$l_1 = \frac{L}{2} \times \frac{V_x - V_{l/2}}{V_0 - V_{l/2}}$$

$$= 9\,750 \times \frac{164.44 - 84}{440 - 84}$$

$$= 2\,203\,(\text{mm})$$

在 l_1 长度内可按构造要求布置箍筋。

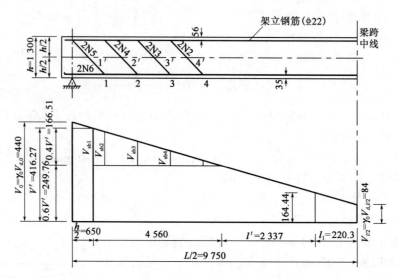

图 4-12　计算剪力分配图(尺寸单位:mm;剪力单位:kN)

同时,根据《公桥规》规定,在支座中心线附近 $h/2 = 650$ mm 范围内,箍筋的间距最大为 100mm。

距支座中心线为 $h/2$ 处的计算剪力值 V',由剪力包络图按比例求得,为:

$$V' = \frac{LV_0 - h(V_0 - V_{l/2})}{L}$$

$$= \frac{19\,500 \times 440 - 1\,300 \times (440 - 84)}{19\,500}$$

$$= 416.27(\text{kN})$$

其中应由混凝土和箍筋承担的剪力计算值至少为 $0.6V' = 249.76$ kN;应由弯起钢筋(包括斜筋)承担的剪力计算值最多为 $0.4V' = 166.51$ kN,设置弯起钢筋区段长度为 4 560mm (图 4-12)。

④箍筋设计

采用直径为 8mm 的双肢箍筋,箍筋截面积 $A_{sv} = nA_{sv1} = 2 \times 50.3 = 100.6(\text{mm}^2)$。

在等截面钢筋混凝土简支梁中,箍筋尽量做到等距离布置。为计算简便,按式(4-2)设计箍筋时,式中的斜截面内纵筋配筋百分率 p 及截面有效高度 h_0 可近似按支座截面和跨中截面的平均值取用,计算如下:

跨中截面:　　$p_{l/2} = 3.06 > 2.5$,取 $p_{l/2} = 2.5, h_0 = 1\,183$ mm

支点截面:　　$p_0 = 0.64, h_0 = 1\,247$ mm

则平均值分别为: $p = \dfrac{2.5 + 0.64}{2} = 1.57, h_0 = \dfrac{1\,183 + 1\,247}{2} = 1\,215(\text{mm})$。

箍筋间距 s_v 计算为:

$$s_v = \frac{\alpha_1^2 \alpha_3^2 0.56 \times 10^{-6} \times (2 + 0.6p)\sqrt{f_{cu,k}} A_{sv} f_{sv} b h_0^2}{V'^2}$$

$$= \frac{1 \times 1.1^2 \times 0.56 \times 10^{-6}(2 + 0.6 \times 1.57)\sqrt{30} \times 100.6 \times 250 \times 200 \times 1215^2}{416.27^2}$$

$$= 468 \text{(mm)}$$

确定箍筋间距 s_v 的实际取值尚应考虑《公桥规》的构造要求。

若箍筋间距取 $s_v = 400\text{mm} \leqslant \frac{1}{2}h = 650\text{mm}$ 及 400mm，满足规范要求。但采用 $\Phi 8$ 双肢箍筋，箍筋配筋率 $\rho_{sv} = \frac{A_{sv}}{bs_v} = \frac{100.6}{200 \times 400} = 0.12\% < 0.14\%$（HPB300 钢筋时），故不满足规范规定。

现取 $s_v = 250\text{mm}$ 计算的箍筋配筋率 $\rho_{sv} = 0.2\% > 0.14\%$，且小于 $\frac{1}{2}h = 650\text{mm}$ 和 400mm。

综合上述计算，在支座中心向跨径长度方向的 1 300mm 范围内，设计箍筋间距 $S_v = 100\text{mm}$，而后至跨中截面统一的箍筋间距取 $s_v = 250\text{mm}$。

⑤弯起钢筋及斜筋设计

设焊接钢筋骨架的架立钢筋（HRB400）为 $\Phi 22$，钢筋重心至梁受压翼板上边缘距离 $a'_s = 56\text{mm}$。

弯起钢筋的弯起角度为 45°，弯起钢筋末端与架立钢筋焊接。为了得到每对弯起钢筋分配的剪力，要由各排弯起钢筋的末端折点应落在前一排弯起钢筋弯起点的构造规定来得到各排弯起钢筋的弯起点计算位置，首先要计算弯起钢筋上、下弯点之间垂直距离 Δh_i（图 4-13）。

现拟弯起 N1～N5 钢筋，将计算的各排弯起钢筋弯起点截面的 Δh_i 以及至支座中心距离 x_i、分配的剪力计算值 V_{sbi}、所需的弯起钢筋面积 A_{sbi} 值列入表 4-3。

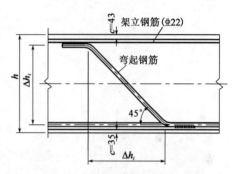

图 4-13 弯起钢筋细节（尺寸单位：mm）

弯起钢筋计算表 表 4-3

弯 起 点	1	2	3	4	5
Δh_i(mm)	1 125	1 090	1 054	1 035	1 017
距支座中心距离 x_i(mm)	1 125	2 215	3 269	4 304	5 321
分配的剪力计算值 V_{sbi}(kN)	166.51	149.17	109.36	70.88	
所需的弯筋面积 A_{sbi}(mm²)	952	852	624	405	
可提供的弯筋面积 A_{sbi}(mm²)	1 609 (2Φ32)	1 609 (2Φ32)	1 609 (2Φ32)	402 (2Φ16)	
弯筋与梁轴交点到支座中心距离 x'(mm)	564	1 690	2 779	3 841	

现将表 4-3 中有关计算举例说明如下。

根据《公桥规》规定，简支梁的第一排弯起钢筋（对支座而言）的末端弯折点应位于支座中心截面处。这时，Δh_1 计算为：

$$\Delta h_1 = 1\,300 - [(35 + 35.8 \times 1.5) + (43 + 25.1 + 35.8 \times 0.5)]$$

$$= 1\,125\text{(mm)}$$

弯筋的弯起角为45°,则第一排弯筋(2N5)的弯起点1距支座中心距离为1 125mm。弯筋与梁纵轴线交点1′距支座中心距离为:

$$1\ 125 - [1\ 300/2 - (35 + 35.8 \times 1.5)] = 564(\text{mm})$$

对于第二排弯起钢筋,可得到:

$$\Delta h_2 = 1\ 300 - [(35 + 35.8 \times 2.5) + (43 + 25.1 + 35.8 \times 0.5)]$$
$$= 1\ 090(\text{mm})$$

弯起钢筋2N4的弯起点2距支点中心距离为$1\ 125 + \Delta h_2 = 1\ 125 + 1\ 090 = 2\ 215(\text{mm})$。分配给第二排弯起钢筋的计算剪力值$V_{sb2}$,由比例关系计算可得到:

$$\frac{4\ 560 + 650 - 1\ 125}{4\ 560} = \frac{V_{sb2}}{166.51}$$

得

$$V_{sb2} = 149.17\text{kN}$$

其中,$0.4V' = 166.51\text{kN}$,$h/2 = 650\text{mm}$;设置弯起钢筋区段长为4 560mm。所需要提供的弯起钢筋截面积A_{sb2}为:

$$A_{sb2} = \frac{1\ 333.33 V_{sb2}}{f_{sd}\sin 45°}$$

$$= \frac{1\ 333.33 \times 149.17}{330 \times 0.707}$$

$$= 852(\text{mm}^2)$$

第二排弯起钢筋与梁轴线交点2′距支座中心距离为:

$$2\ 215 - [1\ 300/2 - (35 + 35.8 \times 2.5)] = 1\ 690(\text{mm})$$

其余各排弯起钢筋的计算方法与第二排弯起钢筋计算方法相同。

由表4-3可见,原拟定弯起N1钢筋的弯起点距支座中心距离为5 321mm,已大于$4\ 560 + h/2 = 4\ 560 + 650 = 5\ 210(\text{mm})$,即在欲设置弯筋区域长度之外,故暂不参加弯起钢筋的计算,图4-14中以截断N1钢筋表示,但在实际工程中,往往不截断而是弯起,以加强钢筋骨架施工时的刚度。

按照计算剪力初步布置弯起钢筋,如图4-14所示。

按照同时满足梁跨间各正截面和斜截面抗弯要求,确定弯起钢筋的弯起点位置。由已知跨中截面弯矩计算值$M_{l/2} = \gamma_0 M_{d,l/2} = 2\ 600\text{kN}\cdot\text{m}$,支点中心处$M_0 = \gamma_0 M_{d,0} = 0$,按式(4-10)作出梁的计算弯矩包络图(图4-14)。在$\frac{1}{4}L$截面处,因$x = 4.875\text{m}$,$L = 19.5\text{m}$,$M_{l/2} = 2\ 200\text{kN}\cdot\text{m}$,则弯矩计算值为:

$$M_{l/4} = 2\ 600 \times \left(1 - \frac{4 \times 4.875^2}{19.5^2}\right) = 1\ 950(\text{kN}\cdot\text{m})$$

与已知值$M_{d,l/4} = 1\ 900\text{kN}\cdot\text{m}$相比,两者相对误差为2.6%,故用式(4-10)来描述简支梁弯矩包络图是可行的。

各排弯起钢筋弯起后,相应正截面抗弯承载力M_{ui}计算如表4-4所示。

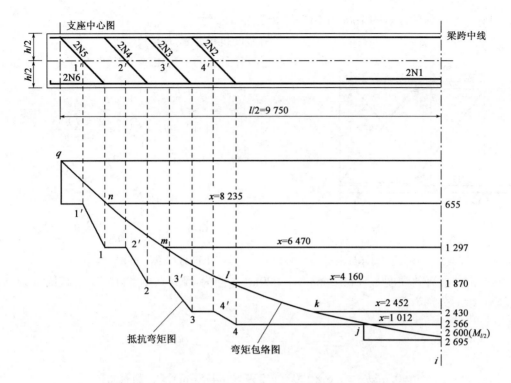

图 4-14 梁的弯矩包络图与抵抗弯矩图(尺寸单位:mm;弯矩单位:kN·m)

钢筋弯起后相应各正截面抗弯承载力   表 4-4

梁 区 段	截面纵筋	有效高度 h_0 (mm)	T形截面类别	受压区高度 x (mm)	抗弯承载力 M_{ui} (kN·m)
支座中心 ~ 1 点	2 ⌀ 32	1247	第一类	24	655.0
1 点 ~ 2 点	4 ⌀ 32	1229	第一类	49	1 279.0
2 点 ~ 3 点	6 ⌀ 32	1211	第一类	73	1 870.0
3 点 ~ 4 点	8 ⌀ 32	1193	第一类	98	2 430.0
4 点 ~ N1 钢筋截断处 j	8 ⌀ 32 + 2 ⌀ 16	1189	第一类	104	2566.0
N1 钢筋截断处 ~ 梁跨中	8 ⌀ 32 + 4 ⌀ 16	1183	第一类	110	2 695.0

将表 4-4 的正截面抗弯承载力 M_{ui} 在图 4-14 上用各平行直线表示出来,它们与弯矩包络图的交点分别为 n、m、…、j,并以各 M_{ui} 值代入式(4-11)中,可求得 n、m、…、j 到跨中截面距离 x 值(图 4-14)。

由弯矩包络图可知图 4-14 所示弯起钢筋弯起点初步位置满足要求。

(3)斜截面抗剪承载力的复核

对于钢筋混凝土简支梁斜截面抗剪承载力的复核,按照《公桥规》关于复核截面位置和复核方法的要求逐一进行。本例以距支座中心 $h/2$ 处斜截面抗剪承载力复核介绍计算方法。

①选定斜截面位置

距支座中心 $h/2$ 处截面的横坐标为 $x = 9\ 750 - 650 = 9\ 100\text{(mm)}$,正截面有效高度 $h_0 = 1\ 247\text{(mm)}$。现取斜截面投影长度 $c' \approx h_0 = 1\ 247\text{(mm)}$,则得到选择的斜截面顶端位置 A

(图 4-15),其横坐标为 $x = 9\,100 - 1\,247 = 7\,853(\text{mm})$。

②斜截面抗剪承载力复核

A 处正截面上的剪力 V_x 及相应的弯矩 M_x 计算如下:

$$V_x = V_{l/2} + (V_0 - V_{l/2}) \frac{2x}{L}$$

$$= 84 + (440 - 84) \times \frac{2 \times 7\,853}{19\,500}$$

$$= 370.74(\text{kN})$$

$$M_x = M_{l/2}\left(1 - \frac{4x^2}{L^2}\right)$$

$$= 2\,600 \times \left(1 - \frac{4 \times 7\,853^2}{19\,500^2}\right)$$

$$= 913.31(\text{kN}\cdot\text{m})$$

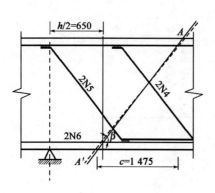

图 4-15 距支座中心 $h/2$ 处斜截面抗剪
承载力计算图式(尺寸单位:mm)

A 处正截面有效高度 $h_0 = 1\,229\text{mm} = 1.229\text{m}$(主筋为 4Φ32),则实际广义剪跨比 m 及斜截面投影长度 c 分别为:

$$m = \frac{M_x}{V_x h_0} = \frac{913.31}{370.74 \times 1.229} = 2.00 < 3$$

$$c = 0.6mh_0 = 0.6 \times 2.0 \times 1.229 = 1.475(\text{m}) > 1.247(\text{m})$$

将要复核的斜截面是图 4-15 中所示 AA' 斜截面(虚线表示),斜角 $\beta = \tan^{-1}\frac{h_0}{c} = \tan^{-1}\frac{1.229}{1.475} \approx 39.8°$。

斜截面内纵向受拉主筋平均配筋率 p 为:

$$p = 100\frac{2A_s}{bh_0} = \frac{2 \times 100 \times 1\,608}{200 \times 1\,247} = 1.28 < 2.5$$

箍筋的配筋率(取 $S_v = 250\text{mm}$ 时)ρ_{sv} 为:

$$\rho_{sv} = \frac{A_{sv}}{bS_v} = \frac{100.6}{200 \times 250} = 0.201\% > \rho_{\min}(=0.14\%)$$

与斜截面相交的弯起钢筋有 2N5(2Φ32)、2N4(2Φ32)。

将以上计算值代入式(4-3),则得到 AA' 斜截面抗剪承载力为:

$$V_u = \alpha_1\alpha_2\alpha_3 0.45 \times 10^{-3}bh_0\sqrt{(2+0.6p)}\sqrt{f_{cu,k}\rho_s f_{sv}} + 0.75 \times 10^{-3}f_{sd}\sum A_{sb}\sin\theta_s$$

$$= 1 \times 1 \times 1.1 \times 0.45 \times 10^{-3} \times 200 \times 1\,247 \times \sqrt{(2+0.6 \times 1.28)} \times \sqrt{30} \times 0.002\,01 \times 250 +$$

$$\quad 0.75 \times 10^{-3} \times 330 \times 2 \times 1\,608 \times 0.707$$

$$= 316.23 + 633.09$$

$$= 949.32(\text{kN}) > V_x = 370.74(\text{kN})$$

故距支座中心为 $h/2$ 处的斜截面抗剪承载力满足设计要求。

思考练习题

1. 钢筋混凝土受弯构件沿斜截面破坏的形态有几种？各有什么特点？
2. 影响钢筋混凝土受弯构件斜截面抗剪承载力的主要因素有哪些？
3. 钢筋混凝土受弯构件斜截面抗弯承载力基本公式的适用范围是什么？公式的上、下限值物理意义是什么？
4. 箍筋有什么作用？配箍率是怎样定义的？它对斜截面抗剪强度有什么影响？
5. 什么叫抵抗弯矩图？它与设计弯矩图应有怎样的关系？什么是钢筋的充分利用点和理论切断点？试述纵向钢筋在支座处锚固有哪些规定？
6. 钢筋混凝土抗剪承载力复核时，如何选择复核截面？
7. 试述受弯构件梁的斜截面破坏形态及各自的特点。
8. 已知如习题 8 图所示的简支梁计算跨径 $L=12.6\text{m}$，两梁中心距为 2.1m，其截面尺寸如图所示，采用 C35 混凝土，HRB400 钢筋，$\gamma_0=1.0$，该截面弯矩为：永久荷载 $M_{Gk}=492.3\text{kN}\cdot\text{m}$，基本可变荷载 $M_{Qk}=401.2\text{kN}\cdot\text{m}$，$\gamma_G=1.2$，$\gamma_Q=1.4$，试进行截面配筋和截面复核。
9. 主梁计算跨径、主梁截面尺寸均与习题 8 相同，C30 混凝土，半跨剪力图如习题图 9 所示，试进行抗剪配筋设计。

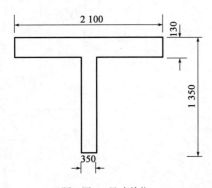

习题 8 图 （尺寸单位:mm）

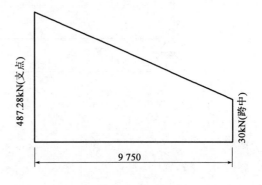

习题 9 图 （尺寸单位:mm）

10. 某均布荷载作用下的简支伸臂梁(如习题 10 图所示，图中荷载已包括自重)，采用 C30 混凝土，HPB300 级钢筋，试设计此梁并进行钢筋布置。

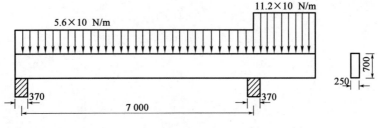

习题10图(尺寸单位:mm)

11. 已知某简支梁，标准跨径 $L_b=20\text{m}$，计算跨径 $L=19.5\text{m}$，梁全长 $L_0=19.96\text{m}$，截面尺寸和剪力图如习题 11 图所示。跨中截面有效高度 $h_0=120\text{cm}$，支点截面有效高度 $h_0'=125\text{cm}$，跨中截面配筋率为 3.17%，支点截面配筋率为 0.6%，主筋采用 HRB400 螺纹钢筋，箍筋为

HPB300 钢筋,混凝土 C30,请进行抗剪配筋设计。

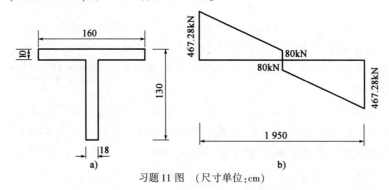

习题 11 图　（尺寸单位:cm）

12. 计算跨径 $L=4.8\mathrm{m}$ 的钢筋混凝土矩形截面简支梁(习题 12 图), $b\times h=20\mathrm{mm}\times 500\mathrm{mm}$,C30 混凝土。Ⅰ类环境条件,安全等级为二级。已知简支梁跨中截面弯矩组合设计值 $M_{\mathrm{d},l/2}=147\mathrm{kN}\cdot\mathrm{m}$,支点处剪力组合设计值 $V_{\mathrm{d},0}=124.8\mathrm{kN}$,跨中处剪力组合设计值 $V_{\mathrm{d},l/2}=25.2\mathrm{kN}$。试求所需的纵向受拉钢筋 A_{s}(HRB400 级钢筋)和仅配置箍筋(HPB300 级)时其布置间距 s_{v},并画出配筋图。

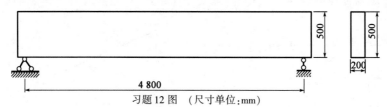

习题 12 图　（尺寸单位:mm）

第 5 章
受扭构件承载力计算

5.1 概 述

工程中出现的弯、斜以及承受偏心荷载作用的桥中,主梁要承担扭矩,本章介绍受扭构件的计算。

1)工程中常见受扭构件

(1)曲线梁(弯梁桥)、斜梁(板);

(2)偏心荷载作用下的梁、板,箱形梁桥;

(3)螺旋楼梯板。

实际上,结构中很少有扭矩单独作用的情况,大多为弯矩、剪力和扭矩同时作用,有时还有轴向力作用。弯、剪、扭共同作用的结构的破坏机理比较复杂,本章从纯扭构件开始,探讨钢筋混凝土受扭构件的设计。

2)受扭构件的受力特点

在曲线梁桥中,即使不考虑活载,仅在恒载作用下,梁(板)的截面上除有弯矩 M、剪力 V 外,还存在着扭矩 T(图 5-1)。

由于扭矩、弯矩和剪力的作用,构件的截面上将产生

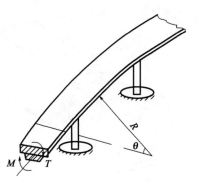

图 5-1 曲线梁示意图

相应的主拉应力。当主拉应力超过混凝土的抗拉强度时,构件便会开裂。构件开裂后,素混凝土极易破坏,因此,必须配置适量的钢筋(纵筋和箍筋)来提高钢筋混凝土构件的抗扭承载能力。

5.2 纯扭构件的破坏特征和承载力计算

1) 钢筋混凝土纯扭构件受力过程分析

从钢筋混凝土受扭构件受力破坏全过程的扭矩 T 和扭转角 θ 的关系曲线(图 5-2)来分析其受力特点。

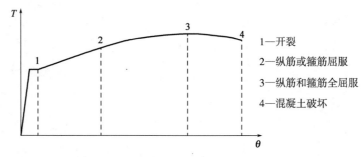

图 5-2　钢筋混凝土受扭构件的 T-θ 曲线

(1)加载初期:截面扭转变形很小,与弹性体材料受扭构件的受力相似。

(2)斜裂缝出现以后:混凝土开裂后退出工作,由于混凝土部分卸载,钢筋应力明显增大,在 T-θ 曲线上出现水平段;当扭转角增加到一定值后,钢筋应变趋于稳定并形成新的受力状态。

(3)荷载继续施加:变形增长较快,裂缝的数量逐步增多,裂缝宽度逐渐加大,这时 T-θ 关系大体还是呈直线变化。

(4)荷载接近极限扭矩:在构件截面长边上的斜裂缝中,有一条发展为临界裂缝,与这条空间斜裂缝相交的部分箍筋(长肢)或部分纵筋将首先屈服,产生较大的塑性变形,这时 T-θ 曲线趋于水平。

(5)荷载达到极限扭矩:和临界斜裂缝相交的箍筋短肢及纵向钢筋相继屈服,混凝土逐步退出工作,构件的抵抗扭矩开始逐步下降,最后在构件的另一长边出现了压区塑性铰或出现两个裂缝间混凝土被压碎的现象时,构件破坏。

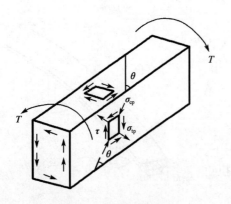

图 5-3　矩形截面纯扭构件

根据钢筋混凝土纯扭构件受力的特点可知,钢筋混凝土构件抗扭计算的两个重要指标是:构件的开裂扭矩和构件的破坏扭矩。

2) 矩形截面纯扭构件的开裂扭矩

钢筋混凝土受扭构件开裂前钢筋中的应力很小,钢筋对开裂扭矩的影响不大,因此,可以忽略钢筋对开裂扭矩的影响,将构件视为纯混凝土受扭构件来计算开裂扭矩。

如图 5-3 所示为矩形截面的纯扭构件。在纯扭矩作用下,匀质弹性材料的矩形截面构件剪应力分布如图 5-4a)所示,裂缝总是沿与构件纵轴成 $\theta=45°$ 方向发展且开裂扭矩即主拉应力 $\sigma_{tp}=\tau=f_t$ 时的扭矩。

假设构件为理想塑性材料的矩形截面构件,截面上某一点应力达到材料的屈服强度时,只意味着局部材料开始进入塑性状态,此时构件仍能继续承担荷载。直到截面上的应力全部达到材料的屈服强度时,构件才能达到其极限承载能力,此时,截面上剪应力的分布如图 5-4b)所示。

现按如图 5-4b)所示理想塑性材料的剪应力分布求其抵抗扭矩,由平衡条件可得到:

$$T = \left\{ 2 \cdot \frac{b}{2} \cdot (h-b) \cdot \frac{b}{4} + 4 \cdot \frac{b}{2} \cdot \frac{b}{2} \cdot \frac{1}{2} \cdot \frac{b}{3} + 2 \cdot \frac{b}{2} \cdot \frac{b}{2} \left[\frac{2}{3} \cdot \frac{b}{2} + \frac{1}{2}(h-b) \right] \right\} \tau_{max}$$

$$= \frac{b^2}{6}(3h-b)\tau_{max} = W_t \tau_{max} \tag{5-1}$$

式中,W_t 称为矩形截面的抗扭塑性抵抗矩,$W_t = \frac{b^2}{6}(3h-b)$。

对于钢筋混凝土构件来说,混凝土既非理想弹性材料,又非理想塑性材料,而是介于两者之间的弹塑性材料。如果按弹性材料的应力分布进行计算,将低估构件的开裂扭矩;而按完全的塑性应力分布进行计算,又会高估构件的开裂扭矩。

综上所述,矩形截面钢筋混凝土受扭构件的开裂扭矩,近似地采用理想塑性材料的剪应力图形进行计算,同时通过试验加以校正,乘以一个折减系数 0.7,开裂扭矩的计算式为:

$$T_{cr} = 0.7 W_t f_{td} \tag{5-2}$$

式中,T_{cr} 为矩形截面纯扭构件的开裂扭矩;f_{td} 为混凝土抗拉强度设计值;W_t 为矩形截面的抗扭塑性抵抗矩。

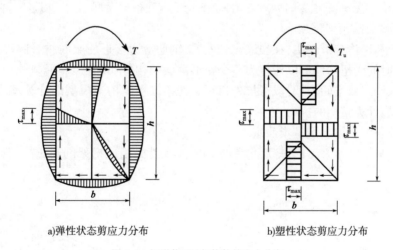

a)弹性状态剪应力分布　　b)塑性状态剪应力分布

图 5-4　矩形截面纯扭构件剪应力分布

3)矩形截面纯扭构件的破坏特征

(1)钢筋混凝土矩形截面受扭构件的破坏类型

①少筋破坏:当抗扭钢筋数量过少时,在构件受扭开裂后,由于钢筋没有足够的能力承受混凝土开裂后卸给它的那部分扭矩,因而构件很快破坏。

②适筋破坏:在正常配筋的条件下,随着外扭矩的不断增加,抗扭箍筋和纵筋首先达到屈

服强度,然后主裂缝迅速开展,最后促使混凝土受压面被压碎,构件破坏。

③超筋破坏:当抗扭钢筋配置过多或混凝土强度过低时,随着外扭矩的增加,构件混凝土先被压碎,从而导致构件破坏,而此时抗扭箍筋和纵筋均未达到屈服强度。

④部分超筋破坏:当抗扭箍筋或纵筋中的一种配置过多时,构件破坏时只有部分纵筋或箍筋屈服,而另一部分抗扭钢筋(箍筋或纵筋)尚未达到屈服强度,破坏具有脆性破坏性质。

(2) 配筋强度比 ζ

纵筋的数量、强度和箍筋的数量、强度的比例(简称配筋强度比,以 ζ 表示)对抗扭承载力有一定的影响。当箍筋用量相对较少时,构件抗扭承载力就由箍筋控制,这时再增加纵筋也不能起到提高抗扭承载力的作用。反之,当纵筋用量很少时,增加箍筋也将不能充分发挥作用。

若将纵筋和箍筋之间的数量比例用钢筋的体积比来表示,则配筋强度比 ζ 的表达式为:

$$\zeta = \frac{f_{sd}A_{st}s_v}{f_{sv}A_{sv1}U_{cor}} \tag{5-3}$$

式中,A_{st}、f_{sd} 分别为对称布置的全部纵筋截面面积及纵筋的抗拉强度设计值;A_{sv1}、f_{sv} 分别为单肢箍筋的截面积和箍筋的抗拉强度设计值;s_v 为箍筋的间距;U_{cor} 为截面核心混凝土部分的周长,计算时可取箍筋内表皮间的距离来得到。

配筋强度比 ζ 可在一定范围内变化,通常限制在 $0.6 \leq \zeta \leq 1.7$。设计时可取 $\zeta = 1.0 \sim 1.2$。

(3) 配筋量对受扭构件破坏形态的影响

即使在配筋强度比 ζ 不变的条件下,纵筋及箍筋的配筋量也会对受扭构件的破坏形态有影响。图 5-5 为 $\zeta=1$ 时箍筋量 $f_{sv}A_{sv1}/s_v$ 和抗扭承载力 T 的关系。对不同的配筋强度比 ζ,少筋和适筋、适筋和超筋的界限位置是不同的。

4) 纯扭构件的承载力计算理论

目前所用的计算理论主要有两种:一种是变角度空间桁架模型,另一种是斜弯曲破坏理论。

(1) 变角度空间桁架模型

试验研究和理论分析表明,在裂缝充分发展、钢筋应力接近屈服时,构件截面核心混凝土退出工作,此时的受扭构件可以假想为一个箱形截面的构件:具有螺旋形裂缝的混凝土外壳、纵筋和箍筋共同组成空间桁架,以抵抗外扭矩。变角度空间桁架模型的计算理论就是以此空间桁架为模型,计算结构的抗力。

①变角度空间桁架模型基本假定

a. 混凝土只承受压力,具有螺旋形裂缝的混凝土外壳组成桁架的斜压杆,其倾角为 α,见图 5-6;

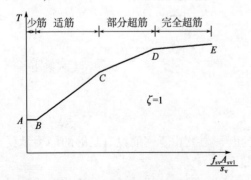

图 5-5 配筋量对抗扭承载力的影响

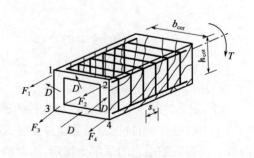

图 5-6 变角度空间桁架模型

b. 纵筋和箍筋只承受拉力,分别构成桁架的弦杆和腹杆;

c. 忽略核心混凝土的抗扭作用和钢筋的销栓作用。

②承载力计算公式

根据破坏状态的受力情况,推导出抗扭极限承载力计算公式为:

$$T_u = 2\frac{A_{sv1}f_{sv}U_{cor}}{s_v}\cot\alpha = 2\sqrt{\zeta} \cdot \frac{A_{sv1}f_{sv}A_{cor}}{s_v} \tag{5-4}$$

由式(5-4)可以看出,构件的抗扭承载力 T_u 主要与钢筋骨架尺寸、箍筋用量及其强度,以及表征纵筋与箍筋的相对用量的参数 ζ 有关,此式为低配筋受扭构件扭矩承载力的计算公式。

(2)斜弯曲破坏理论

对于钢筋混凝土纯扭构件,在截面的长边上出现裂缝,斜截面形成后,扭矩可以分解为垂直于斜截面的力矩(扭矩)和平行于斜截面的力矩(扭矩),斜面长边出现混凝土弯曲受压压碎区,形成斜向弯曲破坏。斜弯曲破坏理论是取截面实际的破坏面为隔离体,导出与纵筋、箍筋用量有关的抗扭承载力计算公式。

①斜弯曲计算理论的基本假定

a. 通过扭曲裂面的纵向钢筋、箍筋在构件破坏时均已达到其屈服强度,见图 5-7;

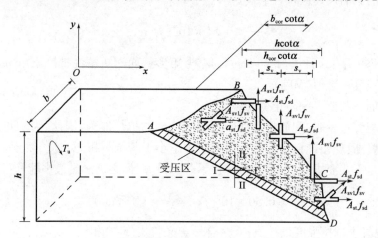

图 5-7 斜弯曲理论计算示意图

b. 受压区高度近似地取为两倍的保护层厚度,假定受压区的合力近似地作用于受压区的形心;

c. 混凝土的抗扭能力忽略不计,扭矩全部由抗扭纵筋和箍筋承担;

d. 抗扭纵筋沿构件核心周边对称、均匀布置,抗扭箍筋沿构件轴线方向等距离布置,且均锚固可靠。

②承载力计算公式

根据上述假定条件,按斜弯曲理论得出扭矩承载力计算式(5-5),它与按变角度空间桁架理论所得的计算式(5-4)完全相同。

$$T_u = 2\sqrt{\zeta} \cdot \frac{A_{sv1}f_{sv}A_{cor}}{s_v} \tag{5-5}$$

(3)两种计算理论的特点

由空间桁架计算模型或斜弯曲理论导出的受扭构件承载力计算公式,充分反映了抗扭钢筋的作用。但试验表明在低配筋时偏于保守,在高配筋时,由于纵向钢筋和箍筋不能同时屈服,计算值又偏高,设计中应尽量避免。

5)《公桥规》对矩形截面纯扭构件的承载力计算规定

(1)承载力计算公式

基于变角度空间桁架的计算模型,并通过受扭构件的试验结果,在保证安全的前提下抗扭能力取试验数据的下限值,得到矩形截面构件抗扭承载力计算公式:

$$\gamma_0 T_d \leq 0.35\beta_a f_{td} W_t + 1.2\sqrt{\zeta}\frac{f_{sv}A_{sv1}A_{cor}}{s_v} \tag{5-6}$$

ζ 为纯扭构件纵向钢筋与箍筋的配筋强度比,按公式(5-3)计算;对钢筋混凝土构件,《公桥规》规定 ζ 值应符合 $0.6 \leq \zeta \leq 1.7$,当 $\zeta > 1.7$ 时,取 $\zeta = 1.7$。

(2)限制条件

①抗扭配筋的上限值——防止超筋破坏

当抗扭钢筋配量过多时,受扭构件可能在抗扭钢筋屈服以前便由于混凝土被压碎而破坏,也就是说,其破坏扭矩取决于混凝土的强度和截面尺寸。因此,《公桥规》规定钢筋混凝土矩形截面纯扭构件的截面尺寸应符合下式要求:

$$\frac{\gamma_0 T_d}{W_t} \leq 0.51 \times 10^{-3}\sqrt{f_{cu,k}} \tag{5-7}$$

式中,T_d 为扭矩组合设计值(kN·mm);W_t 为矩形截面受扭塑性抵抗矩(mm³);$f_{cu,k}$ 为混凝土立方体抗压强度标准值(MPa)。

②抗扭配筋的下限值——防止少筋破坏

当抗扭钢筋配置过少时,配筋将无助于开裂后构件的抗扭能力,因此,为防止纯扭构件在低配筋时混凝土脆断,应使纯扭构件所承担的扭矩不小于其抗裂扭矩。《公桥规》规定钢筋混凝土纯扭构件满足下式要求时,可不进行抗扭承载力计算,但必须满足最小配筋率的要求:

$$\frac{\gamma_0 T_d}{W_t} \leq 0.50 \times 10^{-3} f_{td} \qquad (\text{kN/mm}^2) \tag{5-8}$$

式中,f_{td} 为混凝土抗拉强度设计值;其余符号意义与式(5-7)相同。

《公桥规》规定,纯扭构件的箍筋配筋率应满足 $\rho_{sv} = \dfrac{A_{sv}}{s_v b} \geq 0.055\dfrac{f_{cd}}{f_{sv}}$;纵向受力钢筋配筋率应满足 $\rho_{st} = \dfrac{A_{st}}{bh} \geq 0.08\dfrac{f_{cd}}{f_{sd}}$。

5.3 在弯、剪、扭共同作用下矩形截面构件的承载力计算

1)弯剪扭构件的破坏类型

弯矩、剪力和扭矩共同作用下的钢筋混凝土构件,其受力状态十分复杂。构件的破坏特征及承载能力,与所作用的外部荷载条件和构件的内在因素有关。

(1)外部条件:通常以扭弯比 $\psi\left(\psi = \dfrac{T}{M}\right)$ 以及扭剪比 $\chi\left(\chi = \dfrac{T}{Vb}\right)$ 来表示。

(2)内在因素:系指构件截面形状、尺寸、配筋及材料强度。

当构件的内在因素不变时,其破坏特征仅与扭弯比ψ和扭剪比χ的大小有关;当ψ和χ值不变时,由于构件的内在因素(如截面尺寸)不同,亦可能出现不同类型的破坏形状。

①弯型破坏:受弯剪扭作用的构件,在适筋条件下,扭弯比ψ较小时,即弯矩作用比扭矩显著,即裂缝首先发生在构件的受拉底面;然后裂缝发展到两个侧面,截面顶部纵筋受压,在弯曲受压的基础上承受扭矩引起的拉力,这是互相消弭的有利方面。在三个面上的螺旋形裂缝形成一个扭曲破坏面,第四个面为弯曲受压顶面。构件破坏时体现为先是与螺旋形裂缝相交的纵筋和箍筋受拉达到屈服强度,最终截面上边缘的混凝土受压破坏[图5-8a)]。

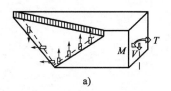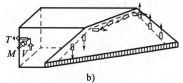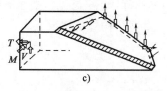

图5-8 弯剪扭构件的破坏类型

②扭型破坏:若扭弯比ψ和扭剪比χ均较大,即扭矩作用显著时,如果构件顶部配筋较少,因为弯矩较小,在构件顶面产生的压应力也较小,受到扭矩作用下的拉力就会抵消弯曲作用下的压应力,且所余的拉力作用较大,这时顶部纵筋先于构件底部纵筋达到受拉屈服强度,破坏面始于构件顶面发展到两个侧面[图5-8b)]。

③剪扭型破坏:当剪力和扭矩都较大,扭剪比又较大时,则裂缝首先出现在构件扭矩产生应力流和剪力流同向的侧面上,然后向顶面和底面发展,这三个面上的螺旋形裂缝形成破坏面,破坏时与螺旋形裂缝相交的钢筋受拉并达到屈服强度,受压区靠近另一侧面[图5-8c)]。

对于弯剪扭共同作用下的构件,除了上述三种破坏形式外,试验表明,如果剪力起显著作用而扭矩较小即扭剪比χ较小时,还会发生与剪压破坏十分相近的受剪破坏形态,实际上已经是剪力在起控制作用了。

2)弯剪扭构件的配筋计算方法

弯剪扭共同作用下的钢筋混凝土构件承载力计算方法,与纯扭构件相同,主要以变角度空间桁架理论和斜弯曲理论为基础的两种计算方法。但是在实际应用中,对于弯扭及弯剪扭共同作用下的构件,按上述两种理论方法计算是非常复杂的。因此需要简化的实用计算方法。

目前我国《混凝土结构设计规范》(GB 50010—2010)规定当构件承受的扭矩小于开裂扭矩的1/4时,可以忽略扭矩的影响,按弯剪共同作用构件计算;当构件承受的剪力小于无腹筋时构件抗剪承载力的1/2时,可忽略剪力的影响,按弯扭共同作用构件计算。对于弯剪扭共同作用构件的配筋计算,采取先按弯矩、剪力和扭矩各自"单独"作用进行配筋计算,然后再把各种相应配筋叠加的截面设计方法。

《公桥规》也采取叠加计算的截面设计简化方法。由于正截面受弯承载力计算方法已如前述,现着重分析剪、扭共同作用下构件的抗扭和抗剪承载力计算问题。

(1)受剪扭构件的承载力计算

《公桥规》在试验研究的基础上,对在剪扭共同作用下矩形截面构件的抗剪和抗扭承载力分别采用了如下的计算公式。

①剪扭构件抗剪承载力按下列公式计算:

$$\gamma_0 V_d \leqslant V_u = 0.5 \times 10^{-4} \alpha_1 \alpha_2 \alpha_3 \frac{10-2\beta_t}{20} bh_0 \sqrt{(2+0.6p)} \sqrt{f_{cu,k}} \rho_{sv} f_{sv} \quad (N) \tag{5-9}$$

$$\beta_t = \frac{1.5}{1+0.5\dfrac{V_d W_t}{T_d bh_0}} \tag{5-10}$$

式中,V_d 为剪扭构件的剪力组合设计值(N);β_t 为剪扭构件混凝土抗扭承载力降低系数,当 $\beta_t<0.5$ 时,取 $\beta_t=0.5$;当 $\beta_t>1.0$ 时,取 $\beta_t=1.0$;W_t 为矩形截面受扭塑性抵抗矩,$W_t=\dfrac{b^2}{6}(3h-b)$;其他符号意义参见斜截面抗剪承载力计算公式。

②剪扭构件抗扭承载力:

$$T_u = \beta_t \left(0.35\beta_a f_{td} + 0.05\frac{N_{po}}{A_o}\right) W_t + 1.2\sqrt{\zeta}\frac{N_\zeta f_{sv} A_{sv1} A_{cor}}{s_v} \tag{5-11}$$

式中,β_t 意义同前;T_d 为剪扭构件的扭矩组合设计值;$\dfrac{N_{po}}{A_o}$ 为考虑预应力影响的参数。

(2)抗剪扭配筋的上下限

①抗剪扭配筋的上限——防止超筋破坏

当构件抗扭钢筋配筋量过大时,构件将由于混凝土首先被压碎而破坏。因此必须规定截面的限制条件,以防止出现这种破坏现象。

在弯剪扭共同作用下,矩形截面构件的截面尺寸必须符合下列条件:

$$\frac{\gamma_0 V_d}{bh_0} + \frac{\gamma_0 T_d}{W_t} \leqslant 0.51 \times 10^{-3} \sqrt{f_{cu,k}} \quad (kN/mm^2) \tag{5-12}$$

式中,V_d 为剪力组合设计值(kN);T_d 为扭矩组合设计值(kN·mm);b 为垂直于弯矩作用平面的矩形或箱形截面腹板总宽度(mm);h_0 为平行于弯矩作用平面的矩形或箱形截面的有效高度(mm);W_t 为截面受扭塑性抵抗矩(mm³);$f_{cu,k}$ 为混凝土立方体抗压强度标准值(MPa)。

②抗剪扭配筋的下限——防止少筋破坏

a. 剪扭构件箍筋配筋率应满足:

$$\rho_{sv} \geqslant \rho_{sv,min} = (2\beta_t - 1)\left(0.055\frac{f_{cd}}{f_{sv}} - c\right) + c \tag{5-13}$$

式中的 β_t 按公式(5-10)计算。对于式中的 c 值,当箍筋采用 R235(Q235)钢筋时取 0.0018;当箍筋采用 HRB335 钢筋时取 0.0012。

b. 纵向受力钢筋配筋率应满足:

$$\rho_{st} \geqslant \rho_{st,min} = \frac{A_{st,min}}{bh} = 0.08(2\beta_t - 1)\frac{f_{cd}}{f_{sd}} \tag{5-14}$$

式中,$A_{st,min}$ 为纯扭构件全部纵向钢筋最小截面面积;h 为矩形截面的长边长度;b 为矩形截面的短边长度;ρ_{st} 为纵向抗扭钢筋配筋率,$\rho_{st}=\dfrac{A_{st}}{bh}$;$A_{st}$ 为全部纵向抗扭钢筋截面积。

c. 矩形截面承受弯剪扭的构件,应当符合下列条件:

$$\frac{\gamma_0 V_d}{bh_0} + \frac{\gamma_0 T_d}{W_t} \leqslant 0.50 \times 10^{-3} f_{td} \quad (kN/mm^2) \tag{5-15}$$

当式(5-15)成立时,可不进行构件的抗扭承载力计算,仅需按构造要求配置钢筋。

(3)在弯矩、剪力和扭矩共同作用下的配筋计算

对于在弯矩、剪力和扭矩共同作用下的构件,其纵向钢筋和箍筋应按下列规定计算。

①抗弯纵向钢筋应按受弯构件正截面承载力计算所需的钢筋截面面积,配置在受拉区边缘。

②按剪扭构件计算纵向钢筋和箍筋,由抗扭所计算得到的纵向抗扭钢筋沿截面四周均匀对称布置,间距不大于300mm,在矩形截面的四角必须配置纵向钢筋。箍筋为按抗剪和抗扭计算所需的面积之和进行布置。

③纵向钢筋的配筋率不应小于受弯构件纵向受力钢筋最小配筋率与受剪扭构件纵向钢筋最小配筋率之和;箍筋配筋率不应小于剪扭构件的箍筋最小配筋率。

5.4　T形和I形截面受扭构件

T形、I形截面可以看作是由简单矩形截面组成的复杂截面(图5-9),在计算其抗裂扭矩、抗扭极限承载力时,可将截面划分为几个矩形截面,并将扭矩 T_d 按各个矩形分块的抗扭塑性抵抗矩按比例分配给各个矩形分块,以求得各个矩形分块所承担的扭矩。

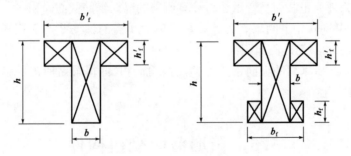

图5-9　T形、I形截面分块示意图

(1)对于肋板部分矩形分块:

$$T_{wd} = \frac{W_{tw}}{W_t} T_d \tag{5-16}$$

(2)对于上翼缘矩形分块:

$$T'_{fd} = \frac{W'_{tf}}{W_t} T_d \tag{5-17}$$

(3)对于下翼缘矩形分块:

$$T_{fd} = \frac{W_{tf}}{W_t} T_d \tag{5-18}$$

上述式中,T_d 为构件截面所承受的扭矩组合设计值;T_{wd} 为肋板所承受的扭矩组合设计值;T'_{fd}、T_{fd} 为上翼缘、下翼缘所承受的扭矩组合设计值。

各个矩形面积划分的原则一般是按截面总高度确定肋板截面,然后再划分受压翼缘和受拉翼缘(图5-9)。

肋板、受压翼缘及受拉翼缘部分的矩形截面受扭塑性抵抗矩计算如下:

肋板：

$$W_{tw} = \frac{b^2}{6}(3h - b) \tag{5-19}$$

受压翼缘：

$$W'_{tf} = \frac{h'^2_f}{2}(b'_f - b) \tag{5-20}$$

受拉翼缘：

$$W_{tf} = \frac{h^2_f}{2}(b_f - b) \tag{5-21}$$

式中，b、h 分别为矩形截面的短边尺寸和长边尺寸；b'_f、h'_f 为 T 形、I 字形截面受压翼缘的宽度和高度；b_f、h_f 为 I 形截面受拉翼缘的宽度和高度。

计算时取用的翼缘宽度应符合 $b'_f \leq b + 6h'_f$ 及 $b_f \leq b + 6h_f$ 的规定。因此，T 形截面总的受扭塑性抵抗矩为：

$$W_t = W_{tw} + W'_{tf} \tag{5-22}$$

I 形截面总的受扭塑性抵抗矩为：

$$W_t = W_{tw} + W'_{tf} + W_{tf} \tag{5-23}$$

对于 T 形截面在弯矩、剪力和扭矩共同作用下构件截面设计可按下列方法进行：

(1) 按受弯构件的正截面受弯承载力计算所需的纵向钢筋截面面积；

(2) 按剪、扭共同作用下的承载力计算承受剪力所需的箍筋截面面积和承受扭矩所需的纵向钢筋截面面积和箍筋截面面积；

(3) 叠加上述二者求得的纵向钢筋和箍筋截面面积，即得最后所需的纵向钢筋截面面积并配置在相应的位置。

5.5 箱形截面受扭构件

在桥梁工程中，箱形截面由于具有抗弯、抗扭刚度大、能承受异号弯矩且结构轻巧等优点，在连续梁桥、曲线梁桥和城市高架桥中得以广泛采用。

试验研究结果表明，箱形梁的抗扭承载力与实心矩形梁相近。

(1) 当箱形梁壁厚与相应计量方向的宽度之比为 $t_2/b \geq 1/4$ 或 $t_1/h \geq 1/4$ 时，其抗扭承载力可按具有相同外形尺寸的带翼缘的矩形截面进行计算（将箱形空洞部分视为实体），如图 5-10 所示。

(2) 当 $1/10 \leq t_2/b < 1/4$ 或 $1/10 \leq t_1/h \leq 1/4$ 时，可近似地将构件截面的抗力乘以一个折减系数 β_a。

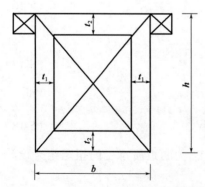

图 5-10 箱形截面构件

由此，箱形截面剪扭构件的抗扭承载力计算公式为：

$$\gamma_0 T_d \leq T_u = 0.35\beta_a\beta_t f_{td} W_t + 1.2\sqrt{\zeta}\frac{f_{sv}A_{sv1}A_{cor}}{S_v} \quad (\text{N} \cdot \text{mm}) \tag{5-24}$$

式中,β_a 为箱形截面有效壁厚折减系数,当 $0.1b \leq t_2 \leq 0.25b$ 或 $0.1h \leq t_1 \leq 0.25h$ 时,取 $\beta_a = 4\dfrac{t_2}{b}$ 或 $\beta_a = 4\dfrac{t_1}{h}$ 两者较小值;当 $t_2 > 0.25b$ 或 $t_1 > 0.25h$ 时,取 $\beta_a = 1.0$。

5.6 构造要求

(1) 由于外荷载扭矩是靠抗扭钢筋的抵抗矩来平衡的,因此在保证必要的保护层的前提下,箍筋与纵筋均应尽可能地布置在构件周边的表面处,以增大抗扭效果。纵向钢筋必须布置在箍筋的内侧,靠箍筋来限制其外鼓。

(2) 根据抗扭强度要求,抗扭纵筋间距不宜大于 300mm,数量至少要有 4 根,布置在矩形截面的四个角隅处,其直径不应小于 8mm;纵筋末端应留有足够的锚固长度。

(3) 架立钢筋和梁肋两侧纵向抗裂分布筋若有可靠的锚固,也可以当抗扭钢筋;在抗弯钢筋一边,可选用较大直径的钢筋来满足抵抗弯矩和扭矩的需要。

(4) 抗扭箍筋必须做成封闭式箍筋(图 5-11),并且将箍筋在角端用 135°弯钩锚固在混凝土核心内,锚固长度约等于 10 倍的箍筋直径。箍筋最大间距根据抗扭要求不宜大于梁高的 1/2 且不大于 400mm,也不宜大于抗剪箍筋的最大间距。箍筋的直径不小于 8mm,且不小于 1/4 主钢筋直径。

(5) 对于由若干个矩形截面组成的复杂截面,如 T 形、L 形、I 形截面的受扭构件,必须将各个矩形截面的抗扭钢筋配成笼状骨架,且使复杂截面内各个矩形单元部分的抗扭钢筋互相交错地牢固联成整体。

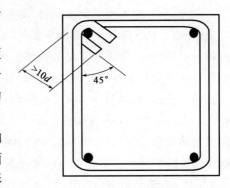

图 5-11 封闭式箍筋示意图

思考练习题

1. 钢筋混凝土纯扭构件有哪几种破坏形态?试说明其发生的条件和破坏特征。
2. 钢筋混凝土纯扭构件的开裂扭矩如何计算?
3. 在受扭承载力计算中,如何避免少筋破坏和超筋破坏?
4. 对受扭箍筋有什么构造要求?抗扭纵筋为什么要沿截面周边对称均匀布置且截面四角必须布置?
5. 钢筋混凝土构件矩形梁,截面尺寸 $b \times h = 300\text{mm} \times 60\text{mm}$,C30 混凝土,HPB300 钢筋。承受的最大弯矩组合设计值 $M_d = 180\text{kN} \cdot \text{m}$,剪力组合设计值 $V_d = 110\text{kN}$,扭矩组合设计值 $T_d = 10\text{kN} \cdot \text{m}$,结构重要性系数 $\gamma_0 = 1.0$。假定 $a_s = 40\text{mm}$,试设计抗弯、剪、扭钢筋。
6. 已知钢筋混凝土矩形截面纯扭构件截面尺寸 $b \times h = 200\text{mm} \times 400\text{mm}$,扭矩设计值 $T_d = 8.5\text{kN} \cdot \text{m}$,C30 混凝土,纵筋 HRB400 级,箍筋 HPB300 级;Ⅰ类环境条件,安全等级为二级,试

求所需钢筋数量。

7. 已知钢筋混凝土受扭构件 $b \times h = 25\text{cm} \times 40\text{cm}, h_0 = 37\text{cm}, d_{he} \times c_{he} = 20\text{cm} \times 35\text{cm}$，C30 混凝土，HPB300 钢筋，承受扭矩 $M_T = 8.5\text{kN} \cdot \text{m}$，剪力 $Q = 51\text{kN}$，试求箍筋和纵向抗扭钢筋，并绘出箍筋的形式。

8. 已知钢筋混凝土受扭构件 $b \times h = 30\text{cm} \times 60\text{cm}$，承受剪力 $Q_{max} = 68\text{kN}, M_{Tmax} = 51\text{kN} \cdot \text{m}$，采用 C15 混凝土，HPB300 钢筋，抗弯的纵向受力钢筋为 4Φ22，一排设置，箍筋为双肢 Φ8@200，试验算已配置的箍筋能否满足强度要求，还需附加多少纵向钢筋，并绘出截面配筋草图。

9. 已知钢筋混凝土矩形截面梁截面尺寸 $b \times h = 200\text{mm} \times 400\text{mm}$，承受弯矩设计值 $M_d = 50\text{kN} \cdot \text{m}$，扭矩设计值 $T_d = 5.0\text{kN} \cdot \text{m}$，剪力设计值 $T_d = 25\text{kN}$；C30 混凝土，纵筋 HRB400 级，箍筋 HPB300 级；Ⅰ类环境条件，安全等级为二级，试求所需钢筋数量。

10. 矩形截面悬臂梁，截面 $b \times h = 250\text{mm} \times 500\text{mm}$；C30 混凝土，纵筋 HRB400 级，箍筋 HPB235 级；该梁在悬臂支座截面处承受的弯矩设计值 $M_d = 100\text{kN} \cdot \text{m}$，扭矩设计值 $T_d = 7.0\text{kN} \cdot \text{m}$，剪力设计值 $T_d = 110\text{kN}$；Ⅰ类环境条件，安全等级为二级；试计算该梁的配筋并绘配筋图。

11. 已知一矩形截面抗扭构件，其截面尺寸及配筋情况如习题图 11 所示，承受剪力设计值 $V_d = 190\text{kN}$。C25 混凝土，截面布置抗扭、剪纵筋 6Φ18，HRB400 级。箍筋用 HPB300 级，箍筋间距 $s_v = 120\text{mm}$，箍筋内表皮至构件表面距离为 35mm，Ⅰ类环境条件，安全等级为二级。试求该构件能承受的最大扭矩设计值。

12. 钢筋混凝土图示雨篷（习题 12 图），板面承受恒载 $g_k = 2.5\text{kN/m}^2$（包含自重），活载 $q_k = 1.6\text{kN/m}^2$，在板端沿板宽度方向每米考虑作用施工活荷载 $p_k = 1.0\text{kN/m}$，它与 q_k 不同时作用。混凝土采用 C30，主筋用 HPB300 级，且取 $h = 80\text{mm}, h_0 = (h - 15)\text{mm}$。试进行板的配筋设计并绘配筋截面图。若上述雨篷梁截面尺寸为 $250\text{mm} \times 400\text{mm}$，计算跨长 3 500mm，净跨长 2 800mm，主筋用 HRB400 级，其他条件与上述基本相同，但只计板面承受荷载 g_k、q_k，且梁两端嵌固。试进行雨篷梁的配筋设计（只计算剪、扭，不计受弯）并绘配筋截面图。

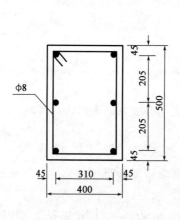

习题 11 图　（尺寸单位:mm）

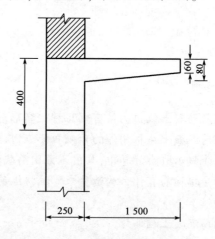

习题 12 图　（尺寸单位:mm）

第6章
轴心受压构件的正截面承载力计算

6.1 概　　述

1）轴心受压构件基本概念

轴向压力的作用点与截面形心重合时的受压构件,称为轴心受压构件。承受以轴向压力为主的受压构件称为"柱"。

在实际结构中,严格的轴心受压构件是很少的。但是,如果轴向力的偏心很小,也可以近似按轴心受压构件计算,例如,钢筋混凝土桁架拱的腹杆。同时,由于轴心受压构件计算简便,故可作为受压构件初步估算截面、复核承载力的手段。

2）钢筋混凝土轴心受压构件的分类

按照箍筋的功能和配置方式的不同可分为两种：

(1)配有纵向钢筋和普通箍筋的轴心受压构件(普通箍筋柱),如图 6-1a)所示；

(2)配有纵向钢筋和螺旋箍筋的轴心受压构件(螺旋箍筋柱),如图 6-1b)所示。

3）箍筋和纵筋的作用

(1)箍筋与纵筋形成钢筋骨架,便于施工,能防止纵向钢筋局部压屈；布置合理的螺旋箍筋能使截面中间部分(核心)混凝土成为约束混凝土,从而提高构件的承载力和延性；

(2)纵向钢筋协助混凝土承受压力,可减小构件截面尺寸；承受可能存在的不大的弯矩；防止构件的突然脆性破坏。

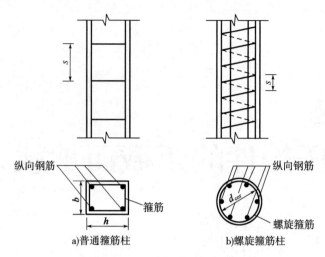

图 6-1 两种钢筋混凝土轴受压构件

6.2 配有纵向钢筋和普通箍筋的轴心受压构件计算

1) 破坏形态

轴心受压柱的破坏按构件的长细比不同，表现出不同的破坏形态。

(1) 短柱（长细比较小）

①试验研究

受压短柱随轴向力 P 增加，混凝土全截面和纵向钢筋均发生压缩变形，试件（图6-2）随之缩短。当轴向力 P 达到破坏荷载的 90% 左右时，柱中部四周混凝土表面出现纵向裂缝，部分混凝土保护层剥落，最后是箍筋间的纵向钢筋发生屈曲，向外鼓出，混凝土被压碎而整个试验柱破坏（图6-3）。它是一种材料破坏，即混凝土压碎破坏。

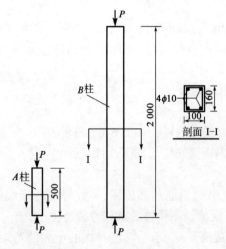

图 6-2 轴心受压构件试件（尺寸单位：mm）

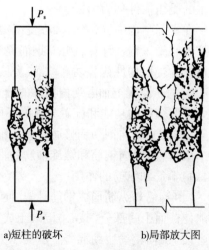

图 6-3 轴心受压短柱的破坏形态

大量试验表明,钢筋混凝土短柱破坏时混凝土的压应变约在 2×10^{-3},混凝土已达到其轴心抗压强度;同时,纵向钢筋采用普通热轧钢筋时,钢筋也能达到抗压屈服强度。

②承载力计算

根据轴向力平衡,就可求得短柱破坏时的轴心力 P_s,它由钢筋和混凝土共同承担:

$$P_s = f_c A_c + f'_s A'_s \qquad (6\text{-}1)$$

(2)长柱(长细比较大)

①试验研究

试件在压力 P 作用下,当压力不大时,也是全截面受压,但随着压力增大,长柱不仅发生压缩变形,同时长柱中部产生较大的横向挠度 u,凹侧压应力较大,凸侧较小。在长柱破坏前,横向挠度增加得很快,破坏突然,属于失稳破坏。破坏时,凹侧的混凝土首先被压碎,混凝土表面有纵向裂缝,纵向钢筋被压弯而向外鼓出,混凝土保护层脱落;凸侧则由受压突然转变为受拉,出现横向裂缝(图6-4)。

②承载力计算

根据大量试验可知,长柱是失稳破坏;长柱的承载力要小于相同截面、相同配筋、相同材料的短柱承载力。因此,可以将短柱的承载力乘以一个折减系数 φ^0 来表示相同截面、配筋和材料的长柱承载力:

$$P_l = \varphi^0 P_s \qquad (6\text{-}2)$$

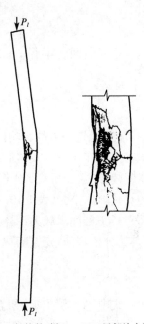

a)长柱的破坏　　b)局部放大图

图6-4　轴心受压长柱的破坏形态

式中,P_s 为短柱破坏时的轴心压力;P_l 为相同截面、配筋和材料的长柱失稳时的轴心压力。

2)稳定系数 φ 的计算

(1)稳定系数的定义

钢筋混凝土轴心受压构件计算中,考虑构件长细比增大的附加效应使构件承载力降低的计算系数称为轴心受压构件的稳定系数,用符号 φ 表示。稳定系数是长柱失稳破坏时的临界承载力 P_l 与短柱压坏时的轴心力 P_s 的比值,表示长柱承载力降低的程度。

(2)稳定系数 φ 值的确定

根据材料力学知识,各种支承条件柱的临界压力计算式为:

$$P_l = \frac{\pi^2 EI}{l_0^2} \qquad (6\text{-}3)$$

式中,EI 为柱截面的抗弯刚度;l_0 为柱的计算长度。

将式(6-3)和式(6-1)代入式(6-2)中,可得到:

$$\varphi^0 = \frac{P_l}{P_s} = \frac{\pi^2 EI}{l_0^2(f_c A + f'_s A'_s)} = \frac{\pi^2 EI}{l_0^2 A(f_c + f'_s \rho')} \qquad (6\text{-}4)$$

$$\rho' = \frac{A'_s}{A}$$

式中,A 为柱截面混凝土面积;A'_s 为纵向钢筋的截面积。

考虑到长柱失稳时截面往往已经开裂,刚度大大降低,为弹性阶段的 30%~50%,所以式(6-4)中的 EI 值要改用柱裂缝出现后的刚度,即用 $\beta_1 E_c I_c$ 来代替式(6-4)中的 EI,β_1 为柱刚度折减系数。于是,可得到:

$$\varphi^0 = \frac{\pi^2 \beta_1 E_c I_c}{l_0^2 A(f_c + f'_s \rho')} = \frac{\pi^2 \beta_1 E_c}{f_c + f'_s \rho'} \cdot \frac{I_c}{A l_0^2} \tag{6-5}$$

柱截面回转半径 $r = \sqrt{I_c/A}$,长细比 $\lambda = l_0/r$,以 φ、f_{cd}、f'_{sd} 分别代替 φ^0、f_c、f'_s,则式(6-5)成为:

$$\varphi = \frac{\pi^2 \beta_1 E_c}{f_{cd} + f'_{sd} \rho'} \cdot \frac{1}{\lambda^2} \tag{6-6}$$

(3)稳定系数的影响因素

由式(6-6)可以看到,当柱的材料和纵筋含筋率一定时,随着长细比 λ 的增加,稳定系数 φ 值就减小。《公桥规》规定,稳定系数 φ 主要与构件的长细比有关,混凝土强度等级及钢筋种类对其影响小,见表 1-10。

3)正截面承载力计算公式及方法

《公桥规》规定配有纵向受力钢筋和普通箍筋的轴心受压构件正截面承载力(图6-5)计算式为:

$$\gamma_0 N_d \leq N_u = 0.9\varphi(f_{cd}A + f'_{sd}A'_s) \tag{6-7}$$

式中,N_d 为轴向力组合设计值;φ 为轴心受压构件稳定系数,按附表 1-10 取用;A 为构件毛截面面积;A'_s 为全部纵向钢筋截面面积;f_{cd} 为混凝土轴心抗压强度设计值;f'_{sd} 为纵向普通钢筋抗压强度设计值。

当纵向钢筋配筋率 $\rho' = \dfrac{A'_s}{A} > 3\%$ 时,式(6-7)中 A 应改用混凝土截面净面积:$A_n = A - A'_s$。

普通箍筋柱的正截面承载力计算分为截面设计和强度复核两种情况。

(1)截面设计

已知截面尺寸,计算长度 l_0,混凝土轴心抗压强度和钢筋抗压强度设计值,轴向压力组合设计值 N_d,求纵向钢筋所需面积 A'_s。

首先计算长细比,由附表 1-10 查得相应的稳定系数 φ。

在式(6-7)中,令 $N_u = \gamma_0 N_d$,γ_0 为结构重要性系数,则可得到:

$$A'_s = \frac{1}{f'_{sd}}\left(\frac{\gamma_0 N_d}{0.9\varphi} - f_{cd}A\right) \tag{6-8}$$

由 A'_s 计算值及构造要求选择并布置钢筋。

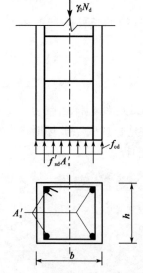

图 6-5 普通箍筋柱正截面承载力计算图式

(2)截面复核

已知截面尺寸,计算长度 l_0,全部纵向钢筋的截面面积 A'_s,混凝土轴心抗压强度和钢筋抗压强度设计值,轴向力组合设计值 N_d,求截面承载力 N_u。

首先,应检查纵向钢筋及箍筋布置构造是否符合要求。

由已知截面尺寸和计算长度 l_0 计算长细比,由附表 1-10 查得相应的稳定系数 φ。

由式(6-7)计算轴心压杆正截面承载力 N_u,且应满足 $N_u > \gamma_0 N_d$。

4)构造要求

(1)混凝土

轴心受压构件的正截面承载力主要由混凝土来提供,故一般多采用 C25~C40 混凝土。

(2)截面尺寸

轴心受压构件截面尺寸不宜过小,因长细比越大,φ 值越小,承载力降低很多,不能充分利用材料强度。构件截面尺寸不宜小于 250mm。

(3)纵向钢筋

①纵向受力钢筋一般采用 HPB300 级、HRB400 级和 HRBF400 级和 HRB500 等热轧钢筋。纵向受力钢筋的直径应不小于 12mm。在构件截面上,纵向受力钢筋至少应有 4 根,并且在截面每一角隅处必须布置一根。

②纵向受力钢筋的净距不应小于 50mm,也不应大于 350mm;对水平浇筑混凝土预制构件,其纵向钢筋的最小净距采用受弯构件的规定要求。纵向钢筋最小混凝土保护层厚度详见附表 1-8。

③对于纵向受力钢筋的配筋率要求,应考虑混凝土徐变及可能存在的较小弯矩。《公桥规》规定了纵向钢筋的最小配筋率 ρ_{min}(%),详见附表 1-9;构件的全部纵向钢筋配筋率不宜超过 5%。一般纵向钢筋的配筋率 ρ' 为 1%~2%。

(4)箍筋

①普通箍筋柱中的箍筋必须做成封闭式,箍筋直径应不小于纵向钢筋直径的 1/4,且不小于 8mm。

②箍筋的间距应不大于纵向受力钢筋直径的 15 倍,且不大于构件截面的较小尺寸(圆形截面采用 0.8 倍直径),并不大于 400mm。

③在纵向钢筋搭接范围内,箍筋的间距应不大于纵向钢筋直径的 10 倍,且不大于 200mm。

④当纵向钢筋截面面积超过混凝土截面面积 3% 时,箍筋间距应不大于纵向钢筋直径的 10 倍,且不大于 200mm。

6.3 配有纵向钢筋和螺旋箍筋的轴心受压构件

1)破坏特性

对于配有纵向钢筋和螺旋箍筋的轴心受压短柱,沿柱高连续缠绕的、间距很密的螺旋箍筋犹如一个套筒,将核心部分的混凝土约束住,有效地限制了核心混凝土的横向变形,从而提高了柱的承载力。

由图 6-6 中所示的螺旋箍筋柱轴压力-混凝土压应变曲线可见,在混凝土压应变 $\varepsilon_c = 2 \times 10^{-3}$ 以前,螺旋箍筋柱的轴力-混凝土压应变变化曲线与普通箍筋柱基本相同。当轴力继续增加,直至混凝土和纵筋的压应变 ε 达到 $3 \times 10^{-3} \sim 3.5 \times 10^{-3}$ 时,纵筋已经开始屈服,箍筋外面的混凝土保护层开始崩裂剥落,混凝土的截面面积减小,轴力略有下降。这时,核心部分混凝土由于受到螺旋箍筋的约束,仍能继续受压,核心混凝土处于三向受压状态,其抗压强度超过了轴心抗压强度f_c,补偿了剥落的外围混凝土所承担的压力,曲线逐渐回升。随着轴力不断增大,螺旋箍筋中的环向拉力也不断增大,直至螺旋箍筋达到屈服,不能再约束核心混凝土横向变形,混凝土被压碎,构件即告破坏。这时,荷载达到第二次峰值,柱的纵向压应变可达到0.01以上。

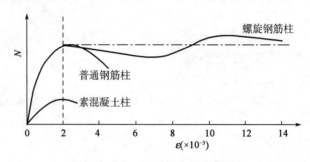

图6-6 轴心受压柱的轴力-应变曲线

由图6-6也可见到,螺旋箍筋柱具有很好的延性,在承载力不降低的情况下,其变形能力比普通箍筋柱提高很多。

2) 正截面承载力计算

(1) 计算公式推导

螺旋箍筋柱的正截面破坏时核心混凝土压碎、纵向钢筋已经屈服,而在破坏之前,柱混凝土保护层早已剥落。

根据图6-7所示螺旋箍筋柱截面受力图式,由平衡条件可得到:

$$N_u = f_{cc}A_{cor} + f_s'A_s' \tag{6-9}$$

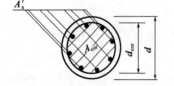

图6-7 螺旋箍筋柱受力计算图式

式中,f_{cc}为处于三向压应力作用下核心混凝土的抗压强度;A_{cor}为核心混凝土面积;A_s'为纵向钢筋面积。

螺旋箍筋对其核心混凝土的约束作用,使混凝土抗压强度提高,根据圆柱体三向受压试验结果,约束混凝土的轴心抗压强度可得到下述近似表达式:

$$f_{cc} = f_c + k\sigma_2 \tag{6-10}$$

式中,σ_2为作用于核心混凝土的径向压应力值。

螺旋箍筋柱破坏,螺旋箍筋达到了屈服强度,它对核心混凝土提供了侧压应力σ_2。现取螺旋箍筋间距s范围内柱体,沿螺旋箍筋的直径切开成脱离体(图6-8),由隔离体的平衡条件可得到:

$$\sigma_2 d_{cor} s = 2f_s A_{s01}$$

整理后为:

$$\sigma_2 = \frac{2f_s A_{s01}}{d_{cor} S} \tag{6-11}$$

式中,A_{s01}为单根螺旋箍筋的截面面积;f_s为螺旋箍筋的抗拉

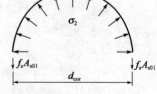

图6-8 螺旋箍筋的受力状态

强度;s 为螺旋箍筋的间距;d_{cor} 为截面核心混凝土的直径,$d_{cor} = d - 2c$;c 为纵向钢筋至柱截面边缘的径向混凝土保护层厚度。

现将间距为 s 的螺旋箍筋,按钢筋体积相等的原则换算成纵向钢筋的面积,称为螺旋箍筋柱的间接钢筋换算截面面积 A_{s0},即:

$$\pi d_{cor} A_{s01} = A_{s0} s \Rightarrow A_{s0} = \frac{\pi d_{cor} A_{s01}}{s} \tag{6-12}$$

将式(6-12)代入式(6-11),则可得到:

$$\sigma_2 = \frac{2f_s A_{s01}}{d_{cor} s} = \frac{2f_s}{d_{cor} s} \cdot \frac{A_{s0} s}{\pi d_{cor}} = \frac{2f_s A_{s0}}{\pi (d_{cor})^2} = \frac{f_s A_{s0}}{2 \frac{\pi (d_{cor})^2}{4}} = \frac{f_s A_{s0}}{2 A_{cor}}$$

将 $\sigma_2 = \dfrac{f_s A_{s0}}{2 A_{cor}}$ 代入式(6-10),可得到:

$$f_{cc} = f_c + \frac{k f_s A_{s0}}{2 A_{cor}} \tag{6-13}$$

将式(6-13)代入式(6-9),整理并考虑实际间接钢筋作用影响,即得到螺旋箍筋柱正截面承载力的计算式:

$$\gamma_0 N_d \leqslant N_u = 0.9(f_{cd} A_{cor} + k f_{sd} A_{s0} + f'_{sd} A'_s) \tag{6-14}$$

式中,k 称为间接钢筋影响系数,混凝土强度等级为 C50 及以下时,取 $k = 2.0$;混凝土强度等级为 C50 ~ C80 时,取 $k = 2.0 \sim 1.70$,中间值直线插入取用。

从式(6-14)可知,螺旋筋仅能间接地提高强度,对柱的稳定性问题毫无帮助,因此长柱和中长柱只能按普通箍筋柱计算,不考虑螺旋筋作用。

(2)《公桥规》对计算公式应用的规定

①为了保证在使用荷载作用下,螺旋箍筋混凝土保护层不致过早剥落,螺旋箍筋柱的承载力计算值[按式(6-14)计算],不应比按式(6-7)计算的普通箍筋柱承载力大 50%,即满足:

$$0.9(f_{cd} A_{cor} + k f_{sd} A_{s0} + f'_{sd} A'_s) \leqslant 1.35 \varphi (f_{cd} A + f'_{sd} A'_s) \tag{6-15}$$

②当遇到下列任意一种情况时,不考虑螺旋箍筋的作用,而按式(6-7)计算构件的承载力。

a. 当构件长细比 $\lambda = \dfrac{l_0}{i} \geqslant 48$($i$ 为截面最小回转半径)时,对圆形截面柱,长细比 $\lambda = \dfrac{l_0}{d} \geqslant 12$($d$ 为圆形截面直径时)。这是由于长细比较大的影响,螺旋箍筋不能发挥其作用。

b. 当按式(6-14)计算承载力小于按式(6-7)计算的承载力时。因为式(6-14)中只考虑了混凝土核心面积,当柱截面外围混凝土较厚时,核心面积相对较小,会出现这种情况,这时就应按式(6-7)进行柱的承载力计算。

c. 当 $A_{s0} < 0.25 A'_s$ 时。螺旋钢筋配置得太少,不能起显著作用。

螺旋箍筋柱的截面设计和复核均依照式(6-14)及其要求来进行。

3)构造要求

(1)螺旋箍筋柱的纵向钢筋应沿圆周均匀分布,其截面积应不小于箍筋圈内核心截面面积的 0.5%。常用的配筋率 $\rho' = A'_s / A_{cor}$ 在 0.8% ~ 1.2% 之间。

(2)构件核心截面面积 A_{cor} 应不小于构件整个截面面积 A 的 2/3。

(3)螺旋箍筋的直径不应小于纵向钢筋直径的 1/4,且不小于 8mm,一般采用 8 ~ 12mm。

为了保证螺旋箍筋的作用,螺旋箍筋的间距 s 应满足:

① s 应不大于核心直径 d_{cor} 的 1/5,即 $s \leq \dfrac{1}{5}d_{cor}$;

② s 应不大于 80mm,且不应小于 40mm,以便施工。

【**例 6-1**】 钢筋混凝土轴心受压圆柱,直径 40cm,柱高 5m,两端固结,C30 混凝土,纵筋 HRB400 为 8⌀20,核心直径 $d_{cor}=35$cm,螺旋形箍筋 Φ20,HPB300,螺距 $s=4$cm,求 N_u。

解:检查是否符合按螺旋箍筋柱公式计算的适用条件:

(1) 长细比:$\lambda = \dfrac{l_0}{d} = \dfrac{0.5 \times 5}{0.4} = 6.25 < 12$,故可以按螺旋箍筋柱计算。

(2) 核心面积:$A_{cor} = \dfrac{\pi d_{cor}^2}{4} = 96\,163\,(\text{mm}^2) > \dfrac{2}{3}A = \dfrac{2}{3} \times \dfrac{\pi}{4} \times 400^2 = 83\,733\,(\text{mm}^2)$。

(3) 纵筋面积:$A'_s = 2\,513\,\text{mm}^2 > 0.5\%A_{cor} = 481\,\text{mm}^2$。

(4) 螺旋箍筋换算截面面积:

$$A_{s0} = \dfrac{\pi d_{cor} A_{s01}}{s} = \dfrac{3.14 \times 350 \times 201.1}{40} = 5525\,(\text{mm}^2) >$$
$$0.25 A'_s = 0.25 \times 2\,513 = 628\,(\text{mm}^2)$$

$$s = 40\,(\text{mm}) < \dfrac{1}{5}d_{cor} = \dfrac{1}{5} \times 350 = 70\,(\text{mm}),且\ s = 40\text{mm} < 80\text{mm},且\ s \nless 40\text{mm}$$

故:

$$N_u = 0.9(f_{cd}A_{cor} + kf_{sd}A_{s0} + f'_{sd}A'_s)$$
$$= 0.9 \times (13.8 \times 96\,163 + 2 \times 250 \times 5\,525 + 330 \times 2\,513)$$
$$= 4\,918.84 \times 10^3\,(\text{N}) = 4\,918.84\,(\text{kN})$$

检查混凝土保护层是否会剥落:

$$N'_u = 0.9\varphi(f_{cd}A + f'_{sd}A'_s)$$
$$= 0.9 \times (13.8 \times \dfrac{\pi}{4} \times 400^2 + 330 \times 2\,513) = 2\,274.42\,(\text{kN})$$

$1.5N'_u = 1.5 \times 2\,274.42 = 3\,411.63\,(\text{kN}) < N_u$,故混凝土保护层会剥落。

所以本柱的承载力取为 3 411.63kN。

由以上可知:间接钢筋(螺旋箍筋)能有效地约束核心混凝土,故能较大地提高柱的承载能力。但与受弯等构件一样,单纯靠增加配筋是不能无限提高构件强度的,条件 $N_u < 1.5N'_u$ 是对采用间接钢筋(螺旋箍筋)提高柱强度的上限值。

思考练习题

1. 试述钢筋混凝土轴心受压短柱和长柱的破坏特征。什么叫作长柱的稳定系数 φ?影响稳定系数 φ 的主要因素有哪些?

2. 钢筋混凝土轴心受压柱在长期荷载作用下为什么承载力有所降低?

3. 为什么轴心受压柱在卸载时混凝土会产生拉应力?

4. 对于轴心受压普通箍筋柱,《公桥规》为什么规定纵向受压钢筋的最大配筋率和最小配筋率? 对于纵向钢筋在截面上的布置以及复合箍筋设置,《公桥规》有什么规定?

5. 试说明钢筋混凝土轴心受压构件中纵筋和箍筋的作用。

6. 配有纵向钢筋和普通箍筋的轴向受压构件与配有纵向钢筋和螺旋箍筋的轴心受压构件的正截面承载力计算有何不同?

7. 试分述螺旋箍与普通箍在钢筋混凝土受压柱中的作用?

8. 某现浇柱截面尺寸定为 $250\text{mm} \times 250\text{mm}$,柱高 $H=4\text{m}$,根据两端支承条件决定其计算长度 $l_0 = 0.7H = 2.8\text{m}$,柱内配有 $4\Phi16$ 的纵筋;构件混凝土 C25,柱承受的纵向力 $N_d = 530.1\text{kN}$,问此柱是否安全?

9. 某现浇圆形截面柱的直径为 $d=400\text{mm}$,承受轴力 $N=176.7\text{kN}$,根据两端支承条件决定其计算长度 $l=0.7H=4.2\text{m}$,混凝土 C25。柱中纵筋及箍筋均采用 HPB300 钢筋。求柱中配筋。

10. 配有纵向钢筋和普通箍筋的轴心受压构件的截面尺寸 $b \times h = 250\text{mm} \times 250\text{mm}$,构件计算长度 $=5\text{m}$;C30 混凝土,HRB400 级钢筋,纵向钢筋面积 $A'_s = 804\text{mm}^2 (4\Phi16)$;Ⅰ类环境条件;安全等级为二级,轴向压力组合设计值 $N_d = 560\text{kN}$,试进行构件承载力校核。

11. 配有纵向钢筋和普通箍筋的轴心受压构件的截面尺寸 $b \times h = 200\text{mm} \times 250\text{mm}$,构件计算长度 $l_0 = 4.3\text{m}$;C30 混凝土,HRB400 级钢筋,纵向钢筋面积 $A'_s = 678\text{mm}^2 (6\Phi12)$,Ⅰ类环境条件,安全等级为二级,试求该构件能承受的最大轴向压力组合设计值 N_d。

12. 配有纵向钢筋和螺旋箍筋的轴心受压构件的截面为圆形,直径 $d = 450\text{mm}$,构件计算长度 $l_0 = 3\text{m}$,C30 混凝土,纵向钢筋采用 HRB400 级钢筋,箍筋采用 HPB300 级钢筋;Ⅱ类环境条件,安全等级为一级,轴向压力组合设计值 $N_d = 550\text{kN}$,试进行构件的截面设计和承载力复核。

第7章
偏心受压构件的正截面承载力计算

偏心受压构件是指当轴向压力 N 的作用点偏离构件的几何形心（有偏心距）时的受力构件，如图 7-1 所示。

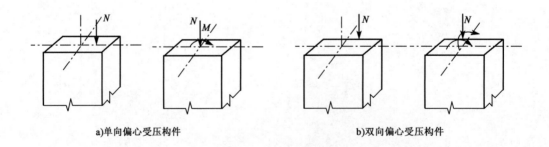

a) 单向偏心受压构件　　　　　　　　b) 双向偏心受压构件

图 7-1　偏心受压构件力的作用位置

截面上同时承受轴向压力和弯矩的构件称为压弯构件，如图 7-2 所示，其正截面承载力有时也按偏心受压构件计算。

偏心受压构件在桥梁工程中广泛应用，如桥墩、桥塔、主拱圈等构件大多是偏心受压构件。常用截面形式有：矩形、圆形、I 形、T 形、箱形，如图 7-3 所示。本章学习单向偏压构件的正截面承载力计算。

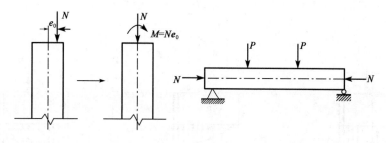

图 7-2　偏心受压构件与压弯构件

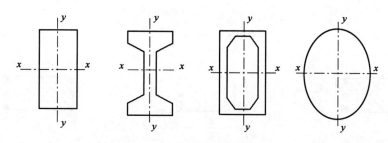

图 7-3　偏心受压构件截面形式

7.1　偏心受压构件正截面受力特点和破坏形态

根据钢筋混凝土偏心受压构件的试验结果,归纳出其正截面破坏形态和受力特点。

1) 偏心受压构件的破坏形态

偏心受压构件随偏心距的大小不同,有以下两种主要的破坏形态。

(1) 受拉破坏——大偏心受压破坏

大偏心受压破坏一般出现在相对偏心距(e_0/h)较大且受拉钢筋配置得不太多的情况下。

破坏特征表现为:截面部分受拉,部分受压,受拉一侧钢筋应力先达到抗拉屈服强度,随后,混凝土被压碎,受压区钢筋达到压屈强度,其破坏形态与双筋截面梁相似。

(2) 受压破坏——小偏心受压破坏

小偏心受压破坏出现在偏心距较小时,或偏心距(e_0/h)较大,但受拉钢筋数量比较多的情况;当偏心距(e_0/h)很小,尤其远离偏心力一侧的钢筋少、靠近偏心力一侧的钢筋过多时的情况。

破坏特征表现为:靠近偏心力一侧的混凝土首先达到极限压应变而压碎,该侧的钢筋达到压屈强度,远离偏心力一侧的钢筋可能受拉也可能受压,一般达不到屈服强度。

小偏心受压短柱截面受力的几种情况如图 7-4 所示。

2) 大小偏心受压破坏的界限

(1) 界限破坏定义:受拉一侧钢筋达到屈服强度时,受压边缘混凝土也刚好达到极限压应变而压碎。

(2) 判别条件:偏心受压构件的破坏形式与受弯构件类似,因此在判断大小偏压的破坏条件时,采用了受弯构件相对界限受压区高度 ξ_b 来判别两种破坏形态。

445

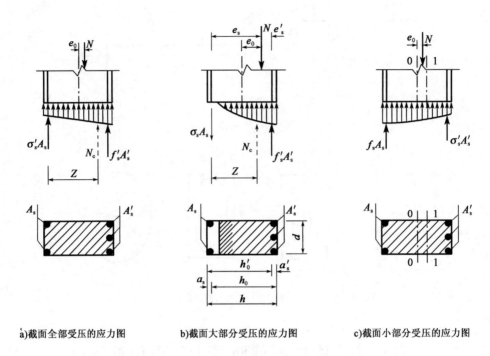

a) 截面全部受压的应力图　　b) 截面大部分受压的应力图　　c) 截面小部分受压的应力图

图 7-4　小偏心受压短柱截面受力的几种情况

当 $\xi \leq \xi_b$ 时，为大偏心受压破坏；

当 $\xi > \xi_b$ 时，为小偏心受压破坏。

3) 偏心受压构件的弯矩-轴力相互关系

根据大量的试验得到偏压构件的弯矩-轴力相互关系，用 M-N 关系曲线表示，如图 7-5 所示。

（1）当截面上的 M-N 组合落在曲线 $abdc$ 上或曲线以外，则构件发生破坏，曲线 ab 为二次抛物线，曲线 cd 是接近于直线的二次函数曲线。

（2）$e_0 = M_d/N_d = \tan\theta$，$\theta$ 越大，e_0 越大，曲线 cb 对应小偏心受压破坏，曲线 ab 对应大偏心受压破坏。

（3）三个特征点：c 点（轴心受压），b 点（大小界限破坏），a 点（受弯构件）。

（4）各部分的特点：大偏心受压的 ab 段（受拉破坏段），轴压力的增加会使抗弯能力增加；小偏心受压的 bcd 段（受压破坏段），轴压力的增加会使其抗弯能力减小。

偏压构件的 M-N 曲线说明了在受力过程中，弯矩和轴力的相互制约关系，在工程中应该正确把握。

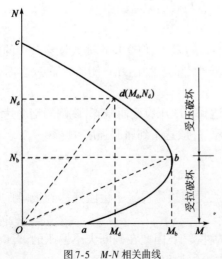

图 7-5　M-N 相关曲线

7.2 偏心受压构件的纵向弯曲

偏心受压构件的长细比大小不同(短柱、长柱),破坏类型也不同。

1)偏心受压构件的破坏特点

(1)短柱

在偏心压力作用下,侧向挠度 u 值很小,偏心压力产生的附加变形较小,可不计其影响。构件的破坏是由于材料强度耗尽而丧失承载力,俗称"材料破坏",如图7-6的 OB 曲线所示。

(2)长柱

在偏心压力作用下,侧向挠度 u 盾较大,偏心压力产生的附加变形不能忽略,附加弯矩影响加速破坏。构件最终还是因为材料强度耗尽而破坏,如图7-6的 OC 曲线所示。

(3)细长柱

长细比很大的柱,当偏心压力达到一定值时,侧向挠度 u 值迅速增加,此时,构件截面上钢筋和混凝土的应变均未达到破坏时的极限值,提前丧失承载力,发生了失稳破坏,如图7-6的 OE 曲线所示。

从以上可知,工程中应尽量控制采用细长柱,而对于长柱则应考虑附加偏心的影响。

2)偏心距增大系数

对于长柱,偏心压力产生的附加弯矩影响,是通过计入附加挠度来考虑的(图7-7),即:

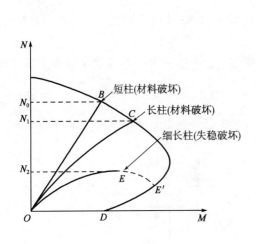
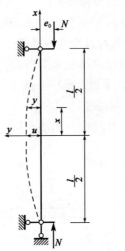
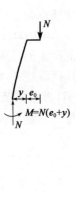

图 7-6 构件长细比的影响   图 7-7 偏心受压构件的受力图式

$$M = N(e_0 + u) = N\frac{e_0 + u}{e_0}e_0$$

令

$$\eta = \frac{e_0 + u}{e_0} = 1 + \frac{u}{e_0} \tag{7-1}$$

则

$$M = N \cdot \eta e_0$$

式中,η 称为偏心受压构件轴向力偏心距增大系数。

《公桥规》规定,偏心距增大系数按下式计算:

$$\eta = 1 + \frac{1}{1\,300 e_0/h_0}(L_0/h)^2 \zeta_1 \zeta_2 \tag{7-2}$$

$$\zeta_1 = 0.2 + 2.7 \frac{e_0}{h_0} \leqslant 1.0 \tag{7-3a}$$

$$\zeta_2 = 1.15 - 0.01 \frac{l_0}{h} \leqslant 1.0 \tag{7-3b}$$

式中,ζ_1 为荷载偏心率对截面曲率的影响系数;ζ_2 为偏心受压构件长细比对截面曲率的影响系数。

《公桥规》规定,$l_0/i > 17.5$、l_0/h(矩形截面)> 5、l_0/d_1(圆形截面)> 4.4 的长柱,应考虑构件在弯矩作用平面内的变形对轴向力偏心距的影响。此时,应将轴向力对截面重心轴的偏心距 e_0 乘以偏心距增大系数 η。

7.3　矩形截面偏心受压构件

1)正截面承载力计算的基本公式

(1)基本假定

①截面应变分布符合平截面假定。

②不考虑混凝土的抗拉强度。

③受压混凝土的极限压应变 $\varepsilon_{cu} = 0.0033 \sim 0.003$。C50 及以下者,$\varepsilon_{cu} = 0.0033$;C80 及以上者,$\varepsilon_{cu} = 0.003$;中间通过线性内插取值。

④混凝土的压应力图形等效为矩形,应力集度为 f_{cd},矩形应力图的高度 x 取按平截面确定的受压区高度 x_c 乘以系数 β,即 $x = \beta x_c$。

(2)计算公式及适用条件

根据矩形截面破坏时的特点和基本假定,得到正截面承载力的计算图式,如图 7-8 所示。

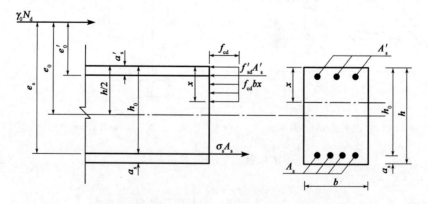

图 7-8　矩形截面偏心受压构件正截面承载力计算图式

沿构件纵轴方向的力平衡,可得:

$$\gamma_0 N_d \leqslant N_u = f_{cd} bx + f'_{sd} A'_s - \sigma_s A_s \tag{7-4}$$

由截面上的力矩平衡,可得:

$$\gamma_0 N_d e_s \leq f_{cd} bx\left(h_0 - \frac{x}{2}\right) + f'_{sd} A'_s (h_0 - a'_s) \tag{7-5}$$

也可以通过对钢筋 A'_s 合力点的力矩之和等于零,得:

$$\gamma_0 N_d e'_s \leq -f_{cd} bx\left(\frac{x}{2} - a'_s\right) + \sigma_s A_s (h_0 - a'_s) \tag{7-6}$$

还可以通过对 $\gamma_0 N_d$ 作用点力矩之和为零,得:

$$f_{cd} bx\left(e_s - h_0 + \frac{x}{2}\right) = \sigma_s A_s e_s - f'_{sd} A'_s e'_s \tag{7-7}$$

式中:

$$e_s = \eta e_0 + h/2 - a_s \tag{7-8}$$

$$e'_s = \eta e_0 - h/2 + a'_s \tag{7-9}$$

$$e_0 = M_d / N_d$$

公式的应用说明如下:

① 钢筋 A_s 的应力 σ_s 取值。

a. 当 $\xi = x/h_0 \leq \xi_b$ 时,构件属于大偏心受压构件,取 $\sigma_s = f_{sd}$;

b. 当 $\xi = x/h_0 > \xi_b$ 时,构件属于小偏心受压构件,σ_s 应按式(7-10)计算,但应满足 $-f'_{sd} \leq \sigma_s \leq f_{sd}$。

$$\sigma_s = \varepsilon_{cu} E_s \left(\frac{\beta h_0}{x} - 1\right) \tag{7-10a}$$

用经验拟合公式:

$$\sigma_s = \frac{f_{sd}}{\xi_b - \beta}(\xi - \beta) \tag{7-10b}$$

② 为了保证构件破坏时,大偏心受压构件截面上的受压钢筋能达到抗压强度设计值 f'_{sd},应满足:

$$x \geq 2a'_s \tag{7-11}$$

当 $x < 2a'_s$ 时,取 $x = 2a'_s$,截面应力分布如图 7-9 所示,受压区混凝土合力作用位置点与受压钢筋承担的压力 $f'_{sd} A'_s$ 作用点重合。

对受压钢筋 A'_s 合力点的力矩之和为零,得:

$$\gamma_0 N_d e'_s \leq f_{sd} A_s (h_0 - a'_s) \tag{7-12}$$

③ 当偏心距很小,离轴向力近的一侧纵向钢筋 A'_s 配置较多,实际重心位置偏向钢筋数量多的一侧,偏心方向将改变(图 7-10),为使远离钢筋 A_s 数量不致过少,尚应符合下列条件:

$$\gamma_0 N_d e' \leq f_{cd} bh\left(h'_0 - \frac{h}{2}\right) + f'_{sd} A_s (h'_0 - a_s) \tag{7-13}$$

$$\begin{cases} h'_0 = h - a'_s \\ e' = h/2 - e_0 - a'_s \end{cases}$$

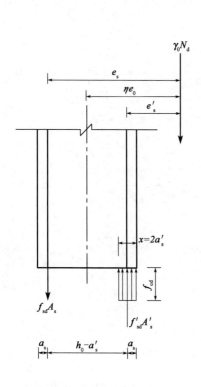
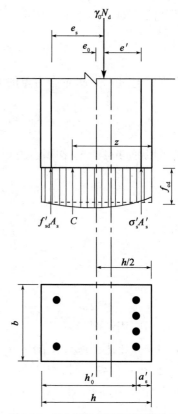

图7-9 当 $x < 2a'_s$ 时，大偏心受压截面计算图式　　图7-10 偏心距很小时截面计算图式

2）矩形截面偏心受压构件非对称配筋的计算方法

构件非对称配筋，是指 $A_s \neq A'_s$。在设计计算时，通常需要根据初始偏心距的大小初步判断构件类型，之后即可采用相应公式计算。

(1) 截面设计

偏压构件的设计要首先确定大、小偏心构件的类型。根据经验，当 $\eta e_0 \geq 0.3 h_0$ 时，可假定为大偏心受压构件；当 $\eta e_0 < 0.3 h_0$ 时，可假定为小偏心构件。

在确定类型后，设计计算步骤如下。

① 当 $\eta e_0 \geq 0.3 h_0$ 时，为大偏心受压构件设计。

a. 第一种情况。

已知：截面尺寸 $b \times h$，计算长度 l_0，荷载效应 N_d 和 M_d，材料强度 f_{cd}、f_{sd} 和 f'_{sd}，初始偏心距 e_0。

求：受压和受拉钢筋数量 A'_s 和 A_s。

解：Ⅰ. 假设 a_s、a'_s，求出 $h_0 = h - a'_s$，确认 $\eta e_0 = 0.3 h_0$。

Ⅱ. 计算 η，短柱 $\eta = 1.0$。计算 e_s、e'_s。

Ⅲ. 画受力图，列基本公式。

Ⅳ. 补充条件 $x = \xi_b h_0$。

Ⅴ. 求 A_s。

取 $\xi = \xi_b$，即 $x = \xi_b h_0$，由式 (7-5) 可得：

$$A'_s = \frac{\gamma_0 N_d e_s - f_{cd} b h_0^2 \xi_b (1 - 0.5\xi_b)}{f'_{sd}(h_0 - a'_s)} \geqslant \rho'_{\min} bh \qquad (7\text{-}14)$$

如果 $0 < A'_s < \rho'_{\min} bh$，$\rho'_{\min} = 0.2\% = 0.002$。

当 A'_s 为负值时，说明截面尺寸偏大，可考虑修改设计，也可以按照 $A'_s = \rho'_{\min} bh$ 选择钢筋并布置 A'_s，然后按 A'_s 为已知情况计算求 A_s。

当 $A'_s \geqslant \rho'_{\min} bh$ 时，将 A'_s 代入基本公式，则所需要的钢筋 A_s 为：

$$A_s = \frac{f_{cd} b h_0 \xi_b + f'_{sd} A'_s - \gamma_0 N_d}{f_{sd}} \geqslant \rho_{\min} bh \qquad (7\text{-}15)$$

式中，$\rho_{\min}$ 为截面受拉一侧钢筋的最小配筋率，按附表1-9选用。

b. 第二种情况。

已知：截面尺寸 $b \times h$，计算长度 l_0，荷载效应 N_d 和 M_d，材料强度 f_{cd}、f_{sd} 和 f'_{sd}，初始偏心距 e_0，受压钢筋面积 A'_s。

求：受拉钢筋面积 A_s。

解：Ⅰ. 假设 a'_s，$h_0 = h - a'_s$。

Ⅱ. 计算 η，短柱 $\eta = 1.0$，确认 $\eta e_0 \geqslant 0.3 h_0$，计算 e_s、e'_s。

Ⅲ. 画受力图，列基本公式，求受压区高度 x：

$$x = h_0 - \sqrt{h_0^2 - \frac{2[\gamma_0 N_d e_s - f'_{sd} A'_s (h_0 - a'_s)]}{f_{cd} b}} \qquad (7\text{-}16)$$

当 $2a'_s < x \leqslant \xi_b h_0$ 时，A_s 为：

$$A_s = \frac{f_{cd} bx + f'_{sd} A'_s - \gamma_0 N_d}{f_{sd}} \qquad (7\text{-}17)$$

当 $x \leqslant \xi_b h_0$，且 $x \leqslant 2a'_s$ 时，取 $x = 2a'_s$，A_s 为：

$$A_s = \frac{\gamma_0 N_d e'_s}{f_{sd}(h_0 - a'_s)} \qquad (7\text{-}18)$$

A_s 应满足最小配筋率的构造要求。

②当 $\eta e_0 < 0.3 h_0$ 时，为小偏心受压构件设计。

a. 第一种情况。

已知：截面尺寸 $b \times h$，计算长度 l_0，荷载效应 N_d 和 M_d，材料强度 f_{cd}、f_{sd} 和 f'_{sd}，初始偏心距 e_0。

求：受压和受拉钢筋数量 A'_s 和 A_s。

解：Ⅰ. 假设 a_s、a'_s，$h_0 = h - a_s$。

Ⅱ. 计算 η，短柱 $\eta = 1.0$，确认 $\eta e_0 < 0.3 h_0$。计算 e_s、e'_s。

Ⅲ. 列基本公式，补充条件。

令 $A_s = \rho'_{\min} bh = 0.002 bh$。

由式(7-6)和式(7-10)可得到以 x 为未知数的方程为：

$$\gamma_0 N_d e'_s = -f_{cd} bx \left(\frac{x}{2} - a'_s\right) + \sigma_s A_s (h_0 - a'_s) \qquad (7\text{-}19)$$

以及

$$\sigma_s = \varepsilon_{cu} E_s \left(\frac{\beta h_0}{x} - 1\right)$$

即得到关于 x 的一元三次方程为：

$$Ax^3 + Bx^2 + Cx + D = 0 \quad (7\text{-}20)$$

$$A = -0.5f_{cd}b \quad (7\text{-}21\text{a})$$

$$B = f_{cd}ba'_s \quad (7\text{-}21\text{b})$$

$$C = \varepsilon_{cu}E_sA_s(a'_s - h_0) - \gamma_0 N_d e'_s \quad (7\text{-}21\text{c})$$

$$D = \beta\varepsilon_{cu}E_sA_s(h_0 - a'_s)h_0 \quad (7\text{-}21\text{d})$$

其中，$e'_s = \eta e_0 - h/2 + a'_s$。

由方程(7-20)求得 x 值后，即可得到相应的相对受压区高度 $\xi = x/h_0$。

当 $h/h_0 > \xi > \xi_b$ 时，以 $\xi = x/h_0$ 代入式(7-10)求得钢筋 A_s 中的应力 σ_s 值。再将钢筋面积 A_s、钢筋应力计算值 σ_s 以及 x 值代入式(7-4)中，即可得所需钢筋面积 A'_s 值，且应满足 $A'_s \geq \rho'_{min}bh$。

当 $\xi \geq h/h_0$ 时，取 $x = h$，则钢筋 A'_s 计算式为：

$$A'_s = \frac{\gamma_0 N_d e_s - f_{cd}bh(h_0 - h/2)}{f'_{sd}(h_0 - a'_s)} \geq \rho'_{min}bh$$

经验公式：

$$\sigma_s = \frac{f_{sd}}{\xi_b - \beta}(\xi - \beta), \quad -f'_{sd} \leq \sigma_s \leq f_{sd} \quad (7\text{-}22)$$

以式(7-22)代入式(7-6)可得到关于 x 的一元二次方程为：

$$Ax^2 + Bx + C = 0 \quad (7\text{-}23)$$

方程中的各系数计算表达式为：

$$A = -0.5f_{cd}bh_0 \quad (7\text{-}24\text{a})$$

$$B = \frac{h_0 - a'_s}{\xi_b - \beta}f_{sd}A_s + f_{cd}bh_0 a'_s \quad (7\text{-}24\text{b})$$

$$C = -\beta\frac{h_0 - a'_s}{\xi_b - \beta}f_{sd}A_s h_0 - \gamma_0 N_d e'_s h_0 \quad (7\text{-}24\text{c})$$

由于式(7-22)中钢筋应力 σ_s 与 ξ 近似为线性关系，因而，利用式(7-22)来求近似解 x，就避免了按式(7-20)来解 x 的一元三次方程的麻烦，这种近似方法适用于构件混凝土强度级别 C50 以下的普通强度混凝土情况。

b. 第二种情况。

已知：截面尺寸 $b \times h$，计算长度 l_0，荷载效应 N_d 和 M_d，材料强度 f_{cd}、f_{sd} 和 f'_{sd}，初始偏心距 e_0，受压钢筋面积 A'_s。

求：受拉钢筋面积 A_s。

先假设 a_s、a'_s，计算 h_0、η、e_s、e'_s 与前相同。

由式(7-5)求截面受压区高度 x，并得到截面相对受压区高度 $\xi = x/h_0$。

当 $h/h_0 > \xi > \xi_b$ 时，为小偏心受压构件。以计算得到的 ξ 值代入式(7-9)，求得钢筋 A_s 的应力 σ_s。由式(7-4)计算得到所需钢筋 A_s 的数量。

当 $\xi \geq h/h_0$ 时，则全截面受压。以 $\xi = h/h_0$ 代入式(7-9)，求得钢筋 A_s 的应力 σ_s，再由式(7-4)可求得钢筋面积 A_{s1}。

为防止钢筋数量 A_s 过少，需满足：

$$A_{s2} \geq \frac{\gamma_0 N_d e'_s - f_{cd}bh\left(h'_0 - \dfrac{h}{2}\right)}{f'_{sd}(h'_0 - a_s)} \tag{7-25}$$

取 A_{s1} 和 A_{s2} 中的较大值,并满足最小配筋率等相关构造要求。

(2) 承载力复核

对已设计好的偏心受压构件求其实际的承载力(N_u),并检验是否满足 $N_u = \gamma_0 N_d$ 的条件,通常应对弯曲平面内的承载力校核,还要对垂直于弯曲平面内的承载力进行校核。对于垂直于弯曲平面内的校核,则是按照轴心受压构件进行校核。

已知:截面尺寸 $b \times h$,计算长度 l_0,荷载效应 N_d 和 M_d,材料强度 f_{cd}、f_{sd}、f'_{sd} 和 γ_0,初始偏心距 e_0,受压和受拉钢筋数量 A'_s 和 A_s。

求:抗压承载力 N_u。

① 弯矩作用平面内的承载力复核

a. 大偏心受压构件承载力复核步骤如下:

(a) 检查复核钢筋的布置数量是否满足构造要求,计算偏心距增大系数 η。

(b) 列出大偏心受压构件计算的基本公式(7-4)和式(7-5),取 $\sigma_s = f_{sd}$,抗力为 N_u。

(c) 求 x:

$$x = h_0 - e_s + \sqrt{(h_0 - e_s)^2 + 2 \times \frac{f_{sd}A_s e_s - f'_{sd}A'_s e'_s}{f_{cd}b}} \tag{7-26}$$

即 $$\xi = \frac{x}{h_0}。$$

(d) 若 $2a'_s \leq x \leq \xi_b h_0$,$N_u = f_{cd}bx + f'_{sd}A'_s - f_{sd}A_s$。

(e) 若 $2a'_s > x$ 时,由式(7-12)求截面承载力:$N_u = \dfrac{f_{sd}A_s(h_0 - a'_s)}{\gamma_0 e'_s}$。

b. 当 $\xi > \xi_b$ 时,为小偏压构件复核,小偏压构件复核步骤如下:

(a) 检查钢筋布置的构造要求。

(b) 列出小偏心受压构件计算的基本公式,抗力为 N_u,由式(7-7)和式(7-10),有:

$$f_{cd}bx\left(e_s - h_0 + \frac{x}{2}\right) = \sigma_s A_s e_s - f'_{sd}A'_s e'_s$$

$$\sigma_s = \varepsilon_{cu} E_s \left(\frac{\beta h_0}{x} - 1\right)。$$

(c) 通过以下一元三次方程,求 x:

$$Ax^3 + Bx^2 + Cx + D = 0 \tag{7-27}$$

式(7-27)中各系数计算表达式为:

$$A = 0.5 f_{cd} b \tag{7-28a}$$

$$B = f_{cd} b (e_s - h_0) \tag{7-28b}$$

$$C = \varepsilon_{cu} E_s A_s e_s + f'_{sd} A'_s e'_s \tag{7-28c}$$

$$D = -\beta \varepsilon_{cu} E_s A_s e_s h_0 \tag{7-28d}$$

式中,$e'_s = \eta e_0 - h/2 + a'_s$。

若钢筋 A_s 中的应力 σ_s 采用 ξ 的线性表达,即式(7-22),则可得到关于 x 的一元二次方

程为:

$$Ax^2 + Bx + C = 0 \quad (7\text{-}29)$$

式(7-29)中各系数计算表达式为:

$$A = 0.5f_{cd}bh_0 \quad (7\text{-}30a)$$

$$B = f_{cd}bh_0(e_s - h_0) - \frac{f_{sd}A_s e_s}{\xi_b - \beta} \quad (7\text{-}30b)$$

$$C = \left(\frac{\beta f_{sd}A_s e_s}{\xi_b - \beta} + f'_{sd}A'_s e'_s\right)h_0 \quad (7\text{-}30c)$$

由式(7-27)或式(7-29),可得到小偏心受压构件截面受压区高度 x 及相应的 ξ 值。

(d) 当 $\xi > \xi_b$ 时,将计算的 ξ 值代入基本公式,可求得钢筋 A_s 的应力 σ_s 值。然后,按照基本公式(7-4),求截面承载力 N_u 并且复核截面承载力。

(e) 当 $\xi > h/h_0$ 时,用 $\xi = h/h_0$ 代入基本公式中求得钢筋 A_s 的应力 σ_s,然后求得截面承载力 N_{u1}。因全截面受压,还需考虑距纵向压力作用点远侧截面边缘破坏的可能性,再由式(7-13)求得截面承载力 N_{u2}。

(f) 构件承载能力 N_u 应取 N_{u1} 和 N_{u2} 中较小值,其意义为既然截面破坏有这种可能性,则截面承载力也可能由其决定。

②垂直于弯矩作用平面内的截面承载力复核

按轴心受压构件复核。

长细比 $l_0/b > 8$,查附录表 1-10 得 φ,则:

$$N_{u\perp} = 0.9\varphi[f_{cd}bh + f'_{sd}(A_s + A'_s)]$$

③构件的承载力

$$N_u = \min\{N_u, N_{u\perp}\}$$

3) 矩形截面偏心受压构件的构造要求

(1) 截面尺寸

$b_{min} \geq 300mm, h/b = 1.5 \sim 3$。为了模板尺寸的模数化,边长宜采用 50mm 的倍数。矩形截面的长边应设在弯矩作用方向。

(2) 纵向钢筋的配筋率

纵向受力钢筋沿截面短边 b 配置。

①截面全部纵向钢筋最小配筋率 $\rho_{min} = 0.5\%$;当混凝土等级大于 C50 时,$\rho_{min} = 0.6\%$。

②一侧纵向钢筋最小配筋率 $\rho_{min} = 0.2\%$。

纵向受力钢筋的常用筋配筋率(全部钢筋截面积与构件截面积之比):

①大偏心受压构件:$\rho = 1\% \sim 3\%$。

②小偏心受压构件:$\rho = 0.5\% \sim 2\%$。

长边 $h \geq 600mm$ 时,应在长边 h 方向设置直径为 $10 \sim 16mm$ 的纵向构造钢筋,必要时相应地设置附加箍筋或复合箍筋。

【例 7-1】 钢筋混凝土偏心受压构件,截面尺寸为 $b \times h = 300mm \times 400mm$,两个方向(弯矩作用方向和垂直于弯矩作用方向)的计算长度均为 $l_0 = 4m$。轴向力组合设计值 $N_d = 232kN$,相应弯矩组合设计值 $M_d = 148kN \cdot m$。预制构件拟采用水平浇筑 C30 混凝土,纵向钢筋为 HRB400 级钢筋,Ⅰ类环境条件,安全等级为一级。试设计和选择钢筋,并进行截面复核。

解: $f_{cd} = 13.8\text{MPa}, f_{sd} = f'_{sd} = 280\text{MPa}, \xi_b = 0.53, \gamma_0 = 1.1$。

1) 截面设计

轴向力计算值 $N = \gamma_0 N_d = 255.2\text{kN}$,弯矩计算值 $M = \gamma_0 M_d = 162.8\text{kN·m}$,可得到偏心距 e_0 为:

$$e_0 = \frac{M}{N} = \frac{162.8 \times 10^6}{255.2 \times 10^3} = 638(\text{mm})$$

弯矩作用平面内的长细比为 $\frac{l_0}{h} = \frac{4\,000}{400} = 10 > 5$,故应考虑偏心距增大系数 η。η 值按式(7-2)计算。设 $a_s = a'_s = 45\text{mm}$,则 $h_0 = h - a_s = 4\,000 - 45 = 355(\text{mm})$。

$$\zeta_1 = 0.2 + 2.7 \frac{e_0}{h_0} = 0.2 + 2.7 \times \frac{638}{355} = 5.05 > 1, \text{取} \zeta_1 = 1.0;$$

$$\zeta_2 = 1.15 - 0.01 \frac{l_0}{h} = 1.15 - 0.010 \times 10 = 1.05 > 1, \text{取} \zeta_2 = 1.0。$$

则

$$\eta = 1 + \frac{1}{1\,300 e_0/h_0}\left(\frac{l_0}{h}\right)^2 \zeta_1 \zeta_2 = 1 + \frac{1}{1\,300 \times \frac{638}{355}} \times 10^2 = 1.04$$

① 大、小偏心受压的初步判定

$\eta e_0 = 1.04 \times 638 = 664(\text{mm}) > 0.3 h_0 [= 0.3 \times 355 = 107(\text{mm})]$,故可先按大偏心受压情况进行设计。$e_s = \eta e_0 + h/2 - a_s = 664 + 400/2 - 45 = 819(\text{mm})$。

② 计算所需的纵向钢筋面积

属于大偏心受压,求钢筋 A_s 和 A'_s 的情况。取 $\xi = \xi_b = 0.53$,由式(7-14)可得到:

$$A'_s = \frac{\gamma_0 N_d e_s - \xi_b (1 - 0.5\xi_b) f_{cd} b h_0^2}{f'_{sd}(h_0 - a'_s)}$$

$$= \frac{1.1 \times 232 \times 10^3 \times 819 - 0.53 \times (1 - 0.5 \times 0.53) \times 13.8 \times 300 \times 355^2}{330 \times (355 - 45)}$$

$$= 56.3(\text{mm}^2) < 240(\text{mm}^2)$$

取 $A'_s = 240\text{mm}^2$。

现选择受压钢筋为 3Φ12,则实际受压钢筋面积 $A'_s = 339\text{mm}^2$,$a'_s = 45\text{mm}$,$\rho' = 0.28\% > 0.2\%$。

由式(7-16)可得到截面受压区高度 x 值为:

$$x = h_0 - \sqrt{h_0^2 - \frac{2[\gamma_0 N_d e_s - f'_{sd} A'_s (h_0 - a'_s)]}{f_{cd} b}}$$

$$= 355 - \sqrt{355^2 - \frac{2[1.1 \times 232 \times 10^3 \times 819 - 330 \times 339 \times (355 - 45)]}{13.8 \times 300}}$$

$$= 151(\text{mm}) < \xi_b h_0 = 0.53 \times 355 = 188(\text{mm})$$

$$> 2a'_s (2 \times 45 = 90\text{mm})$$

取 $\sigma_s = f_{sd}$ 并代入式(7-17)可得到:

$$A_s = \frac{f_{cd} b x + f'_{sd} A'_s - \gamma_0 N_d}{f_{sd}}$$

$$= \frac{13.8 \times 300 \times 151 + 330 \times 339 - 1.1 \times 232 \times 10^3}{330}$$

$$= 1460 (\text{mm}^2) > \rho_{\min} bh = 0.002 \times 300 \times 400 = 240 (\text{mm}^2)$$

现选受拉钢筋为 2 ⊕ 20 + 2 ⊕ 25,$A_s = 1610\text{mm}^2$,$\rho = 1.34\% > 0.2\%$。$\rho + \rho' = 1.62\% > 0.5\%$。

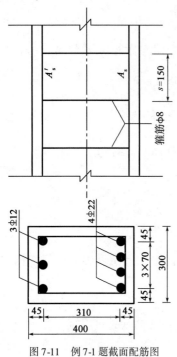

图 7-11 例 7-1 题截面配筋图
(尺寸单位:mm)

设计的纵向钢筋沿截面短边 b 方向布置一排(图 7-11),因偏心压杆采用水平浇筑混凝土预制构件,故纵筋最小净距采用 30mm。设计截面中取 $a_s = a'_s = 45\text{mm}$,钢筋 A_s 的混凝土保护层的厚度为 $(45 - 28.4/2) = 30.4(\text{mm})$,满足规范要求。所需截面最小宽度 $b_{\min} = 2 \times 31 + 3 \times 30 + 2 \times (22.7 + 28.4) = 255(\text{mm}) < b = 300\text{mm}$。

(2) 截面复核

① 垂直于弯矩作用平面的截面复核

因为长细比 $l_0/b = 4000/300 = 13 > 8$,故由附表 1-10 可查得 $\varphi = 0.935$,则:

$$N_u = 0.9\varphi[f_{cd}bh + f'_{sd}(A_s + A'_s)]$$
$$= 0.9 \times 0.935 \times [13.8 \times 300 \times 400 + 330 \times (1610 + 339)]$$
$$= 1934.7 \times 10^3 (\text{N}) = 1934.7 (\text{kN}) > \gamma_0 N_d = 255.2\text{kN}$$

满足设计要求。

② 弯矩作用平面的截面复核

截面实际有效高度 $h_0 = 400 - 45 = 355(\text{mm})$,计算得 $\eta = 1.04$。而 $\eta e_0 = 664\text{mm}$,则:

$$e_s = \eta e_0 + \frac{h}{2} - a_s = 664 + \frac{400}{2} - 45 = 819(\text{mm})$$

$$e'_s = \eta e_0 - \frac{h}{2} + a'_s = 664 - \frac{400}{2} + 45 = 509(\text{mm})$$

假定为大偏心受压,即取 $\sigma_s = f_{sd}$,由式(7-26)可解得混凝土受压区高度 x 为:

$$x = h_0 - e_s + \sqrt{(h_0 - e_s)^2 + 2 \times \frac{f_{sd}A_s e_s - f'_{sd}A'_s e'_s}{f_{cd} b}}$$

$$= 355 - 819 + \sqrt{(355 - 819)^2 + 2 \times \frac{330 \times 1610 \times 819 - 330 \times 339 \times 509}{13.8 \times 300}}$$

$$= 167(\text{mm}) \begin{cases} < \xi_b h_0 = 0.53 \times 355 = 188(\text{mm}) \\ > 2a'_s = 2 \times 45 = 90(\text{mm}) \end{cases}$$

计算表明为大偏心受压。

由式(7-4)可得截面承载力为:

$$N_u = f_{cd}bx + f'_{sd}A'_s - \sigma_s A_s$$
$$= 13.8 \times 300 \times 167 + 330 \times 339 - 330 \times 1610$$
$$= 271.95 \times 10^3 (\text{N}) = 271.95 (\text{kN}) > \gamma_0 N_0 = 255.2\text{kN}$$

满足正截面承载力要求。

4) 矩形截面偏心受压构件对称配筋的计算方法

在实际工程中,偏心受压构件当两侧钢筋数量相差不大时,为使构造简单及便于施工,宜采用对称配筋。装配式偏心受压构件,为了保证安装时不会出错,一般也宜采用对称配筋。

对称配筋是指截面的两侧用相同钢筋等级和数量的配筋,即 $A_s = A'_s, f_{sd} = f'_{sd}, a_s = a'_s$。对称配筋构件的设计原理与非对称配筋设计完全相同。

(1) 截面设计

① 大、小偏心受压构件的判别

假定为大偏心受压,由式(7-4)可得到:

$$\gamma_0 N_d = f_{cd} bx$$

将 $x = \xi h_0$ 代入上式,整理后可得到:

$$\xi = \frac{\gamma_0 N_d}{f_{cd} b h_0} \tag{7-31}$$

当 $\xi \leqslant \xi_b$ 时,按大偏心受压构件设计;
当 $\xi > \xi_b$ 时,按小偏心受压构件设计。

② 大偏心受压构件 ($\xi \leqslant \xi_b$) 的计算

当 $2a'_s \leqslant x \leqslant \xi_b h_0$ 时,直接利用式(7-5)可得到:

$$A_s = A'_s = \frac{\gamma_0 N_d e_s - f_{cd} b h_0^2 \xi (1 - 0.5\xi)}{f'_{sd} (h_0 - a'_s)} \tag{7-32}$$

式中,$e_s = \eta e_0 + \frac{h}{2} - a_s$。当 $x < 2a'_s$ 时,按照式(7-18)求得钢筋。

③ 小偏心受压构件 ($\xi > \xi_b$) 的计算

对称配筋的小偏心受压构件,由于 $A_s = A'_s$,即使在全截面受压情况下,也不会出现远离偏心压力作用点一侧混凝土先破坏的情况。

首先应计算截面受压区高度 x。《公桥规》建议矩形截面对称配筋的小偏心受压构件截面相对受压区高度 ξ 按下式计算:

$$\xi = \frac{\gamma_0 N_d - f_{cd} b h_0 \xi_b}{\dfrac{\gamma_0 N_d e_s - 0.43 f_{cd} b h_0^2}{(\beta - \xi_b)(h_0 - a'_s)} + f_{cd} b h_0} + \xi_b \tag{7-33}$$

求得 ξ 值后,由式(7-32)可求得所需的钢筋面积。

(2) 截面复核

截面复核仍是对偏心受压构件在垂直于弯矩作用平面和弯矩作用平面都进行计算,计算方法与非对称配筋构件方法相同。

【例 7-2】 钢筋混凝土偏心受压构件,截面尺寸为 $b \times h = 400\text{mm} \times 500\text{mm}$,构件在弯矩作用方向和垂直于弯矩作用方向上的计算长度均为 4m。I 类环境条件。轴向力计算值 $N = 600\text{kN}$,弯矩计算值 $M = 360\text{kN} \cdot \text{m}$。C30 混凝土,纵向钢筋采用 HRB400 级钢筋,试求对称配筋时所需钢筋数量并复核截面。

解:$f_{cd} = 13.8\text{MPa}, f_{sd} = f'_{sd} = 330\text{MPa}, \xi_b = 0.53$。

(1) 截面设计

由 $N = 600\text{kN}, M = 300\text{kN} \cdot \text{m}$,可得到偏心距为:

$$e_0 = \frac{M}{N} = \frac{360 \times 10^6}{600 \times 10^3} = 600(\text{mm})$$

在弯矩作用方向,构件长细比 $l_0/h = 4\,000/500 = 8 > 5$。设 $a_s = a'_s = 45\text{mm}, h_0 = h - a_s = 455\text{mm}$,由式(7-2)可计算得到 $\eta = 1.034, \eta e_0 = 620\text{mm}$。

①判别大、小偏心受压

由式(7-31)可得截面相对受压区高度 ξ 为:

$$\xi = \frac{N}{f_{cd}bh_0} = \frac{600 \times 10^3}{13.8 \times 400 \times 455}$$
$$= 0.24 < \xi_b = 0.53$$

故可按大偏心受压构件设计。

②求纵向钢筋面积

由 $\xi = 0.24, h_0 = 450\text{mm}$,得到受压区高度 $x = \xi h_0 = 0.24 \times 455 = 109(\text{mm}) > 2a'_s = 100\text{mm}$。

而

$$e_s = \eta e_0 + \frac{h}{2} - a_s = 620 + \frac{500}{2} - 45 = 825(\text{mm})$$

由式(7-32)可得到所需纵向钢筋面积为:

$$A_s = A'_s = \frac{\gamma_0 N_d e_s - f_{cd}bh_0^2\xi(1 - 0.5\xi)}{f_{sd}(h_0 - a'_s)}$$
$$= \frac{600 \times 10^3 \times 825 - 13.8 \times 400 \times 455^2 \times 0.24 \times (1 - 0.5 \times 0.24)}{330 \times (455 - 45)}$$
$$= 1\,875(\text{mm}^2)$$

选每侧钢筋为 5Φ22,即 $A_s = A'_s = 1\,900\text{mm}^2 > 0.002bh = 0.002 \times 400 \times 500 = 400(\text{mm}^2)$,每侧布置钢筋所需最小宽度 $b_{min} = 2 \times 33.65 + 4 \times 50 + 5 \times 25.1 = 392.8(\text{mm}) < b = 400(\text{mm})$,而 a_s 和 a'_s 取为 45mm。截面布置如图 7-12 所示,构造布置的复合箍筋省略。

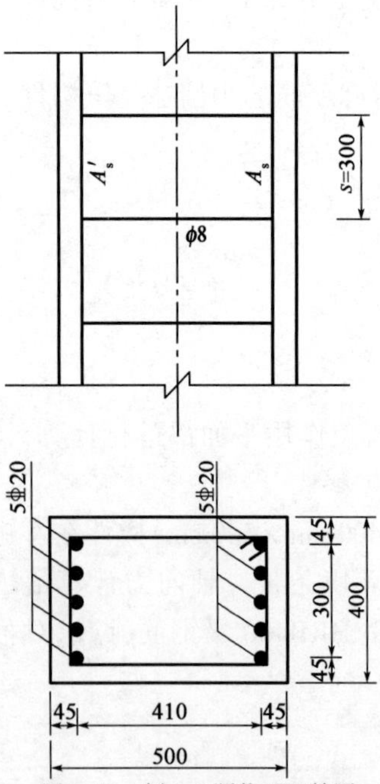

(2)截面复核

①在垂直于弯矩作用平面内的截面抗压承载力

长细比 $l_0/h = 4\,000/400 = 10$,由附录表 1-10 查得 $\varphi = 0.98$,则由式(6-7)可求得 $N_{u1} = 2\,498\text{kN} > N = 600\text{kN}$,满足要求。

②在弯矩作用平面内的截面承载力

由图 7-12 可得到 $a_s = a'_s = 45\text{mm}, A_s = A'_s = 1\,570\text{mm}^2, h_0 = 455\text{mm}$。由式(7-2)求得 $\eta = 1.035$,则 $\eta e_0 = 621\text{mm}$。$e_s = 826\text{mm}, e'_s = \eta e_0 - h/2 + a'_s = 621 - 500/2 + 45 = 416(\text{mm})$。

假定为大偏心受压,即取 $\sigma_s = f_{sd}$,由式(7-26)可解得混凝土受压区高度 x 为:

$$x = h_0 - e_s + \sqrt{(h_0 - e_s)^2 + \frac{2f_{sd}A_s(e_s - e'_s)}{f_{cd}b}} = 455 - 826 +$$
$$\sqrt{(455 - 826)^2 + \frac{2 \times 330 \times 1\,900 \times (826 - 416)}{13.8 \times 400}}$$

图 7-12 例 7-2 题截面配筋图
(尺寸单位:mm)

$$= 110(\text{mm}) \begin{cases} < \xi_b h_0 = 0.56 \times 455 = 255(\text{mm}) \\ > 2a'_s = 2 \times 45 = 90(\text{mm}) \end{cases}$$

故确定为大偏心受压构件。

截面承载力为：

$$N_u = f_{cd}bx = 11.5 \times 400 \times 111 = 510.6 \times 10^3 (\text{N}) = 510.6(\text{kN}) > \gamma_0 N_d = 500(\text{kN})$$

满足要求。

7.4 工字形和 T 形截面偏心受压构件

为了节省混凝土和减轻自重，对于截面尺寸较大的偏心受压构件，一般采用工字形、箱形和 T 形截面，例如大跨径钢筋混凝土拱桥的拱肋、刚架桥的立柱等，常采用这些截面形式。对于工字形、箱形和 T 形截面偏心受压构件的构造要求，也与矩形偏心受压构件相同；工字形、箱形和 T 形截面偏心受压构件的破坏形态、计算方法及原则都与矩形截面偏心受压构件相同，但不允许采用有内折角的箍筋，否则易导致内折角处混凝土崩裂。T 形截面偏压构件箍筋形式如图 7-13 所示。

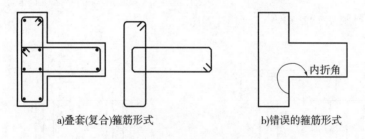

a) 叠套(复合)箍筋形式　　b) 错误的箍筋形式

图 7-13　T 形截面偏压构件箍筋形式

1) 正截面承载力基本计算公式

对于工形截面，在达到承载力极限时，受压区高度可能出现下列情况，如图 7-14 所示，计算公式需针对不同情况采用不同表达式。

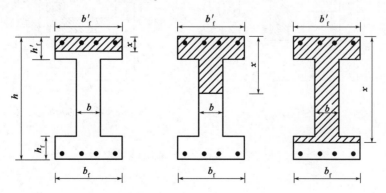

图 7-14　不同受压区高度 x 的工字形截面

(1) 当 $x \leqslant h'_f$ 时,受压区高度位于工字形截面受压翼板内——按照翼板有效宽度为 b'_f、有效高度为 h_0、受压区高度为 x 的矩形截面计算(图 7-15)。

基本计算公式为:

$$\gamma_0 N_d \leqslant N_u = f_{cd} b'_f x + f'_{sd} A'_s - f_{sd} A_s \quad (7\text{-}34)$$

$$\gamma_0 N_d e_s \leqslant f_{cd} b'_f x \left(h_0 - \frac{x}{2} \right) + f'_{sd} A'_s (h_0 - a'_s) \quad (7\text{-}35)$$

$$f_{cd} b'_f x \left(e_s - h_0 + \frac{x}{2} \right) = f_{sd} A_s e_s - f'_{sd} A'_s e'_s \quad (7\text{-}36)$$

式中,$e_s = \eta e_0 + h_0 - y_s$;$e'_s = \eta e_0 - y_s + a'_s$;$y_s$ 为截面形心轴至截面受压边缘距离。

公式的适用条件是:

$$x \leqslant \xi_b h_0$$

及

$$2a'_s \leqslant x \leqslant h'_f \quad (7\text{-}37)$$

式中,h'_f 为截面受压翼板厚度。

当 $x < 2a'_s$ 时,应按式(7-12)来进行计算。

(2) 当 $h'_f < x \leqslant h - h_f$ 时,受压区高度 x 位于肋板内(图 7-16)。

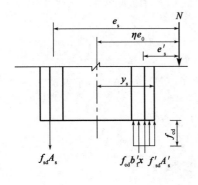
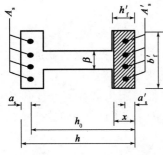

图 7-15 $x \leqslant h'_f$ 时截面计算图式

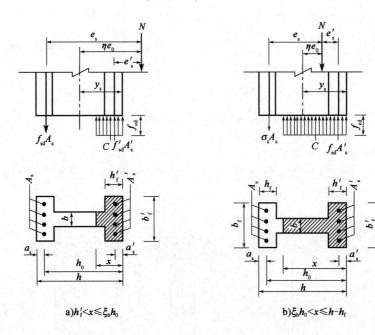

a) $h'_f < x \leqslant \xi_b h_0$ b) $\xi_b h_0 < x \leqslant h - h_f$

图 7-16 $h'_f < x \leqslant h - h_f$ 时截面计算图式

基本计算公式为:

$$\gamma_0 N_d \leqslant N_u = f_{cd} [bx + (b'_f - b) h'_f] + f'_{sd} A'_s - \sigma_s A_s \quad (7\text{-}38)$$

$$\gamma_0 N_d e_s \leq f_{cd}\left[bx\left(h_0 - \frac{x}{2}\right) + (b'_f - b)h'_f\left(h_0 - \frac{h'_f}{2}\right)\right] + f'_{sd}A'_s(h_0 - a'_s) \quad (7\text{-}39)$$

$$f_{cd}bx\left(e_s - h_0 + \frac{x}{2}\right) + f_{cd}(b'_f - b)h'_f\left(e_s - h_0 + \frac{h'_f}{2}\right) = \sigma_s A_s e_s - f'_{sd}A'_s e'_s \quad (7\text{-}40)$$

式中各符号意义同前。

(3) 当 $h - h_f < x \leq h$ 时,受压区高度 x 进入工字形截面受拉或受压较小的翼板内。这时,显然为小偏心受压。

基本计算公式为:

$$\gamma_0 N_d \leq N_u = f_{cd}[bx + (b'_f - b)h'_f + (b_f - b)(x - h + h_f)] + f'_{sd}A'_s - \sigma_s A_s \quad (7\text{-}41)$$

$$Ne_s \leq f_{cd}\left[bx\left(h_0 - \frac{x}{2}\right) + (b'_f - b)h'_f\left(h_0 - \frac{h'_f}{2}\right) + (b_f - b)(x - h + h_f)\right.$$
$$\left.\left(h_f - a_s - \frac{x - h + h_f}{2}\right)\right] + f'_s A'_s (h_0 - a'_s) \quad (7\text{-}42)$$

$$f_{cd}\left[bx\left(e_s - h_0 + \frac{x}{2}\right) + (b'_f - b)h'_f\left(e_s - h_0 + \frac{h'_f}{2}\right) + (b_f - b)(x - h + h_f)\right.$$
$$\left.\left(e_s + a_s - h_f + \frac{x - h + h_f}{2}\right)\right] = \sigma_s A_s e_s - f'_{sd}A'_s e'_s \quad (7\text{-}43)$$

式中, $\sigma_s = \varepsilon_{cu} E_s \left(\dfrac{\beta}{\xi} - 1\right)$。截面计算图式见图 7-17。

(4) 当 $x > h$ 时,则全截面混凝土受压,显然为小偏心受压。这时,取 $x = h$,基本公式为:

$$N \leq N_u = f_{cd}[bh + (b'_f - b)h'_f + (b_f - b)h_f] + f'_{sd}A'_s - \sigma_s A_s \quad (7\text{-}44)$$

$$\gamma_0 N_d e_s \leq f_{cd}\left[bh\left(h_0 - \frac{h}{2}\right) + (b'_f - b)h'_f\left(h_0 - \frac{h'_f}{2}\right) + \right.$$
$$\left.(b_f - b)h_f\left(\frac{h_f}{2} - a_s\right)\right] + f'_{sd}A'_s(h_0 - a'_s) \quad (7\text{-}45)$$

$$f_{cd}\left[bh\left(e_s - h_0 + \frac{h}{2}\right) + (b'_f - b)h'_f\left(e_s - h_0 + \frac{h'_f}{2}\right) + \right.$$
$$\left.(b_f - b)h_f\left(e_s + a_s - \frac{h_f}{2}\right)\right] = \sigma_s A_s e_s - f'_{sd}A'_s e'_s \quad (7\text{-}46)$$

对于 $x > h$ 的小偏心受压构件,还应防止远离偏心压力作用点一侧截面边缘混凝土先压坏的可能性,即应满足:

$$\gamma_0 N_d e'_s \leq f_{cd}\left[bh\left(h'_0 - \frac{h}{2}\right) + (b'_f - b)h'_f\left(\frac{h'_f}{2} - a'_s\right)\right] +$$
$$f_{cd}(b_f - b)h_f\left(h'_0 - \frac{h_f}{2}\right) + f'_{sd}A_s(h'_0 - a_s) \quad (7\text{-}47)$$

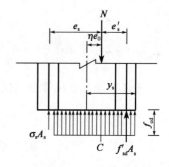
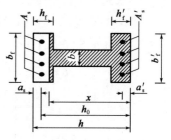

图 7-17 $(h - h_f) < x \leq h$ 时截面的计算图式

式(7-34)~式(7-47)给出了工字形偏心受压构件正截面承载力计算公式。

①当 $h_f = 0, b_f = b$ 时，即 T 形截面承载力计算公式。
②当 $h'_f = h_f = 0, b'_f = b_f = b$ 时，即矩形截面承载力计算公式。

2) 计算方法

在实际工程中，工字形截面偏心受压构件一般采用对称配筋，因此，以下仅介绍对称配筋的工字形截面的计算方法。

对称配筋截面指的是截面对称且钢筋配置对称，对于对称配筋的工字形和箱形截面，即 $b'_f = b_f, h'_f = h_f, A'_s = A_s, f'_{sd} = f_{sd}, a_s = a'_s$。

(1) 截面设计

对于对称配筋截面，可由式(7-38)并且取 $\sigma_s = f_{sd}$，可得到：

$$\xi = \frac{\gamma_0 N_d - f_{cd}(b'_f - b) h'_f}{f_{cd} b h_0} \tag{7-48}$$

当 $\xi \leq \xi_b$ 时，按大偏心受压构件计算；当 $\xi > \xi_b$ 时，按小偏心受压构件计算。

① 当 $\xi \leq \xi_b$ 时

若 $h'_f < x \leq \xi_b h_0$，中和轴位于肋板中，则可将 x 代入式(7-39)，求得钢筋截面面积为：

$$A_s = A'_s = \frac{\gamma_0 N_d e_s - f_{cd}\left[bx\left(h_0 - \frac{x}{2}\right) + (b'_f - b) h'_f\left(h_0 - \frac{h'_f}{2}\right)\right]}{f'_{sd}(h_0 - a'_s)} \tag{7-49}$$

式中，$e_s = \eta e_0 + h/2 - a_s$。

若 $2a'_s \leq x \leq h'_f$，中和轴位于受压翼板内，那么应该重新计算受压区高度 x 为：

$$x = \frac{N}{f_{cd} b'_f} \tag{7-50}$$

则所需钢筋截面为：

$$A_s = A'_s = \frac{\gamma_0 N_d e_s - f_{cd} b'_f x (h_0 - 0.5x)}{f'_{sd}(h_0 - a'_s)} \tag{7-51}$$

当 $x < 2a'_s$ 时，则可按矩形截面方法计算，即按式(7-18)计算所需钢筋 $A_s = A'_s$。

② 当 $\xi > \xi_b$ 时

这时必须重新计算受压区高度 x，然后代入相应公式，求得 $A_s = A'_s$。

计算受压区高度 x 时，采用 $\sigma_s = \varepsilon_{cu} E_s\left(\frac{\beta}{\xi} - 1\right)$ 与相应的基本公式联立求解，例如，当 $h'_f < x \leq h - h_f$ 时，应与式(7-38)和式(7-39)联立求解；当 $h - h_f < x \leq h$ 时，应与式(7-41)和式(7-42)联立求解，将导致关于 x 的一元三次方程的求解。

在设计时，也可以近似采用下式求截面受压区相对高度系数 ξ：

a. 当 $\xi_b h_0 < x \leq h - h_f$ 时：

$$\xi = \frac{\gamma_0 N_d - f_{cd}[(b'_f - b) h'_f + b \xi_b h_0]}{\dfrac{\gamma_0 N_d e_s - f_{cd}\left[(b'_f - b) h'_f\left(h_0 - \dfrac{h'_f}{2}\right) + 0.43 b h_0^2\right]}{(\beta - \xi_b)(h_0 - a'_s)} + f_{cd} b h_0} + \xi_b \tag{7-52}$$

b. 当 $h - h_f < x \leq h$ 时：

$$\xi = \frac{\gamma_0 N_d + f_{cd}[(b_f - b)(h - 2h_f) - b_f \xi_b h_0]}{\dfrac{N e_s + f_{cd}[0.5(b_f - b)(h - 2h_f)(h_0 - a'_s) - 0.43 b_f h_0^2]}{(\beta - \xi_b)(h_0 - a'_s)} + f_{cd} b_f h_0} + \xi_b \tag{7-53}$$

c. 当 $x > h$ 时，取 $x = h_0$。

(2) 截面复核

截面复核方法与矩形截面对称配筋截面复核方法相似，唯计算公式不同。

7.5　圆形截面偏心受压构件

大量圆形截面偏心受压构件试验表明，其破坏特征类似于矩形截面。当 e_0 较大时（$e_0/r \geq 0.86$），首先在远离轴向力 N 一侧的受拉区出现多道明显的横向裂缝，只靠近受拉边缘的一部分钢筋达到屈服强度，具有较大的纵向弯曲变形；当 e_0 较小（$e_0/r \leq 0.35$），远离轴力一侧的横向裂缝出现较晚，数量少且较窄，最后表现为受压区压坏而丧失承载力，受压钢筋也只有靠近受压边缘的一部分达到屈服强度。

圆形截面偏心受压构件与矩形截面的不同之处在于纵向受力钢筋的应力各不相同，视应变大小才能确定，且圆形截面不能截然划分为大小偏心受压界限破坏。

1) 构造要求

(1) 纵向受力钢筋，通常是沿圆周均匀布置，其根数不少于 6 根。

(2) 纵向钢筋的直径不宜小于 12mm。而对于钻孔灌注桩，其截面尺寸较大（桩直径 $D = 800 \sim 1500\text{mm}$），桩内纵向受力钢筋的直径不宜小于 14mm，根数不宜小于 8 根，钢筋间净距不宜小于 50mm，混凝土保护层厚度不小于 $60 \sim 80\text{mm}$。

(3) 箍筋直径不小于 8mm，箍筋间距 $200 \sim 400\text{mm}$。

2) 正截面承载力计算的基本假定

沿周边均匀配筋的圆形偏心受压构件（图 7-18），正截面承载力计算的基本假定为：

(1) 截面变形符合平截面假定。

(2) 构件达到破坏时，受压边缘处混凝土的极限压应变取值为 $\varepsilon_{cu} = 0.0033$。

(3) 受压区混凝土应力分布采用等效矩形应力图。

(4) 不考虑受拉区混凝土参加工作，拉力由钢筋承受。

(5) 将钢筋视为理想的弹塑性体。

对于周边均匀配筋的圆形截面偏心受压构件，由内外力平衡关系来推导出承载力计算公式。

3) 正截面承载力计算的基本公式

由截面上所有水平力平衡条件可知：

$$N_u = D_c + D_s \tag{7-54}$$

由截面上所有力对截面形心轴 y-y 的合力矩平衡条件可知：

$$M_u = M_c + M_s \tag{7-55}$$

式中，D_c 和 D_s 分别为受压区混凝土压应力的合力和所有钢筋的应力合力；M_c 和 M_s 分别为受压区混凝土应力的合力对 y 轴力矩和所有钢筋应力合力对 y 轴的力矩。

(1) 受压区混凝土压应力的合力 D_c 与力矩 M_c

圆形截面偏心受压构件正截面的受压区为弓形，若用 r 表示圆截面的半径，$2\pi\alpha$ 表示受压

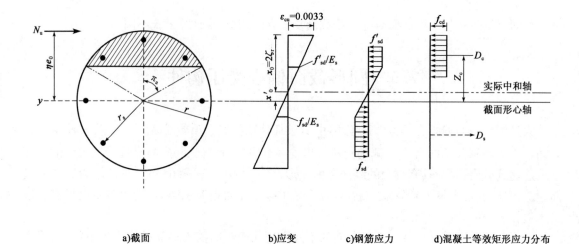

a)截面　　　　b)应变　　　c)钢筋应力　　d)混凝土等效矩形应力分布

图 7-18　圆形截面偏心受压构件计算简图

区对应的圆心角(rad),则截面受压区混凝土面积 A_c 可表示为:

$$A_c = \alpha \left(1 - \frac{\sin 2\pi\alpha}{2\pi\alpha} \right) A \tag{7-56}$$

式中,A 为截面总面积,$A = \pi r^2$。

按照截面受压区等效矩形应力简化,假设受压区混凝土应力相等,均为混凝土抗压强度 f_c,则受压区混凝土的合压力 D_c 及合压力对截面中心产生的力矩 M_c 可表示为:

$$D_c = \alpha f_{cd} A \left(1 - \frac{\sin 2\pi\alpha}{2\pi\alpha} \right) \tag{7-57}$$

$$M_c = \frac{2}{3} f_{cd} A r \frac{\sin^3 \pi\alpha}{\pi} \tag{7-58}$$

(2) 截面钢筋应力的合力 D_s 与力矩 M_s

一般情况下,截面中有部分钢筋的应力达到屈服强度而部分钢筋的应力未达到屈服强度,即靠近受压区或受拉边缘的钢筋可能达到屈服强度,而接近中和轴的钢筋一般达不到屈服强度[图7-19c)]。为简化计算,近似将受拉区和受压区钢筋的应力等效为钢筋强度 f_s 和 f'_s 的均匀分布,纵向受拉普通钢筋截面面积与全部纵向普通钢筋的面积比 α_t 近似表示为:

$$\alpha_t = 1.25 - 2\alpha \geq 0 \tag{7-59}$$

以 A_s 表示钢筋截面总面积,假设 $f_s = f'_s$,截面中钢筋合压力 D_s 以及合力对截面中心产生力矩 M_s 可表示为:

$$D_s = (\alpha - \alpha_t) f_{sd} A_s \tag{7-60}$$

$$M_s = f_{sd} A_s r_s \frac{\sin \pi\alpha + \sin \pi\alpha_t}{\pi} \tag{7-61}$$

将式(7-57)、式(7-58)、式(7-60)、式(7-61)分别代入式(7-54)和式(7-55)中,可得到圆形截面偏心受压构件正截面承载力计算表达式。

$$N_u = \alpha f_{cd} A \left(1 - \frac{\sin 2\pi\alpha}{2\pi\alpha}\right) + (\alpha - \alpha_t) f_{sd} A_s \tag{7-62}$$

$$N_u e_i = \frac{2}{3} f_{cd} A r \frac{\sin^3 \pi\alpha}{\pi} + f_{sd} A_s r_s \frac{\sin \pi\alpha + \sin \pi\alpha_t}{\pi} \tag{7-63}$$

$$\alpha_t = 1.25 - 2\alpha \geq 0$$

$$e_i = \eta e_0$$

式中,A 为圆形截面面积;A_s 为全部纵向普通钢筋截面面积;r 为圆形截面的半径;r_s 为纵向普通钢筋重心所在圆周的半径;e_0 为轴向力对截面重心的偏心距;α 为对应于受压区混凝土截面面积的圆心角(rad)与 2π 的比值;α_t 为纵向受拉普通钢筋截面面积与全部纵向普通钢筋的面积比值,当 α 大于 0.625 时,α_t 为 0;η 为偏心受压区构件轴心力偏心距增大系数。

当采用手算进行圆形截面偏心受压区构件正截面承载力计算时,一般需要对 α 值进行假设并对式(7-62)和式(7-63)采用迭代法来进行计算。

在工程计算中,为了避免圆形截面偏心受压构件正截面承载力迭代法计算的麻烦,对式(7-62)和式(7-63)进行数学处理,即式(7-63)除以式(7-62),可得:

$$\eta \frac{e_0}{r} = \frac{\frac{2}{3} \frac{\sin^3 \pi\alpha}{\pi} + \rho \frac{f_{sd}}{f_{cd}} \frac{r_s}{r} \frac{\sin \pi\alpha + \sin \pi\alpha_t}{\pi}}{\alpha \left(1 - \frac{\sin 2\pi\alpha}{2\pi\alpha}\right) + (\alpha - \alpha_t) \rho \frac{f_{sd}}{f_{cd}}} \tag{7-64}$$

取

$$n_u = \alpha \left(1 - \frac{\sin 2\pi\alpha}{2\pi\alpha}\right) + (\alpha - \alpha_t) \rho \frac{f_{sd}}{f_{cd}}$$

则得到:

$$\eta \frac{e_0}{r} = \frac{\frac{2}{3} \cdot \frac{\sin^3 \pi\alpha}{\pi} + \rho \cdot \frac{f_{sd}}{f_{cd}} \cdot \frac{r_s}{r} \cdot \frac{\sin \pi\alpha + \sin \pi\alpha_t}{\pi}}{n_u} \tag{7-65}$$

式中,ρ 为截面纵向钢筋配筋率,$\rho = \sum_{i=1}^{n} A_{si} / \pi r^2$;$\sum_{i=1}^{n} A_{si}$ 为圆形截面纵向钢筋截面积之和;A_{si} 为单根纵向钢筋截面积;n 为圆形截面上全部纵向钢筋根数。

由式(7-62)可得到圆形截面偏心受压构件正截面承载力计算表达式为:

$$N_u = n_u A f_{cd} \tag{7-66}$$

在式(7-65)中,可将工程中常用的钢筋所在重心所在圆周半径 r_s 与构件圆形截面半径 r 之比(r_s/r)取为代表值,只要给定 $\eta e_0/r$ 和 $(\rho f_{sd}/f_{cd})$ 的值,由式(7-65)可求得相应的 α 和 n_u 值,并可由式(7-66)计算圆形截面偏心受压构件正截面承载力 N_u。

对于混凝土强度等级为 C30~C50、纵向钢筋配筋率在 0.5%~4% 之间,沿周边均匀配置纵向钢筋的圆形截面钢筋混凝土偏心受压构件,《公桥规》采用式(7-65)以及相应的数值计算,给出了由计算表格直接确定或经内插得到计算参数的正截面抗压承载力计算方法,通过查表计算的圆形截面钢筋混凝土偏心受压构件正截面抗压承载力应符合以下要求:

$$\gamma_0 N_d \leq N_u = n_u A f_{cd} \tag{7-67}$$

式中，γ_0 为结构重要性系数；N_d 为构件轴向压力的设计值；n_u 为构件相对抗压承载力，按附表 1-11 确定；A 为构件截面面积；f_{cd} 为混凝土抗压强度设计值。

4）计算方法

（1）截面设计

已知截面尺寸、计算长度、材料强度级别、轴向力计算值 N 及相应的弯矩计算值 M。求所需的纵向钢筋面积 A_s。

首先计算截面偏心距 e_0。判断是否要考虑纵向弯曲对偏心距的影响，通过试算确定 α，计算钢筋面积 A_s，最后选择钢筋并进行截面布置。

（2）截面复核

已知圆形截面直径和实际纵向钢筋面积和布置，构件计算长度，材料强度级别，轴向力计算值 N 及相应的弯矩计算值 M。要求复核截面抗压承载力。

计算截面偏心距 e_0；判断是否要考虑纵向弯曲对偏心距的影响，计算偏心距增大系数 η，并得到参数 $\eta e_0/r$ 的值。

由实际纵向钢筋面积、混凝土和钢筋强度设计值计算得到参数 $\rho f_{sd}/f_{cd}$ 的计算值。

由参数 $\rho f_{sd}/f_{cd}$ 和 $\eta e_0/r$ 计算值，查附表 1-11 得到相应的表格参数 n_u 的值。若不能直接查到时，采用内插法来得到与已知计算参数 $\rho f_{sd}/f_{cd}$ 和 $\eta e_0/r$ 值一致的参数 n_u 的表格值。

由查表得到的 n_u 值代入式（7-67）计算，得到圆形截面钢筋混凝土偏心受压构件正截面抗压承载力 N_u，并满足式（7-67）的要求。

思考练习题

1. 试述偏心受压构件的破坏形态和破坏类型。
2. 钢筋混凝土偏心受压构件截面形式与纵向钢筋布置有什么特点？
3. 现行规范如何考虑偏心受压长柱的纵向弯曲影响，偏心距增大系数 η 与哪些因素有关？
4. 钢筋混凝土矩形截面（非对称配筋）偏心受压构件的截面设计和截面复核中，如何判断是大偏心受压还是小偏心受压？
5. 与非对称布筋的矩形截面偏心受压构件相比，对称布筋设计时的大、小偏心受压的判别方法有何不同之处？
6. 圆形截面偏心受压构件的纵向受力钢筋布置有何特点和要求？箍筋布置有何构造要求？
7.《公桥规》在柱构件设计中引入系数 η 和 φ，它们是分别针对什么受力构件？考虑何种因素？
8. 钢筋混凝土构件设计中，其构造措施的重要性和必要性是什么？
9. 试述钢筋混凝土大偏心受压柱的受力过程与破坏特征。
10. 矩形截面偏心受压构件的截面尺寸为 $b \times h = 300\text{mm} \times 500\text{mm}$，弯矩作用平面内和垂直于弯矩作用平面的计算长度 $l_0 = 6\text{m}$；C30 混凝土和 HRB400 级钢筋；Ⅰ类环境条件，安全等级为二级,轴向力组合设计值 $N_d=265\text{kN}$,相应弯矩组合设计值 $M_d=120\text{kN}\cdot\text{m}$,试按非对称布筋进行截面设计和截面复核。

11. 矩形截面偏心受压构件截面尺寸为 $b\times h=300\text{mm}\times600\text{mm}$,弯矩作用平面内的构件计算长度 $l_0=6\text{m}$,C30 混凝土,HRB400 级钢筋;Ⅰ类环境条件,安全等级为二级;轴向力组合设计值 $N_d=542.8\text{kN}$,相应弯矩组合设计值 $M_d=326.6\text{kN}\cdot\text{m}$,试按非对称布筋进行截面设计。

12. 矩形截面偏心受压构件的截面尺寸为 $b\times h=300\text{mm}\times400\text{mm}$,弯矩作用平面内的构件计算长度 $l_0=4\text{m}$,C30 混凝土;HRB400 级钢筋,Ⅰ类环境条件,安全等级为二级,轴向力组合设计值 $N_d=188\text{kN}$,相应弯矩组合设计值 $M_d=120\text{kN}\cdot\text{m}$,现截面受压区已经配置了 3⌀20 钢筋(单排),$a_s'=40\text{mm}$,试计算所需的受拉钢筋面积 A_s,并选配钢筋。

13. 矩形截面偏心受压构件的截面尺寸为 $b\times h=300\text{mm}\times500\text{mm}$,弯矩作用平面内的构件计算长度 $l_{0x}=3.5\text{m}$,垂直于弯矩作用平面方向的计算长度 $l_{0y}=6\text{m}$。C30 混凝土,HRB300 级钢筋。Ⅰ类环境条件,安全等级为二级。截面钢筋布置如习题 13 图所示,$A_s=339\text{mm}^2$(3⌀12),$A_s'=308\text{mm}^2$(2⌀14)。轴向力组合设计值 $N_d=174\text{kN}$,相应弯矩组合设计值 $M_d=54.8\text{kN}\cdot\text{m}$。试进行截面复核。

14. 矩形截面偏心受压构件的截面尺寸为 $b\times h=250\text{mm}\times300\text{mm}$,弯矩作用平面内和垂直于弯矩作用平面的计算长度均为 $l_0=2.2\text{m}$;C30 混凝土和 HRB300 级钢筋;Ⅰ类环境条件,安全等级为二级,轴向力组合设计值 $N_d=122\text{kN}$,相应弯矩组合设计值 $M_d=58.5\text{kN}\cdot\text{m}$,试按对称布筋进行截面设计和截面复核。

15. 工字形截面偏心受压构件的截面如习题 15 图所示,弯矩作用平面内的计算长度 $l_{0x}=5\text{m}$,垂直于弯矩作用平面方向的计算长度 $l_{0y}=7\text{m}$。C30 混凝土和 HRB300 级钢筋。$A_s=A_s'=1\,257\text{mm}^2$(4⌀20)。Ⅰ类环境条件,安全等级为二级。轴向力组合设计值 $N_d=552\text{kN}$,相应弯矩组合设计值 $M_d=750\text{kN}\cdot\text{m}$,试进行截面复核。

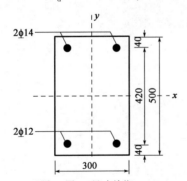

习题 13 图 (尺寸单位:mm)

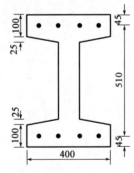

习题 15 图 (尺寸单位:mm)

16. 圆形截面偏心受压构件的截面半径 $r=380\text{mm}$,计算长度 $l_0=8.8\text{m}$。C30 混凝土和 HRB300 级钢筋。Ⅰ类环境条件,安全等级为二级。轴向力组合设计值 $N_d=970\text{kN}$,相应弯矩组合设计值 $M_d=315\text{kN}\cdot\text{m}$,试按查表法进行截面设计和截面复核。

17. 已知钢筋混凝土柱截面尺寸(习题 17 图):$b\times h=400\text{mm}\times550\text{mm}$,$l_0=5.8\text{m}$,内力设计值 $N=900\text{kN}$,$M=300\text{kN}\cdot\text{m}$。混凝土采用 C30,纵筋用 HRB300 级。已知 $(e_0)_{\min}=0.338\,h_0$,且已求得 $\eta=1.16$。试求所需配置的钢筋 $A_s=A_s'$。

18. 矩形截面短柱($\eta=1$)截面尺寸 $b \times h = 450\text{mm} \times 600\text{mm}$,内力设计值 $N = 1\,102\text{kN}$,$M = 385.7\text{kN} \cdot \text{m}$,混凝土 C30,纵筋为 HRB300 级,附加偏心距 $e_a = 20\text{mm}$,$a_s = a_s' = 35\text{mm}$,按对称配筋求纵筋截面面积 $A_s = A_s'$。

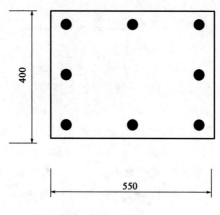

习题 17 图 (尺寸单位:mm)

第8章 受拉构件的承载力计算

8.1 概述

当纵向拉力作用线与构件截面形心轴线相重合时,此构件为轴心受拉构件。当纵向拉力作用线偏离构件截面形心轴线时,或者构件上既作用有拉力,同时又作用有弯矩时,则为偏心受拉构件。桥梁工程中常见的受拉构件有:桁架拱、桁架梁桥中的拉杆、系杆拱桥中的系杆等。

钢筋混凝土受拉构件需配置纵向钢筋和箍筋,箍筋直径应不小于6mm,间距一般为 150~200mm。

8.2 轴心受拉构件的承载力计算

轴心受拉构件开裂以前,混凝土与钢筋共同承担拉力。当构件开裂后,裂缝截面处的混凝土退出工作,拉力由钢筋承担。而当钢筋拉应力达到屈服强度时,构件也达到其极限承载能力。轴心受拉构件的正截面承载力计算式如下:

$$\gamma_0 N_d \leq N_u = f_{sd} A_s \tag{8-1}$$

式中,N_d 为轴向拉力设计值;N_u 为轴心抗拉承载力;f_{sd} 为钢筋抗拉强度设计值;A_s 为截面上全部纵向受拉钢筋截面面积。

对于轴心受拉构件，由已知的拉力设计值进行配筋设计，或者由已知的设计结果求出结构的抗拉力 N_u。

《公桥规》规定：轴心受拉构件及偏心受拉构件的一侧受拉钢筋的配筋率应不小于 $45f_{td}/f_{sd}$，同时不应小于0.2；轴心受拉构件及小偏心受拉构件一侧受拉钢筋的配筋率应按构件毛截面面积计算；大偏心受拉构件的一侧受拉纵筋的配筋率按 $A_s/(bh_0)$ 计算。

8.3 偏心受拉构件的承载力计算

偏心受拉构件的强度计算，按纵向拉力 N_d 的作用位置可分为两种情况：当 N_d 作用在截面钢筋 A_s（离轴向拉力近的钢筋）合力点与 A_s'（离轴向拉力远的一侧的钢筋）合力作用点之间时 $\left(e_0 \leqslant \dfrac{h}{2-a_s}\right)$，为小偏心受拉；当 N_d 作用在截面钢筋 A_s 合力点与 A_s' 合力作用点以外时 $\left(e_0 > \dfrac{h}{2-a_s}\right)$，为大偏心受拉。

1) 小偏心受拉构件的强度计算

小偏心受拉时，构件临破坏前截面混凝土已全部裂通，拉力完全由两侧钢筋承担，因此，小偏心受拉构件的正截面承载力计算图式（图8-1）中，不考虑混凝土的受拉工作；构件破坏时，钢筋 A_s 及 A_s' 的应力均达到抗拉强度设计值 f_{sd}。根据平衡条件，得到基本计算式如下：

$$\gamma_0 N_d e_s \leqslant N_u e_s = f_{sd} A_s' (h_0 - a_s') \tag{8-2}$$

$$\gamma_0 N_d e_s' \leqslant N_u e_s' = f_{sd} A_s (h_0 - a_s) \tag{8-3}$$

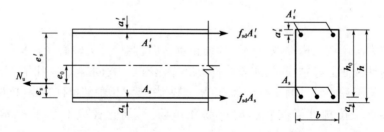

图8-1 小偏心受拉构件正截面承载力计算图式

式(8-2)、式(8-3)中的符号意义详见图8-1。

式(8-2)和式(8-3)中的 e_s 和 e_s' 分别根据下式计算：

$$e_s = \dfrac{h}{2} - e_0 - a_s \tag{8-4}$$

$$e_s' = e_0 + \dfrac{h}{2} - a_s' \tag{8-5}$$

对于偏心拉力的作用，可看成是轴向拉力和弯矩的共同作用，在设计中如有若干组不同的内力组合（M_d、N_d）时，应按最大的轴向拉力组合设计值 N_d 与相应的弯矩组合设计值 M_d 计算钢筋面积。当对称布筋时，离轴向力较远一侧的钢筋 A_s' 的应力可能达不到其抗拉强度设计值,因此,截面设计时,钢筋 A_s 和 A'_s 值均按式(8-2)来求解。

2) 大偏心受拉构件的强度计算

对于正常配筋的矩形截面,大偏心受拉构件的破坏类似于大偏心受压构件,在荷载作用下,截面部分受拉,部分受压,靠近偏心力一侧首先出现裂缝,而离轴向力较远一侧的混凝土仍然受压。随荷载增加,裂缝不断开展,使压区面积减小,破坏时,钢筋 A_s(适当配筋)的应力达到其抗拉强度,然后受压钢筋 A'_s 的应力也达到屈服强度,受压区边缘混凝土的应变达到压应变极限值而破坏。

矩形截面大偏心受拉构件正截面承载力计算图式如图8-2所示,纵向受拉钢筋 A_s 的应力达到其抗拉强度设计值 f_{sd},受压区混凝土应力图形可简化为矩形,其应力为混凝土抗压强度设计值 f_{cd}。受压钢筋 A'_s 的应力可假定达到其抗压强度设计值。

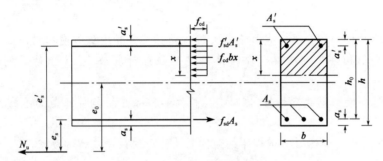

图8-2 大偏心受拉构件计算图式

根据平衡条件,可得到:

$$\gamma_0 N_d \leq N_u = f_{sd}A_s - f'_{sd}A'_s - f_{cd}bx \tag{8-6}$$

$$\gamma_0 N_d e_s \leq N_u e_s = f_{cd}bx\left(h_0 - \frac{x}{2}\right) + f'_{sd}A'_s(h_0 - a'_s) \tag{8-7}$$

$$f_{sd}A_s e_s - f'_{sd}A'_s e'_s = f_{cd}bx\left(e_s + h_0 - \frac{x}{2}\right) \tag{8-8}$$

式中:

$$e_s = e_0 - \frac{h}{2} + a_s \tag{8-9}$$

上述公式的适用条件为:

$$2a'_s \leq x \leq \xi_b h_0$$

式中,ξ_b 为混凝土相对界限受压区高度。

当 $x < 2a'_s$ 时,因受压钢筋离中和轴距离很近,破坏时其应力不能达到抗压强度设计值。此时,可假定混凝土合力中心与受压钢筋 A'_s 重合,即近似地取 $x = 2a'_s$ 进行承载力计算,计算式为:

$$\gamma_0 N_d e'_s \leq N_u = f_{sd}A_s(h_0 - a'_s) \tag{8-10}$$

如果按上式所求得的构件强度比不考虑受压钢筋还小时,则在计算中不考虑受压钢筋 A'_s 的作用。

当已知截面尺寸 $b \times h$、偏心拉力组合值 N_d、偏心距 e_0、需求钢筋面积时,为了能充分发挥材料的强度,宜取 $x = \xi_b h_0$,此时的设计当最为经济。因此,由式(8-6)和式(8-7)可得到:

$$A'_s = \frac{\gamma_0 N_d e_s - f_{cd}bh_0^2 \xi_b(1 - 0.5\xi_b)}{f'_{sd}(h_0 - a'_s)} \tag{8-11}$$

$$A_{s} = \frac{\gamma_0 N_d + f'_{sd} A'_s + f_{cd} b h_0 \xi_b}{f_{sd}} \quad (8\text{-}12)$$

若按式(8-11)求得的 A'_s 过小或为负值,可按最小配筋率或有关构造要求配置 A'_s,然后按式(8-6)~式(8-8)计算 A_s。一般情况下,计算的 x 往往小于 $2a'_s$,这时可按式(8-10)求 A_s。

当为对称配筋的大偏心受拉构件时,由于 $f_{sd} = f'_{sd}$,$A_s = A'_s$,若将上述各值代入式(8-6)后,必然会求得负值 x,亦即属于 $x < 2a'$ 的情况。此时可按式(8-10)求得 A_s 值。

思考练习题

1. 偏心受拉构件如何分类?怎样判别大偏心受拉构件和小偏心受拉构件?

2. 大偏心受拉构件有何破坏特征?试推导其强度计算公式,画出计算图形,写出适用条件。

3. 分析矩形截面双筋受弯构件、偏心受压构件和偏心受拉构件正截面承载力基本计算公式的异同性。

4. 《公桥规》对大、小偏心受拉构件纵向钢筋的最小配筋率有哪些要求?

5. 已知某偏心受拉构件所承受的纵向拉力组合设计值为 $N_d = 700\text{kN}$,弯矩组合设计值 $M_d = 70\text{kN} \cdot \text{m}$,截面尺寸为 $b \times h = 300\text{mm} \times 500\text{mm}$,采用 C30 混凝土和 HRB400 钢筋,结构重要性系数 $\gamma_0 = 1.0$,试进行截面配筋。

6. 某矩形截面偏心受拉构件,采用 C30 混凝土和 HRB400 级钢筋,纵向拉力组合设计值为 $N_d = 150\text{kN}$,弯矩组合设计值为 $M_d = 125\text{kN} \cdot \text{m}$,截面尺寸为 $b \times h = 350\text{mm} \times 600\text{mm}$,结构重要性系数 $\gamma_0 = 1.0$,试对截面进行配筋。

7. 已知矩形截面偏心受拉构件,截面尺寸为 $b \times h = 300\text{mm} \times 500\text{mm}$,采用 C30 混凝土和 HRB400 级钢筋,承受纵向拉力作用,偏心距 $e_0 = 100\text{mm}$,$A_s = 2\,100\text{mm}^2$,$A'_s = 1\,500\text{mm}^2$,$a_s = a'_s = 40\text{mm}$,结构重要性系数 $\gamma_0 = 1.0$,求构件所能承受的纵向力组合设计值。

第 9 章
钢筋混凝土受弯构件的应力、裂缝和变形计算

9.1 概 述

为保证构件的安全、适用、耐久性,《公桥规》规定钢筋混凝土构件必须进行两种状态(承载能力极限状态、正常使用极限状态)、四种状况(持久、短暂、偶然、地震)的计算,前面已介绍了承载力极限状态的计算方法,本章以受弯构件为例,介绍持久状况正常使用极限状态的计算,短期状况构件应力的计算。

1)两种极限状态的区别

(1)承载能力极限状态计算

讨论构件在各种不同受力状态下的承载力计算,承载力满足要求是保证结构安全的首要条件,由此决定了构件的材料、尺寸、配筋及构造。

(2)正常使用极限状态验算

钢筋混凝土构件除了可能由于强度破坏或失稳等原因达到承载能力极限状态外,还可能由于构件变形或裂缝过大等影响构件的适用性及耐久性,而达不到结构正常使用要求。因此,对于所有的钢筋混凝土构件,除要求进行承载力计算外,还要根据使用条件进行正常使用极限状态的验算,以保证构件的正常使用功能要求。

2）钢筋混凝土受弯构件的应力、裂缝和变形计算的内容
（1）施工阶段混凝土和钢筋应力；
（2）使用阶段的变形；
（3）使用阶段的最大裂缝宽度。
3）计算的特点（与承载能力极限状态相比）
（1）计算依据不同。承载能力极限状态是以破坏阶段（Ⅲa）的状态为建立计算图式的基础；而使用阶段一般是指第Ⅱ阶段，即构件带裂缝工作阶段，施工阶段一般在第Ⅰ和第Ⅱ阶段。
（2）影响程度不同。与承载能力极限状态相比，超过正常使用极限状态所造成的后果（如人员伤亡和经济损失）的危害性和严重性相对要小一些、轻一些，因而可适当放宽对其可靠性保证率的要求。
（3）计算内容不同。
①承载能力极限状态。包括截面设计和截面复核，其计算决定了构件设计尺寸、材料、配筋数量及钢筋布置。
②施工和正常使用阶段。施工阶段仅需验算混凝土和钢筋应力不要超过限值，正常使用阶段验算裂缝宽度和变形小于规范规定的各项限值。
（4）荷载效应取值不同。
①承载能力极限状态。汽车荷载应计入冲击系数，作用（或荷载）效应及结构构件的抗力均采用考虑了分项安全系数的设计值；在多种作用（或荷载）效应情况下，应将各效应设计值进行最不利组合，并根据参与组合的作用（或荷载）效应情况，取用不同的效应组合系数。
②正常使用极限状态。汽车荷载应可不计冲击系数，作用（或荷载）效应取用短期效应和长期效应的一种或几种组合。
（5）截面几何参数的计算不同。
与单一材料结构截面几何参数计算不同，考虑钢筋对混凝土截面的贡献，需要采用等效的（或当量的）换算截面计算。

9.2 换算截面

1）定义
将钢筋和混凝土两种材料组成的实际截面换算成为一种拉压性能相同的单一材料组成的当量截面，称换算截面。
2）计算的基本假定
在计算换算截面几何特性时，考虑到在施工阶段是处于整体工作状况，钢筋与混凝土黏结完好，形成一个有机整体；在使用阶段，混凝土可能开裂退出工作的实际情况，计算中采用了一些假定。
（1）平截面假定：梁的正截面在梁受力并发生弯曲变形以后，仍保持为平面。
（2）弹性体假定：压区混凝土应力近似按线性分布。
（3）开裂后受拉区完全不承担拉应力，拉应力完全由钢筋承受。
由上述基本假定可得到计算图式，如图9-1所示。

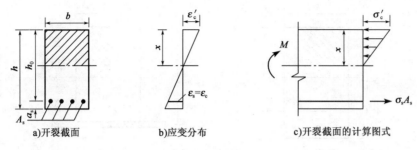

图 9-1 受弯构件的开裂截面

3）换算原则

换算前后合力的大小和作用点的位置不变。

$$A_s\sigma_s = A_{sc}\sigma_c$$

$$A_{sc} = A_s\frac{\sigma_s}{\sigma_c} = A_s\frac{E_s\varepsilon_s}{E_c\varepsilon_c} = \alpha_{Es}A_s\frac{\varepsilon_s}{\varepsilon_c} = \alpha_{Es}A_s$$

式中，A_{sc} 为钢筋换算面积，为钢筋截面面积 A_s 换算成的假想混凝土截面面积；α_{Es} 为钢筋混凝土构件截面的换算系数，等于钢筋的弹性模量与混凝土的弹性模量之比，即 $\alpha_{Es} = \dfrac{E_s}{E_c}$。

4）开裂截面的换算截面几何参数计算

(1) 定义

钢筋混凝土受弯构件受力进入带裂缝工作阶段后，受拉区混凝土退出工作，拉应力由钢筋承受。将受压区的混凝土面积和受拉区钢筋换算的当量截面称为开裂截面的换算截面。

(2) 几何参数计算

① 换算截面面积 A_0

$$A_0 = bx + \alpha_{Es}A_s \tag{9-1}$$

② 换算截面对中和轴的静矩 S_0

a. 受压区：

$$S_{0c} = \frac{1}{2}bx^2 \tag{9-2}$$

b. 受拉区：

$$S_{0t} = \alpha_{Es}A_s(h_0 - x) \tag{9-3}$$

③ 换算截面惯性矩 I_{cr}

$$I_{cr} = \frac{1}{3}bx^3 + \alpha_{Es}A_s(h_0 - x)^2 \tag{9-4}$$

④ 换算截面的受压区高度 x 的计算

a. 矩形截面（图 9-2）。

对于受弯构件，开裂截面的中和轴通过其换算截面的形心轴，即 $S_{0c} = S_{0t}$，可得到：$\dfrac{1}{2}bx^2 = \alpha_{Es}A_s(h_0 - x)$，化简后解得换算截面的受压区高度为：

$$x = \frac{\alpha_{Es}A_s}{b}\left(\sqrt{1 + \frac{2bh_0}{\alpha_{Es}A_s}} - 1\right) \tag{9-5}$$

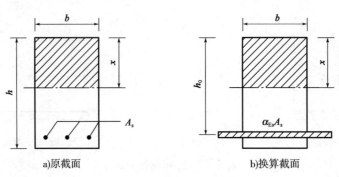

图 9-2　换算截面图

b. T 形截面（图 9-3）。

当 $x \leqslant$ 受压翼板高度 h'_f 时，为第一类 T 形截面，可按宽度为 b'_f 的矩形截面计算开裂截面的换算截面几何特性。

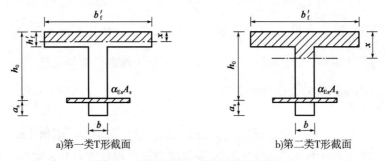

图 9-3　开裂状态下 T 形截面换算计算图式

当受压区高度 $x > h'_f$ 时，表明中和轴位于 T 形截面的肋部，为第二类 T 形截面，则：

$$x = \sqrt{A^2 + B} - A$$

$$A = \frac{\alpha_{Es}A_s + (b'_f - b)h'_f}{b}, \quad B = \frac{2\alpha_{Es}A_s h_0 + (b'_f - b)(h'_f)^2}{b}$$

开裂截面的换算截面对其中和轴的惯性 I_{cr} 为：

$$I_{cr} = \frac{b'_f x^3}{3} - \frac{(b'_f - b)(x - h'_f)^3}{3} + \alpha_{Es}A_s(h_0 - x)^2 \tag{9-6}$$

5) 全截面的换算截面几何参数计算（图 9-4）

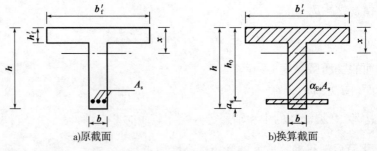

图 9-4　全截面换算示意图

(1) 换算截面概念

混凝土全截面参与工作，混凝土和钢筋面积所组成的换算截面。

(2) 几何参数计算

①换算截面面积

$$A_0 = bh + (b'_f - b)h'_f + (\alpha_{Es} - 1)A_s \tag{9-7}$$

②受压区高度

$$x = \frac{\frac{1}{2}bh^2 + \frac{1}{2}(b'_f - b)(h'_f)^2 + (\alpha_{Es} - 1)A_s h_0}{A_0} \tag{9-8}$$

③换算截面对中和轴的惯性矩

$$I_0 = \frac{1}{12}bh^3 + bh\left(\frac{1}{2}h - x\right)^2 + \frac{1}{12}(b'_f - b)(h'_f)^3 + (b'_f - b)h'_f\left(\frac{h'_f}{2} - x\right)^2 + (\alpha_{Es} - 1)A_s(h_0 - x)^2 \tag{9-9}$$

9.3 应力验算

钢筋混凝土受弯构件在施工阶段的应力验算,只要保证相应阶段的应力不超过规范的限值即可,此阶段的应力采用材料力学方法进行计算。

1) 应力限值

《公桥规》规定,施工阶段的应力限值,应根据可能出现的施工荷载进行内力组合,受弯构件正截面应力应符合下列条件。

(1) 受压区混凝土边缘纤维应力:

$$\sigma^t_{cc} \leq 0.80 f_{ck} \tag{9-10}$$

(2) 受拉钢筋应力:

$$\sigma^t_{si} \leq 0.75 f_{sk} \tag{9-11}$$

式中,f_{ck} 和 f_{sk} 分别为施工阶段相应的混凝土轴心抗压强度标准值和普通钢筋的抗拉强度标准值;σ^t_{si} 为按短暂状况计算时受拉区第 i 层钢筋的应力。

(3) 中性轴处主拉应力:

$$\sigma^t_{tp} = \frac{V^t_k}{bz_0} \leq f'_{tk} \tag{9-12}$$

式中,V^t_k 为由施工荷载标准值产生的剪力值;b 为矩形、T 形、I 形截面的腹板宽度;z_0 为受压合力作用点至受拉钢筋合力作用点的距离,按压应力图形为三角形计算确定;f'_{tk} 为施工阶段混凝土轴心抗拉标准强度。

2) 应力计算

(1) 应力计算原则如下:

施工阶段出现的荷载,对其进行组合后,按材料力学方法计算,应注意截面的几何特性值。

(2) 矩形截面梁正应力计算步骤。

①计算受压区高度 x;

②计算开裂截面的换算截面惯性矩 I_{cr};

③计算截面应力:

a. 受压区混凝土边缘:

$$\sigma^t_{cc} = \frac{M^t_k x}{I_{cr}} \leq 0.80 f_{ck} \tag{9-13}$$

b. 受拉钢筋的面积重心处：
$$\sigma_{si}^{t} = \alpha_{Es} \frac{M_k^t(h_{0i} - x)}{I_{cr}} \leq 0.75 f_{sk} \qquad (9\text{-}14)$$

式中，I_{cr} 为开裂截面换算截面的惯性矩；M_k^t 为由临时的施工荷载标准值产生的弯矩值。

(3) T 形截面梁正应力计算方法同矩形截面梁。

9.4 受弯构件的裂缝和裂缝宽度验算

钢筋混凝土构件在正常使用时允许出现裂缝，但如果裂缝过宽会影响使用功能和耐久性，因此，我们允许其开裂，但同时又要限制其裂缝宽度。

1) 控制裂缝宽度的原因

(1) 外观要求：满足人们心理界限，裂缝宽度不宜超过 0.3mm。

(2) 耐久性要求：防锈蚀，过宽的裂缝导致钢筋锈蚀，严重影响耐久性。

2) 产生裂缝的原因及控制措施

(1) 由作用效应引起的裂缝：由于截面上的弯矩、轴力、扭矩等作用效应，会使混凝土产生裂缝，主要通过验算和采取构造措施加以控制。

(2) 由外加变形或约束变形引起的裂缝：如混凝土收缩、温度变化、基础不均匀沉降等外加变形或约束变形引起开裂，主要通过采取构造措施和相关施工工艺加以控制。

(3) 钢筋锈蚀裂缝：采取保证混凝土保护层厚度和混凝土密实性的措施，严格控制早凝剂的掺入量等。

3) 受弯构件弯曲裂缝宽度计算理论和方法

(1) 第一类是基于试验的理论分析方法。基于试验结果，通过理论分析来建立力学模型，最后得到裂缝宽度计算公式，对公式中一些不易确定的系数，利用试验资料加以确定，主要有黏结滑移理论、无滑移理论以及两种理论的综合。

① 黏结滑移理论。裂缝控制主要取决于钢筋和混凝土之间的黏结性能，裂缝出现后，钢筋和混凝土产生相对滑移，变形不一致导致裂缝开展。

② 无滑移理论。表面裂缝宽度是由钢筋至构件表面的应变梯度控制的，即裂缝宽度随着与钢筋距离的增大而增大，钢筋的保护层厚度是影响裂缝宽度的主要因素。

③ 综合理论。考虑了混凝土保护层厚度对裂缝宽度的影响，也考虑了钢筋和混凝土之间可能出现的滑移。

(2) 第二类是基于试验的数理统计方法。基于大量试验资料，分析影响裂缝宽度的因素，然后利用数理统计方法给出简单、适用而又有一定可靠性的裂缝宽度计算公式。

4) 影响裂缝宽度的主要因素

(1) 钢筋应力 σ_{ss}：钢筋应力为最主要影响因素，最大裂缝宽度与 σ_{ss} 呈线性关系。

(2) 钢筋直径 d：在配筋率 ρ 与钢筋应力大致相同的情况下，最大裂缝宽度 W_{fmax} 随 d 的增加而增大。

(3) 配筋率 ρ：当钢筋直径 d、钢筋应力不变的情况下，W_{fmax} 随 ρ 的增加而减小，当 ρ 接近某一数值时，W_{fmax} 接近不变。

(4) 保护层厚度 c：试验发现，c 越大，W_{fmax} 越大。

(5)钢筋外形:引入系数 c_1 来考虑钢筋外形的影响。带肋钢筋黏结性好;圆钢筋黏结力小,裂缝宽。

(6)荷载作用性质:短、长期重复作用,引入系数 c_2。

(7)构件受力性质的影响:构件受到的作用性质不同,裂缝宽度不同,引入系数 c_3。

5)最大裂缝宽度计算公式

《公桥规》对矩形、T 形和工字形截面的钢筋混凝土受弯构件,规定其最大裂缝宽度 W_{cr}(mm)按下式计算:

$$W_{cr} = C_1 C_2 C_3 \frac{\sigma_{ss}}{E_s} \left(\frac{c+d}{0.36+1.7\rho_{te}} \right) \quad (9-15)$$

式中,C_1 为钢筋表面形状系数,对于光面钢筋,$C_1=1.4$;对于带肋钢筋,$C_1=1.0$;对环氧树脂层带肋钢筋,$C_1=1.15$;C_2 为长期效应影响系数,$C_2 = 1+0.5\frac{M_l}{M_s}$,其中 M_l 和 M_s 分别为按作用准永久组合和作用频遇组合计算的弯矩设计值(或轴力设计值);C_3 为与构件受力性质有关的系数。当为钢筋混凝土板式受弯构件时,$C_3=1.15$;其他受弯构件时,$C_3=1.0$,偏心受拉构件时,$C_3=1.1$;圆形截面偏心受压时,$C_3=0.75$,其他截面偏心受压构件时,$C_3=0.9$;轴心受拉构件时,$C_3=1.2$;d 为纵向受拉钢筋的直径(mm)。当用不同直径的钢筋时,改用换算直径 d_e,$d_e = \frac{\sum n_i d_i^2}{\sum n_i d_i}$,对钢筋混凝土构件,$n_i$ 为受拉区第 i 种普通钢筋的根数,d_i 为受拉区第 i 种普通钢筋的公称直径,当受拉区普通钢筋为单根钢筋时,d_i 取公称直径,当为普通钢筋的束筋时,d_i 取等代直径 d_{se},$d_{se} = \sqrt{n}d$,n 为组成束筋的普通钢筋根数。对于焊接钢筋骨架,d 或 d_e 应乘以 1.3 的系数;ρ_{te} 为纵向受拉钢筋配的有效配筋率,对于矩形、T 形和 I 字形截面受弯构件,$\rho_{te} = \frac{A_s}{A_{te}}$,$A_s$ 为受拉区纵向钢筋截面面积,A_{te} 为有效受拉混凝土截面面积,对于钢筋混凝土受弯构件取 $2a_s b$,a_s 为受拉钢筋重心至受拉区边缘的距离,对于矩形截面,b 为截面宽度,对于翼缘位于受拉区的 T 形、I 字形截面,b 为受拉区有效翼缘宽度,当 $\rho_{te} > 0.1$ 时,取 $\rho_{te} = 0.1$,当 $\rho_{te} < 0.01$ 时,取 $\rho_{te} = 0.01$;σ_{ss} 为钢筋应力(MPa),对于钢筋混凝土受弯构件,$\sigma_{ss} = \frac{M_s}{0.87 A_s h_0}$;其他受力性质构件的 σ_{ss} 计算式参见《公桥规》的有关规定;E_s 为钢筋弹性模量(MPa);h_0 为有效高度。

6)裂缝宽度限值

《公桥规》规定,钢筋混凝土受弯构件在荷载作用下,最大裂缝宽度应满足下列要求:

Ⅰ类和Ⅱ类环境:$[W_{fk}] = 0.2$mm。

Ⅲ类和Ⅳ类环境:$[W_{fk}] = 0.15$mm。

9.5 受弯构件的变形(挠度)验算

过大的变形会影响结构的使用功能和耐久性,需要对结构变形进行控制。但钢筋混凝土结构是由两种性能不同的材料组成的,其挠度计算不同于单一材料的计算方法。本节从研究单一弹性材料的挠度计算方法入手,考虑钢筋混凝土受弯构件在开裂后刚度降低的特点,计算构件开裂后的刚度,从而计算挠度。

1) 为何对钢筋混凝土受弯构件进行变形限制

(1) 挠度过大,影响使用功能。如简支梁跨中挠度过大,将使梁端部转角过大,车辆对该处产生冲击,破坏伸缩缝和桥面;连续梁的挠度过大,将使桥面不平顺,行车时引起颠簸和冲击等问题。

(2) 使相邻构件开裂、压碎。

(3) 考虑行驶时人们心理安全的需要。

(4) 挠度过大,发生振动、动力效应。

2) 变形计算方法

对均质弹性体材料的简支梁,挠度计算可写成如下公式:

$$f = \alpha \frac{ML^2}{B} \tag{9-16}$$

式中,α 为与荷载形式有关的荷载效应系数,又称挠度系数;B 为截面的抗弯刚度,对匀质弹性梁,抗弯刚度 $B = EI$;M 为跨中截面弯矩;L 为计算跨径。

对于给定荷载形式和支承条件的钢筋混凝土构件而言,挠度计算的关键在于确定开裂后截面的刚度。

3) 钢筋混凝土受弯构件抗弯刚度计算

(1) 等效刚度的概念

钢筋混凝土受弯构件各截面的配筋不一样,承受的弯矩也不相等,弯矩小的截面可能不出现弯曲裂缝,其刚度要较弯矩大的开裂截面大得多,因此沿梁长度的抗弯刚度大小不同,为简化计算,把变刚度构件等效为等刚度构件进行计算。

(2) 等效刚度的计算方法(图9-5)。按在两端部弯矩作用下构件转角相等的原则,求得受弯构件的等效刚度 B,也为开裂构件等效截面的抗弯刚度。

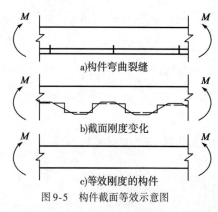

图9-5 构件截面等效示意图

(3)《公桥规》建议的计算公式:

$$B = \frac{B_0}{\left(\frac{M_{cr}}{M_s}\right)^2 + \left[1 - \left(\frac{M_{cr}}{M_s}\right)^2\right]\frac{B_0}{B_{cr}}} \tag{9-17}$$

式中,B 为开裂构件等效截面的抗弯刚度;B_0 为全截面的抗弯刚度,$B_0 = 0.95E_cI_0$;B_{cr} 为开裂截面的抗弯刚度,$B_{cr} = E_cI_{cr}$;E_c 为混凝土的弹性模量;I_0 为全截面换算截面惯性矩;I_{cr} 为开裂截面的换算截面惯性矩;M_s 为按短期效应组合计算的弯矩值;M_{cr} 为开裂弯矩,$M_{cr} = \gamma f_{tk} W_0$;$f_{tk}$ 为混凝土轴心抗拉强度标准值;γ 为构件受拉区混凝土塑性影响系数,$\gamma = 2S_0/W_0$;S_0 为全截面换算截面重心轴以上(或以下)部分面积对重心轴的面积矩;W_0 为全截面换算截面抗裂验算边缘的弹性抵抗矩。

4) 钢筋混凝土受弯构件使用阶段的挠度计算要求

(1) 验算要求

《公桥规》规定:钢筋混凝土受弯构件长期挠度值 - 结构自重产生的长期挠度值 ≤ 挠度限值。

第9章 钢筋混凝土受弯构件的应力、裂缝和变形计算

钢筋混凝土受弯构件使用阶段的挠度计算应考虑长期效应的影响,按荷载短期效应组合计算的挠度值乘以挠度长期增大系数 η_θ。

即
$$f_l = \eta_\theta f_s \tag{9-18}$$

式中,η_θ 为挠度长期增大系数。

当采用 C40 以下混凝土时,$\eta_\theta = 1.60$;当采用 C40 ~ C80 混凝土时,$\eta_\theta = 1.45 \sim 1.35$,中间强度等级可按直线内插取用。

(2)钢筋混凝土受弯构件挠度限值

梁式桥主梁的最大挠度处:$l/600$。

梁式桥主梁的悬臂端:$l_1/300$。

此处,l 为受弯构件的计算跨径,l_1 为悬臂长度。

5)预拱度的设置

(1)概念

预拱度为施工时预设的反向拱度。

(2)设置目的

①为了消除结构重力这个长期荷载引起的变形;

②希望构件在无静荷载作用时保持一定的拱度。

(3)设置条件

当由作用(或荷载)短期效应组合并考虑作用(或荷载)长期效应影响产生的长期挠度不超过计算跨径 L 的 $1/1\,600$ 时不设预拱度,否则设预拱度。

(4)预拱度值

$$\Delta = w_G + \frac{1}{2}w_Q \tag{9-19}$$

式中,Δ 为预拱度值;w_G 为结构重力产生的长期竖向挠度;w_Q 为可变荷载频遇值产生的长期竖向挠度。

需要注意的是,预拱的设置按最大的预拱值沿顺桥向做成平顺的曲线。

6)举例

【例9-1】 钢筋混凝土简支 T 梁梁长 $L_0 = 19.96\text{m}$,计算跨径 $L = 19.50\text{m}$。C30 混凝土,$f_{ck} = 20.1\text{MPa}$,$f_{tk} = 2.01\text{MPa}$,$E_c = 3.0 \times 10^4 \text{MPa}$。I 类环境条件,安全等级为一级。

主梁截面尺寸如图 9-6a)所示。跨中截面主筋为 HRB400 级,钢筋截面积 $A_s = 6\,836\text{mm}^2$($8\Phi32 + 2\Phi16$),$a_s = 111\text{mm}$,$E_s = 2 \times 10^5 \text{MPa}$,$f_{sk} = 400\text{MPa}$。

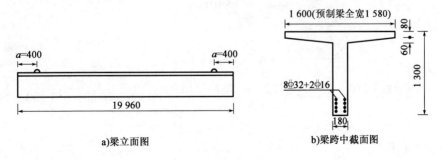

图 9-6 例 9-1 题图(尺寸单位:mm)

简支梁吊装时,其吊点设在距梁端 $a=400\text{mm}$ 处[图 9-6a)],梁自重在跨中截面引起的弯矩 $M_{G1}=505.69\text{kN}\cdot\text{m}$。

T 梁跨中截面使用阶段汽车荷载标准值产生的弯矩为 $M_{Q1}=596.04\text{kN}\cdot\text{m}$(未计入汽车冲击系数),人群荷载标准值产生的弯矩 $M_{Q2}=55.30\text{kN}\cdot\text{m}$,永久作用(恒载)标准值产生的弯矩 $M_G=751\text{kN}\cdot\text{m}$。试进行钢筋混凝土简支 T 梁的验算。

解:(1)施工吊装时的正应力验算

根据如图 9-6a)所示梁的吊点位置及主梁自重(看作均布荷载),可以看到在吊点截面处有最大负弯矩,在梁跨中截面有最大正弯矩,均为正应力验算截面。本例以梁跨中截面正应力验算为例介绍计算方法。

①梁跨中截面的换算截面惯性矩 I_{cr} 计算

在施工阶段,梁受压翼板的宽度为 1 580mm,根据《公桥规》计算得到的翼板有效宽度 $b_f'=1\,500\text{mm}$,而受压翼板平均厚度为 110mm。有效高度 $h_0=h-a_s=1\,300-111=1\,189(\text{mm})$。

$$\alpha_{Es}=\frac{E_s}{E_c}=\frac{2\times 10^5}{3.0\times 10^4}=6.667$$

截面混凝土受压区高度为:

$$\frac{1}{2}\times 1\,500\times x^2=6.667\times 6\,836\times(1\,189-x)$$

得到:
$$x=240.12\text{mm}>h_f'=110\text{mm}$$

故为第二类 T 形截面。

这时,换算截面受压区高度 x:

$$A=\frac{\alpha_{Es}A_s+h_f'(b_f'-b)}{b}$$

$$=\frac{6.667\times 6\,836+110\times(1\,500-180)}{180}=1\,060$$

$$B=\frac{2\alpha_{Es}A_s h_0+(b_f'-b)h_f'^2}{b}$$

$$=\frac{2\times 6.667\times 6\,836\times 1\,189+(1\,500-180)\times 110^2}{180}=690\,838$$

故
$$x=\sqrt{A^2+B}-A$$
$$=\sqrt{1\,060^2+690\,838}-1\,060$$
$$=287(\text{mm})>h_f'=110(\text{mm})$$

开裂截面的换算截面惯性矩 I_{cr} 为:

$$I_{cr}=\frac{b_f'x^3}{3}-\frac{(b_f'-b)(x-h_f')^3}{3}+\alpha_{Es}A_s(h_0-x)^2$$

$$=\frac{1\,500\times 287^3}{3}-\frac{(1\,500-180)\times(287-110)^3}{3}+6.667\times 6\,836\times(1\,189-287)^2$$

$$=4.646\,055\times 10^{10}(\text{mm}^4)$$

②正应力验算

吊装时动力系数为 1.2(起吊时主梁超重),则跨中截面计算弯矩为:$M_k^t=1.2M_{G1}=1.2\times$

$505.69 \times 10^6 = 606.828 \times 10^6 (\text{N} \cdot \text{mm})$。

则受压区混凝土边缘正应力为：

$$\sigma_{cc}^t = \frac{M_k^t x}{I_{cr}} = \frac{606.828 \times 10^6 \times 287}{46\,460.55 \times 10^6}$$

$$= 3.75(\text{MPa}) < 0.8 f_{ck}' = 0.8 \times 16.7 = 13.36(\text{MPa})$$

受拉钢筋的面积重心处的应力为：

$$\sigma_s^t = \alpha_{Es} \frac{M_k^t (h_0 - x)}{I_{cr}} = 6.667 \times \frac{606.828 \times 10^6 \times (1\,189 - 287)}{46\,460.55 \times 10^6}$$

$$= 78.54(\text{MPa}) < 0.75 f_{sk} = 0.75 \times 335 = 251(\text{MPa})$$

最下面一层钢筋（2⌀32）重心距受压边缘高度为：$h_{01} = 1\,300 - \left(\frac{35.8}{2} + 35\right) = 1\,247(\text{mm})$，则钢筋应力为：

$$\sigma_s = \alpha_{Es} \frac{M_k^t}{I_{cr}} (h_{01} - x)$$

$$= 6.667 \times \frac{606.828 \times 10^6}{46\,460.55 \times 10^6} \times (1\,247 - 287)$$

$$= 83.6(\text{MPa}) < 0.75 f_{sk} = 251(\text{MPa})$$

验算结果表明，主梁吊装时混凝土正应力和钢筋拉应力均小于规范限值，可取图9-6a)的吊点位置。

(2) 裂缝宽度 W_{fk} 的验算

①系数 c_1

带肋钢筋 $c_1 = 1.0$。

作用频遇组合弯矩计算值为：

$$M_s = M_G + \psi_{11} \times M_{Q1} + \psi_{12} \times M_{Q2}$$

$$= 751 + 0.7 \times 596.04 + 0.4 \times 55.30$$

$$= 1\,190.35(\text{kN} \cdot \text{m})$$

作用准永久组合弯矩计算值为：

$$M_1 = M_G + \psi_{21} \times M_{Q1} + \psi_{22} \times M_{Q2}$$

$$= 751 + 0.4 \times 596.04 + 0.4 \times 55.30$$

$$= 1\,011.54(\text{kN} \cdot \text{m})$$

系数 $c_2 = 1 + 0.5 \frac{M_1}{M_s} = 1 + 0.5 \frac{1\,011.54}{1\,190.35} = 1.42$；

非板式受弯构件系数 $c_3 = 1.0$。

②钢筋应力 σ_{ss} 的计算

$$\sigma_{ss} = \frac{M_s}{0.87 h_0 A_s} = \frac{1\,190.35 \times 10^6}{0.87 \times 1\,189 \times 6\,836} = 168(\text{MPa})$$

③换算直径 d

因为受拉区采用不同的钢筋直径,d 应取用换算直径 d_e,则可得到:

$$d = d_e = \frac{8 \times 32^2 + 2 \times 16^2}{8 \times 32 + 2 \times 16} = 30.2(\text{mm})$$

对于焊接钢筋骨架:$d = d_e = 1.3 \times 30.2 = 39.26(\text{mm})$。

④纵向受拉钢筋配筋率 ρ_{te}

根据 T 梁尺寸,计算有效受拉的混凝土面积 A_{te} 为:$A_{te} = 2a_s b = 2 \times 111 \times 180 = 39\,960(\text{mm}^2)$

$$\rho_{te} = \frac{A_s}{A_{te}} = \frac{6\,836}{39\,960} = 0.171 > 0.1$$

取 $\rho = 0.02$。

⑤最大裂缝宽度 W_{fk}

$$W_{fk} = c_1 c_2 c_3 \frac{\sigma_{ss}}{E_s}\left(\frac{C+d}{0.36 + 1.7\rho}\right)$$

$$= 1 \times 1.42 \times 1 \times \frac{168}{2 \times 10^5}\left(\frac{35 + 39.26}{0.36 + 1.7 \times 0.1}\right)$$

$$= 0.17(\text{mm}) < [W_f] = 0.2(\text{mm})$$

满足要求。

(3) 梁跨中挠度的验算

在进行梁变形计算时,应取梁与相邻梁横向连接后截面的全宽度受压翼板计算,即为 $b'_{f1} = 1\,600\text{mm}$,而 h'_f 仍为 110mm。

①T 梁换算截面的惯性矩 I_{cr} 和 I_0 计算

对 T 梁的开裂截面:

$$\frac{1}{2} \times 1\,600 \times x^2 = 6.667 \times 6\,836(1\,189 - x)$$

$$x = 233.3\text{mm} > h'_f = 110\text{mm}$$

梁跨中截面为第二类 T 形截面。这时,受压区 x 高度重新计算,即:

$$A = \frac{\alpha_{Es}A_s + h'_f(b'_{f1} - b)}{b}$$

$$= \frac{6.667 \times 6\,836 + 110 \times (1\,600 - 180)}{180}$$

$$= 1\,120.98$$

$$B = \frac{2\alpha_{Es}A_s h_0 + (b'_{f1} - b)h'^2_f}{b}$$

$$= \frac{2 \times 6.667 \times 6\,836 \times 1\,189 + (1\,600 - 180) \times 110^2}{180}$$

$$= 697\,560.0$$

则

$$x = \sqrt{A^2 + B} - A$$

$$= \sqrt{1\,120.98^2 + 697\,560.0} - 1\,120.98$$

$$= 277(\text{mm}) > h'_f = 110(\text{mm})$$

开裂截面的换算截面惯性矩 I_{cr} 为：

$$I_{cr} = \frac{1\,600 \times 277^3}{3} - \frac{(1\,600 - 180) \times (277 - 110)^3}{3} + 6.667 \times 6\,836 \times (1\,189 - 277)^2$$

$$= 47\,038.14 \times 10^6 (\text{mm}^4)$$

T 梁的全截面换算截面面积 A_0 为：

$$A_0 = 180 \times 1\,300 + (1\,600 - 180) \times 110 + (6.667 - 1) \times 6\,836 = 428\,939.6(\text{mm}^2)$$

形心轴到上缘的高度 x 为：

$$x = \frac{\frac{1}{2} \times 180 \times 1\,300^2 + \frac{1}{2} \times (1\,600 - 180) \times 110^2 + (6.667 - 1) \times 6\,836 \times 1\,189}{428\,939.6}$$

$$= 482(\text{mm})$$

全截面换算惯性矩 I_0 的计算为：

$$I_0 = \frac{1}{12}bh^3 + bh\left(\frac{h}{2} - x\right)^2 + \frac{1}{12}(b'_{fl} - b)h'^3_f +$$

$$(b'_{fl} - b)h'_f \cdot \left(x - \frac{h'_f}{2}\right)^2 + (\alpha_{ES} - 1)A_s(h_0 - x)^2$$

$$= \frac{1}{12} \times 180 \times 1\,300^3 + 180 \times 1\,300 \times \left(\frac{1\,300}{2} - 482\right)^2 + \frac{1}{12} \times (1\,600 - 180) \times 110^3 +$$

$$(1\,600 - 180) \times 110 \times \left(482 - \frac{110}{2}\right)^2 + (6.667 - 1) \times 6\,836 \times (1\,189 - 482)^2$$

$$= 8.76 \times 10^{10}(\text{mm}^4)$$

②计算开裂构件的抗弯刚度

全截面抗弯刚度：

$$B_0 = 0.95E_cI_0 = 0.95 \times 3.0 \times 10^4 \times 8.76 \times 10^{10} = 2.49 \times 10^{15}(\text{N} \cdot \text{mm}^2)$$

开裂截面抗弯刚度：

$$B_{cr} = E_cI_{cr} = 3.0 \times 10^4 \times 47\,038.14 \times 10^6 = 1.41 \times 10^{15}(\text{N} \cdot \text{mm}^2)$$

全截面换算截面受拉区边缘的弹性抵抗矩为：

$$W_0 = \frac{I_0}{h - x} = \frac{8.76 \times 10^{10}}{1\,300 - 482} = 1.07 \times 10^8(\text{mm}^3)$$

全截面换算截面的面积矩为：

$$S_0 = \frac{1}{2}b'_{fl}x^2 - \frac{1}{2}(b'_{fl} - b)(x - h'_f)^2$$

$$= \frac{1}{2} \times 1\,600 \times 482^2 - \frac{1}{2} \times (1\,600 - 180) \times (482 - 110)^2$$

$$= 8.76 \times 10^7(\text{mm}^3)$$

塑性影响系数为：

$$\gamma = \frac{2S_0}{W_0} = \frac{2 \times 8.76 \times 10^7}{1.07 \times 10^8} = 1.64$$

开裂弯矩：

$$M_{cr} = \gamma f_{tk} W_0 = 1.64 \times 2.01 \times 1.07 \times 10^8 = 3.527 \times 10^8(\text{N} \cdot \text{mm}) = 352.7(\text{kN} \cdot \text{m})$$

开裂构件的抗弯刚度为：$B = \dfrac{B_0}{\left(\dfrac{M_{cr}}{M_s}\right)^2 + \left[1 - \left(\dfrac{M_{cr}}{M_s}\right)^2\right]\dfrac{B_0}{B_{cr}}}$

$= \dfrac{2.49 \times 10^{15}}{\left(\dfrac{352.7}{1\,223.53}\right)^2 + \left[1 - \left(\dfrac{352.7}{1\,223.53}\right)^2\right] \times \dfrac{2.49 \times 10^{15}}{1.41 \times 10^{15}}}$

$= 1.466 \times 10^{15}(\text{N} \cdot \text{mm}^2)$

③受弯构件跨中截面处的长期挠度值

短期荷载效应组合下跨中截面弯矩标准值：$M_s = 1\,223.53 \text{kN} \cdot \text{m}$，结构自重作用下跨中截面弯矩标准值 $M_G = 751 \text{kN} \cdot \text{m}$。对 C25 混凝土，挠度长期增长系数 $\eta_\theta = 1.60$。

受弯构件在使用阶段的跨中截面的长期挠度值为：

$w_l = \dfrac{5}{48} \times \dfrac{M_s L^2}{B} \times \eta_\theta$

$= \dfrac{5}{48} \times \dfrac{1\,190.35 \times 10^6 \times (19.5 \times 10^3)^2}{1.466 \times 10^{15}} \times 1.60$

$= 51.5(\text{mm})$

在结构自重作用下跨中截面的长期挠度值为：

$w_G = \dfrac{5}{48} \times \dfrac{M_G L^2}{B} \times \eta_\theta$

$= \dfrac{5}{48} \times \dfrac{751 \times 10^6 \times (19.5 \times 10^3)^2}{1.466 \times 10^{15}} \times 1.60$

$= 32.5(\text{mm})$

长期挠度计算值 w_{ll} 为：

$w_{ll} = w_l - w_G = 51.5 - 32.5 = 19(\text{mm}) < \dfrac{L}{600} = \dfrac{19.5 \times 10^3}{600} = 33(\text{mm})$

符合《公桥规》的要求。

(4) 预拱度设置

在荷载短期效应组合并考虑荷载长期效应影响下梁跨中处产生的长期挠度为：$w_l = 51.5(\text{mm}) > \dfrac{L}{1\,600} = \dfrac{19.5 \times 10^3}{1\,600} = 12(\text{mm})$，故跨中截面需设置预拱度。

根据《公桥规》对预拱度设置的规定，梁跨中截面处的预拱度为：

$\Delta = w_G + \dfrac{1}{2} w_{ll} = 32.5 + \dfrac{1}{2} \times 19 = 42.0(\text{mm})$

9.6 混凝土结构的耐久性

1) 结构耐久性的基本概念

(1) 定义

所谓混凝土结构的耐久性，是指混凝土结构在自然环境、使用环境及材料内部因素的作用下,在设计要求的目标使用期内,不需要花费大量资金加固处理而保持安全、使用功能和外观要求的能力。

(2)影响混凝土结构耐久性的主要因素

①混凝土冻融破坏:处于饱水状态(含水率达91.7%的极限值)的混凝土受冻时,毛细孔中同时受到膨胀压力和渗透压力,使混凝土结构产生内部裂缝和损伤,经多次反复,损伤积累到一定程度就引起结构破坏。

②混凝土的碱骨料反应:混凝土集料中某些活性矿物与混凝土微孔中的碱性溶液产生化学反应称为碱骨料反应。碱骨料反应产生碱—硅酸盐凝胶,并吸水膨胀,体积可增大3~4倍,从而引起混凝土剥落、开裂、强度降低,甚至导致结构破坏。

③侵蚀性介质的腐蚀:有些化学介质侵入,造成混凝土中一些成分被溶解、流失,引起裂缝、孔隙、松散破碎;有的化学介质侵入,与混凝土中一些成分反应生成物体积膨胀,引起混凝土结构破坏。

④机械磨损:机械磨损常见于工业地面、公路路面、桥面、飞机跑道等。

⑤混凝土的碳化:混凝土的碳化是指大气中的二氧化碳与混凝土中的碱性物质氢氧化碳发生反应使混凝土的pH值下降,从而使混凝土中钢筋的保护膜受到破坏,引起钢筋锈蚀。

⑥钢筋锈蚀:钢筋锈蚀使混凝土保护层脱落,钢筋有效面积减小,导致承载力下降甚至结构破坏。

2)混凝土结构耐久性设计基本要求

(1)结构混凝土材料耐久性的基本要求应符合《混凝土结构耐久性设计规范》(CB/T 50476)的规定。

(2)对于预应力混凝土构件,混凝土材料中的最大氯离子含量为0.6%,最小水泥用量为350kg/m³,最低混凝土强度等级为C40。

(3)特大桥和大桥的混凝土最大含碱量宜降至1.8kg/m³,当处于Ⅲ类、Ⅳ类或使用除冰盐和滨海环境时,宜使用非碱活性集料。

(4)处于Ⅲ类或Ⅳ类环境的桥梁,当耐久性确实需要时,其主要受拉钢筋宜采用环氧树脂涂层钢筋;预应力钢筋、锚具及连接器应采取专门防护措施。

(5)水位变动区有抗冻要求的结构混凝土,其抗冻等级应符合有关标准的规定。

(6)有抗渗要求的混凝土结构,混凝土的抗渗等级应符合有关标准的要求。

思考练习题

1.对于钢筋混凝土构件,为什么《公桥规》规定必须进行持久状况正常使用极限状态计算和短暂状况应力计算?

2.钢筋混凝土构件持久状况正常使用极限状态验算主要有哪些内容?验算的目的是什么?

3.什么是钢筋混凝土构件的换算截面?将钢筋混凝土开裂截面化为等效的换算截面基本前提是什么?

4. 引起钢筋混凝土构件出现裂缝的主要因素有哪些?

5. 裂缝宽度和挠度验算公式中如何体现作用(或荷载)短期效应组合并考虑长期效应的影响?

6. 在持久状况正常使用极限状态下,为什么要验算受弯构件的挠度?挠度不满足要求如何处理?

7. 什么是钢筋混凝土受弯构件的预拱度?如何确定预拱度的大小?是否任何情况下都要设置预拱度?

8. 影响混凝土结构耐久性的主要因素有哪些?混凝土结构耐久性设计应考虑哪些问题?

9. 提高钢筋混凝土构件刚度的主要措施是什么?

10. 试述钢筋混凝土梁裂缝验算时,应力不均匀系数 φ 的定义和工程意义。

11. 验算钢筋混凝土构件裂缝宽度的工程意义是什么?

12. 已知矩形截面钢筋混凝土简支梁的截面尺寸为 $b \times h = 200\text{mm} \times 500\text{mm}$,$a_s = 40\text{mm}$。C30 混凝土,HRB400 级钢筋。在截面受拉区配有纵向受拉钢筋 $A_s = 603\text{mm}^2$($3\underline{\Phi}16$),永久作用(恒载)产生的弯矩标准值 $M_G = 40\text{kN} \cdot \text{m}$,汽车荷载产生的弯矩标准值为 $M_{Q1} = 15\text{kN} \cdot \text{m}$(未计入汽车冲击系数)。Ⅰ类环境条件,安全等级为一级。若不考虑长期荷载的作用,试求:①构件的最大裂缝宽度;②当主钢筋改为 $2\Phi20$($A_s = 628\text{mm}^2$)时,求梁的最大裂缝宽度。

13. 已知某钢筋混凝土 T 形截面梁计算跨径 $L = 19.6\text{m}$,截面尺寸为 $b'_f = 1\,580\text{mm}$,$h'_f = 110\text{mm}$,$b = 180\text{mm}$,$h = 1\,300\text{mm}$,$h_0 = 1\,180\text{mm}$。C35 混凝土。HRB400 级钢筋,在截面受拉区配有纵向受拉钢筋为($6\underline{\Phi}32 + 6\underline{\Phi}16$),$A_s = 6\,031\text{mm}^2$。永久作用(恒载)产生的弯矩标准值 $M_G = 750\text{kN} \cdot \text{m}$,汽车荷载产生的弯矩标准值为 $M_{Q1} = 710\text{kN} \cdot \text{m}$(未计入汽车冲击系数)。Ⅰ类环境条件,安全等级为二级。试验算此梁跨中挠度,并确定是否应设计预拱度。

第 10 章

局部承压

局部承压是指在构件的表面上,仅有部分面积承受压力的受力状态。桥梁工程中局部承压构件比较多,如后张法预应力混凝土构件端部锚固区,桥梁墩(台)帽直接支承支座的部分,拱上立柱对拱圈的作用,支座反力对梁底混凝土的作用等,都属于局部承压构件。

取局部承压构件的端部区分析,力的传递示意如图 10-1 所示,并有以下特点:

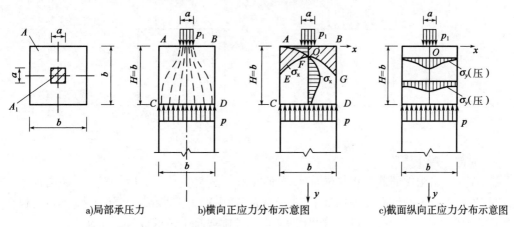

a)局部承压力　b)横向正应力分布示意图　c)截面纵向正应力分布示意图

图 10-1　构件端部的局部承受压区

(1) x 方向正应力 σ_x：在局部承压区的 $AOBGFE$ 部分,σ_x 为压应力,在其余部分为拉应力

[图10-1b)],最大横向拉应力 σ_{xmax} 发生在局部承压区 $ABCD$ 的中点附近。

(2) y 方向正应力 σ_y：在局部承压区内,绝大部分的 σ_y 都是压应力,Oy 轴处的压应力 σ_y 较大,其中又以 O 点处为最大,即等于 p_1。

(3) 剪应力 τ：在局部受压面以外的任何平行于 y 轴的截面上都有剪应力。

总体来说,局部承压构件具有下列特点：

(1) 局部承压构件的混凝土抗压强度比全面积受压时混凝土抗压强度高。

(2) 在局部承压区的中部有横向拉应力 σ_x（图10-1）,这种横向拉应力可使混凝土产生裂缝,裂缝是平行于局部压力方向的。

由局部承压构件特点可知,局部承压构件应该进行抗裂性和强度的计算。

10.1 局部承压的破坏形态和破坏机理

1) 混凝土局部承压的破坏形态

混凝土局部承压的破坏形态主要与 A_l/A（A_l 为局部承压面积,A 为构件截面面积）以及 A_l 在表面上的位置有关。对于 A_l 对称布置于构件端面上的轴心局部承压,其破坏形态主要有三种。

(1) 先开裂后破坏

破坏特征：当构件截面积与局部承压面积比较接近时（一般 $A/A_l<9$）,在50%~90%破坏荷载时,构件某一侧面首先出现纵向裂缝。随着荷载增加,裂缝逐渐延伸,其他侧面也相继出现类似裂缝。最后承压面下的混凝土被冲切出一个楔形体[图10-2a)],构件被劈成数块而发生劈裂破坏。

(2) 一开裂即破坏

破坏特征：当构件截面积与局部承压面积相比较大时（一般 $9<A/A_l<36$）,构件一开裂就破坏,破坏很突然,裂缝从顶面向下发展,裂缝宽度上大下小,局部承压面积外围混凝土被劈成数块,而局部承压面下的混凝土被冲剪成一个楔形体[图10-2b)]。

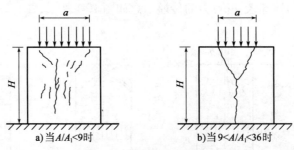

图10-2 局部承压的破坏形态

(3) 局部混凝土下陷

破坏特征：当构件的截面积与局部承压面积相比很大（一般 $A/A_l>36$）时,在构件整体破坏前,局部承压面下的混凝土先局部下陷,沿局部承压面四周的混凝土出现剪切破坏,但此时外围混凝土尚未劈裂,荷载还可以继续增加,直至外围混凝土被劈成数块而最终破坏。

2) 局部承压的工作机理分析

(1) 套箍理论：局部承压区的混凝土可看作是承受侧压力作用的混凝土芯块。当局部荷载作用增大时,受挤压的混凝土向外膨胀,而周围混凝土起着套箍作用阻止其横向膨胀,因此,挤压区混凝土处于三向受压状态,提高了芯块混凝土的抗压强度。当周围混凝土环向拉应力达到抗拉极限强度时,构件即告破坏,如图10-3所示。

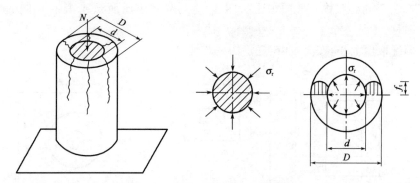

图10-3　套箍理论的局部承压受力模型

(2)剪切理论:在局部荷载作用下,构件端部的受力特征可以比拟为一个带多根拉杆的拱结构。紧靠承压板下面的混凝土,亦即位于拉杆部位的混凝土,承受横向拉力。当局部承压荷载达到开裂荷载时,部分拉杆由于局部承压区中横向拉应力 σ_x 大于混凝土极限抗拉强度 f_t 而断裂,从而产生了局部纵向裂缝,但此时尚未破坏结构[图10-4b)]。随着荷载继续增加,更多的拉杆被拉断,裂缝进一步增多和延伸,内力进一步重分配。当达到破坏荷载时,承压板下的混凝土在剪压作用下形成楔形体,产生剪切滑移面,楔形体的劈裂最终导致拱机构破坏[图10-4c)]。

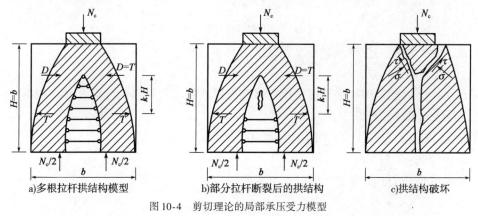

a)多根拉杆拱结构模型　　b)部分拉杆断裂后的拱结构　　c)拱结构破坏

图10-4　剪切理论的局部承压受力模型

两种理论从不同的角度分析了局部承压构件的不同破坏形式,《公桥规》采用了基于剪切理论建立的计算公式。

10.2　混凝土局部承压强度提高系数

1)混凝土局部承压提高系数 β

(1)规定的 β 计算公式

$$\beta = \sqrt{\frac{A_b}{A_1}} \tag{10-1}$$

式中,A_l 为局部承压面积(考虑在钢垫板中沿45°刚性角扩大的面积),当有孔道时(对圆形承压面积而言)不扣除孔道面积;A_b 为局部承压的计算底面积。

(2)局部承压的计算面积 A_b 的确定

①确定方法:"同心对称有效面积法",即 A_b 应与局部承压面积 A_l 具有相同的形心位置,且要求相应对称。

②常用情况取值大小:按图10-5 规定的公式计算 A_b。

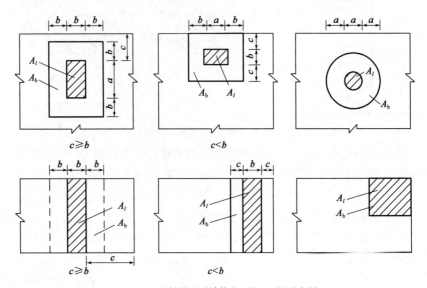

图10-5　局部承压时计算底面积 A_b 的示意图

2)配置间接钢筋的混凝土局部承压强度提高系数 β_{cor}

(1)间接钢筋的形式

局部承压区内配置间接钢筋可采用方格钢筋网或螺旋式钢筋两种形式,如图10-6 所示。

(2)间接钢筋体积配筋率 ρ_v

①定义:指核心面积 A_{cor} 范围内单位体积所含间接钢筋的体积。

②计算公式:

a. 当间接钢筋为方格钢筋网时[图10-6a)]:

$$\rho_v = \frac{n_1 A_{s1} l_1 + n_2 A_{s2} l_2}{A_{cor} s} \tag{10-2}$$

式中,s 为钢筋网片层距;n_1、A_{s1} 分别为单层钢筋网沿 l_1 方向的钢筋根数和单根钢筋截面面积;n_2、A_{s2} 分别为单层钢筋网沿 l_2 方向的钢筋根数和单根钢筋截面面积;A_{cor} 为方格网间接钢筋内表面范围的混凝土核心面积,其重心应与 A_l 的重心重合,计算时按同心、对称原则取值。

此外,钢筋网在两个方向的钢筋截面面积相差不应大于50%,且局部承压区间接钢筋不应少于4层钢筋网。

b. 当间接钢筋为螺旋形钢筋时[图10-6b)]:

$$\rho_v = \frac{4 A_{ss1}}{d_{cor} s} \tag{10-3}$$

式中,A_{ss1}为单根螺旋形钢筋的截面面积;d_{cor}为螺旋形间接钢筋内表面范围内混凝土核心的直径;s为螺旋形钢筋的间距,螺旋形钢筋不应少于4圈。

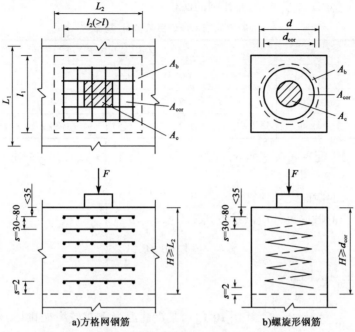

图 10-6 局部承压区内的间接钢筋配筋形式(尺寸单位:mm)

(3)配有间接钢筋的混凝土局部承压强度提高系数β_{cor}的计算公式:

$$\beta_{cor} = \sqrt{\frac{A_{cor}}{A_1}} \geqslant 1 \tag{10-4}$$

式中,A_{cor}为间接钢筋网或螺旋钢筋范围内混凝土核心面积。

10.3 局部承压区的计算

1)局部承压受力构件计算的内容

《公桥规》要求对局部承压构件应进行局部承压区承载能力和截面尺寸验算两项内容计算。

2)局部承压构件的承载力计算

配有间接钢筋的局部承压构件的抗力包括混凝土的抗力和间接钢筋的抗力,《公桥规》给出了下列计算公式:

$$\gamma_0 F_{ld} \leqslant F_u = 0.9(\eta_s \beta f_{cd} + k\rho_v \beta_{cor} f_{sd}) A_{ln} \tag{10-5}$$

式中,F_{ld}为局部受压面积上的局部压力设计值,对后张法预应力混凝土构件的锚头局部受压区,可取1.2倍张拉时的最大压力;η_s为混凝土局部承压修正系数,按表10-1采用;β为混凝土承压强度的提高系数;k为间接钢筋影响系数,混凝土强度等级C50及以下时,取$k = 2.0$;C50~C80取$k = 2.0 \sim 1.70$,中间直接插值取用,见表10-1;ρ_v为间接钢筋的体积配筋率;

β_{cor} 为配置间接钢筋时局部承压承载能力提高系数；f_{sd} 为间接钢筋的抗拉强度设计值；A_{ln} 为当局部受压面有孔洞时，扣除孔洞后的混凝土局部受压面积（计入钢垫板中按 45°刚性角扩大的面积），即 A_{ln} 为局部承压面积 A_l 减去孔洞的面积。

混凝土局部承压计算系数 η_s 与 k 表 10-1

计算系数	混凝土强度等级						
	≤C50	C55	C60	C65	C70	C75	C80
η_s	1.0	0.96	0.92	0.88	0.84	0.80	0.76
k	2.0	1.95	1.90	1.85	1.80	1.75	1.70

计算公式适用于配置间接钢筋的局部承压区，且当 $A_{cor} > A_l$ 时，A_{cor} 的重心与 A_l 的重心相重合。

3) 局部承压区的截面尺寸计算

《公桥规》给出的截面尺寸应满足下列要求：

$$\gamma_0 F_{ld} \leq 1.3 \eta_s \beta f_{cd} A_{ln} \quad (10\text{-}6)$$

式中各符号的意义同上。

4) 举例

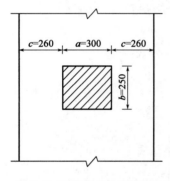

图 10-7 示意图（尺寸单位：mm）

[例 10-1] 某钢筋混凝土双铰桁架拱桥，台帽承受主拱拱脚作用的支承力设计值 $F_{ld} = 2\,500\text{kN}$，其局部承压面积为 250mm × 300mm（已考虑了矩形钢垫板厚度且沿 45°刚性角扩大后的面积），如图 10-7 所示。台帽采用 C30 混凝土，$f_{cd} = 13.8\text{MPa}$。配置方格网的间接钢筋采用 HPB300 级钢筋，$f_{sd} = 250\text{MPa}$。试进行台帽的局部承压计算。

解：由图 10-7 可知，矩形局部承压面积 A_l 的长边边长 $a = 300\text{mm}$，短边边长 $b = 250\text{mm}$，而矩形局部承压面积边缘至台帽边缘的最小距离 $c = 260\text{mm} > b$，故由图 10-5a) 可得到局部承压时的计算底面积 A_b 及边长为：

$$L_1 = 250 + 2 \times 250 = 750(\text{mm})$$

$$L_2 = 300 + 2 \times 250 = 800(\text{mm})$$

$$A_b = 750 \times 800 = 6 \times 10^5 (\text{mm}^2)$$

而局部承压面积 $A_l = 250 \times 300 = 0.75 \times 10^5 (\text{mm}^2)$。

根据图 10-6 的规定，设间接钢筋的网格核心混凝土面积尺寸为 $l_1 = 500\text{mm}$，$l_2 = 600\text{mm}$，$A_{cor} = 500 \times 600 = 3 \times 10^5 (\text{mm}^2) > A_l = 0.75 \times 10^5 (\text{mm}^2)$。

则混凝土局部承压强度提高系数 β 为：

$$\beta = \sqrt{\frac{A_b}{A_l}} = \sqrt{\frac{6 \times 10^5}{0.75 \times 10^5}} = 2.83$$

因采用 C30 级混凝土，由表 10-1 可查得 $\eta_s = 1$，故 $\eta_s\beta = 2.83$。
配置间接钢筋时局部承压强度提高系数 β_{cor} 为：

$$\beta_{cor} = \sqrt{\frac{A_{cor}}{A_l}} = \sqrt{\frac{3 \times 10^5}{0.75 \times 10^5}} = 2$$

设焊接钢筋网片沿 l_1 方向的钢筋根数 $n_1 = 6$（间距为 100mm）；沿 l_2 方向的钢筋数 $n_2 = 7$（间距为 100mm）。而单根钢筋为 $\phi 6$ 的面积 $A_{s1} = A_{s2} = 28.3 \text{mm}^2$。间接钢筋设置情况如图 10-8 所示。

图 10-8 中，顶层钢筋网片距局部承压面为 30mm，各网片的层距 s 取 100mm，间接钢筋网片的设置深度 $H = 580$mm，布置层数为 $m = 5$，则间接钢筋的配筋率 ρ_v 为：

$$\rho_v = \frac{n_1 A_{s1} l_1 + n_2 A_{s2} l_2}{A_{cor} s}$$

$$= \frac{6 \times 28.3 \times 500 + 7 \times 28.3 \times 600}{(500 \times 600) \times 100}$$

$$= 0.0068$$

且

$$\frac{n_1 A_{s1}}{n_2 A_{s2}} = \frac{6 \times 28.3}{7 \times 28.3} = 0.86 > 0.5$$

（1）局部承压区截面尺寸验算

$$1.3\eta_s\beta f_{cd} A_{ln} = 1.3 \times 2.83 \times 13.8 \times 0.75 \times 10^5$$
$$= 3807.765(\text{kN}) > F_{ld}$$
$$= 2500(\text{kN})$$

局部承压区尺寸满足要求。

（2）局部承压承载力计算

$$0.9(\eta_s\beta f_{cd} + R\rho_v\beta_{cor} f_{sd})A_{ln} = 0.9 \times (2.83 \times 13.8 + 2 \times 0.0068 \times 2 \times 250) \times 0.75 \times 10^5$$
$$= 3095.15(\text{kN}) > F_{ld}$$
$$= 2500(\text{kN})$$

局部承压区承载能力满足要求。

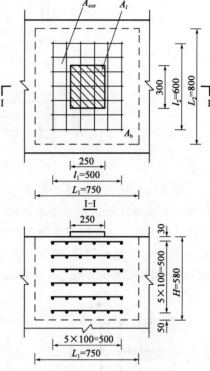

图 10-8 间接钢筋设置情况（尺寸单位：mm）

思考练习题

1. 什么叫作混凝土构件的局部受压？试说明符号 β、β_{cor}、β_b、A_t、A_{cor}、A_{ln} 的意义。
2. 局部承压区段混凝土内设置间接钢筋的作用是什么？常用的间接钢筋有哪几种配筋形式？
3. 对局部承压的计算包括哪些内容？

4. 预应力混凝土简支 T 形梁在支点处设置板式橡胶支座(习题 4 图)。对梁的支承反力设计值为 $F_{ld} = 602.47 \text{kN}$。橡胶板式支座对梁底面的局部承压作用面积为 $200\text{mm} \times 350\text{mm}$，C50 混凝土，梁体在支承位置处未设间接钢筋。试对梁支承处混凝土局部承压进行复核计算。

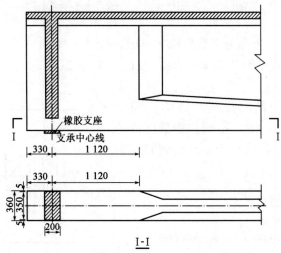

习题 4 图　(尺寸单位:mm)

5. 预应力混凝土梁端锚具布置如习题 5 图所示。梁体混凝土为 C40，张拉预应力钢束时对梁体局部压力设计值 $F_{ld} = 534.68 \text{kN}$。锚具直径 $D = 110\text{mm}$，孔道直径 $d_1 = 50\text{mm}$，锚具下钢板厚度 $t = 20\text{mm}$。端部锚下混凝土中设置有两种间接钢筋:一种是方格钢筋网，共设 4 片，$S = 100\text{mm}$；另一种是螺旋钢筋，螺旋圈直径 $d = 90\text{mm}$，螺距 $S = 30\text{mm}$，埋置深度 $h_1 = 210\text{mm}$。间接钢筋均为 HPB235 级钢筋，直径 8mm。试进行主梁端部锚下混凝土局部承压承载力计算。

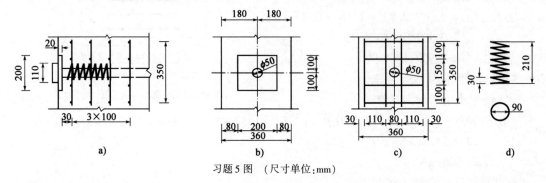

习题 5 图　(尺寸单位:mm)

PART 2 | 第二部分

预应力混凝土结构

第11章 预应力混凝土结构的基本概念及其组成材料

11.1 概 述

1）预应力混凝土结构的基本概念

（1）基本原理

预应力混凝土结构，就是事先人为地在混凝土或钢筋混凝土中引入内部应力，且其数值和分布恰好能将使用荷载产生的内力抵消到一个合适程度的混凝土。这种预先给混凝土引入内部应力的结构，就称为预应力混凝土结构。

以图11-1所示的简支梁为例，承受均布荷载的简支梁施加了偏心压力后，截面应力状态明显改变。

上面的示例说明了两个重要的问题：

①由于预先给混凝土梁施加了预压力 N_y，使混凝土梁在均布荷载 q 作用下，下缘产生的拉应力完全或部分被预压应力所抵消，因而可以避免混凝土出现裂缝或推迟裂缝出现。

②实际结构中，必须针对荷载作用下可能产生的应力状态来施加预应力，且所需施加的预压力 N_y 应当与荷载（或者说弯矩 M）的大小相适应，因此，预应力的大小、位置必须进行设计。

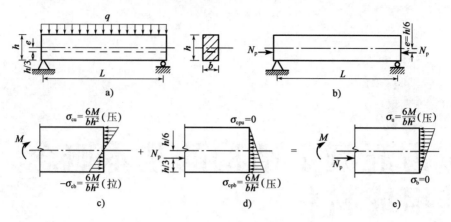

图 11-1 预应力混凝土结构基本原理图

(2) 加筋混凝土结构的分类

加筋混凝土结构是根据预应力度的大小来分类的。

① 预应力度的定义

预应力度是指由预加应力大小确定的消压弯矩 M_0 与外荷载产生的弯矩 M_s 的比值，即

$$\lambda = \frac{M_0}{M_s} \tag{11-1}$$

式中，M_0 为消压弯矩，也就是构件抗裂边缘预压应力抵消到零时的弯矩；M_s 为按作用（或荷载）频遇效应组合计算的弯矩值；λ 为预应力混凝土构件的预应力度。

② 配筋混凝土构件的分类

a. 全预应力混凝土构件：在作用（荷载）频遇效应组合下控制的正截面受拉边缘不允许出现拉应力，即 $\lambda \geqslant 1$。

b. 部分预应力混凝土构件：在作用（荷载）频遇效应组合下控制的正截面受拉边缘出现拉应力或出现不超过规定宽度的裂缝，即 $0 < \lambda < 1$。

c. 钢筋混凝土构件：不预加应力的混凝土构件，即 $\lambda = 0$。

2) 预应力混凝土结构的优缺点

(1) 优点

① 提高了构件的抗裂度和刚度；

② 节约材料，降低造价；

③ 结构质量安全可靠；

④ 增强结构耐久性；

⑤ 能促进桥梁新体系的发展。

(2) 缺点

① 工艺较复杂，对质量要求高；

② 需要有一定的专门设备；

③ 预应力反拱不易控制；

④ 开工费用大，对于跨径较小、构件数量少的工程，成本较高。

了解到预应力混凝土结构的特点，在实际应用中应扬长避短，更好地为工程所用。

11.2 预加应力的方法与设备

1) 预加应力的主要方法

(1) 先张法:施工工艺如图 11-2 所示。

①定义:先张法是先张拉钢筋、后浇筑构件混凝土的方法。先张法所用的预应力钢筋,一般可用高强钢丝、直径较小的钢绞线和小直径的冷拉钢筋。

②特点:靠钢筋和混凝土之间的黏结力来传递和保持预应力。

③优点:先张法生产工艺简单、工序少,效率高、质量容易保证,适宜工厂化大批量生产。

④缺点:需要专门的张拉台座,基建投资较大,且预应力钢筋只能直线配筋,一般用于中小型构件。

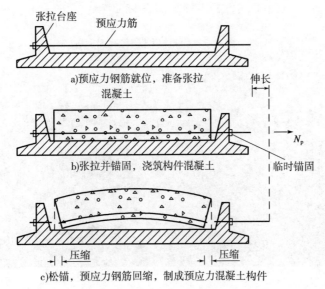

图 11-2 先张法工艺流程示意图

(2) 后张法:施工工艺如图 11-3 所示。

①定义:先浇筑混凝土后张拉钢筋的方法,张拉钢筋时,构件混凝土开始储备应力。

②特点:后张法主要是靠锚具传递和保持预加应力。

③优点:后张法不用加力台座,张拉设备简单,便于现场施工,是生产大型预应力混凝土构件的主要方法,且预应力钢筋可配合荷载的弯矩和剪力变化,布置成合理的曲线形,某预应力混凝土 T 梁的预应力筋布置如图 11-4 所示。

由以上内容可知,两种预加力方法的施工工艺不同,建立和保持预应力的方法也不同。后张法是靠工作锚具来传递和保持预加应力的;先张法则是靠黏结力来传递并保持预加应力的。

2) 锚具

(1) 概念

锚具是在制作预应力混凝土构件时锚固预应力筋的装置。在先张法中,构件制成后锚具可取下重复使用;后张法是靠锚具传递预应力,锚具永久地锚固在构件内。

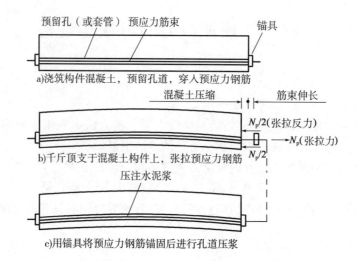

图 11-3　后张法工艺流程示意图

图 11-4　预应力筋的布置

(2) 对锚具的要求
① 锚具受力安全可靠;
② 预应力损失小;
③ 构造简单,制作方便,用钢量少,价格便宜;
④ 施工设备简便,张拉锚固方便迅速。
(3) 锚具的分类
① 依靠摩阻力锚固的锚具,如锥形锚、夹片式群锚、JM 锚;
② 依靠承压锚固的锚具,如镦头锚、钢筋螺纹锚;
③ 依靠黏结力锚固的锚具,如先张法的预应力筋黏结锚固、后张法固定端的钢绞线压花锚具。
(4) 几种常用的锚具
① 锥形锚:如图 11-5 所示。

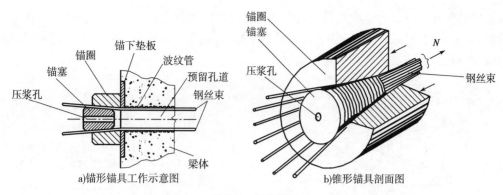

图 11-5　锥形锚具

工作原理:通过顶压锥形的锚塞,将预应力的钢丝卡在锚圈与锚塞之间,当张拉千斤顶放松预应力钢丝后,钢丝向梁内回缩时带动锚塞向锚圈内楔紧,这样预应力钢丝通过摩阻力将预应力传到锚圈,然后由锚圈承压,将预加力传到混凝土构件上。

②镦头锚:如图 11-6 所示。

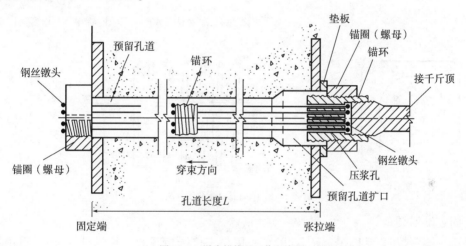

图 11-6　镦头锚锚具工作示意图

工作原理:先将钢丝逐一穿过锚杯的蜂窝眼,然后用专门的镦头机将钢丝端头镦粗,借粗镦头直接承压将钢丝固定于锚杯上。锚杯的外缘车有螺纹,穿束后,在固定端将锚圈(螺母)拧上,即可将钢束锚固于梁端。在张拉端,则先将与千斤顶连接的拉杆旋入锚杯内进行张拉,待锚杯带动钢筋或钢丝伸长到设计需要时,将锚圈沿锚杯外的螺纹旋紧顶在构件表面,再慢慢放松千斤顶,退出拉杆,钢丝束的回缩力就通过锚圈垫板,传到梁体混凝土上而获得锚固。

③钢筋螺纹锚具。

工作原理:借助于钢筋两端的螺纹,在钢筋张拉后直接拧上螺母进行锚固,钢筋的回缩力由螺母经支承垫板承压传递给梁体而获得预应力(图 11-7)。

优点:钢筋螺纹锚具受力明确,锚固可靠,且构造简单,施工方便,并能重复张拉、放松或拆卸,还可以简便地采用套筒接长。

④夹片锚具。

钢绞线夹片锚,它由带锥孔的锚板和夹片组成(图 11-8)。张拉力筋后,用夹片将钢绞线锚固。扁型夹片锚具是为适应扁薄截面构件(如桥面板梁等)预应力钢筋锚固的需要而研制的,简称扁锚。其工作原理与一般夹片锚具体系相同,只是工作锚板、锚下钢垫板和喇叭管,以及形成预留孔道的波纹管等均为扁形而已。

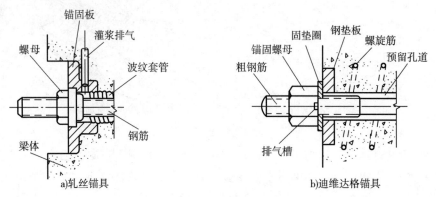

图 11-7 钢筋螺纹锚具

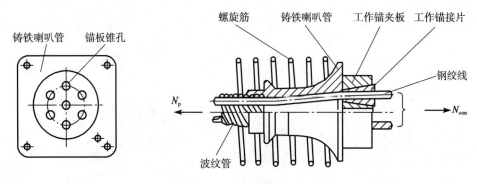

图 11-8 夹片锚具配套示意图

⑤固定端锚具。

采用一端张拉时,其固定端锚具,除可采用与张拉端相同的夹片锚具外,还可采用挤压锚具(图 11-9)和压花锚具,如图 11-10 所示。

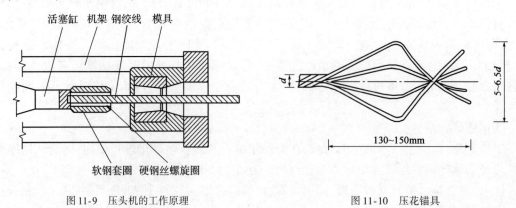

图 11-9 压头机的工作原理　　　　　　图 11-10 压花锚具

(5) 连接器

锚头连接器如图 11-11a)所示,钢绞线束 N_1 锚固后,再用来连接钢绞线束 N_2 的装置。接长连接器如图11-11b)所示,连接两段未张拉的钢绞线束的设备。

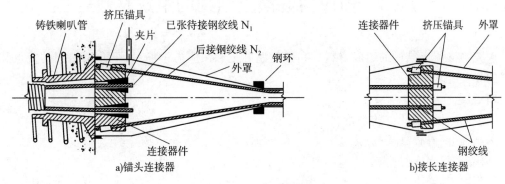

图 11-11　连接器构造

3)千斤顶

千斤顶用于张拉预应力钢筋,工程中千斤顶根据张拉力大小不同有许多类型,大直径的穿心单作用千斤顶如图11-12所示。

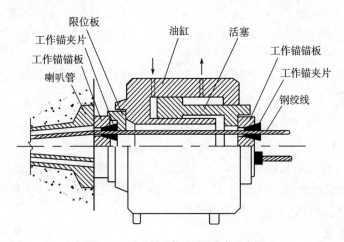

图 11-12　夹片锚张拉千斤顶安装示意图

4)预加应力的其他设备

(1)制孔器:后张法构件中,预留孔道的设备。

①抽拔橡胶管;

②螺旋金属波纹管(简称波纹管)。

(2)穿索机:牵引预应力钢筋穿过孔道的设备。

①液压式,桥梁中常用;

②电动式。

(3)灌孔水泥浆用压浆机:在有黏结的预应力混凝土构件中,张拉钢筋后给孔道压注水泥浆的设备。它主要由灰浆搅拌桶、储浆桶、灰浆泵以及供水系统组成。

(4)张拉台座:先张法预应力混凝土施工中,张拉预应力筋并临时锚固的工作平台。

11.3 预应力混凝土结构的组成材料

预应力混凝土结构的组成材料是高强度混凝土和预应力钢筋,以下介绍它们的特点和对材料的相关要求。

1) 混凝土

(1) 预应力混凝土结构对混凝土的要求

① 高强度,有较高承载力;

② 快硬、早强,满足施工快速的要求;

③ 收缩、徐变小,有效预应力大,效率高。

(2) 强度要求

用于预应力结构的混凝土,必须有高的抗压强度,正因为如此,《公桥规》规定:预应力混凝土构件的混凝土强度等级不应低于 C40。

(3) 混凝土的收缩与徐变

混凝土的收缩和徐变,使预应力混凝土构件缩短,因而引起预应力钢筋预应力下降,通常称此为预应力损失,在预应力混凝土结构设计计算中,必须正确估算这些损失。

① 混凝土的收缩

混凝土的收缩包括两部分:水泥凝结硬化时所产生的收缩变形即凝缩;混凝土干硬出水后产生的收缩变形即干缩。

《公桥规》推荐用的混凝土收缩应变计算式为:

$$\varepsilon_{cs}(t,t_s) = \varepsilon_{cs0} \cdot \beta_s(t-t_s) \tag{11-2}$$

式中,$\varepsilon_{cs}(t,t_s)$ 为收缩开始时龄期 t_s(一般假定为 3~7d),计算考虑的龄期为 t 时的收缩应变;ε_{cs0} 为名义收缩系数,对于强度等级 C20~C50 的混凝土,按表 11-1 取值;β_s 为收缩随时间发展的系数,见《公桥规》中的计算方法。

② 混凝土的徐变

定义:混凝土在应力不变的情况下,随时间增长的塑性变形。

混凝土名义收缩系数 ε_{cs0} 表 11-1

40% ≤ RH ≤ 70%	70% ≤ RH ≤ 90%
0.529×10^{-3}	0.310×10^{-3}

主要影响因素:荷载应力、持荷时间、混凝土的品质与加载龄期,以及构件尺寸和工作的环境等。

徐变系数:为徐变应变与弹性应变的比例系数 ϕ,即:

$$\phi = \frac{\varepsilon_c}{\varepsilon_e} \tag{11-3}$$

式中,ε_c 为徐变应变值;ε_e 为加载(σ_c 作用)时的弹性应变(急变)值;ϕ 为徐变系数,亦称徐变特征值。

《公桥规》推荐用的徐变系数计算式为:

$$\phi(t,t_0) = \phi_0 \cdot \beta_c(t-t_0) \tag{11-4}$$

式中，ϕ_0 为混凝土名义徐变系数；$\beta_c(t-t_0)$ 为加载后徐变随时间发展的系数，按《公桥规》推荐方法计算。

2) 预应力钢材

(1) 预应力混凝土结构对钢筋的基本要求

①强度高，提供高的承受能力；
②有较好的塑性和可焊性，满足施工质量要求；
③与混凝土的黏着力强，尤其对先张法构件而言，提供可靠的锚固；
④应力松弛损失小。

(2) 常用预应力钢筋的种类

①高强度钢丝：直径为 5～9mm，消除应力的光面和螺旋肋钢丝。
②钢绞线：由多根高强钢丝扭结并经消除内应力后的盘卷状钢束。
③精轧螺纹钢筋：钢筋表面轧制有规律的螺纹肋条，可用螺母锚固。

(3) 其他材料

用玻璃纤维、增强塑料、碳纤维增强塑料等制成的预应力钢筋。

思考练习题

1. 为什么要对构件施加预应力？预应力混凝土结构的主要优点是什么？
2. 《公桥规》对预应力混凝土构件如何分类？
3. 什么是先张法？先张法构件按什么样的工序施工？先张法构件如何实现预应力筋的锚固？先张法构件有何优缺点？
4. 预应力钢筋对锚具有哪些要求？从原理上分为哪几类？
5. 为什么预应力混凝土构件所选用的材料都要求有较高的强度？
6. 什么叫预应力度？预应力度是否"预应力越大越好"？
7. 在施加预应力工艺中，何谓先张法与后张法？各有什么特点？后张法构件按什么样的工序施工？后张法构件如何实现预应力筋的锚固？后张法构件有何优缺点？
8. 预应力混凝土构件对混凝土材料有哪些特别的要求？何谓高强混凝土？
9. 什么是张拉控制应力？张拉控制应力是如何确定的？为什么张拉控制应力既不能定得过高，也不能定得过低？
10. 混凝土的收缩、徐变对预应力混凝土构件有何影响？如何配制收缩、徐变小的混凝土？
11. 预应力混凝土结构对所使用的预应力钢筋有何要求？公路桥梁中常用的预应力钢筋有哪些？
12. 理解预应力混凝土结构的三种概念是什么？它们在分析和设计中有何用？
13. 建立预应力的方法有哪些？试述各自适用范围与特点。
14. 预应力混凝土较普通钢筋混凝土结构的优越性一般包括哪些方面？

第12章
预应力混凝土受弯构件的设计与计算

12.1 预应力混凝土受弯构件的受力阶段与计算特点

1) 预应力混凝土受弯构件的受力阶段

钢筋混凝土受弯构件经历了整体工作、带裂缝工作、破坏三个阶段,预应力混凝土受弯构件的受力过程是怎样的呢?研究表明,它从施加预应力开始直到破坏的过程分为三大阶段。

(1) 施工阶段

从施加预应力开始到梁的运输、安装就位为止的全过程称为施工阶段,它又包括预加应力和运输安装两个受力阶段。

① 预加应力阶段

从预加应力开始到传力锚固为止——施加预应力的全过程。对于简支梁,由于 N_p 的偏心作用,构件将产生向上的反拱,形成以梁两端为支点的简支梁,因此梁的一期恒载(自重荷载) G_1 也在施加预加力 N_p 的同时一起参加作用。此阶段的应力分布如图12-1所示。

此阶段的设计要点:

a. 预应力(张拉应力)的控制;

b. 混凝土上、下缘的应力应满足《公桥规》要求;

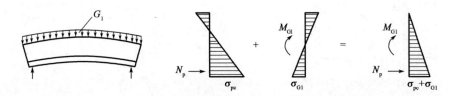

图 12-1 预加应力阶段截面应力分布

c. 端部锚固区局部承受能力满足要求,并且不能出现裂缝。

此阶段的特点:

a. 荷载:有效预应力最大,活载最小,仅有自重作用;

b. 混凝土强度相对较低,没有完全达到设计强度。

②运输安装阶段

在预加应力阶段梁体上缘可能产生拉应力,加上运输安装时的吊梁形式与成桥不同,使梁体悬臂部分在自重作用下产生负弯矩,增加上缘的拉应力而使混凝土开裂,这是不允许的;另外加上起吊放梁的冲击作用,加大了应力效应。按《公桥规》规定计入动力系数,来验算构件支点或吊点截面上缘混凝土的拉应力。

(2)使用阶段

桥梁通车运营使用整个工作阶段——运营阶段。它经历了消压、裂缝出现和带裂缝工作过程。

对全预应力混凝土结构此阶段不允许出现拉应力;对 A 类部分预应力混凝土结构,此阶段允许出现拉应力,不允许开裂;对 B 类部分预应力混凝土结构,此阶段允许出现裂缝,但裂缝宽度不允许超过规定值。

①消压阶段

构件除承受偏心预加力 N_p 和梁的一期恒载 G_1 外,还要承受桥面铺装、人行道、栏杆等后加的二期恒载 G_2 和车辆、人群等活荷载 Q。当梁控制截面由恒载、活载产生的正应力(图 12-2)与偏心预加力 N_p 所产生的应力相等时,下边缘应力为零[图 12-3b)],消掉了储备的预加应力。

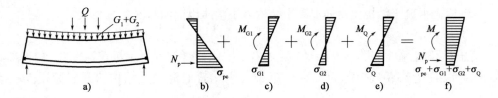

图 12-2 使用阶段各种作用下的截面应力分布

②裂缝出现

在使用荷载作用下,截面下缘经历了压应力—零应力—拉应力—混凝土开裂过程,如图 12-3 所示。

此阶段的设计要点:

a. 混凝土的正应力、主应力、预应力钢筋的应力均应满足要求;

b. 进行构件的正截面、斜截面抗裂性验算;

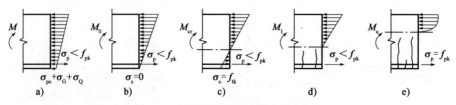

图 12-3 梁使用及破坏阶段的截面应力图

c. 构件维持正常使用的变形验算。

此阶段的特点:

a. 预应力损失大部分已经发生,有效预应力小;

b. 外荷载最大,包括全部使用活载。

③带裂缝工作

当混凝土开裂后,随着荷载增加,裂缝发生发展,混凝土的拉应力由钢筋承受,钢筋应力逐渐增大到屈服强度。

(3) 破坏阶段——极限承载力状态

适筋梁达到极限状态时,受拉区钢筋先屈服,而后受压区混凝土被压碎,构件即告破坏[图 12-3e)]。破坏时截面的应力状态与钢筋混凝土受弯构件相似,计算方法也相同。

特别注意:预应力混凝土结构的极限承载力也是以材料强度耗尽而达到极限。

2) 构件的消压弯矩、开裂弯矩、极限弯矩的计算

(1) 消压弯矩

由外荷载引起的拉应力恰好与控制截面下边缘预存的压应力 σ_{pc} 抵消,控制截面下边缘达到零应力状态时称为消压,此时外荷载弯矩为消压弯矩 M_0。

$$\sigma_{pc} = \frac{M_0}{I_0} \cdot y_\text{下} = \frac{M_0}{W_0} \tag{12-1}$$

$$M_0 = \sigma_{pc} \cdot W_0 \tag{12-2}$$

式中, σ_{pc} 为由永存预加力 N_p 引起的梁下边缘混凝土的有效预压应力;W_0 为换算截面对受拉边的弹性抵抗矩。

"消压"表示预应力已经耗尽,与没有施加预应力的钢筋混凝土结构的初始状态处于同一个起点,预应力混凝土结构比钢筋混凝土结构多了消压以前的状况。

(2) 开裂弯矩

"消压"后随着荷载的增加,下缘混凝土出现拉应力直到混凝土抗拉强度 f_{tk},构件即将出现裂缝,此时理论临界弯矩称为开裂弯矩 M_{cr},它比普通混凝土结构的开裂弯矩 $M_{cr,c}$ 多一个 M_0。

$$M_{cr} = M_0 + M_{cr,c} \tag{12-3}$$

$$M_{cr} = (\sigma_{pc} + \gamma f_{tk}) W_0 \tag{12-4}$$

$$\gamma = 2S_0/W_0 \tag{12-5}$$

式中, $M_{cr,c}$ 相当于同截面钢筋混凝土梁的开裂弯矩;γ 为受拉区混凝土塑性影响系数;S_0 为全截面换算截面重心轴以上(或以下)部分面积对重心轴的面积矩。

(3) 破坏弯矩

构件达到极限状态,钢筋屈服,混凝土压碎时的极限弯矩即破坏弯矩。进一步看出预应力

混凝土结构的"本质"：

①预应力的加施，使混凝土在施工阶段和正常使用阶段均处于弹性范围，可按"材料力学"方法计算——预应力的作用是把混凝土变成弹性材料；

②在正常配筋范围内，预应力混凝土受弯构件的破坏弯矩与同条件普通钢筋梁的破坏弯矩几乎相同——预应力只是用来平衡荷载；

③预先给混凝土储备的压应力，在使用阶段混凝土不开裂或裂缝较细，这样高强钢筋就可以与混凝土一起共同工作——预应力促成高强度钢筋与混凝土的共同工作。

12.2 预应力混凝土受弯构件承载力计算

1) 正截面承载力计算

(1) 计算依据

受拉区配有预应力筋 A_p 和非预应力筋 A_s，受压区有预应力筋 A'_p 和非预应力筋 A'_s 的适筋梁，在破坏时受拉区钢筋都达到屈服强度 f_{pd}、f_{sd}，受压区普通钢筋达到压屈强度 f'_{sd}，预应力筋一般不会压屈，应力为 σ_{pa}。

(2) 计算假定

①平截面假定；

②不计混凝土抗拉作用；

③受压区混凝土矩形应力图的假定。

(3) 计算公式

由基本假定得到受力图式(图 12-4)和基本公式。

$$f_{sd}A_s + f_{pd}A_p = f_{cd}bx + f'_{sd}A'_s + (f'_{pd} - \sigma'_{p0})A'_p \tag{12-6}$$

$$\gamma_0 M_d \leq f_{cd}bx\left(h_0 - \frac{x}{2}\right) + f'_{sd}A'_s(h_0 - a'_s) + (f'_{pd} - \sigma'_{p0})A'_p(h_0 - a'_p) \tag{12-7}$$

式中，σ'_{p0} 为钢筋当其重心水平处混凝土应力为零时的有效预应力，$\sigma'_{p0} = \sigma'_p + \alpha E_p \sigma'_{pc}$。

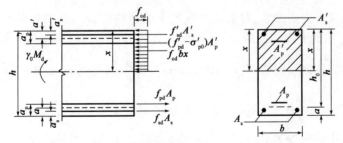

图 12-4 受压区配置预应力钢筋的矩形截面受弯构件正截面承载力计算图

(4) 适用条件

①受压区高度 x 应满足《公桥规》的规定：

$$x \leq \xi_b h_0, \xi_b = \frac{\beta}{1 + \dfrac{0.002}{\varepsilon_{cu}} + \dfrac{f_{pd} - \sigma_{p0}}{\varepsilon_u E_p}} \tag{12-8a}$$

ξ_b 按表 3-2 取值。

②当受压区预应力钢筋受压,即 $f'_{pd} - \sigma'_{p0} > 0$ 时,应满足:
$$x \geq 2a' \tag{12-8b}$$

③当受压区预应力钢筋受拉,即 $f'_{pd} - \sigma'_{p0} < 0$ 时,应满足:
$$x \geq 2a'_s \tag{12-8c}$$

式中,a' 为受压区钢筋 A'_s 和 A'_p 的合力作用点至截面最近边缘的距离;当预应力钢筋 A'_p 中的应力为拉应力时,则以 a'_s 代替 a';a'_p 为钢筋 A'_p 的合力作用点至截面最近边缘的距离。

(5)设计方法

①截面设计:预应力钢筋的设计先按 $A_s = A'_s = A'_p = 0$,初步计算所需 A_p,还要根据抗裂度等要求最后确定预应力钢筋面积,具体见 12.8 节。

②强度复核:步骤与普通钢筋混凝土结构相同,具体见 12.8 节。

③T 形截面计算类似。采用纵向体外预应力钢筋的 T 形截面或 I 形截面,参考《公桥规》的相关规定。

2)斜截面承载力计算

(1)预应力在斜截面受力时的作用

①提高了斜截面抗剪承载力;

②延缓斜裂缝的出现,抑制斜裂缝发展;

③增加了剪压区高度和集料的咬合作用。

(2)斜截面抗剪承载力计算

《公桥规》采用的斜截面抗剪承载力计算公式:
$$\gamma_0 V_d \leq V_{cs} + V_{sb} = V_{pb} + V_{pb,ex} \tag{12-9}$$

式中,V_d 为斜截面受压端正截面上由作用(或荷载)产生的最大剪力组合设计值(kN);V_{cs} 为斜截面内混凝土和箍筋共同的抗剪承载力设计值(kN);V_{pb} 为与斜截面相交的预应力弯起钢筋抗剪承载力设计值(kN);$V_{pb,ex}$ 为与斜截面相交的体外预应力弯起钢筋抗剪承载力计算值(kN),具体计算参见《公桥规》。

$$V_{cs} = \alpha_1 \alpha_2 \alpha_3 0.45 \times 10^{-3} bh_0 \sqrt{(2 + 0.6p)\sqrt{f_{cu,k}} \rho_{sv} f_{sv}} \tag{12-10}$$

式中,α_2 为预应力提高系数,对预应力混凝土受弯构件,$\alpha_2 = 1.25$,但当由钢筋合力引起的截面弯矩与外弯矩的方向相同时,或允许出现裂缝的预应力混凝土受弯构件,取 $\alpha_2 = 1.0$;p 为斜截面内纵向受拉钢筋配筋率,$p = 100\rho$,$\rho = \dfrac{A_p + A_{pb} + A_s}{bh_0}$;当 $p > 2.5$ 时,取 $p = 2.5$;ρ_{sv} 为斜截面内箍筋配筋率,$\rho_{sv} = \dfrac{A_{sv}}{s_v b}$,在实际工程中,预应力混凝土箱梁也有采用腹板内设置竖向预应力钢筋的情况,这时 ρ_{sv} 应换为竖向预应力钢筋的配筋率 ρ_{pv};s_v 为斜截面内竖向预应力钢筋的间距(mm);f_{sv} 为竖向预应力钢筋抗拉强度设计值;A_{sv} 为斜截面内配置在同一截面的竖向预应力钢筋截面面积。

$$V_{pb} = 0.75 \times 10^{-3} f_{pd} \sum A_{pb} \sin\theta_p \tag{12-11}$$

式中,θ_p 为预应力弯起钢筋(在斜截面受压端正截面处)的切线与水平线的夹角;A_{pb} 为斜截面内在同一弯起平面的预应力弯起钢筋的截面面积(mm²);f_{pd} 为预应力钢筋抗拉强度设计值。

注意：预应力混凝土受弯构件抗剪承载力计算,所需满足的公式上、下限值与普通钢筋混凝土受弯构件相同。

(3)斜截面抗弯承载力计算

根据斜截面的受弯破坏形态,仍取斜截面以左部分为脱离体(图12-5),并以受压区混凝土合力作用点 O(转动铰)为中心取矩,由 $\sum M_O = 0$,得到矩形、T形和I形截面的受弯构件斜截面抗弯承载力计算公式为:

$$\gamma_0 M_d \leq f_{sd}A_s Z_s + f_{pd}A_p Z_p + \sum f_{pd}A_{pb}Z_{pb} + \sum f_{sv}A_{sv}Z_{sv} \tag{12-12}$$

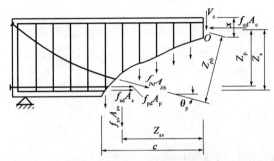

图12-5 斜截面抗弯承载力计算图

式中,M_d 为斜截面受压端正截面的最大弯矩组合设计值;Z_s、Z_p 分别为纵向普通受拉钢筋合力点、纵向预应力受拉钢筋合力点至受压区中心点 O 的距离;Z_{pb} 为与斜截面相交的同一弯起平面内预应力弯起钢筋合力点至受压区中心点 O 的距离;Z_{sv} 为与斜截面相交的同一平面内箍筋合力点至斜截面受压端的水平距离。

计算斜截面抗弯承载力时,其最不利斜截面的位置,需选在预应力钢筋数量变少、箍筋数量与间距的变化处,以及构件混凝土截面腹板厚度的变化处进行。但其斜截面的水平投影长度 c,仍需自下而上,按不同倾斜角度试算确定。最不利的斜截面水平投影长度按下列公式试算确定:

$$\gamma_0 V_d = \sum f_{pd}A_{pb}\sin\theta_p + \sum f_{sv}A_{sv} \tag{12-13}$$

假设最不利斜截面与水平方向的夹角为 α,水平投影长度为 c,则斜截面上箍筋截面积为 $\sum A_{sv} = A_{sv} \cdot \dfrac{c}{s_v}$,代入上式可得到最不利水平投影长度 c 的表达式为:

$$c = \frac{\gamma_0 V_d - \sum f_{pd}A_{pb}\sin\theta_p}{\dfrac{f_{sv} \cdot A_{sv}}{s_v}} \tag{12-14}$$

式中,V_d 为斜截面受压端正截面相应于最大弯矩组合设计值的剪力组合设计值;s_v 为箍筋间距(mm)。

水平投影长度 c 确定后,尚应确定受压区合力作用点的位置 O,以便确定各力臂的长度。由斜截面的受力平衡条件 $\sum H = 0$,可得到:

$$\sum f_{pd}A_{pb}\cos\theta_p + f_{sd}A_s + f_{pd}A_p = f_{cd}A_c \tag{12-15}$$

由此可求出混凝土截面受压区的面积 A_c。因 A_c 是受压区高度 x 的函数,故截面形式确定后,斜截面受压区高度 x 也就不难求得,受压区合力作用点的位置也随之可以确定。

预应力混凝土梁斜截面抗弯承载力的计算比较复杂,通常用构造措施来加以保证。

12.3 张拉控制应力与预应力损失的计算

1) 张拉控制应力的确定

张拉控制应力 σ_{con}：

$$\sigma_{con} = \frac{N_{p.con}}{A_p} \tag{12-16}$$

式中，$N_{p.con}$ 为预应力钢筋锚固前张拉设备所显示的总张拉力；A_p 为预应力钢筋面积。

《公桥规》规定，预应力钢筋端部（锚下）的控制应力 σ_{con} 应符合下列规定：

对于钢丝、钢绞线、体外预应力：

$$\sigma_{con} \leq 0.75 f_{pk} \tag{12-17}$$

对于预应力螺纹钢筋：

$$\sigma_{con} \leq 0.85 f_{pk} \tag{12-18}$$

式中，f_{pk} 为预应力钢筋的抗拉强度标准值。

在实际工程中，当对构件进行超张拉或计入锚圈口摩擦损失时，预应力钢筋最大控制应力值（千斤顶油泵上显示的值）可增加 $0.05 f_{pk}$。

2) 预应力损失及其计算

预应力钢筋张拉后到使用阶段，会产生应力损失，《公桥规》中主要考虑了 6 项损失。

(1) 预应力筋与管道壁间摩擦引起的应力损失——摩擦损失 σ_{l1}

摩擦损失是指后张法张拉钢筋时，预应力钢筋与周围接触的管道之间存在摩擦，引起[图 12-8a)]预应力钢筋中的应力随张拉端距离的增加而逐渐减少的现象。摩擦损失包括两部分：

①由管道的位置偏差引起的，孔壁粗糙及钢筋表面的粗糙等原因，使得预应力钢筋与孔壁摩擦产生损失，也称为管道偏差影响的摩擦损失，其数值较小；

②张拉曲线预应力钢筋，由于孔道的曲率，预应力筋与孔之间产生径向压力所引起的摩擦损失，称为弯道影响摩擦损失，其数值较大。受力示意如图 12-6 所示。

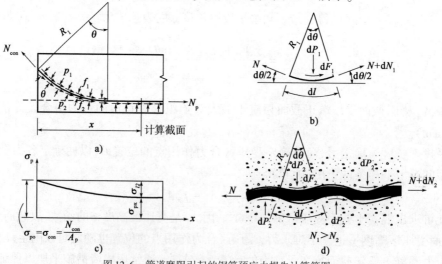

图 12-6 管道摩阻引起的钢筋预应力损失计算简图

摩擦损失 σ_{l1} 计算公式为：

$$\sigma_{l1} = \frac{N_{con} - N_x}{A_p} = \sigma_{con}[1 - e^{-(\mu\theta + kx)}] \quad (12\text{-}19)$$

式中，σ_{con} 为锚下张拉控制应力，$\sigma_{con} = \frac{N_{con}}{A_p}$；$N_{con}$ 为钢筋锚下张拉控制力；A_p 为预应力钢筋的截面面积；x 为从张拉端至计算截面的管道长度在构件纵轴上的投影长度；或为三维空间曲线管道的长度(m)；k 为管道每米长度的局部偏差对摩擦的影响系数，可按附表 2-6(见中文部分)采用；μ 为钢筋与管道壁间的摩擦系数，可按附表 2-6 采用；θ 为从张拉端至计算截面间管道平面曲线的夹角[图 12-6a)]之和，即曲线包角，按绝对值相加，单位以弧度计，如管道为竖平面内和水平面内同时弯曲的三维空间曲线管道，则 θ 可按下式计算：

$$\theta = \sqrt{\theta_H^2 + \theta_V^2} \quad (12\text{-}20)$$

式中，θ_H、θ_V 分别为在同段管道水平面内的弯曲角与竖向平面内的弯曲角。

为减少摩擦损失，一般可采用如下措施：
①采用两端张拉，以减小 θ 值及管道长度 x 值；
②采用超张拉。对于后张法预应力钢筋，其张拉工艺按下列要求进行：

对于钢绞线束：

$0 \rightarrow$ 初应力$(0.1\sigma_{con} \sim 0.15\sigma_{con}) \rightarrow 1.05\sigma_{con}$(持荷 2min)$\rightarrow \sigma_{con}$(锚固)

对于钢丝束：

$0 \rightarrow$ 初应力$(0.1\sigma_{con} \sim 0.15\sigma_{con}) \rightarrow 1.05\sigma_{con}$(持荷 2min)$\rightarrow 0 \rightarrow \sigma_{con}$(锚固)

注意：对于一般夹片式锚具，不宜采用超张拉工艺。

(2)锚具变形、钢筋回缩和接缝压缩引起的应力损失——锚具损失 σ_{l2}

当预应力张拉结束并进行锚固时，锚具受力变形、缝隙被挤密以及钢筋向内回缩，此外，拼装式构件的接缝，在锚固后也将被压密变形引起应力损失 σ_{l2}：

$$\sigma_{l2} = \frac{\sum \Delta l}{l} E_p \quad (12\text{-}21)$$

式中，$\sum \Delta l$ 为张拉端锚具变形、钢筋回缩和接缝压缩值之和(mm)，可根据试验确定，当无可靠资料时，按附录表 2-7 采用；l 为张拉端至锚固端之间的距离(mm)；E_p 为预应力钢筋的弹性模量。

《公桥规》规定：后张法预应力混凝土构件应计算由锚具变形、钢筋回缩等引起反摩阻后的预应力损失。

减小 σ_{l2} 值的方法：
①采用超张拉；
②注意选用 $\sum \Delta l$ 值小的锚具，尤其对于短小构件。

(3)钢筋与台座间的温差引起的应力损失——温差损失 σ_{l3}

先张法构件在采用蒸汽或其他加热方法养护混凝土时，新浇混凝土尚未结硬，钢筋受热膨胀，台座固定不动，即钢筋长度保持不变，钢筋中的应力随温度的增高而降低产生的应力损失 σ_{l3}。

假设张拉时钢筋与台座的温度均为 t_1，混凝土加热养护时的最高温度为 t_2，此时钢筋尚未与混凝土黏结，温度由 t_1 升为 t_2 后钢筋在混凝土中自由变形 Δl_t：

$$\Delta l_t = \alpha \cdot (t_2 - t_1) \cdot l \tag{12-22}$$

式中，α 为钢筋的线膨胀系数，一般可取 $\alpha = 1 \times 10^{-5}$；$l$ 为钢筋的有效长度；t_1 为张拉钢筋时制造场地的温度（℃）；t_2 为混凝土加热养护时已张拉钢筋的最高温度（℃）。

当停止升温养护时，混凝土已与钢筋黏结在一起，钢筋和混凝土将同时随温度变化而共同伸缩，因养护升温所降低的应力已不可恢复，形成温差应力损失 σ_{l3}：

$$\sigma_{l3} = \frac{\Delta l_t}{l} \cdot E_p = \alpha(t_2 - t_1) \cdot E_p \tag{12-23}$$

取预应力钢筋的弹性模量 $E_p = 2 \times 10^5 \text{MPa}$，则有：

$$\sigma_{l3} = 2(t_2 - t_1) \tag{12-24}$$

减小温差损失，常采用二次升温的养护方法，即第一次由常温 t_1 升温至 t_2' 进行养护。初次升温的温度一般控制在20℃以内，待混凝土达到一定强度（例如 7.5～10MPa）能够阻止钢筋在混凝土中自由滑移后，再将温度升至 t_2 进行养护。此时，钢筋将和混凝土一起变形，不会因第二次升温而引起应力损失，故计算 σ_{l3} 的温差只是 $t_2' - t_1$。

如果张拉台座与被养护构件是共同受热、共同变形时，则不应计入此项应力损失。

(4) 混凝土弹性压缩引起的应力损失——弹性压缩损失 σ_{l4}

当混凝土受到预压力产生压缩变形时，对于已经张拉锚固于构件上的预应力钢筋来说，也会产生压缩应变 $\varepsilon_p = \varepsilon_c$，从而产生应力损失，即混凝土弹性压缩损失 σ_{l4}。

① 先张法构件

当预应力钢筋放张时，混凝土的弹性压缩应变引起的应力损失为：

$$\sigma_{l4} = \varepsilon_p \cdot E_p = \varepsilon_c \cdot E_p = \frac{\sigma_{pc}}{E_c} \cdot E_p = \alpha_{Ep} \cdot \sigma_{pc} \tag{12-25}$$

式中，α_{Ep} 为预应力钢筋弹性模量 E_p 与混凝土弹性模量 E_c 的比值；σ_{pc} 为在先张法构件计算截面钢筋重心处，由预加力 N_{p0} 产生的混凝土预压应力，可按 $\sigma_{pc} = \frac{N_{p0}}{A_0} + \frac{N_{p0} e_p^2}{I_0}$ 计算；N_{p0} 为全部钢筋的有效预加力（扣除相应阶段的预应力损失）；A_0、I_0 为构件全截面的换算截面面积和换算截面惯性矩；e_p 为预应力钢筋重心至换算截面重心轴间的距离。

② 后张法构件

后张法构件一般是采用分批张拉锚固，当张拉后批钢筋时所产生的混凝土弹性压缩变形将使先批已张拉并锚固的预应力钢筋产生应力损失，称为分批张拉应力损失，也以 σ_{l4} 表示。《公桥规》规定 σ_{l4} 可按下式计算：

$$\sigma_{l4} = \alpha_{Ep} \sum \Delta \sigma_{pc} \tag{12-26}$$

式中，α_{Ep} 为预应力钢筋弹性模量与混凝土的弹性模量的比值；$\sum \Delta \sigma_{pc}$ 为在计算截面先张拉的钢筋重心处，由后张拉各批钢筋所产生的混凝土法向应力之和。

后张法构件中钢筋在各截面的效应不同（表现为三个不同——钢筋位置不同、每次的张拉力可能不同、计算的截面不同），使各截面的"$\sum \Delta \sigma_{pc}$"不相同，要详细计算，非常麻烦。为简化计算，对简支梁，进行如下近似处理：

a. 取应力控制的截面作为全梁的平均截面进行计算，其余截面不另计算，简支梁可以取 $l/4$ 截面。

b. 假定同一截面（如 $l/4$ 截面）内的所有预应力钢筋，都集中布于其合力作用点处，并假定各批预应力钢筋的张拉力都相等,等于各批钢筋张拉力的平均值。求得各批钢筋张拉时,在先批张拉钢筋重心点处所产生的混凝土正应力为 $\Delta\sigma_{pc}$,即

$$\Delta\sigma_{pc} = \frac{N_p}{m}\left(\frac{1}{A_n} + \frac{e_{pn} \cdot y_i}{I_n}\right) \tag{12-27}$$

式中,N_p 为所有预应力钢筋预加应力(扣除相应阶段的应力损失 σ_{l1} 与 σ_{l2} 后)的合力;m 为张拉预应力钢筋的总批数;e_{pn} 为预应力钢筋预加应力的合力 N_p 至净截面重心轴间的距离;y_i 为先批张拉钢筋重心(假定的全部预应力钢筋重心)处至混凝土净截面重心轴间的距离,故 $y_i \approx e_{pn}$;A_n、I_n 为混凝土梁的净截面面积和净截面惯性矩。

张拉各批钢筋所产生的混凝土正应力 $\Delta\sigma_{pc}$ 之和,就等于由全部(m 批)钢筋的合力 N_p 在其作用点(或全部筋束的重心点)处所产生的混凝土正应力 σ_{pc},即

$$\sum\Delta\sigma_{pc} = m\Delta\sigma_{pc} = \sigma_{pc}$$

或写成

$$\Delta\sigma_{pc} = \frac{\sigma_{pc}}{m} \tag{12-28}$$

c. 进一步假定同一截面上($l/4$ 截面)全部预应力筋重心处混凝土弹性压缩应力损失的总平均值,作为各批钢筋由混凝土弹性压缩引起的应力损失值。第 i 批钢筋的应力损失 $\sigma_{l4(i)}$ 为:

$$\sigma_{l4(i)} = (m-i) \cdot \alpha_{Ep}\Delta\sigma_{pc} \tag{12-29}$$

截面上各批钢筋弹性压缩损失平均值为:

$$\sigma_{l4} = \frac{\sigma_{l4(1)} + \sigma_{l4(m)}}{2} = \frac{m-1}{2} \cdot \alpha_{Ep}\Delta\sigma_{pc} \tag{12-30}$$

对于各批张拉预应力钢筋根数相同的情况,将式(12-29)代入式(12-30),得到分批张拉引起的平均应力损失为:

$$\sigma_{l4} = \frac{m-1}{2m} \cdot \alpha_{Ep}\sigma_{pc} \tag{12-31}$$

式中,σ_{pc} 为计算截面全部钢筋重心处由张拉所有预应力钢筋产生的混凝土法向应力。

减小 σ_{l4} 的措施:采用超张拉,或者提高混凝土强度等级。

(5)钢筋松弛引起的应力损失——松弛损失 σ_{l5}

钢筋长度固定不变,则钢筋中的应力将随时间延长而降低,产生钢筋的应力松弛,图 12-7 为典型的预应力钢筋松弛曲线。

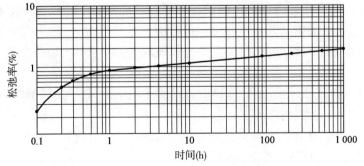

图 12-7 典型的预应力钢筋松弛曲线

由钢筋松弛引起的应力损失终值,按下列规定计算。

对于精轧螺纹钢筋：
一次张拉： $$\sigma_{l5} = 0.05\sigma_{con} \tag{12-32}$$
超张拉： $$\sigma_{l5} = 0.035\sigma_{con} \tag{12-33}$$
对于预应力钢丝、钢绞线：
$$\sigma_{l5} = \psi \cdot \zeta \cdot \left(0.52\frac{\sigma_{pe}}{f_{pk}} - 0.26\right) \cdot \sigma_{pe} \tag{12-34}$$

式中，ψ 为张拉系数，一次张拉时，$\psi = 1.0$；超张拉时，$\psi = 0.9$；ζ 为钢筋松弛系数，Ⅰ级松弛（普通松弛），$\zeta = 1.0$；Ⅱ级松弛（低松弛），$\zeta = 0.3$；σ_{pe} 为传力锚固时的钢筋应力，对后张法构件 $\sigma_{pe} = \sigma_{con} - \sigma_{l1} - \sigma_{l2} - \sigma_{l4}$；对先张法构件 $\sigma_{pe} = \sigma_{con} - \sigma_{l2}$。

《公桥规》还规定，对碳素钢丝、钢绞线，当 $\sigma_{pe}/f_{pk} \leq 0.5$ 时，应力松弛损失值为零。

减小 σ_{l5} 的措施：采用超张拉，或者采用低松弛的预应力钢筋。

（6）混凝土收缩和徐变引起的应力损失——收缩徐变损失 σ_{l6}

混凝土收缩、徐变会使预应力混凝土构件缩短，从而引起应力损失，《公桥规》规定损失按下列方法计算。

①受拉区预应力钢筋的预应力损失 σ_{l6} 为：

$$\sigma_{l6}(t) = \frac{0.9[E_p\varepsilon_{cs}(t,t_0) + \alpha_{Ep}\sigma_{pc}\phi(t,t_0)]}{1 + 15\rho\rho_{ps}} \tag{12-35}$$

式中，$\sigma_{l6}(t)$ 为构件受拉区全部纵向钢筋截面重心处由混凝土收缩、徐变引起的预应力损失；σ_{pc} 为构件受拉区全部纵向钢筋截面重心处由预应力（扣除相应阶段的预应力损失）和结构自重产生的混凝土法向应力(MPa)，对于简支梁，一般可取跨中截面和 $l/4$ 截面的平均值作为全梁各截面的计算值；σ_{pc} 不得大于 $0.5f'_{cu}$，f'_{cu} 为预应力钢筋传力锚固时混凝土立方体抗压强度；E_p 为预应力钢筋的弹性模量；α_{Ep} 为预应力钢筋弹性模量与混凝土弹性模量的比值；ρ 为构件受拉区全部纵向钢筋配筋率；对先张法构件，$\rho = (A_p + A_s)/A_0$；对于后张法构件，$\rho = (A_p + A_s)/A_n$，其中 A_p、A_s 分别为受拉区的预应力钢筋和非预应力钢筋的截面面积，A_0 和 A_n 分别为换算截面面积和净截面面积；$\rho_{ps} = 1 + \frac{e_{ps}^2}{i^2}$，其中 i 为截面回转半径，$i^2 = \frac{I}{A}$，先张法构件取 $I = I_0$，$A = A_0$；后张法构件取 $I = I_n$，$A = A_n$，其中，I_0 和 I_n 分别为换算截面惯性矩和净截面惯性矩；e_{ps} 为构件受拉区预应力钢筋和非预应力钢筋截面重心至构件截面重心轴的距离，$e_{ps} = (A_p e_p + A_s e_s)/(A_p + A_s)$；$e_p$ 为构件受拉区预应力钢筋截面重心至构件截面重心的距离；e_s 为构件受拉区纵向非预应力钢筋截面重心至构件截面重心的距离；$\varepsilon_{cs}(t,t_0)$ 为预应力钢筋传力锚固龄期为 t_0，计算考虑的龄期为 t 时的混凝土收缩应变；$\phi(t,t_0)$ 为加载龄期为 t_0，计算考虑的龄期为 t 时的徐变系数。

对于受压区配置预应力钢筋 A'_p 和非预应力钢筋 A'_s 的构件，其受拉区预应力钢筋的预应力损失也可取 $A'_p = A'_s = 0$，近似地计算。

②受压区配置预应力钢筋 A'_p 和非预应力钢筋 A'_s 的构件，由混凝土收缩、徐变引起构件受压区预应力钢筋的预应力损失 σ'_{l6} 为：

$$\sigma'_{l6} = \frac{0.9[E_p\varepsilon_{cs}(t,t_0) + \alpha_{Ep}\sigma'_{pc}\phi(t,t_0)]}{1 + 15\rho'\rho'_{ps}} \tag{12-36}$$

式中，$\sigma'_{l6}(t)$ 为构件受压区全部纵向钢筋截面重心处由混凝土收缩、徐变引起的预应力损失；σ'_{pc} 为构件受压区全部纵向钢筋截面重心处由预应力(扣除相应阶段的预应力损失)和结构自重产生的混凝土法向应力(MPa)；σ'_{pc} 不得大于 $0.5f'_{cu}$；当 σ'_{pc} 为拉应力时，应取其为零；ρ' 为构件受压区全部纵向钢筋配筋率；对先张法构件，$\rho = (A'_p + A'_s)/A_0$；对于后张法构件，$\rho = (A'_p + A'_s)/A_n$；其中 A'_p、A'_s 分别为受压区的预应力钢筋和非预应力筋的截面面积；$\rho'_{ps} = 1 + e'^2_{ps}/i^2$；$e'_{ps}$ 为构件受压区预应力钢筋和非预应力钢筋截面重心至构件截面重心轴的距离，$e'_{ps} = (A'_p e'_p + A'_s e'_s)/(A'_p + A'_s)$；$e'_p$ 为构件受压区预应力钢筋截面重心至构件截面重心的距离；e'_s 为构件受压区纵向非预应力钢筋截面重心至构件截面重心的距离。

应当指出，混凝土收缩、徐变应力损失与钢筋的松弛应力损失等是相互影响的，目前采用分开单独计算的方法不够完善。

减小 σ_{l6} 的措施：采用高强混凝土和提高施工养护质量。

3) 钢筋的有效预应力计算

预应力钢筋的有效预应力 σ_{pe} 为预应力钢筋控制应力 σ_{con} 扣除相应阶段的应力损失 σ_l 后实际存余的应力值，具体见表 12-1。

各阶段预应力损失值的组合　　　　　　　　　　　　　　表 12-1

预应力损失值的组合	先张法构件	后张法构件
传力锚固时的损失(第一批) σ_{lI}	$\sigma_{l2} + \sigma_{l3} + \sigma_{l4} + 0.5\sigma_{l5}$	$\sigma_{l1} + \sigma_{l2} + \sigma_{l4}$
传力锚固后的损失(第二批) σ_{lII}	$0.5\sigma_{l5} + \sigma_{l6}$	$\sigma_{l5} + \sigma_{l6}$

预应力钢筋的有效预应力 σ_{pe}：

在预加应力阶段，预应力筋中的有效预应力：

$$\sigma_{pe} = \sigma_{pI} = \sigma_{con} - \sigma_{lI} \tag{12-37}$$

在使用阶段，预应力筋中的有效预应力，即永存预应力：

$$\sigma_{pe} = \sigma_{pII} = \sigma_{con} - (\sigma_{lI} + \sigma_{lII}) \tag{12-38}$$

12.4　预应力混凝土受弯构件的应力计算与验算

1) 概述

(1) 应力计算(验算)的目标：使混凝土在施工、使用阶段的应力满足规范要求。

(2) 应力验算的内容：混凝土的法向压应力、钢筋的拉应力、混凝土的主压应力。

注意：① 对预应力混凝土简支结构，只计算预应力引起的主效应；

② 对预应力混凝土连续梁等超静定结构，此外还应计算预应力引起的次效应。

2) 短暂状况的应力验算

(1) 验算内容：在制作、运输及安装等施工阶段，由预应力作用、构件自重和施工荷载等引起的正截面和斜截面的应力。

(2) 计算原理：短暂状况的应力计算，构件属弹性阶段，按材料力学方法计算。

(3) 计算截面：对于简支梁来说，其受力最不利截面往往在支点附近，特别是直线配筋的预应力混凝土等截面简支梁，其支点上缘拉应力常常成为计算的控制力。

(4) 计算公式：

①预加应力阶段的正应力计算

预加应力阶段的正应力按材料力学方法计算，如图12-8所示，主要承受偏心的预加力 N_p 和一期恒载（梁自重荷载）G_1 作用效应 M_{G1}。

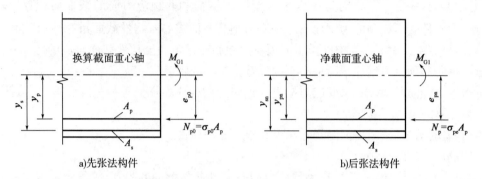

图 12-8 预加力阶段预应力钢筋和非预应力钢筋合力及其偏心矩

预加应力阶段截面上、下缘混凝土的总的正应力 σ_{ct}^t、σ_{cc}^t 为：

先张法构件
$$\left.\begin{array}{l}\sigma_{ct}^t = \dfrac{N_{p0}}{A_0} - \dfrac{N_{p0}e_{p0}}{W_{0u}} + \dfrac{M_{G1}}{W_{0u}} \\[2mm] \sigma_{cc}^t = \dfrac{N_{p0}}{A_0} + \dfrac{N_{p0}e_{p0}}{W_{0b}} - \dfrac{M_{G1}}{W_{0b}}\end{array}\right\} \quad (12\text{-}39)$$

后张法构件
$$\left.\begin{array}{l}\sigma_{ct}^t = \dfrac{N_p}{A_n} - \dfrac{N_p e_{pn}}{W_{nu}} + \dfrac{M_{G1}}{W_{nu}} \\[2mm] \sigma_{cc}^t = \dfrac{N_p}{A_n} + \dfrac{N_p e_{pn}}{W_{nb}} - \dfrac{M_{G1}}{W_{nb}}\end{array}\right\} \quad (12\text{-}40)$$

式中，σ_{p0} 为受拉区预应力钢筋合力点处混凝土法向应力等于零时的预应力钢筋应力；$\sigma_{p0} = \sigma_{con} - \sigma_{l1} + \sigma_{l4}$，其中 σ_{l4} 为受拉区预应力钢筋由混凝土弹性压缩引起的预应力损失；σ_{l1} 为受拉区预应力钢筋传力锚固时的预应力损失；e_{p0} 为预应力钢筋的合力对构件全截面换算截面重心的偏心距；e_{pn} 为预应力钢筋的合力对构件净截面重心的偏心距；A_n 为构件净截面的面积；W_{0u}、W_{0b} 为构件全截面换算截面对上、下缘的截面抵抗矩；W_{nu}、W_{nb} 为构件净截面对上、下缘的截面抵抗矩；N_p 为后张法构件的预应力钢筋的合力，按下式计算：

$$N_p = \sigma_{pe} A_p \quad (12\text{-}41)$$

对于配置曲线预应力钢筋的构件，上式中的 A_p 取为 $A_p + A_{pb}\cos\theta_p$；其中，A_{pb} 为弯起预应力钢筋的截面积，θ_p 为计算截面上弯起的预应力钢筋的切线与构件轴线的夹角，A_p 为受拉区预应力钢筋的截面面积；σ_{pe} 为受拉区预应力钢筋的有效预应力，$\sigma_{pe} = \sigma_{con} - \sigma_{l1}$，$\sigma_{l1}$ 为受拉区预应力钢筋传力锚固时的预应力损失。

②运输、吊装阶段的正应力计算

此阶段构件应力计算方法与预加应力阶段相同。唯应注意的是预加力 N_p 已变小；计算一期恒载作用时产生的弯矩应考虑计算图式的变化，并考虑动力系数。

③施工阶段混凝土的限制应力

《公桥规》要求施工阶段算得的混凝土正应力应符合下列规定：

a. 混凝土压应力 σ_{cc}^t。

$$\sigma_{cc}^t \leq 0.70 f'_{ck} \tag{12-42}$$

式中，f'_{ck} 为制作、运输、安装各施工阶段的混凝土轴心抗压强度标准值。

b. 混凝土拉应力 σ_{ct}^t。

当 $\sigma_{ct}^t \leq 0.70 f'_{tk}$ 时，预拉区应配置配筋率不小于 0.2% 的纵向非预应力钢筋；

当 $\sigma_{ct}^t = 1.15 f'_{tk}$ 时，预拉区应配置配筋率不小于 0.4% 的纵向非预应力钢筋；

当 $0.70 f'_{tk} < \sigma_{ct}^t < 1.15 f'_{tk}$ 时，预拉区应配置的纵向非预应力钢筋配筋率按以上两者直线内插取用，拉应力 σ_{ct}^t 不应超过 $1.15 f'_{tk}$。

3) 运营使用持久状况

(1) 验算内容：混凝土的正应力，主应力，预应力钢筋应力。

(2) 计算特点：预应力损失已全部完成，有效预应力 σ_{pe} 最小，计算时作用取其标准值，汽车荷载应计入冲击系数，预加应力效应考虑在内，荷载分项系数均取为 1.0。

(3) 计算原理：持久状况的应力计算，对全预应力混凝土和 A 类部分预应力混凝土结构而言，仍属构件弹性阶段的计算，采用材料力学的方法进行计算。

(4) 控制截面：对于直线配筋等截面简支梁，一般以跨中为最不利控制截面；但对于曲线配筋的等截面或变截面简支梁，则应根据预应力筋的弯起和混凝土截面变化的情况，一般取跨中、$l/4$、$l/8$、支点截面和截面变化处的截面进行计算。

(5) 计算公式：

① 正应力计算

配有普通钢筋的预应力混凝土构件中（图 12-9），正应力如下。

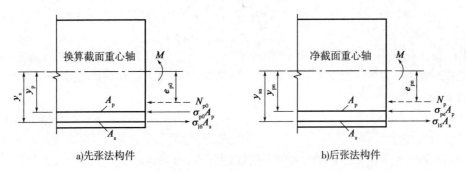

图 12-9 使用阶段预应力钢筋和非预应力钢筋合力及其偏心矩

a. 先张法构件。

先张法构件由作用标准值和预加力在构件截面上缘产生的混凝土法向压应力为：

$$\sigma_{cu} = \sigma_{pt} + \sigma_{kc} = \left(\frac{N_{p0}}{A_0} - \frac{N_{p0} \cdot e_{p0}}{W_{0u}} \right) + \frac{M_{G1}}{W_{0u}} + \frac{M_{G2}}{W_{0u}} + \frac{M_Q}{W_{0u}} \tag{12-43}$$

预应力钢筋中的最大拉应力为：

$$\sigma_{pmax} = \sigma_{pe} + \alpha_{Ep} \left(\frac{M_{G1}}{I_0} + \frac{M_{G2}}{I_0} + \frac{M_Q}{I_0} \right) \cdot y_{p0} \tag{12-44}$$

式中，σ_{kc} 为作用标准值产生的混凝土法向压应力；σ_{pe} 为预应力钢筋的永存预应力，即 $\sigma_{pe} = \sigma_{con} - \sigma_{lI} - \sigma_{lII} = \sigma_{con} - \sigma_l$；$N_{p0}$ 为使用阶段预应力钢筋和非预应力钢筋的合力[图 12-9a)]，按

下式计算：

$$N_{p0} = \sigma_{p0}A_p - \sigma_{l6}A_s \qquad (12\text{-}45)$$

式中，σ_{p0} 为受拉区预应力钢筋合力点处混凝土法向应力等于零时的预应力钢筋应力；$\sigma_{p0} = \sigma_{con} - \sigma_l + \sigma_{l4}$，其中 σ_{l4} 为使用阶段受拉区预应力钢筋由混凝土弹性压缩引起的预应力损失，σ_l 为受拉区预应力钢筋总的预应力损失；σ_{l6} 为受拉区预应力钢筋由混凝土收缩和徐变引起的预应力损失；e_{p0} 为预应力钢筋与非预应力钢筋合力作用点至构件换算截面重心轴的距离，可按下式计算：

$$e_{p0} = \frac{\sigma_{p0}A_p y_p - \sigma_{l6}A_s y_s}{\sigma_{p0}A_p - \sigma_{l6}A_s} \qquad (12\text{-}46)$$

式中，A_s 为受拉区非预应力钢筋的截面面积；y_s 为受拉区非预应力钢筋重心至换算截面重心的距离；W_{0u} 为构件混凝土换算截面对截面上缘的抵抗矩；α_{Ep} 为预应力钢筋与混凝土的弹性模量比；M_{G2} 为由桥面铺装、人行道和栏杆等二期恒载产生的弯矩标准值；M_Q 为由可变荷载标准值组合计算的截面最不利弯矩，汽车荷载考虑冲击系数。

b. 后张法构件。

后张法构件由作用（或荷载）标准值和预应力在构件截面上缘产生的混凝土法向压应力 σ_{cu} 为：

$$\sigma_{cu} = \sigma_{pt} + \sigma_{kc} = \left(\frac{N_p}{A_n} - \frac{N_p \cdot e_{pn}}{W_{nu}}\right) + \frac{M_{G1}}{W_{nu}} + \frac{M_{G2}}{W_{0u}} + \frac{M_Q}{W_{0u}} \qquad (12\text{-}47)$$

预应力钢筋中的最大拉应力为：

$$\sigma_{pmax} = \sigma_{pe} + \alpha_{Ep}\frac{M_{G2} + M_Q}{I_0} \cdot y_{0p} \qquad (12\text{-}48)$$

式中，N_p 为预应力钢筋和非预应力钢筋的合力，按下式计算：

$$N_p = \sigma_{pe}A_p - \sigma_{l6}A_s \qquad (12\text{-}49)$$

式中，σ_{pe} 为受拉区预应力钢筋的有效预应力，$\sigma_{pe} = \sigma_{con} - \sigma_l$；$W_{nu}$ 为构件混凝土净截面对截面上缘的抵抗矩；e_{pn} 为预应力钢筋和非预应力钢筋合力作用点至构件净截面重心轴的距离，按下式计算：

$$e_{pn} = \frac{\sigma_{pe}A_p y_{pn} - \sigma_{l6}A_s y_{sn}}{\sigma_{pe}A_p - \sigma_{l6}A_s} \qquad (12\text{-}50)$$

式中，y_{sn} 为受拉区非预应力钢筋重心至净截面重心的距离；y_{0p} 为计算的预应力钢筋重心到换算截面重心轴的距离。

当截面受压区配置预应力钢筋 A'_p 时，则计算式还需考虑 A'_p 的作用。

② 混凝土主应力计算

预应力混凝土受弯构件由作用（或荷载）标准值和预加力作用产生的混凝土主压应力 σ_{cp} 和主拉应力 σ_{tp} 可按下列公式计算，即：

$$\left.\begin{array}{c}\sigma_{tp}\\ \sigma_{cp}\end{array}\right\} = \frac{\sigma_{cx} + \sigma_{cy}}{2} \mp \sqrt{\left(\frac{\sigma_{cx} - \sigma_{cy}}{2}\right)^2 + \tau^2} \qquad (12\text{-}51)$$

式中，σ_{cx} 为在计算主应力点，由作用标准值和预加力产生的混凝土法向应力。

先张法构件：

$$\sigma_{cx} = \frac{N_{p0}}{A_0} - \frac{N_{p0}e_{p0}}{I_0}y_0 + \frac{M_{G1} + M_{G2} + M_Q}{I_0}y_0 \qquad (12\text{-}52)$$

后张法构件：
$$\sigma_{cx} = \frac{N_p}{A_n} - \frac{N_p e_{pn}}{I_n} y_n + \frac{M_{G1}}{I_n} y_n + \frac{M_{G2} + M_Q}{I_0} y_0 \quad (12\text{-}53)$$

式中，y_0、y_n 分别为计算主应力点至换算截面、净截面重心轴的距离；I_0、I_n 分别为换算截面惯性矩、净截面惯性矩；σ_{cy} 为由竖向预应力钢筋的预加力产生的混凝土竖向压应力，按下式计算：

$$\sigma_{cy} = 0.6 \frac{n \sigma'_{pe} A_{pv}}{b \cdot s_v} \quad (12\text{-}54)$$

式中，n 为同一截面上竖向钢筋的肢数；σ'_{pe} 为竖向预应力钢筋扣除全部预应力损失后的有效预应力；A_{pv} 为单肢竖向预应力钢筋的截面面积；s_v 为竖向预应力钢筋的间距；τ 在计算主应力点，按作用标准值组合计算的剪力产生的混凝土剪应力，对于等高度梁截面上任一点，在作用标准值组合下的剪应力 τ 可按下列公式计算：

先张法构件
$$\tau = \frac{V_{G1} S_0}{b I_0} + \frac{(V_{G2} + V_Q) S_0}{b I_0} \quad (12\text{-}55)$$

后张法构件
$$\tau = \frac{V_{G1} S_n}{b I_n} + \frac{(V_{G2} + V_Q) S_0}{b I_0} - \frac{\sum \sigma''_{pe} A_{pb} \sin\theta_p S_n}{b I_n} \quad (12\text{-}56)$$

式中，V_{G1}、V_{G2} 分别为一期恒载和二期恒载作用引起的剪力标准值；V_Q 为可变作用（或荷载）引起的剪力标准值组合；对于简支梁，V_Q 的计算式为：

$$V_Q = V_{Q1} + V_{Q2} \quad (12\text{-}57)$$

式中，V_{Q1}、V_{Q2} 分别为汽车荷载效应（计冲击系数）、人群荷载效应引起的剪力标准值；S_0、S_n 分别为计算主应力点以上（或以下）部分换算截面面积对截面重心轴、净截面面积对截面重心轴的面积矩；θ_p 为计算截面上预应力弯起钢筋的切线与构件纵轴线的夹角（图 12-10）；b 为计算主应力点处构件腹板的宽度；σ''_{pe} 为纵向预应力弯起钢筋扣除全部预应力损失后的有效预应力；A_{pb} 为计算截面上同一弯起平面内预应力弯起钢筋的截面面积。

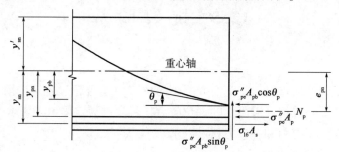

图 12-10 剪力计算图

注意：a. 以上公式中均取压应力为正，拉应力为负；

b. 对连续梁等超静定结构，应计及预加力、温度作用等引起的次效应；

c. 对变高度预应力混凝土连续梁，计算由作用引起的剪应力时，应计算截面上弯矩和轴向力产生的附加剪应力。

③持久状况的钢筋和混凝土的应力限值

对于按全预应力混凝土和 A 类部分预应力混凝土设计的受弯构件，《公桥规》中对持久状况应力计算的限值规定如下。

a. 使用阶段正截面混凝土的最大压应力，应满足：

$$\sigma_{kc} + \sigma_{pt} \leq 0.5 f_{ck} \quad (12\text{-}58)$$

式中，σ_{kc} 为作用标准值产生的混凝土法向压应力；σ_{pt} 为预加力产生的混凝土法向拉应力；f_{ck} 为混凝土轴心抗压强度标准值。

b. 使用阶段预应力钢筋的最大拉应力限值。

《公桥规》规定钢筋的最大拉应力限值为：

对体内预应力钢绞线、钢丝 $\quad\sigma_{pe} + \sigma_p \leq 0.65 f_{pk}$ (12-59a)

对预应力螺纹钢筋 $\quad\sigma_{pe} + \sigma_p \leq 0.75 f_{pk}$ (12-59b)

式中，σ_{pe} 为预应力钢筋扣除全部预应力损失后的有效预应力；σ_p 为作用产生的预应力钢筋应力增量；f_{pk} 为预应力钢筋抗拉强度标准值。

c. 使用阶段混凝土主应力限值。

混凝土的主压应力应满足：

$$\sigma_{cp} \leq 0.6 f_{ck} \quad (12\text{-}60)$$

式中，f_{ck} 为混凝土轴心抗压强度标准值。

计算所得的混凝土主拉应力 σ_{tp} 作为构件斜截面抗剪计算的补充，按下列规定设置箍筋：

a. 在 $\sigma_{tp} \leq 0.5 f_{tk}$ 的区段，箍筋可仅按构造要求配置；

b. 在 $\sigma_{tp} > 0.5 f_{tk}$ 的区段，箍筋的间距 s_v 可按下式计算：

$$s_v = \frac{f_{sk} A_{sv}}{\sigma_{tp} b} \quad (12\text{-}61)$$

式中，f_{sk} 为箍筋的抗拉强度标准值；f_{tk} 为混凝土轴心抗拉强度标准值；A_{sv} 为同一截面内箍筋的总截面面积；b 为矩形截面宽度、T形或I形截面的腹板宽度。

当按上式计算的箍筋用量少于按斜截面抗剪承载力计算的箍筋用量时，构件箍筋按抗剪承载力计算要求配置。

12.5 抗裂验算

1）概述

构件正截面抗裂性验算是以正截面混凝土的拉应力是否超过规定的限值为标准表示的，斜截面的抗裂性验算是以构件混凝土的主拉应力是否超过规定限值来表示的，因此，抗裂验算只要计算控制截面的正应力和主应力就可以了。

2）正截面抗裂性验算要求

(1) 全预应力构件在作用频遇组合下应满足：

预制构件 $\quad\sigma_{st} \leq 0.85 \sigma_{pc}$ (12-62)

分段浇筑或砂浆接缝的纵向分块 $\quad\sigma_{st} \leq 0.80 \sigma_{pc}$ (12-63)

(2) A 类部分预应力构件在作用频遇组合下应满足：

$$\sigma_{st} - \sigma_{pc} \leq 0.7 f_{tk} \quad (12\text{-}64)$$

在准永久组合下应满足：

$$\sigma_{lt} - \sigma_{pc} \leq 0 \quad (12\text{-}65)$$

①由作用频遇组合产生的构件抗裂验算边缘混凝土的法向拉应力 σ_{st}，计算式为：

先张法构件：
$$\sigma_{st} = \frac{M_s}{W} = \frac{M_{G1} + M_{G2} + M_{Qs}}{W_0} \tag{12-66}$$

后张法构件：
$$\sigma_{st} = \frac{M_s}{W} = \frac{M_{G1}}{W_n} + \frac{M_{G2} + M_{Qs}}{W_0} \tag{12-67}$$

式中，σ_{st} 为按作用频遇组合计算的构件抗裂验算边缘混凝土法向拉应力；M_s 为按作用频遇组合计算的弯矩值；M_{Qs} 为按作用频遇组合计算的可变荷载弯矩值，对于简支梁：

$$M_{Qs} = \psi_{11} M_{Q1} + \psi_{12} M_{Q2} = 0.7 M_{Q1} + 1.0 M_{Q2} \tag{12-68}$$

式中，ψ_{11}、ψ_{12} 分别为频遇组合计算中的汽车荷载效应和人群荷载效应的频遇值系数；M_{Q1}、M_{Q2} 分别为汽车荷载效应（不计冲击）和人群荷载效应产生的弯矩标准值；W_0、W_n 分别为构件换算截面和净截面对抗裂验算边缘的弹性抵抗矩。

对于预应力混凝土连续梁和连续刚构，除了考虑直接施加于梁上的荷载如恒载、汽车外，还应考虑间接作用如日照温差、混凝土收缩和徐变的影响。

② 由作用（或荷载）准永久组合下边缘混凝土的正应力计算。

作用准永久组合是永久作用标准值和可变作用准永久值效应相组合。准永久组合下预应力混凝土构件边缘混凝土的正应力 σ_{lt} 计算与频遇组合下的计算基本一致。计算式为：

先张法构件
$$\sigma_{lt} = \frac{M_l}{W} = \frac{M_{G1} + M_{G2} + M_{Ql}}{W_0} \tag{12-69}$$

后张法构件
$$\sigma_{lt} = \frac{M_l}{W} = \frac{M_{G1}}{W_n} + \frac{M_{G2} + M_{Ql}}{W_0} \tag{12-70}$$

式中，σ_{lt} 为按作用准永久组合计算构件抗裂验算边缘混凝土的法向拉应力；M_l 为按作用准永久组合计算的弯矩值；M_{Ql} 为按作用准永久组合计算的可变作用弯矩值，仅考虑汽车、人群等直接作用于构件的荷载产生的弯矩值，可按下式计算：

$$M_{Ql} = \psi_{21} M_{Q1} + \psi_{22} M_{Q2} = 0.4 M_{Q1} + 0.5 M_{Q2} \tag{12-71}$$

式中，M_{Q1}、M_{Q2} 分别为汽车荷载效应（不计冲击）和人群荷载效应产生的弯矩标准值；ψ_{21}、ψ_{22} 分别为作用准永久组合中的汽车荷载效应和人群效应的准永久值系数。

3）斜截面抗裂性验算要求

(1) 全预应力构件，在作用频遇组合下要满足：

预制构件 $\qquad \sigma_{tp} \leq 0.6 f_{tk}$ (12-72)

现浇构件 $\qquad \sigma_{tp} \leq 0.4 f_{tk}$ (12-73)

(2) A 类和 B 类部分预应力构件，在频遇组合下要满足：

预制构件 $\qquad \sigma_{tp} \leq 0.7 f_{tk}$ (12-74)

现浇构件 $\qquad \sigma_{tp} \leq 0.5 f_{tk}$ (12-75)

式中，f_{tk} 为混凝土轴心抗拉强度标准值。

σ_{tp} 计算式为：

$$\sigma_{tp} = \frac{\sigma_{cx} + \sigma_{cy}}{2} - \sqrt{\left(\frac{\sigma_{cx} - \sigma_{cy}}{2}\right)^2 + \tau^2} \tag{12-76}$$

式中的正应力 σ_{cx}、σ_{cy} 和剪应力 τ 的计算方法见式(12-52)~式(12-56)。在计算剪应力 τ 时，剪力 V_Q 取按作用频遇组合计算的可变作用引起的剪力值 V_{Qs}。

4)斜截面抗裂性验算截面的选择

沿构件长度方向应选择剪力和弯矩均较大的截面,或外形有突变的截面;沿截面高度应选择换算截面形心及截面宽度改变处;对先张法构件还应考虑传递长度 l_{cr} 范围内预应力值的实际大小。

12.6 变形计算与预拱度设置

1)预应力混凝土受弯构件在荷载频遇组合下的总挠度 w_s

$$w_s = -\delta_{pe} + w_{Ms} \tag{12-77}$$

式中,δ_{pe} 为永存预加力 N_{pe} 所产生的上挠度;w_{Ms} 为由作用频遇组合计算的弯矩值引起的挠度值。

(1)预加力 N_{pe} 所产生的上挠度 δ_{pe}

δ_{pe} 的计算按照单位力法进行,具体表达式为:

$$\delta_{pe} = \int_0^l \frac{M_{pe} \cdot \overline{M}_x}{B_0} dx \tag{12-78}$$

式中,M_{pe} 为由永存预加力(永存预应力的合力)在任意截面 x 处所引起的弯矩值;$\overline{M}_x$ 为跨中作用单位力时在任意截面 x 处所产生的弯矩值;B_0 为构件抗弯刚度,计算时按实际受力阶段取值。

(2)作用频遇组合计算的弯矩值引起的挠度值 w_{Ms}

《公桥规》规定,对于全预应力构件以及 A 类部分预应力混凝土构件的等高度简支梁、悬臂梁的挠度计算表达式为:

$$w_{Ms} = \frac{\alpha M_s l^2}{0.95 E_c I_0} \tag{12-79}$$

式中,l 为梁的计算跨径;α 为挠度系数,与弯矩图形状和支承的约束条件有关(表 12-2);M_s 为按作用(或荷载)频遇组合计算的弯矩;I_0 为构件全截面的换算截面惯性矩。

梁的最大弯矩 M_{max} 和跨中(或悬臂端)挠度系数 α 表 12-2

荷载图式	弯矩图和最大弯矩 M_{max}	挠度系数 α
均布荷载 q,跨度 l	$\dfrac{ql^2}{8}$	$\dfrac{5}{48}$
部分均布 βl,q	$\dfrac{\beta^2(2-\beta)^2 ql^2}{8}$	$\beta \leqslant \dfrac{1}{2}$ 时:$\dfrac{3-2\beta}{12(2-\beta)^2}$ $\beta > \dfrac{1}{2}$ 时:$\dfrac{4\beta^4 - 10\beta^3 + 9\beta^2 - 2\beta + 0.25}{12\beta^2(\beta-2)^2}$
三角形荷载 q	$\dfrac{ql^2}{15.625}$	$\dfrac{5}{48}$

续上表

荷 载 图 式	弯矩图和最大弯矩 M_{max}	挠度系数 α
简支梁，距左端 βl 处作用集中力 F	$F\beta(1-\beta)l$	$\beta \geq \dfrac{1}{2}$ 时：$\dfrac{4\beta^2 - 8\beta + 1}{-48\beta}$
悬臂梁，距固定端 βl 处作用集中力 F	$F\beta l$	$\dfrac{\beta(3-\beta)}{6}$
悬臂梁，从固定端起长度 βl 作用均布荷载 q	$\dfrac{q\beta^2 l^2}{2}$	$\dfrac{\beta(4-\beta)}{12}$

2）作用频遇组合并考虑长期效应影响的挠度值 w_l

预应力混凝土受弯构件随时间的增长，由于混凝土徐变等原因，使得构件挠度增加。《公桥规》中通过挠度长期增长系数 η_θ 来实现，具体计算式为：

$$w_l = -\eta_{\theta,pe} \cdot \delta_{pe} + \eta_{\theta,Ms} \cdot w_{Ms}$$
$$= -\eta_{\theta,pe} \cdot \delta_{pe} + \eta_{\theta,Ms} \cdot (w_{G1} + w_{G2} + w_{Qs}) \quad (12-80)$$

式中，w_l 考虑长期荷载效应的挠度值；$\eta_{\theta,pe}$ 预加力反拱值考虑长期效应增长系数；计算使用阶段预加力反拱值时，预应力钢筋的预加力应扣除全部预应力损失，并取 $\eta_{\theta,pe} = 2$；$\eta_{\theta,Ms}$ 为短期荷载效应组合考虑长期效应的挠度增长系数，按表 12-3 取值。

作用频遇组合考虑长期效应的挠度增长系数值　　表 12-3

混凝土强度等级	C40 以下	C40	C45	C50	C55	C60	C65	C70	C75	C80
$\eta_{\theta,Ms}$	1.60	1.45	1.44	1.43	1.41	1.40	1.39	1.38	1.36	1.35

预应力混凝土受弯构件在频遇组合、考虑长期效应影响下最大竖向挠度的容许值，与钢筋混凝土梁相同。

3）预拱度的设置

《公桥规》规定：预应力混凝土受弯构件由预加应力产生的长期反拱值大于按作用频遇组合计算的长期挠度时，不设预拱度；当预加应力的长期反拱值小于按作用频遇组合计算的长期挠度时应设预拱度，预拱度值 Δ 按该项荷载的挠度值与预加应力长期反拱值之差采用，即：

$$\Delta = \eta_{\theta,Ms} w_{Ms} - \eta_{\theta,pe} \delta_{pe} \quad (12-81)$$

设置预拱度时，按最大的预拱值沿顺桥向做成平顺的曲线。

12.7　端部锚固区计算

1）后张法构件锚下局部承压计算

后张法构件巨大的预加压力 F_{ed} 是通过锚具及其下面不大的垫板传给混凝土的。混凝土在局部受压后要进行局部抗压承载力计算和局部的抗拉承载力验算,并满足有关构件要求。

(1) 锚下局部承压的强度计算

配有螺旋箍筋的端部抗压能力应满足:

$$\gamma_0 N_{ed} \leq 0.9(\eta_s \beta f_{cd} + k\rho_{st}\beta_{con} f_{sd})A_{ln} \quad (12-82)$$

如果验算不满足要求,还可以:

① 加大局部受压面积;
② 调整锚具位置;
③ 提高混凝土强度等级。

(2) 局部受压时的抗拉承载力验算

为防止局部承压区出现沿构件长度方向的裂缝,对于在局部承压区配有间接钢筋的情况,按下式抗压杆模型验算抗拉承载力:

$$\gamma_0 T_{(\cdot),d} \leq f_{sd} \cdot A_s \quad (12-83)$$

式中,$T_{(\cdot),d}$ 按《公桥规》规定取。

(3) 构造要求

① 锚具下应设置厚度不小于16mm的垫板或采用具有喇叭管的锚具垫板;
② 板下螺旋筋圈数的长度不应小于喇叭管长度;
③ 锚下总体区应配置抵抗横向劈裂力的闭合箍筋,其间距不应大于120mm;
④ 梁端截面应配置抵抗表面劈裂力的抗裂钢筋;
⑤ 梁端平面尺寸由锚具尺寸、锚具间距以及张拉千斤顶的要求等布置而定。

2) 先张法构件预应力钢筋的传递长度与锚固长度

(1) 传递长度 l_{tr} 和锚固长度 l_a

先张法构件通过钢筋与混凝土的黏结力来传递和保持预应力,预应力钢筋在端部外露处应力应变均为零。经过一定长度,由于黏结的作用,预应力钢筋的应力为有效预应力 σ_{pe},这一长度称为预应力钢筋的传递长度 l_{tr}。传递长度范围内钢筋的受力比较复杂,应力是变化的。《公桥规》规定近似按线性规律来变化,如图12-11所示。同理,当构件达到承载能力极限状态时,预应力钢筋应力从端部的零值到设计强度 f_{pd} 为止的这一长度称为锚固长度 l_a,锚固长度保证钢筋达到屈服时不被拔出。

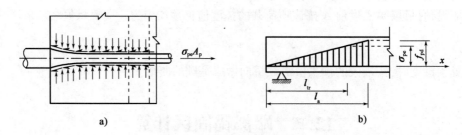

图12-11 先张法预应力筋的锚固

《公桥规》考虑以上各因素后对预应力钢筋的传递长度 l_{tr} 和锚固长度 l_a 的规定取值见附表2-8。同时假定传递长度和锚固长度范围内的预应力钢筋的应力(从零至 σ_{pe} 或 f_{pd})按直线内插计算[图 12-11b)]。因此,在端部锚固长度 l_a 范围内计算斜截面承载力时,预应力筋的应力 σ_{pe} 应根据斜截面所处位置按直线内插求得;在端部预应力传递长度 l_{tr} 范围内进行抗裂性计算时,预应力钢筋的实际应力值也应根据验算截面所处位置按直线内插求得。

(2)其他要求

对先张法构件端部还应采取局部加强措施,对预应力钢筋端部周围混凝土通常采取的加强措施是:单根钢筋时,其端部宜设置长度不小于 150mm 的螺旋筋;当为多根预应力钢筋时,其端部在 10d(预应力筋直径)范围内,设置 3~5 片钢筋网。

12.8 预应力混凝土简支梁设计

1)设计计算内容

以后张法简支梁为例,其设计计算的内容如下:

(1)根据设计要求,选定构件的截面形式与相应尺寸,或者直接对弯矩最大截面,根据截面抗弯要求初步估算构件混凝土截面尺寸;

(2)计算控制截面最大的设计弯矩和剪力;

(3)估算预应力钢筋的数量,并进行合理的布置;

(4)计算主梁截面几何特性;

(5)进行正截面与斜截面承载力计算;

(6)确定钢筋的张拉控制应力,估算各项预应力损失,并计算各阶段相应的有效预应力;

(7)按短暂状况和持久状况进行构件的应力验算;

(8)进行正截面与斜截面的抗裂验算;

(9)主梁的变形计算;

(10)锚固局部承压计算与锚固区设计。

设计中的关键是如何合理地拟定尺寸、估计预应力钢筋数量和正确布置钢筋。设计就是先假设然后计算,计算满足规范各项要求即可行,否则再假设再计算,直到满足要求为止。

2)截面效率指标

荷载效应与截面之间有着对应关系,荷载大,截面也应该大,但荷载效应包括了恒重(自重)和活载,截面尺寸的大小直接影响到总的荷载效应。

通常采用抗弯截面效率指标 ρ 检验其合理性:

$$\rho = \frac{k_u + k_b}{h}$$

式中,k_u 为上核心距;k_b 为下核心距。

ρ 越大,抗弯效率越高,ρ 一般为 0.4~0.55。

3)常用截面形式

图 12-12 表示桥梁工程中常用的一些截面形式,它们分别是空心板、T 梁、I 梁和箱梁等。

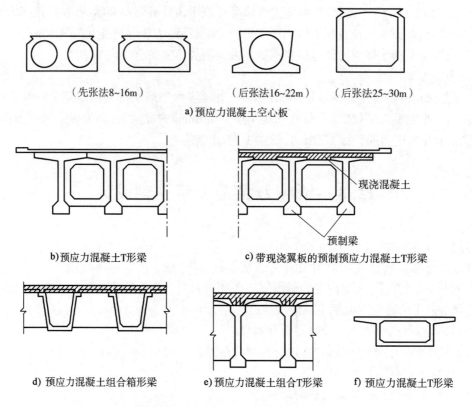

图 12-12 预应力混凝土梁的常用截面形式

4)预应力钢筋数量的估算

预应力钢筋截面积的估算应从以下三方面考虑。

(1)按构件正截面抗裂性要求估算预应力钢筋数量

全预应力混凝土梁按作用频遇组合进行正截面抗裂性验算,计算所得的正截面混凝土法向拉应力应满足限制值的要求,即

$$\frac{M_s}{W} - 0.85 N_{pe}\left(\frac{1}{A} + \frac{e_p}{W}\right) \leqslant 0 \tag{12-84}$$

对上式稍做变化,即可得全预应力混凝土梁满足作用频遇组合抗裂验算所需的有效预加力为:

$$N_{pe} \geqslant \frac{\dfrac{M_s}{W}}{0.85\left(\dfrac{1}{A} + \dfrac{e_p}{W}\right)} \tag{12-85}$$

式中,M_s 为按作用频遇组合计算的弯矩值;N_{pe} 为使用阶段预应力钢筋永存应力的合力;A 为构件混凝土全截面面积;W 为构件全截面对抗裂验算边缘弹性抵抗矩;e_p 为预应力钢筋的合力作用点至截面重心轴的距离。

对于 A 类部分预应力混凝土构件,可以得到类似的计算式,即

$$N_{pe} \geqslant \frac{\dfrac{M_s}{W} - 0.7f_{tk}}{\dfrac{1}{A} + \dfrac{e_p}{W}} \tag{12-86}$$

求得 N_{pe} 的值后,再确定适当的张拉控制应力 σ_{con},并扣除相应的应力损失 σ_l(对于配高强钢丝或钢绞线的后张法构件,σ_l 约为 $0.2\sigma_{con}$),就可以估算出所需要的预应力钢筋的总面积 $A_p = \dfrac{N_{pe}}{(1-0.2)\sigma_{con}}$。

A_p 确定之后,则可按一束预应力钢筋的面积 A_{p1} 算出所需的预应力钢筋束数 n_1 为:

$$n_1 = \frac{A_p}{A_{p1}} \tag{12-87}$$

式中,A_{p1} 为一束预应力钢筋的截面面积。

(2)按构件承载能力极限状态要求估算非预应力钢筋数量

在确定预应力钢筋的数量后,非预应力钢筋根据正截面承载能力极限状态的要求来确定。矩形截面梁、T形截面梁按正截面承载能力极限状态要求估算非预应力钢筋。

(3)最小配筋率的要求

按上述方法估算所得的钢筋数量,还必须满足最小配筋率的要求。《公桥规》规定,预应力混凝土受弯构件的最小配筋率应满足条件:

$$\frac{M_u}{M_{cr}} \geqslant 1.0 \tag{12-88}$$

式中,M_u 为受弯构件正截面抗弯承载力设计值;按式(3-29)或式(3-21)中不等号右边的式子计算;M_{cr} 为受弯构件正截面开裂弯矩值,M_{cr} 的计算式为:

$$M_{cr} = (\sigma_{pc} + \gamma f_{tk})W_0 \tag{12-89}$$

其中,σ_{pc} 为扣除全部预应力损失预应力钢筋和普通钢筋合力 N_{p0} 在构件抗裂边缘产生的混凝土预压应力;W_0 为换算截面抗裂边缘的弹性抵抗矩;γ 为计算参数,按式 $\gamma = 2S_0/W_0$ 计算;S_0 为全截面换算截面重心轴以上(或以下)部分面积对重心轴的面积矩。

5)预应力钢筋的布置

(1)截面核心

矩形、T形截面的核心如图12-13所示,预加力作用在核心内时,截面不会出现拉应力,按照这种思路来布置预应力钢筋。

(2)束界

根据全预应力混凝土构件截面上、下缘混凝土不出现拉应力的原则,按照在最小外荷载作用下和最不利荷载作用下的两种情况,分别确定 N_p 在各个截面上偏心距的极限。绘出如图12-14所示的两条 e_p 的限值线 E_1 和 E_2。只要 N_p 作用点(也即近似为预应力钢筋的截面重心)的位置落在由 E_1 及 E_2 所围成的区域内,就能保证构件在最小外荷载和最不利荷载作用下其上、下缘混凝土均不会出现拉应力。因此,把由 E_1 和 E_2 两条曲线所围成的布置预应力钢筋时的钢筋重心界限,称为束界(或索界)。

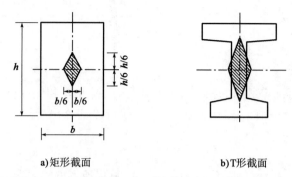

a) 矩形截面　　　　　　　b) T形截面

图 12-13　截面核心区

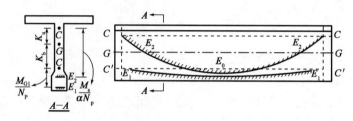

图 12-14　全预应力混凝土简支梁的束界图

在预加应力阶段,保证梁的上缘混凝土不出现拉应力的条件为:

$$\sigma_{ct} = \frac{N_{pI}}{A} - \frac{N_{pI}e_{pI}}{W_u} + \frac{M_{G1}}{W_u} \geqslant 0 \tag{12-90}$$

由此求得到:

$$e_{pI} \leqslant E_1 = K_b + \frac{M_{G1}}{N_{pI}} \tag{12-91}$$

式中,e_{pI} 为预加力的偏心距,合力点位于截面重心轴以下时 e_{pI} 取正值,反之取负值;K_b 为混凝土截面下核心距,$K_b = \dfrac{W_u}{A}$;W_u 为构件全截面对截面上缘的弹性抵抗矩;N_{pI} 为传力锚固时预加力的合力。

同理,在作用频遇组合计算的弯矩值作用下,可以求得预加力合力偏心距 e_{p2} 为:

$$e_{p2} \geqslant E_2 = \frac{M_s}{\alpha N_{pI}} - k_u \tag{12-92}$$

式中,M_s 为按作用(或荷载)频遇组合计算的弯矩值;α 为使用阶段的永存预加力 N_{pe} 与传力锚固时的有效预加力 N_{pI} 之比值,可近似地取 $\alpha = 0.8$;k_u 为混凝土截面上核心距,$k_u = \dfrac{W_b}{A}$;W_b 为构件全截面对截面下缘的弹性抵抗矩。

可以看出:e_{p1}、e_{p2} 分别具有与弯矩 M_{G1} 和弯矩 M_s 相似的变化规律,都可视为沿跨径而变化的抛物线,其上下限值 E_2、E_1 之间的区域就是束筋配置范围,如图 12-14 所示。由此可知,预应力钢筋重心位置(e_p)所应遵循的条件为:

$$\frac{M_s}{\alpha N_{pI}} - k_u \leqslant e_p \leqslant k_b + \frac{M_{G1}}{N_{pI}} \tag{12-93}$$

对于允许出现拉应力或允许出现裂缝的部分预应力混凝土构件,只要根据构件上、下缘混凝土拉应力(包括名义拉应力)的不同限制值作相应的演算,则束界也同样可以确定。

(3)预应力钢筋的布置原则

①预应力钢筋的布置,应使其重心线不超出束界范围。同时,构件端部逐步弯起的预应力钢筋将产生预剪力,这对抵消支点附近较大的外荷载剪力也是非常有利的;而且从构造上来说,预应力钢筋的弯起,可使锚固点分散,有利于锚具的布置。

②预应力钢筋弯起的角度,应与所承受的剪力变化规律相配合。预应力钢筋弯起后所产生的预剪力 V_p 应能抵消作用(或荷载)产生的剪力组合设计值 V_d 的一部分。抵消外剪力后所剩余的称为减余剪力,减余剪力是配置抗剪钢筋的依据。

③预应力钢筋的布置应符合构造要求。

(4)预应力钢筋弯起点的确定

预应力钢筋的弯起点,应从兼顾剪力与弯矩两方面的受力要求来考虑。

①从受剪考虑,理论上应从 $\gamma_0 V_d \geqslant V_{cs}$ 的截面开始起弯,以提供一部分预剪力 V_p 来抵抗作用产生的剪力。因此一般在跨径的三分点到四分点之间开始弯起。

②从受弯考虑,由于预应力钢筋弯起后,其重心线将往上移,使偏心距 e_p 变小,即预加力弯矩 M_p 将变小。应注意预应力钢筋弯起后正截面抗弯承载力的要求。

③预应力钢筋的起弯点尚应考虑满足斜截面抗弯承载力的要求,即保证预应力钢筋弯起后斜截面上的抗弯承载力不低于斜截面顶端所在的正截面的抗弯承载力。

(5)预应力钢筋弯起角度

从理论上讲,预应力钢筋弯起的最佳设计是通过 $N_{pd}\sin\theta_p = V_{G1} + V_{G2} + \dfrac{V_Q}{2}$ 的条件来控制预应力钢筋的弯起角度 θ_p,但对于恒载较大(跨径较大)的梁,按此确定的 θ_p 值显然过大。为此,只能在条件允许的情况下选择较大的 θ_p 值,对于邻近支点的梁段,则可在满足抗弯承载力要求的条件下,预应力钢筋弯起的数量应尽可能多些。但从减小曲线预应力钢筋预拉时摩阻应力损失出发,弯起角度 θ_p 不宜大于 20°,对于弯出梁顶锚固的钢筋,则往往超过 20°,因此 θ_p 常在 25°~30°。θ_p 角较大的预应力钢筋,应注意采取减小摩擦系数值的措施,以减小由此而引起的摩擦应力损失。

(6)预应力钢筋弯起的曲线形状

预应力钢筋弯起的曲线可采用圆弧线、抛物线或悬链线三种形式。公路桥梁中多采用圆弧线。《公桥规》规定,后张法构件预应力构件的曲线形预应力钢筋,其曲率半径应符合下列规定:

①钢丝束、钢绞线束的钢丝直径 $d \leqslant 5mm$ 时,不宜小于 4m;钢丝直径 $d > 5mm$ 时,不宜小于 6m。

②精轧螺纹钢筋直径 $d \leqslant 25mm$ 时,不宜小于 12m;直径 $d > 25mm$ 时,不宜小于 15m。

③对于具有特殊用途的预应力钢筋(如斜拉桥桥塔中围箍用的半圆形预应力钢筋,其半接在 1.5m 左右),因采取特殊的措施,可以不受此限。

(7)预应力钢筋布置的构造要求

①后张法构件

对外形呈曲线形且布置有曲线预应力钢筋的构件(图12-15),其曲线平面内、外管道的最小混凝土保护层厚度,按以下公式计算:

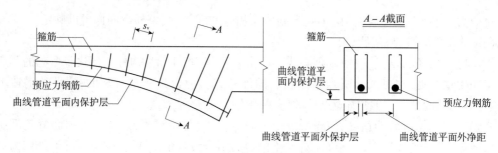

图 12-15 预应力钢筋曲线管道保护层示意图

a. 曲线平面内

$$c_{in} \geq \frac{P_d}{0.266r\sqrt{f'_{cu}}} - \frac{d_s}{2} \quad (12-94)$$

式中：c_{in}——曲线平面内最小混凝土保护层厚度(mm)；

d_s——管道外缘直径(mm)；

P_d——预应力钢筋的张拉力设计值(N)，可取扣除锚圈口摩擦、钢筋回缩及计算截面处管道摩擦损失后的张拉力乘以 1.2；

r——管道曲线半径，按《公桥规》9.3.15 条计算；

f'_{cu}——预应力钢筋张拉时，边长为 150mm 立方体混凝土抗压强度(MPa)。

当按式(12-94)计算的保护层厚度较大时，也可按直线管道设置最小保护层厚度，但应在管道曲线段弯曲平面内设置箍筋(图 12-15)，箍筋单肢的截面面积计算式为：

$$A_{sv1} \geq \frac{P_d s_v}{2r f_{sv}} \quad (12-95)$$

式中：A_{sv1}——箍筋单肢截面面积(mm²)；

s_v——箍筋间距(mm)；

f_{sv}——箍筋抗拉强度设计值(MPa)。

b. 曲线平面外

$$c_{out} \geq \frac{P_d}{0.266\pi r \sqrt{f'_{cu}}} - \frac{d_s}{2} \quad (12-96)$$

式中：c_{out}——曲线平面外最小混凝土保护层厚度(mm)；

P_d、r、f'_{cu} 意义同上。

当计算的保护层厚度小于各类环境的直线管道的保护层厚度，应取相应环境条件的直线管道的保护层厚度。

后张法预应力混凝土构件，其预应力钢筋管道的设置应符合下列规定：

a. 直线管道之间的水平净距不应小于 40mm，且不宜小于管道直径的 0.6 倍；对于预埋的金属或塑料波纹管和铁皮管，在竖直方向可将两管道叠置。

b. 曲线形预应力钢筋管道在曲线平面内相邻管道间的最小净距(图 12-16)计算式为式(12-94)，其中 P_d 和 r 分别为相邻两管道曲线半径较大的一根预应力钢筋的张拉力设计值(N)和曲线半径，c_{in} 为相邻两曲线管道外边缘在曲线平面内净距。当上述计算结果小于其相应直线管道外缘间净距时，应取用直线管道最小外缘间净距。

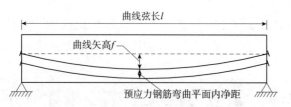

图 12-16　曲线形预应力钢筋弯曲平面内净距

曲线形预应力钢筋管道在曲线平面外相邻外缘间的最小净距计算式为式(12-96),其中 c_{out} 为相邻两曲线外缘在曲线平面外净距(mm)。

c. 管道内径的截面面积不应小于两倍预应力钢筋截面面积。

d. 按计算需要设置预拱度时,预留管道也应同时起拱。

② 先张法构件

a. 先张法预应力混凝土构件宜采用钢绞线、螺旋肋钢丝或刻痕钢丝用作预应力钢筋,当采用光面钢丝作预应力筋时,应采取适当措施,保证钢丝在混凝土中可靠地锚固,防止因钢丝与混凝土间黏结力不足而使钢丝滑动,丧失预应力。

b. 预应力钢绞线之间的净距不应小于其直径的 1.5 倍,且对 2 股、3 股钢绞线不应小于 20mm,对 7 股钢绞线不应小于 25mm。预应力钢丝间净距不应小于 15mm。

c. 对于单根预应力钢筋,其端部应设置长度不小于 150mm 的螺旋筋;对于多根预应力钢筋,在构件端部 10 倍预应力钢筋直径范围内,应设置 3~5 片钢筋网。

d. 普通钢筋和预应力直线形钢筋的最小混凝土保护层厚度(钢筋外缘至混凝土表面的距离)不应小于钢筋公称直径,且应符合附表 1-8 的规定。

(8) 非预应力钢筋的布置

① 箍筋

《公桥规》要求按下列规定配置构造箍筋:

a. 预应力混凝土 T 形、I 形截面梁和箱形截面梁腹板内应分别设置直径不小于 10mm 和 12mm 的箍筋,且应采用带肋钢筋,间距不应大于 200mm;自支座中心起长度不小于一倍梁高范围内,应采用闭合式箍筋,间距不应大于 120mm。

b. 在 T 形、I 形截面梁下部的"马蹄"内,应另设直径不小于 8mm 的闭合式箍筋,间距不应大于 200mm。另外,"马蹄"内还应设直径不小于 12mm 的定位钢筋。

② 水平纵向辅助钢筋

T 形截面预应力混凝土梁,截面上边缘有翼缘、下边缘有"马蹄",它们在梁横向的尺寸都比腹板厚度大,对于预应力混凝土梁,这种钢筋宜采用小直径的钢筋网,紧贴箍筋布置于腹板两侧,以增加与混凝土的黏结力,使裂缝的间距和宽度均减小。

③ 局部加强钢筋

对于局部受力较大的部位,应设置加强钢筋,如"马蹄"中的闭合式箍筋和梁端锚固区的加强钢筋等,除此之外,梁底支座处亦设置钢筋网加强。

④ 架立钢筋与定位钢筋

架立钢筋是用于支撑箍筋的,一般采用直径为 12~20mm 的圆钢筋;定位钢筋系指用于固定预留孔道制孔器位置的钢筋,常做成网格式。

(9)锚具的防护

对于埋入梁体的锚具,在预加应力完成后,其周围应设置构造钢筋与梁体连接,然后浇筑封锚混凝土。封锚混凝土强度等级不应低于构件本身混凝土强度等级的80%,且不低于C30。

12.9 预应力混凝土空心板计算示例

1)设计资料

(1)跨径:标准跨径 $l_k = 13.00\text{m}$;计算跨径 $l = 12.60\text{m}$。

(2)桥面宽:$2.5\text{m} + 4 \times 3.75\text{m} + 2.5\text{m} = 20.0\text{m}$。

(3)设计荷载:公路—Ⅱ级;人群荷载:3.0kN/m^2。

(4)材料:

预应力钢筋采用 $\phi^s 1 \times 7$ 钢绞线,公称直径12.7mm;公称截面积98.7mm^2,$f_{pk} = 1\,860\text{MPa}$,$f_{pd} = 1\,260\text{MPa}$,$E_p = 1.95 \times 10^5 \text{MPa}$。

非预应力钢筋采用 HRB400,$f_{sk} = 400\text{MPa}$,$f_{sd} = 330\text{MPa}$;HPB300,$f_{sk} = 300\text{MPa}$,$f_{sd} = 250\text{MPa}$。

空心板块混凝土采用C50,$f_{ck} = 32.4\text{MPa}$,$f_{cd} = 22.4\text{MPa}$,$f_{tk} = 2.65\text{MPa}$,$f_{td} = 1.83\text{MPa}$;铰缝为C30细集料混凝土;桥面铺装采用C30沥青混凝土,厚度12cm;栏杆及人行道为C30混凝土。

(5)设计要求:

根据《公桥规》要求,按A类预应力混凝土构件设计。

(6)施工方法:

采用先张法施工。

2)空心板尺寸

全桥宽采用20块C40的预制预应力混凝土空心板,每块空心板宽99cm,高62cm,空心板全长12.96m。全桥空心板横断面布置如图12-17所示,每块空心板截面及构造尺寸如图12-18所示。

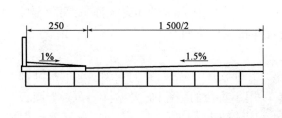

图12-17 桥梁横断面(尺寸单位:cm)

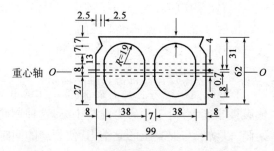

图12-18 空心板截面构造及尺寸(尺寸单位:cm)

3)空心板毛截面几何特性计算

(1)毛截面面积 A

$$A = 99 \times 62 - 2 \times 38 \times 8 - 4 \times \frac{\pi \times 19^2}{2} - 2 \times \left(\frac{1}{2} \times 7 \times 2.5 + 7 \times 2.5 + \frac{1}{2} \times 7 \times 5\right)$$

$$= 3\,174.3(\text{cm}^2)$$

(2) 毛截面重心位置

全截面对 1/2 板高处的静矩：

$$S_{\frac{1}{2}板高} = 2 \times \left[\frac{1}{2} \times 2.5 \times 7 \times \left(24 + \frac{7}{3}\right) + 7 \times 2.5 \times \left(24 + \frac{7}{2}\right) + \frac{1}{2} \times 7 \times 5 \times \left(24 - \frac{7}{3}\right)\right]$$

$$= 2\,181.7(\mathrm{cm}^3)$$

铰缝的面积：

$$A_{铰} = 2 \times \left(\frac{1}{2} \times 2.5 \times 7 + 2.5 \times 7 + \frac{1}{2} \times 5 \times 7\right) = 87.5(\mathrm{cm}^2)$$

则毛截面重心离 1/2 板高的距离为：

$$d = \frac{S_{\frac{1}{2}板高}}{A} = \frac{2\,181.7}{3\,174.3} = 0.687(\mathrm{cm}) \approx 0.7(\mathrm{cm}) = 7(\mathrm{mm})(向下移)$$

铰缝重心对 1/2 板高处的距离为：

$$d_{铰} = \frac{2\,181.7}{87.5} = 24.9(\mathrm{cm})$$

(3) 空心板毛截面对其重心的惯矩 I

由图 12-19，设每个挖空的半圆面积为 A'：

$$A' = \frac{1}{8}\pi d^2 = \frac{1}{8}\pi \times 38^2 = 567.1(\mathrm{cm}^2)$$

半圆重心轴：

$$y = \frac{4d}{6\pi} = \frac{4 \times 38}{6 \times \pi} = 8.06(\mathrm{cm}) = 80.6(\mathrm{mm})$$

半圆对其自身重心轴 O-O 的惯矩为 I'：

$$I' = 0.006\,86 d^4 = 0.006\,86 \times 38^4 = 14\,304(\mathrm{cm}^4)$$

则空心板毛截面对其重心轴的惯矩 I 为：

$$I = \frac{99 \times 62^3}{12} + 99 \times 62 \times 0.7^2 - 2 \times \left(\frac{38 \times 8^3}{12} + 38 \times 8 \times 0.7^2\right) - 4 \times 14\,304 -$$

$$2 \times 567.1 \times \left[(8.06 + 4 + 0.7)^2 + (8.06 + 4 - 0.7)^2\right] - 87.5 \times (24.9 + 0.7)^2$$

$$= 1\,520\,077.25(\mathrm{cm}^4) = 1.520\,1 \times 10^{10}(\mathrm{mm}^4)$$

(不计铰缝对其自身重心轴的惯矩)

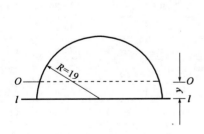

图 12-19 挖空半圆构造(尺寸单位:cm)

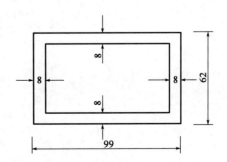

图 12-20 计算 I_T 的空心板截面简化图(尺寸单位:cm)

空心板截面的抗扭刚度可简化为图 12-20 的单箱截面来近似计算：

$$I_\mathrm{T} = \frac{4b^2 h^2}{\frac{2h}{t_1} + \frac{2b}{t_2}} = \frac{4 \times (99-8)^2 \times (62-8)^2}{\frac{2 \times (62-8)}{8} + \frac{2 \times (99-8)}{8}} = 2.6645 \times 10^6 (\mathrm{cm}^4)$$

$$= 2.6645 \times 10^{10} (\mathrm{mm}^4)$$

4）作用效应组合

按承载能力极限状态设计时的基本组合表达式为：

$$\gamma_0 S_\mathrm{ud} = \gamma_0 (1.2 S_\mathrm{Gk} + 1.4 S_\mathrm{Q1k} + 0.8 \times 1.4 S_\mathrm{Qjk})$$

式中，γ_0 为结构重要系数，本桥取 $\gamma_0 = 0.9$；S_ud 为效应组合设计值；S_Gk 为永久作用效应标准值；S_Q1k 为汽车荷载效应（含汽车冲击力）的标准值；S_Qjk 为人群荷载效应的标准值。

按正常使用极限状态设计时，应根据不同的设计要求，采用以下两种效应组合。

作用频遇组合表达式：

$$S_\mathrm{sd} = S_\mathrm{Gk} + 0.7 \times S'_\mathrm{Q1k} + 1.0 \times S_\mathrm{Qjk}$$

式中，S_sd 为作用频遇组合设计值；S_Gk 为永久作用效应标准值；S'_Q1k 为不计冲击的汽车荷载效应标准值；S_Qjk 为人群荷载效应的标准值。

作用准永久组合表达式：

$$S_\mathrm{ld} = S_\mathrm{Gk} + 0.4 \times S'_\mathrm{Q1k} + 0.4 \times S_\mathrm{Qjk}$$

《公桥规》还规定结构构件当需进行弹性阶段截面应力计算时，应采用标准值效应组合，此时效应组合表达式为：

$$S = S_\mathrm{Gk} + S_\mathrm{Q1k} + S_\mathrm{Qjk}$$

式中，S 为标准值效应组合设计值；S_Gk、S_Q1k、S_Qjk 为永久作用效应、汽车荷载效应（计入汽车冲击力）、人群荷载效应的标准值。

根据计算得到的作用效应，按《公桥规》各种组合表达式可求得各效应组合设计值，现将计算汇总于表 12-4 中。

空心板作用效应组合计算汇总表 表 12-4

序号	作用种类		弯矩 M(kN·m)		剪力 V(kN)		
			跨中	1/4	跨中	1/4	支点
作用效应标准值	永久作用效应	g_I	157.49	118.12	0	25.00	50.00
		g_II	65.17	48.88	0	10.34	20.69
		$g = g_\mathrm{I} + g_\mathrm{II}(S_\mathrm{Gk})$	222.66	167.00	0	35.34	70.69
	可变作用效应	车道荷载 不计冲击 S'_Q1k	131.32	98.41	21.52	34.16	108.36
		$\times (1+\mu) S_\mathrm{Q1k}$	172.69	129.42	28.31	44.92	142.51
	人群荷载 S_Qjk		13.40	10.05	1.06	2.39	3.19
承载能力极限状态	基本组合 S_ud	$1.2 S_\mathrm{Gk}$ (1)	267.19	200.40	0	42.41	84.83
		$1.4 S_\mathrm{Q1k}$ (2)	241.77	181.19	39.63	62.89	199.51
		$0.8 \times 1.4 S_\mathrm{Qjk}$ (3)	15.01	11.26	1.19	2.68	3.57
		$S_\mathrm{ud} = (1)+(2)+(3)$	523.97	392.85	40.82	107.98	287.91

续上表

序　号		作　用　种　类		弯矩 M(kN·m)		剪力 V(kN)		
				跨中	1/4	跨中	1/4	支点
正常使用极限状态	作用短期效应组合 S_{sd}	S_{Gk}	(4)	222.66	167.00	0	35.34	70.69
		$0.7 S'_{Q1k}$	(5)	91.92	68.89	15.06	23.91	75.85
		S_{Qjk}	(6)	13.40	10.05	1.06	2.39	3.19
		$S_{sd} = (4)+(5)+(6)$		327.98	245.94	16.12	61.64	149.73
	使用长期效应组合 S_{ld}	S_{Gk}	(7)	222.66	167.00	0	35.34	70.69
		$0.4 S'_{Q1k}$	(8)	52.53	39.36	8.61	.13.66	43.34
		$0.4 S_{Qjk}$	(9)	5.36	4.02	0.42	0.96	1.28
		$S_{ld} = (7)+(8)+(9)$		280.55	210.38	9.03	49.96	115.31
弹性阶段截面应力计算	标准值效应组合 S	S_{Gk}	(10)	222.66	167.00	0	35.34	70.69
		S_{Q1k}	(11)	172.69	129.42	28.31	44.92	142.51
		S_{Qjk}	(12)	13.40	10.05	1.06	2.39	3.19
		$S = (10)+(11)+(12)$		408.75	306.47	29.37	82.65	216.39

5)预应力钢筋数量估算及布置

(1)预应力钢筋数量的估算

采用先张法预应力混凝土空心板,设计时它应满足不同设计状况下规范规定的控制条件要求,例如承载力、抗裂性、裂缝宽度、变形及应力等要求。预应力混凝土桥梁设计时,一般情况下,首先根据结构在正常使用极限状态正截面抗裂性或裂缝宽度限值确定预应力钢筋的数量,再由构件的承载能力极限状态要求确定普通钢筋的数量。本示例以部分预应力 A 类构件设计,首先按正截面抗裂性确定有效加力 N_{pe}。

按《公桥规》6.3.1 条,A 类预应力混凝土构件正截面抗裂性是控制混凝土的法向拉应力,并符合以下条件:

在作用短期效应组合下,应满足 $\sigma_{st} - \sigma_{pc} \leq 0.70 f_{tk}$ 的要求。

式中,σ_{st} 为在作用短期效应组合 M_{sd} 作用下构件抗裂验算边缘混凝土的法向拉应力;σ_{pc} 为构件抗裂验算边缘混凝土的有效预压应力。

在初步设计时,σ_{st} 和 σ_{pc} 可按下列公式近似计算:

$$\sigma_{st} = \frac{M_{sd}}{W}$$

$$\sigma_{pc} = \frac{N_{pc}}{A} + \frac{N_{pc} e_p}{W}$$

式中,A、W 为构件毛截面面积及对毛截面受拉边缘的弹性抵抗矩;e_p 为预应力钢筋重心对毛截面重心轴的偏心矩,$e_p = y - a_p$,a_p 可预先假定。

代入 $\sigma_{st} - \sigma_{pc} \leq 0.70 f_{tk}$ 即可求得满足部分预应力 A 类构件正截面抗裂性要求所需的有效预加力为:

$$N_{pc} = \frac{\dfrac{M_{sd}}{W} - 0.70 f_{tk}}{\dfrac{1}{A} + \dfrac{e_p}{W}}$$

式中,f_{tk}为混凝土抗拉强度标准值。

预应力空心板桥采用C50,$f_{tk}=2.65$,由表12-4得,$M_{sd}=327.98\text{kN}\cdot\text{m}=327.98\times10^6\text{N}\cdot\text{mm}$,空心板毛截面换算面积:$A=3\ 174.3\text{cm}^2=3\ 174.3\times10^2\text{mm}^2$,$W=\dfrac{I}{y_{下}}=\dfrac{1\ 520.1\times10^3\text{cm}^4}{(31-0.70)\text{cm}}=50.17\times10^3\text{cm}^3=50.17\times10^6\text{mm}^3$。

假设$a_p=4\text{cm}$,则$e_p=y_{下}-a_p=31-0.7-3=26.3(\text{cm})=263(\text{mm})$。

代入得:

$$N_{pe}=\dfrac{\dfrac{327.98\times10^6}{50.17\times10^6}-0.7\times2.65}{\dfrac{1}{3\ 174.3\times10^2}+\dfrac{263}{50.17\times10^6}}=557\ 642.2(\text{N})$$

则所需预应力钢筋截面面积A_p为:

$$A_p=\dfrac{N_{pe}}{\sigma_{con}-\sum\sigma_l}$$

式中,σ_{con}为预应力钢筋的张拉控制应力;$\sum\sigma_l$为全部预应力损失值,按张拉控制应力的20%估算。

采用1×7股钢绞线作为预应力钢筋,直径12.7mm,公称截面面积98.7mm²,$f_{pk}=1\ 860\text{MPa}$,$f_{pd}=1\ 260\text{MPa}$,$E_p=1.95\times10^5\text{MPa}$。

按《公桥规》$\sigma_{con}\leq0.75f_{pk}$,现取$\sigma_{con}=0.70f_{pk}$,预应力损失总和近似假定为20%张拉控制应力来估算,则:

$$A_p=\dfrac{N_{pe}}{\sigma_{con}-\sum\sigma_l}=\dfrac{N_{pe}}{\sigma_{con}-0.2\sigma_{con}}=\dfrac{557\ 642.2}{0.8\times0.70\times1\ 860}=535.37(\text{mm}^2)$$

采用7根1×7股钢绞线,即$\Phi^s12.7$钢绞线,单根钢绞线公称面积98.7mm²,则$A_p=7\times98.7=690.9\text{mm}^2$满足要求。

(2)预应力钢筋的布置

预应力空心板选用7根1×7股钢绞线布置在空心板下缘,$a_p=40\text{mm}$,沿空心板跨长直线布置,即沿跨长$a_p=40\text{mm}$保持不变,如图12-21所示。预应力钢筋布置还应满足《公桥规》的构造要求,钢绞线净距不小于25mm,端部设置长度不小于150mm的螺旋钢筋等。

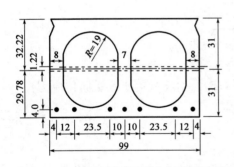

图12-21 空心板跨中截面预应力钢筋的布置
(尺寸单位:cm)

(3)普通钢筋数量的估算及布置

在预应力钢筋数量已经确定的情况下,可由正截面承载能力极限状态要求的条件确定普通钢筋数量,暂不考虑在受压区配置预应力钢筋,也暂不考虑普通钢筋的影响。空心板截面可换算成等效工字形截面来考虑:

由 $$b_kh_k=\dfrac{\pi}{4}\times38^2+8\times38=1\ 438.115(\text{cm}^2)$$

得

$$b_k=\dfrac{1\ 438.115\text{cm}^2}{h_k}$$

$$\frac{1}{12}b_k h_k^3 = \frac{38 \times 8^3}{12} + 2 \times 0.00686 \times 38^4 + 2 \times 567.1 \times (8.06 + 4)^2 = 195\,191.53\,(\text{cm}^4)$$

把 $b_k = \dfrac{1\,438.114\,\text{cm}^2}{h_k}$ 代入 $\dfrac{1}{12}b_k h_k^3 = 195\,191.53\,\text{cm}^4$，求得 $h_k = 40.4\,\text{cm}, b_k = \dfrac{1\,438.114}{40.4} = 35.6\,(\text{cm})$。

则得等效工字形截面的上翼缘板厚度 h'_f：

$$h'_f = y_\text{上} - \frac{h_k}{2} = 31 - \frac{40.4}{2} = 10.8\,(\text{cm})$$

等效工字形截面的下翼缘板厚度 h_f：

$$h_f = y_\text{下} - \frac{h_k}{2} = 31 - \frac{40.4}{2} = 10.8\,(\text{cm})$$

等效工字形截面的肋板厚度：

$$b = b'_f - 2b_k = 99 - 2 \times 35.6 = 27.8\,(\text{cm})$$

等效工字形截面尺寸如图12-22所示。

估算普通钢筋时，可先假定 $x \leqslant h'_f$，则由下式可求得受压区高度 x，设：

$$h_0 = h - a_s = 62 - 4 = 58\,(\text{cm}) = 580\,(\text{mm})$$

$$\gamma_0 M_{ud} \leqslant f_{cd} b'_f x \left(h_0 - \frac{x}{2}\right)$$

由《公桥规》，$\gamma_0 = 0.9, C50, f_{cd} = 22.4\,\text{MPa}$。由表12-4，跨中 $M_{ud} = 523.97\,\text{kN} \cdot \text{m} = 523.97 \times 10^6\,\text{N} \cdot \text{mm}, b'_f = 990\,\text{mm}$，代入上式得：

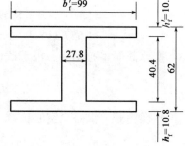

图12-22 空心板换算等效I字形截面
（尺寸单位：cm）

$$0.9 \times 523.97 \times 10^6 \leqslant 22.4 \times 990 \times x \times \left(580 - \frac{x}{2}\right)$$

整理后得：

$$x^2 - 1\,160x + 42.53 \times 10^3 \leqslant 0$$

求得：

$$x = 37.8\,\text{mm} < h'_f = 108\,\text{mm}, 且 x < \xi_b h_0 = 0.4h_0 = 232\,\text{mm}$$

说明中和轴在翼缘板内，可用下式求得普通钢筋面积 A_s：

$$A_s = \frac{f_{cd} b'_f x - f_{pd} A_p}{f_{sd}} = \frac{22.4 \times 990 \times 37.8 - 1\,260 \times 690.9}{280} < 0$$

普通钢筋选用 HRB400，$f_{sd} = 330\,\text{MPa}, E_s = 2 \times 10^5\,\text{MPa}$。

按《公桥规》规定，$A_s \geqslant 0.003 b h_0 = 0.003 \times 278 \times 580 = 483.72\,(\text{mm}^2)$。

普通钢筋采用 5⌀12，$A_s = 5 \times \dfrac{\pi (12)^2}{4} = 565.5\,(\text{mm}^2) > 483.72\,(\text{mm}^2)$。

普通钢筋 5⌀12 布置在空心板下缘一排（截面受拉边缘），沿空心板跨长直线布置，钢筋重心至下缘 40mm 处，即 $a_s = 40\,\text{mm}$。

6）换算截面几何特性计算

由前面计算已知空心板毛截面的几何特性。毛截面面积 $A = 317\,430\,\text{mm}^2$，毛截面重心轴至 1/2 板高的距离 $d = 7\,\text{mm}$（向下），毛截面对其重心轴惯性矩 $I = 15\,201 \times 10^6\,\text{mm}^4$。

(1) 换算截面面积 A_0

$$A_0 = A + (\alpha_{Ep} - 1)A_p + (\alpha_{Es} - 1)A_s$$

$$\alpha_{Ep} = \frac{E_p}{E_c} = \frac{1.95 \times 10^5}{3.45 \times 10^4} = 5.65, A_p = 690.9 \text{mm}^2$$

$$\alpha_{Es} = \frac{E_s}{E_c} = \frac{2 \times 10^5}{3.25 \times 10^4} = 5.79, A_s = 565.5 \text{mm}^2$$

$$A = 317\,430 \text{mm}^2$$

代入得:

$$A_0 = 317\,430 + (5.65 - 1) \times 690.9 + (5.79 - 1) \times 565.5 = 323\,351.4(\text{mm}^2)$$

(2) 换算截面重心位置

所有钢筋换算截面对毛截面重心的静矩为:

$$S_{01} = (\alpha_{Ep} - 1)A_p \times (310 - 7 - 40) + (\alpha_{Es} - 1)A_s \times (310 - 7 - 40)$$
$$= (5.65 - 1) \times 690.9 \times 263 + (5.79 - 1) \times 565.5 \times 263$$
$$= 1\,557\,336.1(\text{mm}^3)$$

换算截面重心至空心板毛截面中心的距离为:

$$d_{01} = \frac{S_{01}}{A_0} = 4.8\text{mm}(向下移)$$

则换算截面重心至空心板截面下缘的距离为:

$$y_{01l} = 310 - 7 - 4.8 = 298.2(\text{mm})$$

换算截面重心至空心板截面上缘的距离为:

$$y_{01u} = 310 + 7 + 4.8 = 321.8(\text{mm})$$

换算截面重心至预应力钢筋中心的距离为:

$$e_{01p} = 298.2 - 40 = 258.2(\text{mm})$$

换算截面重心至普通钢筋中心的距离为:

$$e_{01s} = 298.2 - 40 = 258.2(\text{mm})$$

(3) 换算截面惯性矩 I_0

$$I_0 = I + Ad_{01}^2 + (\alpha_{Ep} - 1)A_p e_{01p}^2 + (\alpha_{Es} - 1)A_s e_{01s}^2$$
$$= 15\,201 \times 10^6 + 317\,430 \times 4.8^2 + (5.65 - 1) \times 690.9 \times 258.2^2 +$$
$$(5.79 - 1) \times 565.5 \times 258.2^2$$
$$= 1.5633 \times 10^{10}(\text{mm}^4)$$

(4) 换算截面弹性抵抗矩

下缘:$W_{01l} = \dfrac{I_0}{y_{01l}} = \dfrac{1.5633 \times 10^{10}}{298.2} = 52.45 \times 10^6(\text{mm}^3)$

上缘:$W_{01u} = \dfrac{I_0}{y_{01u}} = \dfrac{1.5633 \times 10^{10}}{321.8} = 48.57(\text{mm}^3)$

7) 持久状况承载能力极限状态计算

(1) 跨中截面正截面抗弯承载力计算

跨中截面构造尺寸及配筋如图 12-21 所示。预应力钢绞线合力作用点到截面底边的距离 $a_p = 40\text{mm}$,普通钢筋截面底边的距离 $a_s = 40\text{mm}$,则预应力钢筋和普通钢筋的合力作用到截面底边的距离为：

$$a_{ps} = \frac{f_{sd}A_s a_s + f_{pd}A_p a_p}{f_{sd}A_s + f_{pd}A_p} = \frac{330 \times 565.5 \times 40 + 1\,260 \times 690.9 \times 40}{330 \times 565.5 + 1\,260 \times 690.9} = 40(\text{mm})$$

$$h_0 = h - a_{ps} = 620 - 40 = 580(\text{mm})$$

采用换算等效工字形截面来计算，如图 12-22 所示，上翼缘厚度 $h'_f = 108\text{mm}$。
上翼缘工作宽度 $b'_f = 990\text{mm}$，肋宽 $b = 278\text{mm}$。首先按公式 $f_{pd}A_p + f_{sd}A_s \leq f_{cd}b'_f h'_f$ 判断截面类型：

$$f_{pd}A_p + f_{sd}A_s = 1\,260 \times 690.9 + 330 \times 565.5 = 1\,057\,149(\text{N})$$

$$\leq f_{cd}b'_f h'_f = 22.4 \times 990 \times 108 = 2\,395\,008(\text{N})$$

所以属于第一类 T 形，应按宽度 $b'_f = 990\text{mm}$ 的矩形截面来计算其抗弯承载力。由 $\sum x = 0$ 计算混凝土受压区高度 x：

由

$$f_{pd}A_p + f_{sd}A_s = f_{cd}b'_f x$$

得

$$x = \frac{f_{pd}A_p + f_{sd}A_s}{f_{cd}b'_f} = \frac{1\,260 \times 690.9 + 330 \times 565.5}{22.4 \times 990} = 47.7(\text{mm})$$

$$\begin{cases} < \xi_b h_0 = 0.4 \times 580 = 232(\text{mm}) \\ < h'_f = 108(\text{mm}) \end{cases}$$

将 $x = 46.4\text{mm}$ 代入下列公式中计算出跨中截面的抗弯承载力 M_{ud}：

$$M_{ud} = f_{cd}b'_f x \left(h_0 - \frac{x}{2}\right) = 22.4 \times 990 \times 46.4 \times \left(580 - \frac{46.4}{2}\right)$$

$$= 572.92 \times 10^6(\text{N} \cdot \text{mm})$$

$$= 572.92(\text{kN} \cdot \text{m})$$

$$> \gamma_0 M_d = 0.9 \times 523.97 = 471.57(\text{kN} \cdot \text{m})$$

计算结果表明，跨中截面抗弯承载力满足要求。

(2) 斜截面抗剪承载力计算

① 截面抗剪强度上、下限复核

选取距支点 $h/2$ 处截面进行斜截面抗剪承载力计算。截面构造尺寸及配筋如图 12-23 所示。首先进行抗剪强度上、下限复核，按《公桥规》5.2.9 条：

$$\gamma_0 V_d \leq 0.51 \times 10^{-3} \sqrt{f_{cu,k}} b h_0$$

式中，V_d 为验算截面处的剪力组合设计值（kN），由表 12-4 得支点处剪力及跨中截面剪力，内插得到距支点 $h/2 = 310\text{mm}$ 处的截面剪力 V_d：

$$V_d = 287.91 - \frac{310 \times (287.91 - 40.82)}{6\,300} = 275.75(\text{kN})$$

h_0 为截面有效高度，由于本示例预应力筋及普通钢筋都是直线配置，有效高度 h_0 与跨中截面相同，$h_0 = 580\text{mm}$；$f_{cu,k}$ 为边长为 150mm 的混凝土立方体抗压强度，空心板为 C50，则 $f_{cu,k} = 50\text{MPa}$，$f_{td} = 1.83\text{MPa}$；b 为等效工字形截面的腹板宽度，$b = 278\text{mm}$。

代入上述公式：

$$\gamma_0 V_d = 0.9 \times 275.5 = 248.18(\text{kN})$$

$$\gamma_0 V_d \leq 0.51 \times 10^{-3} \sqrt{50} \times 278 \times 580 = 581.47(\text{kN})$$

计算结果表明空心板截面尺寸符合要求。

按《公桥规》第5.2.9条：

$$1.25 \times 0.5 \times 10^{-3} \times \alpha_2 f_{td} b h_0 = 1.25 \times 0.5 \times 10^{-3} \times 1.0 \times 1.83 \times 278 \times 580$$
$$= 184.42(kN)$$

式中，$\alpha_2 = 1.25$ 是按《公桥规》5.2.9条，板式受弯构件可乘以1.25提高系数。

由于 $\gamma_0 V_d = 0.9 \times 275.75 = 248.18(kN) > 1.25 \times 0.5 \times 10^{-3} \times \alpha_2 f_{td} b h_0 = 184.42(kN)$，并对照表12-4中沿跨长各截面的控制剪力组合设计值，在1/4至支点的部分区段内应按计算要求配置抗剪箍筋，其他区段可按构造要求配置箍筋。

为了构造方便和便于施工，空心板不设弯起钢筋，计算剪力全部由混凝土及箍筋承受，则斜截面抗剪承载力按下式计算：

$$\gamma_0 V_d \leq V_{cs}$$

$$V_{cs} = \alpha_1 \alpha_2 \alpha_3 \times 0.45 \times 10^{-3} b h_0 \sqrt{(2 + 0.6P) \sqrt{f_{cu,k}} \rho_{sv} f_{sv}}$$

式中，各系数值按《公桥规》5.2.7条规定取用。α_1 为异号弯矩影响系数，简支梁 $\alpha_1 = 1.0$；α_2 为预应力提高系数，本示例为部分预应力A类构件，偏安全取 $\alpha_2 = 1.0$；α_3 为受压翼缘的影响系数，取 $\alpha_3 = 1.1$；b、h_0 为等效工字形截面的肋宽及有效高度，$b = 278mm$，$h_0 = 580mm$；P 为纵向钢筋的配筋率，$P = 100\rho = 100 \times \dfrac{690.9 + 565.5}{278 \times 580} = 0.78$；$\rho_{sv}$ 为箍筋的配筋率，$\rho_{sv} = \dfrac{A_{sv}}{b s_v}$，

箍筋选用双支箍 $\Phi 10$，$A_{sv} = 2 \times \dfrac{\pi \times 10^2}{4} = 157.08(mm^2)$，则写出箍筋间距 s_v 的计算式为：

$$s_v = \dfrac{\alpha_1^2 \alpha_2^2 \alpha_3^2 \times 0.2 \times 10^{-6}(2 + 0.6P)\sqrt{f_{cu,k}} f_{sv} A_{sv} b h_0^2}{(\gamma_0 V_d)^2}$$

$$= \dfrac{1.0^2 \times 1.0^2 \times 1.1^2 \times 0.2 \times 10^{-6}(2 + 0.6 \times 0.78)\sqrt{50} \times 330 \times 157.8 \times 278 \times 580^2}{(0.9 \times 275.75)^2}$$

$$= 336.5(mm)$$

$f_{cu,k} = 50MPa$。

箍筋选用HRB400，则 $f_{sv} = 330MPa$。

取箍筋间距 $s_v = 150mm$，并按《公桥规》要求，在支座中心向跨中方向不小于1倍梁高范围内，箍筋间距取100mm。

箍筋率：

$$\rho_{sv} = \dfrac{A_{sv}}{b s_v} = \dfrac{157.08}{278 \times 150} = 0.0038 = 0.38\% > \rho_{svmin} = 0.12\%。$$

在组合设计剪力值 $\gamma_0 V_d \leq 1.25 \times 0.5 \times 10^{-3} \times \alpha_2 f_{td} b h_0 = 184.42kN$ 的部分梁段，可只按构造要求配置箍筋，设箍筋仍选用双肢 $\Phi 10$，配筋率 ρ_{sv} 取 ρ_{svmin}，则由此求得构造箍筋间距 s_v'：

$$\dfrac{A_{sv}}{b \rho_{svmin}} = \dfrac{157.08}{278 \times 0.0012} = 470.9(mm)。$$

取 $s_v' = 200mm$。

经比较和综合考虑，箍筋沿空心板跨长布置如图12-23所示。

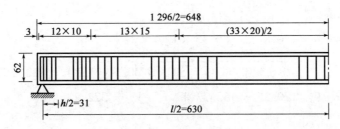

图 12-23 空心板箍筋的布置

②斜截面抗剪承载力计算

由图 12-23 可知,选取以下三个位置进行空心板斜截面抗剪承载力计算:

a. 距支座中心 $h/2 = 310\text{mm}$ 处截面, $x = 5990\text{mm}$;

b. 距跨中位置 $x = 3300\text{mm}$ 处截面(箍筋间距变化处);

c. 距跨中位置 $x = 3300 + 13 \times 150 = 5250(\text{mm})$ 处(箍筋间距变化处)。

计算截面的剪力组合设计值,可按表 12-4 由跨中和支点的设计值内插得到,计算结果见表 12-5。

各计算截面剪力组合设计值　　　表 12-5

截面位置 $x(\text{mm})$	支点 $x = 6300$	$x = 5990$	$x = 5250$	$x = 3300$	跨中 $x = 0$
剪力组合设计值 $V_d(\text{kN})$	287.91	275.75	246.73	170.25	40.82

a. 距支座中心 $h/2 = 310\text{mm}$ 处截面,即 $x = 5990\text{mm}$。

由于预应力筋及普通钢筋是直线配筋,故此截面的有效高度取与跨中近似相同, $h_0 = 580\text{mm}$,其等效工字形截面的肋宽 $b = 278\text{mm}$。由于不设弯起斜筋,因此,斜截面抗剪承载能力按下式计算:

$$V_{cs} = \alpha_1 \alpha_2 \alpha_3 \times 0.45 \times 10^{-3} b h_0 \sqrt{(2 + 0.6P) \sqrt{f_{cu,k}} \rho_{sv} f_{sv}}$$

式中, $\alpha_1 = 1.0, \alpha_2 = 1.0, \alpha_3 = 1.1, b = 278\text{mm}, h_0 = 580\text{mm}, P = 100\rho = 100 \times \dfrac{690.9 + 565.5}{278 \times 580} = 0.78$;

此处,箍筋间距 $s_v = 100\text{mm}, 2\Phi 10, A_{sv} = 157.08\text{mm}^2$,

$$\rho_{sv} = \frac{A_{sv}}{bs_v} = \frac{157.08}{278 \times 100} = 0.00565 = 0.565\% > \rho_{svmin} = 0.12\%$$

$f_{cu,k} = 50\text{MPa}, f_{sv} = 280\text{MPa}$

代入,得:

$$V_{cs} = 1.0 \times 1.0 \times 1.1 \times 0.45 \times 10^{-3} \times 278 \times 580 \times$$
$$\sqrt{(2 + 0.6 \times 0.78) \times \sqrt{50} \times 0.00565 \times 330} = 455.28(\text{kN})$$
$$\gamma_0 V_d = 0.9 \times 275.75 = 248.18(\text{kN}) < V_{cs} = 455.28(\text{kN})$$

抗剪承载力满足要求。

b. 距跨中截面 $x = 3300\text{mm}$ 处。

此处,箍筋间距 $s_v = 200\text{mm}, V_d = 170.25\text{kN}$,

$$\rho_{sv} = \frac{A_{sV}}{bs_v} = \frac{157.08}{278 \times 200} = 0.283\% > \rho_{svmin} = 0.12\%$$

斜截面抗剪承载力：

$$V_{cs} = 1.0 \times 1.0 \times 1.1 \times 0.45 \times 10^{-3} \times 278 \times 580 \times$$

$$\sqrt{(2 + 0.6 \times 0.78) \times \sqrt{50} \times 0.00283 \times 330} = 322.21(\text{kN})$$

$$\gamma_0 V_d = 0.9 \times 170.25 = 153.23(\text{kN}) < V_{cs} = 322.21(\text{kN})$$

截面抗剪承载力满足要求。

c. 距跨中截面距离 $x = 5250$mm 处。

此处，箍筋间距 $s_v = 150$mm，$V_d = 246.73$kN，

$$\rho_{sv} = \frac{A_{sv}}{bs_v} = \frac{157.08}{278 \times 150} = 0.00377 = 0.377\% > \rho_{svmin} = 0.12\%$$

斜截面抗剪承载力：

$$V_{cs} = 1.0 \times 1.0 \times 1.1 \times 0.45 \times 10^{-3} \times 278 \times 580 \times$$

$$\sqrt{(2 + 0.6 \times 0.78)\sqrt{50} \times 0.00377 \times 330} = 371.90(\text{kN})$$

$$\gamma_0 V_d = 0.9 \times 246.73 = 222.06(\text{kN}) < V_{cs} = 371.90\text{kN}$$

计算表明均满足斜截面抗剪承载力要求。

8）预应力损失计算

预应力钢筋采用直径为12.7mm的1×7股钢绞线，$E_p = 1.95 \times 10^5$MPa，$f_{pk} = 1860$MPa，控制应力取 $\sigma_{con} = 0.7 f_{pk} = 0.7 \times 1860 = 1302(\text{MPa})$。

(1) 锚具变形、回缩引起的应力损失 σ_{l2}

预应力钢绞线的有效长度取为张拉台座的长度，设台座长 $L = 50$m，采用一端张拉及夹片式锚具，有顶压时 $\Delta l = 4$mm，则：

$$\sigma_{l2} = \frac{\sum \Delta l}{L} E_p = \frac{4}{50 \times 10^3} \times 1.95 \times 10^5 = 15.6(\text{MPa})$$

(2) 加热养护引起的温度损失 σ_{l3}

先张法预应力混凝土空心板采用加热养护的方法，为减少温度引起的预应力损失，采用分阶段养护措施。设控制预应力钢绞线与台座之间的最大温差 $\Delta t = t_2 - t_1 = 15°C$，则：

$$\sigma_{l3} = 2\Delta t = 2 \times 15 = 30(\text{MPa})$$

(3) 预应力钢绞线由于应力松弛引起的预应力损失 σ_{l5}

$$\sigma_{l5} = \psi \xi (0.52 \frac{\sigma_{pe}}{f_{pk}} - 0.26) \sigma_{pe}$$

式中，ψ 为张拉系数，一次张拉时，$\psi = 1.0$；ξ 为预应力钢绞线松弛系数，低松弛 $\xi = 0.3$；f_{pk} 为预应力钢绞线的抗拉强度标准值，$f_{pk} = 1860$MPa；σ_{pe} 为传力锚固时的钢筋应力，由《公桥规》6.2.6条，对于先张法构件，$\sigma_{pe} = \sigma_{con} - \sigma_{l2} = 1302 - 15.6 = 1286.4(\text{MPa})$。

代入计算式，得：$\sigma_{l5} = 1.0 \times 0.3 \times \left(0.52 \times \frac{1286.4}{1860} - 0.26\right) \times 1286.4 = 38.45(\text{MPa})$。

(4)混凝土弹性压缩引起的预应力损失 σ_{l4}

对于先张法构件:

$$\sigma_{l4} = \alpha_{Ep}\sigma_{pe}$$

式中,α_{Ep} 为预应力钢筋弹性模量与混凝土弹性模量的比值,$\alpha_{Ep} = \dfrac{1.95\times 10^5}{3.45\times 10^4} = 5.65$;$\sigma_{pe}$ 为在计算截面钢筋重心处,由全部钢筋预加力产生的混凝土法向应力(MPa),其值为:

$$\sigma_{pe} = \frac{N_{p0}}{A_0} + \frac{N_{p0}e_{p0}}{I_0}y_0$$

$$N_{p0} = \sigma_{p0}A_p - \sigma_{l6}A_s$$

$$\sigma_{p0} = \sigma_{con} - \sigma'_l$$

式中,σ'_l 为预应力钢筋传力锚固时的全部预应力损失,由《公桥规》6.2.8条,先张法构件传力锚固时的损失为:$\sigma'_l = \sigma_{l2} + \sigma_{l3} + 0.5\sigma_{l5}$,则:

$$\begin{aligned}\sigma_{p0} &= \sigma_{con} - (\sigma_{l2} + \sigma_{l3} + 0.5\sigma_{l5})\\ &= 1\,302 - 15.6 - 30 - 0.5\times 38.45\\ &= 1\,237.18(\text{MPa})\end{aligned}$$

$$N_{p0} = \sigma_{p0}A_p - \sigma_{l6}A_s = 1\,237.18\times 690.9 - 0 = 854.77\times 10^3(\text{N})$$

由前面计算空心板换算截面面积 $A_0 = 323\,351\,\text{mm}^2$,$I_0 = 1.563\,3\times 10^{10}\,\text{mm}^4$,$e_{p0} = 258.2\,\text{mm}$,$y_0 = 257.8\,\text{mm}$,则:

$$\sigma_{pe} = \frac{854.77\times 10^3}{323\,351.4} + \frac{854.77\times 10^3\times 258.2}{1.563\,3\times 10^{10}}\times 258.2 = 6.27(\text{MPa})$$

$$\sigma_{l4} = \alpha_{Ep}\sigma_{pe} = 5.65\times 6.27 = 35.42(\text{MPa})$$

(5)混凝土收缩、徐变引起的预应力损失 σ_{l6}

$$\sigma_{l6} = \frac{0.9[E_p\varepsilon_{cs}(t,t_0) + \alpha_{Ep}\sigma_{pc}\phi(t,t_0)]}{1 + 15\rho\rho_{ps}}$$

式中,ρ 为构件受拉区全部纵向钢筋的含筋率,$\rho = \dfrac{A_p + A_s}{A_0} = \dfrac{690.9 + 565.5}{323\,351.4} = 0.003\,88$;$e_{ps}$ 为构件截面全部纵向钢筋截面重心至构件重心的距离(mm),$e_{ps} = 297.8 - 40 = 257.8$;$i$ 为构件截面回转半径(mm^2),$i^2 = \dfrac{I_0}{A_0} = \dfrac{1.563\,3\times 10^{10}}{323\,351.4} = 48\,340.0$;$\sigma_{pc}$ 为构件受拉区全部纵向钢筋重心处,由预应力(扣除相应阶段的预应力损失)和结构自重产生的混凝土法向压应力,其值为:

$$\sigma_{pc} = \frac{N_{p0}}{A_0} + \frac{N_{p0}\lambda_{p0}}{I_0}y_0$$

N_{p0} 为传力锚固时,预应力钢筋的预加力,其值为:

$$\begin{aligned}N_{p0} &= \sigma_{p0}A_P - \sigma_{l6}A_S = [\sigma_{con} - (\sigma_{l2} + \sigma_{l3} + \sigma_{l4} + 0.5\sigma_{l5})]A_p - 0\\ &= [1\,302 - (15.6 + 30.0 + 35.42 + 0.5\times 38.45)]\times 690.9\\ &= 830\,292.5(\text{N})\end{aligned}$$

$$\lambda_{p0} = \frac{\sigma_{p0}A_Py_P - \sigma_{l6}A_sy_s}{N_{p0}} = \frac{830\,292.5\times 258.2}{830\,292.5}$$

$= 258.2(\text{mm})$ （因为 $y_p = y_s = 258.2\text{mm}$）

y_0 为构件受拉区全部纵向钢筋重心至截面重心的距离,由前面计算 $y_0 = \lambda_{p0} = 258.2\text{mm}$; $\varepsilon_{cs}(t,t_0)$ 为预应力钢筋传力锚固龄期 t_0、计算龄期为 t 时的混凝土收缩应变; $\phi(t,t_0)$ 为加载龄期为 t_0,计算考虑的龄期为 t 时徐变系数。

$$\sigma_{pc} = \frac{N_{p0}}{A_0} + \frac{N_{p0}\lambda_{p0}}{I_0}y_0 = \frac{830\,292.5}{323\,351.4} + \frac{830\,292.5 \times 258.2}{1.563\,3 \times 10^{10}} \times 258.2 = 6.11(\text{MPa})$$

$$\rho_{ps} = 1 + \frac{\lambda_{ps}^2}{i^2} = 1 + \frac{258.2^2}{48\,340} = 2.379$$

$$E_p = 1.95 \times 10^5 \text{MPa}$$

$$\alpha_{Ep} = 5.65$$

考虑自重的影响,由于收缩徐变持续时间较长,采用全部永久作用,空心板跨中截面全部永久作用弯矩 M_{Gk} 由表12-4查得 $M_{Gk} = 222.66\text{kN·m}$,在全部的钢筋重心处由自重产生的拉应力为:

跨中截面:

$$\sigma_t = \frac{M_{Gk}}{I_0}y_0 = \frac{222.66 \times 10^6}{1.5633 \times 10^{10}} \times 258.2 = 3.67(\text{MPa})$$

$l/4$ 截面:

$$\sigma_t = \frac{167.00 \times 10^6}{1.563\,3 \times 10^{10}} \times 258.2 = 2.75(\text{MPa})$$

支点截面:

$$\sigma_t = 0$$

则全部纵向钢筋重心处的压应力为:

跨中:

$$\sigma_{pc} = 6.11 - 3.67 = 2.44(\text{MPa})$$

$l/4$ 截面:

$$\sigma_{pc} = 6.11 - 2.75 = 3.36(\text{MPa})$$

支点截面:

$$\sigma_{pc} = 6.11\text{MPa}$$

《公桥规》6.2.7条规定,σ_{pc} 不得大于传力锚固时混凝土立方体抗压强度 f'_{cu} 的0.5倍,设传力锚固时,混凝土达到C30,则 $f'_{cu} = 30\text{MPa}$, $0.5f'_{cu} = 0.5 \times 30 = 15\text{MPa}$,则跨中、$l/4$ 截面,支点截面全部钢筋重心处的压应力 2.44MPa、3.36MPa、6.11MPa,均小于 $0.5f'_{cu} = 0.5 \times 30 = 15(\text{MPa})$,满足要求。

设传力锚固龄期为7d,计算龄期为混凝土终极值 t_u,设桥梁所在处环境的大气相对湿度为75%。由前面计算,空心板毛截面面积 $A = 3\,174.3 \times 10^2 \text{mm}^2$,空心板与大气接触的周边长度为 u:

$$u = 2 \times 990 + 2 \times 620 + 2\pi \times 380 + 4 \times 80 = 5\,927.6(\text{mm})$$

理论厚度:

$$h = \frac{2A}{u} = \frac{2 \times 3\,174.3 \times 10^2}{5\,927.6} = 107.1(\text{mm})$$

查《公桥规》表6.2.7直线内插得到：
$$\varepsilon_{cs}(t,t_0) = 0.000\ 297$$
$$\phi(t,t_0) = 2.308$$

把各项数值代入 σ_{l6} 计算式中。

跨中：
$$\sigma_{l6}(t) = \frac{0.9 \times (1.95 \times 10^5 \times 0.000\ 297 + 5.65 \times 2.44 \times 2.308)}{1 + 15 \times 0.003\ 88 \times 2.379} = 70.94(\text{MPa})$$

$l/4$ 截面：
$$\sigma_{l6}(t) = \frac{0.9 \times (1.95 \times 10^5 \times 0.000\ 297 + 5.65 \times 3.36 \times 2.308)}{1 + 15 \times 0.003\ 88 \times 2.379} = 80.42(\text{MPa})$$

支点截面：
$$\sigma_{l6}(t) = \frac{0.9 \times (1.95 \times 10^5 \times 0.000\ 297 + 5.65 \times 6.11 \times 2.038)}{1 + 15 \times 0.003\ 88 \times 2.379} = 108.77(\text{MPa})$$

(6) 预应力损失组合

传力锚固时第一批损失 σ_{lI}：
$$\sigma_{lI} = \sigma_{l2} + \sigma_{l3} + \sigma_{l4} + \frac{1}{2}\sigma_{l5} = 15.6 + 30 + 35.42 + \frac{1}{2} \times 38.45 = 100.25(\text{MPa})$$

传力锚固后预应力损失总和 σ_l：
$$\sigma_l = \sigma_{l2} + \sigma_{l3} + \sigma_{l4} + \sigma_{l5} + \sigma_{l6} = 15.6 + 30 + 35.42 + 38.45 + 70.94 = 190.41(\text{MPa})$$

$l/4$ 截面：
$$\sigma_l = 15.6 + 30 + 35.42 + 38.45 + 80.42 = 199.89(\text{MPa})$$

支点截面：
$$\sigma_l = 15.6 + 30 + 35.42 + 38.45 + 108.77 = 228.24(\text{MPa})$$

各截面的有效应力：
$$\sigma_{pe} = \sigma_{con} - \sigma_l$$

跨中截面：
$$\sigma_{pe} = 1\ 302 - 190.41 = 1\ 111.59(\text{MPa})$$

$l/4$ 截面：
$$\sigma_{pe} = 1\ 302 - 199.89 = 1\ 102.11(\text{MPa})$$

支点截面：
$$\sigma_{pe} = 1\ 302 - 228.24 = 1\ 073.76(\text{MPa})$$

9) 正常使用极限状态

(1) 正截面抗裂性验算

正截面抗裂性计算是对构件跨中截面混凝土的拉应力进行验算，并满足《公桥规》6.3条要求。对于本例部分预应力A类构件，应满足两个要求：

在作用短期效应组合下，$\sigma_{st} - \sigma_{pc} \leq 0.7 f_{tk}$；

在荷载长期效应组合下，$\sigma_{lt} - \sigma_{pc} \leq 0$，即不出现拉应力。

式中，σ_{st} 为在作用短期效应组合下，空心板抗裂验算边缘的混凝土法向拉应力，由表12-4，空心板跨中截面弯矩 $M_{sd} = 327.98\text{kN} \cdot \text{m} = 327.98 \times 10^6 \text{N} \cdot \text{mm}$，由前面计算截面下缘弹性抵抗矩 $W_{0l1} = 52.45 \times 10^6 \text{mm}^3$，代入得：

$$\sigma_{st} = \frac{M_{sd}}{W_{0ll}} = \frac{327.98 \times 10^6}{52.45 \times 10^6} = 6.25(\text{MPa})$$

σ_{pc} 为扣除全部预应力损失后的预加力,在构件抗裂验算边缘产生的预应力,其值为:

$$\sigma_{pc} = \frac{N_{p0}}{A_0} = \frac{N_{p0}e_{p0}}{I_0}y_0$$

$$\sigma_{p0} = \sigma_{con} - \sigma_l + \sigma_{l4} = 1\,302 - 190.4 + 35.42 = 1\,147.01(\text{MPa})$$

$$N_{p0} = \sigma_{p0}A_p - \sigma_{l6}A_s = 1\,147.01 \times 690.9 - 70.94 \times 565.5 = 752\,352.64(\text{N})$$

$$e_{p0} = \frac{\sigma_{p0}A_pY_p - \sigma_{l6}A_sY_s}{N_{p0}} = 258.2\,\text{mm}$$

空心板跨中截面下缘的预压应力 σ_{pc} 为:

$$\sigma_{pc} = \frac{N_{p0}}{A_0} + \frac{N_{p0}e_{p0}}{I_0}y_0 = \frac{752\,352.64}{3\,233\,351.4} + \frac{752\,352.64 \times 258.2}{1.563\,3 \times 10^{10}} \times 298.2 = 6.04(\text{MPa})$$

由表 12-4,跨中截面弯矩 $M_{ld} = 280.55\,\text{kN} \cdot \text{m} = 280.55 \times 10^6 \text{N} \cdot \text{mm}$。同样,$W_{0ld} = 52.45 \times 10^6 \text{N} \cdot \text{mm}^3$,代入 α_{lt} 公式,则得:

$$\sigma_{lt} = \frac{M_{ld}}{W_{0ld}} = \frac{280.55 \times 10^6}{52.45 \times 10^6} = 5.34(\text{MPa})$$

由此得:

$$\sigma_{st} - \sigma_{pc} = 6.25 - 6.11 = 0.14(\text{MPa}) < 0.7f_{tk} = 0.7 \times 2.65 = 1.855(\text{MPa})$$

$$\sigma_{lt} - \sigma_{pc} = 5.34 - 6.11 = -0.77(\text{MPa}) < 0$$

符合《公桥规》对 A 类构件的规定。

温差应力按《公桥规》附录 B 计算。本示例桥面铺装厚度 100mm,由《公桥规》4.3.10 条,竖向温度梯度如图 12-24 所示,由于空心板高为 620mm,大于 400mm,取 $a = 300\,\text{mm}$。

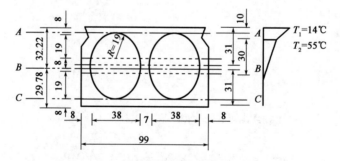

图 12-24 空心板竖向温度梯度(尺寸单位:cm)

对于简支板桥,温差应力:

$$N_t = \sum A_y t_y a_c E_c$$
$$M_t^0 = \sum A_y t_y a_c E_c e_y$$

正温差应力:

$$\sigma_t = \frac{-N_t}{A_0} + \frac{M_t^0}{I_0}y + t_y a_c E_c$$

式中,α_c 为混凝土温度线膨胀系数,$\alpha_c = 0.000\,01$;E_c 为混凝土弹性模量,C50,$E_c = 3.45 \times 10^4 \text{MPa}$;$A_y$ 为截面内的单元面积;t_y 为单元面积 A_y 内温差梯度平均值,均以正值代入;

y 为计算应力点至换算截面重心轴的距离,重心轴以上取正值,以下取负值;A_0、I_0 为换算截面面积和惯矩;e_y 为单位面积 A_y 重心至换算截面重心轴的距离,重心轴以上取正值,以下取负值。

列表计算 A_y、t_y、e_y,计算结果见表 12-6。

温度应力计算表　　　　　　　　　　表 12-6

编号	单元面积 A_y (mm^2)	温度 t_y (℃)	单元面积 A_y 重心至换算截面重心轴的距离 e_y
1	$80 \times 990 = 79\,200$	$\dfrac{14+7.2}{2}=10.6$	$e_y = 321.8 - \dfrac{80(14+2\times 7.2)}{3\times(14+7.2)}=286.5$
2	$(2\times 80+70)\times 20 = 4\,600$	$\dfrac{7.2+5.5}{2}=6.35$	$e_y = 321.8 - \dfrac{80(7.2+2\times 5.5)}{3\times(7.2+5.5)}=232.6$
3	$(2\times 80+70)\times 300 = 69\,000$	$\dfrac{5.5}{2}=2.75$	$e_y = 321.8 - 80 - 20 - \dfrac{1}{3}\times 300 = 122.2$

$$N_t = \sum A_y t_y \alpha_c E_c = (79\,200\times 10.6 + 4\,600\times 6.35 + 69\,000\times 2.75)\times 0.000\,01\times 3.45\times 10^4$$
$$= 365\,175.6(\text{N})$$

$$M_t^0 = -\sum A_y t_y \alpha_c E_c e_y$$
$$= -(79\,200\times 10.6\times 286.5 + 4\,600\times 6.35\times 232.6 + 69\,000\times 2.75\times 122.2)\times$$
$$\quad 0.000\,01\times 3.45\times 10^4$$
$$= -93.92\times 10^6$$

正温差应力:

梁顶:
$$\sigma_t = \dfrac{-N_t}{A_0} + \dfrac{M_t^0}{I_0}y + t_y \alpha_c E_c$$
$$= \dfrac{-365\,175.6}{323\,351.4} + \dfrac{-93.32\times 10^6 \times 321.8}{1.563\,3\times 10^{10}} + 14\times 0.000\,01\times 3.45\times 10^4$$
$$= -1.13 - 1.92 + 4.83$$
$$= 1.78(\text{MPa})$$

梁底:
$$\sigma_t = \dfrac{-365\,175.6}{323\,351.4} + \dfrac{-93.32\times 10^6}{1.563\,3\times 10^{10}}\times(-298.2) + 0$$
$$= -1.13 + 1.78$$
$$= 0.65(\text{MPa})$$

预应力钢筋重心处:
$$\sigma'_t = \dfrac{-365\,175.6}{323\,351.4} + \dfrac{-93.32\times 10^6}{1.563\,3\times 10^{10}}\times(-258.2)$$
$$= -1.13 + 1.54$$
$$= 0.41(\text{MPa})$$

普通钢筋重心处:
$$\sigma'_t = \dfrac{-365\,175.6}{323\,351.4} + \dfrac{-87.913\,9\times 10^6}{1.563\,3\times 10^{10}}\times(-258.2)$$

$$= -1.13 + 1.54$$
$$= 0.41(\text{MPa})$$

预应力钢筋温差应力：
$$\sigma_t = \alpha_{Ep}\sigma'_t = 5.65 \times 0.41 = 2.32(\text{MPa})$$

普通钢筋温差应力：
$$\sigma_t = \alpha_{Ep}\sigma'_t = 5.79 \times 0.41 = 2.37(\text{MPa})$$

反温差应力：

按《公桥规》4.2.10条，反温差为正温差乘以 -0.5，则得反温差应力：

梁顶： $\sigma_t = 1.78 \times (-0.5) = -0.89(\text{MPa})$

梁底： $\sigma_t = 0.65 \times (-0.5) = -0.32(\text{MPa})$

预应力钢绞线反温差应力： $\sigma_t = 2.32 \times (-0.5) = -1.16(\text{MPa})$

普通钢筋反温差应力： $\sigma_t = 2.39 \times (-0.5) = -1.20(\text{MPa})$

以上正值表示压应力，负值表示拉应力。

设温差频遇值系数为0.8，则考虑温差应力，在作用短期效应组合下，梁底总拉应力为：
$$\sigma_{st} = 6.25 + 0.8 \times 0.31 = 6.5(\text{MPa})$$

则 $\sigma_{st} - \sigma_{pc} = 6.50 - 6.11 = 0.39(\text{MPa}) < 0.7 f_{tk} = 0.7 \times 2.65 = 1.855(\text{MPa})$，满足部分预应力A类构件条件。

在长期效应组合下，梁底的总拉应力为：
$$\sigma_{lt} = 5.34 + 0.8 \times 0.32 = 5.59(\text{MPa})$$

则 $\sigma_{lt} - \sigma_{pc} = 5.54 - 6.11 = -0.77(\text{MPa}) < 0$，符合A类预应力混凝土条件。

上述计算结果表明，在短期效应组合及长期效应组合下，并考虑温差应力，正截面抗裂性均满足要求。

(2) 斜截面抗裂性验算

部分预应力A类构件斜截面抗裂性验算是主拉应力控制，采用作用的短期效应组合，并考虑温差作用。温差作用效应可利用正截面抗裂计算中温差应力计算及表12-6、图12-24，并选用支点截面，分别计算支点截面 A-A 纤维（空洞顶面）、B-B 纤维（空心板换算截面重心轴）、C-C 纤维（空洞底面）处主拉应力，对于部分预应力A类构件应满足：
$$\sigma_{tp} \leq 0.7 f_{tk}$$

式中，f_{tk} 为混凝土的抗拉强度标准值，C50，$f_{tk} = 2.65\text{MPa}$；σ_{tp} 为由作用短期效应组合和预应力引起的混凝土主拉应力，并考虑温差作用。

先计算温差应力，由表12-6和图12-24：

① 正温差应力

A-A 纤维：
$$\sigma_t = \frac{-N_t}{A_0} + \frac{M_t^0}{I_0}y + t_y \alpha_c E_c$$
$$= \frac{-365\,175.6}{323\,351.4} + \frac{-93.32 \times 10^6}{1.563\,3 \times 10^{10}} \times (321.8 - 80) + 7.2 \times 0.000\,01 \times 3.45 \times 10^4$$
$$= -1.13 + (-1.43) + 2.48$$
$$= -0.08(\text{MPa})$$

B-B 纤维：

$$\sigma_t = \frac{-365\,175.6}{323\,351.4} + \frac{-93.32 \times 10^6}{1.5633 \times 10^{10}} \times 0 + 1.42 \times 0.00001 \times 3.45 \times 10^4$$

$$= -1.13 + 0.49$$

$$= -0.64(\text{MPa})$$

C-C 纤维：

$$\sigma_t = \frac{-365\,175.6}{323\,351.4} + \frac{-93.32 \times 10^6}{1.5633 \times 10^{10}} \times [-(298.2 - 80)] + 0$$

$$= -1.13 + 1.30$$

$$= 0.17(\text{MPa})$$

② 反温差应力

为正温差应力乘以 -0.5。

A-A 纤维： $\sigma_t = (-0.08) \times (-0.5) = 0.04(\text{MPa})$

B-B 纤维： $\sigma_t = (-0.64) \times (-0.5) = 0.32(\text{MPa})$

C-C 纤维： $\sigma_t = 0.17 \times (-0.5) = -0.08(\text{MPa})$

以上正值表示压应力，负值表示拉应力。

③ 主拉应力 σ_{tp}

a. A-A 纤维（空洞顶面）。

$$\sigma_{tp} = \frac{\sigma_{cx}}{2} - \sqrt{\left(\frac{\sigma_{cx}}{2}\right)^2 + \tau^2}$$

$$\tau = \frac{V_d S_{01A}}{b I_0}$$

式中，V_d 为支点截面短期组合效应剪力设计值，由表 12-4，$V_d = 149.73 \text{kN} = 149.73 \times 10^3 \text{N}$；$b$ 为计算主拉应力处截面腹板总宽，取 $b = 70 + 2 \times 80 = 230(\text{mm})$；$I_0$ 为计算主拉应力截面抗弯惯矩，$I_0 = 1.5633 \times 10^{10} \text{mm}^4$；$S_{01A}$ 为空心板 A-A 纤维以上截面对空心板换算截面重心轴的静矩，$S_{01A} = 990 \times 80 \times (321.8 - 80/2) = 22.32 \times 10^{10}(\text{mm}^3)$，则：

$$\tau = \frac{V_d S_{01A}}{b I_0} = \frac{149.73 \times 10^3 \times 22.32 \times 10^6}{230 \times 1.5633 \times 10^{10}} = 0.93(\text{MPa})$$

$$\sigma_{sx} = \sigma_{pc} + \frac{M_s y_0}{I_0} + \psi_{ij} \sigma_t$$

$$N_{p0} = \sigma_{p0} A_p - \sigma_{l6} A_s$$

$$\sigma_{p0} = \sigma_{con} - \sigma_l + \sigma_{l4} = 1\,302 - 228.2 + 35.42 = 1\,038.38(\text{MPa})$$

$$N_{p0} = \sigma_{p0} A_p - \sigma_{l6} A_s = 1\,038.38 \times 690.9 - 108.77 \times 565.5 = 655\,907.3(\text{N})$$

$$e_{p0} = \frac{\sigma_{p0} A_p Y_p - \sigma_{l6} A_s Y_s}{N_{p0}} = \frac{1\,038.38 \times 690.9 \times 258.2 - 108.77 \times 565.5 \times 258.2}{655\,907.3}$$

$$= 258.2(\text{mm})$$

$$\sigma_{pc} = \frac{N_{p0}}{A_0} - \frac{N_{p0} e_{p0}}{I_0} y_0 = -0.64 \text{MPa}$$

$$\sigma_{sx} = \sigma_{pc} + \frac{M_s y_0}{I_0} + \psi_{ij} \sigma_t = -0.64 + 0 + 0.8 \times (-0.08) = -0.70(\text{MPa})$$

y_0 是指 A-A 纤维到重心轴的距离,$y_0 = 322.2 - 80 = 242.2(\text{mm})$。

式中,M_s 为竖向荷载产生的弯矩,在支点 $M_s = 0$;ψ_{ij} 为温差频遇系数,取 $\psi_{ij} = 0.8$。

计入反温差效应,则:

$$\sigma_{cx} = -0.64 + 0.8 \times 0.04 = -0.61(\text{MPa})$$

主拉应力:

$$\sigma_{tp} = \frac{\sigma_{cx}}{2} - \sqrt{\left(\frac{\sigma_{cx}}{2}\right)^2 + \tau^2} = -1.34(\text{MPa})$$

计入反温差应力:

$$\sigma_{tp} = \frac{-0.61}{2} - \sqrt{\left(\frac{-0.61}{2}\right)^2 + (0.93)^2} = -1.28(\text{MPa})$$

负值表示拉应力。

预应力混凝土 A 类构件,在短期效应组合下,预制构件应符合 $\sigma_{tp} = -1.34\text{MPa} < 1.68\text{MPa}$。现 A-A 纤维处 $\sigma_{tp} \leq 0.7 f_{tk} = 0.7 \times 2.65 = 1.855(\text{MPa})$(计入正温差影响),$\sigma_{tp} = -1.28\text{MPa}$(计入反温差影响),符合要求。

b. B-B 纤维(空心板换算截面重心处)(图 12-24)。

$$\tau = \frac{V_d S_{01B}}{b I_0}$$

式中,S_{01B} 为 B-B 纤维以上截面对重心轴的静矩。

$$S_{01B} = 990 \times 321.8 \times \frac{321.8}{2} - 2 \times \frac{\pi \times 380^2}{8}(321.8 - 80 - 190 + 80.6) -$$

$$2 \times 380 \times (321.8 - 80 - 90) \times \frac{321.8 - 80 - 190}{2}$$

$$= 35.29 \times 10^6(\text{mm}^3)(\text{铰缝未扣除})$$

$$\tau = \frac{V_d S_{01B}}{b I_0} = \frac{149.73 \times 10^3 \times 35.29 \times 10^6}{230 \times 1.5633 \times 10^{10}} = 1.47(\text{MPa})$$

$$\sigma_{cx} = \sigma_{pc} + \frac{M_s y_0}{I_0} + \psi_{ij} \sigma_t$$

式中,y_0 为 B-B 纤维至重心轴距离,$y_0 = 0$。

$$\sigma_{pc} = \frac{655\,907.3}{323\,351.4} - \frac{655\,907.3 \times 258.2}{1.5633 \times 10^{10}} \times 0 = 2.03(\text{MPa})$$

同样,$M_s = 0$,$\psi_{ij} = 0.8$。

$$\sigma_{cx} = 2.03 + 0.8 \times (-0.6) = 1.55(\text{MPa})$$

$$\sigma_{cx} = 2.03 + 0.8 \times 0.3 = 2.27(\text{MPa})$$

$$\sigma_{tp} = \frac{\sigma_{cx}}{2} - \sqrt{\left(\frac{\sigma_{cx}}{2}\right)^2 + \tau^2} = -0.88\text{MPa}(\text{考虑正温差影响})$$

$$\sigma_{tp} = \frac{\sigma_{cx}}{2} - \sqrt{\left(\frac{\sigma_{cx}}{2}\right)^2 + \tau^2} = -0.68\text{MPa}(\text{考虑负温差影响})$$

B-B 纤维处:

$$\sigma_{tp} = -0.88\text{MPa}$$

$$\sigma_{tp} = -0.68 \text{MPa}$$

负值为拉应力,均小于 $0.7f_{tp} = 0.7 \times 2.65 = 1.855(\text{MPa})$,符合《公桥规》对部分预应力 A 类构件斜截面抗裂性要求。

c. $C\text{-}C$ 纤维(空洞底面)。

$$\sigma_{tp} = \frac{\sigma_{cx}}{2} - \sqrt{\left(\frac{\sigma_{cx}}{2}\right)^2 + \tau^2}$$

$$\tau = \frac{V_d S_{01C}}{bI_0}$$

式中,S_{01C} 为 $C\text{-}C$ 纤维以下截面对空心板重心轴的静矩。

$S_{01C} = 990 \times 80 \times (298.2 - 80/2) + (5.65 - 1) \times 690.9 \times 258.2 +$
$\quad (5.79 - 1) \times 565.5 \times 258.2 = 21.98 \times 10^6 (\text{mm}^3)$
$\tau = V_d S_{01C}/(bI_0) = 149.73 \times 10^3 \times 21.98 \times 10^6/(230 \times 1.5633 \times 10^{10}) = 0.92(\text{MPa})$
$\sigma_{pc} = N_{p0}/A_0 + (N_{p0}e_{p0}/I_0) \times y_0$
$\quad = 655\,907.3/323\,351.4 + 655\,907.3 \times 258.2 \times 218.2/(1.5633 \times 10^{10})$
$\quad = 2.03 + 2.36 = 4.39(\text{MPa})$

y_0 为 $C\text{-}C$ 纤维至重心轴距离,$y_0 = 297.8 - 80 = 217.8(\text{mm})$。

$\sigma_{cx} = \sigma_{pc} + (M_s y_0)/I_0 + \psi_{ij}\sigma_t$
$\quad = 4.39 + 0 + 0.8 \times 0.17 = 4.53(\text{MPa})$(计入正温差应力)
$\sigma_{cx} = 4.39 + 0 + 0.8 \times (-0.08) = 4.33(\text{MPa})$(计入反温差应力)

$\sigma_{tp} = \sigma_{cx}/2 - \sqrt{\left(\frac{\sigma_{cx}}{2}\right)^2 + \tau^2}$

$\quad = 4.53/2 - \sqrt{\left(\frac{4.53}{2}\right)^2 + 0.92^2} = -0.18(\text{MPa})$(计入正温差应力)

$\sigma_{tp} = 4.33/2 - \sqrt{\left(\frac{4.33}{2}\right)^2 + 0.92^2} = 0.19(\text{MPa})$(计入反温差应力)

负值为拉应力。

$C\text{-}C$ 纤维处的主拉应力 $\sigma_{tp} = 0.18\text{MPa} < 0.7f_{tp} = 0.7 \times 2.65 = 1.855(\text{MPa})$。
$\sigma_{tp} = 0.19\text{MPa} < 0.7f_{tp} = 0.7 \times 2.65 = 1.855(\text{MPa})$。

上述计算结果表明,满足《公桥规》对部分预应力 A 类构件斜截面抗裂性要求。

10) 变形计算

(1) 正常使用阶段的挠度计算

使用阶段的挠度值,按短期荷载效应组合计算,并考虑挠度长期增长系数 η_θ,对于 C50 混凝土,$\eta_\theta = 1.60$,对于部分预应力 A 类构件,使用阶段的挠度计算时,抗弯刚度 $B_0 = 0.95E_cI_0$。取跨中截面尺寸及配筋情况确定 B_0:

$$B_0 = 0.95E_cI_0 = 0.95 \times 3.45 \times 10^4 \times 1.5633 \times 10^{10} = 5.124 \times 10^{14}(\text{mm}^2)$$

短期荷载组合使用下的挠度值,可简化为按等效均布荷载作用情况计算:

$$f_s = \frac{5}{48}\frac{l^2 M_s}{B_0} = \frac{5 \times 12\,600^2 \times 327.98 \times 10^6}{48 \times 5.124 \times 10^{14}} = 10.5(\text{mm})$$

自重产生的挠度值按等效均布荷载作用情况计算：

$$f_G = \frac{5}{48}\frac{l^2 M_{Gk}}{B_0} = \frac{5 \times 12\,600^2 \times 222.66 \times 10^6}{48 \times 5.124 \times 10^{14}} = 7.2(\text{mm})$$

M_s、M_{Gk} 值查表 12-4 得。

消除自重产生的挠度，并考虑长期影响系数 η_θ 后，正常使用阶段的挠度值为：

$$f_l = \eta_\theta(f_s - f_G) = 1.6 \times (10.5 - 7.2) = 5.28(\text{mm}) < l/600 = 12\,600/600 = 21(\text{mm})$$

计算结果表明，使用阶段的挠度值满足《公桥规》要求。

(2) 预加力引起的反拱度计算及预拱度的设置

① 预加力引起的反拱度计算

当空心板放松预应力钢绞线时，跨中产生反拱度，设这时空心板混凝土强度达到 C30。预加力产生的反拱度计算按跨中截面尺寸及配筋计算，并考虑反拱长期增长系数 $\eta_\theta = 2.0$。

先计算此时的抗弯刚度：$B'_0 = 0.95 E'_c I'_0$。

放松预应力钢绞线时，设空心板混凝土强度达到 C30，这时 $E'_c = 3.0 \times 10^4 \text{MPa}$，则：

$$\alpha'_{Ep} = \frac{E_p}{E'_c} = \frac{1.95 \times 10^5}{3.0 \times 10^4} = 6.5, A_p = 690.9 \text{mm}^2$$

$$\alpha'_{Es} = \frac{E_s}{E'_c} = \frac{2.0 \times 10^5}{3.0 \times 10^4} = 6.7, A_s = 565.5 \text{mm}^2$$

换算截面面积：

$$A'_0 = 317\,430 + (6.5 - 1) \times 690.9 + (6.7 - 1) \times 565.5 = 324\,453(\text{mm}^2)$$

所有钢筋换算面积对毛截面重心的静矩为：

$$S'_{01} = (\alpha'_{Ep} - 1)A_p(310 - 7 - 40) + (\alpha'_{Es} - 1)A_s(310 - 7 - 40)$$
$$= (6.5 - 1) \times 690.9 \times 263 + (6.7 - 1) \times 565.5 \times 263$$
$$= 1\,847\,128(\text{mm}^2)$$

换算截面重心至毛截面重心的距离为：

$$d'_{01} = \frac{S'_{01}}{A'_0} = \frac{1\,847\,128}{324\,453} = 5.7(\text{mm})(\text{向下移})$$

则换算截面重心至空心板下缘的距离：

$$y'_{01l} = 310 - 7 - 5.7 = 297.3(\text{mm})$$

换算截面重心至空心板上缘的距离：

$$y'_{01u} = 310 + 7 + 5.7 = 322.7(\text{mm})$$

预应力钢绞线至换算截面重心的距离：

$$e'_{01p} = 297.3 - 40 = 257.3(\text{mm})$$

普通钢绞线至换算截面重心的距离：

$$e'_{01s} = 297.3 - 40 = 257.3(\text{mm})$$

换算截面惯矩：

$$I'_0 = 15\,201 \times 10^6 + 317\,430 \times 5.7^2 + (6.5 - 1) \times 690.9 \times 257.8^2 +$$
$$(6.7 - 1) \times 565.5 \times 257.8^2$$
$$= 1.567\,6 \times 10^{10}(\text{mm}^4)$$

换算截面的弹性抵抗矩：

下缘：$W'_{011} = I'_0/y'_{011} = 1.5676 \times 10^{10}/297.3 = 52.495 \times 10^6 (\text{mm}^3)$

上缘：$W'_{01u} = I'_0/y'_{01u} = 1.5676 \times 10^{10}/322.7 = 48.5776 \times 10^6 (\text{mm}^3)$

空心板换算截面几何特性汇总于表12-7。

空心板截面几何特性汇总表　　　表12-7

项　目	符号	单位	C30, $\alpha'_{Ep}=6.5$	C50, $\alpha_{Ep}=5.65$
换算截面面积	A'_0	mm²	324 453	323 351.4
换算截面重心至截面下缘距离	y'_{011}	mm	297.3	298.2
换算截面重心至截面上缘距离	y'_{01u}	mm	322.7	298.2
预应力钢筋至截面重心轴距离	e'_{01u}	mm	257.3	258.2
普通钢筋至截面重心轴距离	e'_{01s}	mm	257.3	258.2
换算截面惯矩	I'_0	mm⁴	1.5676×10^{10}	1.5633×10^{10}
换算截面弹性抵抗矩	W'_{011}	mm³	52.495×10^6	52.450×10^6
换算截面弹性抵抗矩	W'_{01u}	mm³	48.5776×10^6	48.57×10^6

由前面计算得扣除预应力损失后的预加力为：$N_{p0} = 752352.64$ N。

$$M_{p0} = 752352.64 \times 257.3 = 193.58 \times 10^6 (\text{N} \cdot \text{mm})$$

则由预加力产生的跨中反拱度，并乘以长期增长系数 $\eta_\theta = 2.0$ 后，得：

$$f_p = 2.0 \times \frac{5l^2 M_{p0}}{48 \times 0.95 E'_c I'_c}$$

$$= 2.0 \times \frac{5 \times 193.586 \times 10^6 \times 12600^2}{48 \times 0.95 \times 3.0 \times 10^4 \times 1.5676 \times 10^{10}}$$

$$= 14.33 (\text{mm})$$

②预拱度的设置

由《公桥规》6.5.5条，当预加应力的长期反拱值 f_p 小于按荷载短期效应组合计算的长期挠度 f_{sl} 时，应设置预拱度，其值按该荷载的挠度值与预加应力长期反拱值之差采用。

本例 $f_p = 14.33\text{mm} < f_{sl} = 1.6 \times 10.5 = 16.8 (\text{mm})$，说明结构刚度不够，应采取措施提高刚度以满足要求，也可以考虑成桥时桥面铺装层参与工作，增加结构的刚度。

11) 持久状态应力验算

持久状态应力验算应计算使用阶段正截面混凝土的法向压应力 σ_{kc}、预应力钢筋的拉应力 σ_p 及斜截面的主压应力 σ_{cp}。计算时作用取标准值，不计分项系数，汽车荷载考虑冲击系数并考虑温差应力。

(1) 跨中截面混凝土法向压应力 σ_{kc} 验算

跨中截面的有效预应力：

$$\sigma_p = \sigma_{con} - \sigma_l = 1302 - 190.41 = 1111.59 (\text{MPa})$$

跨中截面的有效预加力：

$$N_p = \sigma_p A_p = 1111.59 \times 690.9 = 767997.5 (\text{N})$$

由表12-4得标准值效应组合 $M_s = 408.75 \text{kN} \cdot \text{m} = 408.75 \times 10^6 \text{N} \cdot \text{mm}$，则：

$$\sigma_{kc} = \frac{N_p}{A_0} - \frac{N_p e_p}{W_{01u}} + \frac{M_s}{W_{01u}} + \sigma_t$$

$$= \frac{767\,997.5}{323\,351.4} - \frac{767\,997.5}{48.57 \times 10^6} \times 258.2 + \frac{408.75 \times 10^6}{48.7 \times 10^6} + 1.78$$

$$= 8.50(\text{MPa}) < 0.5 f_{ck} = 0.5 \times 32.4 = 16.2(\text{MPa})$$

(2)跨中截面预应力钢绞线拉应力 σ_p 验算

$$\sigma_p = \sigma_{pe} + \sigma_{Ep}\sigma_{kt} \leqslant 0.65 f_{pk}$$

式中,σ_{kt} 为按荷载效应标准值计算的预应力钢绞线重心处混凝土法向应力。

$$\sigma_{kt} = 408.75 \times 10^6 \times 258.2/(1.5633 \times 10^{10}) = 6.75(\text{MPa})$$

有效预应力:

$$\sigma_{pe} = \sigma_{con} - \sigma_l = 1\,302 - 190.41 = 1\,111.59(\text{MPa})$$

考虑温差应力,则预应力钢绞线中的拉应力为:

$$\sigma_p = \sigma_{pe} + \sigma_{Ep}\sigma_{kt} + \sigma_t$$

$$= 1\,111.59 + 5.65 \times 6.75 + 1.16$$

$$= 1\,150.88(\text{MPa}) < 0.65 f_{pk} = 0.65 \times 1\,860 = 1\,209(\text{MPa})$$

(3)斜截面主应力验算

斜截面主应力选取支点截面的 A-A 纤维(空洞顶面)、B-B 纤维(空心板中心轴)、C-C 纤维(空洞底面)在标准值效应组合和预加力作用下产生的主压应力 σ_{cp} 和主拉应力 σ_{tp} 计算,并满足 $\sigma_{cp} \leqslant 0.6 f_{ck} = 0.6 \times 32.4 = 19.44(\text{MPa})$ 的要求。

$$\sigma_{cp} = \frac{\sigma_{cxk}}{2} + \sqrt{\left(\frac{\sigma_{cxk}}{2}\right)^2 + \tau_k^2}$$

$$\sigma_{tp} = \frac{\sigma_{cxk}}{2} - \sqrt{\left(\frac{\sigma_{cxk}}{2}\right)^2 + \tau_k^2}$$

$$\sigma_{cxk} = \sigma_{pc} + \frac{M_k y_0}{I_0} + \sigma_t$$

$$\tau_k = \frac{V_d S_{01}}{b I_0}$$

① A-A 纤维(空洞顶面)

$$\tau_k = \frac{V_d S_{01A}}{b I_0} = \frac{216.39 \times 10^3 \times 22.32 \times 10^6}{230 \times 1.5633 \times 10^{10}} = 1.35(\text{MPa})$$

式中,V_d 为支点截面标准值效应组合设计值,由表12-4,$V_d = 216.391\text{kN} = 216.391 \times 10^3 \text{N}$;$b$ 为腹板宽度,$b = 230\text{mm}$;S_{01A} 为 A-A 纤维以上截面对空心板重心轴的静矩,$S_{01A} = 22.32 \times 10^6 \text{mm}^3$。

$$\sigma_{cxk} = \sigma_{pc} + \frac{M_k y_0}{I_0} + \sigma_t$$

$$= 0.64 + 0 + (-0.08) = 0.72(\text{MPa})$$

式中,σ_{pc} 为预加力产生在 A-A 纤维处的正应力,$\sigma_{pc} = 0.64\text{MPa}$;$M_k$ 为竖向荷载产生的截面弯矩,支点截面 $M_k = 0$;σ_t 为 A-A 纤维处正温差应力,$\sigma_t = -0.08\text{MPa}$,反温差应力 $\sigma_t = 0.04\text{MPa}$,不再赘述。

A-A 纤维处的主应力为(计入正温差应力):

$$\sigma_{cp} = \frac{-0.72}{2} + \sqrt{\left(\frac{-0.72}{2}\right)^2 + 1.35^2} = 1.04(\text{MPa})$$

$$\sigma_{tp} = \frac{-0.72}{2} - \sqrt{\left(\frac{-0.72}{2}\right)^2 + 1.35^2} = -1.76(\text{MPa})$$

计入反温差应力时:

$$\sigma_{cxk} = 0.64 + 0 + 0.04 = -0.60(\text{MPa})$$

则:

$$\sigma_{cp} = \frac{-0.60}{2} + \sqrt{\left(\frac{-0.6}{2}\right)^2 + 1.35^2} = 1.08(\text{MPa})$$

$$\sigma_{tp} = \frac{-0.60}{2} - \sqrt{\left(\frac{-0.6}{2}\right)^2 + 1.35^2} = 1.68(\text{MPa})$$

C50 混凝土主应力限值为:$0.6f_{ck} = 0.6 \times 32.4 = 19.44(\text{MPa})$。

$\sigma_{cpmax} = 1.08\text{MPa} < 16.08\text{MPa}$,符合《公桥规》要求。

② B-B 纤维

$$\tau_k = \frac{V_d S_{01B}}{b I_0} = \frac{216.39 \times 10^3 \times 35.29 \times 10^6}{230 \times 1.5633 \times 10^{10}} = 2.12(\text{MPa})$$

式中,S_{01B} 为 B-B 纤维以上截面对空心板重心轴的静矩,$S_{01B} = 35.29 \times 10^6 \text{mm}^3$。

由前面计算得 $\sigma_{pc} = 2.03\text{MPa}$,$\sigma_t = -0.60\text{MPa}$(计入正温差),$\sigma_t = 0.3\text{MPa}$(计入反温差),则:

$$\sigma_{cxk} = \sigma_{pc} + \frac{M_k y_0}{I_0} + \sigma_t$$

$$= 2.03 + 0 + (-0.6) = 1.43(\text{MPa})(计入正温差应力)$$

$$\sigma_{cxk} = \sigma_{pc} + \frac{M_k y_0}{I_0} + \sigma_t$$

$$= 2.03 + 0 + 0.3 = 2.33(\text{MPa})(计入反温差应力)$$

B-B 纤维处的主应力为(计入正温差应力):

$$\sigma_{cp} = \frac{1.43}{2} + \sqrt{\left(\frac{1.43}{2}\right)^2 + 2.12^2} = 2.95(\text{MPa})$$

$$\sigma_{tp} = \frac{1.43}{2} - \sqrt{\left(\frac{1.43}{2}\right)^2 + 2.12^2} = -1.52(\text{MPa})$$

计入反温差应力:

$$\sigma_{cp} = \frac{2.33}{2} + \sqrt{\left(\frac{2.33}{2}\right)^2 + 2.12^2} = 3.58(\text{MPa})$$

$$\sigma_{tp} = \frac{2.33}{2} - \sqrt{\left(\frac{2.33}{2}\right)^2 + 2.12^2} = -1.25(\text{MPa})$$

混凝土主压应力限值为 16.08MPa > 3.58MPa，符合《公桥规》要求。

③C-C 纤维

$$\tau_k = \frac{V_d S_{01C}}{b I_0} = \frac{216.39 \times 10^3 \times 21.98 \times 10^6}{230 \times 1.5633 \times 10^{10}} = 1.33(\text{MPa})$$

式中，S_{01C} 为 C-C 纤维以下截面对空心板重心轴的静矩，$S_{01C} = 21.98 \times 10^6 \text{mm}^3$。

同样，由前面得 $\sigma_{pc} = 4.39\text{MPa}$，$\sigma_t = 0.18\text{MPa}$（正温差应力），$\sigma_t = -0.19\text{MPa}$（反温差应力），则：

$$\sigma_{cxk} = \sigma_{pc} + \frac{M_k y_0}{I_0} + \sigma_t$$
$$= 4.39 + 0 + 0.18 = 4.57(\text{MPa})(\text{计入正温差应力})$$

$$\sigma_{cxk} = \sigma_{pc} + \frac{M_k y_0}{I_0} + \sigma_t$$
$$= 4.39 + 0 + (-0.19) = 4.2(\text{MPa})(\text{计入反温差应力})$$

C-C 纤维处的主应力为（计入正温差应力）：

$$\sigma_{cp} = \frac{4.57}{2} + \sqrt{\left(\frac{4.57}{2}\right)^2 + 1.33^2} = 4.93(\text{MPa})$$

$$\sigma_{tp} = \frac{4.57}{2} - \sqrt{\left(\frac{4.57}{2}\right)^2 + 1.33^2} = -0.36(\text{MPa})$$

计入反温差应力时：

$$\sigma_{cp} = \frac{4.2}{2} + \sqrt{\left(\frac{4.2}{2}\right)^2 + 1.33^2} = 4.59(\text{MPa})$$

$$\sigma_{tp} = \frac{4.2}{2} - \sqrt{\left(\frac{4.2}{2}\right)^2 + 1.33^2} = -0.39(\text{MPa})$$

混凝土主压应力为 $\sigma_{cp} = 4.93\text{MPa} < 16.08\text{MPa}$，符合《公桥规》要求。

计算结果表明使用阶段正截面混凝土法向力、预应力钢筋拉应力和斜截面主压应力均满足规范要求。

以上主拉应力最大值发生在 A-A 纤维处为 1.76MPa，按《公桥规》7.1.6 条：在 $\sigma_{tp} \leq 0.5 f_{tk} = 0.5 \times 2.65 = 1.33(\text{MPa})$ 区段，箍筋可按构造设置；在 $\sigma_{tp} > 0.5 f_{tk} = 1.33\text{MPa}$ 区段，箍筋间距 s_v 按下式计算：

$$s_v = \frac{f_{sk} A_{sv}}{\sigma_{tp} b}$$

式中，f_{sk} 为箍筋抗拉强度标准值，由前箍筋采用 HRB335，其 $f_{sk} = 335\text{MPa}$；A_{sv} 为同一截面

内箍筋的总截面面积,由前箍筋为双肢 2Φ10,$A_{sv} = 157.08\text{mm}^2$;$b$ 为腹板宽度,$b = 230\text{mm}$。

则箍筋间距计算如下:

$$s_v = \frac{f_{sk}A_{sv}}{\sigma_{tp}b} = \frac{335 \times 157.08}{1.76 \times 230} = 130(\text{mm})$$

采用 $s_v = 100\text{mm}$。

此时配箍率:

$$\rho_{sv} = \frac{A_{sv}}{s_v b} = \frac{157.08}{100 \times 230} = 0.0068 = 0.68\%$$

按《公桥规》9.3.12,对于 HRB400,$\rho_{sv} \geq 0.11\%$,满足要求。支点附近箍筋间距 100mm,其他截面适当加大,需按计算决定,箍筋布置图如图 12-23 所示,既满足斜截面抗剪要求,也满足主拉应力计算要求,箍筋间距也满足不大于板高的一半即 $h/2 = 310\text{mm}$ 以及不大于 400mm 的构造要求。

12) 短暂状态应力验算

预应力混凝土受弯构件短暂状态计算时,应计算构件在制造、运输及安装等施工阶段,由预加力(扣除相应的应力损失)、构件自重及其他施工荷载引起的截面应力,并满足《公桥规》要求。为此,对本示例应计算在放松预应力钢绞线时预制空心板的板底压应力和板顶拉应力。

设预制空心板当混凝土强度达到 C30 时,放松预应力钢绞线,这时,空心板处于初始预加力及空心板自重作用下,计算空心板板顶(上缘)、板底(下缘)法向应力。

C30 混凝土,$E'_c = 3 \times 10^4 \text{MPa}$,$f'_{ck} = 20.1\text{MPa}$,$f'_{tk} = 2.01\text{MPa}$,$E_p = 1.95 \times 10^5 \text{MPa}$,$\alpha'_{Ep} = \frac{E_p}{E'_c} = \frac{1.95 \times 10^5}{3.0 \times 10^4} = 6.5$,$\alpha'_{Es} = \frac{2.0 \times 10^5}{3.0 \times 10^4} = 6.7$,由此计算空心板截面几何特性,见表 12-7。

放松预应力钢绞线时,空心板截面法向应力计算取跨中、$l/4$、支点三个截面,计算如下。

(1) 跨中截面

① 由预加力产生的混凝土法向应力(由《公桥规》6.1.6 条):

$$\begin{aligned}\text{板底压应力 } \sigma_\text{下} \\ \text{板顶拉应力 } \sigma_\text{上}\end{aligned} = \frac{N_{p0}}{A_0} \pm \frac{N_{p0}e_{p0}}{I_0} \times \begin{aligned}y_{01l} \\ y_{01u}\end{aligned}$$

式中,N_{p0} 为先张法预应力钢筋和普通钢筋的合力,其值为:

$$N_{p0} = \sigma_{p0}A_p - \sigma_{l6}A_s$$

$$\sigma_{p0} = \sigma_{con} - \sigma_l + \sigma_{l4}$$

式中,σ_l 为放松预应力钢绞线时预应力损失值,由《公桥规》6.2.8 条,对先张法构件,有:

$$\sigma_l = \sigma_{l1} = \sigma_{l2} + \sigma_{l3} + \sigma_{l4} + 0.5\sigma_{l5}$$

则 $\sigma_{p0} = \sigma_{con} - \sigma_{l1} + \sigma_{l4} = \sigma_{con} - (\sigma_{l2} + \sigma_{l3} + \sigma_{l4} + 0.5\sigma_{l5}) + \sigma_{l4}$

$= 1302 - 15.6 - 30 - 0.5 \times 38.45$

$= 1237.18(\text{MPa})$

$N_{p0} = \sigma_{p0}A_p - \sigma_{l6}A_s = 1237.18 \times 690.9 - 70.94 \times 565.5 = 814651.09(\text{N})$

$$e_{p0} = \frac{\sigma_{p0}A_p y_p - \sigma_{l6}A_s y_s}{N_{p0}}$$

$$= \frac{1\,237.18 \times 690.9 \times 257.3 - 70.94 \times 565.5 \times 257.3}{814\,651.09} = 257.3(\text{mm})$$

下缘应力 $\sigma_{下}$
上缘应力 $\sigma_{上}$ $= \frac{N_{p0}}{A_0} \pm \frac{N_{p0}e_{p0}}{I_0} \times \frac{y_{01l}}{y_{01u}}$

$$= \frac{814\,651.09}{324\,453} \pm \frac{814\,651.09 \times 257.3}{1.567\,6 \times 10^{10}} \times \frac{297.3}{322.7}$$

$$= 2.51 \pm \frac{3.97}{4.31} = \frac{6.48}{-1.80}(\text{MPa})$$

②由板自重产生的板截面上、下缘应力

由表12-4,空心板跨中截面板自重弯矩 $M_{G1} = 157.49\text{kN} \cdot \text{m} = 157.49 \times 10^6 10\text{N} \cdot \text{mm}$,则由板自重产生的截面法向应力为:

下缘应力
上缘应力 $\frac{\sigma_{下}}{\sigma_{上}} = \frac{M_{G1}}{I_0} \times \frac{y_{01l}}{y_{01u}} = \frac{157.49 \times 10^6}{1.567\,6 \times 10^{10}} \times \frac{-297.3}{322.7} = \frac{-2.99}{3.24}(\text{MPa})$

放松预应力钢绞线时,由预加力及板自重共同作用,空心板上下缘产生的法向应力为:

下缘应力 $\sigma_{下} = 6.48 - 2.99 = 3.49(\text{MPa})$

上缘应力 $\sigma_{上} = -1.80 + 3.24 = 1.44(\text{MPa})$

截面上下缘均为压应力,且小于 $0.7f'_{ck} = 0.7 \times 20.1 = 14.07(\text{MPa})$,符合《公桥规》要求。

(2) $l/4$ 截面

$$\sigma_{p0} = \sigma_{con} - \sigma_{lI} + \sigma_{l4} = \sigma_{con} - (\sigma_{l2} + \sigma_{l3} + \sigma_{l4} + 0.5\sigma_{l5}) + \sigma_{l4}$$

$$= \sigma_{con} - \sigma_{l2} - \sigma_{l3} - 0.5\sigma_{l5}$$

$$= 1\,302 - 15.6 - 30 - 0.5 \times 38.45$$

$$= 1\,237.18(\text{MPa})$$

$$N_{p0} = \sigma_{p0}A_p - \sigma_{l6}A_s = 1\,237.1 \times 690.9 - 80.42 \times 565.5 = 809\,234.88(\text{N})$$

$$e_{p0} = \frac{\sigma_{p0}A_p y_p - \sigma_{l6}A_s y_s}{N_{p0}} = 257.3(\text{mm})$$

下缘应力 $\sigma_{下}$
上缘应力 $\sigma_{上}$ $= \frac{N_{p0}}{A_0} \pm \frac{N_{p0}e_{p0}}{I_0} \times \frac{y_{01l}}{y_{01u}}$

$$= \frac{809\,234.88}{324\,453} \pm \frac{809\,234.88 \times 257.3}{1.567\,6 \times 10^{10}} \times \frac{297.3}{322.7}$$

$$= 2.49 \pm \frac{3.94}{4.28} = \frac{6.43}{-1.79}(\text{MPa})$$

由表12-4,$l/4$ 截面板自重弯矩 $M_{G1} = 118.12\text{kN} \cdot \text{m} = 118.12 \times 10^6\text{N} \cdot \text{mm}$,则由板自重$l/4$截面产生的上下缘应力为:

下缘应力
上缘应力 $\frac{\sigma_{下}}{\sigma_{上}} = \frac{M_{G1}}{I_0} \times \frac{y_{01l}}{y_{01u}} = \frac{118.12 \times 10^6}{1.567\,6 \times 10^{10}} \times \frac{-297.3}{322.7} = \frac{-2.24}{2.43}(\text{MPa})$

放松预应力钢绞线时,由预加力及板自重共同作用下板上下缘产生的法向应力为:

下缘应力 $\quad\sigma_{下} = 6.43 - 2.24 = 4.19(\text{MPa})$
上缘应力 $\quad\sigma_{上} = -1.79 + 2.43 = 0.64(\text{MPa})$

截面上下缘均为压应力，且小于 $0.7f'_{ck} = 0.7 \times 20.1 = 14.07(\text{MPa})$，符合《公桥规》要求。

(3) 支点截面

预加力产生的支点截面上下缘的法向应力为：

下缘应力
上缘应力
$$\frac{\sigma_{下}}{\sigma_{上}} = \frac{N_{p0}}{A_0} \pm \frac{N_{p0}e_{p0}}{I_0} \times \frac{y_{0l1}}{y_{0lu}}$$

$$\sigma_{p0} = \sigma_{con} - \sigma_{lI} + \sigma_{l4} = \sigma_{con} - (\sigma_{l2} + \sigma_{l3} + \sigma_{l4} + 0.5\sigma_{l5}) + \sigma_{l4}$$

$$= \sigma_{con} - \sigma_{l2} - \sigma_{l3} - 0.5\sigma_{l5}$$

$$= 1\,302 - 15.6 - 30 - 0.5 \times 38.45$$

$$= 1\,237.18(\text{MPa})$$

$$N_{p0} = \sigma_{p0}A_p - \sigma_{l6}A_s = 1\,237.18 \times 690.9 - 108.77 \times 565.6 = 793\,247.35(\text{N})$$

$$e_{p0} = \frac{\sigma_{p0}A_p y_p - \sigma_{l6}A_s y_s}{N_{p0}} = 257.3\,\text{mm}$$

下缘应力 $\sigma_{下}$
上缘应力 $\sigma_{上}$
$$\frac{\sigma_{下}}{\sigma_{上}} = \frac{N_{p0}}{A_0} \pm \frac{N_{p0}e_{p0}}{I_0} \times \frac{y_{0l1}}{y_{0lu}}$$

$$= \frac{793\,247.35}{324\,453} \pm \frac{793\,247.35 \times 257.3}{1.567\,6 \times 10^{10}} \times \frac{297.3}{322.7}$$

$$= 2.44 \pm \frac{3.86}{4.19} = \frac{6.30}{-1.75}(\text{MPa})$$

板自重在支点截面产生的弯矩为0，因此，支点截面跨中法向应力为：

下缘应力 $\quad\sigma_{下} = 6.30\,\text{MPa}$
上缘应力 $\quad\sigma_{上} = -1.75\,\text{MPa}$

下缘应力 $\sigma_{下} = 6.30\,\text{MPa} < 0.7f'_{ck} = 0.7 \times 20.1 = 14.07(\text{MPa})$。跨中、$l/4$、支点截面在放松预应力钢绞线时板上下缘应力计算结果汇总于表12-8。

短暂状态空心板截面正应力汇总表 表12-8

截面位置		跨中截面		$l/4$ 截面		支点截面	
		$\sigma_{下}$	$\sigma_{上}$	$\sigma_{下}$	$\sigma_{上}$	$\sigma_{下}$	$\sigma_{上}$
作用种类	预加力	-1.8	6.48	-1.79	6.43	-1.75	6.3
	板自重	3.24	-2.99	2.43	-2.24	0	0
总应力值(MPa)		1.44	3.49	0.64	4.19	-1.57	6.30
压应力极限 $0.7f'_{ck}=14.07\,\text{MPa}$		14.07	14.07	14.07	14.07	—	14.07

注：表中负值为拉应力，正值为压应力，压应力均满足《公桥规》要求。

由上述计算，在放松预应力钢绞线时，支点截面上缘拉应力为：

$$\sigma_{\pm} = 1.75\text{MPa} \quad \begin{matrix} > 0.7f'_{tk} = 0.7 \times 2.01 = 1.407(\text{MPa}) \\ < 1.15f'_{tk} = 1.15 \times 2.01 = 2.312(\text{MPa}) \end{matrix}$$

按《公桥规》7.2.8 条,预拉区(截面上缘)应配置纵向钢筋,并应按以下原则配置:

当 $\sigma_{\pm} \leqslant 0.7f'_{tk}$ 时,预拉区应配置其配率筋不小于 0.2% 的纵向钢筋;

当 $\sigma_{\pm} = 1.15f'_{tk}$ 时,预拉区应配置其配率筋不小于 0.4% 的纵向钢筋;

当 $0.7f'_{tk} < \sigma_{\pm} < 1.15f'_{tk}$ 时,预拉区应配置的纵向钢筋配筋率按以上两者直线内插取得。

上述配筋率为 $\dfrac{A'_s}{A}$,A'_s 为预拉区普通钢筋截面积,A 为截面毛截面面积,$A = 317\,430\text{mm}^2$,两者内插得到 $\sigma_{\pm} = 1.75\text{MPa}$ 时的纵向钢筋配筋率为 0.002 76,则 $A'_s = 0.002\,76 \times 317\,430 = 876.1(\text{mm}^2)$。

预拉区的纵向钢筋宜采用带肋钢筋,其直径不宜大于 14mm,现采用 HRB400 钢筋,8⌀12,则 $A'_s = 8 \times \dfrac{\pi \times 12^2}{4} = 904.8(\text{mm}^2)$,大于 876.1mm^2,满足要求,布置在空心板支点截面上边缘,如图 12-25 所示。

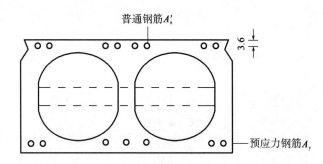

图 12-25 空心板支点截面钢筋布置图(尺寸单位:cm)

为了防止支点截面上缘拉应力过大,还可采用降低支点截面预压力的方法,即支点附近设置套管,使预应力钢绞线与混凝土局部隔离,以不传递预压力。设支点截面附近仅有 5 根钢绞线传递预压力,另 2 根隔离,则此时空心板上缘拉应力将减为 $\sigma_{\pm} = 1.75 \times \dfrac{5}{7} = 1.25(\text{MPa}) < 0.7f'_{tk} = 1.407\text{MPa}$,按《公桥规》要求,预拉区需配置配筋率不小于 0.2% 的纵向普通钢筋,其值为 $0.002 \times 317\,430 = 634.9(\text{mm}^2)$,则可采用 6⌀12 钢筋,$A'_s = 6 \times \dfrac{\pi \times 12^2}{4} = 678.6(\text{mm}^2)$。

13) 最小配筋率复核

按《公桥规》9.1.12 条,预应力混凝土受弯构件最小配筋率应满足下列要求:

$$\frac{M_{ud}}{M_{cr}} \geqslant 1.0$$

式中,M_{ud} 为受弯构件正截面承载力设计值,由计算得 $M_{ud} = 567.86\text{kN}\cdot\text{m}$;$M_{cr}$ 为受弯构件正截面开裂弯矩值,$M_{cr} = (\sigma_{pc} + rf_{tk})W_0$,其中 $r = \dfrac{2S_0}{W_0}$;σ_{pc} 为扣除全部预应力钢筋和普通钢筋合力 N_{p0} 在构件抗裂边缘产生的混凝土预压应力,已得 $\sigma_{pc} = 6.11\text{MPa}$;$S_0$ 为换算截面重心

轴以上部分对重心轴的静矩,其值为:

$$S_0 = 990 \times 322.2 \times \frac{321.8}{2} - 2 \times 380 \times \frac{(321.8 - 80 - 192)^2}{2} - 2 \times \frac{\pi \times 380^2}{8} \times$$

$$[80.6 + (322.2 - 80 - 190)] = 35\,290\,870(\text{mm}^3)$$

W_0 为换算截面抗裂边缘的弹性抵抗矩,由计算得 $W_0 = W_{0 \text{下}} = 52.45 \times 10^6 \text{mm}^3$;$f_{tk}$ 为混凝土轴心抗拉标准值,C50,$f_{tk} = 2.65\text{MPa}$,

$$r = \frac{2S_0}{W_0} = \frac{2 \times 35\,290\,870}{52.495 \times 10^6} = 1.345$$

代入 M_{cr} 计算式,得:

$$M_{cr} = (\sigma_{pc} + rf_{tk})W_0 = (6.01 + 1.345 \times 2.65) \times 52.495 \times 10^6$$

$$= 502.60 \times 10^6 (\text{N} \cdot \text{mm}) = 502.60 \text{kN} \cdot \text{m}$$

$$\frac{M_{ud}}{M_{cr}} = \frac{567.86}{502.60} = 1.13 > 1.0$$

满足《公桥规》要求。

按《公桥规》9.3.5 条,部分预应力受弯构件中普通受拉钢筋的截面面积不应小于 $0.003bh_0$。其中普通受拉钢筋:

$$A_s = 565.5 \text{mm}^2 > 0.003bh_0 (0.003bh_0 = 0.003 \times 178 \times 580 = 483.7 \text{mm}^2)$$

这里 b 采用空心板等效工字形截面的肋宽 $b = 278\text{mm}$,计算结果满足《公桥规》要求。

思考练习题

1. 何谓预应力钢筋的有效预应力?对先张法、后张法构件,各阶段的预应力损失应如何组合?
2. 试写出预应力混凝土矩形截面梁正截面强度计算的基本公式、适用条件、强度复核的步骤。
3. 对预应力混凝土受弯构件的相对界限高度 ξ_{pb} 与普通钢筋混凝土的 ξ_b 有何不同?
4. 预应力混凝土受弯构件的斜截面强度与普通钢筋混凝土构件的有什么相同、不同之处?
5. 斜截面抗剪强度计算公式中 α_1、α_2、α_3 三个系数的物理意义是什么?
6. 解释下列名词术语:截面核心;截面上下核心距;净截面;毛截面;换算截面。
7. 试证明预应力混凝土梁的消压弯矩 M_0,$M_0 = N_y(K_s + e_{yi})$,N_y 为预应力筋永存应力,K_s 为截面上核心距,e_{yi} 为预应力筋合力至换算截面重心轴的距离。
8. 预应力混凝土简支梁设计的主要内容有哪些?基本步骤是什么?
9. 预应力混凝土受弯构件在施工阶段和使用阶段的受力有何特点?
10. 预应力混凝土梁的优越性是什么?决定预应力混凝土梁破坏弯矩的主要因素是什么?

11. 何谓预应力损失？何谓张拉控制应力？张拉控制应力的高低对构件有何影响？

12.《公桥规》中考虑的预应力损失主要有哪些？引起各项预应力损失的主要原因是什么？如何减小各项预应力损失？

13. 何谓预应力钢筋的松弛？钢筋松弛损失计算有何特点？

14. 在构件的受压区配置预应力钢筋对构件的受力特性有何影响？

15. 为什么要进行构件的应力计算？应力计算包括哪些计算项目？如何选择应力计算的截面？

16. 预应力混凝土为什么要进行抗裂性计算？构件的抗裂性主要通过什么来控制？斜截面抗裂性计算中如何选择计算截面？抗裂性计算与应力计算有何异同点？

17. 什么是截面抗弯效率指标？何谓束界？预应力钢筋的布置原则是什么？如何确定预应力钢筋的弯起点？如何确定预应力钢筋的弯起角度？预应力钢筋弯起的曲线形状主要有哪些？

18. 何谓预应力钢筋的传递长度？何谓预应力钢筋的锚固长度？

19. 预应力混凝土构件的挠度由哪些部分组成？何谓上拱度？何谓预拱度？《公桥规》中如何考虑荷载长期作用的影响？《公桥规》中如何设置预拱度？

20. 计算习题 20 图所示后张法预应力混凝土梁截面的净截面及换算截面几何特性。已知预留孔道直径为 50mm^2，每束高强钢丝束为 $24\phi^P 5$，混凝土强度等级为 C50。高强钢丝和混凝土的弹性模量分别为 $E_p = 2.05 \times 10^6 \text{MPa}$ 和 $E_c = 3.45 \times 10^4 \text{MPa}$。

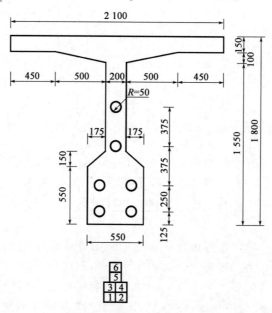

习题 20 图 （尺寸单位:mm）

21. 后张法预应力混凝土等截面简支 T 形截面梁如习题 21 图所示，主梁预制长度为 $L = 14.60\text{m}$。主梁采用 C50 混凝土，配置了 3 束预应力钢筋（每束 6 根 7 股 $\phi^s 15.2$ 钢绞线），夹片式锚具，有预压张拉，预埋金属波纹管成孔，孔洞直径 $d = 67\text{mm}$。主梁各计算控制截面的几何特性如下表所示，全部预应力钢筋截面积 $A_p = 2\,919\text{mm}^2$。当混凝土达到设计强度后，分批张拉各预应力钢绞线,两端张拉。先张拉预应力钢绞线束 N1,再同时张拉 N2、N3。张拉控制应力为 $\sigma_{con} = 0.75 f_{pk} = 0.75 \times 1\,860 = 1\,395(\text{MPa})$。主梁所处桥位的环境年平均相对湿度为 80%。主梁的自重集度为 $G_1 = 7.85\,\text{kN/m}$。试计算预应力损失。

截面几何特征表 习题 21 表

截面 项目	跨中截面	$l/4$ 截面	距支座 $h/2$ 截面	支点处截面
N1 钢束曲线与 x 轴夹角 α 值(rad)	0	0.087	0.175	0.192
净截面面积 $A_n(\text{mm}^2)$	304 173	304 173	304 173	304 173
净截面对自身重心轴的惯性矩 $I_n(\text{mm}^4)$	$29.822\,326 \times 10^9$	$29.936\,277 \times 10^9$	$29.786\,814 \times 10^9$	
净截面重心轴距主梁上边缘距离 $y_{nu}(\text{mm})$	352	354	357	
预应力钢束重心轴距净截面重心的距离 $e_{pn}(\text{mm})$	418	362	264	

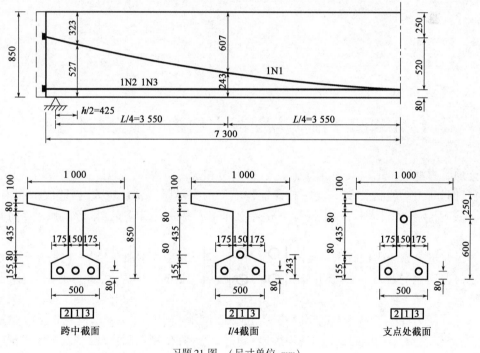

习题 21 图 (尺寸单位:mm)

第13章
部分预应力混凝土受弯构件

13.1 部分预应力混凝土结构的概念与特点

全预应力混凝土结构虽有抗裂刚度大、抗疲劳和抗渗漏等特点，但在工程实践中也发现一些缺点，例如，主梁的反拱大，以至于桥面铺装施工的实际厚度变化较大，易造成桥面损坏，影响行车顺适；预加力过大时，锚下混凝土横向拉应变超出极限拉应变，出现沿预应力钢筋方向不能恢复的裂缝。

为了改进全预应力混凝土结构的不足，出现了部分预应力混凝土结构，它是介于全预应力混凝土与普通钢筋混凝土之间的结构，根据要求施加适量的预应力，配置普通钢筋，以保证承载力要求。

部分预应力混凝土结构不仅充分发挥预应力钢筋的作用，而且注意利用非预应力钢筋的作用，进一步改善预应力混凝土的使用性能；同时，允许在使用期间出现裂缝，扩大了预应力混凝土结构的应用范围，促进了预应力混凝土结构设计思想的发展，使设计人员可以根据结构使用要求来选择预应力度的高低，进行合理的结构设计。

13.2 部分预应力混凝土结构分类与受力特性

1) 部分预应力混凝土结构分类

《公桥规》按预应力度 λ 的不同，将预应力混凝土结构分为两类：全预应力混凝土结构和部分预应力混凝土结构。部分预应力混凝土结构又分为 A 类构件和 B 类构件两类。

(1) A 类预应力混凝土构件

在使用阶段控制截面的正截面允许出现拉应力，拉应力较小，不会出现裂缝，"拉而无裂"。A 类构件的设计与全预应力混凝土结构完全相同。

(2) B 类预应力混凝土构件

在使用阶段控制截面的正截面允许裂缝出现，裂缝宽度在限制范围内，"裂而有限"。

本章主要介绍 B 类预应力混凝土构件。

2) 部分预应力混凝土构件的荷载—挠度曲线

试验研究表明不同预应力度条件下梁的荷载—挠度曲线如图 13-1 所示。

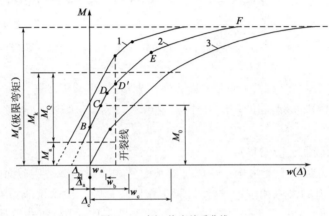

图 13-1 弯矩-挠度关系曲线

从图 13-1 中可以看出，部分预应力混凝土梁的受力特性表现在：

(1) 在荷载作用较小时，部分预应力混凝土梁(曲线 2)受力特性与全预应力混凝土梁(曲线 1)相似：在自重与有效预加力 N_{pe}（扣除相应的预应力损失）作用下，它具有反拱度 Δ_b，但其值较全预应力混凝土梁的反拱度 Δ_a 小——线弹性特性。

(2) 随着外加荷载作用增加，弯矩 M 达到 B 点，这时表示外荷载作用下产生的梁下挠度与预应力反拱度相等，两者正好相互抵消，此时梁的挠度为 0——零挠度。

(3) 当荷载作用继续增加，达到曲线 2 的 C 点时，外荷载作用产生的梁底混凝土拉应力正好与有效预压应力 σ_{pc} 互相抵消，使梁底受拉边缘的混凝土应力为 0，此时相应的外荷载作用产生的弯矩，称为消压弯矩 M_0——零应力状态。

(4) 梁的截面下边缘消压后，如继续加载至 D 点，混凝土的边缘拉应力达到极限抗拉强度。受拉区混凝土进入塑性阶段，构件的刚度下降，达到 D' 点时表示构件即将出现裂缝，此时相应的弯矩就称为部分预应力混凝土构件的抗裂弯矩 M_{pcr}，它就相当于相应的钢筋混凝土梁截面的抗裂弯矩 M_{cr}，即 $M_{cr} = M_{pcr} - M_0$——出现裂缝阶段。

(5) 外荷载作用加大，从 D' 点开始，裂缝开展，刚度继续下降，挠度迅速增加。到达 E 点时，受拉钢筋屈服。E 点以后裂缝进一步扩展，刚度进一步降低，挠度增加速度更快，直到 F 点，这时构件达到极限承载能力状态而破坏——破坏阶段。

3) 实现部分预应力的可行方法

(1) 全部采用高强钢筋，将其中的一部分张拉到最大容许应力，保留其余一部分作为非预应力钢筋，这样可以节省锚具，减少张拉工作量。

(2) 将全部预应力钢筋都张拉到一个较低的应力水平。

(3) 采用张拉的预应力钢筋与普通钢筋的混合配筋。构件中的预应力筋可以平衡一部分荷载,提高抗裂度,减少挠度,并提供部分或大部分的承载力;非预应力钢筋则可分散裂缝,提高承载能力和破坏时的延性,以及加强结构中难以配置预应力钢筋的那些部分。

对于 B 类预应力混凝土构件,工程上采用第三种配筋方法(混合配筋)最多,由于采用了预应力高强钢筋与非预应力普通钢筋的混合配筋,既具有两种配筋的优点,又基本排除了两者的缺点。

13.3　B 类部分预应力混凝土受弯构件的计算

部分预应力混凝土结构与全预应力混凝土结构的设计计算有许多的相同点,但也有一些不同。

(1) 相同点:持久状态正截面和斜截面承载力计算,局部承压计算,预应力损失估算。

(2) 不同点:使用阶段截面的正应力计算,裂缝宽度计算(验算),变形计算,疲劳计算,截面配筋计算,构造要求。

1) 使用阶段的截面正应力计算

B 类部分预应力混凝土受弯构在使用阶段截面已经开裂。开裂后截面的中性轴位置和几何特性取决于预加力的大小和位置,这使计算工作比较复杂。但从预应力混凝土梁的弯矩-挠度曲线可以明确看出,梁开裂后仍具有一个良好的弹性工作阶段,即开裂弹性阶段。因此,部分预应力混凝土梁开裂后使用阶段的应力计算,仍采用弹性分析方法计算。

①部分预应力混凝土梁截面开裂后的应力状态,与钢筋混凝土大偏心受压构件很相似。钢筋混凝土大偏心受压构件截面开裂后,可以用钢筋混凝土结构在使用荷载阶段处于弹性受力的特点求解钢筋应力与混凝土的应力。但应该注意到,当外力为零时,钢筋混凝土大偏心受压构件截面混凝土应力均等于零(称为"零应力"状态)。

②B 类预应力混凝土受弯构件截面上由作用产生的弯矩 M_k,虽然可以用等效的偏心压力来代替,但是偏心压力所产生的应力效应,并不能直接用上述钢筋混凝土大偏心受压构件求解应力的方法来求解,这是因为部分预应力混凝土构件尚存在着预加力的作用,所以,即使截面上没有作用,但是由于预加力的作用,梁的截面上已经存在着由预加力所引起的混凝土正应力。

③鉴于钢筋混凝土大偏心受压构件求解截面应力的公式是在"零应力"状态下建立的,如果能把这个预加力引起的截面应力的特点加以考虑,从计算方法上进行某些处理,将截面上由预加力引起的混凝土压应力退压成"零压力"状态,暂时先消除预加力的影响,就可以借助大偏心受压构件的计算方法来求解截面上钢筋和混凝土的应力。

④预应力混凝土受弯构件开裂截面的应力计算,就是把在作用弯矩 M_k 和预应力钢筋及非预应力钢筋合力 N_p 共同作用下的受弯构件,转化为轴向力作用点距截面重心轴 e_{0N} 的钢筋混凝土偏心受压构件进行计算。对后张法预应力连续梁等超静定结构,上述外弯矩 M_k 还应计入由预加力引起的次弯矩 M_{p2}。

如图 13-2 所示,对后张法 T 形截面 B 类受弯构件截面弯矩超过开裂弯矩时($M > M_{cr}$),将开裂截面正应力状态分解成如下几个阶段来分析。

（1）有效预加力 N_{pe} 作用（获得零应力以前的应力状态）

在有效预加力 N_{pe} 单独作用下的应变图，如图 13-2b)所示线①。此时，受拉区和受压区预应力钢筋中的拉应力为不计梁自重作用的有效预拉力：

$$\sigma_{pe} = \sigma_{con} - \sigma_l \tag{13-1}$$

$$\sigma'_{pe} = \sigma'_{con} - \sigma'_l \tag{13-2}$$

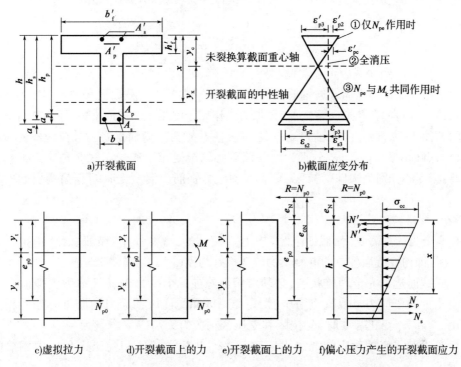

图 13-2 大偏心受压等效过程

在配有普通钢筋的预应力混凝土构件中，由于混凝土的收缩和徐变，使普通钢筋产生与预加力相反的内力，从而减少了受拉区混凝土的法向预压应力。为了简化计算，在以下的计算中，普通钢筋的应力均近似取为混凝土收缩和徐变引起的预应力损失值（图 13-3）。此时，在截面全部预应力钢筋和普通钢筋的合力 N_p 作用下的截面下缘和上缘预应力钢筋重心处混凝土的预压应力 σ_{pc} 和 σ'_{pc} 计算式为：

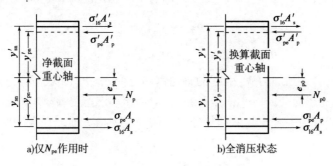

图 13-3 预应力钢筋和普通钢筋合力及其偏心距

$$\sigma_{pc} = \frac{N_p}{A_n} + \frac{N_p e_{pn}}{I_n} y_{pn} \tag{13-3a}$$

$$\sigma'_{pc} = \frac{N_p}{A_n} - \frac{N_p e_{pn}}{I_n} y'_{pn} \tag{13-3b}$$

$$N_p = \sigma_{pe} A_p + \sigma'_{pe} A'_p - \sigma_{l6} A_s - \sigma'_{l6} A'_s \tag{13-4}$$

式中，σ_{pe} 和 σ'_{pe} 分别按式(13-1)和式(13-2)计算；e_{pn} 按式(13-5)计算。

$$e_{pn} = \frac{\sigma_{pe} A_p y_{pn} - \sigma'_{pe} A'_p y'_{pn} - \sigma_{l6} A_s y_{sn} + \sigma'_{l6} A'_s y'_{sn}}{\sigma_{pe} A_p + \sigma'_{pe} A'_p - \sigma_{l6} A_s - \sigma'_{l6} A'_s} \tag{13-5}$$

式中，e_{pn} 为 N_p 到净截面重心轴偏心距；A_n 为净截面面积，即为扣除管道等削弱部分后的混凝土全部截面面积与纵向普通钢筋截面面积换算成混凝土的截面面积之和；I_n 为净截面惯性矩，对后张法而言，在孔道灌浆之前，预加力是作用于混凝土净截面上的，所以在计算中应取用净截面的面积、惯性矩、中心轴和偏心矩；y_{pn}、y'_{pn} 分别为净截面重心至受拉区或受压区预应力钢筋合力点处的距离；y_{sn}、y'_{sn} 分别为净截面重心轴至受拉区或受压区非预应力钢筋合力点处的距离；σ_{l6}、σ'_{l6} 分别为构件受拉区、受压区预应力钢筋由混凝土收缩和徐变引起的预应力损失。

(2) 全消压状态(获得虚拟零应力状态)

这是一个为计算需要(使截面呈"零应力"状态)的"虚拟作用"阶段。在"虚拟作用"下，全截面消压，即构件截面各点的混凝土的应变恰好为零(图13-2中线②)。

为了使截面达到完全消压状态，必须对截面施加一个拉力 N_{p0}（又称为虚拟荷载），使之消除混凝土的预压应力。混凝土消压后，在受拉区和受压区预应力钢筋重心处混凝土应变值分别由 ε_{pc} 和 ε'_{pc} 变化为零时，受拉区和受压区预应力钢筋应变增量为 $(-\varepsilon_{p2})$ 和 $(-\varepsilon'_{p2})$，其绝对值等于其重心处对应的混凝土应变 ε_{pc}、ε'_{pc}。故受拉区和受压区预应力钢筋的拉应力增量为：

$$\sigma_{p2} = E_p \cdot (-\varepsilon_{p2}) = E_p \varepsilon_{pc} = \alpha_{Ep} \sigma_{pc} \tag{13-6}$$

$$\sigma'_{p2} = E_p \cdot (-\varepsilon'_{p2}) = E_p \varepsilon'_{pc} = \alpha_{Ep} \sigma'_{pc} \tag{13-7}$$

式中，α_{Ep} 为预应力钢筋弹性模量与混凝土弹性模量的比值。

这里假定受拉和受压区预应力钢筋为同一类钢筋，同时 $\alpha_{Ep} = E_p / E_c$。

在全消压状态下，受拉区和受压区预应力钢筋中的总拉应力 σ_{p0}、σ'_{p0} 分别为：

$$\sigma_{p0} = \sigma_{pe} + \sigma_{p2} = \sigma_{con} - \sigma_l + \alpha_{Ep} \sigma_{pc} \tag{13-8}$$

$$\sigma'_{p0} = \sigma'_{pe} + \sigma'_{p2} = \sigma'_{con} - \sigma'_l + \alpha_{Ep} \sigma'_{pc} \tag{13-9}$$

全消压状态下，有效预加力 N_{pe} 引起的普通钢筋应变 ε_{s1}、ε'_{s1} 消失。但由于混凝土收缩徐变变形引起的普通钢筋压应变 ε_{s2}、ε'_{s2} 依然存在，所以消压状态下受拉、受压区普通钢筋依然存在压应力，其值均近似取为混凝土收缩和徐变引起的预应力损失值。存在的压力 N_{s2}、N'_{s2} 为：

$$N_{s2} = -\sigma_{l6} A_s \tag{13-10}$$

$$N'_{s2} = -\sigma'_{l6} A'_s \tag{13-11}$$

全消压状态下，预应力钢筋和普通钢筋合力 N_{p0} 就为：

$$N_{p0} = \sigma_{p0}A_p - \sigma_{l6}A_s + \sigma'_{p0}A'_p - \sigma'_{l6}A'_s \tag{13-12}$$

N_{p0} 作用点距截面受压边缘距离[图13-2c)]为：

$$e_{p0} = \frac{\sigma_{p0}A_p y_{pn} - \sigma'_{p0}A'_p y'_{pn} - \sigma_{l6}A_s y_{sn} + \sigma'_{l6}A'_s y'_{sn}}{\sigma_{p0}A_p - \sigma_{l6}A_s + \sigma'_{p0}A'_p - \sigma'_{l6}A'_s} \tag{13-13}$$

对于先张法预应力混凝土构件，由式（13-12）计算 N_{p0} 时，应取：

$$\sigma_{p0} = \sigma_{con} - \sigma_l + \sigma_{l4} \tag{13-14}$$

$$\sigma'_{p0} = \sigma'_{con} - \sigma'_l + \sigma'_{l4} \tag{13-15}$$

(3) 开裂截面应力计算中虚拟作用的处理（对虚拟状态处理）

虚拟拉力（N_{p0}）是为了计算处理而虚设的，因此，最终应消除其影响，必须在预应力钢筋和普通钢筋合力作用点处施加一个与 N_{p0} 大小相等、方向相反的作用力（$-N_{p0}$）[图13-2d)]。此时，作用于构件开裂截面的弯矩值 M_k 和偏心压力 N_{p0}，可用一个等效的偏心压力 R 作用于构件开裂后的换算截面上，并由此可求得 R 的大小及距截面上边缘的距离 e_N（图13-4）。

现取 $R = N_{p0}$，由隔离体对 N_{p0} 作用点的力矩平衡：

$$M_k = R(h_{ps} + e_N) = N_{p0}(h_{ps} + e_N)$$

故

$$e_N = \frac{M_k}{N_{p0}} - h_{ps} \tag{13-16}$$

这样，承受预加力合力 N_p 和弯矩 M_k 作用的预应力混凝土 B 类受弯构件就转化为承受距受压边缘的偏心距为 e_N 的压力 $R = N_{p0}$ 的钢筋混凝土偏心受压构件（图13-4）。

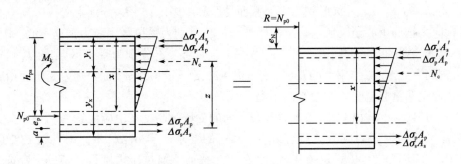

图 13-4　等效的钢筋混凝土偏心受压构件

(4) 按钢筋混凝土结构大偏心受压构件计算梁开裂截面的受压区高度（建立大偏压构件状态）

开裂后的 B 类预应力混凝土受弯构件，按钢筋混凝土偏心受压构件计算时，采用以下假定：

①截面变形符合平截面假定；
②受压混凝土正应力分布取三角形；
③不考虑受拉区混凝土参加工作，拉力全部由钢筋承担。

假定开裂截面的中性轴位于肋板内（图13-5），按内外力对偏心压力 N_{p0} 作用点取矩为0，整理后得到开裂截面受压区高度 x 的计算方程：

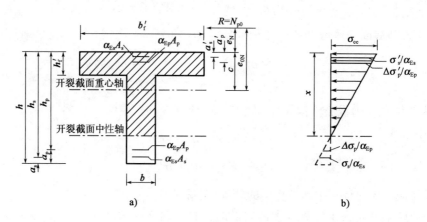

图 13-5 开裂截面及应力图

$$Ax^3 + Bx^2 + Cx + D = 0 \tag{13-17}$$

$$A = b \tag{13-18}$$

$$B = 3be_N \tag{13-19}$$

$$C = 3b_0 h'_f(2e_N + h'_f) + 6\alpha_{Ep}(A_p g_p + A'_p g'_p) + 6\alpha_{Es}(A_s g_s + A'_s g'_s) \tag{13-20}$$

$$D = b_0 h'^2_f(3e_N + 2h'_f) - 6\alpha_{Ep}(A_p h_p g_p + A'_p a'_p g'_p) - 6\alpha_{Es}(A_s h_s g_s + A'_s a'_s g'_s) \tag{13-21}$$

式中，b 为 T 形和工字形截面的肋板宽度或矩形截面的宽度；e_N 为 N_{p0} 作用点至截面受压边缘的距离，N_{p0} 位于截面外为正，位于截面内为负；b_0 为 T 形和工字形截面受压翼缘宽度与肋板宽度之差，$b_0 = b'_f - b$；h'_f 为 T 形和工字形截面受压翼缘厚度；h_p、h_s 分别为受拉区预应力钢筋重心、非预应力钢筋重心至受压区边缘的距离；g_p、g_s 分别为受拉区预应力钢筋重心、非预应力钢筋重心至 N_{p0} 作用点的距离，$g_p = h_p + e_N$，$g_s = h_s + e_N$；g'_p、g'_s 分别为受压区预应力钢筋重心、非预应力钢筋重心至 N_{p0} 作用点的距离，$g'_p = a'_p + e_N$，$g'_s = a'_s + e_N$；a'_p、a'_s 分别为受压区预应力钢筋重心、普通钢筋重心至受压区边缘的距离。

使用式(13-17)~式(13-21)来求解开裂截面的受压区高度 x 中应注意：

① 受压区普通钢筋的应力应符合 $\alpha_{Es}\sigma_{cc} \leqslant f'_{sd}$ 的要求。当 $\alpha_{Es}\sigma_{cc} > f'_{sd}$ 时，式(13-20)和式(13-21)中的 A'_s 应以 $f'_{sd}A'_s/(\alpha_{Es}\sigma_{cc})$ 代替，此处 f'_{sd} 为普通钢筋抗压强度设计值，σ_{cc} 为受压区普通合力点处混凝土压应力，可按式(13-22)计算，但式中的 c 改用该钢筋合力点至开裂截面重心轴的距离。

② 当受压区预应力钢筋为拉应力时，即 $\alpha_{Es}\sigma_{cc} - \sigma'_{p0} < 0$ 时，式(13-20)和式(13-21)中含有 A'_p 项前面的正号应改为负号，此处 σ_{cc} 为受压区预应力钢筋合力点处的混凝土压应力。

③ 当受压区未设预应力钢筋或非预应力钢筋时，式(13-20)和式(13-21)中的 A'_p 项或 A'_s 项令其等于 0。

式(13-17)适用于 T 形和工字形截面的 B 类预应力混凝土受弯构件；对于矩形截面，令有关计算式中的 h'_f 等于零即可。

(5)开裂截面混凝土压应力

《公桥规》依照消压分析法提出允许开裂的 B 类预应力混凝土受弯构件，由作用标准值产生的开裂截面混凝土压应力计算式为：

$$\sigma_{cc} = \frac{N_{p0}}{A_{cr}} + \frac{N_{p0}e_{0N}c}{I_{cr}} \leqslant 0.5f_{ck} \quad (13\text{-}22)$$

$$e_{0N} = e_N + c \quad (13\text{-}23)$$

$$e_N = \frac{M_k}{N_{p0}} - h_{ps} \quad (13\text{-}24)$$

$$h_{ps} = \frac{\sigma_{p0}A_p h_p - \sigma_{l6}A_s h_s + \sigma'_{p0}A'_p a'_p - \sigma'_{l6}A'_s a'_s}{N_{p0}} \quad (13\text{-}25)$$

式中，N_{p0}为混凝土法向预应力等于0时预应力钢筋和普通钢筋的合力，先张法构件和后张法构件均按公式(13-22)计算，其中的σ_{p0}、σ'_{p0}为构件受拉区、受压区预应力钢筋合力点处混凝土法向应力等于零时预应力钢筋的应力，先张法构件按式(13-24)和式(13-25)计算；后张法构件按式(13-8)和式(13-9)计算；c为截面受压区边缘至开裂换算截面重心轴的距离；e_{0N}为N_{p0}作用点至开裂截面重心轴距离；e_N为N_{p0}作用点至截面受弯区边缘的距离，N_{p0}位于截面之外为正，N_{p0}位于截面之内为负；h_{ps}为预应力钢筋与普通钢筋合力点至截面受压区边缘的距离；h_p、a'_p分别为截面受拉区、受压区的预应力钢筋合力点至截面受压区边缘距离；h_s、a'_s分别为截面受拉区、受压区的非预应力钢筋合力点至截面受压区边缘距离；A_{cr}为开裂截面换算截面面积；I_{cr}为开裂截面换算截面惯性矩。

(6) 开裂截面预应力钢筋的应力

开裂截面预应力钢筋的应力增量为：

$$\Delta\sigma_p = \alpha_{Ep}\left[\frac{N_{p0}}{A_{cr}} - \frac{N_{p0}e_{0N}(h_p - c)}{I_{cr}}\right] \quad (13\text{-}26)$$

开裂截面受拉区预应力钢筋总拉应力为：

$$\sigma_p = \sigma_{p0} + \Delta\sigma_p$$

式中，σ_{p0}为构件受拉区预应力钢筋合力点处混凝土法向应力等于零时预应力钢筋的应力，后张法构件按式(13-8)计算，先张法构件按式(13-14)计算；其他符号意义与式(13-19)~式(13-22)相同。

使用阶段开裂截面受拉预应力钢筋的计算总拉应力应满足：

对体内钢绞线、钢丝：

$$\sigma_{p0} + \Delta\sigma_p \leqslant 0.65f_{pk} \quad (13\text{-}27)$$

对预应力螺纹钢筋：

$$\sigma_{p0} + \Delta\sigma_p \leqslant 0.80f_{pk} \quad (13\text{-}28)$$

预应力混凝土受弯构件受拉区的普通钢筋，其使用阶段的应力很小，可不必验算。

2) 裂缝宽度计算

B类部分预应力受弯构件，在正常使用阶段允许出现裂缝，因此，控制裂缝宽度，使之不超过规范限值，就成为B类构件计算的一项重要内容。

(1) 裂缝宽度计算

对B类受弯构件预应力混凝土，《公桥规》采用的最大裂缝宽度计算式为：

$$W_{tk} = C_1 C_2 C_3 \frac{\sigma_{ss}}{E_s}\frac{c+d}{0.36+1.4\rho_{te}}(\text{mm}) \quad (13\text{-}29)$$

式中,ρ_{te} 为纵向受拉钢筋有效配筋率,$\rho_{te} = \dfrac{A_s}{A_{te}}$,当 $\rho > 0.1$ 时,取 $\rho = 0.1$;当 $\rho < 0.01$ 时,取 $\rho = 0.01$;C_1 为钢筋表面形状系数,对光面钢筋,$C_1 = 1.4$;对带肋钢筋,$C_1 = 1.0$;C_2 为作用(或荷载)长期效应影响系数,$C_2 = 1 + 0.5\dfrac{N_l}{N_s}$,其中 N_l 和 N_s 分别为按作用(或荷载)长期效应组合和短期效应组合计算的内力值(弯矩或轴向力);C_3 为与构件受力性质有关的系数,当为钢筋混凝土板式受弯构件时,$C_3 = 1.15$,其他受弯构件 $C_3 = 1.0$;d 为纵向受拉钢筋直径(mm),当用不同直径的钢筋时,d 改为换算直径 d_e,$d_e = \dfrac{\sum n_i d_i^2}{\sum n_i d_i}$;对于混合配筋的预应力混凝土构件,《公桥规》规定,预应力钢筋为由多根钢丝或钢绞线组成的钢丝束或钢绞线束,d_i 为普通钢筋公称直径、钢丝束或钢绞线束的等代直径 d_{pe},$d_{pe} = \sqrt{n}d$,此处,n 为钢丝束中钢丝根数或钢绞线束中钢绞线根数,d 为单根钢丝或钢绞线的公称直径;σ_{ss} 为由作用短期效应组合并考虑长期效应影响引起的开裂截面纵向受拉钢筋的应力,可近似按下式计算:

$$\sigma_{ss} = \frac{M_s - N_{p0}(z - e_p)}{(A_p + A_s)z} \tag{13-30}$$

$$z = \left[0.87 - 0.12(1 - \gamma'_f)\left(\frac{h_0}{e}\right)^2\right]h_0 \tag{13-31}$$

$$\gamma'_f = \frac{(b'_f - b)h'_f}{bh_0} \tag{13-32}$$

$$e = e_p + \frac{M_s}{N_{p0}} \tag{13-33}$$

式中,M_s 为按作用短期效应组合计算的弯矩值;N_{p0} 为混凝土法向应力等于零时预应力钢筋和非预应力钢筋的合力,按式(13-12)计算;z 为受拉区纵向预应力钢筋和非预应力钢筋合力点至截面受压区合力点的距离(图 13-4);γ'_f 为受压翼缘截面积与肋板有效截面积的比值;b'_f、h'_f 为受压翼缘的宽度和厚度,当 $h'_f > 0.2h_0$ 时,取 $h'_f = 0.2h_0$;e_p 为混凝土法向应力等于零时,预应力钢筋和非预应力钢筋的合力 N_{p0} 的作用点至受拉区预应力钢筋和普通钢筋合力点的距离(图 13-4)。

应当指出,对于超静定 B 类预应力混凝土受弯构件,在计算 σ_{ss} 时,尚应考虑由预加力 N_p 产生的次弯矩 M_{p2} 的影响,故式(13-30)和式(13-33)应改写为:

$$\sigma_{ss} = \frac{M_s \pm M_{p2} - N_{p0}(z - e_p)}{(A_p + A_s)z} \tag{13-34}$$

$$e = e_p + \frac{M_s \pm M_{p2}}{N_{p0}} \tag{13-35}$$

式中,当 M_{p2} 与 M_s 方向相同时取正值,相反时取负值。

(2) 裂缝宽度的限值

《公桥规》规定,B 类预应力混凝土构件计算的最大裂缝宽度不应超过下列规定限值:

① 采用预应力螺纹钢筋的预应力混凝土构件,Ⅰ类和Ⅱ类环境条件下为 0.20mm;Ⅲ类和Ⅳ类环境条件下 0.15mm。

② 采用钢丝或钢绞线的预应力混凝土构件,Ⅰ类、Ⅱ类、Ⅲ类和Ⅳ类环境条件下为

0.10mm；Ⅱ类环境条件下不得进行带裂缝的 B 类构件设计。

3）变形计算

《公桥规》规定 B 类预应力混凝土受弯构件的变形计算,应采用作用短期效应组合并考虑长期效应组合的影响,计算原理与全预应力混凝土受弯构件的相同。《公桥规》规定允许开裂的预应力混凝土 B 类构件的抗弯刚度按作用短期效应组合 M_s 分段取用：

在开裂弯矩 M_{cr} 作用下：
$$B_0 = 0.95 E_c I_0 \tag{13-36}$$

在 $M_s - M_{cr}$ 作用下：
$$B_{cr} = E_c I_{cr} \tag{13-37}$$

式中，I_0、I_{cr} 分别为构件全截面换算截面惯性矩和开裂截面换算截面的惯性矩。

4）疲劳计算

部分预应力混凝土受弯构件的疲劳破坏通常是发生在高应力区钢筋断裂,如果构件没有裂缝是不会发生疲劳破坏的,因此构件正截面的疲劳校核主要是验算受拉区钢筋的应力,斜截面的疲劳验算主要是控制箍筋的应力。

受拉钢筋的疲劳按计算应力变化幅 $\Delta\sigma_m = \sigma_{max}^p - \sigma_{min}^p$ 校核,其允许值 $[\Delta\sigma_p]$ 应由试验确定,当缺少该项试验数据时,可参照表 13-1 采用。

钢筋应力变化幅度容许值　　　　　　　　　　　　　表 13-1

钢筋种类	光面圆钢筋	规则变形钢筋	光面预应力钢丝	钢绞线	高强钢筋
$[\Delta\sigma_p]$ (MPa)	250	150	200	200	80

13.4　允许开裂的部分预应力混凝土受弯构件的设计

1）按预应力度 λ 法进行配筋

以 λ 表示预应力度,即：
$$\lambda = M_0 / M_s$$

式中,M_0 被称为消压弯矩,对受弯构件是指其下边缘混凝土的预压应力恰被抵消为零时弯矩,M_0 的表达式为：
$$M_0 = \sigma_{pc} W_0$$

且
$$\sigma_{pc} = \frac{N_{pe}}{A}\left(1 + \frac{e_p \cdot y_x}{i^2}\right)$$

由上述式子整理可得到：
$$N_{pe} = \frac{\lambda M_s}{W_0} \cdot \frac{A}{1 + \dfrac{e_p \cdot y_x}{i^2}} \tag{13-38}$$

$$A_p = \frac{N_{pe}}{\sigma_{con} - \sigma_l} \tag{13-39}$$

式中,σ_{con} 为预应力钢筋的张拉控制应力；σ_l 为预应力总损失值；估算时,对先张法构件可取张拉控制应力的 20%～30%；对后张法构件除摩擦损失外可取 15%～25% 的张拉控制应力。

当由式(13-39)确定了预应力钢筋面积 A_p 后,则由受弯构件正截面承载能力来求所需非预应力钢筋 A_s。例如对于仅在受拉区配置预应力钢筋 A_p 和非预应力钢筋 A_s 的单筋矩形截面,截面宽度为 b,高度为 h,由基本静力平衡方程可得到联立方程:

$$f_{cd}bx = f_{pd}A_p + f_{sd}A_s \tag{13-40}$$

$$\gamma_0 M_d = f_{pd}A_p\left(h - a_p - \frac{x}{2}\right) + f_{sd}A_s\left(h - a_s - \frac{x}{2}\right) \tag{13-41}$$

式中,M_d 为弯矩组合设计值。

由联立方程式(13-40)和式(13-41)可求得受压区高度 x 和非预应力钢筋面积 A_s,求得的 x 应满足 $x \leqslant \xi_b h_0$。

计算步骤(以矩形截面受弯构件为例):

(1) 计算混凝土毛截面的几何特征 A、I、W 和 y_x。

(2) 假定预应力钢筋的合力作用点位置 a_p,求得偏心距 e_p;假定预应力钢筋和普通钢筋合力作用点位置 a,计算有效高度 h_0。

(3) 选择预应力度,一般 $\lambda = 0.6 \sim 0.8$。

(4) 由式(13-39)求得所需预应力钢筋面积,解联立方程式(13-40)和式(13-41)求得相应的非预应力钢筋的面积。

(5) 选择预应力钢筋和普通钢筋并布置在截面上,按正截面抗弯承载能力要求,进行截面复核。

2) 截面配筋设计的名义拉应力法

根据部分预应力混凝土构件在使用阶段的裂缝宽度的限制要求,计算控制截面上预应力钢筋和普通钢筋所需面积的方法。

具体方法:把带裂缝的构件假设为未开裂的构件,用一般材料力学公式算出构件受拉边缘的最大拉应力,因为构件已开裂,所以这个拉应力必定大于混凝土的弯曲抗拉强度,成为名义上的概念,故称为"名义拉应力"。在大量试验的基础上,可以定出对应于不同裂缝宽度限值的容许名义拉应力。

在使用荷载阶段,按匀质未开裂混凝土截面,用材料力学公式计算,仅由使用荷载作用引起的截面边缘最大拉应力 σ_{st} 与相应位置混凝土截面边缘所受的有效预应力 σ_{pc} 叠加,其结果就是相应截面混凝土边缘所受的总的拉应力,也就是名义拉应力。这个名义拉应力值是有限制的,应满足:

$$\sigma_{st} - \sigma_{pc} \leqslant [\sigma_{ct}] \tag{13-42}$$

式中,$[\sigma_{ct}]$ 为混凝土的容许名义拉应力,按式(13-44)、式(13-45)的要求选用。

采用名义拉应力法进行配筋估算步骤为:

(1) 进行混凝土截面几何特性的计算。

(2) 计算 σ_{st}:

$$\sigma_{st} = \frac{M_s}{W} \tag{13-43}$$

式中,M_s 为由荷载短期效应组合产生的弯矩值;W 为截面受拉边缘的弹性抵抗矩,计算时可按毛截面计算。

(3) 确定混凝土的容许名义拉应力 $[\sigma_{ct}]$,根据构件的使用要求及环境条件,根据裂缝宽

度限值规定,$[\sigma_{ct}]$可用以下简单公式计算:

后张法 $$[\sigma_{ct}] = \beta[\sigma'_{ct}] + 4\rho \leqslant \frac{f_{cu,k}}{4} \quad (13-44)$$

先张法 $$[\sigma_{ct}] = \beta[\sigma'_{ct}] + 3\rho \leqslant \frac{f_{cu,k}}{4} \quad (13-45)$$

式中,$[\sigma'_{ct}]$为混凝土基本容许名义拉应力,可查表13-2得到,它仅与预加应力方式、混凝土强度级别及裂缝宽度三个因素有关;β称为构件的高度修正系数,可从表13-3中查到;ρ为受弯构件受拉区非预应力钢筋的配筋率$[\rho = A_s/(bh_0)]$,对于每1%的ρ值,先张法构件容许值$[\sigma'_{ct}]$提高3MPa,后张法构件容许值$[\sigma'_{ct}]$提高4MPa;$f_{cu,k}$为混凝土强度级别。

混凝土基本容许名义拉应力$[\sigma'_{ct}]$(MPa) 表13-2

构件名称	裂缝宽度限值(mm)	混凝土强度等级			构件名称	裂缝宽度限值(mm)	混凝土强度等级		
		C30	C40	≥C50			C30	C40	≥C50
先张法构件	0.1	—	4.6	5.5	后张法构件	0.1	3.2	4.1	5.0
	0.15	—	5.3	6.2		0.15	3.5	4.6	5.6
	0.20	—	6.0	6.9		0.20	2.8	5.1	6.2

注:仅适用于C60以下混凝土。

混凝土容许名义拉应力的构件高度修正系数 表13-3

构件高度(mm)	≤200	400	600	800	≥1 000
修正系数	1.1	1.0	0.9	0.8	0.7

(4)求所需的有效预加力N_{pe}及相应的预应力钢筋面积A_p:

根据式(13-42)计算梁受拉边缘混凝土所需要的有效预压应力σ_{pc},即:

$$\sigma_{pc} \geqslant \sigma_{st} - [\sigma_{ct}] \quad (13-46)$$

$$\sigma_{pc} = N_{pe}\left(\frac{1}{A} + \frac{e_p}{W}\right) \quad (13-47)$$

由式(13-46)和式(13-47),并用式(14-43)表示σ_{st},可得到:

$$N_{pe} \geqslant \frac{\dfrac{M_s}{W} - [\sigma_{ct}]}{\dfrac{1}{A} + \dfrac{e_p}{W}} \quad (13-48)$$

相应所需要的预应力钢筋面积:

$$A_p = \frac{N_{pe}}{\sigma_{con} - \Sigma \sigma_l} \quad (13-49)$$

式中,e_p为预应力钢筋对截面(未开裂)重心轴的偏心距,$e_p = y - a_p$;y、a_p分别为截面重心轴和预应力钢筋重心至截面受拉边缘的距离,a_p可预先假定;A为构件截面面积,可采用毛截面;$\Sigma \sigma_l$为预应力损失总值,估算时对先张法构件可取$(0.2 \sim 0.3)\sigma_{con}$,对后张法构件可取$(0.25 \sim 0.35)\sigma_{con}$。

(5)根据受弯构件正截面承载力要求,计算所需的非预应力钢筋面积A_s。

(6)按正截面强度计算,检查受压区高度x是否满足$x \leqslant \xi_b h_0$,防止超筋破坏。

3)构造要求

考虑到部分预应力混凝土受弯构件的特点,《公桥规》和中国土木工程学会编制的《部分预应力混凝土结构设计建议》(以下简称《建议》)都提出了有关构造要求,简述如下:

(1)部分预应力混凝土梁应采用混合配筋。位于受拉区边缘的非预应力钢筋宜采用直径较小的带肋钢筋,以较密的间距布置。

(2)采用混合配筋得受弯构件,《建议》中建议,其普通钢筋数量,应根据预应力度的大小按下列原则配置:

①当预应力度较高($\lambda > 0.7$)时,为保证构件的安全和延性。宜采用较小直径及较小间距,按最小配筋率 $\rho_s = A_s/A_{he} = 0.2\% \sim 0.3\%$ 设置非预应力钢筋,其中 A_s 为非预应力钢筋面积,A_{he} 为受拉区混凝土面积。

②当预应力度中等($0.4 \leq \lambda \leq 0.7$)时,由于非预应力钢筋的数量相对增多,因此,钢筋的直径,特别是最外排的直径应予以加大。

③当预应力度较低($\lambda < 0.4$)时,非预应力钢筋的数量已超过了预应力钢筋数量,构件受力性能已接近钢筋混凝土构件,故可按钢筋混凝土梁的构造规定配置非预应力钢筋。

(3)普通钢筋宜采用 HRB335 级、HRB400 级热轧钢筋。

(4)截面配筋率。

①最小配筋率。《建议》对部分预应力混凝土受弯构件的最小配筋率要求满足:

$$\frac{M_u}{M_{cr}} > 1.25 \tag{13-50}$$

式中,M_u 为受弯构件正截面抗弯承载能力;M_{cr} 为构件开裂弯矩。

满足条件的部分预应力混凝土构件,从开裂到破坏将具有一定的安全储备,可避免一旦开裂即发生破坏的现象。

②最大配筋率。满足受压区高度 $x \leq \xi_b h_0$。

思考练习题

1. 按截面混凝土应力控制条件,部分预应力混凝土结构可分为几类?各有什么不同?
2. 部分预应力混凝土受弯结构受力特性与全预应力混凝土受弯结构的受力特性主要有哪些不同?部分预应力混凝土结构主要应用范围是什么?
3. 按预应力度法进行部分预应力混凝土结构截面配筋设计的主要步骤有哪些?
4. 部分预应力混凝土结构抗疲劳的主要特点有哪些?
5. 在混合配筋的预应力混凝土结构中,非预应力钢筋的作用是什么?

第14章 无黏结预应力混凝土受弯构件计算

无黏结预应力混凝土梁,是指配置的主筋为无黏结预应力钢筋的后张法预应力混凝土梁。而无黏结预应力钢筋,是指由单根或多根高强钢丝、钢绞线或粗钢筋,沿其全长涂有专用防腐油脂涂料层和外包层,使之与周围混凝土不建立黏结力,张拉时可沿纵向发生相对滑动的预应力钢筋。

无黏结预应力混凝土构件一般采用类似于普通钢筋混凝土构件的方法进行施工,无黏结筋像普通钢筋一样进行敷设,然后浇筑混凝土,待混凝土达到规定的强度后,进行预应力钢筋的张拉和锚固,省去了传统的后张法预应力混凝土的预埋管道、穿束、压浆等工艺,节省了施工设备,简化了施工工艺,缩短了工期,故综合经济性较好。

14.1 无黏结预应力混凝土受弯构件的受力性能

1) 无黏结预应力混凝土受弯构件的基本概念及分类

无黏结预应力混凝土梁,一般分为纯无黏结预应力混凝土梁和无黏结部分预应力混凝土梁。前者是指受力主筋全部采用无黏结预应力钢筋,后者是指其受力主筋采用无黏结预应力钢筋与适当数量非预应力有黏结钢筋的混合配筋梁。

2) 无黏结预应力混凝土受弯构件的受力性能及破坏特征

(1) 纯无黏结预应力混凝土梁与有黏结预应力混凝土梁受力性能及破坏特征对比分析

①裂缝发展及破坏形态。由图 14-1 可以看出,两种梁在荷载作用下裂缝发展的整个过程,试验表明,纯无黏结预应力混凝土梁在试验荷载作用下随着裂缝宽度与高度的急剧增加,受压混凝土压碎而引起的破坏,具有明显的脆性破坏。

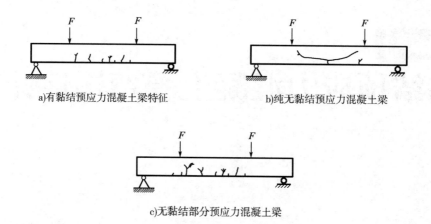

图 14-1　有黏结与无黏结混凝土梁的裂缝形态

②荷载-跨中挠度曲线。由图 14-2 可见,有黏结预应力混凝土梁的荷载-挠度曲线具有三直线形式,而纯无黏结预应力混凝土梁的曲线不仅没有第三阶段,连第二阶段也没有明显的直线段。

③梁最大弯矩截面上钢筋应力随荷载变化的规律。由图 14-3 可以清楚地看到,无黏结预应力筋的应力增量,总是低于有黏结预应力筋的应力增量,而且随着荷载的增大,这个差距也会越来越大。在梁的最大弯矩截面处,无黏结筋的应力增量比有黏结筋少。从开始受力直到破坏,无黏结筋预应力钢筋承受的应力比有黏结筋的应力要低。

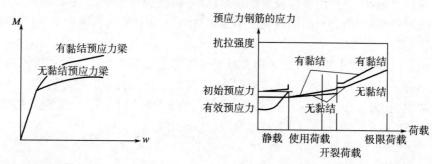

图 14-2　黏结力对梁挠度影响示意图　　图 14-3　黏结力对预应力钢筋应力影响示意图

在构件混凝土开裂之前,由荷载作用引起无黏结筋的应力增量,可以通过纵向变形协调条件,即无黏结筋的总伸长应与沿其整个长度周围混凝土的总伸长相等的条件来求得。

设无黏结筋梁任一截面上的弯矩为 M,则 M 对该截面上任一点引起 y 处的混凝土应变为:

$$\varepsilon_c = \frac{\sigma_c}{E_c} = \frac{M}{E_c I_c} y \tag{14-1}$$

这时,沿无黏结筋全长,构件混凝土的总伸长为:

$$\Delta = \int \varepsilon_c \, dx = \int \frac{M}{E_c I_c} y \, dx \tag{14-2}$$

无黏结筋长度为 l,则无黏结筋的应变增量为:

$$\frac{\Delta}{l} = \int \frac{M}{E_c I_c l} y \, dx \tag{14-3}$$

无黏结筋相应的应力增量为:

$$\Delta \sigma_p = E_p \frac{\Delta}{l} = \frac{E_p}{E_c l} \int \frac{M}{I_c} y \, dx \tag{14-4a}$$

令 $\alpha_{Ep} = \dfrac{E_p}{E_c}$,则可得到:

$$\Delta \sigma_p = \frac{\alpha_{Ep}}{l} \int \frac{M}{I_c} y \, dx \tag{14-4b}$$

现以承受均布荷载的直线无黏结筋配筋方式的矩形截面简支梁(图 14-4)为例,来比较无黏结筋在弯矩作用下的应力增量。

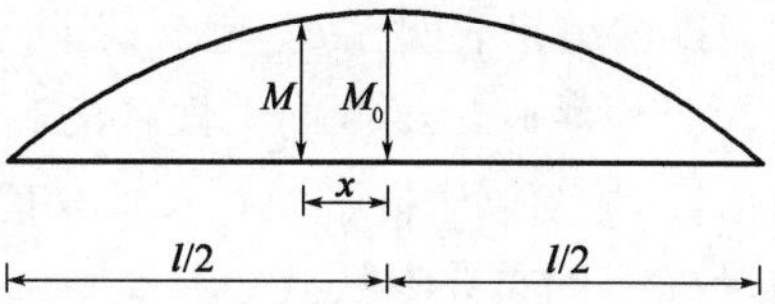

设梁跨中截面弯矩为 M_0,预应力筋在跨中截面处的偏心矩为 e,则距跨中截面为 x 处的弯矩(M)为:

$$M = M_0 \left[1 - 4 \left(\frac{x}{l} \right)^2 \right] \tag{14-5}$$

图 14-4 无黏结钢筋混凝土简支梁

由式(14-1)可得无黏结筋的应力增量为:

$$\Delta \sigma_p = \frac{\alpha_{Ep}}{l I_c} \int_{-l/2}^{l/2} M_0 \left[1 - 4 \left(\frac{x}{l} \right)^2 \right] e \, dx$$

$$= \frac{2}{3} \frac{\alpha_{Ep} M_0 e}{I_c} \tag{14-6}$$

式中,$\dfrac{\alpha_{Ep} M_0 e}{I_c}$ 恰为跨中截面处有黏结筋的应力增量。

因此,相同弯矩下无黏结筋的应力增量要比有黏结筋的应力增量小,在直线布筋的情况下,无黏结筋的应力增量是有黏结筋的 2/3。

用同样的方法也可以得到抛物线形式布置的无黏结筋应力增量(简支梁跨中截面处)是有黏结筋的应力增量的 8/15。

综上所述,纯无黏结筋梁的抗弯强度较有黏结筋梁要低;在荷载作用下,裂缝少且发展迅速,破坏呈明显脆性。这些不足,可采用附加有黏结非预应力钢筋的方法改变,即采用混合配筋的无黏结部分预应力混凝土梁,以获得较好的结构性能。

(2)纯无黏结部分预应力混凝土梁与有黏结预应力混凝土梁受力性能及破坏特征的对比分析

根据大量研究报道总结出以下几点:

①无黏结部分预应力混凝土梁的弯矩—挠度曲线(图 14-5)和有黏结部分预应力混凝土梁一样,也具有三

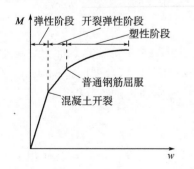

图 14-5 无黏结部分预应力混凝土梁跨中截面的弯矩—挠度曲线

直线的形状。

②无黏结部分预应力混凝土梁的裂缝,由于受到非预应力钢筋的约束,其根数及裂缝间距与配有同样钢筋的普通混凝土梁非常接近[图14-1c)]。

③在一般情况下,无黏结部分预应力混凝土梁,先是普通钢筋屈服,裂缝向上延伸直到受压区边缘混凝土达到极限压应变时,梁才呈现弯曲破坏。

④无黏结部分预应力钢筋,虽仍具有沿全长应力相等(忽略摩擦的影响)和在梁破坏时极限应力不超过条件屈服强度 $\sigma_{0.2}$ 的特点,但极限应力的量值较纯无黏结梁要大得多。

⑤无黏结部分预应力钢筋,在梁到达破坏时的应力增量 $\Delta\sigma_y$,与梁的综合配筋指标 β_0 有密切的关系,其表达式为:

$$\beta_0 = \beta_p + \beta_s = \frac{A_p \sigma_{pe}}{bh_p f_{cd}} + \frac{A_s f_{sd}}{bh_s f_{cd}} \tag{14-7}$$

式中,A_p、σ_{pe} 分别为无黏结预应力钢筋的截面面积和有效预应力;A_s、f_{sd} 分别为有黏结非预应力钢筋的截面面积和抗拉强度设计值;b 为梁的宽度;h_p 为无黏结预应力钢筋截面重心至截面受压边缘的距离;f_{cd} 为混凝土的抗压强度设计值。

试验结果表明,$\Delta\sigma_p$ 与 β_0 之间呈较好的直线关系,$\Delta\sigma_p$ 随 β_0 值的下降而增加。事实上,β_0 可近似反映出梁截面中性轴的高低和梁正截面破坏时的转动能力,而无黏结钢筋的极限应力增量,是与梁中性轴位置及转动能力密切相关的。因此 β_0 是确定梁的无黏结钢筋极限应力增量 $\Delta\sigma_p$ 的重要参数。

⑥对于无黏结部分预应力混凝土梁,在三分点荷载作用下,跨高比对 $\Delta\sigma_p$ 无明显影响;在跨中集中荷载作用下,跨高比对 $\Delta\sigma_p$ 有一定的影响。当跨高比不同,β_0 值相近时,$\Delta\sigma_p$ 随跨高比的增加而降低;当跨高比相同且 β_0 值相近时,跨中集中荷载作用下的 $\Delta\sigma_p$ 低于三分点荷载作用下的 $\Delta\sigma_p$ 值。

14.2 无黏结部分预应力混凝土受弯构件的计算

与有黏结预应力混凝土梁一样,无黏结部分预应力混凝土梁也要进行截面极限承载能力计算、施工阶段和使用阶段的应力验算,以及变形、裂缝最大宽度验算。这些内容的计算方法可参照《公桥规》中有关条文或相关设计规范来进行。这里,仅介绍关于无黏结预应力钢筋的计算方法。

1) 无黏结预应力钢筋的极限应力 σ_{pu}

无黏结预应力钢筋的极限应力值 σ_{pu} 是无黏结部分预应力混凝土梁抗弯承载能力计算中的关键值。然而影响 σ_{pu} 值的因素较多,例如无黏结筋的有效预应力、综合配筋指标 β_0、构件的高跨比、加载条件等,因此受弯破坏时无黏结筋的极限应力 σ_{pu} 常通过试验分析得到。

σ_{pu} 的一般表达式为:

$$\sigma_{pu} = \sigma_{pe} + \Delta\sigma_p \tag{14-8}$$

式中,σ_{pe} 为无黏结预应力筋的有效预应力;$\Delta\sigma_p$ 为无黏结预应力筋在极限荷载作用下的应力增量。

《无黏结预应力混凝土结构技术规程》规定,采用碳素钢丝、钢绞线作为无黏结预应力筋

的受弯构件,在承载能力极限状态下无黏结筋的应力设计值按下列公式计算:

跨高比≤35 的构件:

$$\sigma_{pu} = \frac{1}{\gamma_s}[\sigma_{pe} + (500 - 770\beta_0)] \tag{14-9a}$$

跨高比>35 的构件:

$$\sigma_{pu} = \frac{1}{\gamma_s}[\sigma_{pe} + (250 - 380\beta_0)] \tag{14-9b}$$

式中,β_0 为综合配筋指标,$\beta_0 \leq 0.45$;γ_s 为材料分项系数,取 1.2。同时,σ_{pu} 不应取大于无黏结预应力钢筋的抗拉强度设计值,不应小于无黏结预应力钢筋的有效应力 σ_{pe}。

中国土木工程学会编制的《部分预应力混凝土结构建议》(以下简称《建议》),建议受弯构件无黏结预应力筋的极限应力按公式(14-3)计算,而其中无黏结预应力钢筋在极限荷载下的应力增量 $\Delta\sigma_p$,对于采用高强钢丝、钢绞线的无黏结预应力筋,可按照表 14-1 查用。

无黏结预应力钢筋在极限荷载下的应力增量 $\Delta\sigma_p$(MPa)　　　表 14-1

配筋指标 $\beta_p + \beta_s$	l/h_p		配筋指标 $\beta_p + \beta_s$	l/h_p	
	10	20		10	20
0.05	500	500	0.20	350	300
0.10	500	500	0.25	250	200
0.15	450	400			

注:l 为梁的总长度;h_p 为无黏结预应力筋截面重心至混凝土受压边缘的距离;$\beta_0 = \beta_p + \beta_s$,按式(14-7)计算。

当按式(14-8)计算得到 $\sigma_{pu} > \sigma_{0.2}$ 时,《建议》规定 $\sigma_{pu} = \sigma_{0.2}$,即取抗拉强度设计值,且要求式中的 σ_{pe} 的值不宜低于 $0.6\sigma_{0.2}$。

2) 无黏结预应力钢筋的摩擦损失计算

无黏结预应钢筋的预应力损失值的估算,参照第 12 章中后张法预应力混凝土构件的相应公式计算。

在由摩擦引起预应力损失的估算中,合理采用无黏结预应力筋与孔壁之间摩擦系数 k、μ 是很重要的。无黏结预应力钢筋,与周围混凝土之间的摩擦系数,随着所用涂料和外包材料、制作工艺不同以及截面形式的差异,摩擦系数亦变化较大。表 14-2 为我国《无黏结预应力混凝土结构技术规程》建议的无黏结筋的摩擦系数。

无黏结预应力筋的摩擦系数　　　表 14-2

无黏结预应力筋种类	k	μ	无黏结预应力筋种类	k	μ
$7\phi^P 5$ 钢丝	0.003 5	0.10	$\phi^S 15.2$ 钢铰线	0.004 0	0.12

3) 无黏结部分预应力混凝土梁正截面承载力计算

无黏结部分预应力混凝土梁正截面承载力计算方法,与有黏结部分预应力混凝土梁的方法相同,但是对于无黏结钢筋,这时取其极限应力 σ_{pu},而不是抗拉强度设计值 f_{sp}。

以仅在受拉区布置无黏结预应力筋 A_p 和非预应力筋 A_s 的矩形截面梁为例(图 14-6),由隔离体平衡并参照《公桥规》的计算原则,可以得到正截面承载力计算式为:

$$\sigma_{pu}A_p + f_{sd}A_s = f_{cd}bx \tag{14-10}$$

$$\gamma_0 M_d \leq M_u = f_{cd}bx\left(h_0 - \frac{x}{2}\right) \tag{14-11}$$

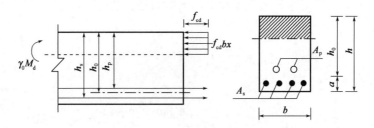

图 14-6　无黏结部分预应力混凝土梁承载力示意图

式(14-11)中截面有效高度：

$$h_0 = h - a$$

而 a 值可以按照下式计算：

$$a = \frac{\sigma_{pu}A_p a_p + f_{sd}A_s a_s}{\sigma_{pu}A_p + f_{sd}A_s} \tag{14-12}$$

式中，a_p、a_s 分别为无黏结筋和非预应力钢筋截面重心至截面受拉边缘距离。

计算的受压区高度 x 仍应满足条件：

$$x \leqslant \xi_b h_0$$

但中国土木工程学会的《建议》，从考虑适当提高部分预应力混凝土受弯构件延性的要求出发，建议：

对一般构件： $\qquad x \leqslant 0.4 h_p$

对延性较高要求的构件： $\qquad x \leqslant 0.3 h_p$

h_p 的意义及计算见表 14-1 的说明。

14.3　无黏结部分预应力混凝土受弯构件的截面设计

在部分预应力混凝土结构设计计算中，为了满足构件使用性能的要求，设计时须选择施加一个合适的预应力量值，并由此可确定预应力钢筋的数量和张拉控制应力，也可在混合配筋中确定预应力钢筋与非预应力钢筋的配置比例，这就是在第 11 章和第 13 章中介绍的预应力度 λ。利用预应力度的概念可以进行无黏结部分预应力混凝土梁的钢筋估算。

除了预应力度法外，在无黏结部分预应力混凝土梁的钢筋估算中还常用部分预应力比率法。

部分预应力比率以预应力钢筋提供的极限抵抗弯矩$(M_u)_p$与全部受拉钢筋（混合配筋时）提供的抵抗弯矩$(M_u)_{p+s}$之比来表示。当部分预应力比率以预应力钢筋和非预应力钢筋特征来表示时，可写成：

$$PPR = \frac{A_p \sigma_{pu}}{A_p \sigma_{pu} + A_s f_{sd}} \tag{14-13}$$

式中，A_p、σ_{pu} 分别为预应力钢筋的截面积和极限拉应力；A_s、f_{sd} 分别为非预应力钢筋的截面积和抗拉强度设计值。

下面以矩形截面为例来介绍无黏结部分预应力混凝土梁钢筋估算的 PPR 法。

1) 按正截面承载能力要求计算预应力钢筋和普通钢筋数量

由图 14-6 可以看到,由于:

$$(A_p\sigma_{pu} + A_s f_{sd})h_0 = A_p\sigma_{pu}h_p + A_s f_{sd}h_s$$

$$h_0 = \frac{A_p\sigma_{pu}h_p + A_s f_{sd}h_s}{A_p\sigma_{pu} + A_s f_{sd}}$$

故
$$h_0 = (\text{PPR})h_p + [1 - (\text{PPR})]h_s \tag{14-14}$$

由式(14-11),取 $M_u = \gamma_0 M_d$,则可以得到:

$$\gamma_0 M_d = f_{cd} b h_0^2 \xi (1 - 0.5\xi)$$

解得:
$$\xi = 1 - \sqrt{1 - \frac{2\gamma_0 M_d}{b h_0^2 f_{cd}}} \tag{14-15}$$

由式(14-10)以及式(14-13)可得到:

$$f_{cd} b x = \sigma_{pu} A_p + f_{sd} A_s$$

$$= \frac{\sigma_{pu} A_p}{\text{PPR}}$$

代入式(14-11),取 $M_u = \gamma_0 M_d$,则可得到:

$$\gamma_0 M_d = \frac{A_p \sigma_{pu}}{\text{PPR}} h_0 (1 - 0.5\xi)$$

则有:
$$A_p = \frac{(\text{PPR})\gamma_0 M_d}{(1 - 0.5\xi) h_0 \sigma_{pu}} \tag{14-16}$$

由公式(14-13)得:
$$A_s = [1 - \text{PPR}] \frac{A_p \sigma_{pu}}{(\text{PPR}) f_{sd}} \tag{14-17}$$

以上是采用预应力比率法估算 A_p 和 A_s 的计算式,具体步骤可概括为:

(1) 假定 PPR,一般可在 0.7~0.95 范围内选择;

(2) 假设预应力钢筋和非预应力钢筋重心到梁截面下缘距离分别为 a_p 和 a_s,计算 $h_p = h - a_p, h_s = h - a_s$,按式(14-14)求 h_0;

(3) 按式(14-15)求得 ξ 值,检查是否满足 $\xi \leq \xi_b$,若不满足,应该调整 PPR 值或修改截面尺寸,重新计算;

(4) 假设无黏结钢筋极限应力 σ_{pu},根据已设计的桥梁经验,可取 $\sigma_{pu} = (0.6 \sim 0.8) f_{pd}$;

(5) 由式(14-16)和式(14-17)分别计算 A_p、A_s 用量。

2) 按施工和使用阶段的要求进行估算

这时,主要进行无黏结预应力钢筋面积 A_p 的估算。对于部分预应力混凝土 A 类构件,可参照第 13 章介绍的方法进行;对于部分预应力混凝土 B 类构件,按使用阶段要求估算时,可取允许名义拉应力来估算 A_p 值。

由上述两方面考虑选择合适的 A_p 和 A_s 值,同时应满足:

$$\frac{A_s f_{sd}}{A_s f_{sd} + A_p \sigma_{pu}} > 0.25$$

14.4 无黏结部分预应力混凝土受弯构件的构造

无黏结部分预应力混凝土受弯构件的构造是,在有黏结部分预应力混凝土构造要求的基础上,提出如下要求:

(1)无黏结钢筋应尽量采用碳素钢丝、钢绞线和热处理钢筋,相应地,混凝土强度等级不宜低于 C40。非预应力钢筋宜选用热轧 HRB335、HRB400 钢筋,钢筋直径宜选用 12mm 或 14mm,不应超过 20mm。

(2)采用混合配筋的受弯构件,应将非预应力钢筋布置靠近截面受拉边缘,并有足够的满足规范要求的混凝土保护层厚度,而将无黏结预应力钢筋布置在非预应力钢筋位置的上方,即使得 $h_p < h_s$,以增大无黏结预应力筋的混凝土保护层厚度,而且一旦裂缝出现,可由非预应力钢筋控制裂缝宽度。

(3)非预应力钢筋的最小数量,应满足:$\dfrac{A_s f_{sd}}{A_s f_{sd} + A_p \sigma_{pu}} > 0.25$,而 f_{sd} 的取用值不得大于 400MPa。

(4)为了保证无黏结预应力混凝土梁的耐久性,首先需要保证无黏结预应力钢筋在构件使用中不锈蚀。因此,必须沿预应力钢筋全长的表面上涂刷防腐蚀材料。防腐处理的方法可采用涂沥青、沥青玛蹄脂、黄油石蜡等材料。目前,工厂生产的无黏结预应力钢筋,多采用化学稳定性好、对周围外包层及混凝土无腐蚀作用、防水性能优良、润滑性能好的专用油脂来处理。在施工中,对涂刷的材料层应特别加以保护,例如可以采用外加塑料套管或缠绕防水塑料纸袋等,并保证形成保护层与预应力钢筋间有一定的空隙,使无黏结预应力钢筋能相对滑动。

(5)无黏结预应力钢筋的锚具,必须按照《预应力筋用锚具、夹具和连接器》(GB/T 14370—2007)的规定程序进行试验验收,合格者方可使用。结构上的锚具应采取可靠的长期防腐蚀保护措施,可采用后浇混凝土封闭等的方法。

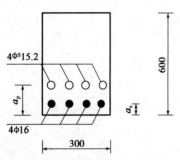

图 14-7 梁截面(尺寸单位:mm)

【例 14-1】 无黏结部分预应力混凝土矩形截面梁,其截面尺寸为 300mm×600mm(图 14-7),无黏结预应力钢筋采用钢绞线 $4\phi^s15.2$,面积 $A_p = 560\text{mm}^2$;非预应力钢筋采用 HRB335 钢筋,$4\phi16$ 的面积 $A_s = 804\text{mm}^2$;C40 混凝土。无黏结预应力钢筋的有效预应力为 $\sigma_{pe} = 720\text{MPa}$。梁计算跨径为 8m,全长为 8.5m。构件处于一般大气环境中,结构重要性等级为二级。梁跨中截面作用的弯矩为梁恒载产生的弯矩 $M_{G1} = 36\text{kN·m}$;二期恒载产生的弯矩 $M_{G2} = 34\text{kN·m}$;汽车荷载产生的弯矩 $M_{Q1} = 144.4\text{kN·m}$。

试进行该梁的正截面承载能力计算。

解:根据《公桥规》规定,梁跨中截面按承载能力极限状态设计时作用效应值基本组合值计算为:

$$M_d = 1.2 \times (36 + 34) + 1.4 \times 144.4 = 286.2(\text{kN·m})$$

结构重要性等级系数取 $\gamma_0 = 1.0$。

(1) 无黏结预应力钢筋的极限拉应力 σ_{pu}

现采用《建议》的方法进行计算。

由图 14-7 所示截面布置的 $a_p = 82\text{mm}$，$a_s = 40\text{mm}$，可求得 $h_p = h - a_p = 518\text{mm}$，$h_s = h - a_s = 560\text{mm}$，则：

$$\frac{l}{h_p} = \frac{8\,500}{518} = 16.4$$

而综合配筋指标 β_0 按式(14-7)计算，即：

$$\begin{aligned}\beta_0 &= \beta_p + \beta_s \\ &= \frac{A_p \sigma_{pe}}{b h_p f_{cd}} + \frac{A_s f_{sd}}{b h_s f_{cd}} \\ &= \frac{560 \times 720}{300 \times 518 \times 18.4} + \frac{804 \times 280}{300 \times 560 \times 18.4} \\ &= 0.214 \end{aligned}$$

由 $l/h_p = 16.4$ 及 $\beta_0 = 0.214$，查表 14-1 得到应力增量 $\Delta\sigma_p = 290\text{MPa}$。

由式(14-7)可求得无黏结预应力钢筋的极限拉应力 σ_{pu} 为：

$$\begin{aligned}\sigma_{pu} &= \sigma_{pe} + \Delta\sigma_p = 720 + 290 \\ &= 1\,010(\text{MPa}) < \sigma_{0.2} = f_{pd} = 1\,260\text{MPa}\end{aligned}$$

而

$$\frac{A_s f_{sd}}{A_s f_{sd} + A_p \sigma_{pu}} = \frac{280 \times 804}{280 \times 804 + 1\,010 \times 560} = 0.285 > 0.25$$

符合非预应力钢筋最小含量要求。

(2) 求截面受压高度 x

由式(14-10)，可得到：

$$\begin{aligned}x &= \frac{\sigma_{pu} A_p + f_{sd} A_s}{f_{cd} b} \\ &= \frac{1\,010 \times 560 + 280 \times 804}{18.4 \times 300} \\ &= 143(\text{mm}) < 0.3 h_p = 0.3 \times 518 = 155(\text{mm})(按《建议》的延性要求) \\ &\qquad\qquad < \xi_b h_0 = 0.4 \times 512 = 208(\text{mm})\end{aligned}$$

满足《公桥规》要求。

(3) 正截面承载力复核

由式(14-12)可得到：

$$\begin{aligned}a &= \frac{\sigma_{pu} A_p a_p + f_{sd} A_s a_s}{\sigma_{pu} A_p + f_{sd} A_S} \\ &= \frac{1\,010 \times 560 \times 82 + 280 \times 804 \times 40}{1\,010 \times 560 + 280 \times 804} = 70(\text{mm})\end{aligned}$$

则截面有效高度：$h_0 = h - a = 600 - 70 = 530(\text{mm})$

由式(14-11)可求得梁跨中截面的抗弯承载能力为：

$$M_u = f_{cd} b x \left(h_0 - \frac{x}{2}\right)$$

$$= 18.4 \times 300 \times 143 \times \left(530 - \frac{143}{2}\right)$$
$$= 361.92(\text{kN} \cdot \text{m}) > \gamma_0 M_d = 286.2(\text{kN} \cdot \text{m})$$

满足设计要求。

思考练习题

1. 纯无黏结预应力混凝土梁相比有黏结梁而言，受力性能上有何明显不同，可采取什么措施进行改进？

2. 无黏结预应力混凝土梁在进行截面承载力计算时，其预应力钢筋的应力如何取值？为什么？

3. 普通钢筋对于改善无黏结预应力混凝土梁的力学性能有何作用？

PART 3 | 第三部分
圬工结构

第15章 圬工结构的基本概念与材料

15.1 圬工结构的基本概念

圬工结构是以砖、石及混凝土预制块作为块材,通过砂浆按照一定的砌筑规则所建成的结构。

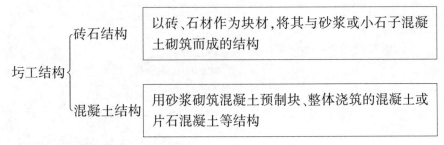

1) 圬工结构的受力特点与工程应用

圬工是一种抗压强度高而抗拉和抗剪强度较低的建筑材料,广泛用于桥涵的基础及受压为主的拱桥、桥墩和挡土墙中。

2) 圬工结构的优缺点

(1) 优点

①圬工结构中所用的天然石料、砂等原材料分布广,易于就地取材,造价低廉。

②有较强的耐久性、耐火性及稳定性,维修养护费用低。
③施工简便,不需特殊设备,施工技术易于掌握。
④具有较强的抗冲击性能及较大的承载性能。
⑤与钢筋混凝土结构相比,可节约水泥和钢材,砌体砌筑时不需模板,可以节省木材。

(2)缺点

①自重大:由于砌体强度不高,特别是抗拉、抗剪强度很低,为保证足够的抗力,故构件截面尺寸大,材料用量多,造成结构自重大,跨越能力低。

②施工周期长、机械化程度低:全部用人工砌筑,施工时间较长。

③抗拉、抗弯强度很低,抗震能力差。

15.2 材料种类

桥涵圬工结构的材料主要有石材、混凝土、砂浆和小石子混凝土。

1)石材

(1)概念

石材是无明显风化的天然岩石经过人工开采和加工后的外形规则的建筑用材。

(2)优点

抗压强度高、抗冻与抗气性能好。

(3)应用

广泛用于建造桥梁基础、墩台、挡土墙等。桥涵结构所用石材应选择质地坚硬、均匀、无裂纹且不易风化的石料。

(4)类型

根据开采方法、形状、尺寸的不同,可分为片石、块石、细料石、半细料石、粗料石五类,各类的开采方法和技术要求见表15-1。

各类石材的技术要求 表15-1

类 型	开采方法	技术要求
片石	由爆破或楔劈法开采的不规则石块	a. 形状不受限制; b. 厚度不得小于150mm; c. 卵形和薄片不得采用
块石	按岩石层理放炮或楔劈而成的石材	a. 形状大致方正,上下面大致平整; b. 厚度为200~300mm,宽度为厚度的1.0~1.5倍,长度为厚度的1.5~3.0倍; c. 块石一般不修凿,但应敲去尖角突出部分
细料石	由岩层或大块石材开劈并经修凿而成	a. 要求外形方正,成六面体,表面凹陷深度不大于10mm; b. 厚度为200~300mm,宽度为厚度的1.0~1.5倍,长度为厚度的2.5~4.0倍
半细料石	同细料石	同细料石,表面凹陷深度不大于15mm
粗料石	同细料石	同细料石,表面凹陷深度不大于20mm

(5)强度

桥涵结构中所用石材的强度等级有:(桥涵)MU30、MU40、MU50、MU60、MU80、MU100和MU120,其中符号MU表示石材强度等级,后面的数字是边长70mm含水饱和试件立方体的抗压强度,以MPa为计量单位。

2)混凝土

(1)强度

用在圬工结构中的混凝土强度等级有C15、C20、C25、C30、C35和C40。

(2)成形方式

混凝土预制块和整体浇筑的混凝土。

①混凝土预制块的特点

a. 预先制作成一定形状,具有足够强度的块材,施工方便快捷;

b. 由于混凝土预制块形状和尺寸统一,故砌体表面整齐美观。

②整体浇筑混凝土的特点

a. 混凝土收缩变形较大,施工期间容易产生收缩裂缝或温度裂缝;

b. 浇筑时耗费木材较多,工期相对预制构件长。

③小石子混凝土的特点

a. 小石子混凝土是由胶结料(水泥)、粗集料(细卵石或碎石,粒径小于20mm)、细粒料(砂)加水拌和而成,可节省水泥和砂,在一定条件下是水泥砂浆的代用品;

b. 与同强度等级砂浆来砌筑的片石和块石砌体的抗压极限强度相比,小石子混凝土的抗压极限强度更高。

3)砂浆

(1)定义

砂浆是由一定比例的胶结料(水泥、石灰等)、细集料(砂)及水配制而成的砌筑材料。

(2)作用

①将块材黏结成整体。

②铺砌时抹平块材不平的表面,使块材在砌体受压时能比较均匀地受力。

③砂浆填满了块材间隙,减少了砌体的透气性,提高密实度、保温性与抗冻性。

(3)类型

①无塑性掺料的水泥砂浆:由一定比例的水泥和砂加水配制而成的砂浆,强度较高。

②有塑性掺料的混合砂浆:由一定比例的水泥、石灰和砂加水配制而成的砂浆,又称水泥石灰砂浆。

③石灰(石膏、黏土)砂浆:胶结料为石灰(石膏、黏土)的砂浆,强度较低。

(4)强度等级

桥涵结构用的砂浆强度等级有M5、M7.5、M10、M15和M20,其中符号M表示砂浆强度等级,后面的数字是边长70.7mm标准立方体试块的抗压强度,以MPa为计量单位。

(5)砌体对砂浆的基本要求

有较高的强度、可塑性和保水性。

①砂浆应满足砌体强度、耐久性的要求,并与块材间有良好的黏结力。

②砂浆的可塑性应保证砂浆在砌筑时能很容易且较均匀地铺开,以提高砌体强度和施工

效率。

③砂浆应具有足够的保水性。

4)砌体种类

工程上根据块材的不同,将砌体分为以下几种:

(1)片石砌体。用片石作为块材,用小石子作为填充料,按照一定砌筑规则砌筑而成。

(2)块石砌体。用方正的块石作块材,错缝砌筑成规则的砌体。

(3)半细料石砌体。用料石作块材,控制表面凹陷深度和砌缝宽度砌筑而成。

(4)细料石砌体。用细料石作块材,控制表面凹陷深度和砌缝宽度砌筑而成。

(5)混凝土预制块砌体。用混凝土预制块材,严格控制砌缝宽度和凹陷深度砌筑而成。

在桥涵工程中,应根据结构的重要程度、尺寸大小、工程环境施工条件以及材料供应情况来选择砌体的种类。

15.3　砌体的强度与变形

1)砌体的抗压强度

(1)砌体的受压破坏特征

砌体轴心受压时从荷载作用开始到破坏,大致分为下列三个阶段:

①第Ⅰ阶段为整体工作阶段:从开始加载到砌体内出现裂缝。作用荷载为砌体极限荷载的50%~70%。

②第Ⅱ阶段为带裂缝工作阶段:随荷载继续增大,单块块材内裂缝不断发展,并逐渐连接起来形成连续的裂缝。

③第Ⅲ阶段为破坏阶段:当荷载稍微增加,裂缝急剧发展,并连成几条贯通的裂缝,将砌体分成若干受压柱,各受压柱受力极不均匀,最后,柱被压碎或丧失稳定导致砌体的破坏。

(2)受压时的应力状态

砌体受压的一个重要特征是单块材料先开裂,在受压破坏时,砌体的抗压强度低于所使用的块材的抗压强度。这主要是因为砌体即使承受轴向均匀压力,砌体中的块材实际上不是均匀受压,而是处于复杂应力状态。

(3)影响砌体抗压强度的主要因素

①块材的强度:砌体的开裂是从单个薄弱块材出现裂缝而引起的,随荷载的逐渐增大,块材强度直接影响砌体强度。

②块材形状和尺寸:块材形状规则,砌体抗压强度高,砌体强度随块材厚度的增大而增加。

③砂浆的物理力学性能:砂浆的强度等级、砂浆的可塑性和流动性、砂浆的弹性模量。

④砌缝厚度:砂浆水平砌缝越厚,砌体强度越低,以10~12mm为宜。

⑤砌筑质量:砌体均匀密实,强度高,反之,砌体强度低。砌体抗压强度见附表3-4~附表3-6。

2)砌体的抗拉、抗弯、抗剪强度

试验表明,在多数情况下,砌体的受拉、受弯及受剪破坏一般发生于砂浆与块材的连接面上。因此,砌体的抗拉、抗弯与抗剪强度取决于砌缝强度,亦即取决于砌缝间块材与砂浆的黏结强度。通常的破坏形式有：

(1) 轴向受拉的破坏形式

①沿砌体齿缝截面发生破坏，破坏面呈齿状，如图 15-1a) 所示。

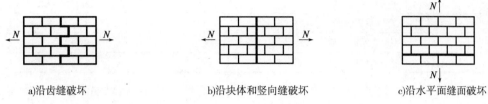

图 15-1　轴心受拉砌体的破坏形式

②沿竖向砌缝和块材破坏，如图 15-1b) 所示。
③沿通缝截面发生破坏，如图 15-1c) 所示。

(2) 弯曲受拉的破坏形式

①通缝截面发生破坏，如图 15-2a) 所示。
②齿缝截面发生破坏，如图 15-2b) 所示。

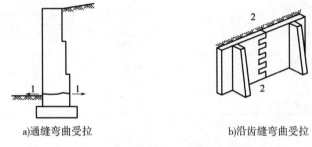

图 15-2　弯曲抗拉砌体的破坏形式

(3) 受剪的破坏形式

①通缝截面受剪破坏，如图 15-3a) 所示。
②沿砌体齿缝截面发生破坏，破坏面呈齿状，如图 15-3b) 所示。

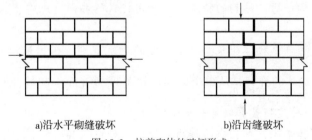

图 15-3　抗剪砌体的破坏形式

3) 砌体的其他性能

(1) 弹性模量

砌体不是弹性材料，但在工程计算中需要能反映砌体的受力性能取值明确的弹性模量，试验表明，当受压应力上限为抗压强度平均值的 40%～50% 时，割线模量接近为弹性模量。

《公路圬工桥涵设计规范》(JTG D61—2005)（以下简称《圬工规范》）规定，按不同等级的砂浆，以砌体弹性模量与砌体抗压强度成正比的关系确定弹性模量。

(2)线膨胀系数、收缩变形与摩擦系数

①线膨胀系数。考虑超静定的圬工结构在温度变化下结构的受力,按表15-2取值。

砌体的线膨胀系数　　　　　　　　　　　　　表15-2

砌 体 种 类	线膨胀系数($10^{-6}/℃$)
混凝土	10
混凝土预制块砌体	9
细料石、半细料石、粗料石、块石、片石砌体	8

②收缩变形。一般通过砌体收缩试验确定干缩变形的大小,如对混凝土预制块砌体,其28d的干缩变形约为0.2mm/m。

③摩擦系数。砌体摩擦系数的大小,取决于接触砌体摩擦面的材料种类和干湿情况等。砌体的摩擦系数μ_f可参照表15-3取值。

砌体的摩擦系数μ_f　　　　　　　　　　　　表15-3

材 料 种 类	摩 擦 面 情 况	
	干　燥	潮　湿
砌体沿砌体或混凝土滑动	0.70	0.60
木材沿砌体滑动	0.60	0.50
钢沿砌体滑动	0.45	0.35
砌体沿砂或卵石滑动	0.60	0.50
砌体沿粉土滑动	0.55	0.40
砌体沿黏性土滑动	0.50	0.30

思考练习题

1. 什么是圬工结构?圬工结构的特点以及所用材料的共同特点是什么?对圬工材料的选择有哪些要求?
2. 什么是砌体?为什么砌体砌筑要满足一定的砌筑规则?根据选用的块材的不同,常用的砌体有哪几类?
3. 石材是怎样分类的?有哪几类?
4. 石料强度等级、砂浆强度等级是如何确定的?
5. 什么是砂浆?其在砌体结构中的作用是什么?砂浆按其胶结料的不同分为哪几种?
6. 什么是小石子混凝土、片石混凝土?为什么有时用小石子混凝土代替砂浆?
7. 对砌体所用的砂浆、石材及混凝土材料有哪些基本要求?
8. 为什么砌体的抗压强度低于所使用的块材的抗压强度?
9. 为什么砌体中的块材实际上处于复杂应力状态?
10. 试述影响砌体抗压强度的主要因素及其原因。
11. 试述砌体的受拉、受弯及受剪破坏的破坏形式。

第 16 章
圬工结构构件的承载力计算

16.1 计算原则

圬工结构构件只要进行承载力计算,其正常使用极限状态的要求,则采取相应的构造措施保证。承载力计算原则为:

(1)《圬工规范》对公路桥梁圬工结构采用以概率理论为基础的极限状态设计方法,以分项安全系数的设计表达式进行计算。

(2)桥涵圬工构件按承载能力极限状态设计时的表达式为:

$$\gamma_0 S \leqslant R(f_d, a_d) \tag{16-1}$$

式中,γ_0 为桥梁结构的重要性系数,对应于规定的一级、二级和三级设计安全等级,分别取用 1.1、1.0 和 0.9;S 为作用效应组合设计值;$R(\cdot)$ 为构件承载力设计值函数;f_d 为材料强度设计值;a_d 为几何参数设计值,可采用几何参数标准值 a_k,即设计文件规定值。

16.2 受压构件的承载力计算

1)砌体受压短柱的实验研究

(1)砌体破坏过程

国内外试验研究资料表明,短柱砌体的受力破坏有以下特点:

①当构件承受轴心压力时,砌体截面上产生均匀的压应力,如图 16-1a)所示;构件破坏时,正截面所能承受的最大压力即砌体的抗压承载力。

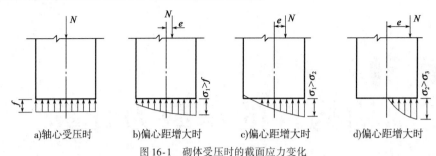

a)轴心受压时　　b)偏心距增大时　　c)偏心距增大时　　d)偏心距增大时

图 16-1　砌体受压时的截面应力变化

②当构件承受偏心压力时,砌体截面上产生的压应力是不均匀的,压应力分布随着偏心距 e 的变化而变化,砌体表现出弹塑性性能。当偏心距不大时,整个截面受压,应力图形呈曲线分布,这时破坏将发生在压应力较大一侧,破坏时该侧压应变比轴心受压时的均匀应变略高,而边缘压应力也比轴心抗压强度略大,如图 16-1b)所示。

③随着偏心距 e 的增大,在远离偏心压力作用的截面边缘,由受压逐步过渡到受拉,但只要在受压边压碎之前受拉边的拉应力尚未达到通缝的抗拉强度,则截面的受拉边就不会开裂,直至破坏为止,仍是全截面受力,如图 16-1c)所示。

④若偏心距 e 再大时,砌体受拉区出现沿截面通缝的水平裂缝。已开裂的截面脱离工作,实际受压区面积减小,受压区应力的合力将与作用的偏心压力保持平衡,这种平衡随裂缝的不断展开被打破而达到新的平衡,如图 16-1d)所示,剩余截面的压应力进一步加大,并出现竖向裂缝,最后由于受压区的承载能力耗尽而破坏。破坏时,虽然砌体受压一侧的受压变形比轴心受压构件高,但由于压应力分布不均匀的加剧和受压区面积的减少,构件的承载力将随偏心距的增大而降低。

(2)纵向力的偏心影响系数

偏心受压圬工构件的承载力,随偏心距的增大而降低,工程应用时,是在材料力学偏心距影响系数计算形式的基础上,根据我国大量试验资料的统计分析,《圬工规范》最终采用的纵向力的偏心影响系数为:

$$\alpha_x = \frac{1 - (e_x/x)^m}{1 + (e_x/i_y)^2} \qquad (16\text{-}2)$$

$$\alpha_y = \frac{1 - (e_y/y)^m}{1 + (e_y/i_x)^2} \qquad (16\text{-}3)$$

图 16-2　砌体构件偏心受压

式中,α_x、α_y 分别为 x 方向和 y 方向的纵向力偏心影响系数;x、y 分别为 x 方向、y 方向截面重心至偏心方向的截面边缘的距离,如图 16-2 所示;e_x、e_y 为轴向力在 x 方向和 y 方向的截面偏心距,$e_x = M_{yd}/N_d$、$e_y = M_{xd}/N_d$,其值不应超过规定值,其中 M_{yd} 和 M_{xd} 分别为绕 x 轴和 y 轴的弯矩设计值,N_d 为轴向力设计值,如图 16-2 所示;m 为截面形状系数,对于圆形截面取 2.5,对于 T 形或 U 形截面取 3.5,对于箱形截面或矩形截面(包括两端设有曲线形或圆弧形的矩形墩身截面)取 8.0；i_x、i_y 为弯曲平面内的截面回转半径，$i_x = \sqrt{I_x/A}$ 和 $i_y = \sqrt{I_y/A}$。

式(16-2)和式(16-3)是半理论半经验公式，它满足轴心受压和偏心受压两个受力边界条件，即当 $e_x = 0$ 和 $e_y = 0$ 时，$\alpha_x = 1$，$\alpha_y = 1$，构件为轴心受压，承载力不受偏心距的影响；当 e_x 和 e_y 有一个不等于 0 时，构件为单向偏心受压；当 e_x 和 e_y 都不等于 0 时，构件为双向偏心受压。

2) 砌体受压长柱的受力特点

(1) 砌体受压长柱的受力特点

① 细长柱在承受轴心压力时，由于材料不均匀和各种偶然因素的影响，轴向力不可能完全作用在砌体截面中心，会产生一定的初始偏心。

② 在偏心压力作用下，会产生附加偏心距 u，实际的偏心距已将达到 $e + u$，使构件承载力大大降低。

③ 在砌体构件中，水平砂浆缝削弱了砌体的整体性，故纵向弯曲现象较钢筋混凝土更为明显，构件的长细比越大，这种纵向弯曲的影响就越大(图 16-3)。

由上可知，对于细长构件，不论是轴心受压还是偏心受压，构件长细比 λ 的变化将影响砌体的承载能力，所以，应根据砌体构件长细比 λ 的大小、砂浆的强度等考虑纵向弯曲对砌体构件承载能力的影响。

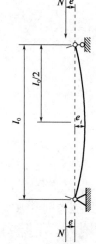

图 16-3 偏心受压构件的附加偏心距

(2) 偏心受压构件的纵向弯曲系数计算

《圬工规范》建议砌体偏心受压构件的纵向弯曲系数的计算式为：

x 方向
$$\phi_x = \frac{1}{1 + \alpha\lambda_x(\lambda_x - 3)\left[1 + 1.33\left(\dfrac{e_x}{i_y}\right)^2\right]} \quad (16-4)$$

y 方向
$$\phi_y = \frac{1}{1 + \alpha\lambda_y(\lambda_y - 3)\left[1 + 1.33\left(\dfrac{e_y}{i_y}\right)^2\right]} \quad (16-5)$$

3) 砌体受压构件的承载力计算

(1) 承载力计算公式

根据以上对砌体受压短柱和长柱的分析，对砌体受压构件，可以采用一个综合系数 φ 来考虑纵向弯曲和轴向力的偏心距对受压构件承载力的影响。《圬工规范》规定：在受压偏心距限值范围内，砌体受压构件的承载力按下式计算：

$$\gamma_0 N_d \leqslant N_u = \varphi A f_{cd} \quad (16-6)$$

式中，N_d 为轴向力设计值；A 为构件的截面面积，对于组合截面按强度比换算；φ 为构件轴向力的偏心距 e 和长细比 λ 对受压构件承载力的影响系数。

式(16-6)适用于砌体轴心受压和偏心受压。

(2) 影响系数 φ 的确定

砌体偏心受压构件承载力影响系数 φ 同时考虑了偏心距和构件的长细比影响。《圬工规范》给出的砌体偏心受压构件承载力影响系数表达式如下：

$$\varphi = \cfrac{1}{\cfrac{1}{\varphi_x} + \cfrac{1}{\varphi_y} - 1} \tag{16-7}$$

$$\varphi_x = \alpha_x \phi_x = \cfrac{1 - \left(\cfrac{e_x}{x}\right)^m}{1 + \left(\cfrac{e_x}{i_y}\right)^2} \cdot \cfrac{1}{1 + \alpha\lambda_x(\lambda_x - 3)[1 + 1.33(e_x/i_y)^2]} \tag{16-8}$$

$$\varphi_y = \alpha_y \phi_y = \cfrac{1 - \left(\cfrac{e_y}{y}\right)^m}{1 + \left(\cfrac{e_y}{i_x}\right)^2} \cdot \cfrac{1}{1 + \alpha\lambda_y(\lambda_y - 3)[1 + 1.33(e_y/i_x)^2]} \tag{16-9}$$

式中,φ_x、φ_y 分别为 x 方向和 y 方向偏心受压构件承载力影响系数。

4) 混凝土受压构件的承载力计算

根据试验分析,混凝土偏心受压构件进入塑性状态,可以认为受压区的法向应力图形为矩形,受压应力的合力作用点与轴心力作用点重合。

因此,《圬工规范》规定,对混凝土偏心受压构件在表 16-1 规定的受压构件偏心距限值范围内,进行受压承载力计算时,假定受压区的法向应力图形为矩形,其应力取混凝土抗压强度设计值。受压承载力按式(16-10)计算,即:

$$\gamma_0 N_d \leqslant N_u = \varphi A_c f_{cd} \tag{16-10}$$

式中,N_d 为轴向力设计值;φ 为弯曲平面内轴心受压构件弯曲系数,按表 16-2 采用;f_{cd} 为混凝土构件抗压强度设计值,按附表 3-3(见英文部分)的规定采用;A_c 为混凝土受压区面积。

受压构件偏心距限值　　　　表 16-1

作用组合	偏心距限值 e	作用组合	偏心距限值 e
基本组合	≤0.6s	偶然组合	≤0.7s

注:s 为截面或者换算截面重心轴至偏心方向截面边缘的距离。

以下分别介绍单向偏心受压构件和双向偏心受压构件的计算。

混凝土轴向受压构件弯曲系数　　　　表 16-2

l_0/b	<4	4	6	8	10	12	14	16	18	20	22	24	26	28	30
l_0/i	<14	14	21	28	35	42	49	56	63	70	76	83	90	97	104
φ	1.00	0.98	0.96	0.91	0.86	0.82	0.77	0.72	0.68	0.63	0.59	0.55	0.51	0.47	0.44

(1) 单向偏心受压

由图 16-4 可知,受压区高度 h_c 应按下列条件确定:

$$h_c = h - 2e \tag{16-11}$$

矩形截面的受压承载力可按下列公式计算:

$$\gamma_0 N_d \leqslant N_u = \varphi f_{cd} b (h - 2e) \tag{16-12}$$

式中,e_c 为受压区混凝土法向应力合力作用点至截面重心的距离;e 为轴向力的偏心距;b 为矩形截面宽度;h 为矩形截面高度。

当构件弯曲平面外长细比大于弯曲平面内长细比时,尚应按轴心受压验算其承载力。

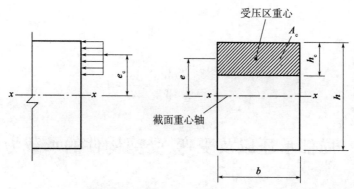

图 16-4 混凝土构件偏心受压(单向偏心)

(2)双向偏心受压

试验表明,双向偏心受压构件破坏形态比较复杂,计算方法尚不成熟,一般采用近似的计算公式。

受压区高度和宽度应按下列条件确定(图 16-5):

$$h_c = h - 2e_y \tag{16-13}$$

$$b_c = b - 2e_x \tag{16-14}$$

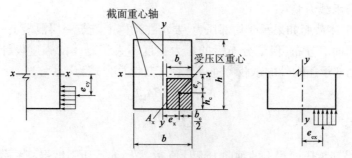

图 16-5 混凝土构件双向偏心受压

矩形截面的偏心受压承载力可按下列公式计算:

$$\gamma_0 N_d \leq N_u = \varphi f_{cd}(h - 2e_y)(b - 2e_x) \tag{16-15}$$

图和式中,φ 为轴心受压构件弯曲系数,按表 16-2 采用;e_{cy} 为受压区混凝土法向应力合力作用点在 y 轴方向至截面重心距离;e_{cx} 为受压区混凝土法向应力合力作用点在 x 轴方向至截面重心距离;e_x、e_y 分别为轴向力沿 x 轴方向和沿 y 轴方向的偏心距。

5)偏心距验算

试验结果表明,若轴向力作用点的偏心距较大,圬工构件的承载力会显著降低。为了保证结构的正常使用状态和截面的稳定性,应该对轴向力作用的偏心距 e 有所限制。如偏心距超出限制值应采用配筋混凝土结构设计。

根据试验结果并参考国内外有关规范,《圬工规范》建议砌体和混凝土的单向和双向偏心受压构件,其偏心距 e 的限值应符合表 16-1 的规定。

当轴向力的偏心距 e 超过表 16-1 规定的偏心距限值时,构件承载力应按下列公式计算:

单向偏心：
$$\gamma_0 N_d \leq N_u = \varphi \frac{A f_{tmd}}{\frac{Ae}{W} - 1} \tag{16-16}$$

双向偏心：
$$\gamma_0 N_d \leq N_u = \varphi \frac{A f_{tmd}}{\frac{Ae_x}{W_y} + \frac{Ae_y}{W_x} - 1} \tag{16-17}$$

16.3　局部承压以及受弯、受剪构件的承载力计算

1）局部承压承载力计算

（1）概念

轴向力仅作用于构件的部分截面上的受压构件。

（2）受力特点

①受压局部范围内砌体抗压强度有较大程度的提高,一般认为这是由于存在"套箍强化"和"应力扩散"的作用。

②劈裂破坏。

③在实际工程中,在局部承压面下可能会出现构件局部压碎的现象。

（3）局部承压承载力验算

桥涵结构的砌体截面如果承受局部压力,要求在砌体上浇筑一层混凝土,在混凝土上面的压力以45°扩散角向下分布,保证分布后的压力强度不大于砌体的强度设计值。故《圬工规范》仅对混凝土局部承压构件的承载力进行计算。其计算公式为：

$$\gamma_0 N_d \leq 0.9 \beta A_1 f_{cd} \tag{16-18}$$

$$\beta = \sqrt{\frac{A_b}{A_1}} \tag{16-19}$$

式中,N_d 为局部承压面积上的轴向力设计值；β 为局部承压强度提高系数；A_1 为局部承压面积；A_b 为局部承压计算底面积,根据底面积重心与局部受压面积重心相重合的原则计算；f_{cd} 为混凝土轴心抗压强度设计值。

2）受弯构件的承载力计算

受弯构件在弯矩作用下砌体可能沿通缝截面或齿缝截面产生弯曲受拉而破坏。对受弯构件正截面的承载力,要求截面的受拉边缘最大计算拉应力必须小于弯曲抗拉强度设计值。

考虑到结构的设计安全等级,计入桥梁结构重要性系数,《圬工规范》规定按下式计算：

$$\gamma_0 M_d \leq W f_{tmd} \tag{16-20}$$

式中,M_d 为弯矩设计值；W 为截面受拉边缘的弹性抵抗矩,对于组合截面应按弹性模量比换算为换算截面受拉边缘弹性抵抗矩；f_{tmd} 为构件受拉边缘的弯曲抗拉强度设计值。

3）受剪构件承载力计算

在受剪构件中,除水平剪力外,还作用有垂直压力。砌体构件的受剪试验表明,砌体沿水平向缝的抗剪承载能力为砌体沿通缝的抗剪承载能力及作用在截面上的压力所产生的摩擦力

的总和。因此，对砌体沿通缝或沿阶梯形截面破坏时的受剪承载力计算时，《圬工规范》规定砌体构件或混凝土构件直接受剪时，按下式计算：

$$\gamma_0 V_d \leqslant V_u = A f_{vd} + \frac{1}{1.4} \mu_f N_k \tag{16-21}$$

式中，V_d 为剪力设计值；A 为受剪截面面积；f_{vd} 为砌体或混凝土抗剪强度设计值；μ_f 为摩擦系数，采用 $\mu_f = 0.7$；N_k 为与受剪截面的压力标准值。

思考练习题

1. 简述圬工结构的计算原则。
2. 为什么不直接采用材料力学公式计算砌体受压构件？
3. 构件长细比对砌体的承载能力有何影响？《圬工规范》是怎样考虑此影响的？
4. 砌体受压构件包括哪些计算内容？
5. 为什么对圬工受压构件要进行偏心矩验算？
6. 空腹式无铰拱桥的拱上横墙为矩形截面，厚度 $h = 500\text{mm}$，宽度 $b = 8.5\text{m}$。拱上横墙采用 M7.5 水泥砂浆，MU40 块石砌筑。横墙沿宽度方向的单位长度上作用的基本弯矩设计值 $M_y = 13.59\text{kN} \cdot \text{m}$，轴向力设计值 $N_d = 234.86\text{kN}$，横墙的计算长度 $l_0 = 4.34\text{m}$，结构安全等级为一级，试复核该横墙的承载力。
7. 主孔净跨径为 30m 的等截面悬链线空腹式无铰石拱桥，安全等级为一级，主拱圈厚度 $h = 80\text{mm}$，宽度 $b = 8.5\text{m}$，矢跨比 1/5，拱轴长度 $s = 33.876\text{m}$，主拱圈采用 M10 水泥砂浆，MU60 块石砌筑。作用效应基本组合下得到在拱顶截面单位宽度作用的弯矩设计值 $M_{d(y)} = 142.689\text{kN} \cdot \text{m/m}$，轴向力设计值 $N_d = 1\,083.064\text{N/m}$；相应拱的水平推力设计值的拱跨 1/4 处的弯矩设计值 $M_{d(y)} = 86.671\text{kN} \cdot \text{m/m}$，拱的轴向力设计值为 $N_d = 935.482\text{kN/m}$。试复核拱顶截面处的承载力以及该拱的整体承载力。

附 表

混凝土强度标准值和设计值（单位：MPa）　　　　附表 1-1

强度种类		符号	混凝土强度等级											
			C20	C25	C30	C35	C40	C45	C50	C55	C60	C65	C70	C75	C80
强度标准值	轴心抗压	f_{ck}	13.4	16.7	20.1	23.4	26.8	29.6	32.4	35.5	38.5	41.5	44.5	47.4	50.2
	轴心抗拉	f_{tk}	1.54	1.78	2.01	2.20	2.40	2.51	2.65	2.74	2.85	2.93	3.0	3.05	3.10
强度设计值	轴心抗压	f_{cd}	9.2	11.5	13.8	16.1	18.4	20.5	22.4	24.4	26.5	28.5	30.5	32.4	34.6
	轴心抗拉	f_{td}	1.06	1.23	1.39	1.52	1.65	1.74	1.83	1.89	1.96	2.02	2.07	2.10	2.14

注：计算现浇钢筋混凝土轴心受压和偏心受压构件时，如截面的长边或直径小于 300mm，表中混凝土强度设计值应乘以系数 0.8；当构件质量（混凝土成形、截面和轴线尺寸等）确有保证时，可不受此限。

混凝土的弹性模量（单位：10^4 MPa）　　　　附表 1-2

混凝土强度等级	C20	C25	C30	C35	C40	C45	C50	C55	C60	C65	C70	C75	C80
E_c	2.55	2.80	3.00	3.15	3.25	3.35	3.45	3.55	3.60	3.65	3.70	3.75	3.80

注：① 混凝土剪变模量 G_c 按表中数值的 0.4 倍采用。

② 对高强混凝土，当采用引气剂及较高砂率的泵送混凝土且无实测数据时，表中 C50～C80 的 E_c 值应乘折减系数 0.95。

普通钢筋强度标准值和设计值（单位：MPa） 附表1-3

钢筋种类	直径 d (mm)	符号	抗拉强度标准值 f_{sk}	抗拉强度设计值 f_{sd}	抗压强度设计值 f'_{sd}
HPB235（R235）	8~20	Φ	235	195	195
HRB335	6~50	Φ	335	280	280
HRB400	6~50	Φ	400	330	330
RRB400（KL400）	8~40	Φ^R	400	330	330

注：①表中 d 系指国家标准中的钢筋公称直径。
②钢筋混凝土轴心受拉和小偏心受拉构件的钢筋抗拉强度设计值大于330MPa时，仍应取用330MPa。
③构件中有不同种类钢筋时，每种钢筋应采用各自的强度设计值。

普通钢筋的弹性模量（单位：10^5MPa） 附表1-4

钢筋种类	弹性模量 E_s	钢筋种类	弹性模量 E_s
HPB235（R235）	2.1	HRB335、HRB400、RRB400（KL400）	2.0

钢筋混凝土受弯构件单筋矩形截面承载力计算用表 附表1-5

ξ	A_0	ζ_0	ξ	A_0	ζ_0
0.01	0.010	0.995	0.34	0.282	0.830
0.02	0.020	0.990	0.35	0.289	0.825
0.03	0.030	0.985	0.36	0.295	0.820
0.04	0.039	0.980	0.37	0.301	0.815
0.05	0.048	0.975	0.38	0.309	0.810
0.06	0.058	0.970	0.39	0.314	0.805
0.07	0.067	0.965	0.40	0.320	0.800
0.08	0.077	0.960	0.41	0.326	0.795
0.09	0.085	0.955	0.42	0.332	0.790
0.10	0.095	0.950	0.43	0.337	0.785
0.11	0.104	0.945	0.44	0.343	0.780
0.12	0.113	0.940	0.45	0.349	0.775
0.13	0.121	0.935	0.46	0.354	0.770
0.14	0.130	0.930	0.47	0.359	0.765
0.15	0.139	0.925	0.48	0.365	0.760
0.16	0.147	0.920	0.49	0.370	0.755
0.17	0.155	0.915	0.50	0.375	0.750
0.18	0.164	0.910	0.51	0.380	0.745
0.19	0.172	0.905	0.52	0.385	0.740
0.20	0.180	0.900	0.53	0.390	0.735
0.21	0.188	0.895	0.54	0.394	0.730
0.22	0.196	0.890	0.55	0.399	0.725
0.23	0.203	0.885	0.56	0.403	0.720
0.24	0.211	0.880	0.57	0.408	0.715
0.25	0.219	0.875	0.58	0.412	0.710
0.26	0.226	0.870	0.59	0.416	0.705
0.27	0.234	0.865	0.60	0.420	0.700
0.28	0.241	0.860	0.61	0.424	0.695
0.29	0.248	0.855	0.62	0.428	0.690
0.30	0.255	0.850	0.63	0.432	0.685
0.31	0.262	0.845	0.64	0.435	0.680
0.32	0.269	0.840	0.65	0.439	0.675
0.33	0.275	0.835			

普通钢筋截面面积、质量表 附表1-6

公称直径(mm)	在下列钢筋根数时的截面面积(mm²)									质量(kg/m)	带肋钢筋	
	1	2	3	4	5	6	7	8	9		公称直径(mm)	外径(mm)
6	28.3	57	85	113	141	170	198	226	254	0.222	6	7.0
8	50.3	101	151	201	251	302	352	402	452	0.395	8	9.3
10	78.5	157	236	314	393	471	550	628	707	0.617	10	11.6
12	113.1	226	339	452	566	679	792	905	1 018	0.888	12	13.9
14	153.9	308	462	616	770	924	1 078	1 232	1 385	1.21	14	16.2
16	201.1	402	603	804	1 005	1 206	1 407	1 608	1 810	1.58	16	18.4
18	254.5	509	763	1 018	1 272	1 527	1 781	2 036	2 290	2.00	18	20.5
20	314.2	628	942	1 256	1 570	1 884	2 200	2 513	2 827	2.47	20	22.7
22	380.1	760	1 140	1 520	1 900	2 281	2 661	3 041	3 421	2.98	22	25.1
25	490.9	982	1 473	1 964	2 454	2 945	3 436	3 927	4 418	3.85	25	28.4
28	615.8	1 232	1 847	2 463	3 079	3 695	4 310	4 926	5 542	4.83	28	31.6
32	804.2	1 608	2 413	3 217	4 021	4 826	5 630	6 434	7 238	6.31	32	35.8

在钢筋间距一定时板每米宽度内钢筋截面积(单位:mm²) 附表1-7

钢筋间距(mm)	钢筋直径(mm)									
	6	8	10	12	14	16	18	20	22	24
70	404	718	1 122	1 616	2 199	2 873	3 636	4 487	5 430	6 463
75	377	670	1 047	1 508	2 052	2 681	3 393	4 188	5 081	6 032
80	353	628	982	1 414	1 924	2 514	3 181	3 926	4 751	5 655
85	333	591	924	1 331	1 811	2 366	2 994	3 695	4 472	5 322
90	314	559	873	1 257	1 711	2 234	2 828	3 490	4 223	5 027
95	298	529	827	1 190	1 620	2 117	2 679	3 306	4 001	4 762
100	283	503	785	1 131	1 539	2 011	2 545	3 141	3 801	4 524
105	269	479	748	1 077	1 466	1 915	2 424	2 991	3 620	4 309
110	257	457	714	1 028	1 399	1 828	2 314	2 855	3 455	4 113
115	246	437	683	984	1 339	1 749	2 213	2 731	3 305	3 934
120	236	419	654	942	1 283	1 676	2 121	2 617	3 167	3 770
125	226	402	628	905	1 232	1 609	2 036	2 513	3 041	3 619
130	217	387	604	870	1 184	1 547	1 958	2 416	2 924	3 480
135	209	372	582	838	1 140	1 490	1 885	2 327	2 816	3 351
140	202	359	561	808	1 100	1 436	1 818	2 244	2 715	3 231
145	195	347	542	780	1 062	1 387	1 755	2 166	2 621	3 120
150	189	335	524	754	1 026	1 341	1 697	2 084	2 534	3 016
155	182	324	507	730	993	1 297	1 642	2 027	2 452	2 919
160	177	314	491	707	962	1 257	1 590	1 964	2 376	2 828
165	171	305	476	685	933	1 219	1 542	1 904	2 304	2 741

续上表

钢筋间距(mm)	钢筋直径(mm)									
	6	8	10	12	14	16	18	20	22	24
170	166	296	462	665	905	1 183	1 497	1 848	2 236	2 661
175	162	287	449	646	876	1 149	1 454	1 795	2 172	2 585
180	157	279	436	628	855	1 117	1 414	1 746	2 112	2 513
185	153	272	425	611	832	1 087	1 376	1 694	2 035	2 445
190	149	265	413	595	810	1 058	1 339	1 654	2 001	2 381
195	145	258	403	580	789	1 031	1 305	1 611	1 949	2 320
200	141	251	393	565	769	1 005	1 272	1 572	1 901	2 262

普通钢筋和预应力直线形钢筋最小混凝土保护层厚度(单位:mm)　　附表1-8

序　号	构件类别	环境条件		
		Ⅰ	Ⅱ	Ⅲ、Ⅳ
1	基础、桩基承台 (1)基坑底面有垫层或侧面有模板(受力钢筋) (2)基坑底面无垫层或侧面无模板	40 60	50 75	60 85
2	墩台身、挡土结构、涵洞、梁、板、拱圈、拱上建筑(受力钢筋)	30	40	45
3	人行道构件、栏杆(受力钢筋)	20	25	30
4	箍筋	20	25	30
5	缘石、中央分隔带、护栏等行车道构件	30	40	45
6	收缩、温度、分布、防裂等表层钢筋	15	20	25

注:①对于环氧树脂涂层钢筋,可按环境类别Ⅰ取用。
　②后张法预应力混凝土锚具,其最小混凝土保护层厚度,Ⅰ、Ⅱ及Ⅲ(Ⅳ)环境类别,分别为40mm、45mm及50mm。
　③先张法预应力钢筋端部应加保护,不得外露。
　④Ⅰ类环境是指非严寒或寒冷地区的大气环境,与无侵蚀性的水或土接触的环境条件。
　　Ⅱ类环境是指严寒地区的大气环境,与无侵蚀性的水或土接触的环境;使用除冰盐环境;滨海环境条件。
　　Ⅲ类环境是指海水环境。
　　Ⅳ类环境是受人为或自然侵蚀性物质影响的环境。

钢筋混凝土构件中纵向受力钢筋的最小配筋率(单位:%)　　附表1-9

受力类型		最小配筋百分率
受压构件	全部纵向钢筋	0.5
	一侧纵向钢筋	0.2
受弯构件、偏心受拉构件及轴心受拉构件的一侧受拉钢筋		0.2 和 $45f_{td}/f_{sd}$ 中较大值
受扭构件		$0.08f_{cd}/f_{sv}$(纯扭时),$0.08(2\beta_t-1)f_{cd}/f_{sv}$(剪扭时)

注:①受压构件全部纵向钢筋最小配筋百分率,当混凝土强度等级为C50及以上时不应小于0.6。
　②当大偏心受拉构件的受压区配置按计算需要的受压钢筋时,其最小配筋百分率不应小于0.2。
　③轴心受压构件、偏心受压构件全部纵向钢筋的配筋率和一侧纵向钢筋(包括大偏心受拉构件的受压钢筋)的配筋百分率应按构件的毛截面面积计算;轴心受拉构件及小偏心受拉构件一侧受拉钢筋的配筋百分率应按构件毛截面面积计算;受弯构件、大偏心受拉构件的一侧受拉钢筋的配筋百分率为 $100A_s/(bh_0)$,其中 A_s 为受拉钢筋截面面积,b 为腹板宽度(箱形截面为各腹板宽度之和),h_0 为有效高度。
　④当钢筋沿构件截面周边布置时,"一侧的受压钢筋"或"一侧的受拉钢筋"是指受力方向两个对边中的一边布置的纵向钢筋。
　⑤对受扭构件,其纵向受力钢筋的最小配筋率为 $A_{st,min}/(bh)$,$A_{st,min}$ 为纯扭构件全部纵向钢筋最小截面积,h 为矩形截面基本单元长边长度,b 为短边长度,f_{sv} 为箍筋抗拉强度设计值。

钢筋混凝土轴心受压构件的稳定系数 φ

附表 1-10

l_0/b	≤8	10	12	14	16	18	20	22	24	26	28
l_0/d	≤7	8.5	10.5	12	14	15.5	17	19	21	22.5	24
l_0/r	≤28	35	42	48	55	62	69	76	83	90	97
φ	1.0	0.98	0.95	0.92	0.87	0.81	0.75	0.70	0.65	0.60	0.56
l_0/b	30	32	34	36	38	40	42	44	46	48	50
l_0/d	26	28	29.5	31	33	34.5	36.5	38	40	41.5	43
l_0/r	104	111	118	125	132	139	146	153	160	167	174
φ	0.52	0.48	0.44	0.40	0.36	0.32	0.29	0.26	0.23	0.21	0.19

注:①表中 l_0 为构件计算长度,b 为矩形截面短边尺寸,d 为圆形截面直径,r 为截面最小回转半径。
②构件计算长度 l_0 的确定,两端固定为 $0.5l$;一端固定、一端为不移动的铰为 $0.7l$;两端均匀不移动的铰为 l;一端固定、一端自由为 $2l$。

圆形截面钢筋混凝土偏心受压构件正截面相对抗压承载力 η_u

附表 1-11

$\eta\dfrac{e_0}{r}$	$\rho f_{sd}/f_{cd}$										
	0.06	0.09	0.12	0.15	0.18	0.21	0.24	0.27	0.3	0.4	0.5
0.01	1.0487	1.0783	1.1079	1.1375	1.1671	1.1968	1.2264	1.2561	1.2857	1.3846	1.4835

预应力钢筋抗拉强度标准值

附表 2-1

钢筋种类		符号	直径 d(mm)	抗拉强度标准值 f_{pk}(MPa)
钢绞线	1×2 (二股)	ϕ^S	8.0、10.0	1470、1570、1720、1860
			12.0	1470、1570、1720
	1×3 (三股)		8.6、10.8	1470、1570、1720、1860
			12.9	1470、1570、1720
	1×7 (七股)		9.5、11.1、12.7	1860
			15.2	1720、1860
消除应力钢丝	光面钢丝	ϕ^P	4、5	1470、1570、1670、1770
			6	1570、1670
	螺旋肋钢丝	ϕ^H	7、8、9	1470、1570
	刻痕钢丝	ϕ^I	5、7	1470、1570
精轧螺纹钢筋		JL	40	540
			18、25、32	540、785、930

注:表中 d 系指国家标准中的钢绞线、钢丝和精轧螺纹钢筋的公称直径。

预应力钢筋抗拉、抗压强度设计值

附表 2-2

钢筋种类	抗拉强度标准值 f_{pk}(MPa)	抗拉强度设计值 f_{pd}(MPa)	抗压强度设计值 f'_{pd}(MPa)
钢绞线 1×2（二股） 1×3（三股） 1×7（七股）	1470	1000	390
	1570	1070	
	1720	1170	
	1860	1260	

续上表

钢筋种类	抗拉强度标准值f_{pk}(MPa)	抗拉强度设计值f_{pd}(MPa)	抗压强度设计值f'_{pd}(MPa)
消除应力钢丝 螺旋肋钢丝	1 470	1 000	410
	1 570	1 070	
	1 670	1 140	
	1 770	1 200	
刻痕钢丝	1 470	1 000	410
	1 570	1 070	
精轧螺纹钢筋	540	450	400
	785	650	
	930	770	

预应力钢筋的弹性模量（单位：10^5 MPa）　　　　　附表 2-3

预应力钢筋种类	E_p
精轧螺纹钢筋	2.0
消除应力钢丝、螺旋肋钢丝、刻痕钢丝	2.05
钢绞线	1.95

钢绞线公称直径、截面面积及理论质量　　　　　附表 2-4

钢绞线种类	公称直径(mm)	公称截面面积(mm²)	每1000m的钢绞线理论质量(kg)
1×2	8	25.1	197
	10	39.3	309
	12	56.5	444
1×3	8.6	37.7	296
	10.8	58.9	462
	12.9	84.8	666
1×7 标准型	9.5	54.8	430
	11.1	74.2	582
	12.7	98.7	775
	15.2	140	1101
1×7 模拔型	12.7	112	890
	15.2	165	1295

钢丝公称直径、公称截面面积及理论质量　　　　　附表 2-5

公称直径(mm)	公称横截面面积(mm²)	理论质量参考值(kg/m)
4.0	12.57	0.099
5.0	19.63	0.154
6.0	28.27	0.222
7.0	38.48	0.302
8.0	50.26	0.394
9.0	63.62	0.499

系数 k 和 μ 值   附表 2-6

管道成型方式	k	μ 钢绞线、钢丝束	μ 精轧螺纹钢筋
预埋金属波纹管	0.001 5	0.20~0.25	0.50
预埋塑料波纹管	0.001 5	0.14~0.17	—
预埋铁皮管	0.003 0	0.35	0.40
预埋钢管	0.001 0	0.25	—
抽心成型	0.001 5	0.55	0.60

锚具变形、钢筋回缩和接缝压缩值（单位：mm）   附表 2-7

锚具、接缝类型		Δl
钢丝束的钢制锥形锚具		6
夹片式锚具	有顶压时	4
夹片式锚具	无顶压时	6
带螺母锚具的螺母缝隙		1
墩头锚具		1
每块后加垫板的缝隙		1
水泥砂浆接缝		1
环氧树脂砂浆接缝		1

预应力钢筋的预应力传递长度 l_{tr} 与锚固长度 l_a（单位：mm）   附表 2-8

项次	钢筋种类	混凝土强度等级	传递长度 l_{tr}	锚固长度 l_a
1	钢绞线 1×2、1×3 $\sigma_{pe}=1\,000\text{MPa}$ $f_{pd}=1\,170\text{MPa}$	C30	$75d$	—
		C35	$68d$	—
		C40	$63d$	$115d$
		C45	$60d$	$110d$
		C50	$57d$	$105d$
		C55	$55d$	$100d$
		C60	$55d$	$95d$
		≥C65	$55d$	$90d$
	钢绞线 1×7 $\sigma_{pe}=1\,000\text{MPa}$ $f_{pd}=1\,260\text{MPa}$	C30	$80d$	—
		C35	$73d$	—
		C40	$67d$	$130d$
		C45	$64d$	$125d$
		C50	$60d$	$120d$
		C55	$58d$	$115d$
		C60	$58d$	$110d$
		≥C65	$58d$	$105d$

续上表

项 次	钢筋种类	混凝土强度等级	传递长度 l_{tr}	锚固长度 l_a
2	螺旋肋钢丝 $\sigma_{pe}=1\,000\mathrm{MPa}$ $f_{pd}=1\,200\mathrm{MPa}$	C30	70d	—
		C35	64d	—
		C40	58d	95d
		C45	56d	90d
		C50	53d	85d
		C55	51d	83d
		C60	51d	80d
		≥C65	51d	80d
3	刻痕钢丝 $\sigma_{pe}=1\,000\mathrm{MPa}$ $f_{pd}=1\,070\mathrm{MPa}$	C30	89d	—
		C35	81d	—
		C40	75d	125d
		C45	71d	115d
		C50	68d	110d
		C55	65d	105d
		C60	65d	103d
		≥C65	65d	100d

注：①预应力钢筋的预应力传递长度 l_{tr} 按有效预应力值 σ_{pe} 查表；锚固长度 l_a 按抗拉强度设计值 f_{pd} 查表。
②预应力传递长度应根据预应力钢筋放松时混凝土立方体抗压强度 f'_{cu} 确定，当 f'_{cu} 在表列混凝土强度等级之间时，预应力传递长度按直线内插取用。
③当采用骤然放松预应力钢筋的施工工艺时，锚固长度的起点及预应力传递长度的起点应从离构件末端 $0.25l_{tr}$ 处开始，l_{tr} 为预应力钢筋的预应力传递长度。
④当预应力钢筋的抗拉强度设计值 f_{pd} 或有效预应力值 σ_{pe} 与表值不同时，其锚固长度或预应力传递长度应根据表值按比例增减。

石材强度设计值（单位：MPa）　　　　　　　　　　　　　　　　　附表 3-1

强度类别	强度等级						
	MU120	MU100	MU80	MU60	MU50	MU40	MU30
抗压 f_{cd}	31.78	26.49	21.19	15.89	13.24	10.59	7.95
弯曲抗拉 f_{tmd}	2.18	1.82	1.45	1.09	0.91	0.73	0.55

石材强度等级的换算系数　　　　　　　　　　　　　　　　　　　附表 3-2

立方体试件边长(mm)	200	150	100	70	50
换算系数	1.43	1.28	1.14	1.00	0.86

混凝土强度设计值（单位：MPa）　　　　　　　　　　　　　　　　附表 3-3

强度类别	强度等级					
	C40	C35	C30	C25	C20	C15
轴心抗压 f_{cd}	15.64	13.69	11.73	9.78	7.82	5.87
弯曲抗拉 f_{tmd}	1.24	1.14	1.04	0.92	0.80	0.66
直接抗剪 f_{vd}	2.48	2.28	2.09	1.85	1.59	1.32

混凝土预制块砂浆砌体抗压强度设计值 f_{cd}（单位：MPa） 附表3-4

砌块强度等级	砂浆强度等级					砂浆强度
	M20	M15	M10	M7.5	M5	0
C40	8.25	7.04	5.84	5.24	4.64	2.06
C35	7.71	6.59	5.47	4.90	4.34	1.93
C30	7.14	6.10	5.06	4.54	4.02	1.79
C25	6.52	5.57	4.62	4.14	3.67	1.63
C20	5.83	4.98	4.13	3.70	3.28	1.46
C15	5.05	4.31	3.58	3.21	2.84	1.26

块石砂浆砌体的抗压强度设计值 f_{cd}（单位：MPa） 附表3-5

砌块强度等级	砂浆强度等级					砂浆强度
	M20	M15	M10	M7.5	M5	0
MU120	8.42	7.19	5.96	5.35	4.73	2.10
MU100	7.68	6.56	5.44	4.88	4.32	1.92
MU80	6.87	5.87	4.87	4.37	3.86	1.72
MU60	5.95	5.08	4.22	3.78	3.35	1.49
MU50	5.43	4.64	3.85	3.45	3.05	1.36
MU40	4.86	4.15	3.44	3.09	2.73	1.21
MU30	4.21	3.59	2.98	2.67	2.37	1.05

注：对各类石砌体，应按表中数值分别乘以下列系数：细料石砌体1.5；半细料石砌体1.3；粗料石砌体1.2；干砌块石可采用砂浆强度为零时的抗压强度设计值。

片石砂浆砌体的抗压强度设计值 f_{cd}（单位：MPa） 附表3-6

砌块强度等级	砂浆强度等级					砂浆强度
	M20	M15	M10	M7.5	M5	0
MU120	1.97	1.68	1.39	1.25	1.11	0.33
MU100	1.80	1.54	1.27	1.14	1.01	0.30
MU80	1.61	1.37	1.14	1.02	0.90	0.27
MU60	1.39	1.19	0.99	0.88	0.78	0.23
MU50	1.27	1.09	0.90	0.81	0.71	0.21
MU40	1.14	0.97	0.81	0.72	0.64	0.19
MU30	0.98	0.84	0.70	0.63	0.55	0.16

注：干砌片石砌体可采用砂浆强度为零时的抗压强度设计值。

砂浆砌体轴心抗拉、弯曲抗拉和直接抗剪强度设计值(单位:MPa)　　附表3-7

强度类别	破坏特征	砌体种类	砂浆强度等级				
			M20	M15	M10	M7.5	M5
轴心抗拉 f_{td}	齿缝	规则砌块砌体	0.104	0.090	0.073	0.063	0.052
		片石砌体	0.096	0.083	0.068	0.059	0.048
弯曲抗拉 f_{tmd}	齿缝	规则砌块砌体	0.122	0.105	0.086	0.074	0.061
		片石砌体	0.145	0.125	0.102	0.089	0.072
	通缝	规则砌块砌体	0.084	0.073	0.059	0.051	0.042
直接抗剪 f_{vd}	—	规则砌块砌体	0.104	0.090	0.073	0.063	0.052
		片石砌体	0.241	0.208	0.170	0.147	0.120

注：①砌体龄期为28d。
②规则块材砌体包括：块石砌体、粗料石砌体、半细料石砌体、细料石砌体、混凝土预制块砌体。
③规则块材砌体在齿缝方向受剪时，系通过砌块和灰缝剪破。

小石子混凝土砌块石砌体轴心抗压强度设计值 f_{cd}(单位:MPa)　　附表3-8

石材强度等级	小石子混凝土强度等级					
	C40	C35	C30	C25	C20	C15
MU120	13.86	12.69	11.49	10.25	8.95	7.59
MU100	12.65	11.59	10.49	9.35	8.17	6.93
MU80	11.32	10.36	9.38	8.37	7.31	6.19
MU60	9.80	9.98	8.12	7.24	6.33	5.36
MU50	8.95	8.19	7.42	6.61	5.78	4.90
MU40	—	—	6.63	5.92	5.17	4.38
MU30	—	—	—	—	4.48	3.79

注：砌块为粗料石时，轴心抗压强度为表值乘1.2；砌块为细料石时、半细料石时，轴心抗压强度为表值乘1.4。

小石子混凝土砌片石砌体轴心抗压强度设计值 f_{cd}(单位:MPa)　　附表3-9

石材强度等级	小石子混凝土强度等级			
	C30	C25	C20	C15
MU120	6.94	6.51	5.99	5.36
MU100	5.30	5.00	4.63	4.17
MU80	3.94	3.74	3.49	3.17
MU60	3.23	3.09	2.91	2.67
MU50	2.88	2.77	2.62	2.43
MU40	2.50	2.42	2.31	2.16
MU30	—	—	1.95	1.85

小石子混凝土砌块石、片石砌体的轴心抗拉、弯曲抗拉和直接抗剪强度设计值（单位：MPa）

附表 3-10

强度类别	破坏特征	砌体种类	小石子混凝土强度等级					
			C40	C35	C30	C25	C20	C15
轴心抗拉 f_{td}	齿缝	块石砌体	0.285	0.267	0.247	0.226	0.202	0.175
		片石砌体	0.425	0.398	0.368	0.336	0.301	0.260
弯曲抗拉 f_{tmd}	齿缝	块石砌体	0.335	0.313	0.290	0.265	0.237	0.205
		片石砌体	0.493	0.461	0.427	0.387	0.349	0.300
	通缝	块石砌体	0.232	0.217	0.201	0.183	0.164	0.142
直接抗剪 f_{vd}	—	块石砌体	0.285	0.267	0.247	0.226	0.202	0.175
		片石砌体	0.425	0.398	0.368	0.336	0.301	0.260

注：对其他规则砌块砌体强度值为表内块石砌体强度值乘以下列系数：粗料石砌体 0.7；细料石、半细料石砌体 0.35。

各类砌体受压弹性模量 E_m（单位：MPa）

附表 3-11

砌体种类	砂浆强度等级				
	M20	M15	M10	M7.5	M5
混凝土预制块砌体	$1700f_{cd}$	$1700f_{cd}$	$1700f_{cd}$	$1600f_{cd}$	$1500f_{cd}$
粗料石、块及片石砌体	7 300	7 300	7 300	5 650	4 000
细料石、半细料石砌体	22 000	22 000	22 000	17 000	12 000
小石子混凝土砌体	$2100f_{cd}$				

注：f_{cd} 为砌体轴心抗压强度设计值。

参 考 文 献

[1] 叶见曙. 结构设计原理[M]. 2版. 北京:人民交通出版社,2005.

[2] 刘立新. 混凝土结构基本原理[M]. 武汉:武汉理工大学出版社,2004.

[3] 中华人民共和国行业标准. JTG D60—2015 公路桥涵设计通用规范[S]. 北京:人民交通出版社股份有限公司,2015.

[4] 中华人民共和国行业标准. JTG 3362—2018 公路钢筋混凝土及预应力混凝土桥涵设计规范[S]. 北京:人民交通出版社股份有限公司,2018.

[5] 中华人民共和国行业标准. JTG D61—2005 公路圬工桥涵设计规范[S]. 北京:人民交通出版社,2005.

[6] 中华人民共和国国家标准. GB 50153—2008 工程结构可靠度设计统一标准[S]. 北京:中国建筑工业出版社,2008.

[7] 中华人民共和国国家标准. GB/T 50283—2008 公路工程结构可靠度设计统一标准[S]. 北京:中国计划出版社,2008.

[8] CEB-FIP Specification.

[9] McComac Jack C, Nelson James K. Design of Reinforced Concrete[M]. John Wiley & Sons, Inc, 2006.